Introduction to Structural [illegible]

This textbook provides fundamental concepts, and a comprehensive analysis of indeterminate structures by both force and displacement methods. Major coverage includes the analysis of beams, rigid-jointed plane frames, and pin-jointed plane frames by various force and displacement methods, followed by the analysis of multi-storey frames using approximate methods, influence lines for indeterminate structures, and two-hinged arches. Each chapter contains an introduction, methodology, necessary derivations/equations, and examples.

Features:

- Discusses advanced levels of structural analysis with a focus on indeterminate structures.
- Covers approximate methods for the analysis of multi-storey frames, two-hinged arches, and influence lines for indeterminate beams.
- Separately discusses both flexibility and stiffness matrix methods for beams, rigid joint plane frames, and pin joint plane frames.
- Step-by-step procedure for solving problems in each method.
- Explains the problems with neat coloured free-body diagrams, shear force and bending moment diagrams, and probable elastic curves.
- Includes review questions and answers for numerical problems and examples.

This book is aimed at undergraduate and senior undergraduate students in structural and civil engineering.

Introduction to Structural Analysis
Indeterminate Structures

Meesala Chakradhara Rao

CRC Press
Taylor & Francis Group
Boca Raton London New York

CRC Press is an imprint of the
Taylor & Francis Group, an **informa** business

Designed cover image: © www.alamy.com

First edition published 2026
by CRC Press
2385 NW Executive Center Drive, Suite 320, Boca Raton FL 33431

and by CRC Press
4 Park Square, Milton Park, Abingdon, Oxon, OX14 4RN

CRC Press is an imprint of Taylor & Francis Group, LLC

ISBN: 978-1-032-91378-0 (hbk)
ISBN: 978-1-032-88910-8 (pbk)
ISBN: 978-1-003-56295-5 (ebk)

DOI: 10.1201/9781003562955

Typeset in Times
by codeMantra

Access the Student Resources: www.routledge.com/9781032889108

In memory of my Parents

Sri Appala Naidu and Smt. Satyavathi

Contents

Foreword xi
Preface xii
Acknowledgements xiii
About the Author xiv

Chapter 1 Methods of Analysis 1

1.1 Introduction 1
1.2 Force Method of Analysis 2
1.3 Displacement Method of Analysis 3
1.4 Equilibrium Equations 4
1.5 Degree of Static Indeterminacy 4
1.5.1 Degree of Static Indeterminacy of Beams 4
1.5.2 Degree of Static Indeterminacy of Rigid-Jointed Plane Frames 4
1.5.3 Degree of Static Indeterminacy of Pin-Jointed Frames 6
1.6 Stability and Unstability of Structures 6
1.7 Kinematic Indeterminacy of Structures 11
1.7.1 Rigid-Jointed Structures 11
1.7.2 Pin-Jointed Plane Frames (Trusses) 12
1.8 Problems for Practice 14
1.9 Answer Key 17
Further Reading 17

Chapter 2 Three-Moment Theorem 18

2.1 Introduction 18
2.2 Derivation of Clapeyron's Theorem or Three-Moment Equation (Without Yielding of Supports) 18
2.3 Derivation of Generalised Three Moments Equation (With Yielding of Supports) 19
2.3.1 Analysis Procedure 20
2.4 Application to Continuous Beams with Extreme Boundaries Simply Supported 21
2.5 Application to the Continuous Beams with Extreme Boundaries Not Simply Supported and with Yielding of Supports 23
2.6 Problems for Practice 31
2.7 Answer Key 31
Further Reading 31

Chapter 3 Consistent Deformation Method 32

3.1 Introduction 32
3.2 Formulation of Consistent Deformation Method 34
3.3 Application to Indeterminate Beams 36
3.4 Application to Rigid-Jointed Plane Frames 50
3.5 Application to Pin-Jointed Plane Frames 65
3.5.1 Introduction 65
3.6 Problems for Practice 71
3.7 Answer Key 74

Chapter 4 Strain Energy Method: Theorem of Least Work 75

4.1 Introduction 75
4.2 Proof of the Theorem of Least Work 75
4.3 Application to Beams 77
4.4 Application to Rigid-Jointed Plane Frames 89

4.5 Application to Pin-Jointed Plane Frames (Trusses) 106
4.6 Problems for Practice 115
4.7 Answer Key 117
Further Reading 117

Chapter 5 Column Analogy Method 118

5.1 Introduction 118
5.2 Development of the Method 118
5.3 Sign Convention 119
5.4 Application to Beams 119
5.5 Application to Symmetrical Rigid-Jointed Plane Frames 126
5.6 Column Analogy for Box Frames 134
5.7 Application to Unsymmetrical Rigid-Jointed Plane Frames 139
5.8 Stiffness Factors of Non-prismatic Members 142
5.9 Problems for Practice 144
5.10 Answer Key 145
Further Reading 145

Chapter 6 Slope–Deflection Method 146

6.1 Introduction 146
6.2 Formulation of SD Equations 146
6.2.1 End Moments Due to the Rotational Displacement at C (θ_C) 146
6.2.2 End Moments Due to the Rotational Displacement at D (θ_D) 147
6.2.3 End Moments Due to the Relative Settlement of Supports between C and D (φ_{CD}) 147
6.2.4 Fixed End Moments Due to the External Loads 147
6.2.5 SD Equations 148
6.3 Modified SD Equations 148
6.3.1 Extreme Support is Hinge or Roller 148
6.3.2 Guided Support at Far End D 149
6.4 Analysis Procedure 149
6.5 Numerical Problems 151
6.5.1 Applications to the Beams Without Yielding of Supports 151
6.5.2 Application to the Beams with Yielding of Supports 160
6.5.3 Application to the Rigid-Jointed Plane Frames (without Side Sway) 168
6.5.4 Application to the Rigid-Jointed Plane Frames with Lateral Sway 175
6.6 Problems for Practice 197
6.7 Answer Key 201
Further Reading 201

Chapter 7 Moment Distribution Method 202

7.1 Introduction 202
7.2 Sign Convention 202
7.3 Flexural (Rotational) Stiffness of a Beam Element 202
7.3.1 Case 1: When the Far End Is Fixed and the Near End Is Subjected to a Clockwise Moment, M_O 202
7.3.2 Case 2: When Far End is Pinned (Simply Supported) and the Near End is Subjected to a Moment, M_O 203
7.3.3 Case 3: When the Far End is Sliding/Guided Support 204
7.4 Joint Stiffness Factor and Distribution Factor 204
7.5 Check on the Moment Distribution 205
7.6 Numerical Problems 206
7.6.1 Application to Beams 207

7.6.2 Application to Rigid Joint Plane Frames (Non-Sway) 228
7.6.3 Application to Rigid Joint Plane Frames (Sway) 238
7.7 Problems for Practice 262
7.8 Answer key 264
Further Reading 265

Chapter 8 Kani's Method 266

8.1 Introduction 266
8.2 Beams Subjected to Only Rotational Degrees of Freedom 266
8.2.1 Kani's Analysis Procedure Formulation When Far End Fixed 266
8.2.2 Far End is Hinged/Pinned Support 266
8.2.3 Far End is Guided-Fixed Support 267
8.3 Numerical Examples on Beams 269
8.4 Rigid Joint Plane Frames (Non-Sway) 285
8.5 Rigid Joint Plane Frames (Side Sway) 293
8.5.1 All Columns Are of Equal Height in a Storey 295
8.5.2 Columns with Unequal Heights 304
8.6 Problems for Practice 308
8.7 Answer Key 311
Further Reading 312

Chapter 9 Matrix Methods: Flexibility Matrix Method 313

9.1 Introduction 313
9.2 Flexibility Matrix Equation 313
9.2.1 Procedure to Solve the Problems 314
9.3 Application to Beams 314
9.3.1 Problems without Yielding of Supports 314
9.3.2 Problems with Yielding of Supports 326
9.4 Applications to Rigid-Jointed Plane Frames 330
9.5 Applications to Pin-Jointed Plane Frames (Trusses) 348
9.6 Problems for Practice 356
9.7 Answer Key 358
Further Reading 358

Chapter 10 Matrix Methods: Stiffness Matrix Method 359

10.1 Introduction 359
10.2 Formulation of Stiffness Matrix Equation or Force Displacement Relationship 360
10.2.1 Procedure to Solve the Problems 362
10.2.2 Development of Stiffness Matrix for a General Beam Element 362
10.3 Application to Beams 364
10.4 Applications to Rigid-Jointed Plane Frames 379
10.4.1 Frames Subjected to No Lateral Sway 379
10.4.2 Frame Subjected to Lateral Sway 388
10.5 Applications to Pin-Jointed Plane Frames 403
10.6 Problems for Practice 410
10.7 Answer Key 413
Further Reading 414

Chapter 11 Analysis of Multi-Storey Frames by Approximate Methods (online only: www.routledge.com/9781032889108)

11.1 Introduction
11.2 Substitute Frame Method
11.3 Multi-Storey Frames Subjected to Lateral Loads

11.3.1 Portal Frame Method
11.3.2 Cantilever Method
11.4 Problems for Practice
11.5 Answer Key

Chapter 12 Influence Lines for Indeterminate Beams (online only: www.routledge.com/9781032889108)

12.1 Introduction
12.2 Muller-Breslau Principle
12.3 Quantitative Influence Lines
12.4 Numerical Examples
12.5 Problems for Practice
12.6 Answer Key

Chapter 13 Two-Hinged Arches (online only: www.routledge.com/9781032889108)

13.1 Introduction
13.2 Two Hinged Symmetrical Arches Subjected to Gravity Loading
13.3 Analysis of a Parabolic Arch
13.4 Analysis of a Segmental Arch
13.5 Effect of Temperature, Yielding of Supports and Rib Shortening
13.6 Influence Lines
13.7 Problems for Practice
13.8 Answer Key

Index 415

Foreword

Statistically Indeterminate structures are those that require more than just equations of equilibrium to solve for reactions and internal forces. Analysis of *Statistically Indeterminate* structures, therefore, forms the backbone of modern civil engineering structural analysis, offering solutions to complex problems where the behaviour of structures cannot be determined using simple static equations. Unlike determinate structures, indeterminate systems demand a deeper understanding of structural behaviour and responses under various loading conditions.

In this book, the author, Dr Chakradhara Rao, has meticulously developed a comprehensive and accessible text integrating fundamental concepts with detailed formulations of various analytical methods. This book delves into a wide array of techniques for analysing indeterminate structures, including the force and displacement methods. Each chapter is enriched with numerous examples, complete with detailed free-body diagrams, shear force and bending moment diagrams, probable deflected shapes, and problem-solving exercises with answers. This structured approach facilitates effective learning and enhances comprehension for the reader.

With Dr Chakradhara Rao's extensive experience and expertise in structural engineering, this book offers both authoritative guidance and an engaging approach to understanding structural systems. The first ten chapters focus on the force and displacement methods for analysing beams, pin-jointed plane frames, and rigid-jointed plane frames. The final three chapters address advanced topics, including the analysis of multi-storey frames under lateral loads, influence lines for indeterminate structures, and two-hinged arches.

As you progress through this textbook, I am confident you will find it an indispensable tool for mastering the complexities of indeterminate structural analysis. Whether you are a student aiming to build a strong foundation or a professional seeking to refine your expertise, this book will undoubtedly serve as a reliable guide.

I commend the author for successfully creating a comprehensive and detailed resource on the analysis of *Indeterminate Structures*. I am certain this book will prove invaluable not only to undergraduate and postgraduate students but also to practising engineers.

January 13, 2025
The University of British Columbia
Vancouver, Canada

Nemkumar Banthia PhD, PEng, FACI, FICI, FCSCE, FCAE, FNAE, FRSC
University Killam Professor, Distinguished University Scholar and Sr. Canada Research Chair

Preface

The primary objective of the structural engineer is to ensure that structures safely transfer all the loads imposed on them to the sub-soil. To achieve this objective, the structural engineer must thoroughly understand the behaviour of the structure, or its structural response, under various loads. This entails understanding the force–displacement relationship under the action of a system of forces, as it characterises the structural response. This understanding is particularly important for indeterminate structures, as the majority of the structures in the real world are indeterminate. However, most textbooks available in the market primarily focus on the analysis of determinate structures, with only limited coverage of indeterminate structures. As a result, these textbooks often become bulky but fail to provide a deep explanation of the fundamental theoretical concepts, along with a diverse range of solved problems. Instead, they attempt to cover an entire course on the analysis of determinate and indeterminate structures using both force and displacement methods, leading to a lack of depth.

Additionally, many existing books rarely include probable elastic curves or deflected shapes in their worked examples. These visualisations are essential for readers to understand the behaviour of structures under load. Furthermore, a significant number of textbooks do not provide answers to the exercise problems at the end of each chapter, which hinders self-assessment and learning.

The proposed book aims to address these gaps by offering a detailed explanation of both force methods and displacement methods, along with an extensive range of solved examples. Each solved problem is accompanied by detailed free-body diagrams, shear force and bending moment diagrams, and probable deflected shapes. These features set this book apart from existing textbooks. Moreover, answers to practice problems at the end of each chapter are included to facilitate better learning and understanding.

The book contains a total of 13 chapters. Of these, ten chapters are dedicated to the analysis of beams, pin-jointed plane frames, and rigid-jointed plane frames using various force and displacement methods. The remaining three chapters cover the analysis of multi-storey frames subjected to lateral loads, influence lines for indeterminate structures, and two-hinged arches.

The introductory chapter provides an overview of indeterminate structures and the force and displacement methods. It also explains static and kinematic indeterminacy, including their degrees, with numerous examples involving beams, rigid-jointed plane frames, and pin-jointed plane frames. Chapters 2 through 5 focus on the analysis of indeterminate structures using force methods:

Chapter 2 introduces the theorem of three moments, including its formulation, and discusses its applications to indeterminate beams with and without yielding of supports. **Chapter 3** covers the consistent deformation method, including its formulation and applications to indeterminate beams, rigid-jointed plane frames, and pin-jointed plane frames (trusses) with and without settlement of supports. **Chapter 4** explains strain energy principles and their applications to indeterminate beams, frames, and trusses. **Chapter 5** details the concept and development of the column analogy method, with its application to prismatic and non-prismatic propped cantilevers, fixed beams, and symmetrical and unsymmetrical rigid-jointed plane frames.

Chapters 6 through 8 cover the analysis of indeterminate structures using displacement methods: **Chapter 6** develops the concept of slope–deflection, including the formulation of slope–deflection equations and their applications to indeterminate beams and rigid-jointed plane frames. **Chapter 7** introduces the moment distribution method, its formulation, and applications to various problems involving beams and rigid-jointed plane frames with and without lateral sway. **Chapter 8** explains Kani's method, including the formulation of Kani's equations for different conditions. Applications of these concepts are discussed for indeterminate beams and rigid-jointed frames with columns of equal and unequal heights and various end conditions.

Chapters 9 and 10 focus on matrix methods of analysis: **Chapter 9** explains the flexibility method, including the formulation of flexibility equations and their applications to beams, frames, and trusses. **Chapter 10** discusses the concepts of stiffness method, including its derivation and applications to beams, rigid-jointed plane frames, and pin-jointed plane frames. The remaining chapters cover specialised topics. **Chapter 11** describes the analysis of multi-storey frames subjected to gravity and lateral loads using different approximate methods. **Chapter 12** introduces Muller–Breslau's principle and its application to constructing influence lines for indeterminate beams. **Chapter 13** focuses on the analysis of parabolic and circular two-hinged arches.

Each chapter begins with a detailed introduction, methodology, and necessary derivations or equations. It includes numerous examples of varying complexity, accompanied by clear, diagrams for better understanding. For each numerical problem, the book presents neat, free-body diagrams, shear force diagrams, bending moment diagrams, and probable elastic curves. At the end of each chapter, review questions are provided to test the reader's understanding. Practice problems with answers are included to help readers verify whether they have adopted the correct approach. Standard expressions are strategically presented for easy reference. This comprehensive structure reflects the author's intention to provide a clear, thorough, and practical resource for students and practitioners alike.

Meesala Chakradhara Rao

Acknowledgements

The completion of this book, *Introduction to Structural Analysis: Indeterminate Structures*, has been a rewarding journey, and I would like to express my heartfelt gratitude to all those who contributed to its realisation.

First and foremost, I would like to express my sincere gratitude to the authors whose sources on structural analysis I referred to and through which I have immensely expanded my knowledge. I personally thank the reviewers for their constructive comments and suggestions in bringing this book into final shape. I would also like to express sincere gratitude to the contributions of the editorial team of the Taylor and Francis publication particularly Dr. Gagandeep Singh (Senior Publisher, STEM), Jahnavi Vaid (Edotorial Assistant), and Assunta Petrone (Project Manager), whose meticulous reviews and valuable suggestions helped refine the manuscript. Your commitment to excellence has been truly remarkable. The author also wish to thank all those who helped directly or indirectly and made this book possible.

The author express his sincere gratitude to Dr. Nemkumar Banthia, University Killam Professor, Distinguished University Scholar and Sr. Canada Research Chair, The University of British Columbia, Vancouver, Canada for giving the foreword of this book. I am especially grateful to my students, whose thought-provoking discussions, questions, and challenges inspired the depth and breadth of this book. I am deeply thankful to my mentors and colleagues of Civil Engineering Department, Guru Ghasidas Vishwavidyalaya, Bilaspur, and former colleague Dr. Ch. Naga Satish Kumar, Professor, Civil Engineering Department, Bapatla Engineering College, Andhra Pradesh whose guidance and encouragement have been instrumental in shaping the concepts and ideas presented in this book. Their expertise and feedback have significantly enhanced the quality of this work.

I would like to place on record the contribution of Mr. Madhumangal Singh Koshe, Technical Assistant, Civil Engineering Department in typing the manuscript and preparing the diagrams.

I am grateful to my wife Bhavana for her understanding, diligent encouragement and support, and patience during the long period I spent writing the manuscript. A word of appreciation is also due to my children Charan and Nithin, who abstained from disturbing me during the preparation of this manuscript. Your belief in my efforts has been a source of strength and inspiration.

I would like to thank the administration of Guru Ghasidas Vishwavidyalaya (Central University), Bilaspur, for their support and encouragement during the completion of this book.

To the readers of this book, I hope it provides the clarity and insights you seek in mastering the intricacies of indeterminate structural analysis. I would be grateful if any errors in this book are brought to my attention. Your feedback and thoughts are welcome and will be invaluable for future editions.

Meesala Chakradhara Rao

About the Author

Meesala Chakradhara Rao holds a Ph.D. from IIT Kharagpur, India, and is currently working as a Professor at Guru Ghasidas Vishwavidyalaya (Central University), Bilaspur (Chhattisgarh), India. He obtained his M.Tech. from NIT Surathkal (formerly Karnataka Regional Engineering College) with a specialisation in Structural Engineering; graduated from Acharya Nagarjuna University, Guntur, Andhra Pradesh; and holds a Diploma in Civil Engineering from SBTET Hyderabad, India. He has over 25 years of teaching, research, and consultancy experience. He previously worked as a Professor at Bapatla Engineering College, Bapatla, and GITAM University, Visakhapatnam, Andhra Pradesh, India. His research areas include sustainable construction materials, microstructural characterisation of concrete, and static and impact testing of concrete structures. He has published more than 40 papers in national and international reputed journals and conferences. He has also authored a book: *Systematic Approach of Characterisation and Behaviour of Recycled Aggregate Concrete*.

1 Methods of Analysis

1.1 INTRODUCTION

Indeterminate structures are widely used due to their advantages over determinate structures. Most reinforced concrete buildings are considered indeterminate structures because of the monolithic construction of beams, columns, and slabs. Structures are generally classified as either statically determinate or statically indeterminate. In the former, structures can be analysed using only static equilibrium equations. In the latter, support reactions or internal forces cannot be determined solely using these equations because the number of unknowns exceeds the number of equilibrium equations. To solve for the unknown reactions and internal forces in indeterminate structures, compatibility conditions must be used in addition to the equilibrium equations. When analysing an indeterminate structure, it is necessary to satisfy:

1. **Equilibrium equations:** These ensure the structure is in equilibrium (i.e., the reactions keep the structure at rest).
2. **Compatibility conditions**: These ensure the structure remains continuous, without breaks or overlaps.
3. **Force–displacement equations**: These describe the relationship between forces and displacements.

Although the analysis of indeterminate structures is more time-consuming and complex compared to determinate structures, they are often chosen for design due to their numerous advantages. These benefits can be illustrated through the comparisons presented in Table 1.1.

In contrast to the merits, there are few demerits associated with indeterminate structures, particularly when the structures are subjected to differential settlements and temperature stresses. These are explained in Table 1.2.

In general, the following two methods are used for analysing indeterminate structures, depending on how the above three conditions are satisfied.

i. Force method
ii. Displacement method

TABLE 1.1
Merits of Indeterminate Structures in Comparison to Determinate Structures

	Determinate Structure	Indeterminate Structure
Deflection	The maximum deflection under a given load and span is considerably larger than that of the indeterminate structures. For example, a simply supported beam of span L with uniform flexural rigidity subjected to a uniformly distributed load of w/m run, the maximum deflection is $5wL^4/384EI$, as shown in Figure 1. 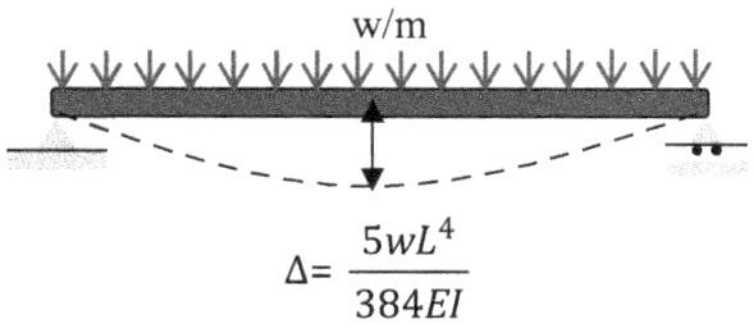	The maximum deflection under a given load and span is considerably smaller than that of the determinate structures. Consider a fixed beam of the same span, loading and flexural rigidity; the maximum deflection is $wL^4/384EI$. This is five times lower than that of the simply supported beam. 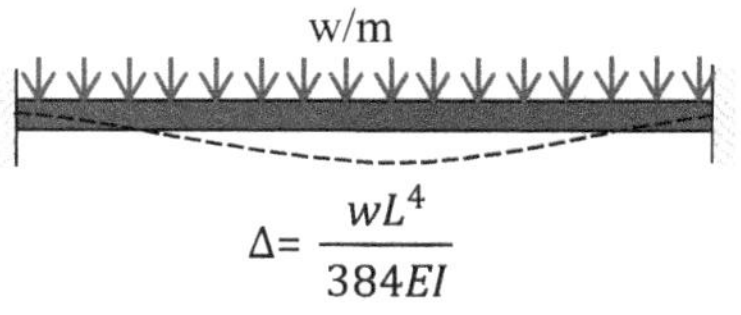
Stress	The maximum bending moment in the simply supported beam when subjected to a uniformly distributed load of w/m run is $wL^2/8$. 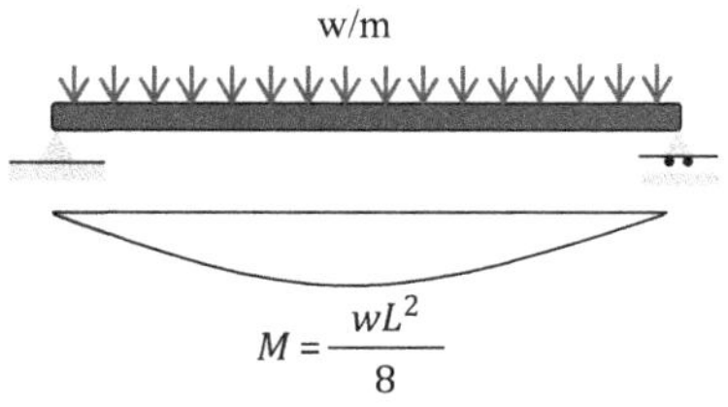	Whereas in the same beam with fixed supports, the maximum bending moment at the supports is $wL^2/12$, and at the centre, it is $wL^2/24$. As a result, the stresses are lower and the required cross-sectional dimensions and material are reduced. 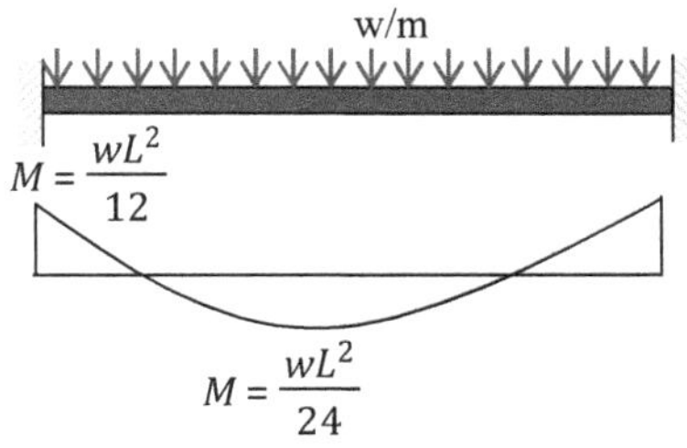

(Continued)

DOI: 10.1201/9781003562955-1

TABLE 1.1 (*Continued*)
Merits of Indeterminate Structures in Comparison to Determinate Structures

	Determinate Structure	Indeterminate Structure
Stability in case of overload	In the case of determinate structures, when subjected to overloading, the supports cannot develop the horizontal forces and moment reactions required to prevent collapse. These structures lack the ability to redistribute loads when overloaded. This behaviour can be illustrated using a simply supported beam subjected to a uniformly distributed load, as shown Figure 5. As the load increases, a plastic hinge forms at the centre of the beam, resulting in large deflections that ultimately lead to collapse, since the supports are unable to provide the necessary horizontal forces and moment reactions to maintain structural stability. 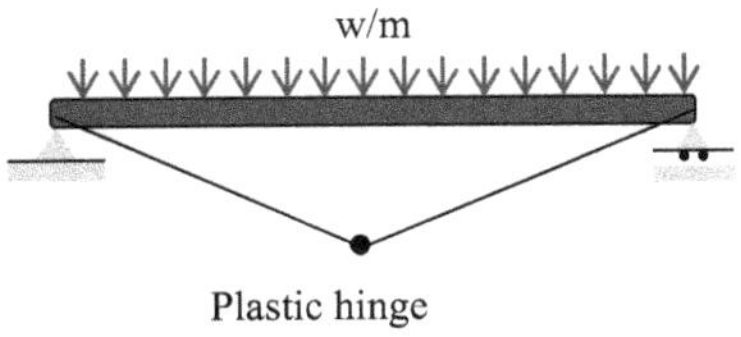	In the case of overloading or faulty design, the supports in indeterminate structures can develop horizontal forces and moment reactions that help stabilise the beam and prevent collapse. Indeterminate structures have the ability to redistribute loads to their redundant supports when subjected to overloading. This behaviour can be illustrated with the following example: Consider a fixed beam, as shown Figure 6, subjected to a uniformly distributed load. As the load increases, the material begins to yield. Initially, a plastic hinge forms at the supports, followed by another at the centre. Although this results in large deflections, the beam does not collapse because the supports generate the necessary horizontal forces and moment reactions to maintain structural stability. 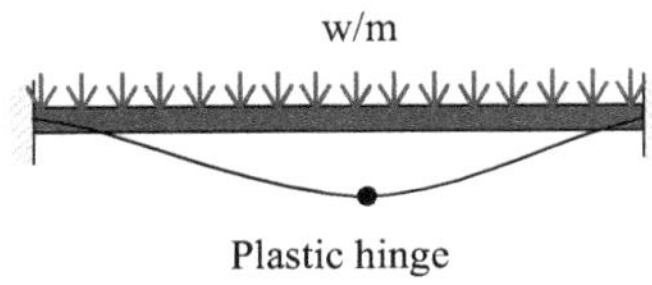

TABLE 1.2
Demerits of Indeterminate Structures in Comparison to Determinate Structures

	Determinate Structure	Indeterminate Structure
Yielding of supports	Support settlements or yielding of supports have no effect on the stresses in determinate structures. 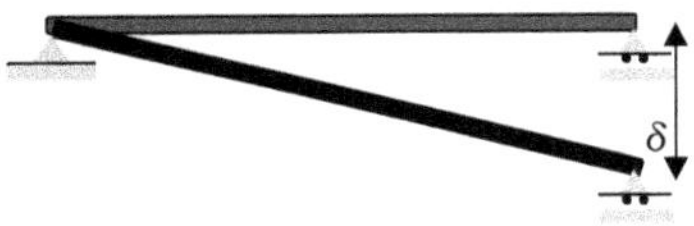	In contrast, for indeterminate structures, the yielding or differential settlement of supports induces additional stresses, which must be considered during the design process. For example, in a fixed beam subjected to support settlement, as shown in Figure 8, additional moments are induced at the supports, calculated as $M = 6EI\delta/L^2$, where E is the modulus of elasticity, I is the moment of inertia, δ is the settlement, and L is the length of the beam. 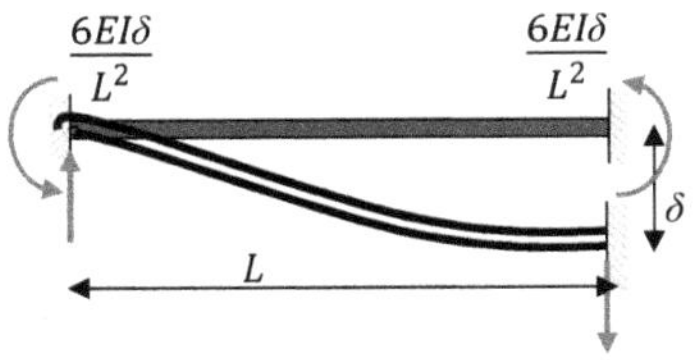
Temperatures, fabrication errors	No additional stresses due to temperatures or fabrication errors are developed in determinate structures. 	In case of indeterminate structures, additional stresses are induced due to temperature changes or fabrication errors, and these must be considered in the design of the structures. 

1.2 FORCE METHOD OF ANALYSIS

In the force method, the basic unknowns are the reactions or internal forces. First, the degree of static indeterminacy of the structure is determined, and the redundants (which may be support reactions or internal forces) are chosen. The basic structure, or primary structure, is obtained by releasing or removing the identified redundants, resulting in a statically determinate structure. Compatibility equations are then written using the force–displacement relationship for the applied loads and redundant forces. By solving these equations, the redundant forces are calculated. After determining the redundant forces, the remaining unknown reactions or forces can be calculated using simple static equilibrium equations. Since compatibility is the basis of this method, it is sometimes referred to as the compatibility method. This procedure is explained below with an example.

Consider a propped cantilever beam, as shown in Figure 1.1a. Calculate the degree of static indeterminacy, $I_s = 4 - 3 = 1$.

We can select any one reaction (M_A, R_A, R_B) as redundant. Let us choose R_A as the redundant and temporarily remove its effect from the original beam. The resulting beam becomes statically determinate and is referred to as the primary or basic beam (Figure 1.1b). As a result, the load W causes the beam to deflect downward, and point B deflects by an amount Δ_{BL} (Figure 1.1b). Now, on the primary beam, remove the external load W and apply the redundant reaction R_B at B (by the principle of superposition), which causes the beam to deflect upward, and point B deflects by an amount Δ_{BR} (Figure 1.1c). From Figure 1.1b and c, the compatibility condition can be written as,

$$\Delta_{BL} - \Delta_{BR} = 0 \tag{1.1}$$

Let us now apply a unit force in place of R_B on the primary beam and find the flexibility coefficient δ_{BB} (Figure 1.1d). Here, the first subscript indicates the location where the deflection is measured, and the second subscript indicates the unit load position. Therefore, δ_{BB} represents the deflection at B due to a unit load at B. Assuming the material is linearly elastic, the force and displacement are directly proportional; thus, we can write as

$$\Delta_{BR} = \delta_{BB} \cdot R_B \tag{1.1a}$$

∴ The compatibility equation 1.1 can be written as

$$\Delta_{BL} - R_B \times \delta_{BB} = 0$$

or

$$R_B = \frac{\Delta_{BL}}{\delta_{BB}} \tag{1.1b}$$

where Δ_{BL} and δ_{BB} can be calculated using the methods studied earlier in basic structural analysis. By solving equation 1.1b, the redundant reaction R_B can be determined. With the known value of R_B and using the static equilibrium equations, the remaining unknown reactions can be determined. This method is also sometimes referred to as the Method of Consistent Deformation or the Flexibility Method. Some of the force methods are

i. Consistent deformation method
ii. Energy method
iii. Column analogy method
iv. Flexibility matrix method
v. Theorem of three moments

These methods are discussed in detail with numerical examples in the subsequent chapters.

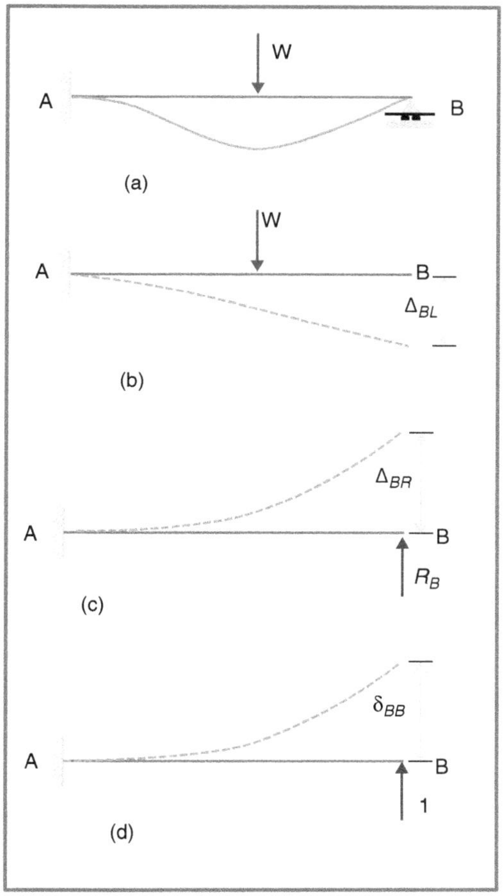

FIGURE 1.1 (a) Original beam, (b) primary beam, (c) primary beam with redundant reactions, (d) primary beam with unit load.

1.3 DISPLACEMENT METHOD OF ANALYSIS

The displacement method works in the opposite way to the force method. It first requires satisfying the equations of equilibrium for a given structure. In this method, the primary unknowns are the displacements. Therefore, the degree of kinematic indeterminacy must be determined first. The displacements are then expressed in terms of forces by writing the force–displacement relations, and the equations are subsequently formulated to satisfy the equilibrium requirements of the structure. These equations, with displacements as unknowns, are solved to determine the displacements. Once the displacements are calculated, the forces can be determined using the compatibility conditions and force–displacement relations. Because the condition of equilibrium is used to find the displacements, the method is also called the equilibrium method. It is sometimes also referred to as the stiffness matrix method. All displacement methods generally follow the same procedure. The detailed procedure adopted in each of the displacement methods is explained in the subsequent chapters.

Some of the displacement methods are

i. Slope deflection method
ii. Moment distribution method
iii. Kani's method
iv. Stiffness matrix method

1.4 EQUILIBRIUM EQUATIONS

It may be recalled from statics that if the structure is in equilibrium, every part of it must also be in equilibrium and must satisfy the equilibrium equations. In the case of three-dimensional space structures (with reference to the coordinate axes *X*, *Y*, *Z*), the following six equilibrium equations must be satisfied.

$$\sum F_X = 0, \quad \sum F_Y = 0, \quad \sum F_Z = 0$$
$$\sum M_X = 0, \quad \sum M_Y = 0, \quad \sum M_Z = 0 \tag{1.2}$$

In the case of two-dimensional (plane) structures, where the members and loads lie in the same plane, the structure must satisfy the following three equilibrium equations.

$$\sum F_X = 0, \quad \sum F_Y = 0, \quad \sum M_Z = 0 \tag{1.3}$$

1.5 DEGREE OF STATIC INDETERMINACY

Before analysing any structure using the force method, it is essential to determine its stability and degree of determinacy . The static equilibrium equations provide the necessary and sufficient conditions for a structure's equilibrium. As studied in basic structural analysis, a structure that can be analysed using static equilibrium equations alone is called a statically determinate structure. On the other hand, statically indeterminate structures, also referred to as hyperstatic structures, cannot be analysed using static equilibrium equations alone. In such structures, the number of unknowns exceeds the number of equilibrium equations. Therefore, additional equations based on displacement compatibility must be introduced to provide the sufficient number of equations needed to solve for all the unknowns. The number of extra equations required to solve the problem is known as the degree of static indeterminacy, which can be classified as either external or internal. Consequently, the total degree of static indeterminacy is the sum of both internal and external indeterminacies. If I_{se} is the degree of external indeterminacy and I_{si} is the degree of internal indeterminacy, then the total degree of indeterminacy is given by

$$I_s = I_{se} + I_{si} \tag{1.4}$$

The external indeterminacy of the structure is primarily related to its support system. A structure is said to be externally indeterminate if the number of support reactions exceeds the number of equilibrium equations. As discussed in Section 1.4, there are six equilibrium equations in the case of space structures (3-D) and three equations of equilibrium in the case of plane (2-D) structures, which must be satisfied for equilibrium. If *R* is the number of support reactions, the external degree of indeterminacy in the case of space structures (3-D) is

$$I_{se} = R - 6 \tag{1.5}$$

and in the case of coplanar structures (2-D), the external degree of indeterminacy is

$$I_{se} = R - 3 \tag{1.6}$$

The internal indeterminacy primarily depends on the arrangement of members. All beams are statically internally determinate. The degree of internal indeterminacy of rigid-jointed frames and pin-jointed frames is discussed separately in the subsequent sections.

1.5.1 Degree of Static Indeterminacy of Beams

The degree of static indeterminacy

$$I_s = I_{se} + I_{si} = I_{se} = R - 3$$

(OR)

In the case of coplanar structures, for each free body diagram, at most three equilibrium equations can be used. Therefore, for '*n*' bodies/parts and *R* total reactions, the degree of static indeterminacy is given by

$$I_s = R - 3n$$

Let us consider a few examples shown in Figure 1.2 to illustrate how to find the degree of static indeterminacy.

1.5.2 Degree of Static Indeterminacy of Rigid-Jointed Plane Frames

The rigid-jointed plane frames may be internally determinate or indeterminate. The internal indeterminacy mainly depends on how the members are interconnected in the frame. A rigid-jointed plane frame is said to be statically internally determinate if the members form an open configuration. This means that if you start from any point in the structure and move along any path without retracing your steps, you cannot return to the initial or starting point. In other words, the structure does not have any closed loops or cells. If the structure does not have an open configuration, it is statically indeterminate, meaning the structure has closed loops or cells. By making a sufficient number of cuts, the structure can be converted into an open configuration, i.e., the structure can be made determinate. At each

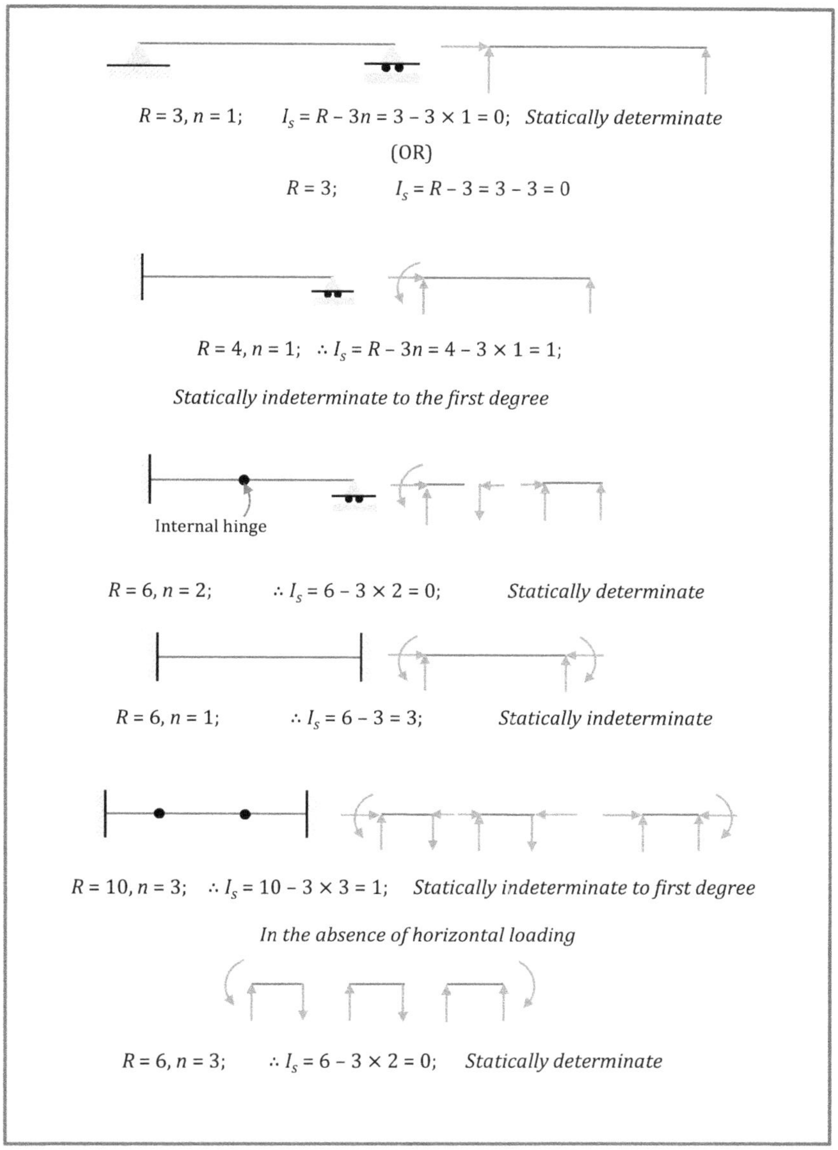

FIGURE 1.2 Typical beams.

(Continued)

cut, there will be three and six internal forces in the case of 2-D and 3-D structures, respectively, and let the degree of internal indeterminacy be I_{si}, then,

$$I_{si} = 3 \times c \quad \text{for 2-D structures}$$

$$I_{si} = 6 \times c \quad \text{for 3-D structures}$$

where '*c*' is the number of cuts or openings required to make the structure an open configuration. As discussed previously, the external indeterminacy is mainly related to the supports and their arrangement. In the case of rigid-jointed plane frames, three equilibrium equations apply, and in the case of space frames, six equilibrium equations apply. If '*R*' is the total number of reactions, then the degree of external indeterminacy is

$$I_{se} = R - 3 \quad \text{for plane frames}$$

$$I_{se} = R - 6 \quad \text{for space frames}$$

The total degree of static indeterminacy is the sum of both internal and external degrees of indeterminacy. The following examples in Figure 1.3 show how the degree of indeterminacy is calculated.

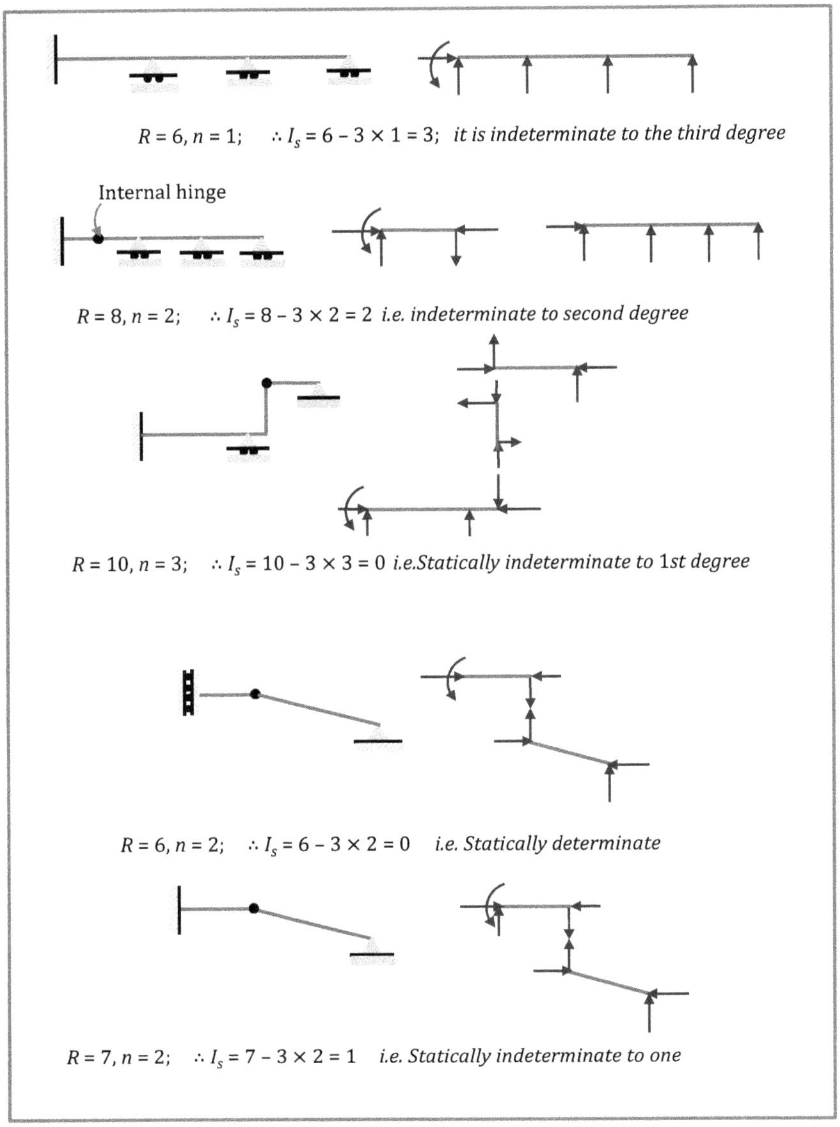

FIGURE 1.2 (*Continued*) Typical beams.

1.5.3 Degree of Static Indeterminacy of Pin-Jointed Frames

In the case of pin-jointed plane frames, each member has one unknown member force, and at each joint, two equilibrium equations are available. If there are R unknown reactions, then the static indeterminacy of the pin-jointed plane frame is given by

$$I_s = (m + R) - 2j$$

where

m is the number of members,
R is the total number of unknown reactions, and
j is the total number of joints.

In the case of pin-jointed space frames, at each joint, three equilibrium equations are available, then the static indeterminacy is given by

$$I_s = (m + R) - 3j$$

Determination of indeterminacy of pin-jointed plane frames is illustrated in the following examples shown in Figure 1.4.

1.6 STABILITY AND UNSTABILITY OF STRUCTURES

Every structural engineer must ensure that, before analysing any structure, it not only satisfies the equilibrium equations but also that the structure and its members are

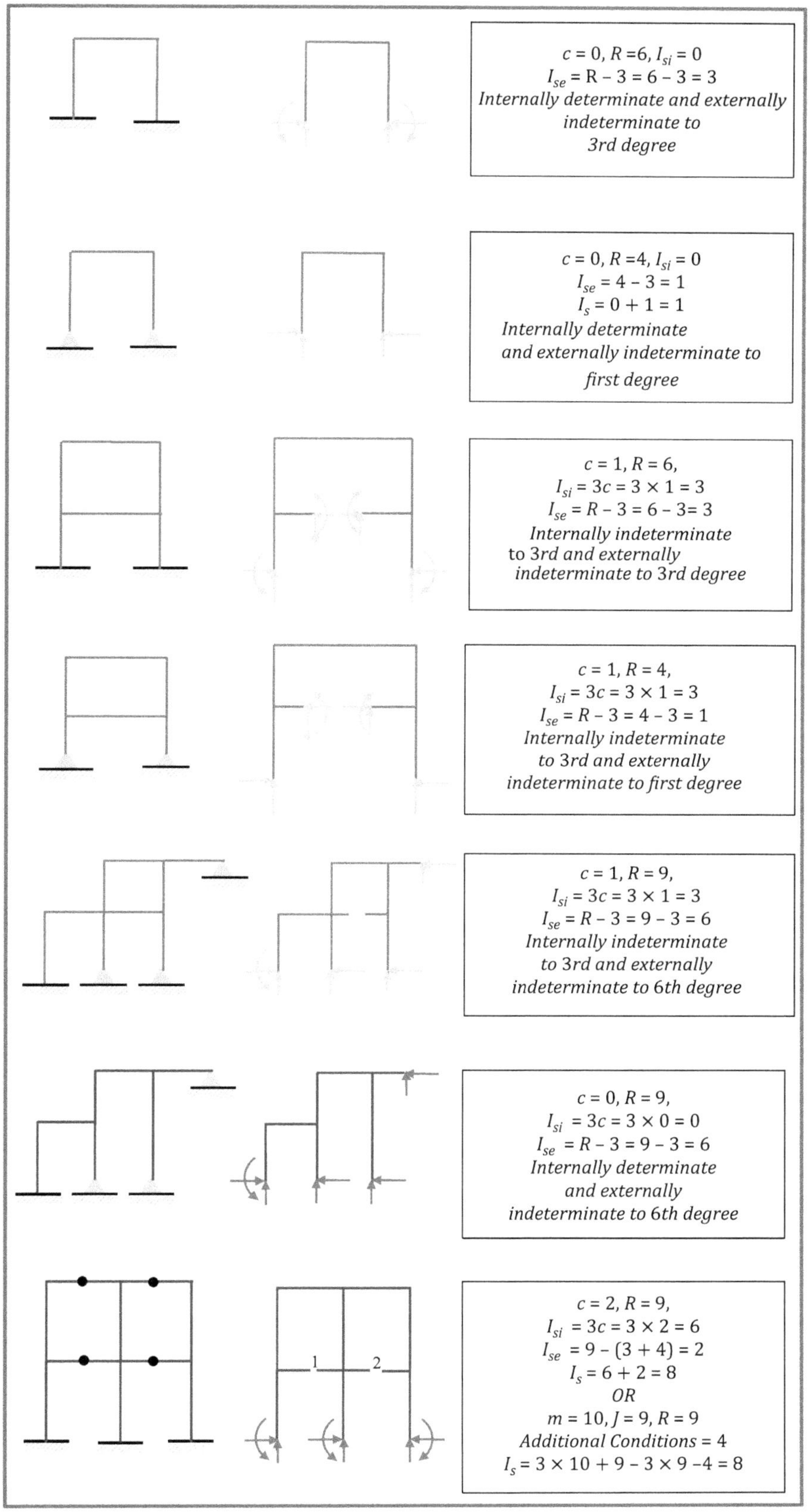

FIGURE 1.3 Typical rigid-jointed plane frames.

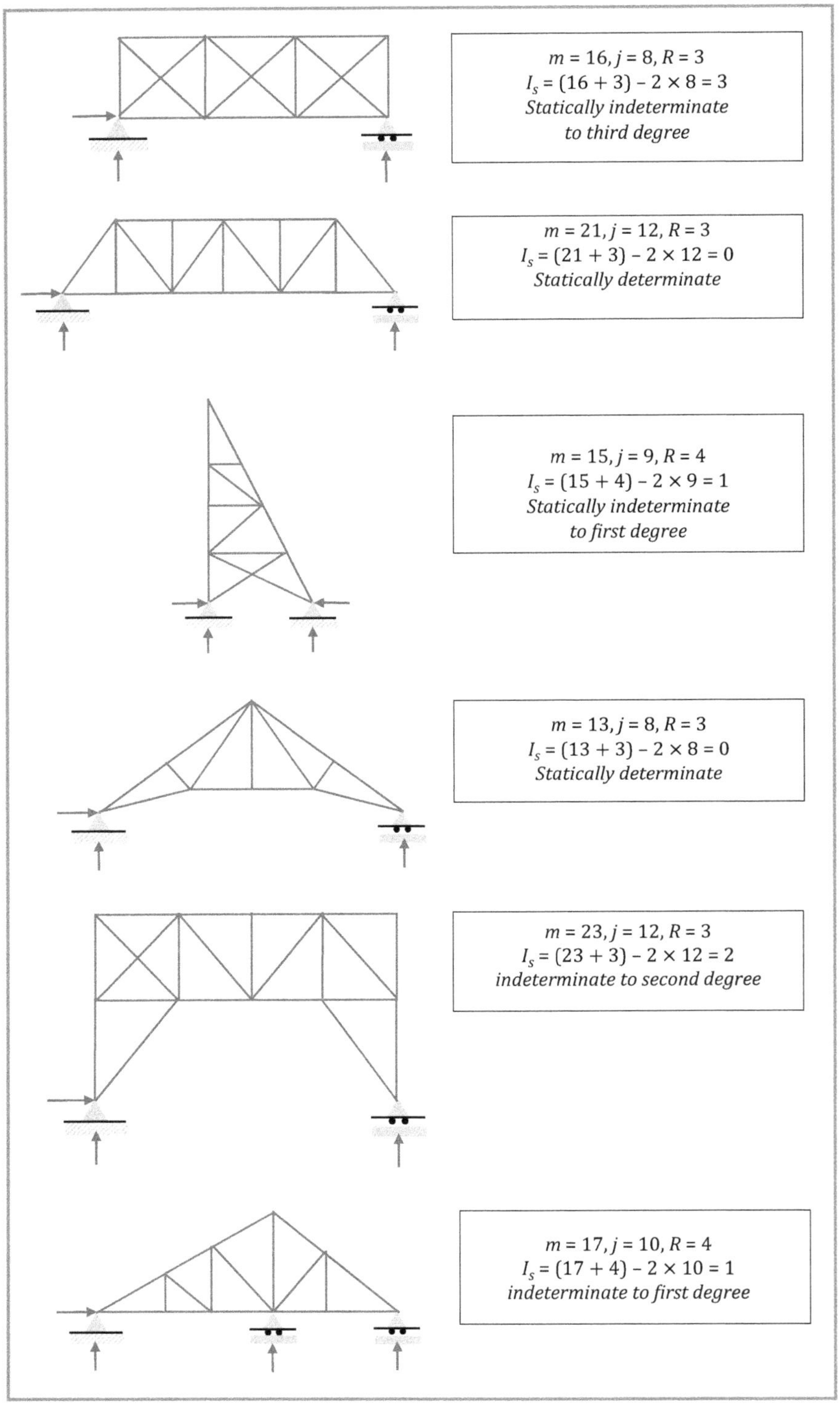

FIGURE 1.4 Typical pin-jointed plane frames.

properly supported to ensure stability against all kinds of loads. The stability can be either external or internal. As discussed previously, external stability is primarily due to the number of supports and their arrangement. If the structure or its members do not have sufficient supports – i.e., if the reactions are less than the number of equations of equilibrium – the structure is externally unstable. This indicates that the structure is partially constrained.

Consider the following example, Figure 1.5; in this case, $\sum F_Y = 0$ and $\sum M = 0$ are satisfied, but $\sum F_X = 0$ is not satisfied for the given loading conditions, and hence, the member is unstable.

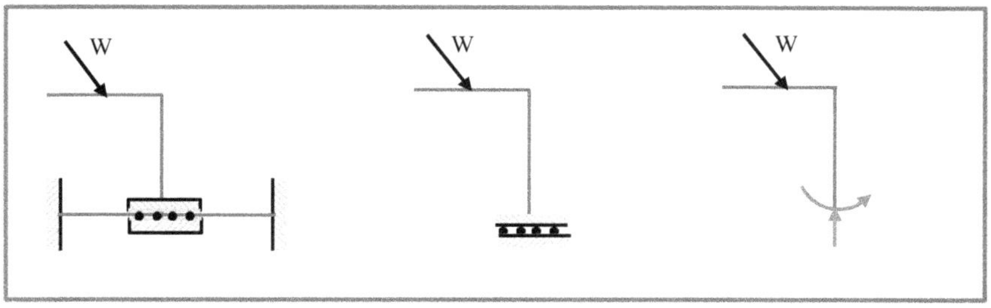

FIGURE 1.5 Unstable structure.

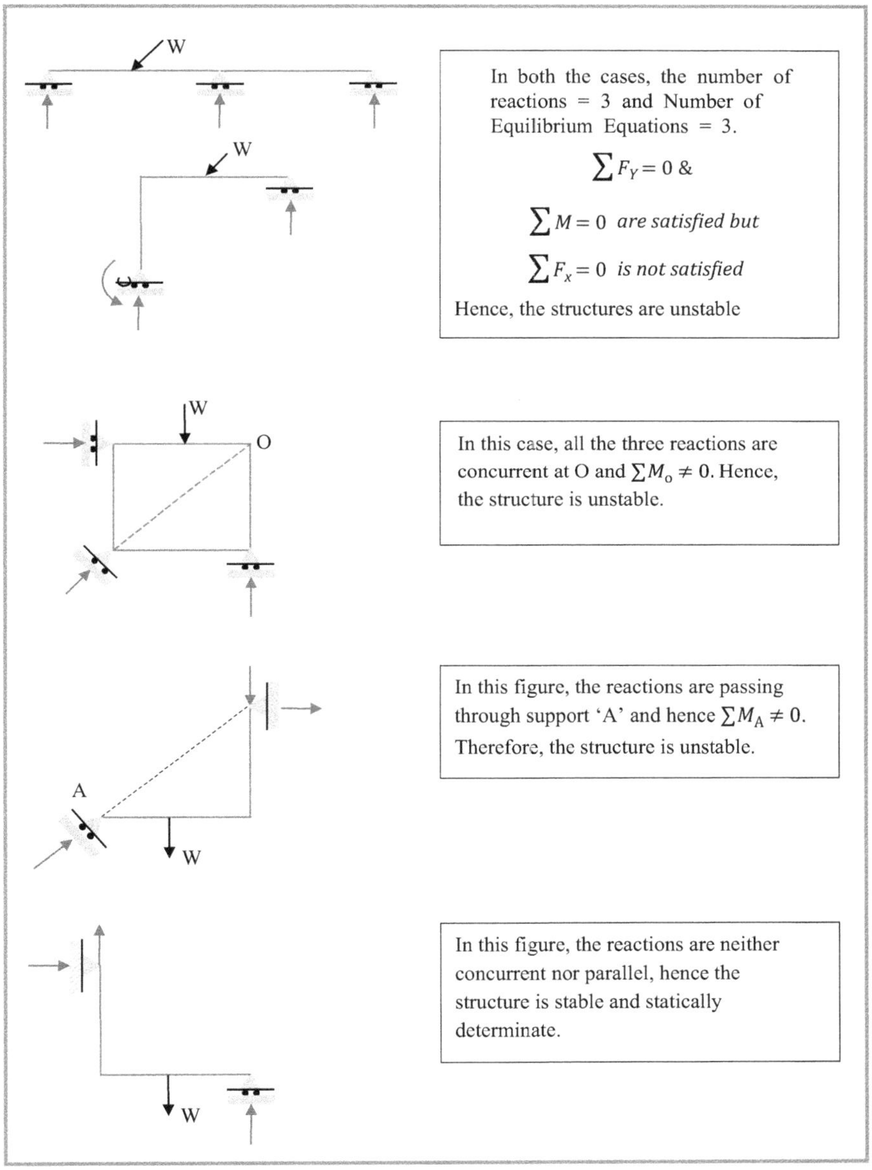

FIGURE 1.6 Typical beams and frames.

In some situations, where the structure is supported by a sufficient number of supports – i.e., the number of reactions is equal to the number of equilibrium equations – but the supports are not properly constrained or arranged, the structure is unstable due to improper constraints. This situation arises in the following two cases:

i. If the support reactions are parallel.
ii. If the support reactions are concurrent.

These are demonstrated in the following examples, shown in Figure 1.6.

A structure is internally stable if it can maintain its geometry when subjected to external forces that tend to deform it. On the other hand, if the members form a mechanism – i.e., if the geometry changes significantly without generating appreciable internal member forces – then the structure is called internally unstable. This can be observed locally through inspection. In general, the overall stability of a structure, without distinguishing between internal and external instabilities, can be assessed using criteria based on the member components and support reactions.

Let R be the number of reactions, 'm' be the number of members, and 'j' be the number of joints; then for rigid-jointed plane frames,

If $(3m+R)<3j$ $\rightarrow$ Unstable
If $(3m+R)=3j$ $\rightarrow$ Stable and determinate
If $(3m+R)>3j$ $\rightarrow$ Stable and statically indeterminate

For rigid-jointed space frames,

If $(6m+R)<6j$ $\rightarrow$ Unstable
If $(6m+R)=6j$ $\rightarrow$ Stable and determinate
If $(6m+R)>6j$ $\rightarrow$ Stable and indeterminate

Pin-jointed plane frames,

If $(m+R)<2j$ $\rightarrow$ Unstable
If $(m+R)=2j$ $\rightarrow$ Stable and determinate
If $(m+R)>2j$ $\rightarrow$ Stable and indeterminate

For pin-jointed space frames,

If $(m+R)<3j$ $\rightarrow$ Unstable
If $(m+R)=3j$ $\rightarrow$ Stable and statically determinate
If $(m+R)>3j$ $\rightarrow$ Stable and statically indeterminate

The examples in Figure 1.7 demonstrates how the trusses becomes unstable.

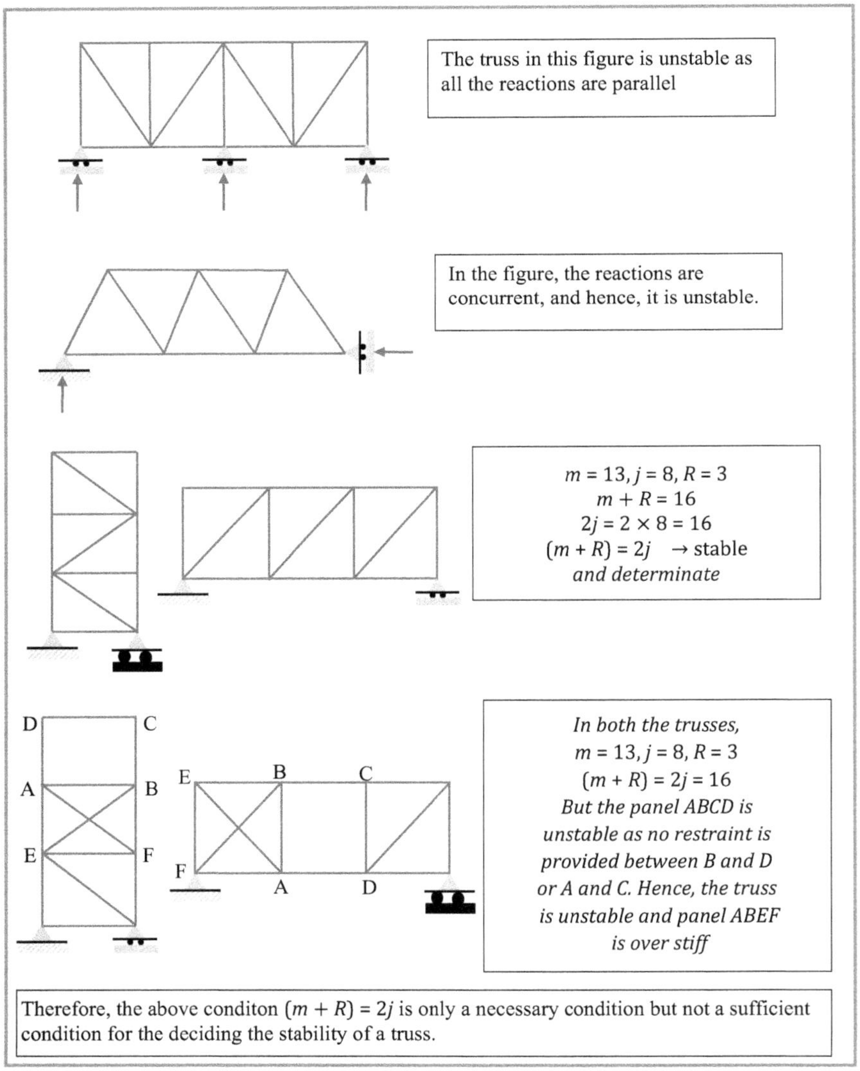

FIGURE 1.7 Typical trusses in an unstable condition.

1.7 KINEMATIC INDETERMINACY OF STRUCTURES

When a structure is subjected to loads, the joints experience displacements (translations and rotations), and these displacements are called the degrees of freedom (DOF). If these displacement components cannot be determined using compatibility conditions alone, the structure is said to be kinematically indeterminate. In such structures, additional equations based on equilibrium must be considered to determine all unknown joint displacements. The number of these additional equations required to find the unknown joint displacements is called the degree of kinematic indeterminacy. Therefore, the degree of kinematic indeterminacy, or the number of joint DOF, is simply the number of independent joint displacements (translations and rotations) in a structure.

1.7.1 Rigid-Jointed Structures

In general, in the case of rigid-jointed plane frames or beams, each joint has three independent displacements (two linear and one rotational), while for rigid-jointed space frames, there are six independent displacements (three linear and three rotational). Furthermore, the nodal degrees of freedom may be restricted based on the assumptions considered in the behaviour of the structure. For example, in the case of beams, the deformations due to axial forces are generally neglected, and only deformations due to bending are considered. In this case, the degrees of freedom at each joint are reduced to two: one rotation and one transverse displacement. Additionally, the displacements at the nodes may be restricted by the supports. If j is the number of joints, and k is the number of constraints imposed by the supports, then the degree of kinematic indeterminacy is given by

$$I_K = 2j - k \quad \rightarrow \text{For beams}$$

$$I_K = 3j - k \quad \rightarrow \text{For plane frames}$$

$$I_K = 6j - k \quad \rightarrow \text{For space frames}$$

Consider the rigid-jointed plane frame, shown in Figure 1.3, in which j=4, k=6 (constraints imposed by supports, i.e., each reaction imposes one constraint), then, $I_K = 3\times 4 - 6 = 6$.

In general, the members are assumed to be inextensible in the analysis of rigid-jointed frames. This assumption imposes additional conditions of compatibility, i.e., constraints on the joint displacements. In Figure 1.8, the inextensibility of the beam provides one condition of compatibility, which results in the horizontal displacements of joints B and C being the same $(\Delta_1 = \Delta_4)$. Similarly, the inextensibility of the columns provides two additional conditions of compatibility, i.e., the column joints B and C have zero vertical displacement $(\Delta_2 = \Delta_5 = 0)$. Therefore, the assumption of inextensibility provides a total of three additional conditions of compatibility (Figure 1.9). Hence,

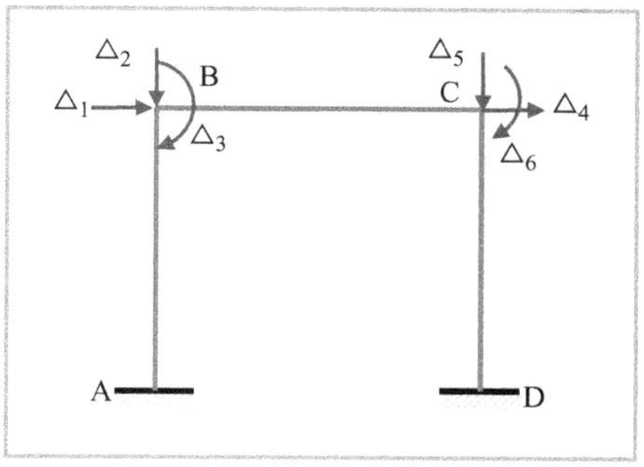

FIGURE 1.8 Rigid-jointed plane frame.

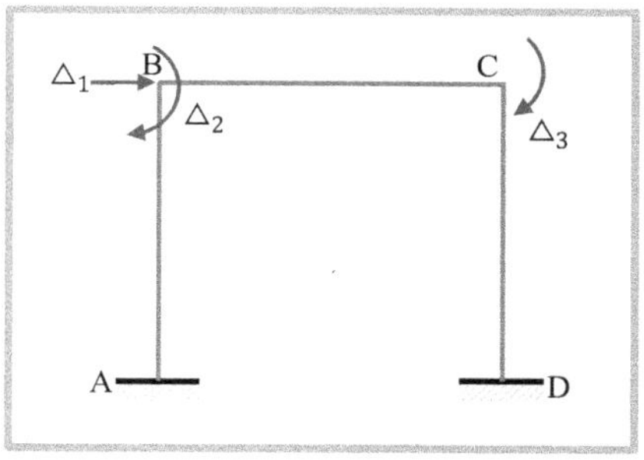

FIGURE 1.9 Rigid-jointed plane frame with inextensibility of members.

the compatibility conditions due to the inextensibility of the members are m=3.

The degree of kinematic indeterminacy, $I_K = 3j - (k+m)$.

$$I_K = 3\times 4 - (6+3) = 3$$

In case of space frames, if 'm' is the additional conditions of compatibility, the degree of kinematic indeterminacy is $I_K = 6j - (k+m)$.

It is clear from the above that the number of additional conditions of compatibility due to the assumption of the inextensibility of members is equal to the number of members (m). This holds true only for unbraced building frames. In general, for a rigid-jointed plane frame with B number of bays and 'S' number of storeys, and fixed column bases, the total number of joints is equal to $S(B+1)$, excluding the supports. We know that each joint has three DOF; therefore, the degree of kinematic indeterminacy is

$$I_K = S(B+1)\times 3 \quad \rightarrow \text{If members are extensible}$$

If the members are inextensible, each joint has only one rotational degree of freedom, and each floor has an independent horizontal displacement. Then, the degree of kinematic indeterminacy is

$$I_K = S(B+1)\times 1 + S = S(B+2)$$

If the column bases are not fixed ends, the DOF will increase accordingly.

Examples in Figure 1.10 demonstrate the calculation of the degree of kinematic indeterminacy.

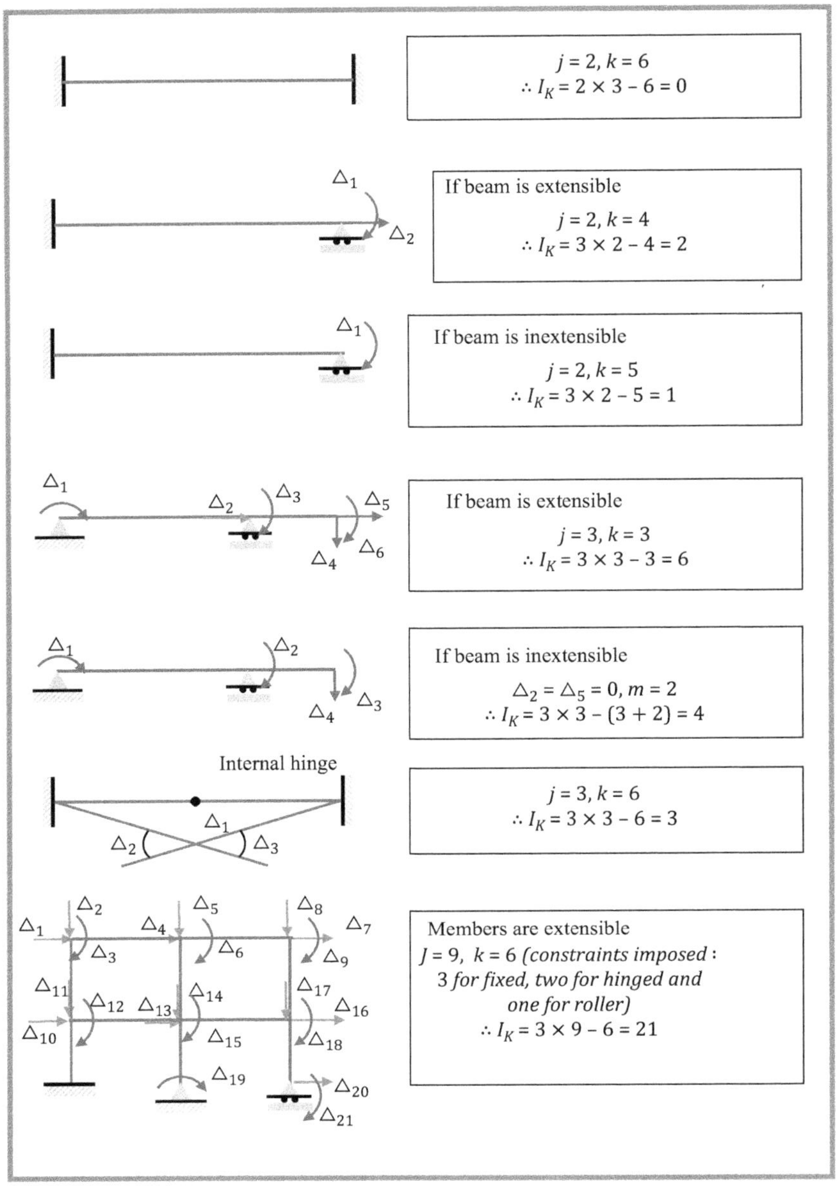

FIGURE 1.10 Degree of kinematic indeterminacy of rigid-jointed beams and frames.

(Continued)

1.7.2 Pin-Jointed Plane Frames (Trusses)

In the case of pin-jointed plane frames, each joint can move in both orthogonal directions (*X* and *Y*), and hence, each joint has two independent displacement (linear) components. Similarly, in the case of space frames, each joint has three independent displacement components (*X*, *Y*, *Z*). Therefore, if there are '*j*' joints, the number of unknown displacements equals to 2*j* for plane frames and 3*j* for space frames. If '*k*' is the number of constraints imposed by the supports, then the degree of kinematic indeterminacy is given by

$$I_K = 2j - k \quad \rightarrow \text{For plane frames}$$

$$I_K = 3j - k \quad \rightarrow \text{For space frames}$$

However, if the number of constraints imposed by the supports is equal to the number of support reactions (*R*), the above equation can be simply written as

$$I_K = 2j - R \quad \rightarrow \text{For plane frames}$$

$$I_K = 3j - R \quad \rightarrow \text{For space frames}$$

The kinematic indeterminacy of some typical trusses is shown in Figure 1.11.

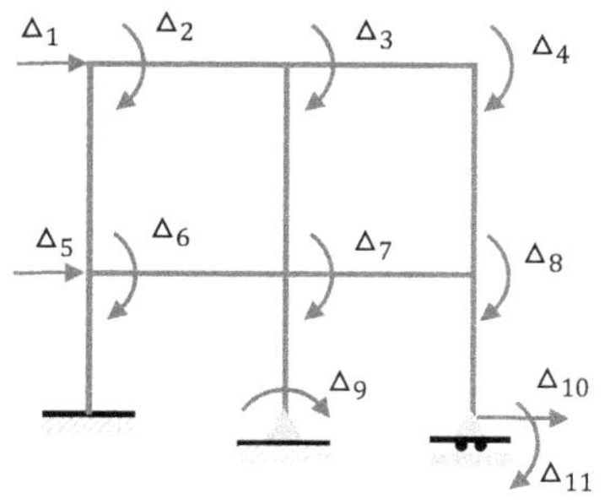

If members are inextensible,

In the previous figure, $\Delta_1 = \Delta_4 = \Delta_7$

$\Delta_{10} = \Delta_{13} = \Delta_{16}$

$\Delta_2 = \Delta_5 = \Delta_8 = \Delta_{11} = \Delta_{14} = \Delta_{17} = 0$

$\therefore I_K = 3 \times 9 - (6+10) = 11$

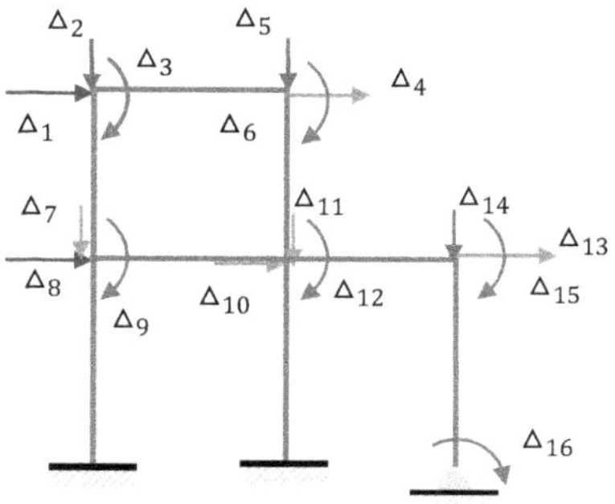

If members are extensible,

$j = 8, k = 8$

$\therefore I_K = 3 \times 8 - 8 = 16$

if members are inextensible, m = 8

$\Delta_1 = \Delta_4; \ \Delta_8 = \Delta_{10} = \Delta_{13}$

$\Delta_2 = \Delta_5 = \Delta_7 = \Delta_{11} = \Delta_{14} = 0$

$\therefore I_K = 3 \times 8 - (8+8) = 8$

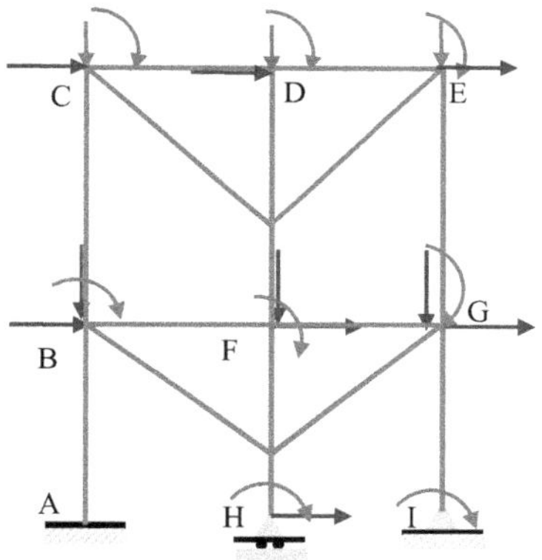

if members are extensible

$j = 9, k = 6$

$\therefore I_K = 3 \times 9 - 6 = 21$

if the members are inextensible both horizontal and vertical displacements of the members are zero. only the rotational displacements are developed at the joints.

$\therefore I_K = 9 - 1 = 8$

FIGURE 1.10 (*Continued*) Degree of kinematic indeterminacy of rigid-jointed beams and frames.

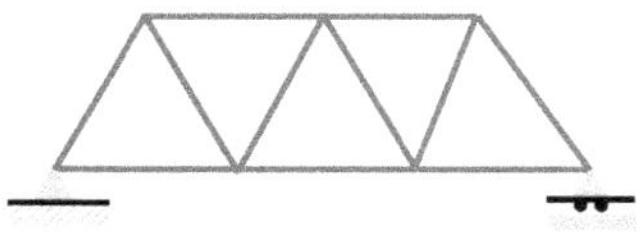

Here, $j = 7, R = 3$

$\therefore I_K = 2 \times 7 - 3 = 11$

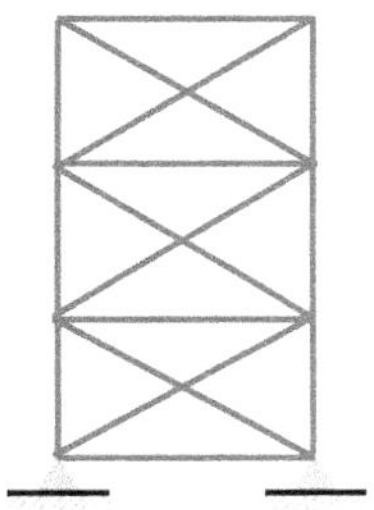

$j = 8, \quad R = 4$

$\therefore I_K = 2 \times 8 - 4 = 12$

FIGURE 1.11 Degree of kinematic indeterminacy of pin-jointed plane frames.

1.8 PROBLEMS FOR PRACTICE

1.8.1 Identify each of the structures shown in Figure 1.12 as stable, unstable, determinate, or indeterminate. If indeterminate, find the degree of static indeterminacy.

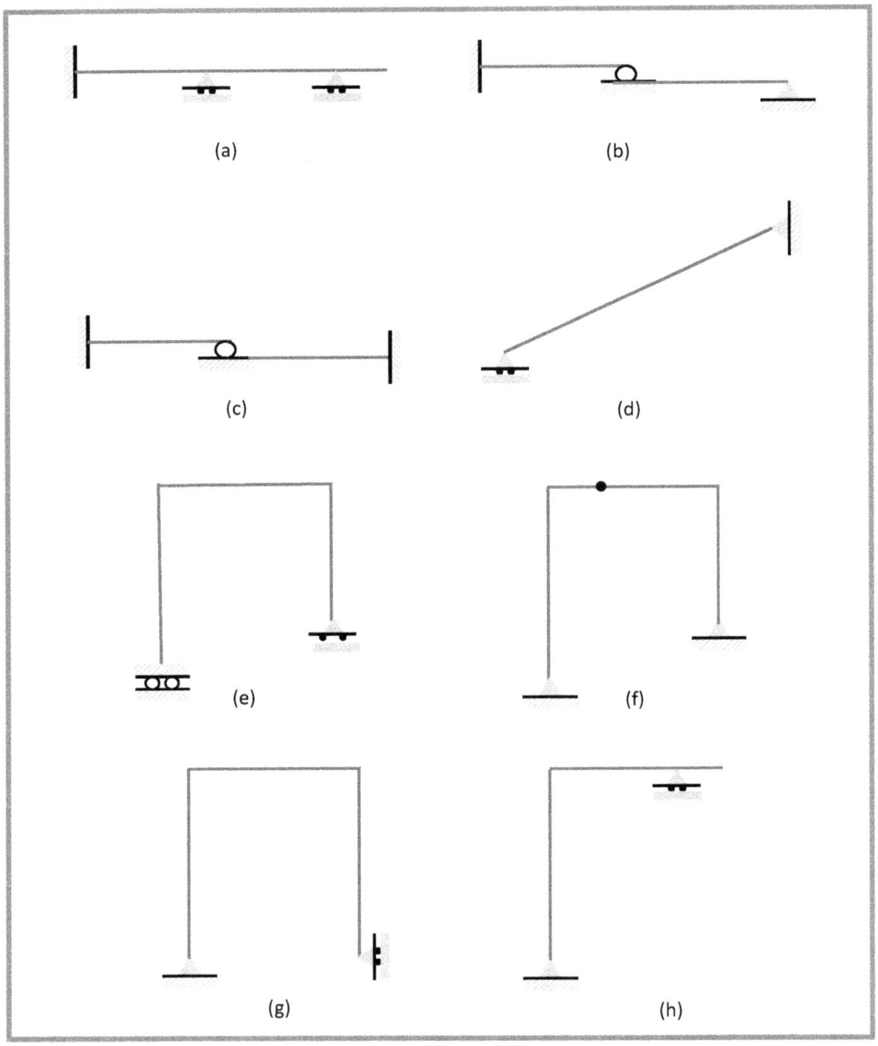

FIGURE 1.12 Typical beams and rigid-jointed plane frames.

(Continued)

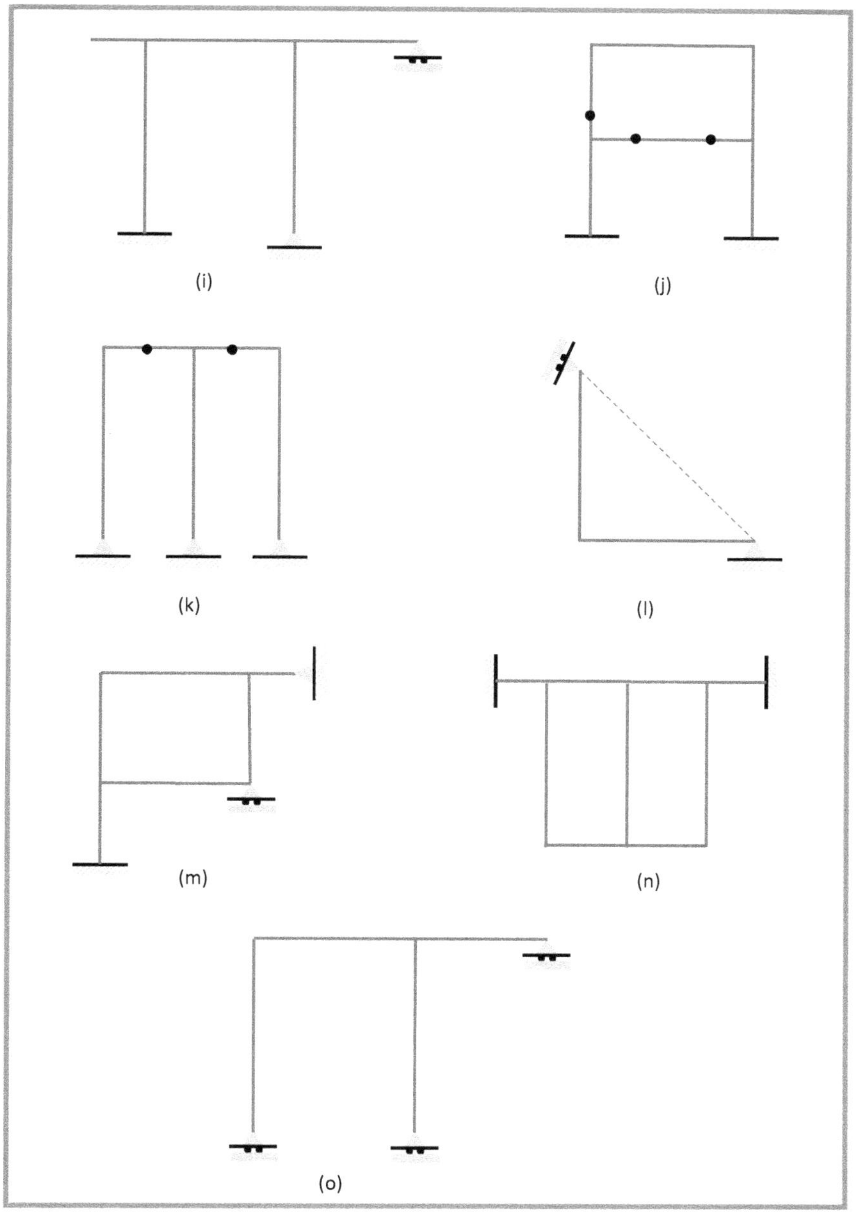

FIGURE 1.12 (*Continued*) Typical beams and rigid-jointed plane frames.

1.8.2 Classify each of the pin-jointed frames shown in Figure 1.13 as stable, unstable, determinate, or indeterminate. If indeterminate, find the degree of static indeterminacy.

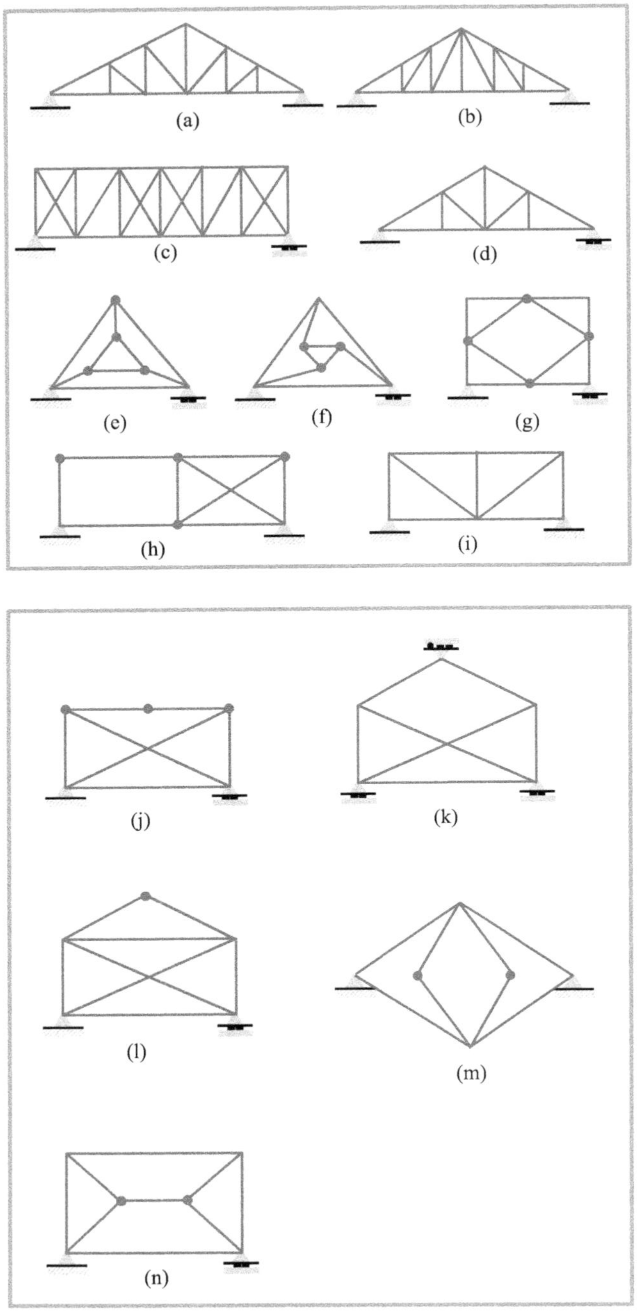

FIGURE 1.13 Typical pin-jointed plane frames for static indeterminacy.

1.8.3 Find the kinematic indeterminacy of the rigid-jointed frames shown in Figure 1.14. Assume members are inextensible.

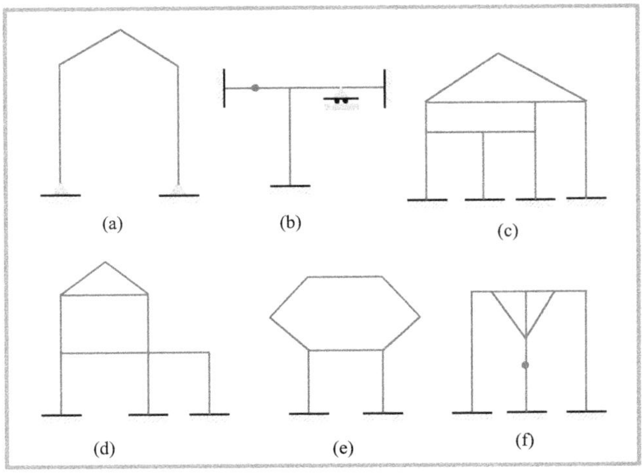

FIGURE 1.14 Typical rigid-jointed plane frames.

1.8.4 Find the static indeterminacy of the pin-jointed plane frames shown in Figure 1.15.

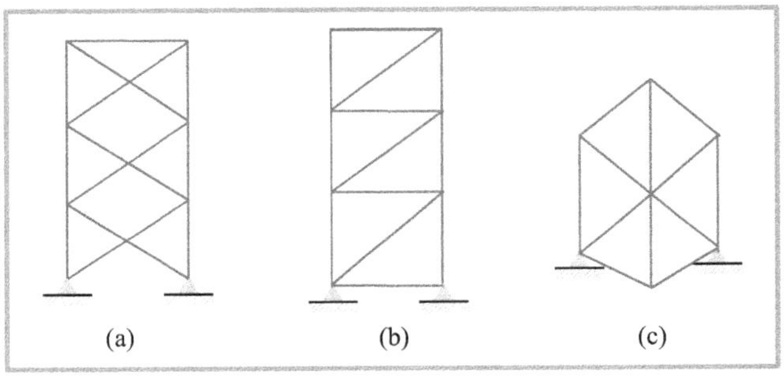

FIGURE 1.15 Typical pin-jointed plane frames.

1.8.5 For Figure 1.13, Q. No. 1.8.2, identify the trusses that are stable, and if stable, determine the kinematic indeterminacy.

1.9 ANSWER KEY

Q.No.	Answer	Q.No.	Answer
1.8.1(a)	Indeterminate, $I_s=2$	1.8.2(a)	Externally indeterminate, $I_s=1$
1.8.1(b)	Unstable	1.8.2(b)	Externally indeterminate, $I_s=1$
1.8.1(c)	Indeterminate, $I_s=1$	1.8.2(c)	Internally indeterminate, $I_s=4$
1.8.1(d)	Determinate	1.8.2(d)	Determinate
1.8.1(e)	Unstable	1.8.2(e)	Unstable
1.8.1(f)	Determinate	1.8.2(f)	Stable and determinate
1.8.1(g)	Unstable	1.8.2(g)	Unstable
1.8.1(h)	Determinate	1.8.2(h)	Unstable
1.8.1(i)	Indeterminate, $I_s=3$	1.8.2(i)	Stable and externally indeterminate, $I_s=1$
1.8.1(j)	Indeterminate, $I_s=3$	1.8.2(j)	Unstable
1.8.1(k)	Indeterminate, $I_s=1$	1.8.2(k)	Unstable
1.8.1(l)	Unstable	1.8.2(l)	Internally indeterminate, $I_s=1$
1.8.1(m)	Indeterminate, $I_s=6$	1.8.2(m)	Unstable
1.8.1(n)	Indeterminate, $I_s=9$	1.8.2(n)	Unstable
1.8.1(o)	Unstable	1.8.4(a)	Externally indeterminate, $I_s=1$
1.8.3(a)	$I_K=11$ (general); $I_K=8$ (members inextensible)	1.8.4(b)	Externally indeterminate, $I_s=1$
1.8.3(b)	$I_K=9$ (general); $I_K=5$ (members inextensible)	1.8.4(c)	Externally and internally indeterminate to first degree, $I_s=2$
1.8.3(c)	$I_K=21$ (general); $I_K=10$ (Members inextensible)	1.8.5(a)	$I_K=20$
1.8.3(d)	$I_K=18$ (general); $I_K=9$ (Members inextensible)	1.8.5(b)	$I_K=20$
1.8.3(e)	$I_K=18$ (general); $I_K=10$ (members inextensible)	1.8.5(c)	$I_K=25$
1.8.3(f)	$I_K=22$ (general); $I_K=11$ (members inextensible)	1.8.5(d)	$I_K=13$
		1.8.5(f)	$I_K=11$
		1.8.5(i)	$I_K=8$
		1.8.5(l)	$I_K=7$

FURTHER READING

Abdul Salam, S. *Structural Analysis and Mechanics*, Zagazig University, Ash Sharqiyah, Egypt.

Derucher, K. Putcha, C., and Kim, U. (2013). *Indeterminate Structural Analysis*, The Edwin Mellen Press, New York, ISBN 978-0773444706.

Hibbeler, R. C. (2022). *Structural Analysis*, Pearson Education (Singapore) Pte. Ltd, Delhi, ISBN 978-9354497841.

Khalfallah, S. (2018). *Structural Analysis 2: Statically Indeterminate Structures*, John Wiley & Sons, Inc., Hoboken, NJ, ISBN 978-1786303394.

Pandit, G. S. and Gupta, S. P. (2008). *Structural Analysis: A Matrix Approach*, McGraw Hill Education, New Delhi, ISBN 978-0070667358.

Sahoo, D. R. (2023). *Indeterminate Structural Analysis*, PHI Learning Pvt Ltd, ISBN 9789354432958.

Shah, H. J and Junarkar, S. B. (2015). *Mechanics of Structures*, Vol. II, (24 E) Charotar Publishing House, Anand, Gujarat, ISBN 978-9385039027.

Vijayanand, M., Muthu, K. U., Narendra, H., and Janardhana, M. (2019). *Indeterminate Structural Analysis*, Dreamtech Press Publisher, Delhi, ISBN 978-9389307177.

Williams, A. (2009). *Structural Analysis in Theory and Practices*, Elsevier.

2 Three-Moment Theorem

2.1 INTRODUCTION

The theorem of three moments, or three-moment equation, was developed in 1857 by the French engineer Clapeyron. This equation establishes the relationship between the loads acting on the spans and the internal moments at three adjacent supports in a continuous beam. It can be applied to two adjacent spans at a time, and successively to the remaining spans of the beam, generating a set of simultaneous equations. If there are 'n' number of spans in a continuous beam, $(n-1)$ equations can be generated. By solving these simultaneous equations, the unknown internal support moments can be determined. Since the equation involves three moments, it is referred to as the three-moment equation. Clapeyron first derived the equation for continuous beams without yielding of supports and later extended it to beams with yielding supports.

2.2 DERIVATION OF CLAPEYRON'S THEOREM OR THREE-MOMENT EQUATION (WITHOUT YIELDING OF SUPPORTS)

Consider a continuous beam shown in Figure 2.1a. Since the three-moment equation relates the moments at three successive supports to the applied loading on adjacent spans, let us consider two adjacent spans of a continuous beam as shown in Figure 2.1b. Denote the supports as left 'L', right 'R', and centre 'C', with the corresponding end moments denoted as M_L, M_R, and M_C, respectively (assumed to act in the positive direction, i.e., as sagging moments). Let I_L, L_L, and I_R, L_R represent the second moment of area and span lengths for the left and right sides, respectively. At support C, the tangent to the elastic curve is horizontal, i.e., the slope is zero, as the beam is continuous at C. In other words, support C is considered rigid. Therefore, the compatibility equation can be written as

$$\theta_{CL} + \theta_{CR} = 0 \tag{2.1}$$

where θ_{CL} and θ_{CR} are the slopes at the left and right of support C.

The slopes may be calculated using any of the methods studied earlier. Let us use the moment–area theorem.

$$\theta_{CL} = \frac{\text{The tangential devaition of } C \text{ from } L}{L_L} = \frac{t_{CL}}{L_L}$$

$$\theta_{CL} = \frac{\text{Moment of area of } \dfrac{M}{EI} \text{ diagram between } C \text{ and } L \text{ about } L}{L_L}$$

$$\theta_{CL} = \frac{1}{L_L}\left[\frac{A_L x_L}{EI_L} + \frac{1}{2}L_L\frac{M_L}{EI_L}\times\frac{L_L}{3} + \frac{1}{2}L_L\frac{M_C}{EI_L}\times\frac{2L_L}{3}\right]$$

$$\theta_{CL} = \frac{A_L x_L}{EI_L L_L} + \frac{M_L L_L}{6EI_L} + \frac{M_C L_L}{3EI_L} \tag{2.2}$$

$$\text{Similarly, } \theta_{CR} = \frac{\text{The tangential devaition of } C \text{ from } R}{L_R} = \frac{t_{CR}}{L_R}$$

$$\theta_{CR} = \frac{\text{Moment of area of } \dfrac{M}{EI} \text{ diagram between } C \text{ and } R \text{ about } R}{L_R}$$

$$\theta_{CR} = \frac{1}{L_R}\left[\frac{A_R x_R}{EI_R} + \frac{1}{2}L_R\frac{M_R}{EI_R}\times\frac{L_R}{3} + \frac{1}{2}L_R\frac{M_C}{EI_R}\times\frac{2L_R}{3}\right]$$

$$\theta_{CR} = \frac{A_R x_R}{EI_R L_R} + \frac{M_R L_R}{6EI_R} + \frac{M_C L_R}{3EI_R} \tag{2.3}$$

where

A_L and A_R are the areas of bending moment diagrams due to external loads on the left and right side spans of simply supported ends.

x_L and x_R are the centroidal distances of A_L and A_R from the left and right side supports, respectively.

By substituting equations 2.2 and 2.3 into equation 2.1, we obtain

$$\frac{A_L x_L}{EI_L L_L} + \frac{M_L L_L}{6EI_L} + \frac{M_C L_L}{3EI_L} + \frac{A_R x_R}{EI_R L_R} + \frac{M_R L_R}{6EI_R} + \frac{M_C L_R}{3EI_R} = 0$$

$$\Rightarrow M_L\left(\frac{L_L}{I_L}\right) + 2M_C\left(\frac{L_L}{I_L} + \frac{L_R}{I_R}\right) + M_R\left(\frac{L_R}{I_R}\right) = -\left[\frac{6A_L x_L}{I_L L_L} + \frac{6A_R x_R}{I_R L_R}\right] \tag{2.4}$$

In general equation 2.4 can be written as

$$M_L\left(\frac{L_L}{I_L}\right) + 2M_C\left(\frac{L_L}{I_L} + \frac{L_R}{I_R}\right) + M_R\left(\frac{L_R}{I_R}\right) = -\left[\sum\frac{6A_L x_L}{I_L L_L} + \sum\frac{6A_R x_R}{I_R L_R}\right] \tag{2.5}$$

Equations 2.4 and 2.5 are called the three-moment equations. They show the relationship between the three moments M_L, M_C, and M_R and the applied loads on two

DOI: 10.1201/9781003562955-2

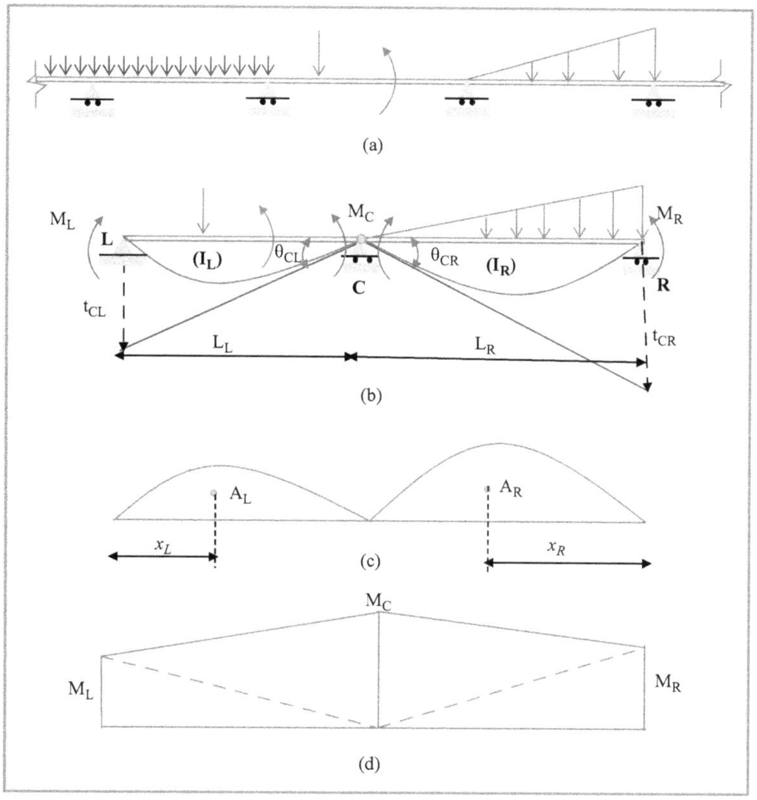

FIGURE 2.1 (a) Continuous beam, (b) intermediate two spans of a continuous beam, (c) moment diagram due to loads on simply supported beams, and (d) moment diagram due to end moments.

adjacent spans. In equation 2.5, the summation symbol Σ is used because different types of loads may act on each span.

2.3 DERIVATION OF GENERALISED THREE MOMENTS EQUATION (WITH YIELDING OF SUPPORTS)

In the previous section, we derived Clapeyron's equation for continuous beams with unyielding of supports. Due to the nature of the soil conditions and foundations, the supports may sometimes settle unevenly over time. These unequal settlements, referred to as yielding of supports, cause additional stresses in indeterminate structures. Hence, it is essential to consider these effects in the analysis of such structures. In 1860, Otto Mohr expanded Clapeyron's equation (equation 2.4) for continuous beams to account for these yieldings of supports. He incorporated the effect of chord rotations due to uneven settlements into the compatibility condition in the derivation of Clapeyron's theorem. This modified equation is a generalised form of the three-moment equation. The derivation of the three-moment equation with yielding of supports is presented below.

Consider two adjacent spans of a continuous beam subjected to loading, as shown in Figure 2.2. Let the supports L, C, and R settle by δ_L, δ_C, and δ_R, respectively, as shown in Figure 2.2, with the settlements taken as negative since they are downward. After the settlement of the supports, the chord of the beam is shown in the figure as $L^1C^1R^1$. The slopes of the tangents drawn to the elastic curve at C^1 are θ_{CL} and θ_{CR}, as shown in the figure.

From the Figure 2.2, the chord rotations with respect to the horizontal line are given by

$$\alpha_{CL} = \frac{\delta_C - \delta_L}{L_L} \tag{2.6a}$$

$$\alpha_{CR} = \frac{\delta_C - \delta_R}{L_R} \tag{2.6b}$$

The rotations β_{CL} and β_{CR} due to the external loads and end moments are already calculated in equations 2.2 and 2.3 and are equal to

$$\beta_{CL} = \frac{A_L x_L}{EI_L L_L} + \frac{M_L L_L}{6EI_L} + \frac{M_C L_L}{3EI_L} \tag{2.7a}$$

$$\beta_{CR} = \frac{A_R x_R}{EI_R L_R} + \frac{M_R L_R}{6EI_R} + \frac{M_C L_R}{3EI_R} \tag{2.7b}$$

From Figure 2.2, we can write

$$\theta_{CL} = \beta_{CL} - \alpha_{CL} \tag{2.8a}$$

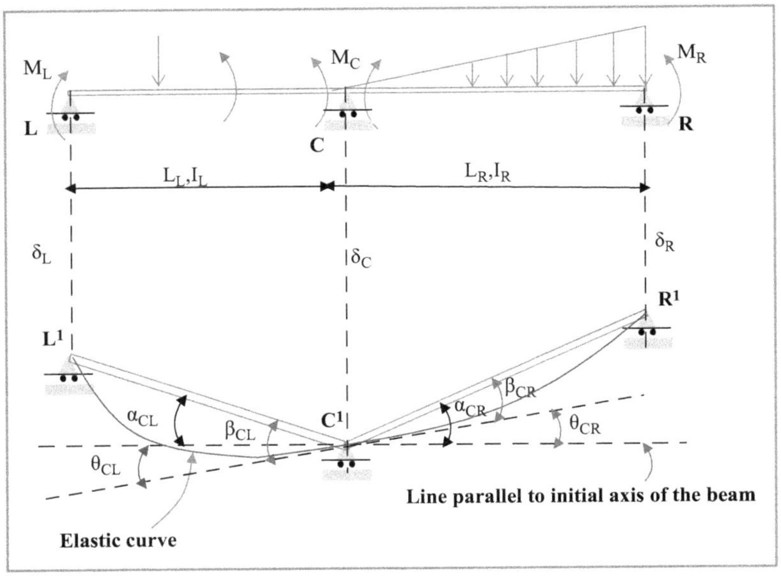

FIGURE 2.2 Two adjacent spans of a continuous beam with yielding of supports.

$$\theta_{CR} = \alpha_{CR} - \beta_{CR} \tag{2.8b}$$

Furthermore, from Figure 2.2, it can be noticed that

$$\theta_{CL} = \theta_{CR} \tag{2.9}$$

Substituting equations 2.6–2.8 into equation 2.9, we obtain

$$\beta_{CL} - \alpha_{CL} = \alpha_{CR} - \beta_{CR}$$

$$=> \left[\frac{A_L x_L}{EI_L L_L} + \frac{M_L L_L}{6EI_L} + \frac{M_C L_L}{3EI_L}\right] - \left[\frac{\delta_C - \delta_L}{L_L}\right]$$

$$= \left[\frac{\delta_C - \delta_R}{L_R}\right] - \left[\frac{A_R x_R}{EI_R L_R} + \frac{M_R L_R}{6EI_R} + \frac{M_C L_R}{3EI_R}\right]$$

$$=> \frac{M_L L_L}{6EI_L} + \frac{2M_C L_L}{3EI_L} + \frac{M_R L_R}{6EI_R}$$

$$= -\frac{A_R x_R}{EI_R L_R} - \frac{A_L x_L}{EI_L L_L} + \left[\frac{\delta_C - \delta_L}{L_L} + \frac{\delta_C - \delta_R}{L_R}\right]$$

$$=> M_L\left(\frac{L_L}{I_L}\right) + 2M_C\left(\frac{L_L}{I_L} + \frac{L_R}{I_R}\right) + M_R\left(\frac{L_R}{I_R}\right)$$

$$= -\left[\frac{6A_L x_L}{I_L L_L} + \frac{6A_R x_R}{I_R L_R}\right] + 6E\left[\frac{\delta_C - \delta_L}{L_L} + \frac{\delta_C - \delta_R}{L_R}\right] \tag{2.10}$$

Equation 2.10 is called the general three-moment equation or Clapeyron's theorem, which relates three successive support moments to the external loads and yielding of supports.

2.3.1 Analysis Procedure

- **Calculate the static indeterminacy** of a given beam and check the extreme boundary conditions. If the extreme supports are simply supported, select the moments at the interior supports as unknowns in the positive direction (sagging moments). If the extreme supports are fixed, select the moments at both the fixed and interior supports as unknowns.
- **Assume each span behaves as a simply supported** beam and draw the free bending moment diagrams due to the applied loads for each span.
- **Calculate the areas of free-bending moment diagrams** and their centroidal distances from either end of each span.
- **If the extreme ends are fixed**, assume an imaginary span of length L^1, either to the left of the fixed support (if the left end support is fixed) or to the right (if the extreme right support is fixed), with infinite flexural rigidity (i.e., $L'/I = L'/\alpha = 0$). Since the imaginary span has infinite flexural rigidity, it does not generate an M/EI diagram, and the elastic curve at point A remains horizontal.
- **Next, apply the three-moment equation** to the leftmost two adjacent spans and then proceed successively with the remaining spans of the continuous beam to obtain a system of simultaneous equations. It should be noted that the number of three-moment equations is equal to the degree of static indeterminacy.
- **Solve the simultaneous equations** to find the unknown support moments. After determining these support moments, use the free body diagram and the static equilibrium equations, to calculate the support reactions.

2.4 APPLICATION TO CONTINUOUS BEAMS WITH EXTREME BOUNDARIES SIMPLY SUPPORTED

Example 2.1

Analyse the two-span continuous beam, shown in Figure 2.3, using Clapeyron's theorem. Also draw the shear force and bending moment diagrams.

Solution:

The static indeterminacy, $I_s = 3 - 2 = 1$, i.e., the beam is statically indeterminate to the first degree. Since supports *A* and *C* are simply supported, the moments at *A* and *C* are zero, and there is only one unknown moment at *B*, i.e., M_B. Therefore, only one three-moment equation is required to solve for M_B. The other given data are provided below. Let us assume each span behaves as a simply supported beam, and draw the free bending moment diagram due to external loads, as shown in Figure 2.4.

Apply Clapeyron's equation 2.5 for spans *AB* and *BC*.

$$M_L\left(\frac{L_L}{I_L}\right) + 2M_C\left(\frac{L_L}{I_L} + \frac{L_R}{I_R}\right) + M_R\left(\frac{L_R}{I_R}\right) = \sum\frac{-6A_Lx_L}{I_LL_L} - \sum\frac{6A_Rx_R}{I_RL_R}$$

$$M_L = M_A = 0;\quad M_C = M_B;\quad M_R = M_C = 0$$

$$L_L = 5\text{ m}\quad L_R = 5\text{ m}$$

$$I_L = 2I\quad I_R = I$$

$$x_L = \frac{5}{2}\text{ m};\quad x_R = \frac{5}{2}\text{ m}$$

$$A_L = \frac{2}{3}\times 5\times 156.25 = 520.83\text{ kNm}^2;$$

$$A_R = \frac{1}{2}\times 5\times 37.5 = 93.75\text{ kNm}^2$$

Substitute these values into equation 2.5.

$$0 + 2M_B\left(\frac{5}{2I} + \frac{5}{I}\right) + 0 = \frac{-6\times 93.75\times 2.5}{(I)\times 5} - \frac{6\times 520.83\times 2.5}{(2I)\times 5}$$

$$15M_B = -281.25 - 781.25$$

$$M_B = -70.83\text{ kNm}$$

(–ve sign indicates that the moment M_B is acting in the opposite direction to the assumed direction.)

The free body diagram, shear force diagram, and bending moment diagrams are shown in Figure 2.5a–c, respectively.

Example 2.2

Analyse the continuous beam shown in Figure 2.6 using Clapeyron's theorem. Assume *EI* is constant.

Solution:

By inspection, the support *A* is simply supported (M_A=0), and the moment at *D* due to the overhang portion is M_D=–10 kNm (–ve sign indicates the tension at the top). For convenience, we can ignore the overhanging portion in the analysis and take it into

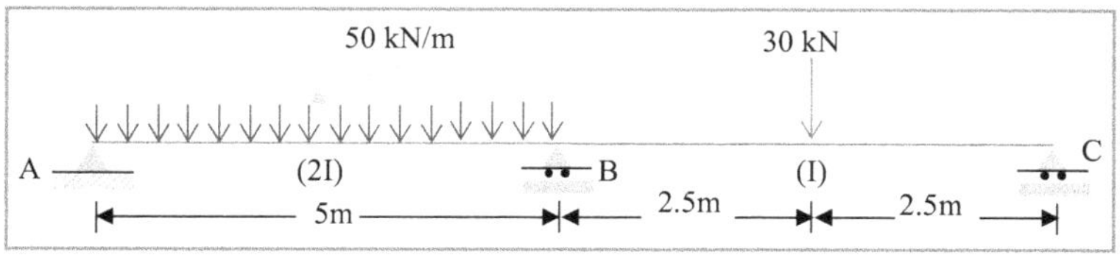

FIGURE 2.3 Two-span continuous beam.

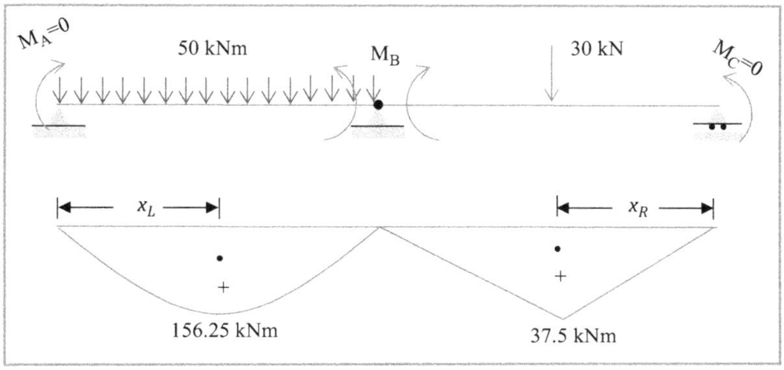

FIGURE 2.4 Free bending moment diagram.

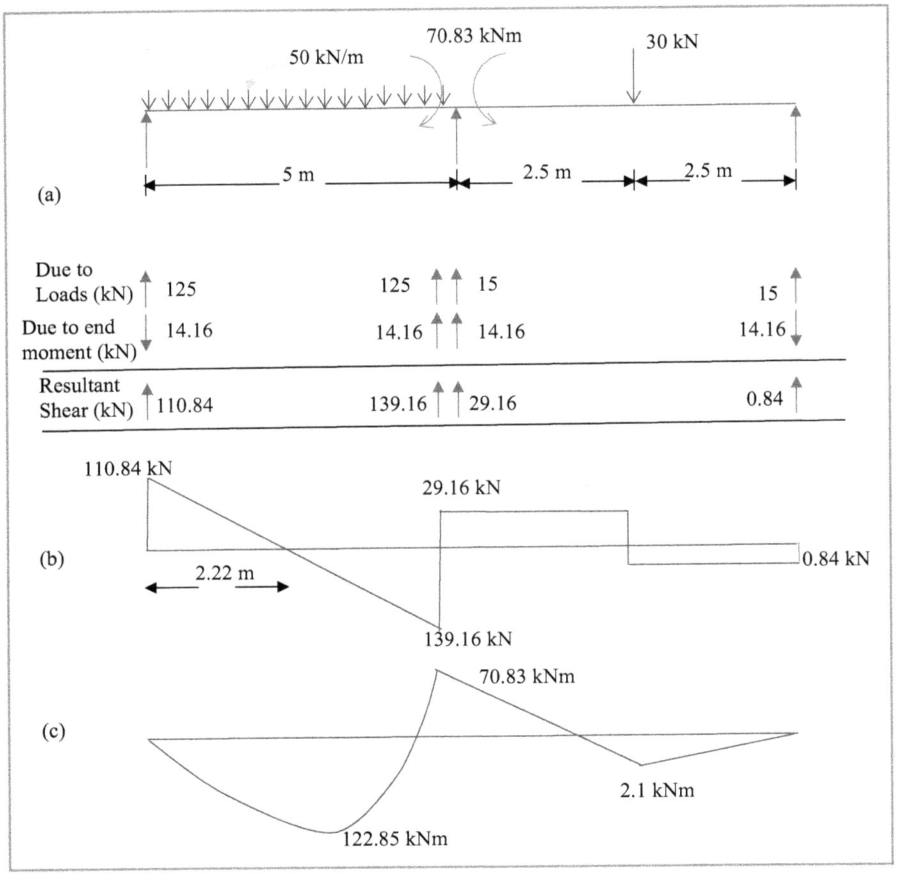

FIGURE 2.5 (a) Free body diagram, (b) shear force diagram and (c) bending moment diagram.

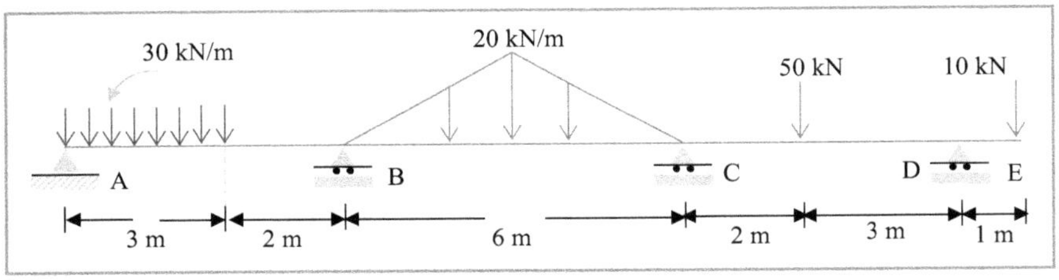

FIGURE 2.6 Continuous beam *ABCDE*.

consideration when calculating the internal forces and drawing the shear force and bending moment diagrams. Therefore, the only redundant moments are at supports *B* and *C*, labelled as M_B and M_C respectively. Hence, we need to apply Clapeyron's equation twice: once for spans *AB* and *BC*, and then again for spans *BC* and *CD*. The free bending moment diagrams are shown in Figure 2.7.

$$A_L = \int_0^3 (63x - 15x^2)dx + \left(\frac{1}{2}\times 2\times 54\right)$$

$$= \left[63\frac{x^2}{2} - 15\frac{x^3}{3}\right]_0^3 + 54 = 202.5 \text{ kNm}^2$$

$$A_L x_L = \int_0^3 (63x - 15x^2)\cdot x dx + \left[\frac{1}{2}\times 2\times 54\times\left(3+\frac{2}{3}\right)\right]$$

$$= \left[63\frac{x^3}{3} - 15\frac{x^4}{4}\right]_0^3 + 198 = 461.25 \text{ kNm}^3$$

$$A_R = 2\int_0^3 30x - \frac{10x^3}{9}\, dx = 2\ 30\frac{x^2}{2} - \frac{10}{9}\frac{x^4}{4}\Big|_0^3 = 225 \text{ kNm}^2$$

(for span BC)

$$A_3 = \frac{1}{2}\times 60\times 5 = 150 \text{ kNm}^2 \text{ (for span CD)}$$

$$x_R = \frac{5+3}{3} = 2.67 \text{ m}$$

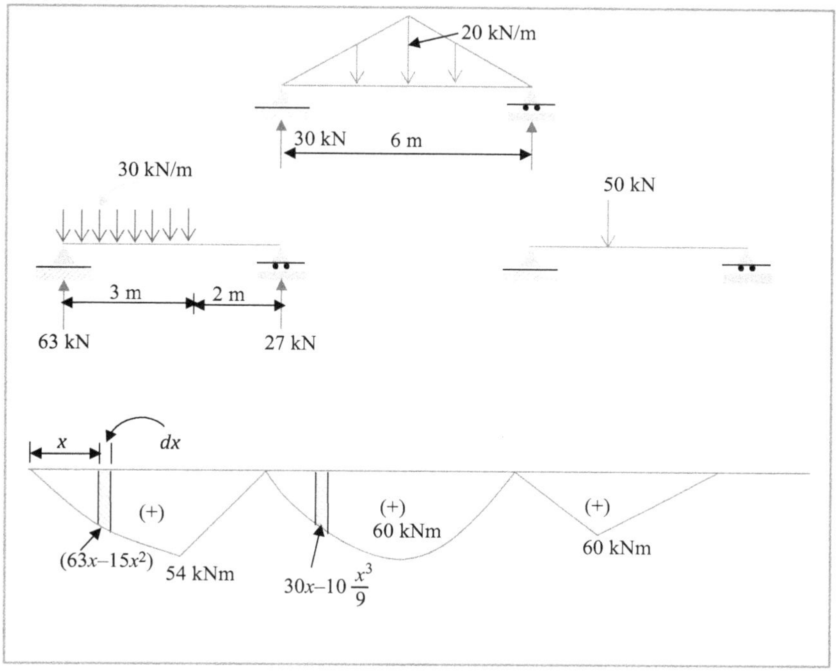

FIGURE 2.7 Free bending moment diagrams.

Apply Clapeyron's equation 2.5 to spans *AB* and *BC*.

$$0+2M_B\left(\frac{5}{I}+\frac{6}{I}\right)+M_C\left(\frac{6}{I}\right)=\frac{-6\times 461.25}{5\times(I)}-\frac{6\times 225\times 3}{6\times(I)}$$

$$\Rightarrow 22M_B+6M_C=-1{,}228.5 \qquad (2.11)$$

Apply Clapeyron's equation to spans *BC* and *CD*.

$$M_B\left(\frac{6}{I}\right)+2M_C\left(\frac{6}{I}+\frac{5}{I}\right)-10\left(\frac{5}{I}\right)$$

$$=\frac{-6\times 225\times 3}{6\times(I)}-\frac{6\times 150\times 2.67}{5\times(I)}$$

$$\Rightarrow 6M_B+22M_C=-1{,}105.6 \qquad (2.12)$$

Combine equations 2.11 and 2.12 in matrix form and solve for the unknowns.

$$\begin{bmatrix} 22 & 6 \\ 6 & 22 \end{bmatrix}\begin{bmatrix} M_B \\ M_C \end{bmatrix}=\begin{bmatrix} -1{,}228.5 \\ -1{,}105.6 \end{bmatrix}$$

$$\begin{bmatrix} M_B \\ M_C \end{bmatrix}=\frac{1}{22^2-36}\begin{bmatrix} 22 & -6 \\ -6 & 22 \end{bmatrix}\begin{bmatrix} -1{,}228.5 \\ -1{,}105.6 \end{bmatrix}$$

$$=\begin{bmatrix} -45.52 \\ -37.84 \end{bmatrix}\text{kNm}$$

The support reactions can be calculated using the static equilibrium equations and the free-body diagram shown in Figure 2.8a. The shear force and bending moment diagrams are shown in Figure 2.8b and c, respectively.

2.5 APPLICATION TO THE CONTINUOUS BEAMS WITH EXTREME BOUNDARIES NOT SIMPLY SUPPORTED AND WITH YIELDING OF SUPPORTS

Example 2.3

Analyse the two-span continuous beam with one end fixed and the other extreme end simply supported, as shown in Figure 2.9. Draw the shear force and bending moment diagrams. Assume *EI* is constant.

Solution:

The beam is statically indeterminate to the second degree. Therefore, two three-moment equations are required to solve for the unknowns. As discussed in the previous section, since the end *A* is fixed, the elastic curve at *A* remains horizontal. This condition can be achieved by assuming an imaginary span AA^1 of length L^1 to the left of support *A*, with infinite flexural rigidity (i.e., $L'/I = L'/\alpha = 0$) (Figure 2.10). Because the span AA^1 has infinite flexural rigidity, it does not yield any M/EI diagram and has no elastic curve at *A*, which remains horizontal.

Apply the three-moment equation separately for beam A^1AB and *ABC*. By inspection, the moment at *C* is zero due to the simply supported end. Let the unknown moments at *A* and *B* be M_A and M_B, and they are assumed to act in the +ve direction (sagging bending

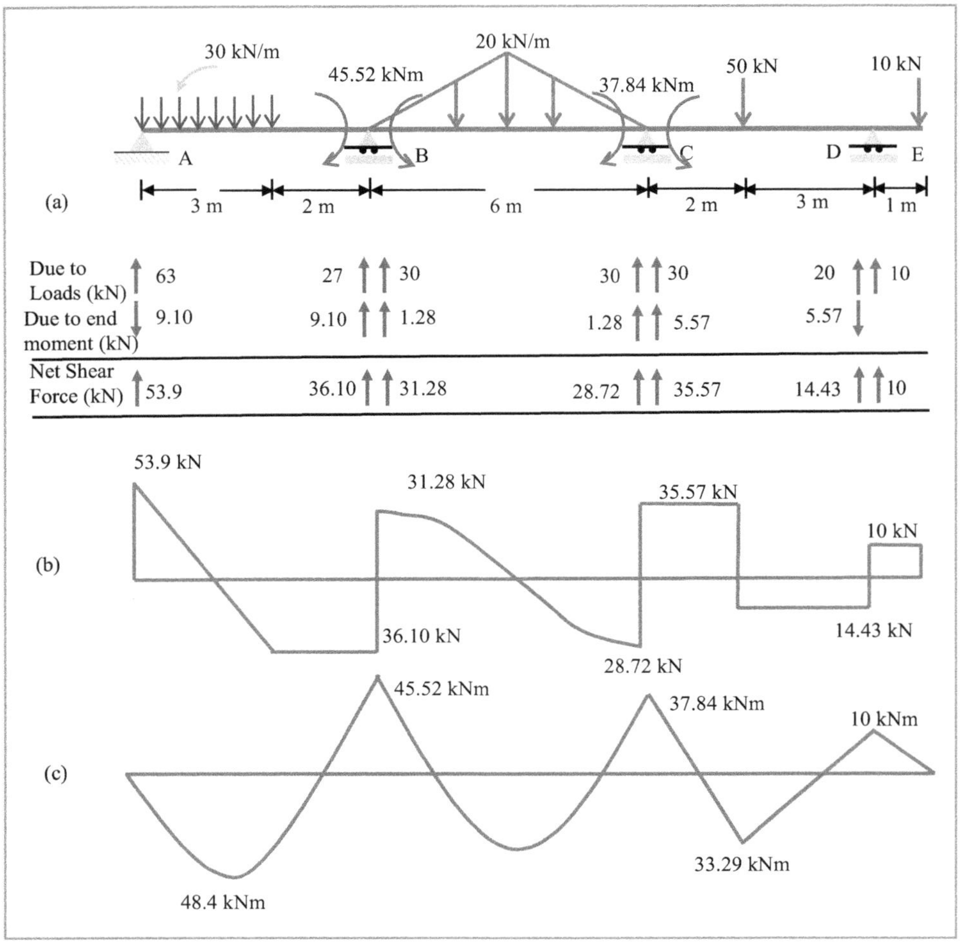

FIGURE 2.8 (a) Free body diagram, (b) shear force diagram and (c) bending moment diagram.

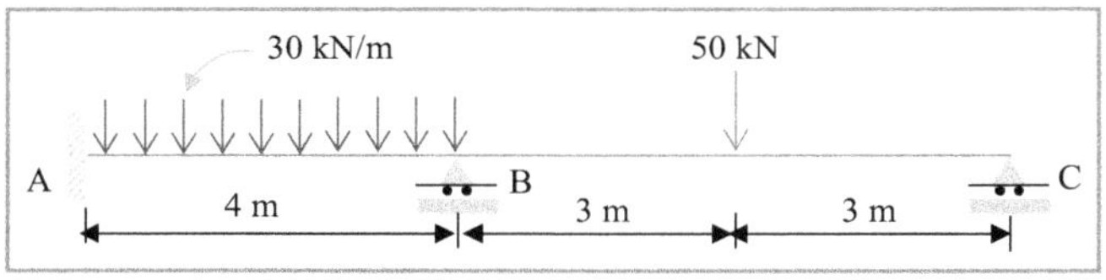

FIGURE 2.9 Two-span continuous beam.

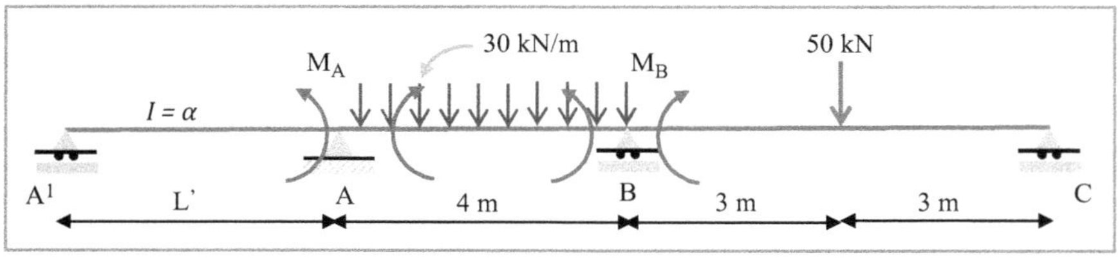

FIGURE 2.10 Continuous beam with imaginary span AA^1.

moments), as shown in Figure 2.10. The free bending moment diagram due to the applied loads on each span is shown in Figure 2.11.

Apply the three-moment equation to spans $A'A$ and AB.

$$M'_A\left(\frac{L'}{\alpha}\right)+2M_A\left(\frac{L'}{\alpha}+\frac{4}{I}\right)+M_B\left(\frac{4}{I}\right)=-0-\frac{6\times160\times2}{4\times I}$$

$$\Rightarrow 8M_A+4M_B=-480$$

$$\Rightarrow 2M_A+M_B=-120 \tag{2.13}$$

Applying the three-moment equation to spans AB and BC.

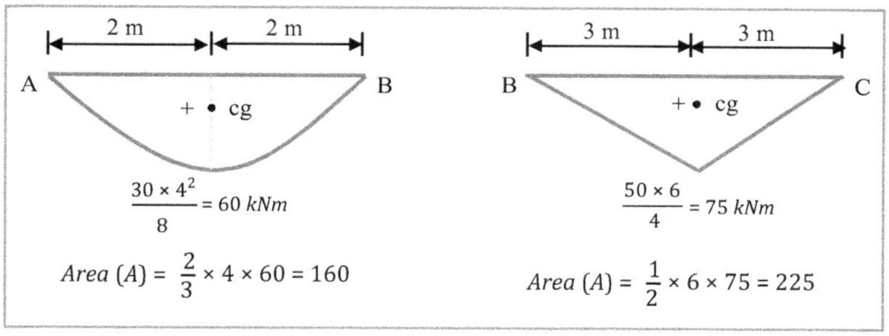

FIGURE 2.11 Free bending moment diagrams.

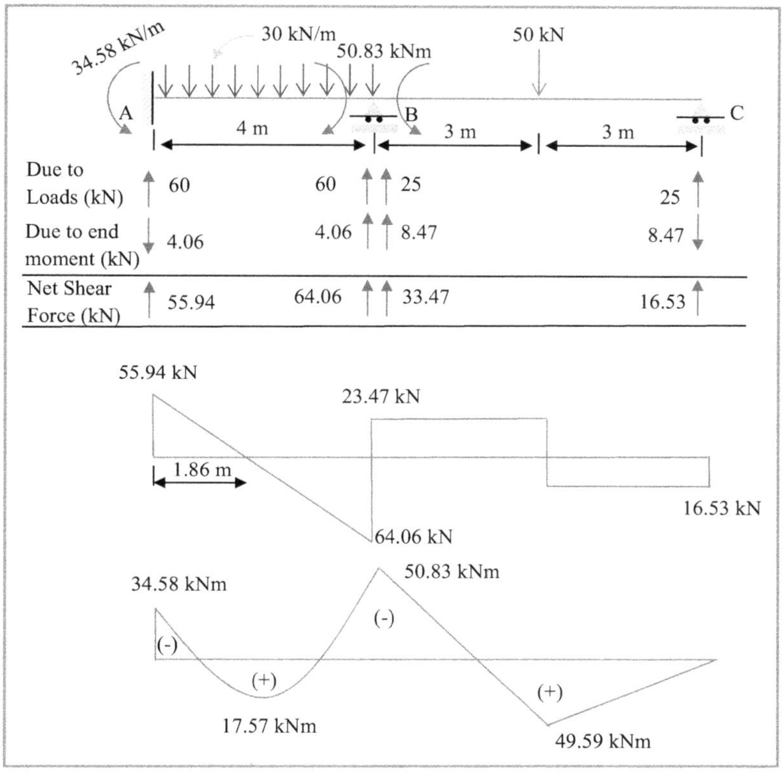

FIGURE 2.12 (a) Free body diagram, (b) shear force diagram and (c) bending moment diagram.

$$M_A\left(\frac{4}{I}\right)+2M_B\left(\frac{4}{I}+\frac{6}{I}\right)+M_C\left(\frac{6}{I}\right)=\frac{-6\times160\times2}{4\times(I)}-\frac{6\times225\times3}{6\times(I)}$$

$$\Rightarrow 4M_A+20M_B=-1{,}155 \qquad (2.14)$$

Write equations 2.13 and 2.14 in matrix form and solve for the unknowns.

$$\begin{bmatrix} 2 & 1 \\ 4 & 20 \end{bmatrix}\begin{bmatrix} M_A \\ M_B \end{bmatrix}=\begin{bmatrix} -120 \\ -1{,}155 \end{bmatrix}$$

$$\begin{bmatrix} M_A \\ M_B \end{bmatrix}=\frac{1}{40-4}\begin{bmatrix} 20 & -1 \\ -4 & 2 \end{bmatrix}\begin{bmatrix} -120 \\ -1{,}155 \end{bmatrix}$$

$$=\begin{bmatrix} -34.58 \\ -50.83 \end{bmatrix} \text{kNm}$$

SUPPORT REACTIONS

Using the static equilibrium equations and the free body diagram, the support reactions are calculated and are shown in Figure 2.12a. The shear force and bending moment diagrams are shown in Figure 2.12b and c, respectively.

Example 2.4

Analyse the three-span continuous beam with an internal hinge shown in Figure 2.13 using the theorem of three moments. Assume *EI* is constant.

Solution:

The static indeterminacy of the beam is 1. By inspection, the moments at *A* and *D* are zero. As the spans *AE* and *EB* are determinate (due to the internal hinge at *E*), the moment at *B* is a known value, which can be determined from the free body diagram shown in Figure 2.14. Therefore, assume the moment at *C* (M_C) is an unknown, acting in the +ve direction (Figure 2.14).

The free bending moment diagram due to the loads is shown in Figure 2.15.

Apply the theorem of three-moment equation 2.5 to spans *BC* and *CD*.

$$\frac{-270\times 4}{I}+2M_C\left(\frac{4}{I}+\frac{6}{I}\right)+0=\frac{-6\times 200\times 2}{4\times(I)}-\frac{6\times 540\times 3}{6\times(I)}$$

$$\Rightarrow 20M_C=-1{,}140$$

$$\Rightarrow M_C=-57\text{ kNm}$$

(Note: The –ve sign indicates the moment produces tension at the top.)

The support reactions are shown in the free body diagram in Figure 2.16a. The shear force and bending moment diagrams are shown in Figure 2.16b and c, respectively.

Example 2.5

Re-analyse the beam shown in Example 2.2 (Figure 2.6) using the three-moment equation, if supports *B* and *C* settle down by 5 and 10 mm, respectively, without any external load, as shown in Figure 2.17. Assume the flexural rigidity $EI=135\times 10^3$ kNm^2.

Solution:

By inspection, the moments at *A* and *D* are zero. Let the moments at *B* and *C* be M_B and M_C, respectively, acting in the +ve direction, as shown in Figure 2.18. The chord rotations in the simply supported spans

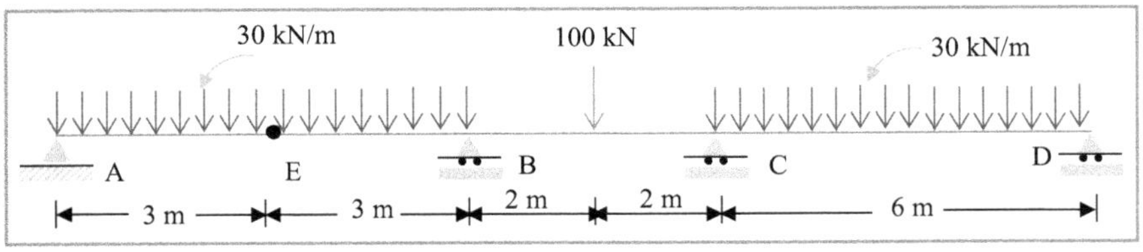

FIGURE 2.13 Three-span continuous beam.

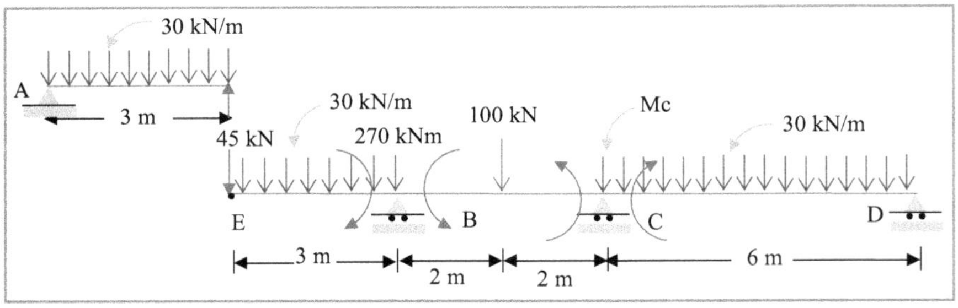

FIGURE 2.14 Free body diagrams.

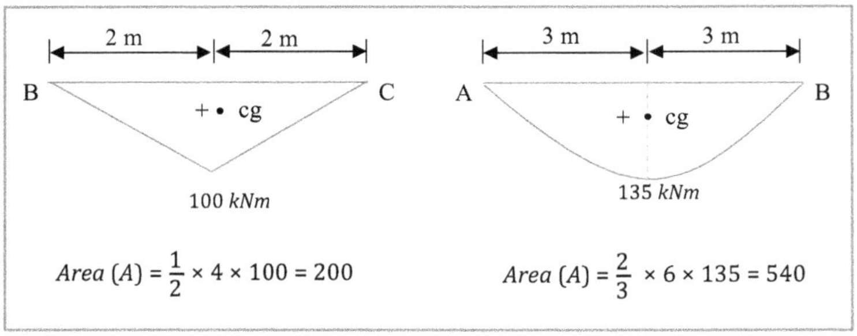

FIGURE 2.15 Free bending moment diagrams.

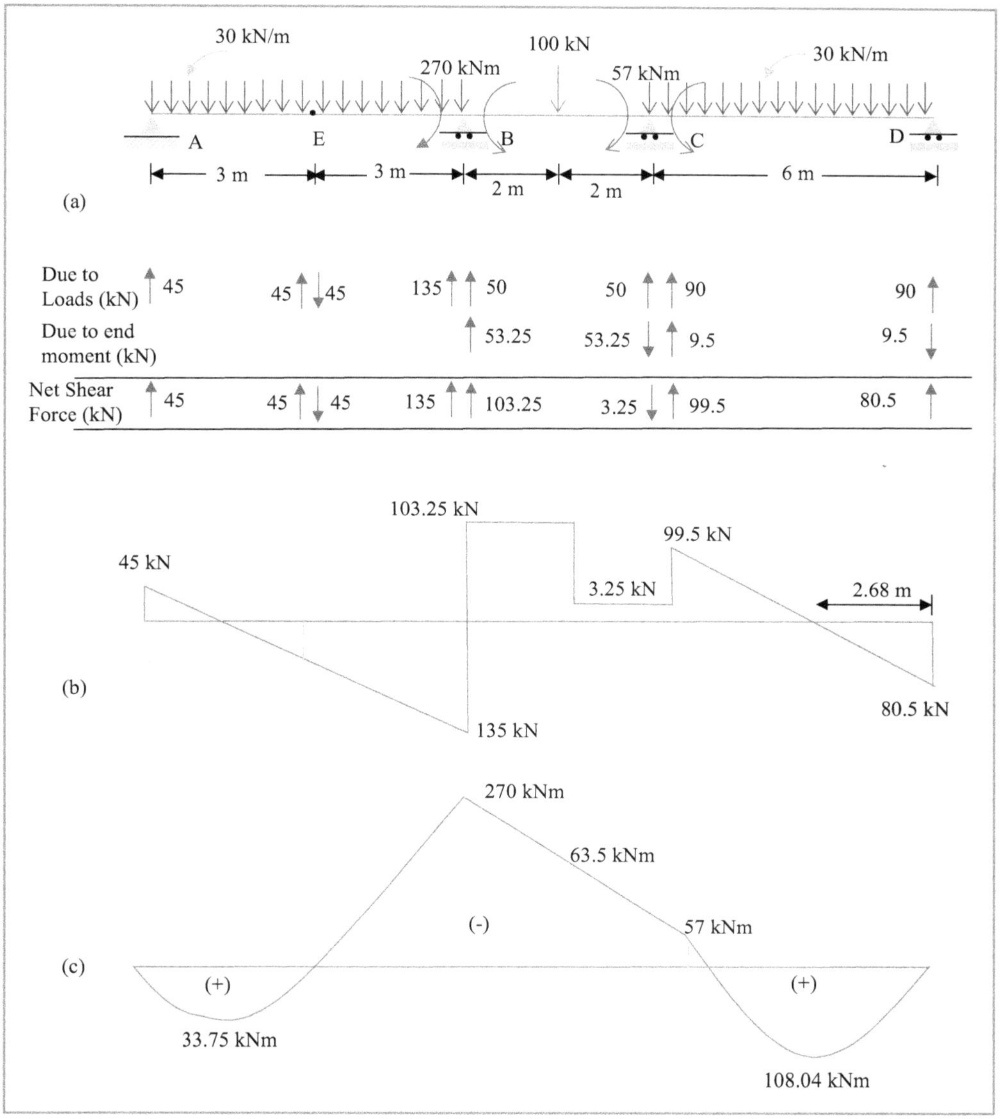

FIGURE 2.16 (a) Free body diagram, (b) shear force diagram and (c) bending moment diagram.

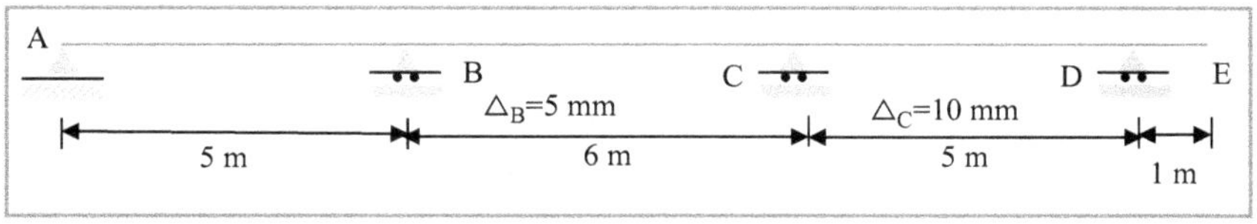

FIGURE 2.17 Three-span continuous beam.

corresponding to $\Delta_B = 5\,\text{mm}$ and $\Delta_C = 10\,\text{mm}$ are also shown in Figure 2.18.

Apply the three-moment equation to spans *AB* and *BC*.

$$0 + 2M_B\left(\frac{5}{I} + \frac{6}{I}\right) + M_C\left(\frac{6}{I}\right) = 0 + 6E\left[\frac{5\times10^{-3}}{5} + \frac{(5-10)\times10^{-3}}{6}\right]$$

$$\Rightarrow 22M_B + 6M_C = 135 \qquad (2.15)$$

Apply the three-moment equation to spans *BC* and *CD*.

$$M_B\left(\frac{6}{I}\right) + 2M_C\left(\frac{6}{I} + \frac{5}{I}\right) + 0 = 0 + 6E\left[\frac{(10-5)\times10^{-3}}{6} + \frac{10\times10^{-3}}{5}\right]$$

$$\Rightarrow 6M_B + 22M_C = 2{,}295 \qquad (2.16)$$

Write equations 2.15 and 2.16 in matrix form to solve for M_B and M_C, as shown below.

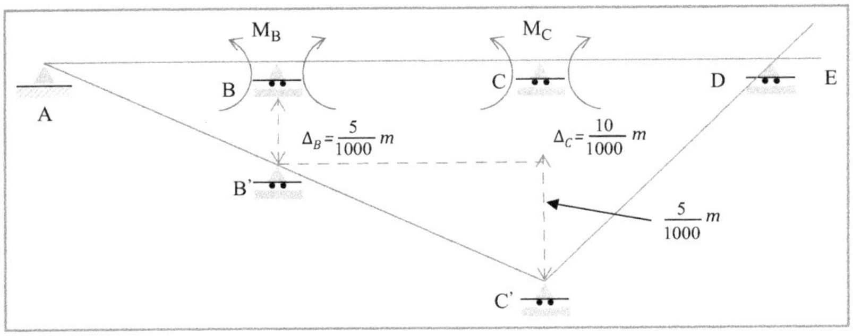

FIGURE 2.18 Redundant moments and chord rotations on simply supported spans.

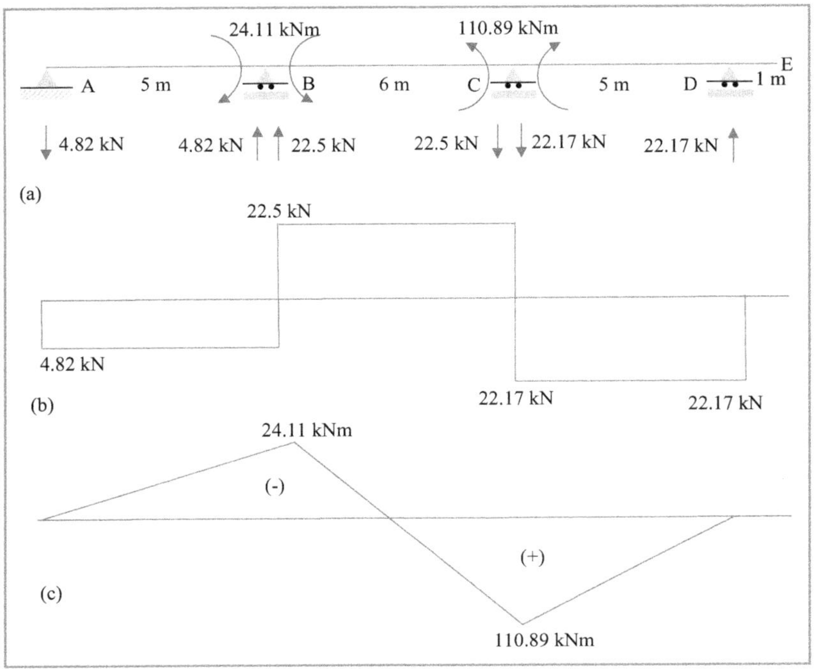

FIGURE 2.19 (a) Free body diagram, (b) shear force diagram, and (c) bending moment diagrams.

$$\begin{bmatrix} 22 & 6 \\ 6 & 22 \end{bmatrix}\begin{bmatrix} M_B \\ M_C \end{bmatrix}=\begin{bmatrix} 135 \\ 2,295 \end{bmatrix}$$

$$\begin{bmatrix} M_B \\ M_C \end{bmatrix}=\frac{1}{22^2-6^2}\begin{bmatrix} 22 & -6 \\ -6 & 22 \end{bmatrix}\begin{bmatrix} 135 \\ 2,295 \end{bmatrix}$$

$$=\begin{bmatrix} -24.11 \\ 110.89 \end{bmatrix}\text{ kNm}$$

Using the known values of the moments at B and C, the other support reactions can be determined using the static equilibrium equations and the free body diagram. The free body diagram, shear force, and bending moment diagrams are shown in Figure 2.19.

Example 2.6

A continuous beam $ABCD$ is supported on springs at supports B and C, as shown in Figure 2.20. The stiffness of the springs K_B is $EI/30$ and K_C is $EI/20$. The loads are also shown in the figure. Analyse the beam $ABCD$ and draw the shear force and bending moment diagrams. Assume EI is constant.

Solution:

By inspection, the bending moments at A and D are zero. The moments at C and D need to be calculated. The beam is supported by springs at B and C; therefore, the supports B and C settle down. Let the reactions at B and C be R_B and R_C, respectively. Then the support settlement at B is R_B/K_B, and at C is R_C/K_C. Thus, $\delta_A = 0 = \delta_D$

$$\delta_B = \frac{30R_B}{EI} \quad \text{and} \quad \delta_C = \frac{20R_C}{EI}$$

The simply supported spans with the redundant moments are shown in Figure 2.21a and the free bending moment diagrams are shown in Figure 2.21b.

Apply the three-moment equation to spans AB and BC.

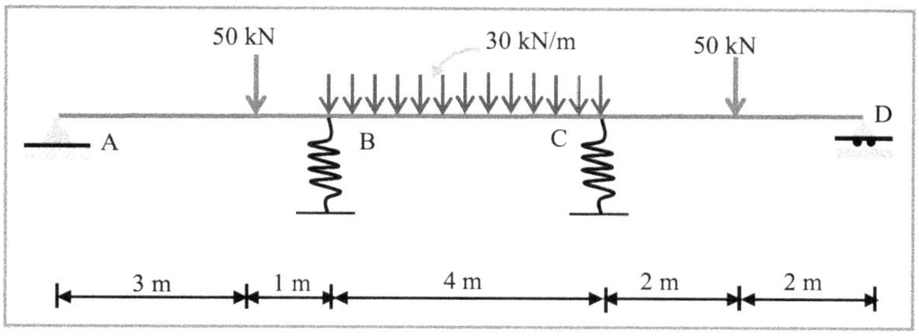

FIGURE 2.20 Continuous beam *ABCD*.

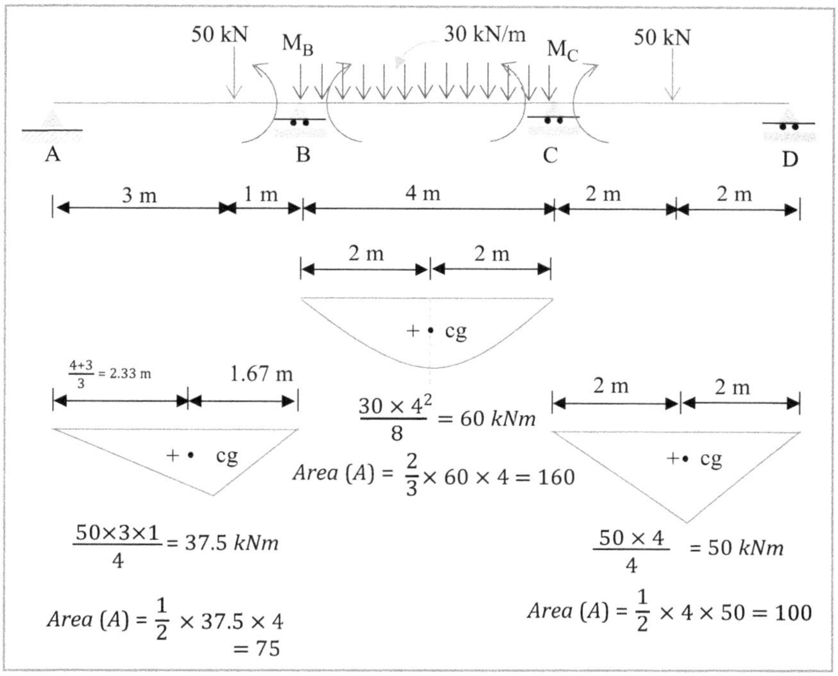

FIGURE 2.21 (a) Simply supported spans and (b) free bending moment diagrams.

$$0 + 2M_B\left(\frac{4}{I} + \frac{4}{I}\right) + M_C\left(\frac{4}{I}\right) = \frac{-6 \times 75 \times 2.33}{4 \times (I)} - \frac{6 \times 160 \times 2}{4 \times (I)}$$

$$-6E\left[\frac{\frac{30R_B}{EI}}{4} + \frac{\frac{30R_B}{EI} - \frac{20R_C}{EI}}{4}\right]$$

$$\Rightarrow 16M_B + 4M_C = -742.12 - 90R_B + 30R_C \quad (2.17)$$

Apply the three-moment equation to spans *BC* and *CD*.

$$M_B\left(\frac{4}{I}\right) + 2M_C\left(\frac{4}{I} + \frac{4}{I}\right) + 0 = \frac{-6 \times 160 \times 2}{4 \times (I)} - \frac{6 \times 100 \times 2}{4 \times (I)}$$

$$-6E\left[\frac{\frac{20R_C}{EI} - \frac{30R_B}{EI}}{4} + \frac{\frac{20R_C}{EI}}{4}\right]$$

$$4M_B + 16M_C = -780 - 60R_C + 45R_B \quad (2.18)$$

Now let us express M_B and M_C in terms of R_B and R_C, using the free body diagrams shown in Figure 2.22. Assume all reactions are in upward direction.

Using static equilibrium equations, we can write

$$\begin{aligned}
R_A &= 12.5 + 0.25M_B \\
R_{BL} &= \frac{150 - M_B}{4} = 37.5 - 0.25M_B \\
R_{BR} &= \frac{240 + M_C - M_B}{4} = 60 + 0.25M_C - 0.25M_B \\
\therefore R_B &= R_{BL} + R_{BR} = 97.5 - 0.5M_B + 0.25M_C \\
R_{CL} &= 60 + 0.25M_B - 0.25M_C \\
R_{CR} &= \frac{100 - M_C}{4} = 25 - 0.25M_C \\
\therefore R_C &= R_{CL} + R_{CR} = 85 + 0.25M_B - 0.5M_C \\
R_D &= 25 + 0.25M_C
\end{aligned} \quad (2.19)$$

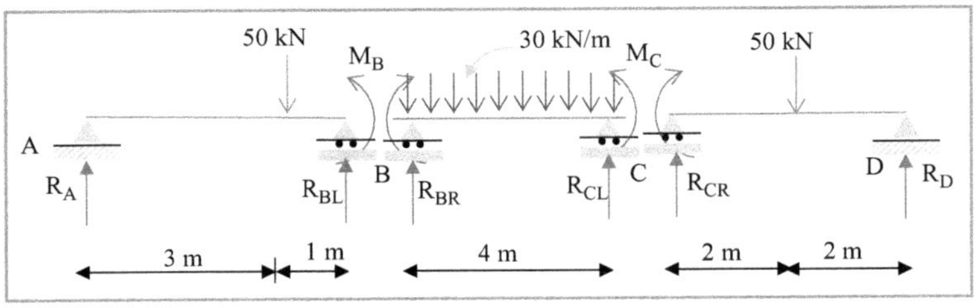

FIGURE 2.22 Free body diagrams.

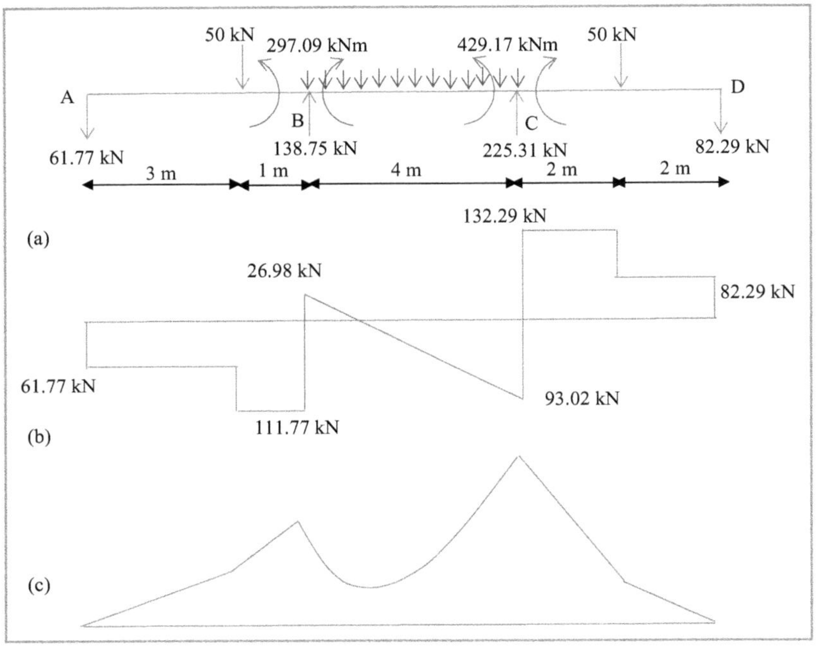

FIGURE 2.23 (a) Free body diagram, (b) shear force diagram, and (c) bending moment diagram.

Substituting equation 2.19 in equations 2.17 and 2.18, we get

$$16M_B + 4M_C = -742.12 - 90(97.5 - 0.5M_B + 0.25M_C)$$
$$+30(85 + 0.25M_B - 0.5M_C)$$

$$\Rightarrow -36.5\ M_B + 41.5\ M_C = -6{,}967.12 \quad (2.17a)$$

Similarly, equation 2.19 becomes

$$4M_B + 16\ M_C = -780 - 60(85 + 0.25M_B - 0.5M_C)$$
$$+45(97.5 - 0.5M_B + 0.25M_C)$$

$$\Rightarrow 41.5\ M_B - 25.5\ M_C = -1{,}492.5 \quad (2.18a)$$

By solving equations 2.17a and 2.18(a, we get

$$M_B = -297.09 \text{ kNm}$$

$$M_C = -429.17 \text{ kNm}$$

From equation 2.19, the reactions are

$$R_A = -61.77 \text{ kN } (\downarrow)$$

$$R_{BL} = 111.77 \text{ kN } (\uparrow)$$

$$R_{BR} = 26.98 \text{ kN } (\uparrow)$$

$$R_B = 138.75 \text{ kN } (\uparrow)$$

$$R_{CL} = 93.02 \text{ kN } (\uparrow)$$

$$R_{CR} = 132.29 \text{ kN } (\uparrow)$$

$$R_C = 225.31 \text{ kN } (\uparrow)$$

$$R_D = -82.29 \text{ kN } (\downarrow)$$

The free body diagram, shear force, and bending moment diagrams are shown in Figure 2.23.

2.6 PROBLEMS FOR PRACTICE

2.6.1 A continuous beam *ABCD* is shown in Figure 2.24. Analyse the beam using the theorem of the three-moment equation. Assume *EI* is constant.

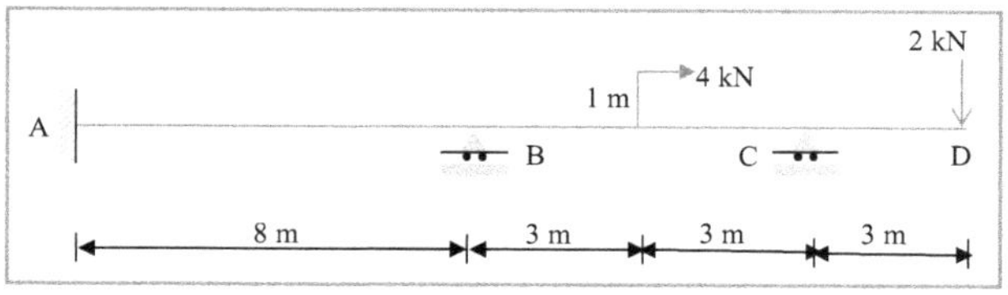

FIGURE 2.24

2.6.2 A continuous beam *ABCD* is shown in Figure 2.25. The loading is also shown in the figure. In addition to the external loads, the support *B* settles down by 10 mm; find the support reactions. Assume $E = 2\times10^5$ N/mm^2; and $I = 1.32\times10^8$ mm^4.

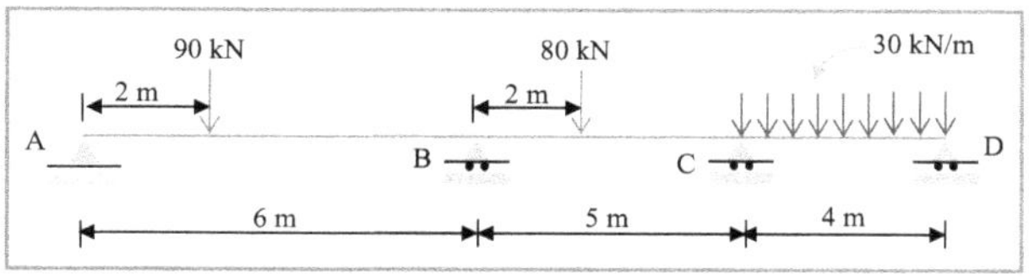

FIGURE 2.25

2.6.3 Analyse the continuous beam shown in Figure 2.26, using the three-moment equation. Assume *EI* is constant. Draw the SF and BM diagrams.

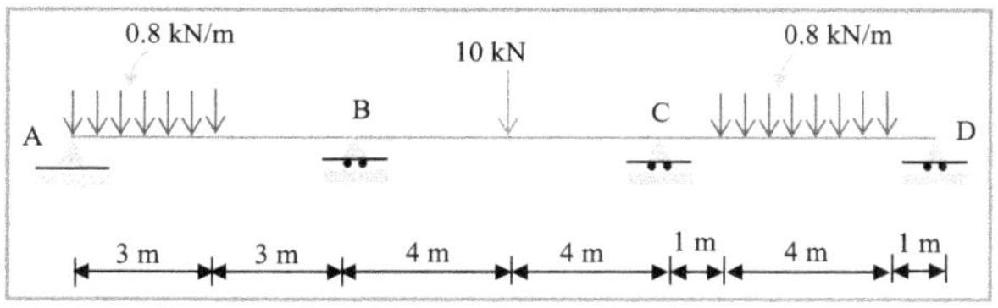

FIGURE 2.26

2.6.4 Analyse the continuous beam *ABCDE* shown in Figure 2.27 using the theorem of three moments. Assume *EI* is constant. Draw the shear force and bending moment diagrams, as well as the probable elastic curve.

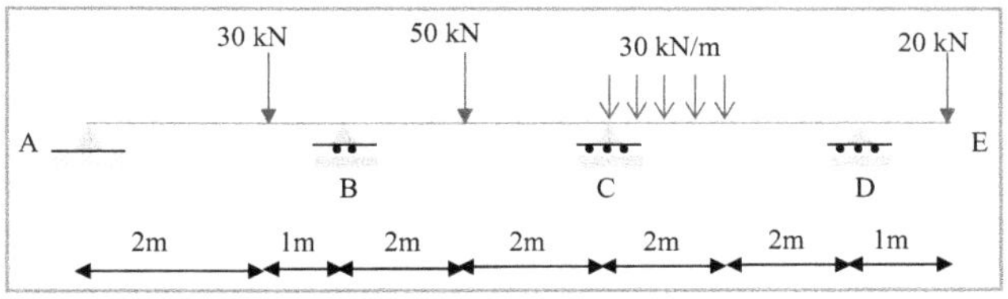

FIGURE 2.27

2.6.5 Analyse the fixed beam shown in Figure 2.28 using the three-moment equation. Draw the shear force and bending moment diagrams.

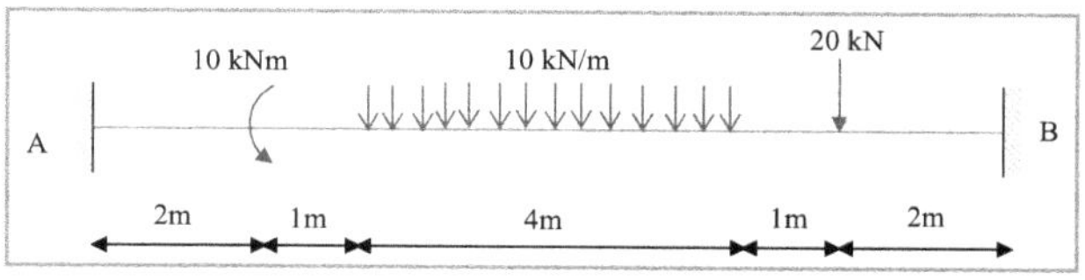

FIGURE 2.28

2.6.6 Analyse the continuous beam shown in Figure 2.29 using the theorem of three moments if the supports *B* and *C* settle down by 5 and 10 mm, respectively. Assume $EI = 85{,}000$ kNm^2.

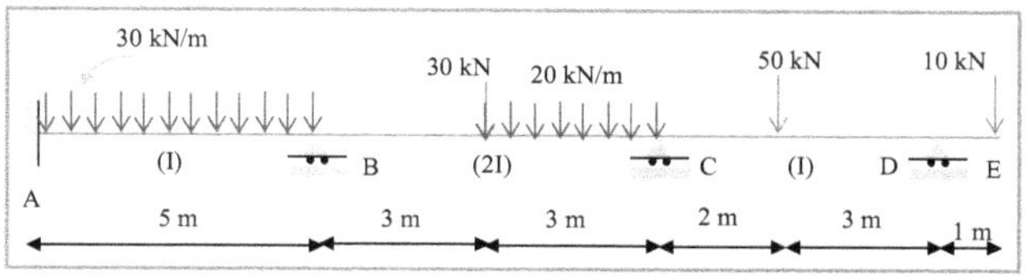

FIGURE 2.29

2.7 ANSWER KEY

Q.No.	Fig. No.	Answer at Key Points
2.6.1	2.24	$M_A = -0.875$ kNm; $M_B = 1.75$ kNm; $R_A = 0.328$ kNm
2.6.2	2.25	$M_B = -35.87$ kNm; $M_C = -71.56$ kNm
2.6.3	2.26	$M_B = -6.99$ kNm; $M_C = -7.89$ kNm
2.6.4	2.27	$M_A = -21.33$ kNm; $M_B = -25.32$ kNm
2.6.5	2.28	$M_A = -50.53$ kNm; $M_B = 70.13$ kNm; $R_A = 23.04$ kN; $R_B = 36.96$ kN
2.6.6	2.29	$M_A = -152.53$ kNm; $M_B = -47.92$ kNm; $M_B = 97.14$ kNm

Note: The answers are provided based on the analysis using structural analysis software. The results from the exact solution may differ slightly (in decimals), but these differences are generally negligible.

FURTHER READING

Khalfallah, S. (2018). *Structural Analysis 2: Statically Indeterminate Structures*, John Wiley & Sons, Inc, Hoboken, NJ, ISBN 978-1786303394.

Menon, D. (2023). *Structural Analysis*, Narosa Publishing House Pvt Ltd, New Delhi, ISBN 978-8184877342.

Reddy, C. S. (2011). *Basic Structural Analysis*, 3rd ed., Tata McGraw-Hill Education Pvt Ltd, New Delhi, ISBN 978-0070702769.

Wang, C. K. (1953). *Statically Indeterminate Structures*, McGraw Hill Higher Education, ISBN 978-0070681309.

3 Consistent Deformation Method

3.1 INTRODUCTION

The force method was discussed in Section 1.2 of Chapter 1. The consistent deformation method is one of the basic force methods of analysis for indeterminate structures. This method was first introduced by J. Maxwell in 1864, and later by Otto Mohr and Heinrich Muller Breslau in 1874. This method is useful for solving all kinds of indeterminate problems, such as beams, rigid-jointed frames, and pin-jointed plane frames. As discussed in Section 1.2 of Chapter 1, the main basis of this method is the compatibility of displacements. In this method, we first choose the redundant reactions/forces based on the degree of static indeterminacy. We then write the compatibility conditions/equations and apply the force–displacement relations to solve for the redundant reactions/forces. Once the redundant forces are known, the structure becomes determinate and can be analysed completely using the conditions of equilibrium. The selection of redundants is completely flexible. A redundant force can either be an external support reaction or an internal member force. Removing it from the structure will not lead to instability. This is explained in Figures 3.1–3.3. In Figure 3.1b, we select the moment at B as redundant by introducing an internal hinge at B, which causes the span BC to become unstable. Therefore, the moment at A is chosen as redundant, ensuring that the basic beam becomes stable.

Similarly, in Figure 3.2, the vertical reaction at D is selected as redundant, which results in all the reactions in the basic structure being concurrent at A, making it unstable (Figure 3.2b). Therefore, instead of selecting the vertical reaction at D as redundant, we will select the horizontal reaction as redundant, ensuring that the basic structure is stable (Figure 3.2c). Figure 3.3b is determinate but unstable due to $\Sigma F_y \neq 0$.

Furthermore, it is recommended that the redundants be chosen to minimise computational effort and maximise computational accuracy. The computational effort required to calculate the response of a given structure using this method can vary significantly depending on the choice of the basic determinate structure. In this regard, the following two cases should be considered:

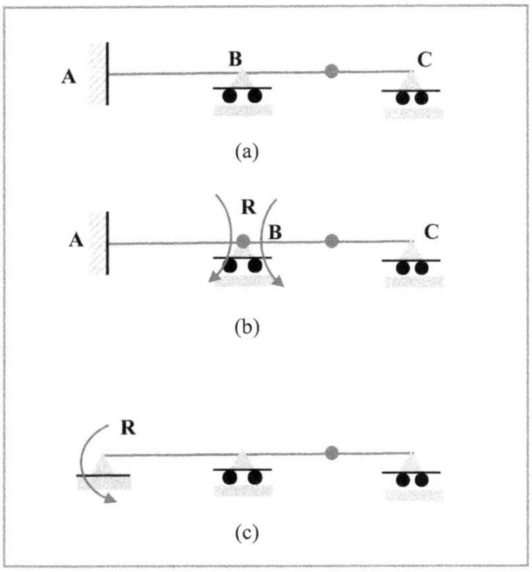

FIGURE 3.1 (a) Basic or determinate, (b) unstable, and (c) stable beams.

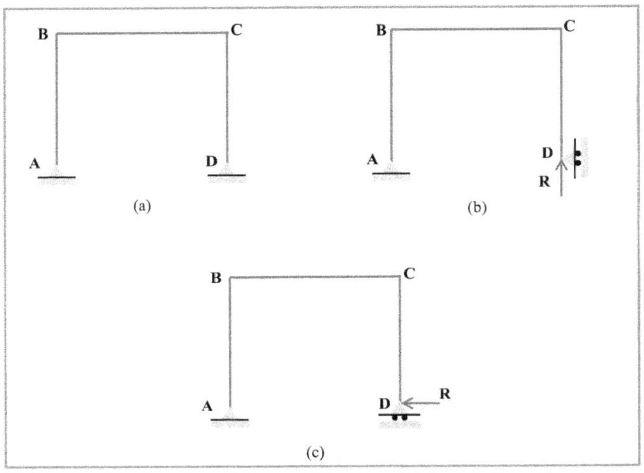

FIGURE 3.2 (a) Basic or determinate structure, (b) unstable, and (c) stable frames.

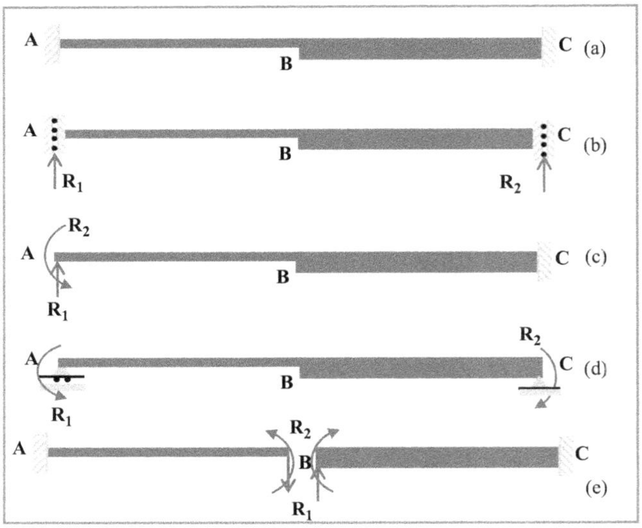

FIGURE 3.3 (a) Fixed beam indeterminate to the second degree, (b) basic or determinate beam but unstable, and (c–e) basic determinate beams that are stable.

DOI: 10.1201/9781003562955-3

a. Select the redundant forces in such a way that, in the resulting basic/determinate structure, the displacements can be easily calculated (i.e., converting it into a simple structure).
b. Select the redundants in such a way that in the resulting basic or determinate structure, the maximum number of flexibility coefficients are equal to zero.

The above cases are illustrated in Figure 3.4. Consider a fixed beam shown in Figure 3.4a. The beam is non-prismatic and statically indeterminate to the second degree. Three basic determinate structures are shown in Figure 3.4b–d. Among these three structures, the beam shown in Figure 3.4d requires the least computational effort, as the resulting basic determinate structure consists of two uniform cantilever beams.

Another example is shown in Figure 3.5a, where the three-span continuous beam has a degree of static indeterminacy of 2. The two basic determinate structures are shown in Figure 3.5b and f. In the first basic determinate structure shown in Figure 3.5b, the vertical support reactions at B and C are chosen as redundants. For this basic determinate structure, the bending moment diagram due to loads (M_L diagram) is shown in Figure 3.5c, and the m_1 and m_2 diagrams due to unit loads corresponding to R_1 and R_2, respectively, are shown in Figure 3.5d and e.

Similarly, the second basic determinate structure is shown in Figure 3.5f, where the internal moments at supports B and C are selected as redundants. The M_L diagram due to loads for this basic structure is shown in Figure 3.5g, and the m_1 and m_2 diagrams due to unit loads corresponding to R_1 and R_2 are shown in Figure 3.5h and i, respectively. Compared to the first case of the basic determinate structure, the second basic structure requires less computational effort for the calculation of deflections, as more flexibility coefficients in the second case are zero, whereas, in the first case, all flexibility coefficients are non-zeros.

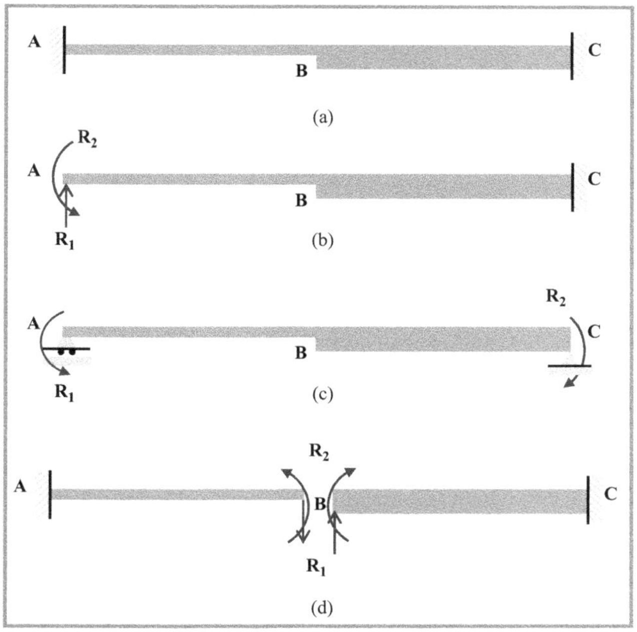

FIGURE 3.4 (a) Fixed beam indeterminate to the second degree, (b) basic or determinate beam with reactions at A as redundants, (c) basic beam with moments at A and C as redundants, and (d) basic beam with internal forces at B as redundants.

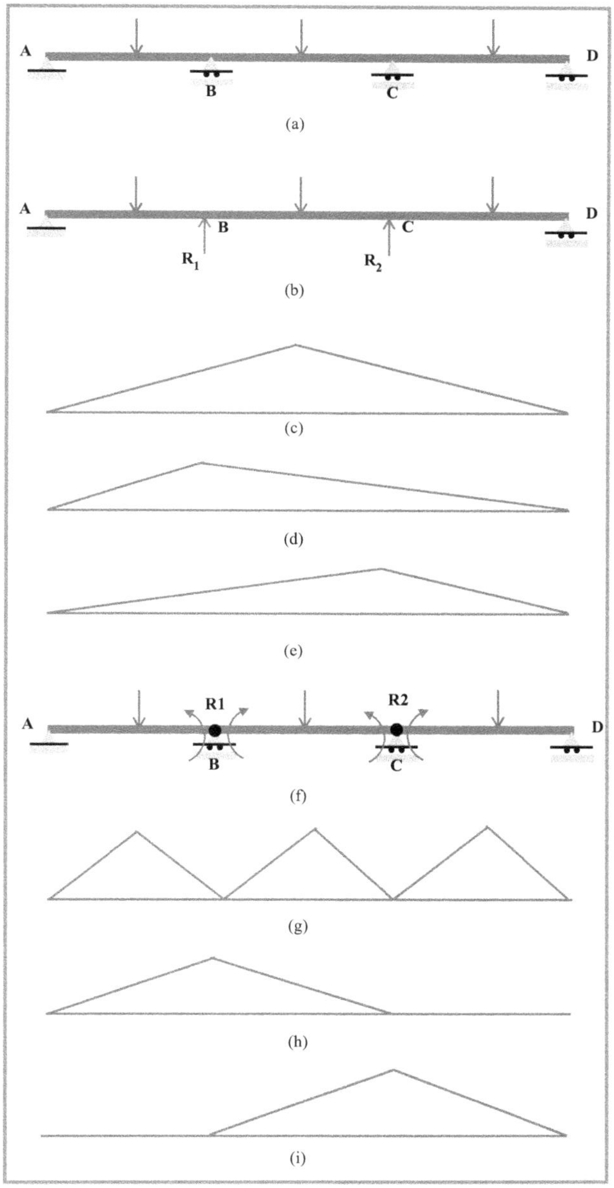

FIGURE 3.5 (a) Indeterminate beam, (b) determinate beam with Rb and Rc as redundants, (c) M_L diagram, (d) m_1 diagram, (e) m_2 diagram, (f) determinate beam with internal moments at B and C as redundants, (g) M_L diagram, (h) m_1 diagram, and (i) m_2 diagram.

3.2 FORMULATION OF CONSISTENT DEFORMATION METHOD

For convenience, let us first consider a fixed beam subjected to some arbitrary loading, as shown in Figure 3.6, which is statically indeterminate to the second degree. Once the procedure is understood, we can easily extend the same approach to structures with an *n* degree of static indeterminacy.

Step 1: **Identification of the degree of static indeterminacy and selection of redundant forces**: The beam is statically indeterminate to the second degree (neglecting the horizontal reactions). This means that there are two reactive forces that can be removed without jeopardising the stability of the structure. Let us select the vertical reaction V_B ($= R_1$) and moment M_B ($= R_2$) at *B* as the redundant reactive forces (Figure 3.6a).

Step 2: **Basic, primary, or determinate structure**: The structure that remains after the removal of the redundant reactions from the original or indeterminate structure is called the primary, basic, or determinate structure. Let us remove the selected redundant forces from the original structure to obtain the primary structure, as shown in Figure 3.6b.

Step 3: **Principle of superposition**: Using the principle of superposition, draw the indeterminate structure and express it as the summation of a series of primary structures with external loads and redundants with unit loads. Also, draw the deflected shape of the determinate structure and indicate the deflections with suitable notations at the locations of the redundants (Figure 3.6b–d).

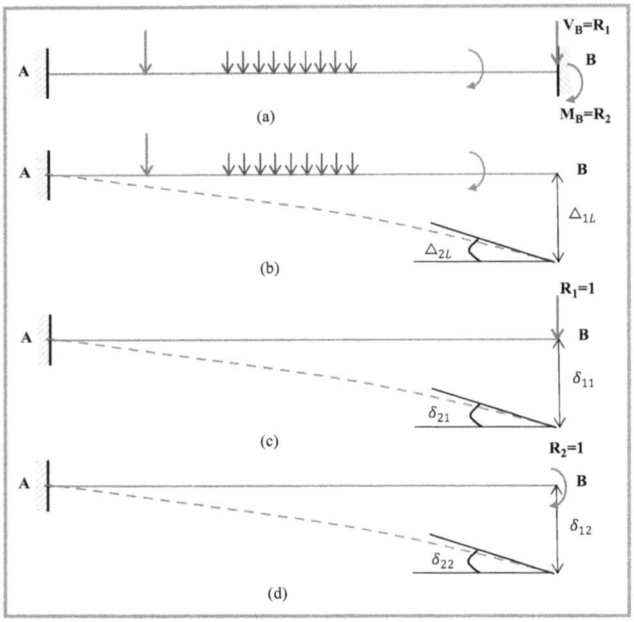

FIGURE 3.6 (a) Indeterminate fixed beam, (b) Primary beam with external loads, (c) primary beam with unit load at R_1 and (d) primary beam unit couple at R_2.

Step 4: **Compatibility equations**: Write the compatibility equations in terms of the displacements due to loads, flexibility coefficients, and redundants corresponding to the locations of the chosen redundants.

$$\Delta_1 = \Delta_{1L} + R_1\delta_{11} + R_2\delta_{12} = 0$$
$$\Delta_2 = \Delta_{2L} + R_1\delta_{21} + R_2\delta_{22} = 0 \qquad (3.1)$$

Step 5: **Displacements and flexibility coefficients**: Calculate the displacements due to loads and the flexibility coefficients due to unit loads, either using the unit load method, any other suitable method, or standard expressions (refer to Tables 3.1–3.3). If the unit load method is applied, then

$$\Delta_{1L} = \int M_L m_1 \frac{dx}{EI}$$
$$\Delta_{2L} = \int M_L m_2 \frac{dx}{EI}$$
$$\delta_{11} = \int m_1 m_1 \frac{dx}{EI} \qquad (3.2)$$
$$\delta_{12} = \delta_{21} = \int m_1 m_2 \frac{dx}{EI}$$
$$\delta_{22} = \int m_2 m_2 \frac{dx}{EI}$$

where M_L, m_1, m_2 are the bending moment diagrams due to external loads and unit loads at redundants on the primary structure.

Step 6: **Solve for the unknown redundants**: Substitute the displacements and flexibility coefficients calculated in Step 5 into the compatibility equations (Step 4) and solve for the unknown redundants. If the numerical value of a redundant is negative, it indicates that the direction of the redundant is opposite to that of the unit load.

Step 7: **Remaining unknown reactions**: With the known values of the redundants, draw the free body diagram and apply the equilibrium equations to determine the remaining unknown reactive forces. After determining all the reactive forces, we can draw the shear force and bending moment diagrams, as well as the probable deflected shape or elastic curve.

The above procedure is illustrated in the following examples.

TABLE 3.1
Slopes and Deflections in Cantilever Beams

Type of Beam	Slope at Free End	Deflection at Free End
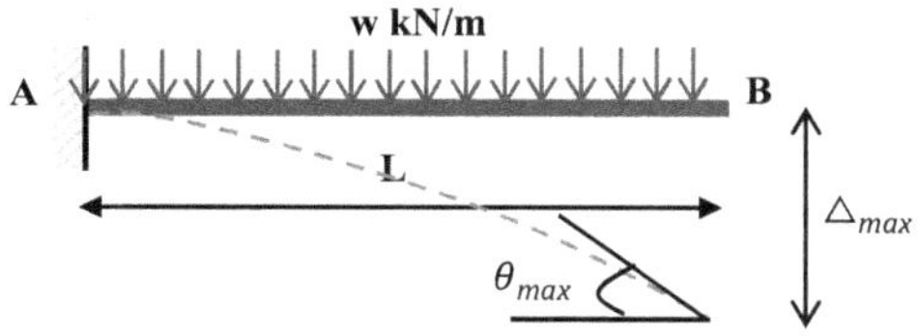	$\theta_{\max} = \dfrac{wL^3}{6EI}$	$\Delta_{\max} = \dfrac{wL^4}{8EI}$
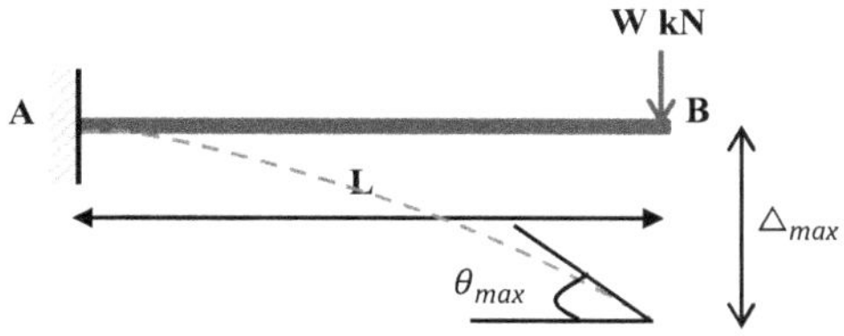	$\theta_{\max} = \dfrac{WL^2}{2EI}$	$\Delta_{\max} = \dfrac{WL^3}{3EI}$
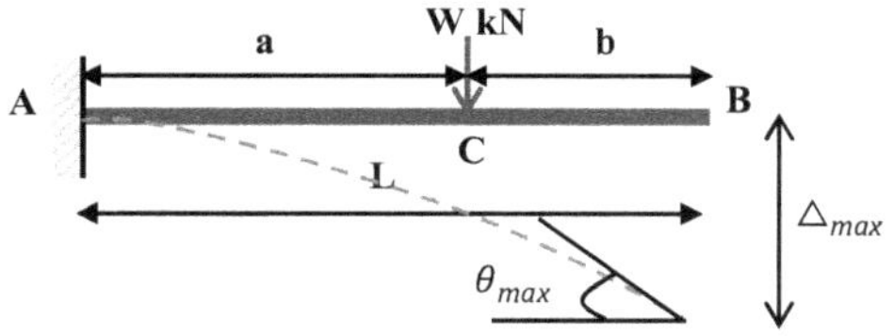	$\theta_{\max} = \dfrac{Wa^2}{2EI}$	$\Delta_C = \dfrac{Wa^3}{3EI}$ $\Delta_{\max} = \dfrac{Wa^3}{3EI} + \dfrac{Wa^2 b}{2EI}$
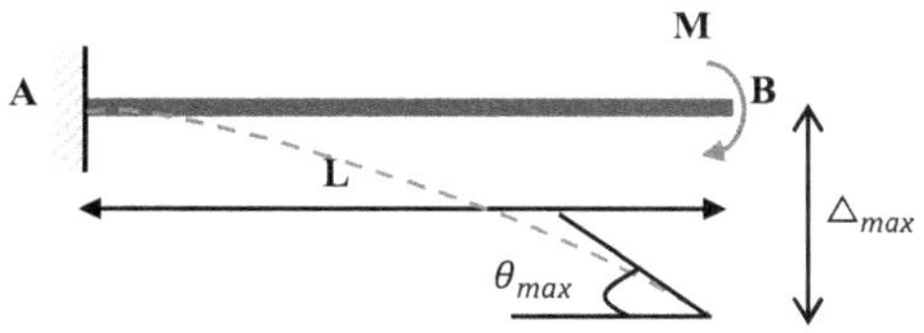	$\theta_{\max} = \dfrac{ML}{EI}$	$\Delta_{\max} = \dfrac{ML^2}{2EI}$
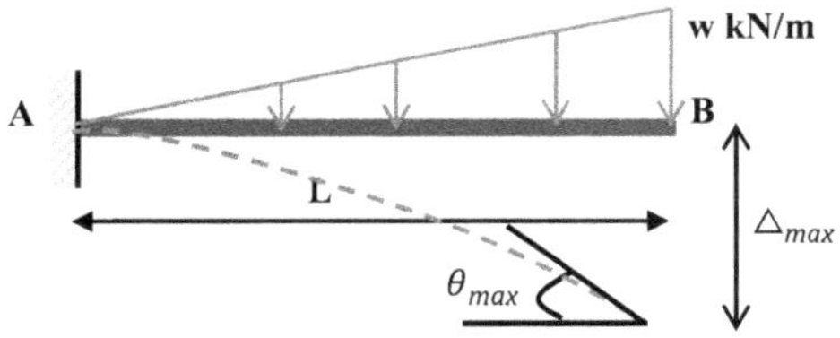	$\theta_{\max} = \dfrac{wL^3}{8EI}$	$\Delta_{\max} = \dfrac{11wL^4}{120EI}$
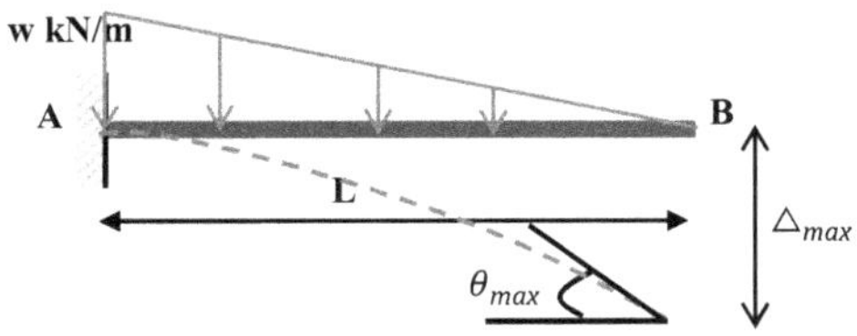	$\theta_{\max} = \dfrac{wL^3}{24EI}$	$\Delta_{\max} = \dfrac{wL^4}{30EI}$
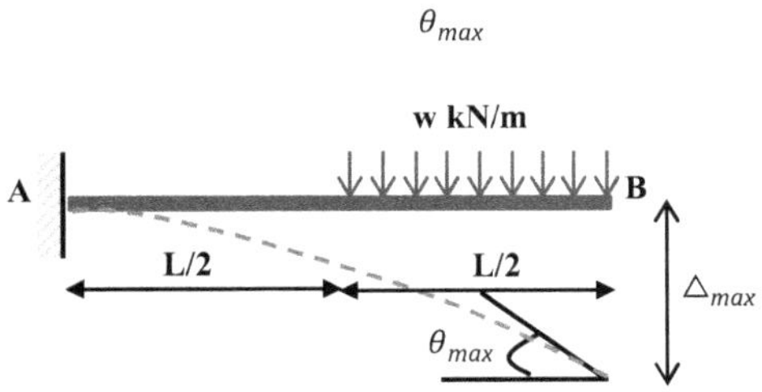	$\theta_{\max} = \dfrac{7wL^3}{48EI}$	$\Delta_{\max} = \dfrac{41wL^4}{384EI}$

TABLE 3.2
Slopes and Deflections in Simply Supported Beams

Type of Beam	Slope	Maximum Deflection
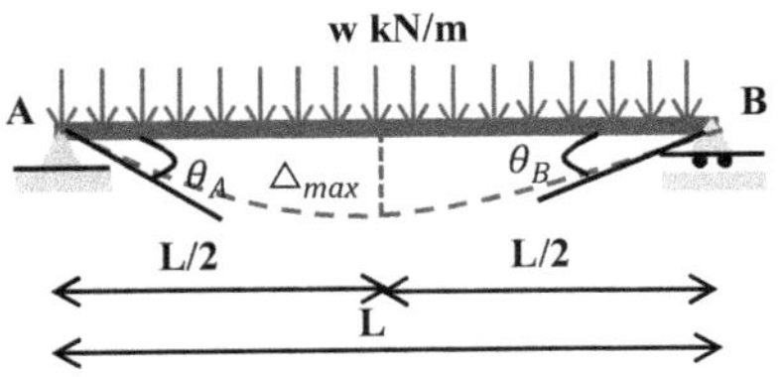	$\theta_A = \theta_B = \dfrac{wL^3}{24EI}$	$\Delta_{\max} = \dfrac{5wL^4}{384EI}$
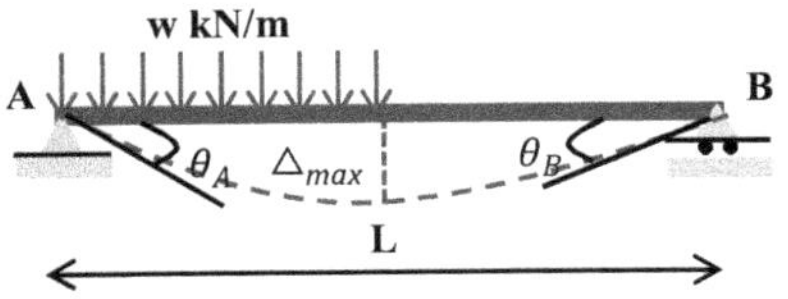	$\theta_A = \dfrac{3wL^3}{128EI}$	$\Delta_{\max} = \dfrac{wL^4}{152.44EI}$
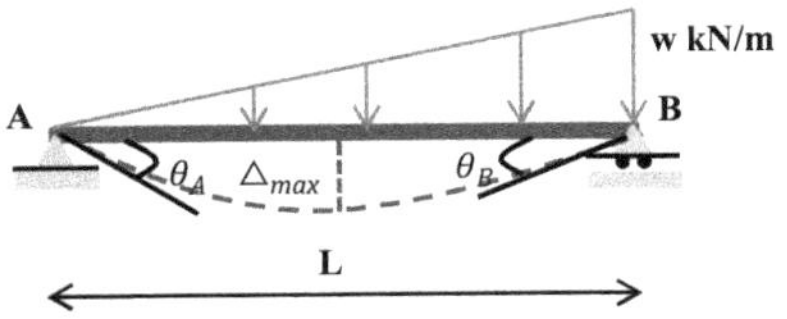	$\theta_A = \dfrac{7wL^3}{180EI}$ $\theta_B = \dfrac{wL^3}{45EI}$	$\Delta_{\max} = \dfrac{wL^4}{153.37EI}$
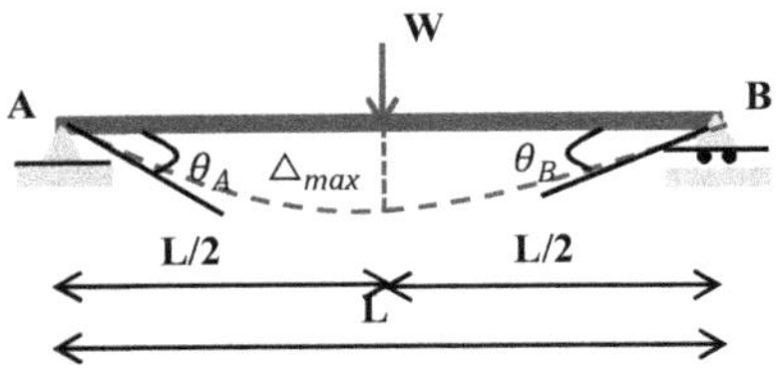	$\theta_A = \theta_B = \dfrac{WL^2}{16EI}$	$\Delta_{\max} = \dfrac{WL^3}{48EI}$
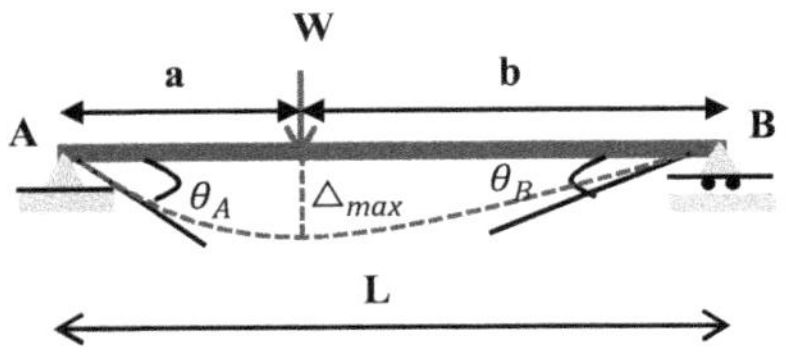	$\theta_A = \dfrac{Wab(L+b)}{6EIL}$ $\theta_B = \dfrac{Wab(L+a)}{6EIL}$	$\Delta_{\max} = \dfrac{Wb\left(L^2-b^2\right)^{3/2}}{9\sqrt{3}EIL}$ $\Delta_{\text{centre}} = \dfrac{Wb\left(3L^2-4b^2\right)}{48EI}$ if $a > b$
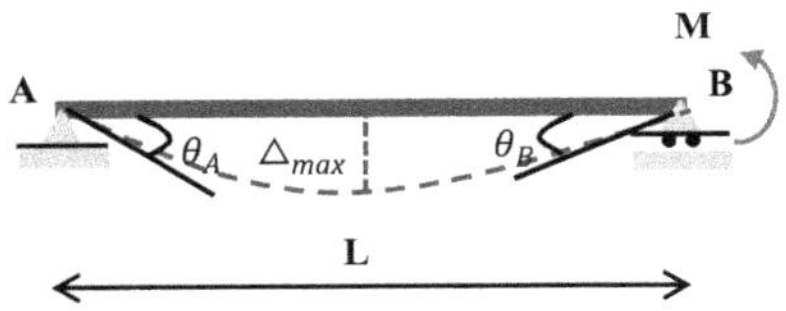	$\theta_A = \dfrac{ML}{6EI}$ $\theta_B = \dfrac{ML}{3EI}$	$\Delta_{\max} = \dfrac{ML^2}{9\sqrt{3}EI}$

3.3 APPLICATION TO INDETERMINATE BEAMS

Example 3.1

Analyse the propped cantilever beam shown in Figure 3.7 using the consistent deformation method. Draw the shear force and bending moment diagrams.

Solution:

The beam is statically indeterminate to the first degree. We can choose either the moment at A (M_A) or the reaction at B (R_B) as the redundant force. Let us choose R_B (assumed to act downwards) as the redundant force in this case. The basic beam with external loads, redundant R_B, and a unit load at R_B=1 is shown in Figure 3.8a–c.

The compatibility equation is

$$\Delta_B = \Delta_{BL} + \Delta_{BR}$$
$$\Delta_B = \Delta_{BL} + \delta_{BB} \cdot R_B \tag{3.3}$$

where

Δ_{BL} = Displacement at B due to external loads on basic beam.

$$\Delta_B = \int \frac{M_{(x)} \cdot m_{(x)}}{EI} \cdot dx$$

where $M_{(x)} \rightarrow BM$ due to external loads on basic beam.

$m_{(x)} \rightarrow BM$ due to unit loads.

TABLE 3.3
Evaluation of $\int M_L m \frac{dx}{EI}$

	M_L, L	M_L, L	M_L, M_L, L	L, M_L
m, L	$M_L \cdot m \cdot L$	$\frac{1}{2} M_L \cdot m \cdot L$	$\frac{L}{2}(M_{L_1} + M_{L_2}) \cdot m$	$\frac{2}{3} M_L \cdot L \cdot m$
m, L	$\frac{1}{2} M_L \cdot m \cdot L$	$\frac{1}{3} M_L \cdot m \cdot L$	$\frac{L}{6}[m(2M_{L_1} + M_{L_2})]$	$\frac{1}{4} M_L \cdot L \cdot m$
m, m, L	$\frac{L}{2}(m_1 + m_2) \cdot M_L$	$\frac{L}{6}[M_L(2m_1 + m_2)]$	$\frac{L}{6}\left[\begin{matrix} M_{L_1}(2m_1 + m_2) + \\ M_{L_2}(m_1 + 2m_2) \end{matrix}\right]$	$\frac{L}{12}[M_L(3m_1 + 5m_2)]$
m, L	$\frac{1}{2} M_L \cdot m \cdot L$	$\frac{1}{6} M_L \cdot m \cdot L$	$\frac{L}{6}[m(M_{L_1} + 2M_{L_2})]$	$\frac{5}{12} M_L \cdot L \cdot m$

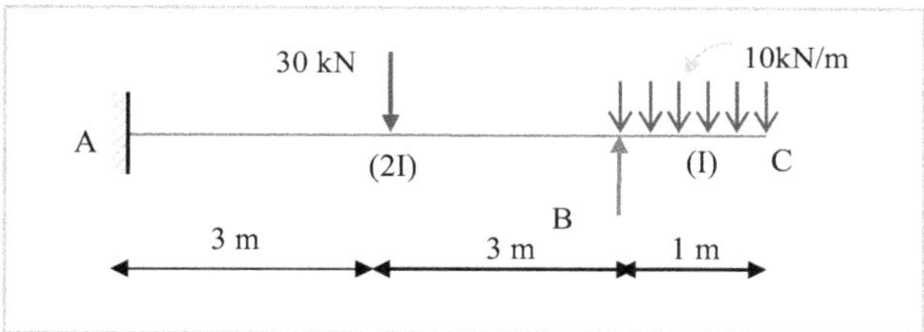

FIGURE 3.7 Propped cantilever beam.

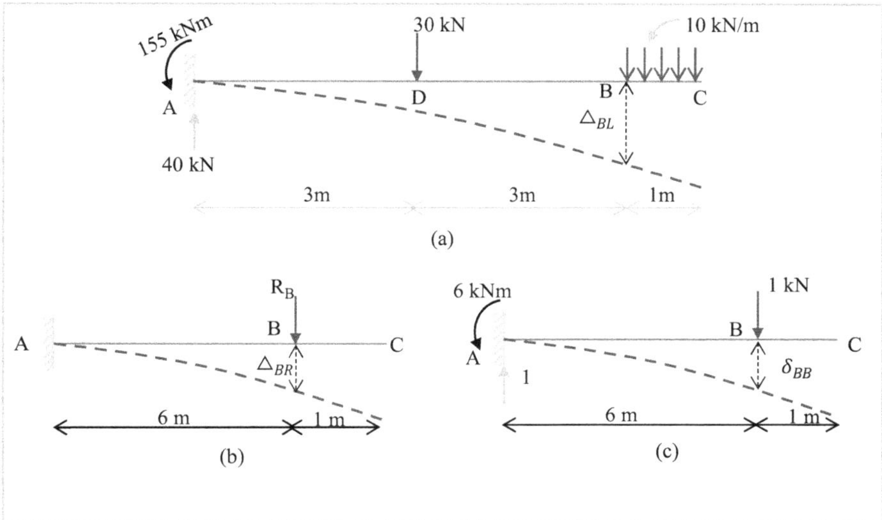

FIGURE 3.8 (a) Primary beams with loads, (b) redundant, and (c) unit load.

δ_{BB} = flexibility coefficient i.e., displacement at B due to unit load at B on the basic beam.

$$\delta_{BB} = \int m_{(x)} \cdot m_{(x)} \cdot \frac{dx}{EI}$$

$$\Delta_{BL} = \int_0^1 \frac{-10 \times x^2}{2}(0)\frac{dx}{EI} + \int_0^3 \left[-10\times 1\times\left(x+\frac{1}{2}\right)\right][(-x)]\frac{dx}{2EI}$$

$$+\int_0^3 (40x-155)(x-6)\frac{dx}{2EI}$$

$$= \left|\frac{10x^3}{3}+\frac{5x^2}{2}\right|_0^3 \frac{1}{2EI} + \left|\frac{40x^3}{3}-\frac{395x^2}{2}+930x\right|_0^3 \frac{1}{2EI}$$

$$= \frac{56.25}{EI}+\frac{686.25}{EI} = \frac{742.5}{EI}$$

$$\delta_{BB} = \int m_{(x)} \cdot m_{(x)} \cdot \frac{dx}{EI} = \int_0^3 (-x)(-x)\frac{dx}{2EI} + \int_0^3 (x-6)(x-6)\frac{dx}{2EI}$$

$$= \frac{1}{2EI}\left|\frac{x^3}{3}\right|_0^3 + \frac{1}{2EI}\left|\frac{x^3}{3}-\frac{12x^2}{2}+36x\right|_0^3$$

$$= \frac{4.5}{EI}+\frac{31.5}{EI} = \frac{36}{EI}$$

From equation 3.1, $0 = \frac{742.5}{EI} + \frac{36}{EI} \cdot R_B = 0$

$R_B = -20.63$ kN $(\uparrow)$ i.e., R_B act upward

The other support reactions can be determined using the static equilibrium equations, with the help of the free body diagram shown in Figure 3.9a. The shear force and bending moment diagrams, along with the deflected shape, are shown in Figure 3.9b–d, respectively.

Example 3.2

Find the support reactions for a two-span continuous beam shown in Figure 3.10 using the consistent deformation method. Draw the shear force and bending moment diagrams. Also, draw the probable deflected shape. Assume *EI* is constant.

Solution:

The beam is statically indeterminate to the second degree. We can choose either of the two unknown reactions as redundants. Let us choose the moment at *A* (M_A) (assumed to act in an anticlockwise direction) and the moment at *B* (M_B) (assumed to act as a sagging moment) as redundant (Figure 3.11a). The primary beam with external loads, and unit loads at M_A and M_B=1 is shown in Figure 3.11b–d, respectively.

The compatibility equations are

$$\Delta_1 = \Delta_{1L} + \delta_{11} \cdot M_A + \delta_{12} \cdot M_B$$

$$0 = \Delta_{1L} + \delta_{11} \cdot M_A + \delta_{12} \cdot M_B \quad (3.4)$$

$$\Delta_2 = \Delta_{2L} + \delta_{21} \cdot M_A + \delta_{22} \cdot M_B$$

$$0 = \Delta_{2L} + \delta_{21} \cdot M_A + \delta_{22} \cdot M_B \quad (3.5)$$

where

Δ_{1L}, Δ_{2L} are the rotations at *A* and *B* due to external loads on the primary beam.

δ_{11}, δ_{12}, and δ_{21}, δ_{22} are the flexibility coefficients due to unit moments at *A* and *B*, respectively

The rotations and flexibility coefficients are calculated using the standard expressions given in Table 3.1, as below.

$$\Delta_{1L} = \frac{ML}{24EI} = \frac{10\times 4}{24EI} = \frac{5}{3EI}$$

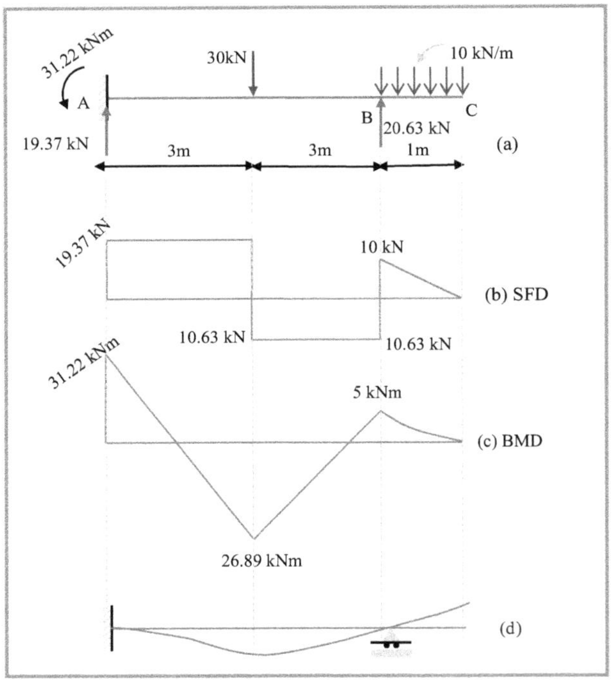

FIGURE 3.9 (a) Free body diagram, (b) shear force diagram, (c) bending moment diagram, and (d) probable deflected shape.

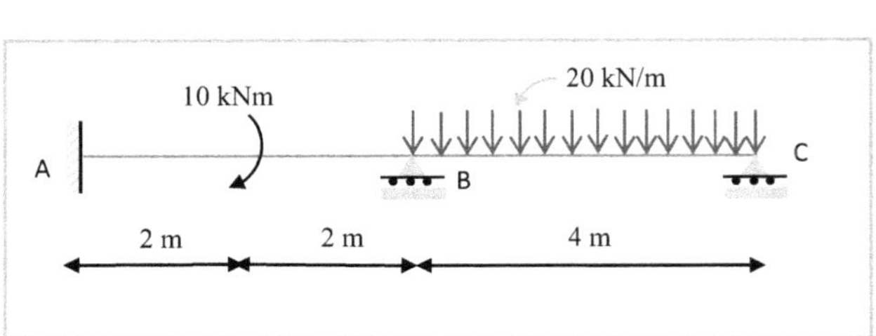

FIGURE 3.10 Two-span continuous beam.

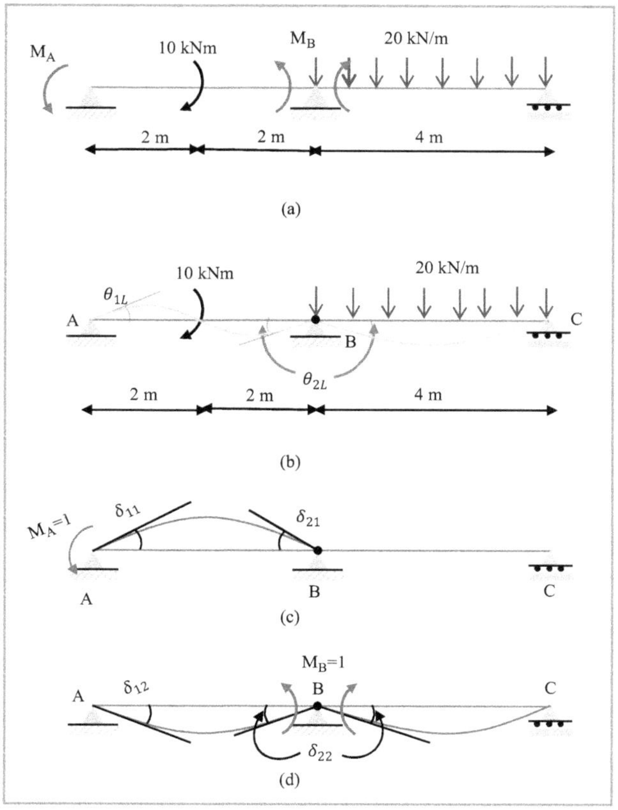

FIGURE 3.11 (a) Primary beam with external loads and redundants, (b) primary beam with external loads, (c) primary beam with unit couple at A, and (d) primary beam with unit couple B.

$$\Delta_{2L} = \frac{ML}{24EI} + \frac{wL^3}{24EI} = \frac{10 \times 4}{24EI} + \frac{20 \times 4^3}{24EI} = \frac{5}{3EI} + \frac{160}{3EI} = \frac{165}{3EI}$$

$$\delta_{11} = \frac{L}{3EI} = \frac{4}{3EI}$$

$$\delta_{21} = \frac{-L}{6EI} = \frac{-4}{6EI} = \frac{-2}{3EI}$$

$$\delta_{12} = \frac{-L}{6EI} = \frac{-4}{6EI} = \frac{-2}{3EI}$$

$(\delta_{12} = \delta_{21}$ from Maxwell's reciprocal theorem$)$

$$\delta_{22} = \frac{L}{3EI} + \frac{L}{3EI} = \frac{4}{3EI} + \frac{4}{3EI} = \frac{8}{3EI}$$

Substituting these values in equations 3.4 and 3.5, we get

$$\frac{5}{3EI} + \frac{4}{3EI} \times M_A + \left(\frac{-2}{3EI}\right) \times M_B = 0$$

$$4M_A - 2M_B = -5 \Rightarrow 2M_A - M_B = \frac{-5}{2} \quad (3.4a)$$

$$\frac{165}{3EI} + \left(\frac{-2}{3EI}\right) \times M_A + \frac{8}{3EI} \times M_B = 0$$

$$-M_A + 4M_B = \frac{-165}{2} \quad (3.5a)$$

By solving equations 3.4a and 3.5a, we get

$$M_B = -23.92 \text{ kNm}$$

The –ve sign indicates that M_B is acting opposite to the assumed direction.

From equation 3.4a, $M_A = -13.21$ kNm

Using the known values of M_A and M_B, and with the help of the static equilibrium equations, the other support reactions can be determined, as shown in Figure 3.12a. The shear force and bending moment diagrams, along with the probable elastic curve, are presented in Figure 3.12b–d, respectively.

Example 3.3

Analyse the continuous beam *ABCDE* shown in Figure 3.13 using the method of consistent deformation. Assume *EI* is constant. Draw the shear force and bending moment diagrams, as well as the probable elastic curve.

Solution:

The degree of static indeterminacy, $I_s = 4 - 2 = 2$. Select the reactions at A and B as redundants, i.e., R_A and R_D, respectively (assumed to act in the downward direction). The primary beam with external loads and the selected redundants is shown in Figure 3.14.

The primary beam with external loads, along with the bending moment diagrams, is shown in Figure 3.15a–c. The basic beam with a unit load at A (R_A= 1), along with the bending moment diagram, is shown in Figure 3.16a and b, respectively. Similarly, the primary beam with a unit load at D (R_D= 1), along with the bending moment diagram, is shown in Figure 3.16c and d, respectively.

Therefore, the compatibility conditions can be written as

$$\Delta_1 = \Delta_{1L} + \delta_{11} \cdot R_A + \delta_{12} \cdot R_D \quad (3.6)$$

$$\Delta_2 = \Delta_{2L} + \delta_{21} \cdot R_A + \delta_{22} \cdot R_D \quad (3.7)$$

where Δ_{1L} and Δ_{2L} are the displacements due to external loads at the redundant locations in the primary beam, and δ represents the flexibility coefficients due to unit loads at the redundant locations in the primary beam. These values can be determined using the unit load method and by applying the area–multiplication technique to the bending moment diagrams. (Negative sign can be neglected, as the terms are products.)

$$\Delta_{1L} = \int M_L \cdot m_1 \cdot \frac{dx}{EI}$$

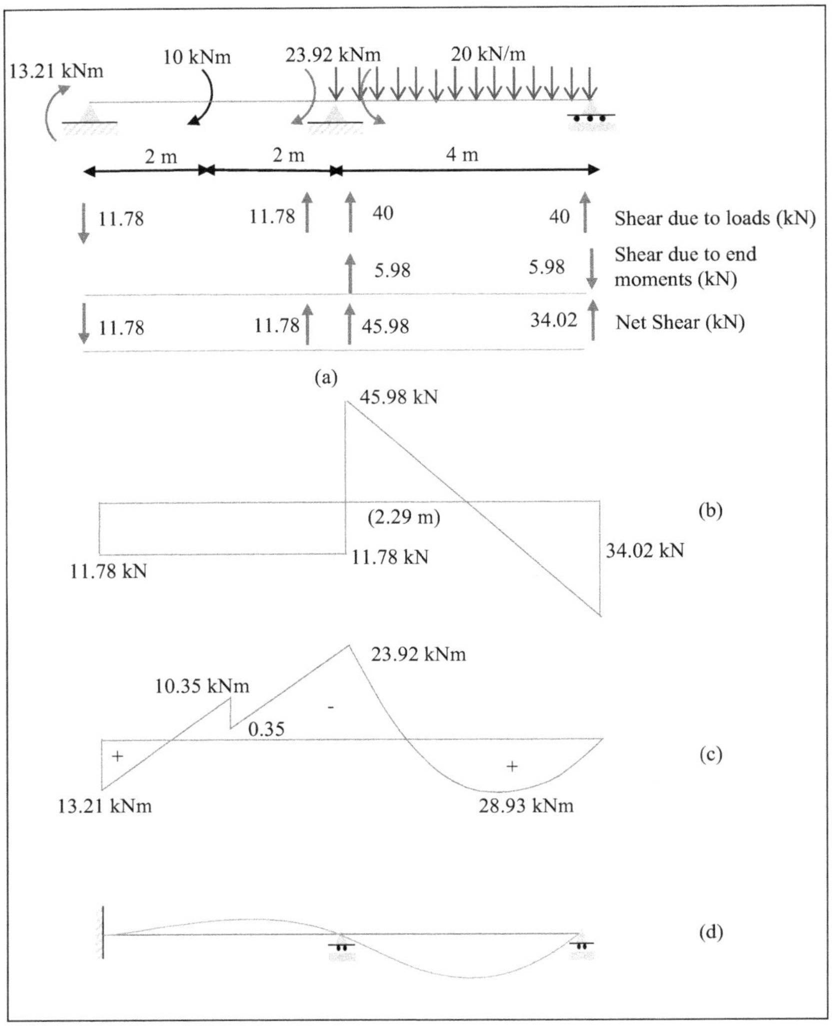

FIGURE 3.12 (a) Free body diagram, (b) shear force diagram, (c) bending moment diagram, and (d) probable deflected shape.

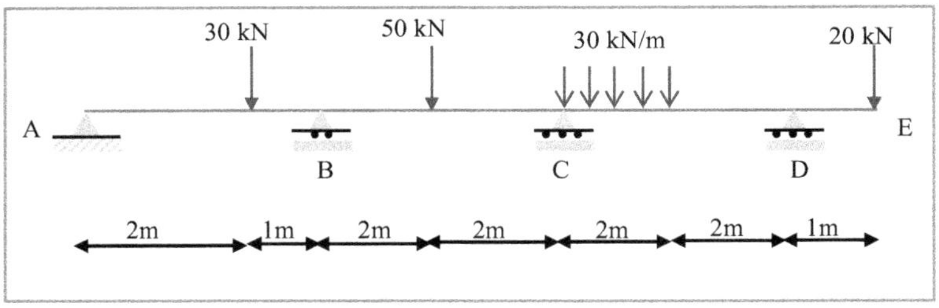

FIGURE 3.13 Continuous beam *ABCDE*.

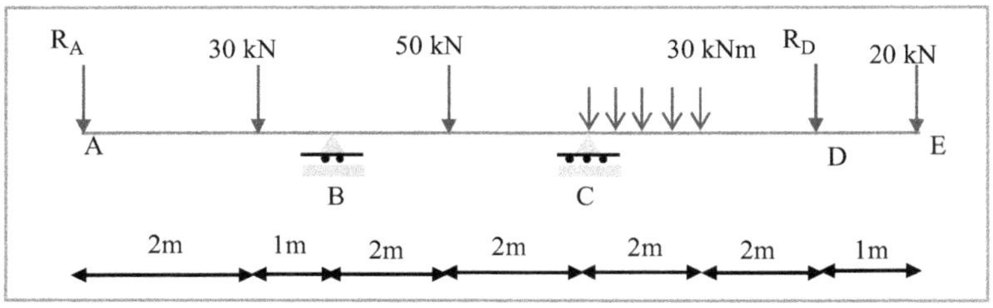

FIGURE 3.14 Primary beam with external loads and redundants.

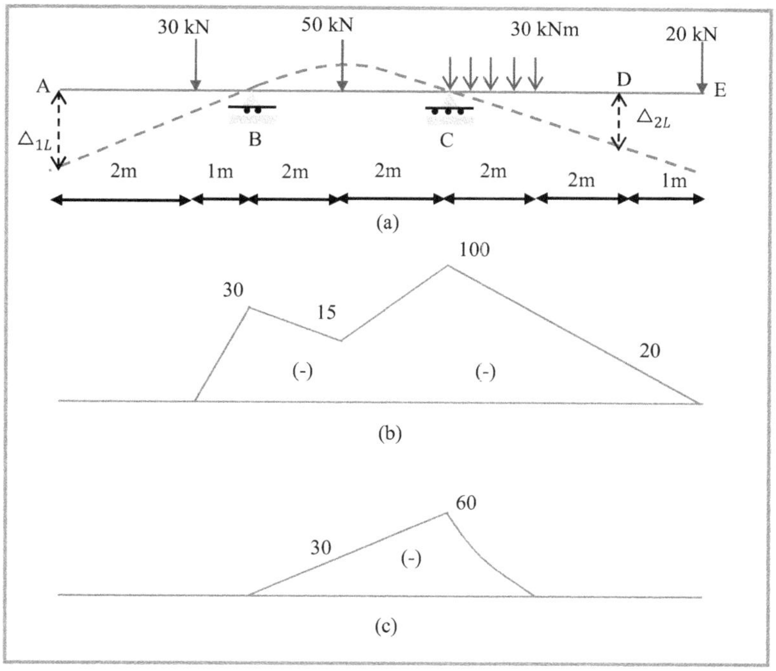

FIGURE 3.15 (a) Primary beam with external loads, (b) bending moment diagram due to concentrated loads, and (c) bending moment diagram due to *udl*.

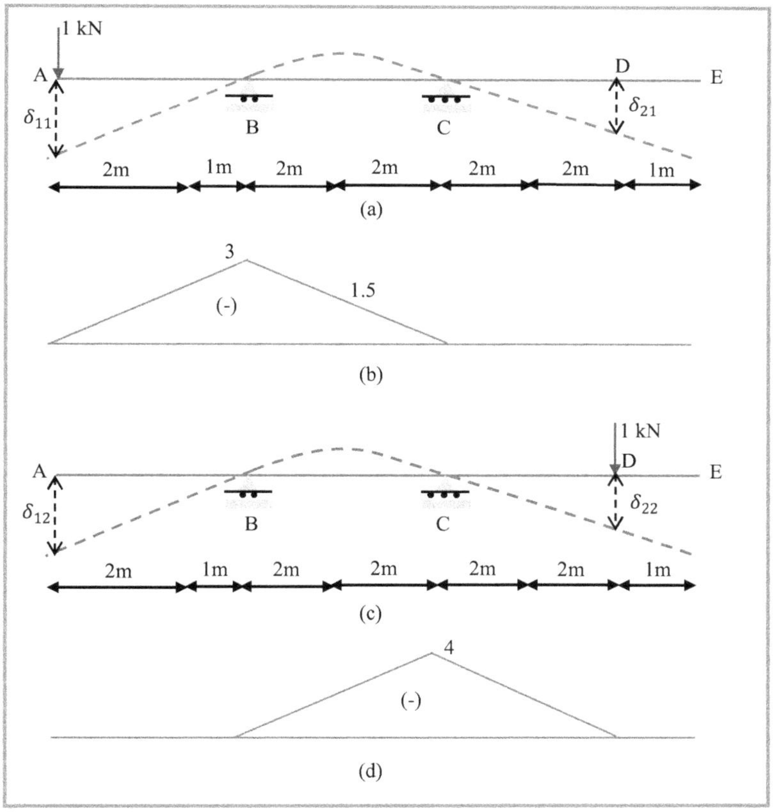

FIGURE 3.16 (a) Primary beam with unit load at A, (b) bending moment diagram due to unit load at A, (c) primary beam with unit load at D, and (d) bending moment diagram due to unit load at D.

$$\Delta_{1L} = \frac{1}{EI}\begin{bmatrix} \frac{1}{6}[30(2\times3+2)] + \frac{2}{6}[30(2\times3+1.5)+15(3+3)] \\ +\frac{2}{6}[15(3+0)+100(1.5)] + \frac{4}{6}(60\times3) \end{bmatrix}$$

$$\Delta_{1L} = \frac{1}{EI}[40+105+65+120] = \frac{330}{EI}$$

$$\Delta_{2L} = \frac{1}{EI}\begin{bmatrix} \frac{2}{6}[30(0+2)+15(2\times2+0)] + \frac{2}{6}\begin{bmatrix}15(2\times2+4) \\ +100(2\times4+2)\end{bmatrix} \\ +\frac{4}{6}[100(2\times4+0)+20(0+4)] + \frac{4}{6}[60(2\times4+0)] \\ +\frac{1}{3}\times60\times2\times3.5 \end{bmatrix}$$

$$\Delta_{2L} = \frac{1}{EI}[40+373.33+586.67+320+140] = \frac{1,460}{EI}$$

$$\delta_{11} = \frac{1}{EI}\left[\frac{3}{6}(3(2\times3+0)) + \frac{4}{6}(3(2\times3+0))\right] = \frac{21}{EI}$$

$$\delta_{12} = \delta_{21} = \frac{1}{EI}\left[\frac{4}{6}(3(2\times0+4))\right] = \frac{8}{EI}$$

$$\delta_{22} = \frac{1}{EI}\left[\frac{4}{6}(4(2\times4+0))\times2\right] = \frac{42.67}{EI}$$

From the compatibility equations, we can write

$$0 = \frac{330}{EI} + \frac{21}{EI}\times R_A + \frac{8}{EI}\times R_D \tag{3.6a}$$

$$0 = \frac{1,460}{EI} + \frac{8}{EI}\times R_A + \frac{42.67}{EI}\times R_D \tag{3.7a}$$

By solving the equations 3.6a and 3.7a, we get $R_A = -2.89$ kN and $R_D = -33.67$ kN

The other support reactions can be determined using the free body diagram shown in Figure 3.17a and by applying the equilibrium equations. The shear force and bending moment diagrams, along with the probable deflected shape, are shown in Figure 3.17b–d, respectively.

Example 3.4

A two-span continuous beam is shown in Figure 3.18. Construct the shear force and BM diagrams after analysing it using the consistent deformation method.

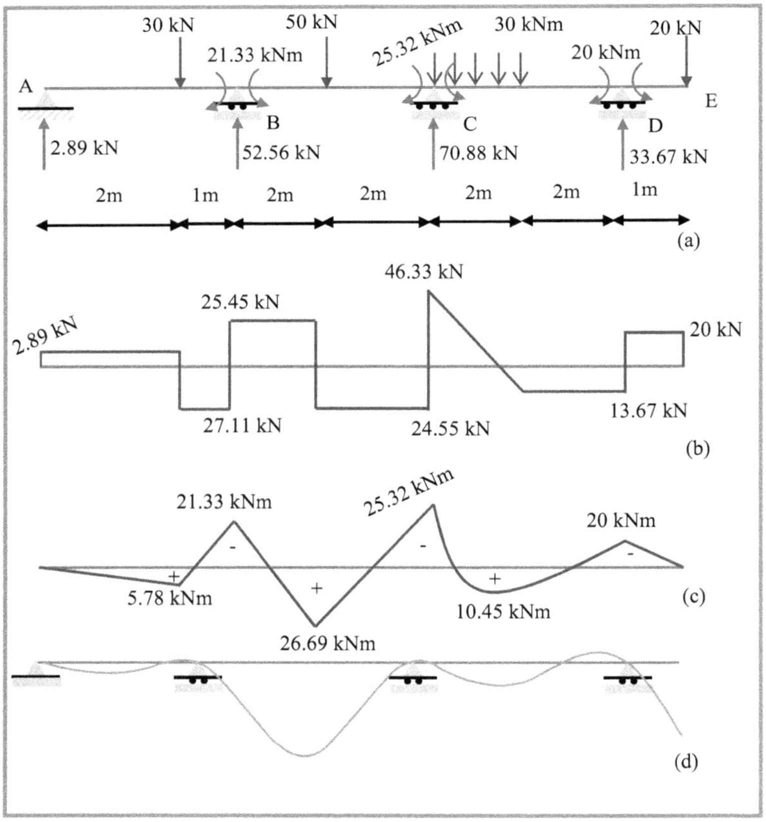

FIGURE 3.17 (a) Free body diagram, (b) shear force diagram, (c) bending moment diagram, and (d) probable deflected shape.

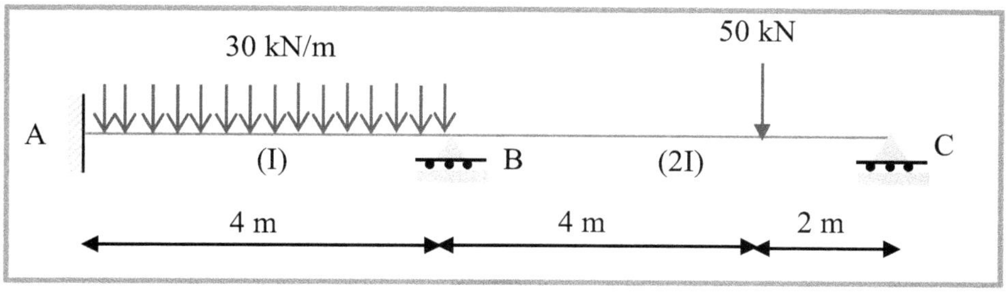

FIGURE 3.18 Two-span continuous beam.

Solution:

The static indeterminacy, $I_s = 5 - 3 = 2$. Out of the total 05 unknown reactions, select any two (except H_A) as redundant reactions. Let us assume R_B and R_C as redundant reactions, which are assumed to act in the downward direction. Remove the selected redundant to obtain the determinate or primary beam, as shown in Figure 3.19a. The primary beam with external loads and the associated bending moment diagrams is shown in Figure 3.19a–c.

Remove the external loads on the primary beam and apply a unit force in place of the redundant reactions R_B and R_C successively. Then, draw the bending moment diagrams due to the unit loads, as shown in Figure 3.20a–d.

$$\Delta_B = \Delta_{BL} + \delta_{11} \cdot R_B + \delta_{12} \cdot R_C = 0 \tag{3.8}$$

$$\Delta_C = \Delta_{CL} + \delta_{21} \cdot R_B + \delta_{22} \cdot R_C = 0 \tag{3.9}$$

The Δ_L and the flexibility coefficients are calculated using the unit load method and by applying the area–moment multiplication technique.

$$\Delta_{1L} = \int_{AB} M_L \cdot m_1 \cdot \frac{dx}{EI} + \int_{BC} M_L \cdot m_1 \cdot \frac{dx}{EI}$$

$$= \frac{1}{EI}\frac{4}{6} \times 4(2\times 400 + 200) + \frac{1}{3} \times 4 \times 240 \times 3 + \frac{1}{2EI}(0)$$

$$= \frac{2{,}666.67}{EI} + \frac{960}{EI} = \frac{3{,}626.67}{EI}$$

$$\Delta_{2L} = \int M_L \cdot m_2 \cdot \frac{dx}{EI} = \frac{14}{6EI}\begin{bmatrix} [10(800+200) + 6(400+400)] \\ +\frac{1}{3} \times 4 \times 240 \times 9 \end{bmatrix}$$

$$+ \frac{1}{2EI}[6(200\times 2 + 0) + 2(0 + 200)] \times \frac{4}{6}$$

$$= \frac{12{,}746.67}{EI} + \frac{933.33}{EI} = \frac{13{,}680}{EI}$$

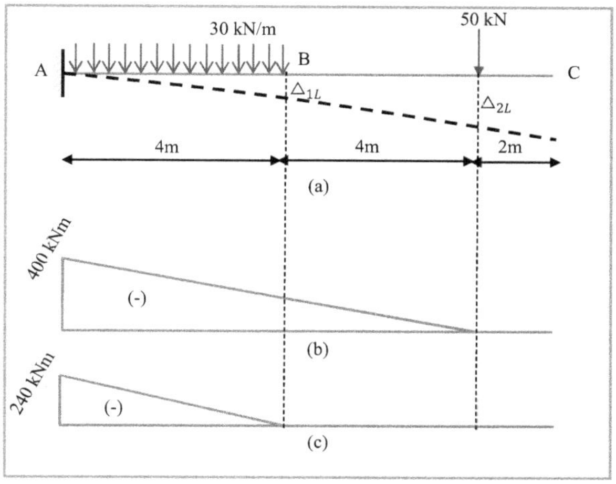

FIGURE 3.19 (a) Basic beam external loads, (b) bending moment diagram due to 50 kN load, and (c) bending moment diagram due to *udl*.

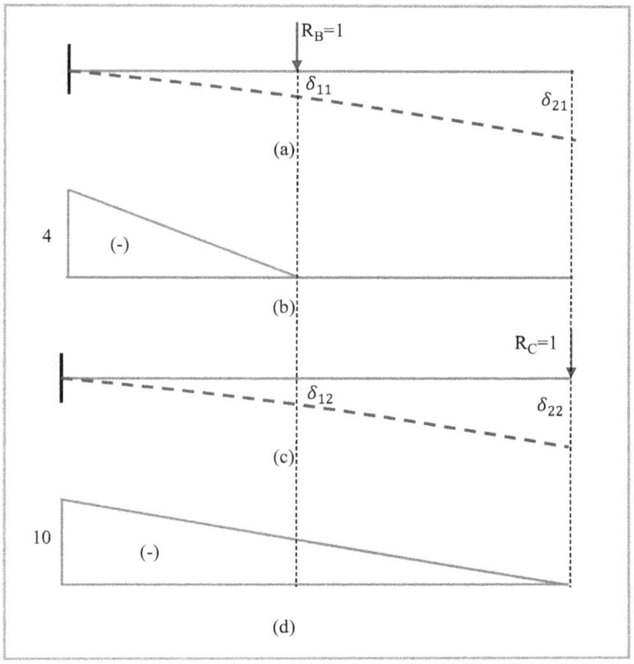

FIGURE 3.20 (a) Basic beam with unit load at *B*, (b) bending moment diagram due to unit load at *B*, (c) basic beam with unit load at *C*, and (d) bending moment diagram due to unit load at *C*.

FLEXIBILITY COEFFICIENTS

$$\delta_{11} = \int m_1 \cdot m_1 \cdot \frac{dx}{EI} = \left(\frac{1}{EI} \times \frac{1}{2} \times 4 \times 4 \times 2.67\right) = \frac{21.33}{EI}$$

$$\delta_{12} = \delta_{21} = \int m_1 \cdot m_2 \cdot \frac{dx}{EI} = \frac{1}{EI} \times \frac{4 \times 4}{6}(2 \times 10 + 6) + 0 = \frac{69.33}{EI}$$

$$\delta_{22} = \int m_2 \cdot m_2 \cdot \frac{dx}{EI} = \frac{1}{EI} \times \frac{4}{6}\left[10(2 \times 10 + 6) + 6(2 \times 6 + 10)\right]$$

$$+\left[\frac{1}{2EI}\left(\frac{1}{2} \times 6 \times 6 \times \frac{2}{3} \times 6\right)\right]$$

$$= \frac{261.33}{EI} + \frac{36}{EI} = \frac{297.33}{EI}$$

Substituting the above values into equations 3.8 and 3.9, we can write the compatibility equations as follows,

$$0 = \frac{33{,}626.67}{EI} + \frac{21.33}{EI} \times R_B + \frac{69.33}{EI} \times R_C \quad (3.8a)$$

$$0 = \frac{13{,}680}{EI} + \frac{69.33}{EI} \times R_B + \frac{297.33}{EI} \times R_C \quad (3.9a)$$

By solving equations 3.8a and 3.9a, we get

$$R_B = -84.59 \text{ kN} \quad \text{and} \quad R_C = -26.28 \text{ kN}$$

ALTERNATE SOLUTION

Select M_A and M_B as redundant reactions. The primary beam with external loads and unit moments at *A* and *B* is shown in Figure 3.21a–c.

$$\Delta_{1L} = \frac{wL^3}{24EI} = \frac{30 \times 4^3}{24EI} = \frac{80}{EI}$$

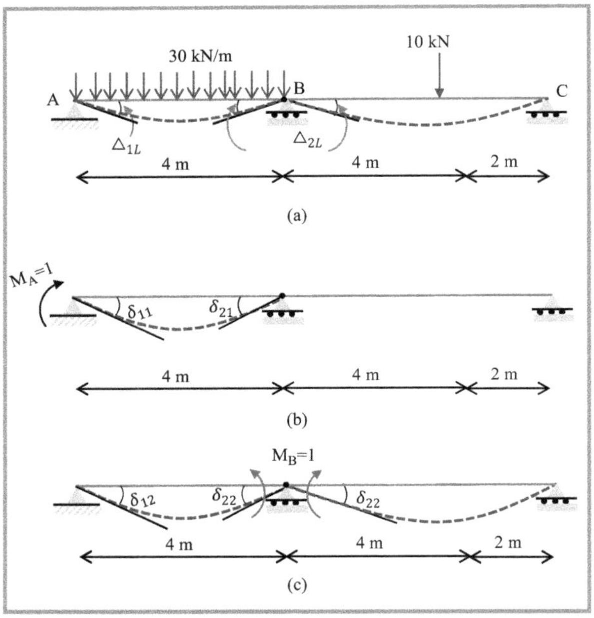

FIGURE 3.21 (a) Determinate beam with external loads, (b) determinate beam with unit moment at *A*, and (c) determinate beam with unit moment at *B*.

$$\Delta_{2L} = \frac{wL^3}{24EI} + \frac{W \times a \times b(L+b)}{6EIL} = \frac{80}{EI} + \frac{50 \times 2 \times 4}{6E(2I)6}(6+2) = \frac{124.44}{EI}$$

$$\delta_{11} = \frac{L}{3EI} = \frac{4}{3EI} = \frac{1.33}{EI}$$

$$\delta_{12} = \delta_{21} = \frac{L}{6EI} = \frac{4}{6EI} = \frac{0.667}{EI}$$

$$\delta_{22} = \frac{L}{3EI} + \frac{L}{3EI} = \frac{4}{3EI} + \frac{6}{3E(2I)} = \frac{2.33}{EI}$$

Substituting the above values into equations 3.8 and 3.9, the compatibility equations become,

$$0 = \frac{80}{EI} + \frac{1.33}{EI} \times M_A + \frac{0.667}{EI} \times M_B \quad (3.8b)$$

$$0 = \frac{124.44}{EI} + \frac{0.667}{EI} \times M_A + \frac{2.33}{EI} \times M_B \quad (3.9b)$$

By solving the equations 3.8b and 3.9b, we get $M_A = -39.01$ kNm and $M_B = 42.31$ kNm

With the known values of M_A and M_B and using the static equilibrium equations, the remaining unknown reactions can be determined (Figure 3.22a). The shear force and bending moment diagrams, along with the tentative deflected shape, are shown in Figure 3.22b–d, respectively.

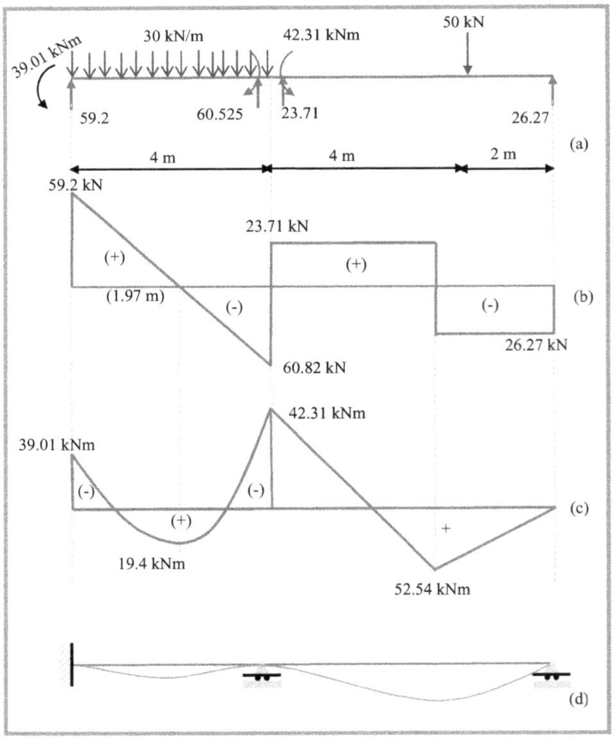

FIGURE 3.22 (a) Free body diagram, (b) shear force diagram, (c) bending moment diagram, and (d) tentative deflected shape.

Example 3.5

Analyse Problem No. 3.4 using the consistent deformation method if support 'A' rotates clockwise by 0.001 radian and support 'B' settles down by 10 mm in the absence of external loads. Assume $E = 2\times10^5$ N/mm^2; $I = 6.75\times10^8$ mm^4.

Solution:

The problem can be solved in two ways:

METHOD 1

The degree of static indeterminacy is 2. Let us select R_B and R_C as redundant reactions, which are assumed to act in the downward direction. Therefore, in compatibility equations 3.8 and 3.9, the values of $\Delta_B = 0.01$ m and $\Delta_C = 0$

The Δ_L or Δ_Y values can be determined due to the rotation of support A (Figure 3.23), i.e., using the chord rotations.

$$\Delta_{1y} = 0.001\times4 = 0.004 \text{ m}$$

$$\Delta_{2y} = 0.001\times10 = 0.01 \text{ m}$$

The flexibility coefficients remain the same as calculated in Q.No. 3.4. Substitute the values in the compatibility equations 3.8 and 3.9; the compatibility equations become

$$0.01 = 0.004 + \frac{21.33}{EI}\times R_B + \frac{69.33}{EI}\times R_C$$

$$0 = 0.01 + \frac{69.33}{EI}\times R_B + \frac{297.33}{EI}\times R_C$$

By solving these equations, we get $R_B = 217.81$ kN and $R_C = -55.32$ kN

METHOD 2

Instead of selecting the reactions R_B and R_C as redundant reactions, we can directly select the reactions corresponding to the locations of yielding of the supports, i.e., the moment at A (M_A, corresponding to the rotation of support A) and the reaction at B (R_B, corresponding to the yielding of support B), as shown in Figure 3.24. Therefore, $\theta_A = \Delta_1 = 0.001$ rad and $\Delta_B = \Delta_2 = 0.01$ m. Since there are no external loads on the beams, the $\Delta_{1L} = 0$ and $\Delta_{2L} = 0$. The primary beam with a unit couple at A and a unit load at B, along with their associated bending moment diagrams, are shown in Figure 3.25a–d.

$$\delta_{11} = \int m_1\cdot m_1\cdot\frac{dx}{EI} = \frac{1}{EI}\left[\frac{4}{6}\left[\left(2\times1+\frac{6}{10}\right)+\frac{6}{10}\left(\frac{2\times6}{10}+1\right)\right]\right]$$

$$+\frac{1}{2EI}\left[\frac{1}{2}\times\frac{6}{10}\times6\times\frac{2}{3}\times\frac{6}{10}\right]$$

$$=\frac{2.61}{EI}+\frac{0.36}{EI}=\frac{2.97}{EI}$$

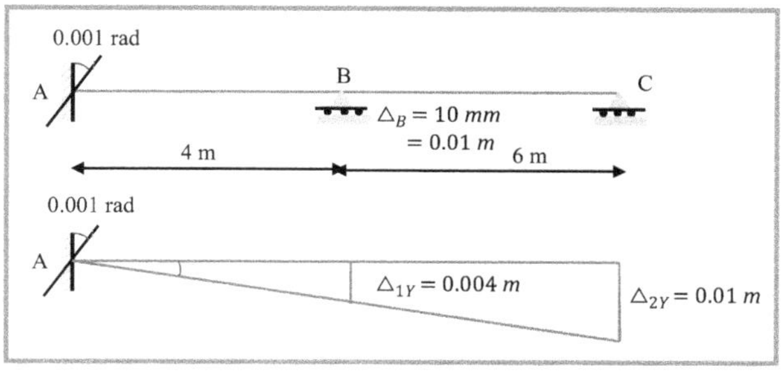

FIGURE 3.23 Beam with yielding of supports.

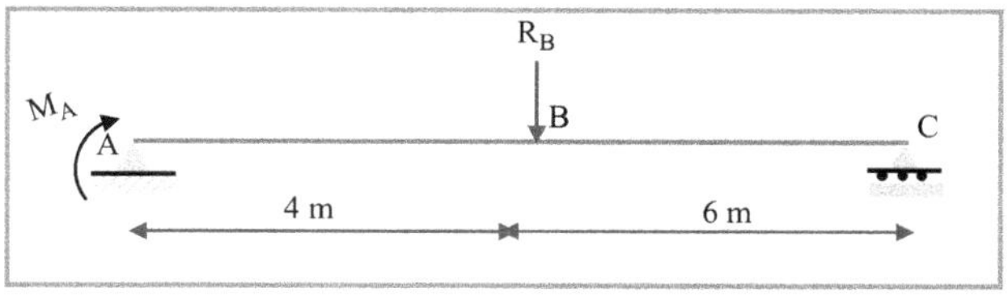

FIGURE 3.24 Primary beam with redundant reaction.

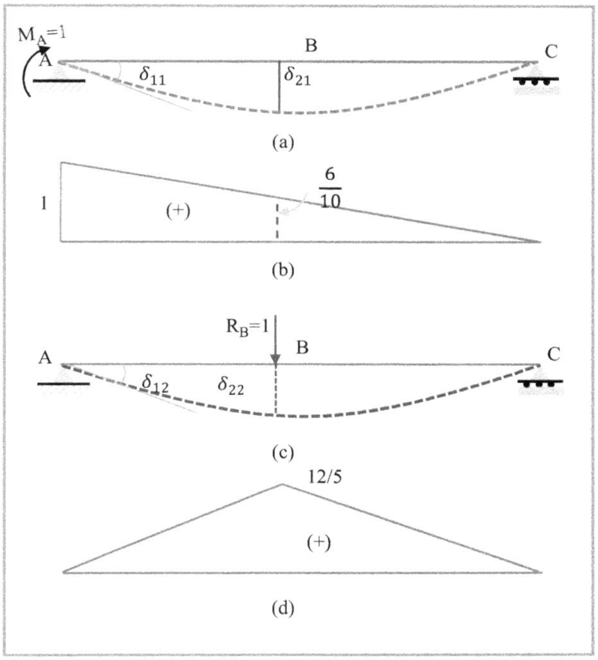

FIGURE 3.25 (a) Primary beam with unit couple at A, (b) Bending moment diagram due to unit couple at A, (c) Primary beam with unit load at B, and (d) Bending moment diagram due to unit load at B.

$$\delta_{12} = \delta_{21} = \int m_1\cdot m_2\cdot\frac{dx}{EI} = \frac{1}{EI}\left[\frac{4}{6}\left[\begin{array}{c}\left(2\times0+\frac{12}{5}\right)\\ +\frac{6}{10}\left(2\times\frac{12}{5}+0\right)\end{array}\right]\right]$$

$$+\frac{1}{2EI}\left[\frac{6}{6}\times\frac{12}{5}\left(2\times\frac{6}{10}+0\right)\right]$$

$$=\frac{3.52}{EI}+\frac{1.44}{EI}=\frac{4.96}{EI}$$

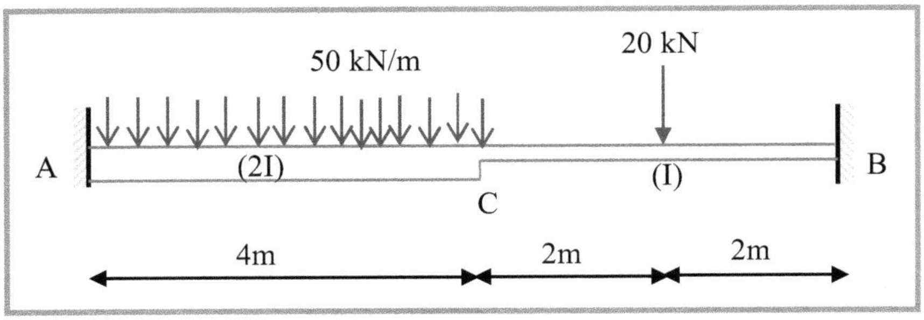

FIGURE 3.26 Non-prismatic beam *ACB*.

$$\delta_{22} = \int m_2 \cdot m_2 \cdot \frac{dx}{EI} = \frac{1}{EI}\left[\frac{4}{6}\times\frac{12}{5}\left(2\times\frac{12}{5}\right)\right]$$

$$+\frac{1}{2EI}\left[\frac{6}{6}\times\frac{12}{5}\times\left(2\times\frac{12}{5}+0\right)\right]$$

$$=\frac{7.68}{EI}+\frac{5.76}{EI}=\frac{13.44}{EI}$$

Substituting the above values into equations 3.8 and 3.9, we can write these equations as

$$0.001 = \frac{2.97}{EI}\times M_A + \frac{4.96}{EI}\times R_B$$

$$0.01 = \frac{4.96}{EI}\times M_A + \frac{13.44}{EI}\times R_B$$

By solving the above equations, we obtain $M_A = -318.85$ kNm and $R_B = 218.15$ kN (approximately the same as the above solution).

Example 3.6

Analyse the non-prismatic beam shown in Figure 3.26 using the consistent deformation method. Also, construct the shear force and bending moment diagrams.

Solution:

The static indeterminacy, $I_s=5-3=2$. Select R_B and M_B as redundant reactions, and assume they act in the downward direction and clockwise direction respectively. Remove the selected redundant reactions to obtain the determinate beam, and the effect of external loads is shown in Figure 3.27a–c. Apply unit loads in place of the redundant reactions (R_B=1 and M_B=1) successively in the determinate beam, and their associated bending moment diagrams are shown in Figure 3.28a–d.

From the figure, the compatibility equations can be written as

$$\theta_B = \theta_{BL} + \delta_{11}\cdot M_B + \delta_{12}\cdot R_B = 0 \qquad (3.10)$$

$$\Delta_B = \Delta_{BL} + \delta_{21}\cdot M_B + \delta_{22}\cdot R_B = 0 \qquad (3.11)$$

Calculate θ_{BL}, Δ_{BL}, and the flexibility coefficients using the unit load method and the area–moment multiplication technique as below.

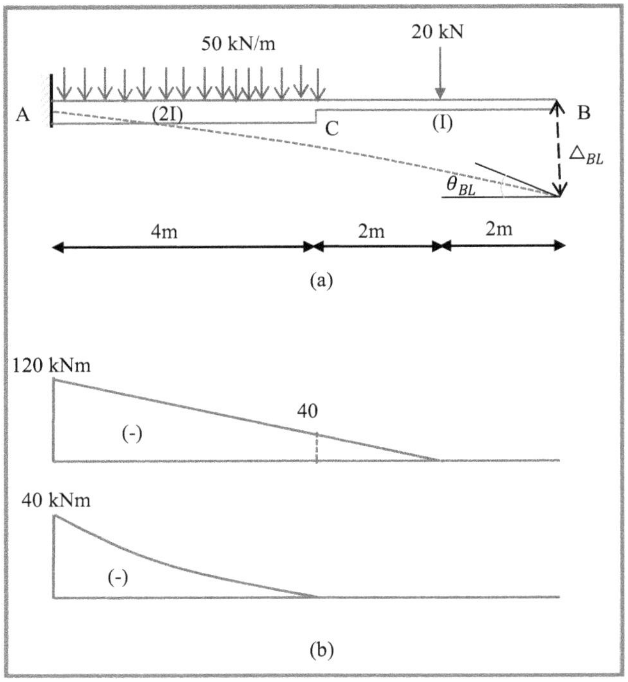

FIGURE 3.27 (a) Basic beam with external loads, and (b) *BMD* due to loads.

$$\theta_{BL} = \int M_L \cdot m_1 \cdot \frac{dx}{EI} = \left[\frac{4}{2}(120+40)\times 1 + \frac{1}{3}\times 4\times 400\times 1\right]$$

$$\frac{1}{2EI}+\frac{1}{EI}\left[\frac{1}{2}\times 2\times 40\times 1\right]$$

$$=\frac{426.67}{EI}+\frac{40}{EI}=\frac{466.67}{EI}$$

$$\Delta_{BL} = \int M_L \cdot m_2 \cdot \frac{dx}{EI} = \begin{bmatrix}\frac{4}{2}(120+40)(8-1.67)\\ +\frac{1}{3}\times 4\times 400(8-1.00)\end{bmatrix}\frac{1}{2EI}$$

$$+\frac{1}{EI}\left[\frac{1}{2}\times 40\times 2\times(2+1.33)\right]$$

$$=\frac{2879.47}{EI}+\frac{133.2}{EI}=\frac{3012.67}{EI}$$

$$\delta_{11} = \int m_1 \cdot m_1 \cdot \frac{dx}{EI} = \frac{1\times 4\times 1}{2EI}+\frac{1\times 4\times 1}{EI}=\frac{6}{EI}$$

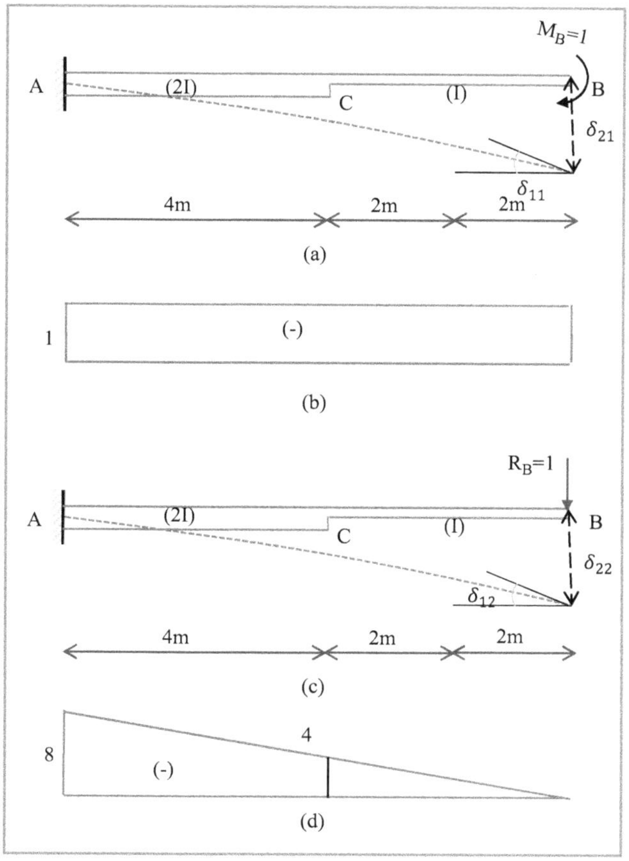

FIGURE 3.28 (a) Basic beam with unit couple at M_B=1, (b) m_1 diagram due to unit couple at M_B = 1, (c) Basic beam with unit load at R_B=1, and (d) m_2 diagram due to unit load at R_B=1.

$$\delta_{12} = \delta_{21} = \int m_1 \cdot m_2 \cdot \frac{dx}{EI} = \frac{1}{2EI}\left[\frac{4}{2}(4+8)\times 1\right]$$

$$+\left[\frac{1}{2}\times 4\times 4\times 1\right]\frac{1}{EI}$$

$$= \frac{12}{EI} + \frac{8}{EI} = \frac{20}{EI}$$

$$\delta_{22} = \int m_2 \cdot m_2 \cdot \frac{dx}{EI} = \frac{1}{2EI}\left[\frac{4}{2}(4+8)\times(8-1.77)\right]$$

$$+\left[\frac{1}{2}\times 4\times 4\times \frac{2}{3}\times 4\right]\frac{1}{EI}$$

$$= \frac{74.76}{EI} + \frac{21.33}{EI} = \frac{96.09}{EI}$$

Substituting the above values into equations 3.10 and 3.11, the equations become

$$0.0 = \frac{466.67}{EI} + \frac{6}{EI}\times M_B + \frac{20}{EI}\times R_B$$

$$0.0 = \frac{3012.67}{EI} + \frac{20}{EI}\times M_B + \frac{96.09}{EI}\times R_B$$

By solving the above equations, we obtain $M_B = 87.30$ kNm and $R_B = -49.52$ kN

By knowing the values of M_B and R_B, and using the static equilibrium equations with the help of the free body diagram, the reactions or end shears (Figure 3.29a) can be determined. The shear force and bending moment diagrams, along with the probable deflected shape, are shown in Figure 3.29b–d, respectively.

Example 3.7

A beam *AB* is supported on beam *CD*, as shown in Figure 3.30. The flexural rigidity of beam *AB* is twice that of beam *CD*. Determine the deflection at *C* or *B*.

Solution:

The beam is statically indeterminate to the first degree. Let us select the reactions at either R_B or R_C (with $R_B = R_C$) as redundant reactions. The primary beam with external loads and the redundant reaction is shown in Figure 3.31. The effect of external loads and a unit load is shown in Figure 3.32.

The compatibility condition is

$$\Delta_B = \Delta_C \tag{3.12}$$

where Δ_B is the net downward deflection at *B* in the beam *AB*.

$$\Delta_B = \Delta_{BL} - \Delta_{R_B}$$

$$= \Delta_{BL} - R_B \cdot \delta_{BB} \tag{3.12a}$$

$$= \frac{10\times 2^4}{8E(2I)} - R_B \cdot \frac{2^3}{3E(2I)} = \frac{10}{EI} - \frac{4R_B}{3EI}$$

$$\Delta_C = \text{deflection at 'C' in the beam } CD \text{ due to } R_C = \frac{R_C\times 1^3}{3EI}$$

From equation 3.12,

$$\frac{10}{EI} - \frac{4R_B}{3EI} = \frac{R_C}{3EI} \quad \text{but } R_B = R_C$$

$$\frac{5R_B}{3EI} = \frac{10}{EI} \Rightarrow R_B = 6\text{ kN}$$

$$\Delta_B = \Delta_C = \frac{6\times 1}{3EI} = \frac{2}{EI}$$

Example 3.8

Analyse the beam shown in Figure 3.33 with springs using the consistent deformation method.

Solution:

The degree of static indeterminacy, I_s=2. Select the moment at *B* (M_B) in the rotational spring and the vertical reaction at *B* (R_B) in the linear spring as redundant reactions. Release the moment and reaction at *B*

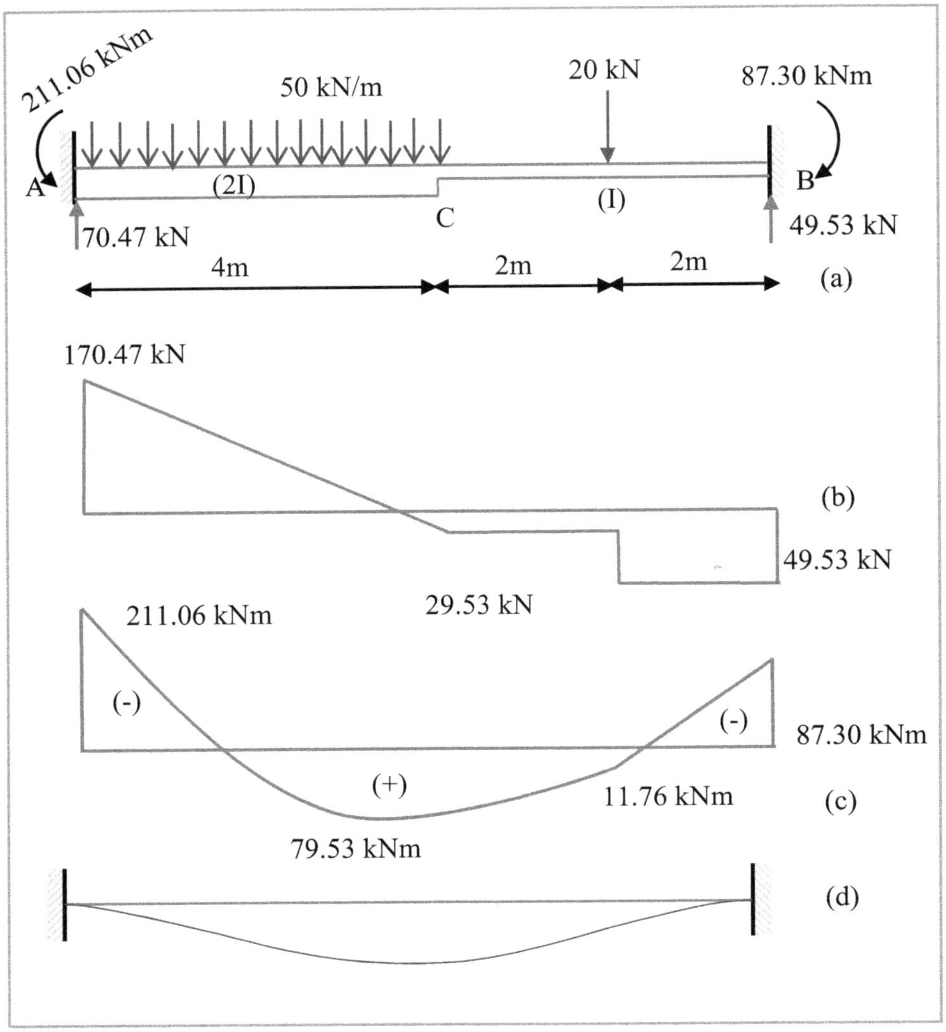

FIGURE 3.29 (a) Free body diagram, (b) shear force diagram, (c) bending moment diagram, and (d) tentative deflected shape.

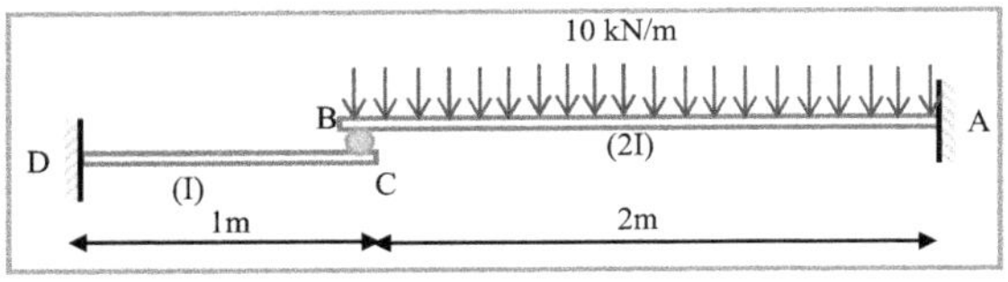

FIGURE 3.30 Compound beam *ABCD*.

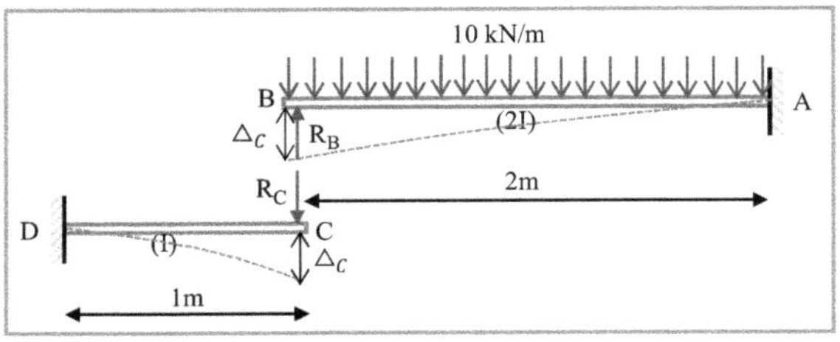

FIGURE 3.31 Primary beam with external loads and redundant.

to obtain the primary beam (Figure 3.34a). Apply a unit force and a unit moment at *B* successively on the primary beam (Figure 3.34b and c).

The compatibility equations require that the relative displacement between the cut ends of each spring is zero. Therefore,

$$\Delta_{1L} + \Delta_{1R_B} + \Delta_{1M_B} + \Delta_{S_1} = 0$$

$$\Rightarrow \Delta_{1L} + \delta_{11} \cdot R_B + \delta_{12} \cdot M_B + \Delta_{S_1} = 0 \tag{3.13}$$

$$\Delta_{2L} + \Delta_{2R_B} + \Delta_{2M_B} + \Delta_{S_2} = 0$$

$$\Delta_{2L} + \delta_{21} \cdot R_B + \delta_{22} \cdot M_B + \Delta_{S_2} = 0 \tag{3.14}$$

$$\Delta_{1L} = \frac{W(L/2)^3}{3EI} + \frac{W(L/2)^2}{2EI} \times \frac{L}{2} = \frac{5}{48} \times \frac{WL^3}{EI}$$

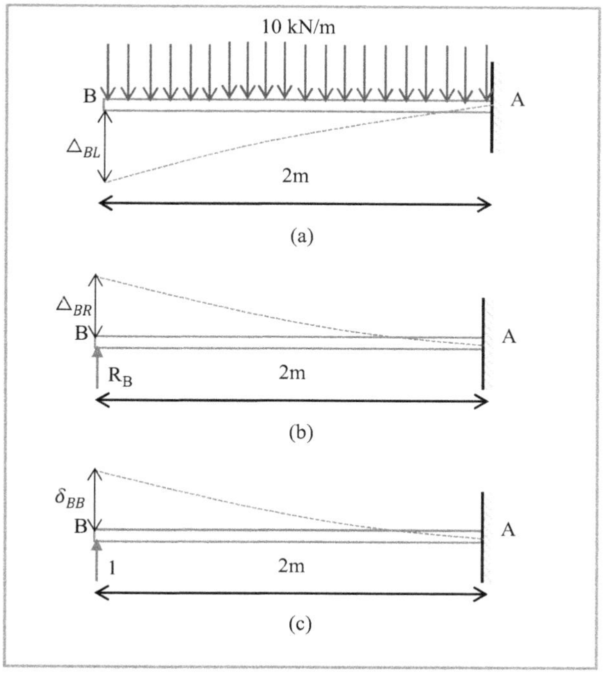

FIGURE 3.32 Primary beam analysis.

$$\Delta_{2L} = \frac{W(L/2)^2}{2EI} = \frac{WL^2}{8EI}$$

$$\delta_{11} = \frac{L^3}{3EI}$$

$$\delta_{12} = \delta_{21} = \frac{L^2}{2EI}$$

$$\delta_{22} = \frac{L}{EI}$$

FOR SPRINGS

Compression in the spring due to a unit force $= \frac{1}{K_1}$

Similarly, the rotation in the spring due to a unit moment $= \frac{1}{K_2}$

Substituting the above values into equations 3.13 and 3.14, the compatibility equations become

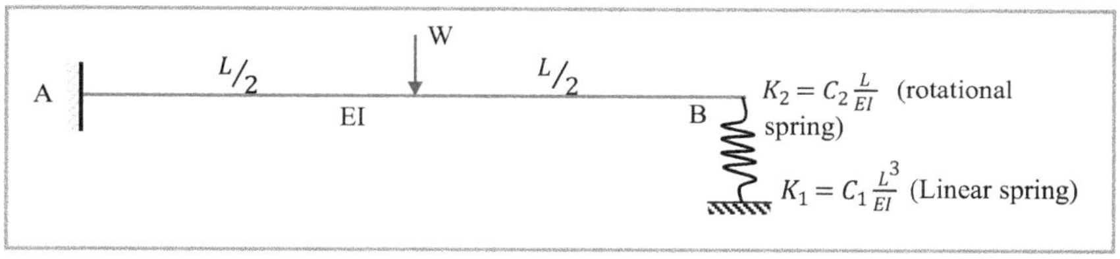

FIGURE 3.33 Beam *AB* with spring.

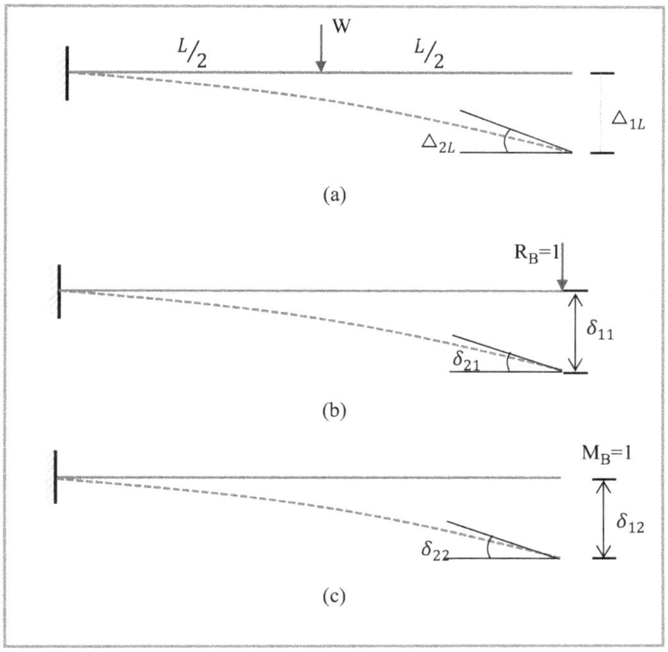

FIGURE 3.34 (a) Basic beam with external load, (b) Basic beam with unit load at $R_B = 1$, and (c) Basic beam with unit moment at $M_{B=}$ 1.

$$\begin{bmatrix} \dfrac{5}{48}\dfrac{WL^3}{EI} \\ \dfrac{WL^2}{8EI} \end{bmatrix} + \frac{1}{EI}\begin{bmatrix} L^3/3 & L^2/2 \\ L^2/2 & L \end{bmatrix}\begin{bmatrix} R_B \\ M_B \end{bmatrix}$$

$$+\begin{bmatrix} 1/K_1 & 0 \\ 0 & 1/K_2 \end{bmatrix}\begin{bmatrix} R_B \\ M_B \end{bmatrix} = \begin{bmatrix} 0 \\ 0 \end{bmatrix}$$

$$\begin{bmatrix} R_B \\ M_B \end{bmatrix} = EI\begin{bmatrix} \left(\dfrac{1}{3}+\dfrac{1}{c_1}\right)L^3 & L^2/2 \\ L^2/2 & \left(1+\dfrac{1}{c_2}\right)L \end{bmatrix}^{-1}\begin{bmatrix} \dfrac{-5}{48}\dfrac{WL^3}{EI} \\ -\dfrac{WL^2}{8EI} \end{bmatrix}$$

$$= \frac{EI}{L^4\left(\dfrac{1}{3}+\dfrac{1}{c_1}\right)\left(1+\dfrac{1}{c_2}\right)-\dfrac{L^4}{4}}$$

$$\begin{bmatrix} \left(1+\dfrac{1}{c_2}\right)L & -L^2/2 \\ -L^2/2 & \left(\dfrac{1}{3}+\dfrac{1}{c_1}\right)L^3 \end{bmatrix}$$

$$\begin{bmatrix} \dfrac{-5}{48}L \\ \dfrac{-1}{8} \end{bmatrix}\frac{WL^2}{EI}$$

By simplifying,

$$R_B = \frac{-W}{12}\frac{5\left(1+\dfrac{1}{c_2}\right)-3}{4\left(1+\dfrac{1}{c_2}\right)\left(\dfrac{1}{3}+\dfrac{1}{c_1}\right)-1};$$

$$M_B = \frac{-WL}{12}\frac{12\left(\dfrac{1}{3}+\dfrac{1}{c_1}\right)-5}{4\left(1+\dfrac{1}{c_2}\right)\left(\dfrac{1}{3}+\dfrac{1}{c_1}\right)-1}$$

For the given values of C_1 and C_2, the support reactions can be determined. Hence, the shear force and bending moment diagrams, along with the deflected shape can be obtained.

3.4 APPLICATION TO RIGID-JOINTED PLANE FRAMES

The primary internal forces in the various members of the rigid-jointed plane frames due to external loads are axial force, shear force, and bending moment. In general, in the case of rigid-jointed plane frames, the members are assumed to be inextensible, as the deformations due to axial forces are less significant when compared to those caused by bending or flexure. Furthermore, the deformations due to shear are also negligible in comparison to flexure. In other words, the members of the frame are considered to have infinite axial and shear rigidities. The analysis procedure for the rigid-jointed plane frames is the same as the procedure adopted for beams, as outlined in Section 3.2.

After determining the static indeterminacy of the frame, select the external support reactions as the primary unknowns (redundants) and draw the primary or basic structure, which is determinate. Then, draw the deflected shapes due to external loads and unit loads on the primary structure, indicating the displacements due to loads as Δ_{1L}, $\Delta_{2L}, \Delta_{3L}, \ldots \Delta_{iL}, \ldots$ and the flexibility coefficients due to unit forces as δ_{jk} in the deflected shapes.

Using these displacements and flexibility coefficients, write the compatibility equations at the chosen redundant locations. Then, find the deflections and flexibility coefficients using the unit load method and apply the area–moment multiplication technique. Substitute these values into the compatibility equations and solve for the unknown reactions/redundants.

Once the redundants are obtained, use the static equilibrium equations and the free body diagrams to find the internal forces, such as axial forces, shear forces, and bending moments, in the members of the frame, as well as the remaining support reactions. Finally, draw the axial force, shear force, and bending moment diagrams, along with the probable elastic or deflected shape. The above procedure is demonstrated in Examples 3.9–3.14.

Example 3.9

Analyse the rigid-jointed plane frame shown in Figure 3.35 using the consistent deformation method. Draw the shear force and bending moment diagrams. Assume *EI* is constant. Also, draw the probable deflected shape.

Solution:

The degree of static indeterminacy, $I_s = 5-3 = 2$.

Let us select the moments at *A* (M_A) and *B* (M_B) as redundant reactions, assumed to produce tension inside. Release the redundants M_A (R_1) and M_B (R_2) from the original structure to obtain the basic or determinate structure, and apply the principle of superposition (Figure 3.36).

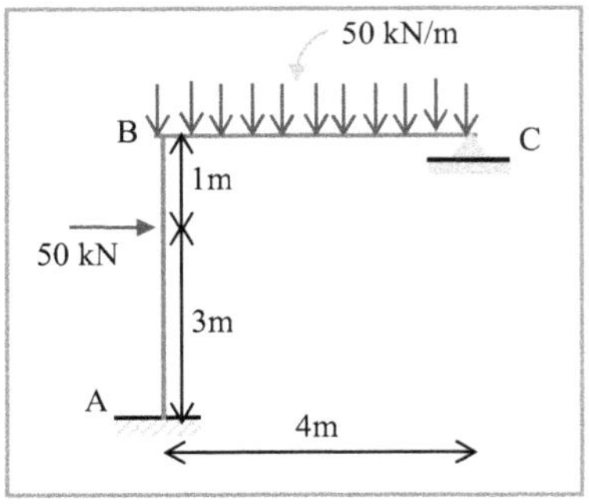

FIGURE 3.35 Rigid-jointed frame *ABC*.

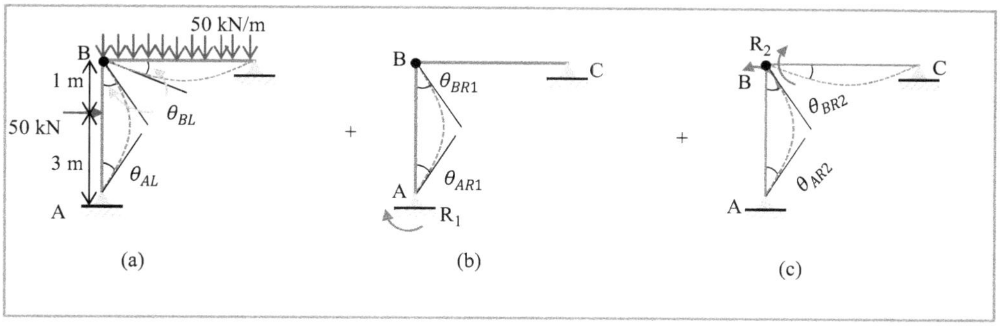

FIGURE3.36 Basic frame with loads and redundants. (a) Basic frame with external loads. (b) Basic frame with redundant R_1. (c) Basic frame with redundant R_2.

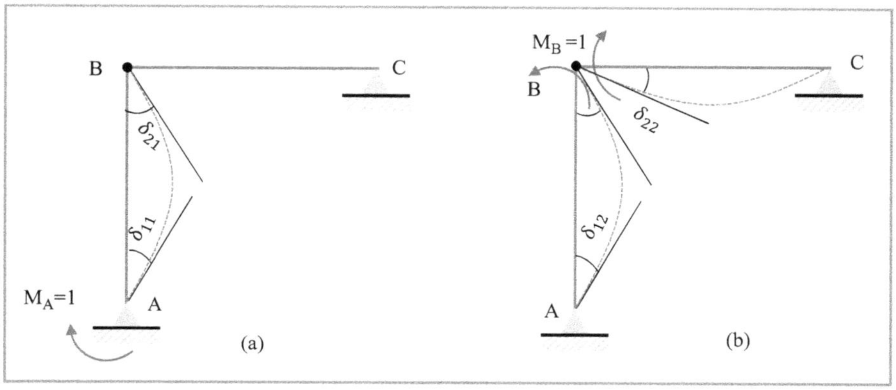

FIGURE 3.37 Basic frame with (a) a unit moment at A and (b) a unit moment at B.

The compatibility equations are

$$\theta_A = \theta_{AL} + \theta_{AR_1} + \theta_{AR_2} = 0$$

$$\Rightarrow \theta_{AL} + \delta_{11} \cdot R_1 + \delta_{12} \cdot R_2 = 0 \quad (3.15)$$

$$\theta_B = \theta_{BL} + \theta_{BR_1} + \theta_{BR_2} = 0$$

$$\Rightarrow \theta_{BL} + \delta_{21} \cdot R_1 + \delta_{22} \cdot R_2 = 0 \quad (3.16)$$

where θ_{AL} and θ_{BL} are the rotations at A and B due to external loads in the basic structure, and δ_{11}, δ_{12}, δ_{21}, δ_{22} are the flexibility coefficients due to unit loads applied in place of the redundants on the basic frame, as shown in Figure 3.37.

$$\delta_{11} = \frac{1 \times L}{3EI} = \frac{4}{3EI}$$

$$\delta_{12} = \delta_{21} = \frac{1 \times L}{6EI} = \frac{4}{6EI}$$

$$\delta_{22} = \left(\frac{L}{3EI}\right)_{AB} + \left(\frac{L}{3EI}\right)_{BC} = \frac{4}{3EI} + \frac{4}{3EI} = \frac{8}{3EI}$$

$$\theta_{AL} = \frac{Wab}{6EIL}(L+b) = \frac{50 \times 3 \times 1}{6EI \times 4}(4+1) = \frac{31.25}{EI}$$

$$\theta_{BL} = \frac{Wab}{6EIL}(L+a) + \frac{wL^3}{24EI} = \frac{43.75}{EI} + \frac{133.33}{EI} = \frac{177.08}{EI}$$

Substituting the above values into equations 3.15 and 3.16, the equations become

$$\frac{31.25}{EI} + \frac{4}{3EI}R_1 + \frac{4}{6EI}R_2 = 0$$

$$\Rightarrow 187.5 + 8R_1 + 4R_2 = 0 \quad (3.15a)$$

$$\frac{177.08}{EI} + \frac{4}{6EI}R_1 + \frac{8}{3EI}R_2 = 0$$

$$\Rightarrow 1{,}062.48 + 4R_1 + 16R_2 = 0 \quad (3.16a)$$

By solving equations 3.15a and 3.16a, we obtain

$$R_1 = 11.16 \text{ kNm} = M_A$$

$$R_2 = -69.20 \text{ kNm} = M_B$$

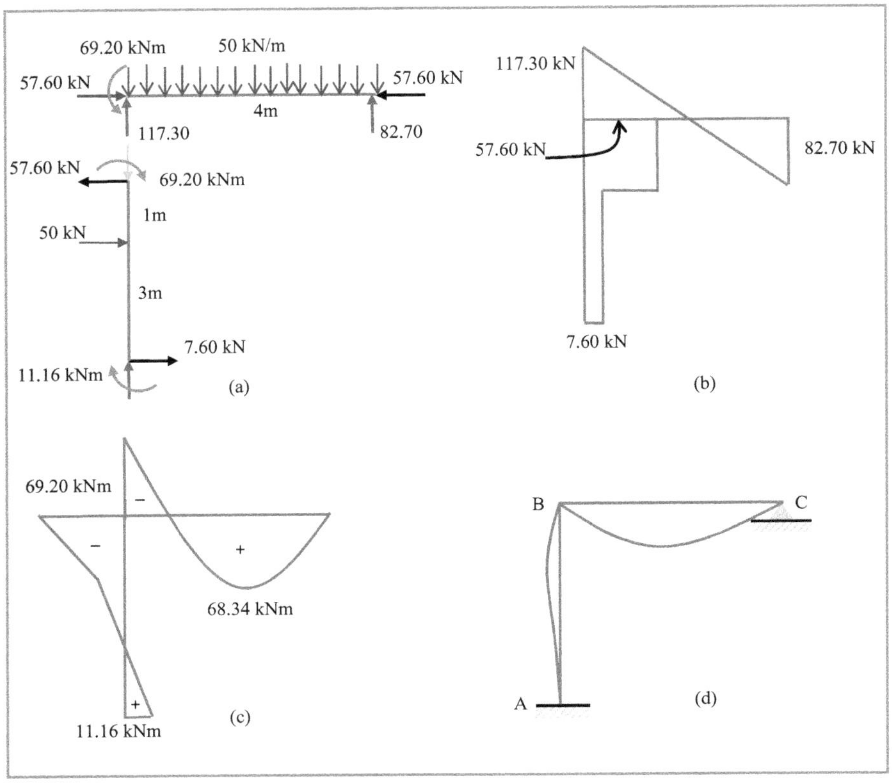

FIGURE 3.38 (a) Free body diagram, (b) shear force diagram, (c) bending moment diagram, and (d) probable elastic curve.

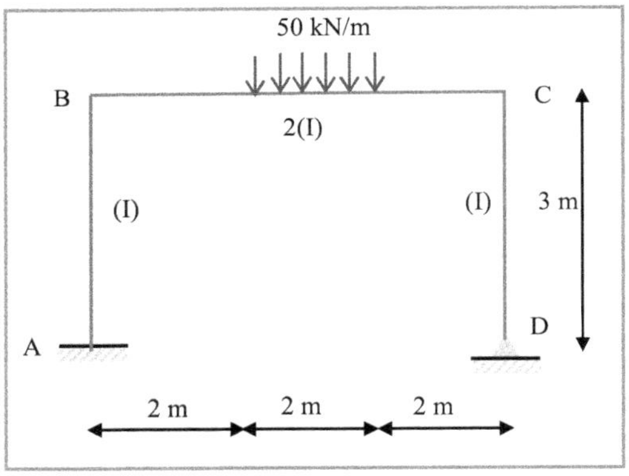

FIGURE 3.39 Rigid-jointed plane frame *ABCD*.

SUPPORT REACTIONS

With the known values of M_A and M_B, and using the static equilibrium equations along with the free body diagram, the support reactions can be calculated, as shown in Figure 3.38a. The shear force and bending moment diagrams, as well as the probable elastic curve, are shown in Figure 3.38b–d, respectively.

Example 3.10

Analyse the rigid-jointed plane frame shown in Figure 3.39 using the consistent deformation method. Draw the SFD, BMD, and the probable deflected shape.

Solution:

The degree of static indeterminacy, $I_s = 5 - 3 = 2$. Let us assume $R_1(V_D)$ and $R_2(H_D)$ as redundant reactions. Release R_1 and R_2 from the original structure to get the primary structure, as shown in Figure 3.40a, and apply the principle of superposition, as illustrated in Figure 3.40b and c.

The compatibility conditions are

$$\Delta_{1L} + \Delta_{1R_1} + \Delta_{1R_2} = 0$$

$$\Rightarrow \Delta_{1L} + \delta_{11} \cdot R_1 + \delta_{12} \cdot R_2 = 0 \tag{3.17}$$

$$\Delta_{2L} + \Delta_{2R_1} + \Delta_{2R_2} = 0$$

$$\Rightarrow \Delta_{2L} + \delta_{21} \cdot R_1 + \delta_{22} \cdot R_2 = 0 \tag{3.18}$$

where Δ_{1L} and Δ_{2L} are the displacements due to the external loads in the determinate structure,and δ_{11}, δ_{12}, δ_{21}, δ_{22} are the flexibility coefficients due to unit loads applied at the positions of redundant reactions in the basic frame, as shown in Figure 3.41a and b.

The M_L diagram due to external loads and the m_1 and m_2 diagrams due to unit loads are shown in Figure 3.42a–c, respectively, for the calculation of displacements and flexibility coefficients.

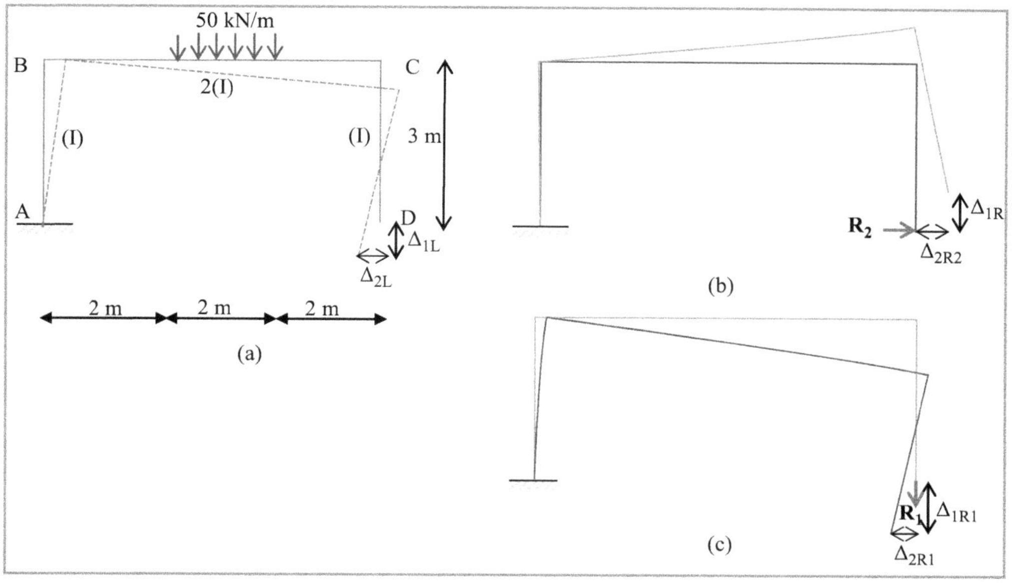

FIGURE 3.40 (a) Primary structure with external loads, (b) Primary structure with R_1, and, (c) Primary structure with R_2.

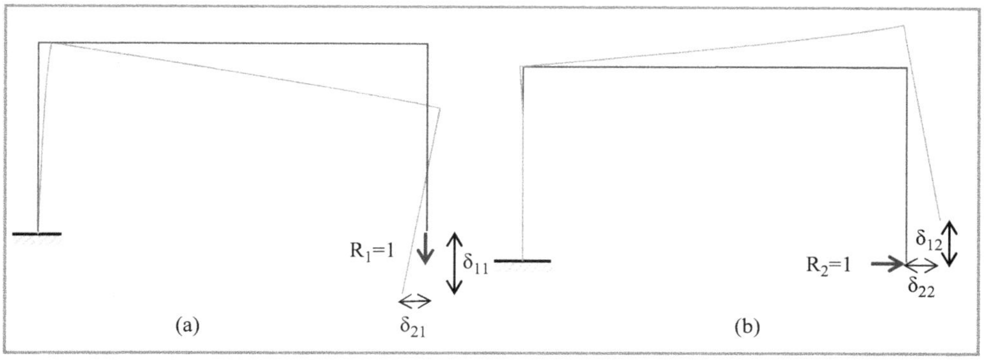

FIGURE 3.41 (a) Primary structure with unit load at R_1=1, (b) Primary structure with unit load at R_2=1

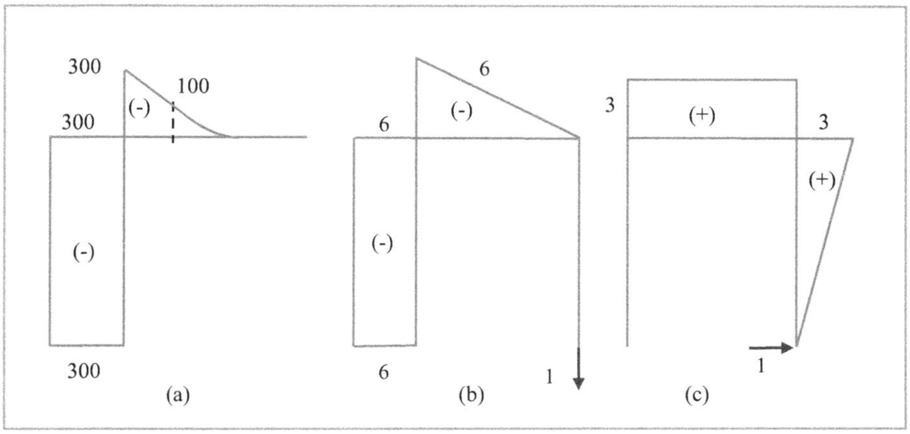

FIGURE 3.42 (a) M_L diagram due to external loads, (b) m_1 diagram due to unit load at $R_1=1$, and (c) m_2 diagram due to unit load at $R_2=1$.

The values can be calculated using the unit load method and by applying the area–moment multiplication technique.

$$\Delta_{1L} = \int M_L \cdot m_1 \cdot \frac{dx}{EI}$$

$$= \frac{1}{2EI}\left[\left(\frac{1}{3}\times 2\times 100 + \frac{2}{2}(100+300)\right)(4.93^*)\right]$$

$$+\frac{1}{EI}[300\times 3\times 6]$$

$$= \frac{6{,}550.33}{EI}$$

$$\Delta_{2L} = \int M_L \cdot m_2 \cdot \frac{dx}{EI}$$

$$= \frac{1}{2EI}\left[\left(\frac{1}{3}\times 2\times(-100) + \frac{2}{2}(-100-300)\right)(3)\right]$$

$$+\frac{1}{EI}[-300\times 3\times 1.5]$$

$$= \frac{-2{,}050}{EI}$$

$$^*cg \text{ from left} = \frac{\frac{1}{3}\times 2\times 100\left(2+\frac{2}{4}\right) + \left(\frac{1}{2}\times 2\times 200\times\frac{2}{3}\right) + (100\times 2\times 1)}{\frac{1}{3}\times 2\times 100 + \frac{1}{2}\times 2\times 200 + 100\times 2} = 1.07 \text{ m}$$

$$\delta_{11} = \int m_1 \cdot m_1 \cdot \frac{dx}{EI} = \frac{1}{2EI}\left[\frac{1}{2}\times 6\times 6\times 4\right] + \frac{1}{EI}[6\times 3\times 6] = \frac{144}{EI}$$

$$\delta_{22} = \int m_2 \cdot m_2 \cdot \frac{dx}{EI} = \frac{1}{EI}\left[\frac{1}{2}\times 3\times 3\times 2\right] + \frac{1}{2EI}[6\times 3\times 3]$$

$$+\frac{1}{EI}\left[\frac{1}{2}\times 3\times 3\times 2\right] = \frac{45}{EI}$$

$$\delta_{12} = \delta_{21} = \int m_1 \cdot m_2 \cdot \frac{dx}{EI} = \frac{1}{2EI}\left[\frac{1}{2}\times 6\times(-6)\times 3\right]$$

$$+\frac{1}{EI}[(-6)\times 3\times 1.5] = \frac{-54}{EI}$$

By substituting the above values into the compatibility conditions, equations 3.17 and 3.18 can be written as

$$\frac{6{,}550.3}{EI} + \frac{144}{EI}R_1 - \frac{54}{EI}R_2 = 0 \quad (3.17a)$$

$$\frac{-2{,}050}{EI} - \frac{54}{EI}R_1 + \frac{45}{EI}R_2 = 0 \quad (3.18a)$$

By solving the above equations, we get

$$R_1 = -51.64 \text{ kN} = V_D, \quad R_2 = -16.42 \text{ kN} = H_D$$

SUPPORT REACTIONS

After determining the reactions $V_D(R_1)$ and $H_D\,(R_2)$, the remaining support reactions can be calculated using the static equilibrium equations and the free body diagram, as shown in Figure 3.43a. The shear force and bending moment diagrams, along with the probable deflected shape, are presented in Figure 3.43b–d, respectively.

Example 3.11

Analyse the rigid-jointed plane frame shown in Figure 3.44 using the consistent deformation method. Assume EI is constant. Draw the SFD, BMD, and probable deflected shape.

Solution:

The degree of static indeterminacy is $I_s = 5 - 3 = 2$. Select $V_D(R_1)$ and $H_D(R_2)$ at D as redundant reactions, assumed to act downwards and towards the left, respectively. The primary structure can be obtained by releasing the redundant reactions R_1 and R_2 from the original structure and applying the principle of superposition, as shown in Figure 3.45.

The compatibility conditions can be written as,

$$\Delta_{VD} = \Delta_{1L} + \Delta_{1R_1} + \Delta_{1R_2} = 0$$

$$\Rightarrow \Delta_{1L} + \delta_{11}\cdot R_1 + \delta_{12}\cdot R_2 = 0 \quad (3.19)$$

$$\Delta_{HD} = \Delta_{2L} + \Delta_{2R_1} + \Delta_{2R_2} = 0$$

$$\Rightarrow \Delta_{2L} + \delta_{21}\cdot R_1 + \delta_{22}\cdot R_2 = 0 \quad (3.20)$$

where $\delta_{11}, \delta_{12}, \delta_{21}, \delta_{22}$ are the flexibility coefficients due to unit loads applied at R_1 and R_2 on the primary structure, as shown in Figure 3.45a and b, respectively. These can be calculated using the unit load method and the area–moment multiplication technique. The M_L diagram due to external loads and the m_1 and m_2 diagrams due to unit loads applied at R_1 and R_2, are shown in Figure 3.46c–e, respectively.

$$\Delta_{1L} = \int M_L \cdot m_1 \cdot \frac{dx}{EI}$$

$$= \frac{1}{EI}\left[\begin{array}{l}\left(\frac{1}{3}\times 4\times 266.67\times 4\right) + \\ \left[266.67\times 2 + \frac{2}{2}(266.67+326.67)\right]\times 5\end{array}\right]$$

$$= \frac{1}{EI}(1{,}422.24 + 5{,}633.4) = \frac{7{,}055.64}{EI}$$

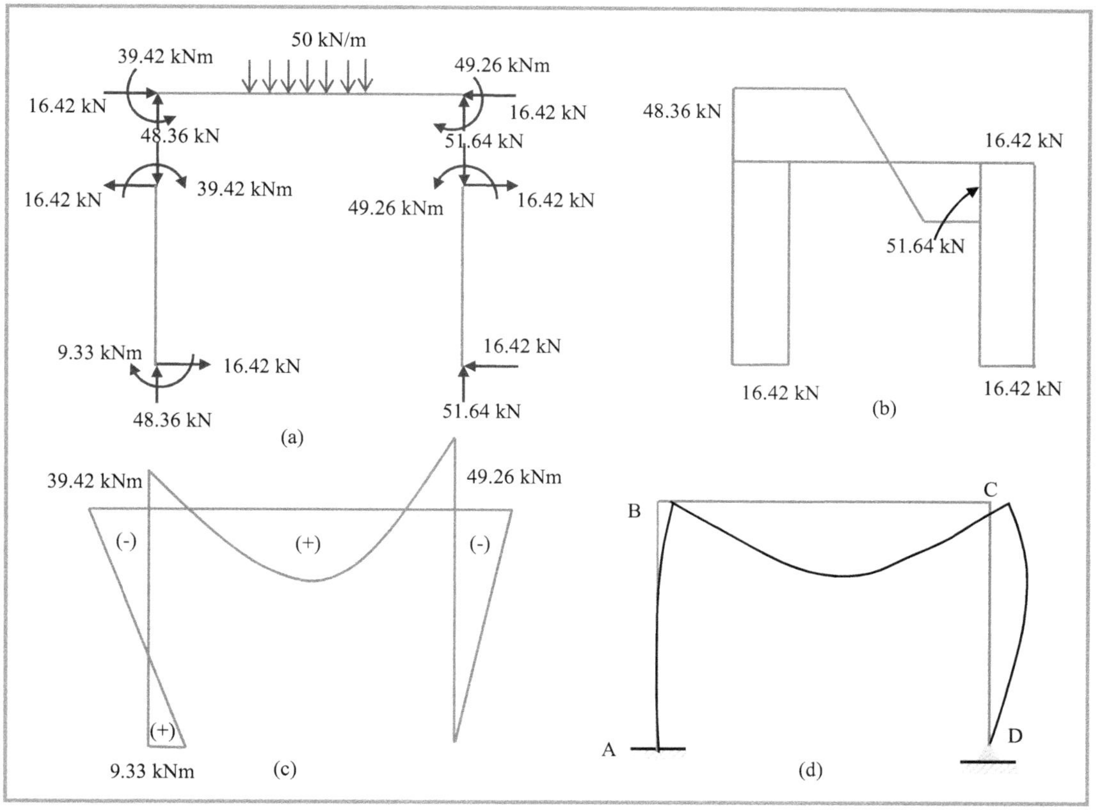

FIGURE 3.43 (a) Free body diagram, (b) shear force diagram, (c) bending moment diagram, and (d) probable elastic curve.

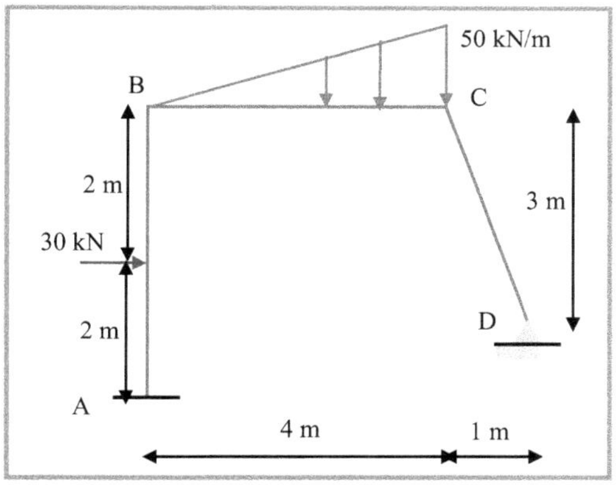

FIGURE 3.44 Indeterminate rigid-jointed plane frame.

$$\Delta_{2L} = \int M_L \cdot m_2 \cdot \frac{dx}{EI}$$

$$= \frac{1}{EI}\left[\begin{array}{l}\frac{1}{3}\times 4\times 266.67\times 3+\\ \left[(266.67\times 2)+\frac{2}{2}(266.67+326.67)\right](0.93)\end{array}\right]$$

$$= \frac{1}{EI}(1{,}066.68+1{,}047.81) = \frac{2{,}114.49}{EI}$$

$$\delta_{11} = \int m_1 \cdot m_1 \cdot \frac{dx}{EI}$$

$$= \frac{1}{EI}\left[\begin{array}{l}\left(\frac{1}{2}\times 3.16\times 1\times 0.67\right)+(5\times 4\times 5)\\ +\frac{4}{6}\left[5\times(2\times 5+1)+1(2\times 1+5)\right]\end{array}\right] = \frac{142.39}{EI}$$

$$\delta_{12} = \delta_{21} = \int m_1 \cdot m_2 \cdot \frac{dx}{EI}$$

$$= \frac{1}{EI}\left[\frac{3.16}{6}(2\times 3+0)+\frac{4}{2}(1+5)\times 3+(5\times 4\times 1)\right]$$

$$= \frac{59.16}{EI}$$

$$\delta_{22} = \int m_2 \cdot m_2 \cdot \frac{dx}{EI}$$

$$= \frac{1}{EI}\left[\begin{array}{l}\frac{3.16}{6}(3\times 2\times 3)+(3\times 4\times 3)+\\ \left(\frac{1}{2}\times 3\times 3\times 2\right)+\left(\frac{1}{2}\times 1\times 1\times\frac{2}{3}\times 1\right)\end{array}\right]$$

$$= \frac{54.81}{EI}$$

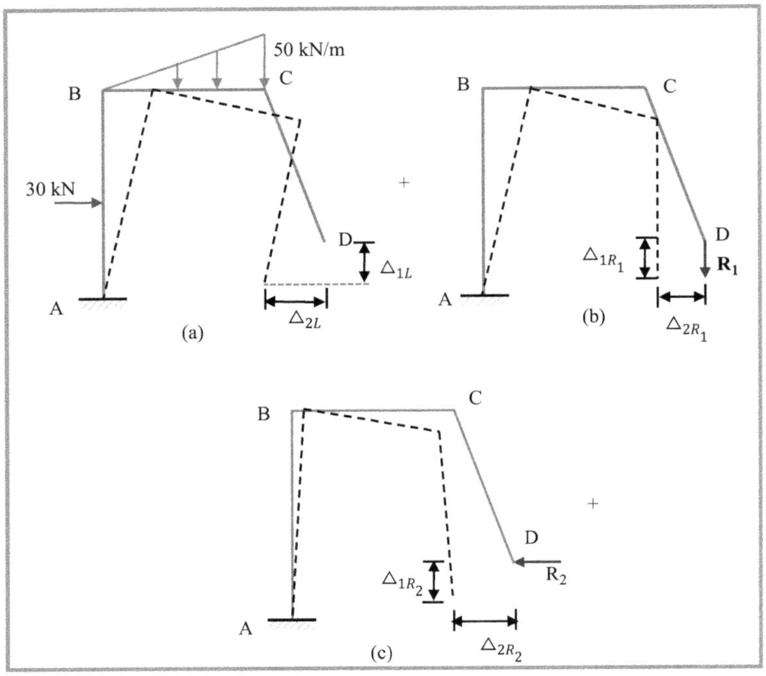

FIGURE 3.45 Primary frame with (a) external loads, (b) Redundant R_1, and (c) Redundant R_2.

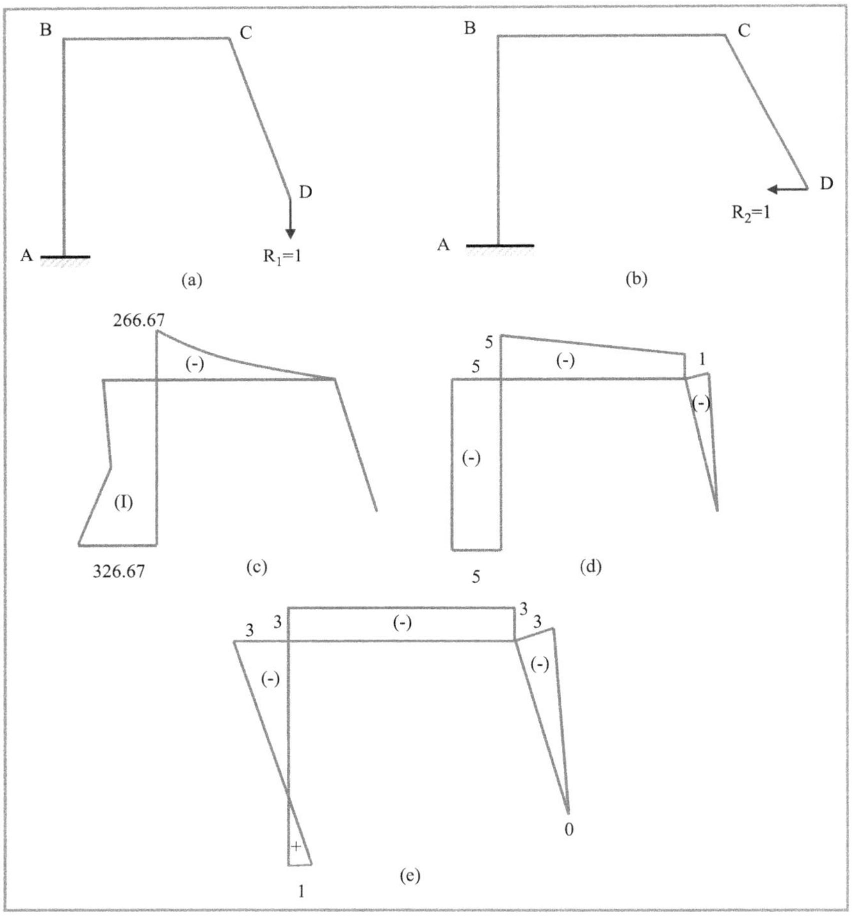

FIGURE 3.46 (a) Unit laod at R_1=1, (b) Unit load at R_2=1, (c) M_L diagram, (d) m_1 diagram, and (e) m_2 diagram.

Substituting the above values into equations 3.19 and 3.20, we can write

$$\frac{7{,}055.64}{EI}+\frac{142.39}{EI}R_1+\frac{59.16}{EI}R_2=0 \qquad (3.19a)$$

$$\frac{2{,}114.49}{EI}+\frac{59.16}{EI}R_1+\frac{54.81}{EI}R_2=0 \qquad (3.20a)$$

By solving the above equations, we obtain

$$R_1=-60.77\text{ kN}=V_D \quad \text{and} \quad R_2=27.02\text{ kN}=H_D$$

SUPPORT REACTIONS

The remaining support reactions can be calculated using the static equilibrium equations, with the help of the free body diagram shown in Figure 3.47a. The shear force and bending moment diagrams, along with the probable deflected shape, are shown in Figure 3.47b–d, respectively.

Example 3.12

Analyse the symmetrical rigid-jointed plane frame shown in Figure 3.48, using the consistent deformation method. Draw the SFD, BMD, and probable deflected shape.

Solution:

The given frame, *ABCD*, with fixed supports, is statically indeterminate to the third degree in general. Any three reactions could be chosen as redundant reactions; for example, the reactions at *D* could be selected as redundant. The primary structure would then become a cantilever frame at *A* (as is typically done). However, since the given frame is symmetrical with respect to both the loading and the structure, the degree of static indeterminacy can be reduced to two by taking advantage of symmetry. Therefore, only half of the frame (ABB^1) is sufficient for the analysis, with appropriate modifications. Specifically, a guided-fixed support is placed at the middle of *BC* (i.e., at B^1) to account for vertical deflection without rotation, due to symmetry at the centre, as shown in Figure 3.49. The absence of a vertical reaction at B^1 in Figure 3.49 indicates that, due to symmetry, the shear force at the midpoint of *BC* is zero, as shown in Figure 3.48.

Let us choose the reactions at B^1, i.e., $H_{B^1}(R_1)$ and $M_{B'}(R_2)$ as redundant reactions. The primary frame with external loads and redundant reactions is shown in Figure 3.50.

The compatibility equations can be written as

$$\Delta_{B'H}=\Delta_{1L}+\Delta_{1R_1}+\Delta_{1R_2}=0$$

$$\Rightarrow \Delta_{1L}+\delta_{11}\cdot R_1+\delta_{12}\cdot R_2=0 \qquad (3.19)$$

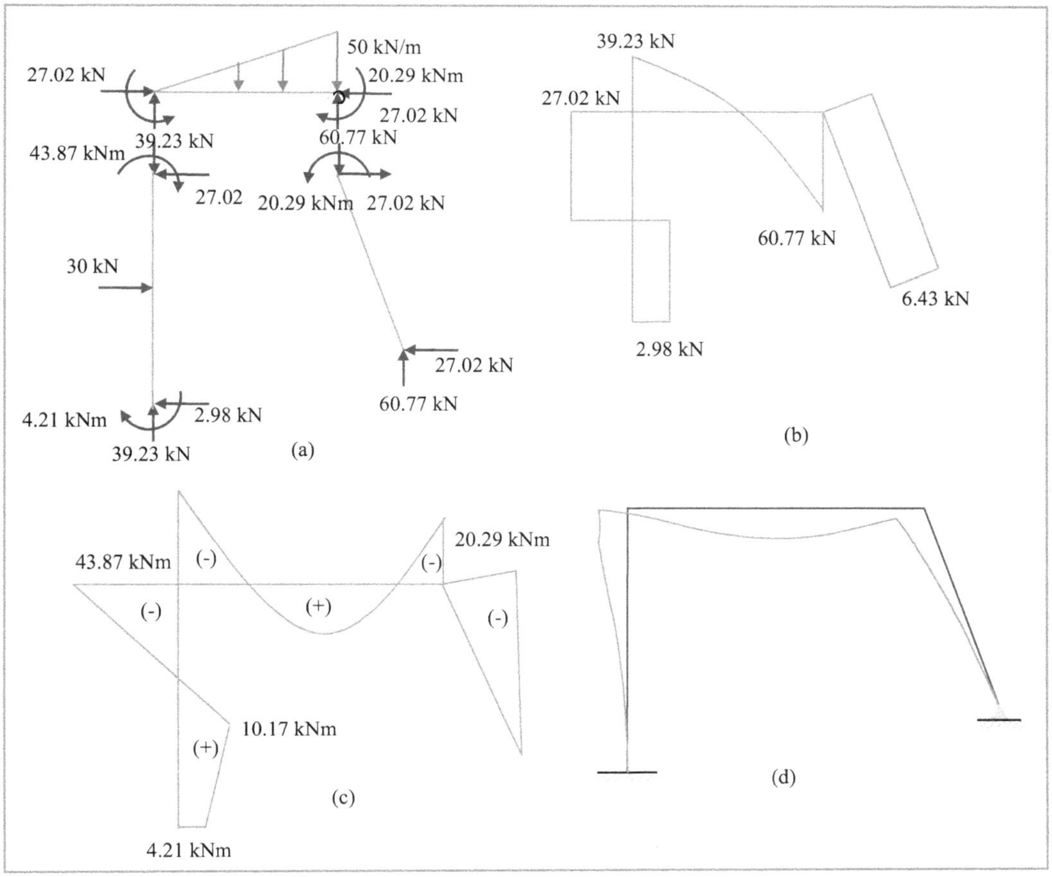

FIGURE 3.47 (a) Free body diagram, (b) shear force diagram, (c) bending moment diagram, and (d) probable deflected curve.

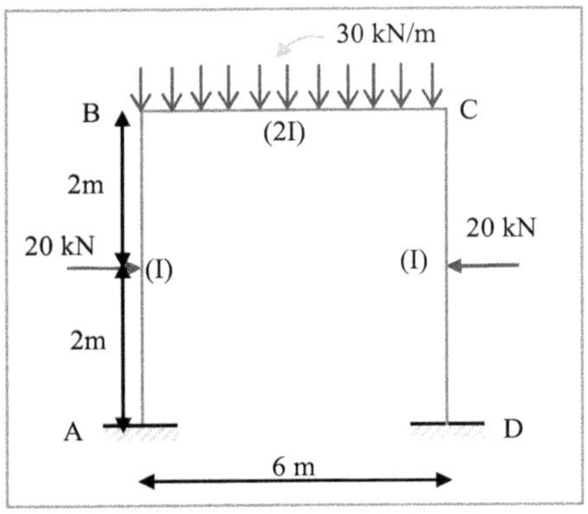

FIGURE 3.48 Symmetrical frame.

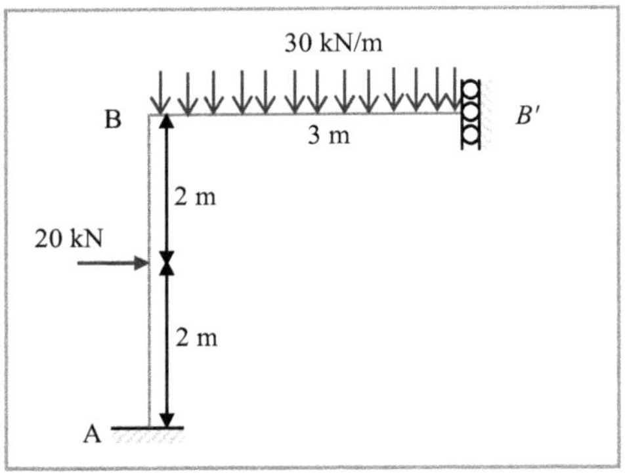

FIGURE 3.49 Half of the portion of frame ABB^1.

$$\Delta_{B^1V} = \Delta_{2L} + \Delta_{2R_1} + \Delta_{2R_2} = 0$$

$$\Rightarrow \Delta_{2L} + \delta_{21} \cdot R_1 + \delta_{22} \cdot R_2 = 0 \quad (3.20)$$

where δ_{11}, δ_{12}, δ_{21}, δ_{22} are the flexibility coefficients due to the unit loads applied at $R_1 = 1$ and $R_2 = 1$ (couple), on the primary structure, as shown in Figure 3.51.

These can be calculated using the unit load method and area–moment multiplication technique. The M_L diagram due to loads, and m_1 and m_2 diagrams due to the unit load at R_1 and unit couple at R_2, respectively, are shown in Figure 3.52.

$$\Delta_{1L} = \int M_L \cdot m_1 \cdot \frac{dx}{EI} = \frac{1}{EI}\left[\left(135\times 4 + \frac{1}{2}\times 2\times 40\right)\times 2.09\right]$$

$$= \frac{1{,}212.2}{EI}$$

$$\Delta_{2L} = \int M_L \cdot m_2 \cdot \frac{dx}{EI}$$

$$= \frac{1}{2EI}\left[\frac{1}{3}\times 3\times 135\times 1\right] + \frac{1}{EI}\left[\left(135\times 4 + \frac{1}{2}\times 2\times 40\right)\times 1\right]$$

$$= \frac{647.5}{EI}$$

$$\delta_{11} = \int m_1 \cdot m_1 \cdot \frac{dx}{EI} = \frac{1}{EI}\left[\frac{1}{2}\times 4\times 4\times 2.66\right] = \frac{21.33}{EI}$$

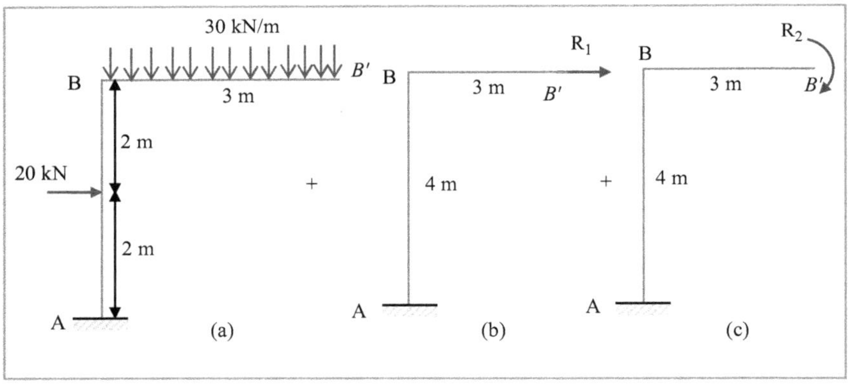

FIGURE 3.50 Primary frame with (a) external loads, (b) redundant R_1, and (c) redundant R_2.

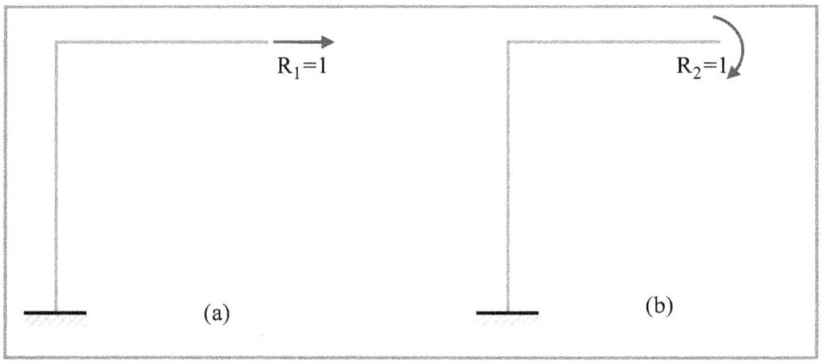

FIGURE 3.51 Primary structure with (a) unit load at R_1 and (b) unit couple at R_2.

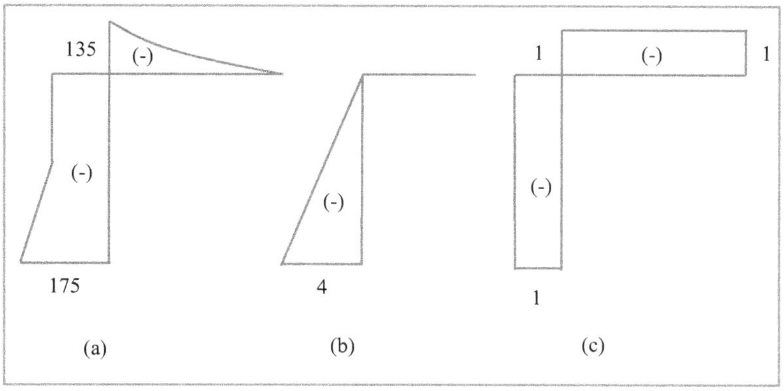

FIGURE 3.52 Effect of external loads and unit loads on the primary structure. (a) M_L diagram, (b) m_1 diagram, and (c) m_2 diagram.

$$\delta_{12} = \delta_{21} = \int m_1 \cdot m_2 \cdot \frac{dx}{EI} = \frac{1}{EI}\left[\frac{1}{2} \times 4 \times 4 \times 1\right] = \frac{8}{EI}$$

$$\delta_{22} = \int m_2 \cdot m_2 \cdot \frac{dx}{EI} = \frac{1}{2EI}[1 \times 3 \times 1] + \frac{1}{EI}[1 \times 4 \times 1] = \frac{5.5}{EI}$$

Substituting the above values into equations 3.19 and 3.20, the equations become

$$\frac{1{,}212.2}{EI} + \frac{21.33}{EI} R_1 + \frac{8}{EI} R_2 = 0 \quad (3.21)$$

$$\frac{647.5}{EI} + \frac{8}{EI} R_1 + \frac{5.5}{EI} R_2 = 0 \quad (3.22)$$

Solving the above equations, we obtain

$$R_1 = -27.93 \text{ kN} \quad \text{and} \quad R_2 = -77.22 \text{ kN}$$

SUPPORT REACTIONS AND INTERNAL FORCES

With the help of the free body diagrams and using static equilibrium equations, the remaining reactions can be determined, as shown in Figure 3.53a and b. The shear force and bending moment diagrams, along with the probable deflected shape, are shown in Figure 3.53c–e, respectively.

ALTERNATE SOLUTION

Without disturbing the symmetry in the primary structure, we can choose the reactions at *A* and *D* as redundant reactions simultaneously, as shown in Figure 3.54. The vertical reactions are statically determinate, i.e., $V_A = V_D = 90$ kN (due to symmetry). To solve this problem, let us take $H_A = H_D = R_1$ and $M_A = M_D = R_2$ as redundant reactions. The primary structure with this choice of redundant reactions is a symmetrical frame with the roller supports at *A* and *D*. While applying the unit load in place of redundant reactions, we must apply the unit load simultaneously at both ends.

The compatibility conditions can be written as

$$\Delta_{H_D} = \Delta_{H_A} = \Delta_{1L} + \Delta_{1R_1} + \Delta_{1R_2} = 0$$

$$\Rightarrow \Delta_{1L} + \delta_{11} \cdot R_1 + \delta_{12} \cdot R_2 = 0 \quad (3.23)$$

$$\theta_A = \theta_D = \Delta_{2L} + \Delta_{2R_1} + \Delta_{2R_2} = 0$$

$$\Rightarrow \Delta_{2L} + \delta_{21} \cdot R_1 + \delta_{22} \cdot R_2 = 0 \quad (3.24)$$

where δ_{11}, δ_{12}, δ_{21}, δ_{22} are the flexibility coefficients due to unit loads applied at R_1 =1 and R_2 =1 (couple) on the primary structure as shown in Figure 3.55. The M_L diagram due to external loads, m_1 diagram due to unit load at *A* and *D*, and m_2 diagram due to unit couple at *A* and *D* on the primary structure, is shown in Figure 3.55. Using the unit load method and the area–moment multiplication technique, the displacements due to loads and the flexibility coefficients can be calculated. After obtaining the values of the displacements and the flexibility coefficients, substitute them into equations 3.23 and 3.24 and solve the equations to obtain R_1 and R_2 values. The calculation is left to the reader.

Example 3.13

Analyse the rigid-jointed plane frame, shown in Figure 3.56, using the consistent deformation method, if support *A* slides towards the left by 10 mm and support *D* rotates by 0.001 rad in a clockwise direction. Assume $EI = 138.375 \times 10^3$ kNm^2 and is constant.

Solution:

The degree of static indeterminacy is $I_s = 5 - 3 = 2$. Let us select the moment at *D* ($M_D = R_1$) and the horizontal reaction at *A* ($H_A = R_2$) as redundant reactions. The primary structure with external loads and the selected redundant reactions is shown in Figure 3.57.

The compatibility equations are

$$\Delta_1 = \Delta_{1L} + \Delta_{1R_1} + \Delta_{1R_2}$$

$$0.01 = \Delta_{1L} + \delta_{11} \cdot R_1 + \delta_{12} \cdot R_2 = 0 \quad (3.25)$$

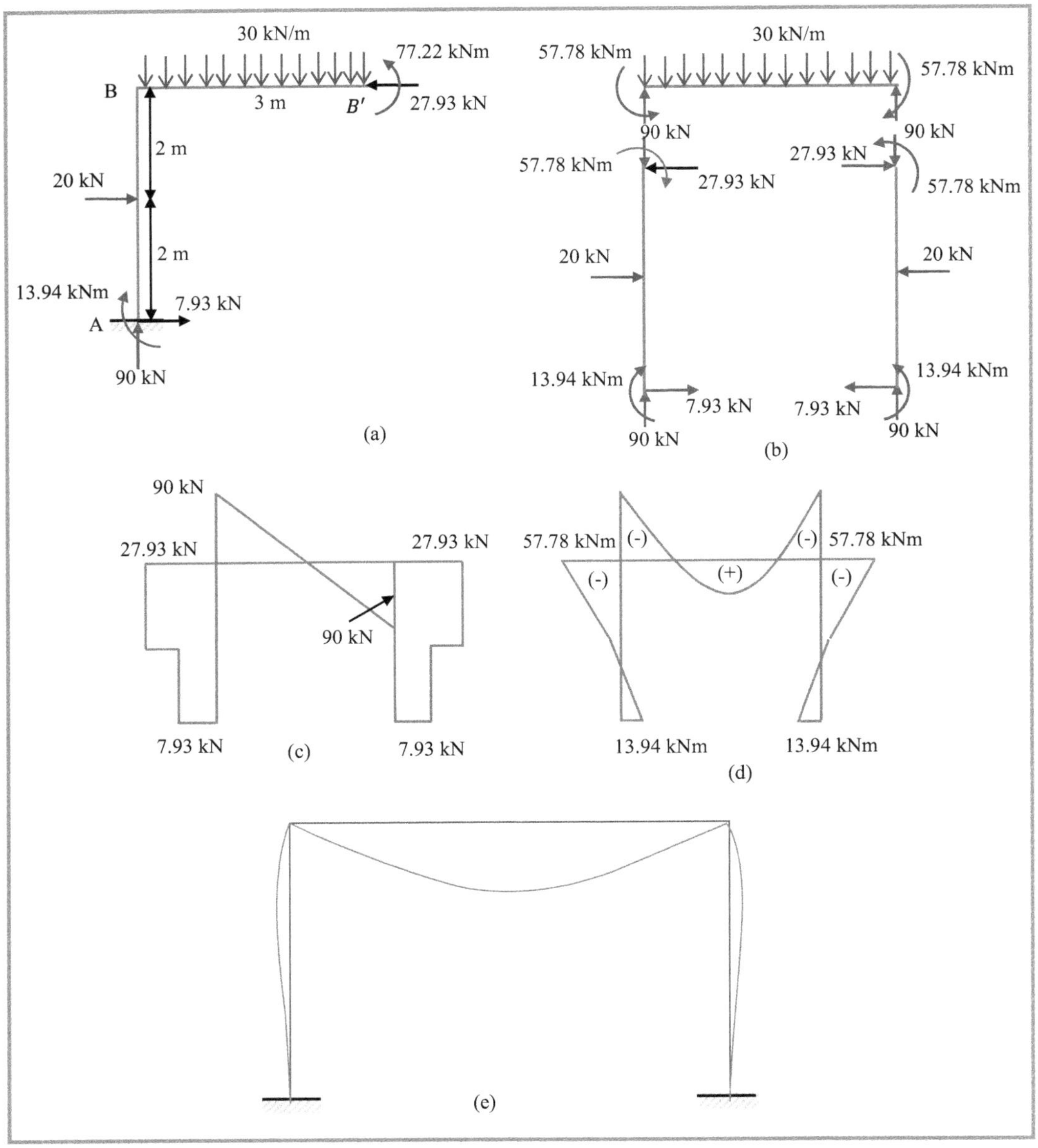

FIGURE 3.53 (a and b) Free body diagrams, (c) shear force diagram, (d) bending moment diagram, and (e) probable deflected curve.

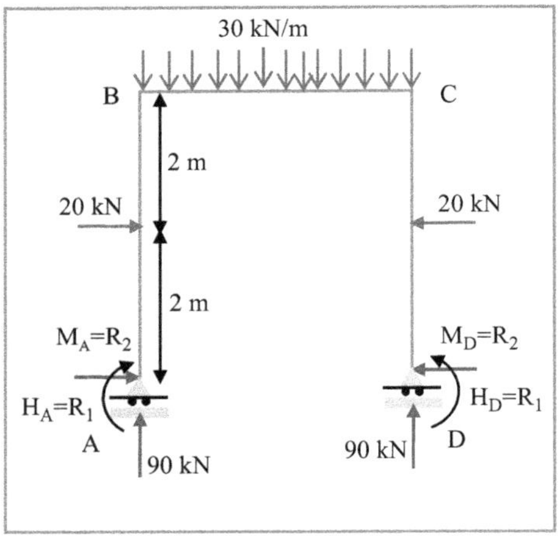

FIGURE 3.54 Symmetrical frame with roller supports at *A* and *D*.

$$\Delta_2 = \Delta_{2L} + \Delta_{2R_1} + \Delta_{2R_2}$$

$$0.002 = \Delta_{2L} + \delta_{21} \cdot R_1 + \delta_{22} \cdot R_2 = 0 \qquad (3.26)$$

where δ_{11}, δ_{12}, δ_{21}, δ_{22} are the flexibility coefficients due to unit loads applied at R_1 and R_2, as shown in Figure 3.58. These can be calculated using the unit load method and the area–moment multiplication technique, as below with the help of the M_L, m_1, and m_2 diagrams, as shown in Figure 3.58.

$$\Delta_{1L} = \int M_L \cdot m_1 \cdot \frac{dx}{EI}; \quad \delta_{11} = \int m_1 \cdot m_1 \cdot \frac{dx}{EI}$$

$$\Delta_{2L} = \int M_L \cdot m_2 \cdot \frac{dx}{EI}; \quad \delta_{22} = \int m_2 \cdot m_2 \cdot \frac{dx}{EI}$$

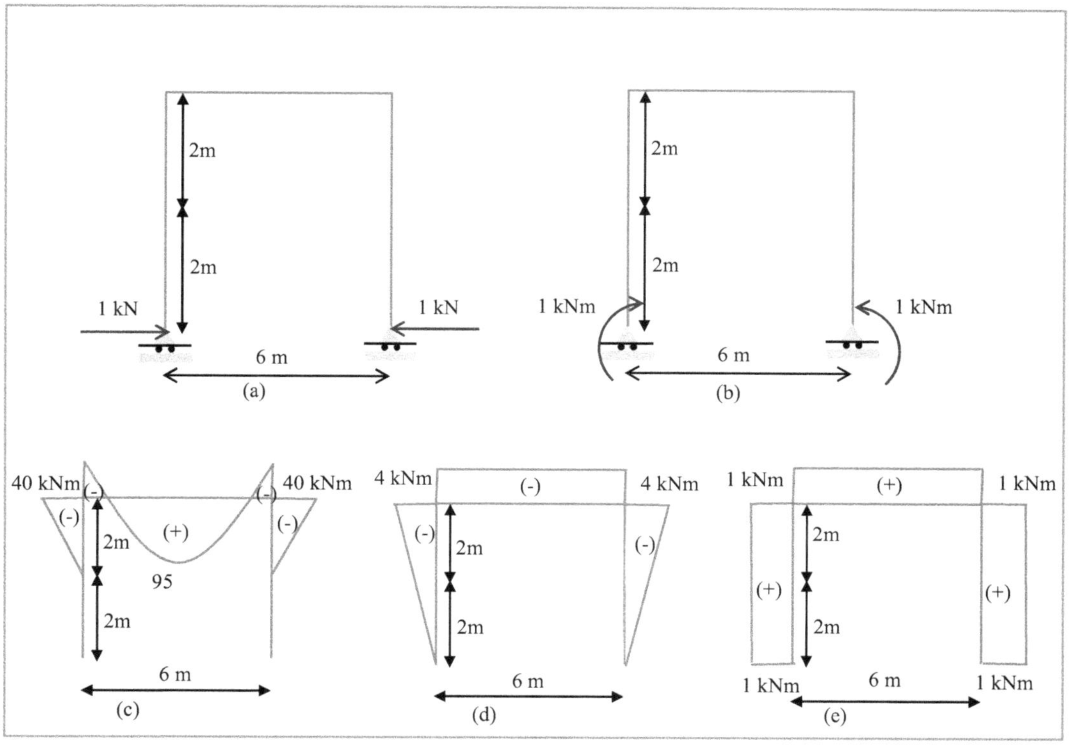

FIGURE 3.55 (a) Unit loads at A and D, (b) unit couples at A and D, (c) M_L diagram, (d) m_1 diagram, and (e) m_2 diagram.

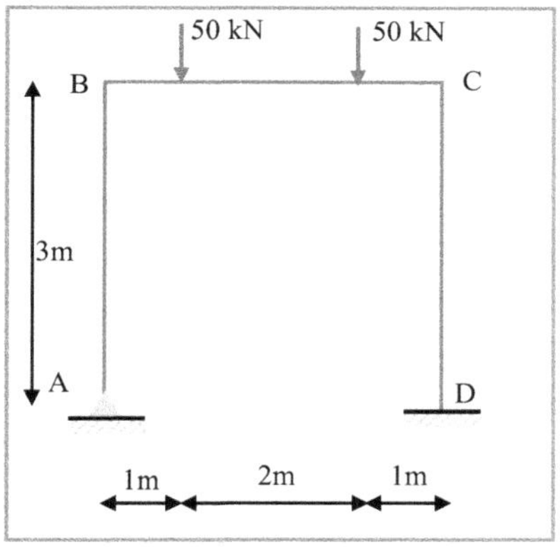

FIGURE 3.56 Indeterminate rigid-jointed plane frame.

$$\Delta_{1L} = \frac{1}{EI}\left[\frac{-50}{2}(2+4)\times 0.5\right] = \frac{-75}{EI}$$

$$\Delta_{2L} = \frac{1}{EI}\left[\frac{50}{2}(4+2)\times 3\right] = \frac{450}{EI}$$

$$\delta_{11} = \frac{1}{EI}\left[\frac{4}{6}(1\times 2\times 1)+1\times 3\times 1\right] = \frac{4.33}{EI}$$

$$\delta_{12} = \delta_{21} = \frac{1}{EI}\left[\frac{1}{2}\times(-1)\times 4\times 3+1\times(-3)\times 1.5\right] = \frac{-10.5}{EI}$$

$$\delta_{22} = \frac{1}{EI}\left[\frac{3}{6}\times(3\times 2\times 3)\times 2+3\times 4\times 3\right] = \frac{54}{EI}$$

Substitute the above values into equations 3.25 and 3.26; thus,

$$0.001 = \frac{-75}{EI} + \frac{4.33}{EI}R_1 - \frac{10.5}{EI}R_2$$

$$213.37 = 4.33R_1 - 10.5R_2 \qquad (3.25a)$$

$$0.01 = \frac{450}{EI} - \frac{10.5}{EI}R_1 + \frac{54}{EI}R_2$$

$$933.75 = -10.5R_1 + 54R_2 \qquad (3.26a)$$

By solving the above equations, we obtain

$$R_1 = 172.58\text{ kN} = M_D \quad \text{and} \quad R_2 = 50.85\text{ kN} = H_D$$

The remaining support reactions, shear force and bending moment diagrams, along with the probable deflected shape, are shown in Figure 3.59a–d, respectively.

Example 3.14

Analyse the rigid-jointed plane frame shown in Figure 3.60 using the consistent deformation method, if the support

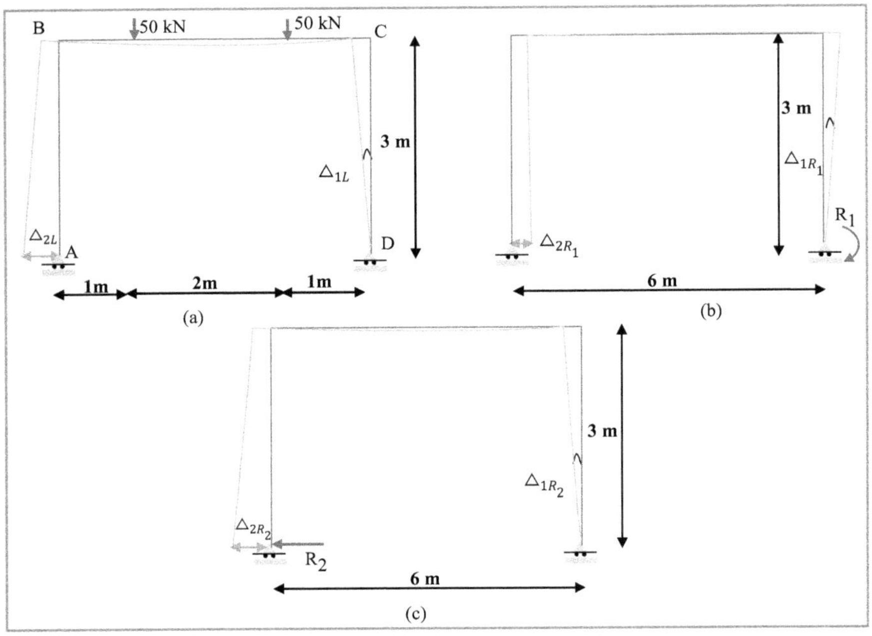

FIGURE 3.57 Basic frame with (a) external loads (b) redundant R_1, and (c) redundant R_2.

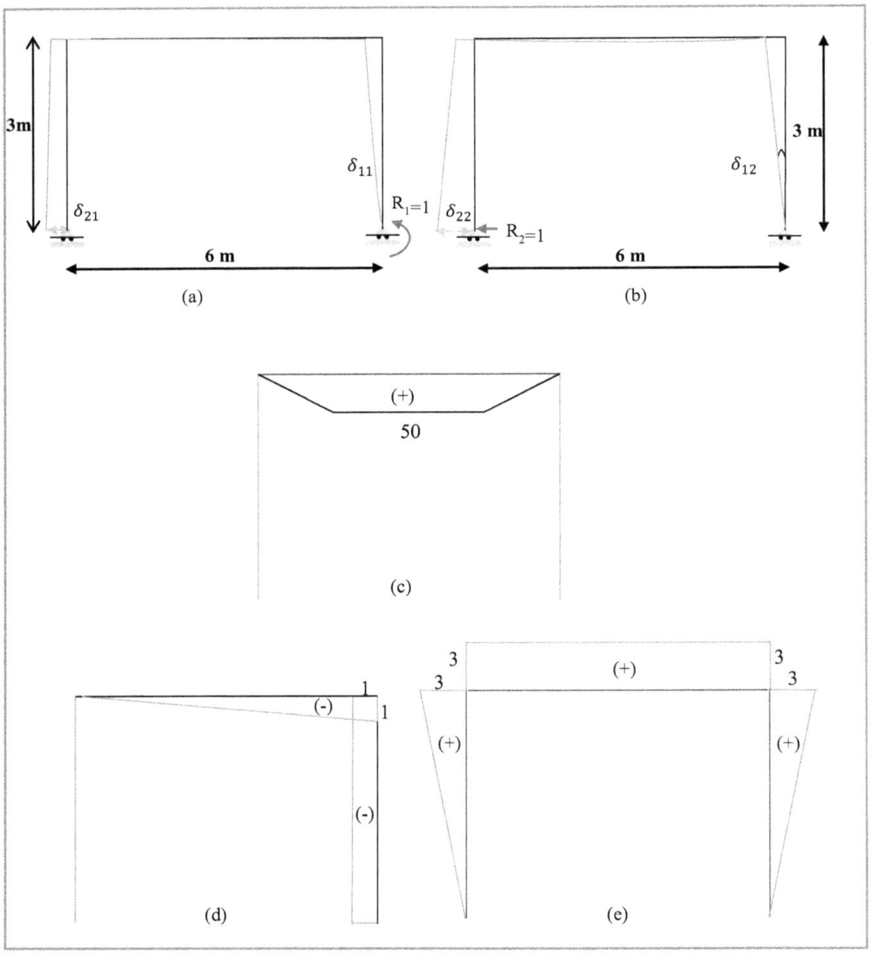

FIGURE 3.58 Basic frame with (a) $R_1 = 1$, (b) $R_2 = 1$, (c) M_L diagram due to external loads, (d) m_1 diagram due to a unit couple at R_1, and (e) m_2 diagram due to a unit load at R_2.

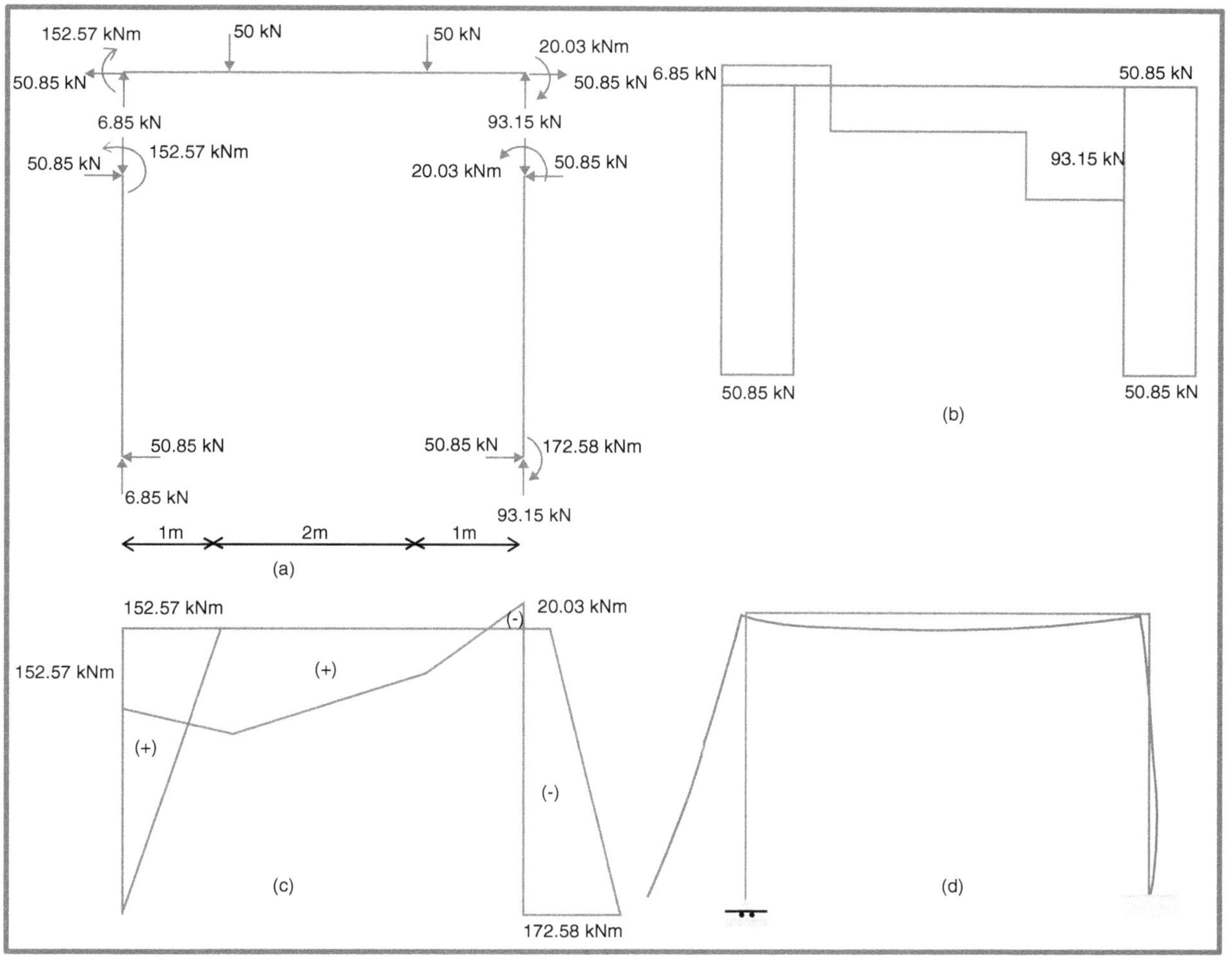

FIGURE 3.59 (a) Free body diagram, (b) shear force diagram, (c) bending moment diagram, and (d) probable deflected curve.

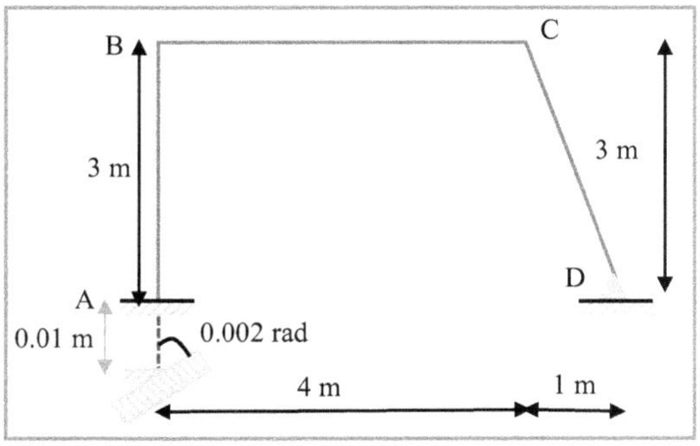

FIGURE 3.60 Indeterminate rigid-jointed plane frame.

A settles down by 10 mm and rotates clockwise by 0.002 rad. Assume, $EI = 141.75 \times 10^3$ kNm^2.

Solution:

The frame is statically indeterminate to the second degree. Let us select the reactions at *D* ($V_D = R_1$ and $H_D = R_2$) as redundant reactions. The primary structure is shown as a cantilever with a fixed end at *A*, as shown in Figure 3.61 with yielding conditions.

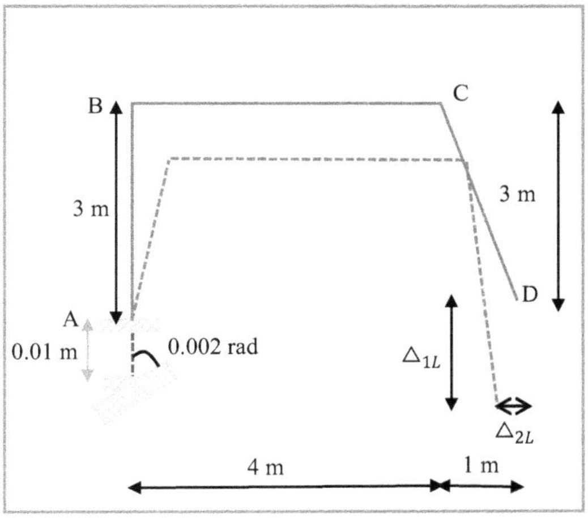

FIGURE 3.61 Primary structure with the effect of yielding of supports.

Apply the unit forces in place of R_1 and R_2 successively in the primary structure, as shown in Figure 3.62. The m_1 and m_2 diagrams, respectively, due to unit loads at R_1 and R_2 are also shown in Figure 3.62.

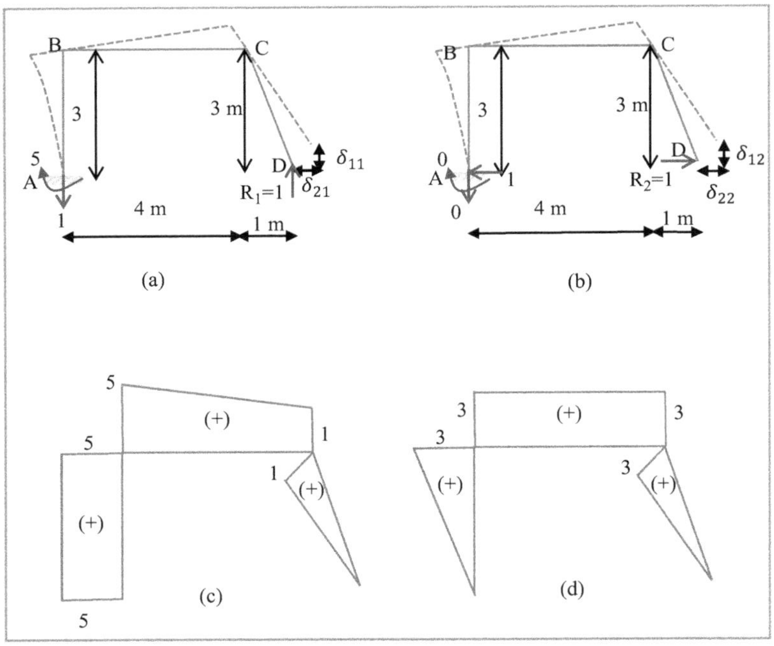

FIGURE 3.62 Primary structure with (a) unit load at R_1, (b) unit load at R_2, (c) m_1 diagram, and (d) m_2 diagram.

The yielding effects at support *A* occur at the non-redundant locations; therefore, to account for these effects, apply the principle of virtual forces, as shown in Figures 3.61 and 3.62.

$$1\times\Delta_{1L}+5\times0.002+1\times0.01=0 \Rightarrow \Delta_{1L}=-0.02\text{ m}$$

$$1\times\Delta_{2L}+0\times0.002+0\times0.01=0 \Rightarrow \Delta_{2L}=0\text{ m}$$

Therefore, the compatibility equations can be written as

$$\Delta_1=\Delta_{1L}+\delta_{11}\cdot R_1+\delta_{12}\cdot R_2=0 \qquad (3.27)$$

$$\Delta_2=\Delta_{2L}+\delta_{21}\cdot R_1+\delta_{22}\cdot R_2=0 \qquad (3.28)$$

where δ_{11}, δ_{12}, δ_{21}, δ_{22} are the flexibility coefficients, and these can be calculated using the unit load method and the area–moment multiplication technique.

$$\delta_{11}=\int m_1\cdot m_1\cdot\frac{dx}{EI}$$

$$=\frac{1}{EI}\left[\begin{aligned}&\left(\frac{1}{2}\times1\times3.16\times0.67\right)\\&+\frac{4}{6}\big(5(2\times5+1)+1(2\times1+5)\big)+(5\times3\times5)\end{aligned}\right]$$

$$=\frac{117.38}{EI}$$

$$\delta_{12}=\delta_{21}=\int m_1\cdot m_2\cdot\frac{dx}{EI}$$

$$=\frac{1}{EI}\left[\begin{aligned}&\frac{3.16}{6}(1\times2\times3+0)\\&+\frac{4}{2}(1+5)\times3+5\times3\times1.5\end{aligned}\right]$$

$$=\frac{61.66}{EI}$$

$$\delta_{22}=\int m_2\cdot m_2\cdot\frac{dx}{EI}$$

$$=\frac{1}{EI}\left[\begin{aligned}&\frac{3.16}{6}(3\times2\times3+0)+(3\times4\times3)\\&+\frac{3}{6}(3\times2\times3+0)\end{aligned}\right]$$

$$=\frac{54.48}{EI}$$

Substituting the above values into equations 3.27 and 3.28, we can write them as

$$-0.02+\frac{117.38}{EI}R_1+\frac{61.66}{EI}R_2=0 \qquad (3.27a)$$

$$0+\frac{61.66}{EI}R_1+\frac{54.48}{EI}R_2=0 \qquad (3.28a)$$

By solving the above equations, we obtain

$$R_1=59.56\text{ kN}=V_D \quad\text{and}\quad R_2=-67.41\text{ kN}=H_D$$

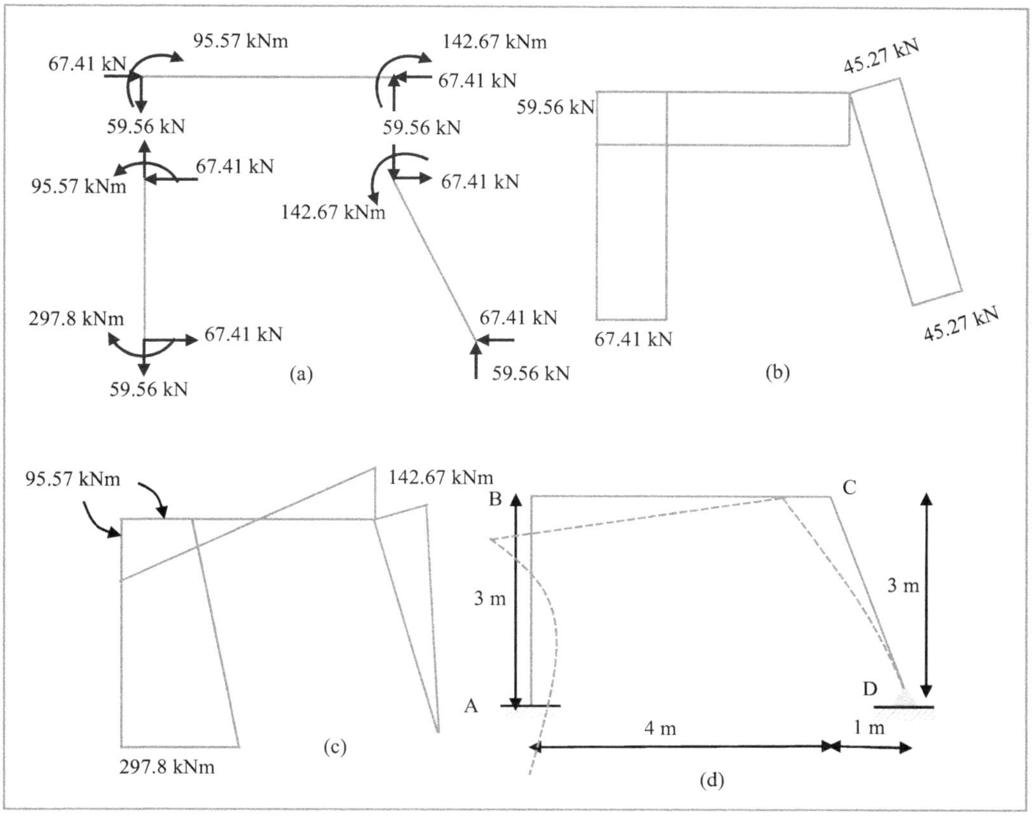

FIGURE 3.63 (a) Free body diagram, (b) shear force diagram, (c) bending moment diagram, and (d) probable deflected curve.

SUPPORT REACTIONS AND INTERNAL FORCES

With the known values of V_D and H_D, and using the free body diagram and the static equilibrium equations, the remaining support reactions can be determined, as shown in Figure 3.63a. The shear force and bending moment diagrams, along with the probable deflected shape, are shown in Figure 3.63b–d, respectively.

3.5 APPLICATION TO PIN-JOINTED PLANE FRAMES

3.5.1 Introduction

As discussed in Chapter 1, the static indeterminacy of a pin-jointed plane frame may be due to either the supports, redundant members, or both. If the redundancy is due to the external supports, the analysis of trusses follows the same procedure as that used for the analysis of beams and rigid-jointed plane frames, as outlined in Section 3.2. However, if the redundancy is due to the members, the member must be cut at a section, and two equal and opposite axial redundant forces are applied to represent the internal force of that member. The compatibility condition is that the relative displacement (axial) between the two ends at the cut section, caused by the external loads and redundant forces, must be zero. The application of this method is illustrated in the following examples.

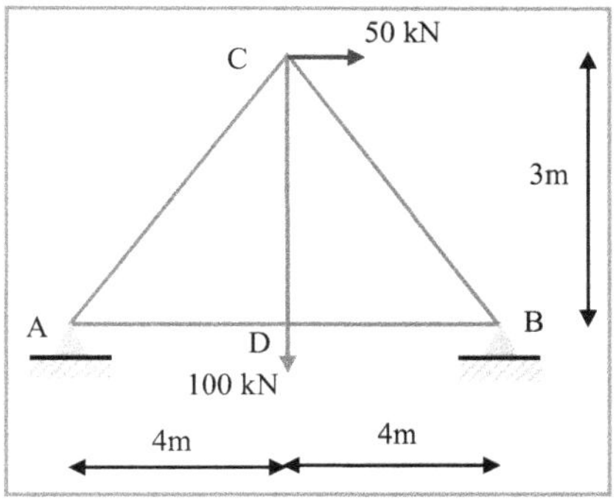

FIGURE 3.64 Indeterminate pin-jointed plane frame.

Example 3.15

Calculate the support reactions and member forces of the pin-jointed plane frame, shown in Figure 3.64, using the consistent deformation method. Assume AE is constant.

Solution:

The degree of static indeterminacy,

$I_s = (m+R) - 2j = (5+4) - 2\times 4 = 1$, i.e., to the first degree, and the truss is internally determinate. Let us select the horizontal reaction at B ($H_B = R_1 \rightarrow$) as the redundant.

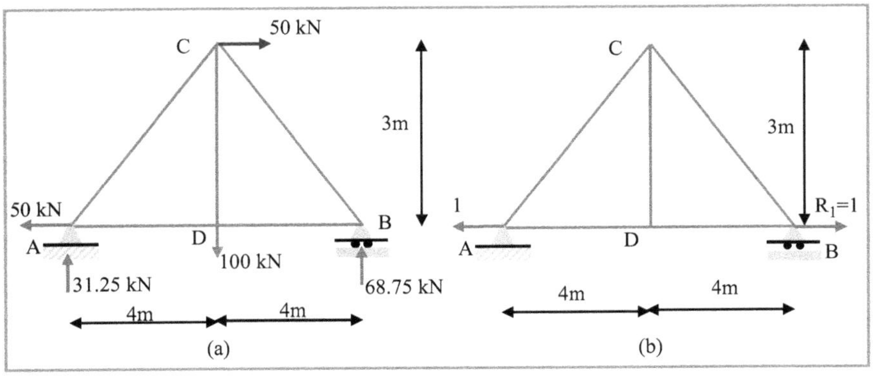

FIGURE 3.65 Primary truss with (a) external loads and (b) unit force at redundant ($R_1 = 1$).

Using the principle of superposition, the primary truss with external loads and a unit load at redundant is shown in Figure 3.65.

The compatibility condition can be written as

$$\Delta_1 = \Delta_{1L} + \Delta_{1R} = 0$$

$$\Delta_1 = \Delta_{1L} + \delta_{11} \cdot R_1 = 0 \qquad (3.29)$$

where δ_{11} is the flexibility coefficient due to the unit load applied at $R_1 = 1$.

Δ_{1L} is the horizontal displacement at D due to external loads.

These can be calculated using the unit load method, as follows.

$$\Delta_{1L} = \sum F_1 f_1 \frac{L}{AE}$$

$$\delta_{11} = \sum f_1 f_1 \frac{L}{AE}$$

where F_1 and f_1 are the forces in the members due to external loads and the unit load applied at R_1 in the primary truss. These forces can be calculated using the method of joints, presented in Table 3.4.

From Table 3.4,

$$\Delta_{1L} = \frac{733.36}{AE}$$

TABLE 3.4

		Member Forces (kN)			
Member	L (m)	F_1 Due to Loads	f_1 Due to Unit Load at $R_1 = 1$	$\sum F_1 f_1 L$ (kNm)	$\sum f_1 f_1 L$ (kNm)
AC	−5	−52.08	0	0	0
AD	4	91.67	1.0	366.68	4
CD	3	100	0	0	0
BD	4	91.67	1.0	366.68	4
BC	5	−114.58	0	0	0
$\sum =$				733.36	8

$$\delta_{11} = \frac{8}{AE}$$

∴From equation 3.29, we can write

$$\frac{733.36}{AE} + \frac{8}{AE} R_1 = 0$$

$$R_1 = -91.67 \text{ kN } (\leftarrow)$$

$$\text{The final member forces} = F + R_1 f_1$$

$$F_{AC} = -52.08 - 91.67 \times 0 = -52.08 \text{ kN}$$

$$F_{AD} = 91.67 - 91.67 \times 1 = 0 \text{ kN}$$

$$F_{CD} = 100 - 0 \times (-91.67) = 100 \text{ kN}$$

$$F_{BD} = 91.67 + (-91.67) \times 1 = 0 \text{ kN}$$

$$F_{BC} = -114.58 + 0 \times (-91.67) = -114.58 \text{ kN}$$

(Note: The +ve sign indicates that the bar force is in tension, and the −ve sign indicates that it is in compression.)

Example 3.16

Analyse the pin-jointed plane frame shown in Figure 3.66, using the consistent deformation method. The cross-sectional areas of the members are shown in parentheses in mm². Assume $E = 2.01 \times 10^5$ N/mm².

Solution:

The truss is internally indeterminate to the first degree and externally determinate, as indicated by $I_s = (m + R) - 2j = (6 + 3) - 2 \times 4 = 1$. Therefore, we

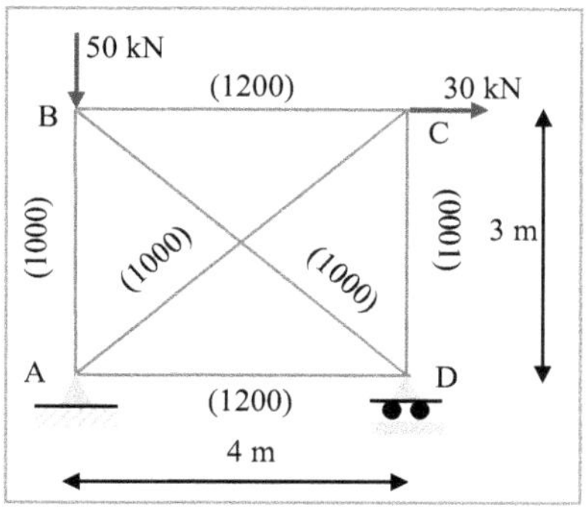

FIGURE 3.66 Indeterminate pin-jointed plane frame.

select the force in member *BD* as the redundant force. The primary truss can be obtained by cutting the member *BD* and showing the internal force in it as R_1, as shown in Figure 3.67. Using the principle of superposition, we can write the compatibility condition such that the relative displacement of the ends of the member *BD* due to loads and due to the redundant force must be zero.

$$\Delta_{BD} = (\Delta_{BD})_L + (\Delta_{BD})_R = 0$$

$$= (\Delta_{BD})_L + \delta_{11} \cdot R_1 = 0 \qquad (3.30)$$

$$= \Delta_{1L} + \delta_{11} \cdot R_1 = 0$$

where Δ_{1L} is the displacement in member *BD* due to loads, and δ_{11} is the flexibility coefficient of *BD* due to a unit load applied at $R_1 = 1$, as shown in Figure 3.68. The member forces due to external loads and due to the unit load at $R_1 = 1$ in the primary truss can be calculated conveniently using the method of joints (details not shown here), and the results are presented in Table 3.5. The Δ_{1L} and δ_{11} can be calculated using the unit load method, as shown in Table 3.5.

$$\Delta_{1L} = \sum_{i=1}^{6} F_i \cdot f_i \cdot \frac{L_i}{A_i E}$$

$$\delta_{11} = \sum_{i=1}^{6} f_i \cdot f_i \cdot \frac{L_i}{A_i E}$$

From Table 3.5 and using equation 3.30, we can write

$$1.512 \times 10^{-3} + 78.218 \times 10^{-6} R_1 = 0$$

$$\therefore R_1 = -19.33 \text{ kN}$$

$$\text{The final member force} = F_i + R_1 f_i$$

$$F_{AB} = -50 - 19.33(-0.6) = -38.40 \text{ kN}$$

$$F_{BC} = 0 - 0.8(-19.33) = 15.46 \text{ kN}$$

$$F_{CD} = -22.5 - 0.6(-19.33) = -10.90 \text{ kN}$$

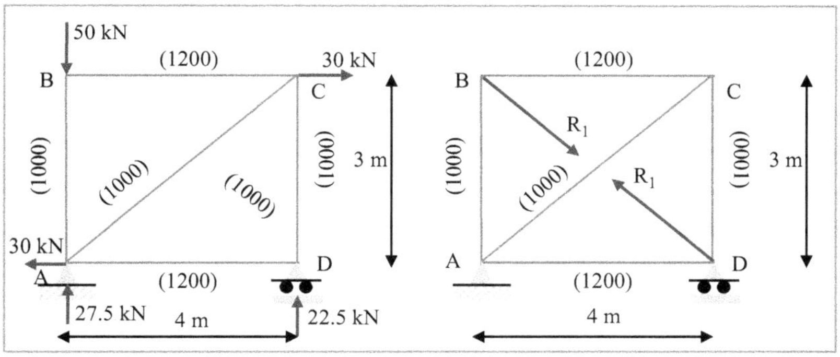

FIGURE 3.67 Primary truss with (a) external loads and (b) redundant R_1.

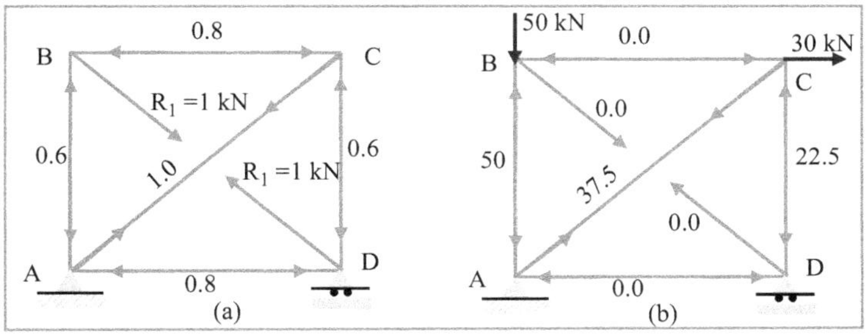

FIGURE 3.68 The member forces due to (a) unit load at $R_1 = 1$ and (b) external loads in the primary truss.

TABLE 3.5

Member	L_i (m)	A_iE_i (kN)	Member Forces (kN) F_i Due to Loads	f_i Due to Redundant $R_1=1$	$F_i \cdot f_i \dfrac{L_i}{A_iE}$ (kNm)	$f_i^2 \cdot \dfrac{L_i}{A_iE}$ (kNm)
AB	3.0	210×10^3	−50.0	−0.6	0.428×10^{-3}	5.142×10^{-6}
BC	4.0	252×10^3	0	−0.8	0.0	10.158×10^{-6}
CD	3.0	210×10^3	−22.5	−0.6	0.192×10^{-3}	5.142×10^{-6}
AD	4.0	252×10^3	0	−0.8	0.0	10.158×10^{-6}
BD	5.0	210×10^3	0	1.0	0.0	23.809×10^{-6}
AC	5.0	210×10^3	37.5	1.0	0.892×10^{-3}	23.809×10^{-6}
					$\Sigma=1.512\times10^{-3}$	$\Sigma=78.218\times10^{-6}$

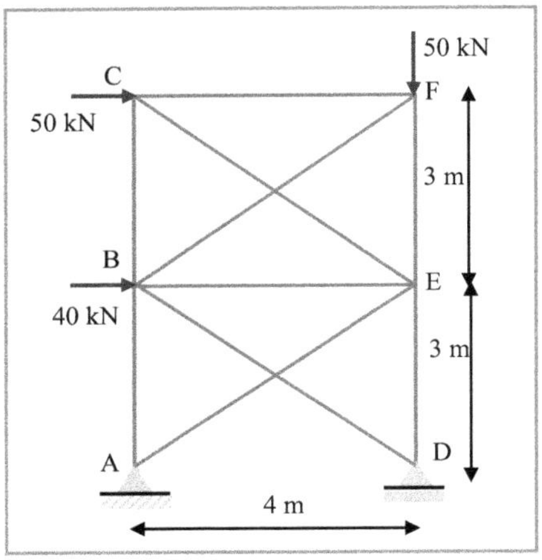

FIGURE 3.69 Indeterminate pin-jointed plane frame.

$$F_{AD} = 0-0.8(-19.33) = 15.46 \text{ kN}$$

$$F_{BD} = 0-19.33\times1 = -19.33 \text{ kN}$$

$$F_{AC} = 37.5+1(-19.33) = 18.17 \text{ kN}$$

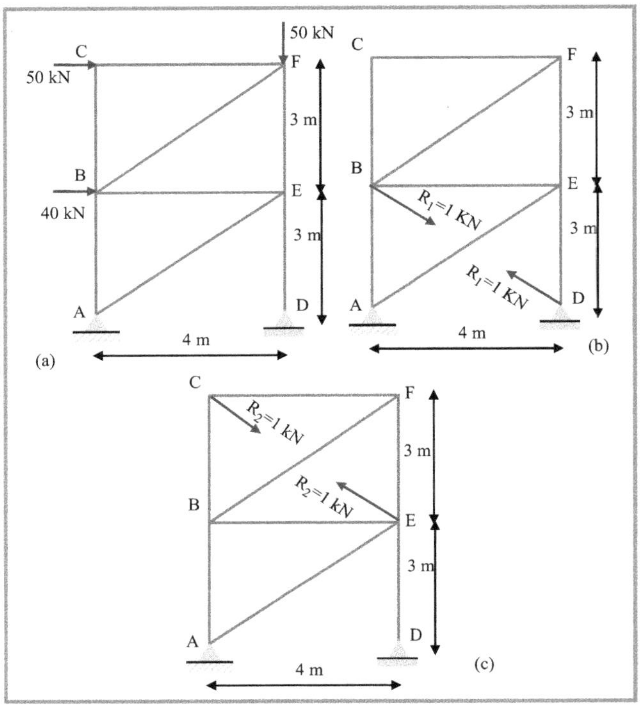

FIGURE 3.70 Determinate pin-jointed plane frame with (a) external loads, (b) unit load at $R_1=1$, and (c) unit load at $R_2=1$.

Example 3.17

Find the member forces of the truss shown in Figure 3.69, using the consistent deformation method. Assume the cross-sectional area is constant for all members.

Solution:

The static indeterminacy, $I_s = (m+R)-2j$

$$I_s = (10+4)-6\times2 = 2$$

Therefore, the truss is statically internally indeterminate to the second degree but externally determinate. The forces in the members *BD* ($F_{BD}=R_1$) and *CE* ($F_{CE}=R_2$) are selected as redundant member forces. The basic truss with external loads, and redundant member forces ($R_1=1$ and $R_2=1$) is shown in Figure 3.70. These primary trusses can be analysed using the method of joints (calculations not shown here). The results are summarised conveniently in Table 3.6.

Using the principle of superposition, the compatibility equations can be written as,

$$\Delta_1 = \Delta_{1L} + \delta_{11}\cdot R_1 + \delta_{12}\cdot R_2 = 0 \tag{3.31}$$

$$\Delta_2 = \Delta_{2L} + \delta_{21}\cdot R_1 + \delta_{22}\cdot R_2 = 0 \tag{3.32}$$

where δ_{11}, δ_{12}, δ_{21}, δ_{22} are the flexibility coefficients due to unit loads applied at $R_1=1$ and $R_2=1$, as shown in Figure 3.70. Δ_{1L} and Δ_{2L} are the displacements due to external loads at the redundant locations in the primary truss. These can be calculated using the unit load method, as shown in Table 3.6.

TABLE 3.6

Member	L (m)	Member Forces (kN) Due to Loads F_{iL}	$R_1 = 1$ kN f_{i1}	$R_2 = 1$ kN f_{i2}	$F_i \cdot f_{i1} L$	$F_i \cdot f_{i2} L$	$f_{i1}^2 \cdot L$	$f_{i1} \cdot f_{i2} L$	$f_{i2} \cdot f_{i2} L$
AB	3	37.5	–0.6	0	–67.5	0	1.08	0	0
BC	3	0	0	-0.6	0	0	0	0	1.08
BD	5	0	1	0	0	0	5	0	0
BE	4	–90	–0.8	–0.8	288	288	2.56	2.56	2.56
CF	4	–50	0	–0.8	0	160	0	0	2.56
AE	5	112.5	1	0	562.5	0	5	0	0
BF	5	62.5	0	1	0	312.5	0	0	5
CE	5	0	0	1	0	0	0	0	5
FE	3	–87.5	0	–0.6	0	157.5	0	0	1.08
ED	3	–155	–0.6	0	279	0	1.08	0	0
				Σ=	1,062	918	14.72	2.56	17.28

From the table, we obtain

$$\Delta_{1L} = \sum_{i=1}^{n} F_{iL} \cdot f_{i1} \left(\frac{L}{AE} \right)_i = \frac{1,062}{AE}$$

$$\Delta_{2L} = \sum_{i=1}^{n} F_{iL} \cdot f_{i2} \left(\frac{L}{AE} \right)_i = \frac{918}{AE}$$

$$\delta_{11} = \sum f_{i1} \cdot f_{i1} \left(\frac{L}{AE} \right)_i = \frac{14.72}{AE}$$

$$\delta_{12} = \delta_{21} = \sum f_{i1} \cdot f_{i2} \left(\frac{L}{AE} \right)_i = \frac{2.56}{AE}$$

$$\delta_{22} = \sum f_{i2} \cdot f_{i2} \left(\frac{L}{AE} \right)_i = \frac{17.28}{AE}$$

By substituting the above into the compatibility equations 3.31 and 3.32,

$$\frac{1,062}{AE} + \frac{14.72}{AE} R_1 + \frac{2.56}{AE} R_2 = 0$$

$$\frac{918}{AE} + \frac{2.56}{AE} R_1 + \frac{17.28}{AE} R_2 = 0$$

By solving the above equations, we obtain

$$R_1 = -64.57 \text{ kN} = F_{BD} \quad \text{and} \quad R_2 = -43.56 \text{ kN} = F_{CE}$$

FINAL MEMBER FORCES

The final member forces can be calculated using equation 3.33, and the results are summarised in Table 3.7.

$$F_i = F_{iL} + R_1 f_{i1} + R_2 f_{i2} \tag{3.33}$$

TABLE 3.7

Members	Force (kN)
AB	76.24
BC	26.14
BD	–64.57
BE	–3.50
CF	–15.15
AE	47.93
BF	18.94
CE	–43.56
FE	–61.36
ED	–116.26

(Note: in the above table, the –ve sign indicates compression, and the +ve sign indicates tension.)

Example 3.18

Analyse the indeterminate truss shown in Figure 3.69 (Q.No. 3.17). If the member *CE* is too long by 3 mm and *BD* is too short by 2 mm in the absence of external loads. Assume $AE = 200 \times 10^3$ kN.

Solution:

The truss has the same geometry as that shown in Example 3.17. Since the member *BD* is shortened and *CE* is elongated, let us choose the forces in these members *BD* and *CE* as the redundant forces, and designate them as R_1 and R_2, respectively.

Since no external loads are acting on the truss, $\Delta_{1L} = \Delta_{2L} = 0$

$$\left(\begin{array}{l} \text{no relative displacements between the ends} \\ \text{of the cut members } CE \text{ and } BD \end{array} \right)$$

As the member *BD* is shortened, it requires tensile force to restore its length to fix; therefore, assume it is positive. In contrast, as member *CE* is elongated, it requires a compressive force to restore its length to fix; therefore, assume it is negative.

Hence, $\Delta_1 = +0.002$ m and $\Delta_2 = -0.003$ m

The compatibility equations can be written as

$$\Delta_1 = \delta_{11} \cdot R_1 + \delta_{12} \cdot R_2 = 0 \quad (3.34)$$

$$\Delta_2 = \delta_{21} \cdot R_1 + \delta_{22} \cdot R_2 = 0 \quad (3.35)$$

where δ_{11}, δ_{12}, δ_{21}, δ_{22} are already calculated in Example 3.17, as shown below.

$$\delta_{11} = \sum f_{1i} \cdot f_{1i}\left(\frac{L}{AE}\right)_i = \frac{14.72}{AE}$$

$$\delta_{12} = \delta_{21} = \sum f_{1i} \cdot f_{2i}\left(\frac{L}{AE}\right)_i = \frac{2.56}{AE}$$

$$\delta_{22} = \sum f_{2i} \cdot f_{2i}\left(\frac{L}{AE}\right)_i = \frac{17.28}{AE}$$

Substituting the above into equations 3.34 and 3.35; thus,

$$0.002 = \frac{13.72}{AE}R_1 + \frac{2.56}{AE}R_2$$

$$\Rightarrow 400 = 13.72R_1 + 2.56\ R_2 \quad (3.34a)$$

$$-0.003 = \frac{2.56}{AE}R_1 + \frac{17.28}{AE}R_2$$

$$\Rightarrow -600 = 2.56R_1 + 17.28R_2 \quad (3.34b)$$

By solving the above equations, we obtain

$$R_1 = +36.64 \text{ kN} = F_{BD} \quad \text{and} \quad R_2 = -40.15 \text{ kN} = F_{CE}$$

FINAL MEMBER FORCES

The final member forces can be calculated using equation 3.36, and the results are presented in Table 3.8.

$$F_i = R_1 f_{i1} + R_2 f_{i2} \quad (3.36)$$

TABLE 3.8

Members	Force (kN)
AB	−21.98
BC	+24.09
BD	+36.64
BE	+2.81
CF	+32.12
AE	+36.64
BF	−40.15
CE	−40.15
FE	+24.09
ED	−21.98

Example 3.19

For the truss shown in Figure 3.69, Q.No. 3.17, if the members *BC*, *CF*, and *FE* are subjected to a temperature rise of 30°C, find the forces in all the members. Assume, $\alpha = 11\times10^{-6}/°\text{C}$, $E = 2.1\times10^5$ N/mm^2. The cross-sectional area of the members *AB*=*BC*=*CF*=*BE*=*FE*=*DE*=1,200 mm², and the cross-sectional area of the members *CE*=*BF*=*AE*=*BD*=1,000 mm².

Solution:

As calculated in Q.No. 3.17, the static indeterminacy is 2. Select the forces in the members *BD* ($F_{BD}=R_1$) and *CE* ($F_{CE}=R_2$) as the redundant member forces. Since the truss is subjected to no external forces, $\Delta_{1L} = \Delta_{2L} = 0$.

Since the members *BC*, *CF*, and *FE* are subjected to a temperature rise, these members become elongated, thereby inducing compressive forces.

The free elongation of these members is

$$\Delta_{EF} = \Delta_{BC} = L\alpha\Delta T = 3{,}000\times11\times10^{-6}\times30 = 0.99 \text{ mm}$$

$$\Delta_{CF} = 4{,}000\times11\times10^{-6}\times30 = 1.32 \text{ mm}$$

The member forces due to the application of unit loads at R_1=1 kN and R_2=1 kN, are already calculated in Q.No. 3.17 and presented in Table 3.6. The displacement of the members *BD* and *CE* due to the members *BC*, *CF*, and *EF* subjected to temperature rise can be calculated using the unit load method as follows:

$$\Delta_{BD}\times1 = \sum_{i=1}^{n} f_{i1}\cdot\delta_t$$

$$\Delta_{CF}\times1 = \sum_{i=1}^{n} f_{i2}\cdot\delta_t$$

where f_{i1} is the forces in the members due to the unit load at R_1=1.

f_{i2} is the forces in the members due to the unit load at R_2=1.

δ_t is the displacement due to temperature = $L \propto \Delta T$

$$\Delta_{BD} = 0(0.99)+0(1.32)+0(0.99) = 0$$

$$\Delta_{CF} = -0.6(0.99)-0.8(1.32)-0.6(0.99) = -2.244 \text{ mm}$$

The compatibility equations are

$$\Delta_1 = \Delta_{BD} = \Delta_{1L} + \delta_{11}.R_1 + \delta_{12}.R_2 = 0 \quad (3.37)$$

$$\Delta_2 = \Delta_{CF} = \Delta_{2L} + \delta_{21}\cdot R_1 + \delta_{22}\cdot R_2 = -2.244 \quad (3.38)$$

where δ_{11}, δ_{12}, δ_{21}, δ_{22} are already calculated and presented in Table 3.6 in Q.No. 3.17, and the same are reproduced in Table 3.9.

TABLE 3.9

Member	L (mm)	A (mm²)	Member Forces (kN) $R_1 = 1$ kN	Member Forces (kN) $R_2 = 1$ kN	$f_{i1}^2\left(\frac{L}{AE}\right)_i$	$f_{i2}^2\left(\frac{L}{AE}\right)_i$	$f_{i1}f_{i2}\frac{L}{AE}$	Final Member Forces (kN)
AB	3,000	1,200	−0.6	0	0.0042	0	0	−2.98
BC	3,000	1,200	0	−0.6	0	0.00428	0	19.53
BD	5,000	1,000	+1	0	0.0238	0	0	4.98
BE	4,000	1,200	−0.8	−0.8	0.01015	0.01015	0.01015	22.06
CF	4,000	1,200	0	−0.8	0	0.01015	0	26.04
AE	5,000	1,000	1.0	0	0.0238	0	0	4.98
BF	5,000	1,000	0	1.0	0	0.0238	0	−32.55
CE	5,000	1,000	0	1.0	0	0.0238	0	−32.55
FE	3,000	1,200	0	−0.6	0	0.00428	0	19.53
ED	3,000	1,200	−0.6	0	0.00428	0	0	−2.98
$\sum =$					0.06634	0.0765	0.01015	

Substituting the above into equations 3.37 and 3.38, thus

$$0.06634\ R_1 + 0.01015\ R_2 = 0 \qquad (3.37a)$$

$$0.01015\ R_1 + 0.0765\ R_2 = -2.244 \qquad (3.38a)$$

By solving the above equations, we obtain

$$R_1 = 4.98\ \text{kN} = F_{BD} \quad \text{and} \quad R_2 = -32.55\ \text{kN} = F_{CE}$$

FINAL MEMBER END FORCES

The final member forces are calculated using equation 3.39, and the results are summarised in the last column of Table 3.9.

$$F_i = R_1 f_{i1} + R_2 f_{i2} \qquad (3.39)$$

3.6 PROBLEMS FOR PRACTICE

3.6.1 Analyse the beams shown in Figure 3.71a–e using the consistent deformation method. Construct the shear force and bending moment diagrams. Also, plot the probable deflected shapes.

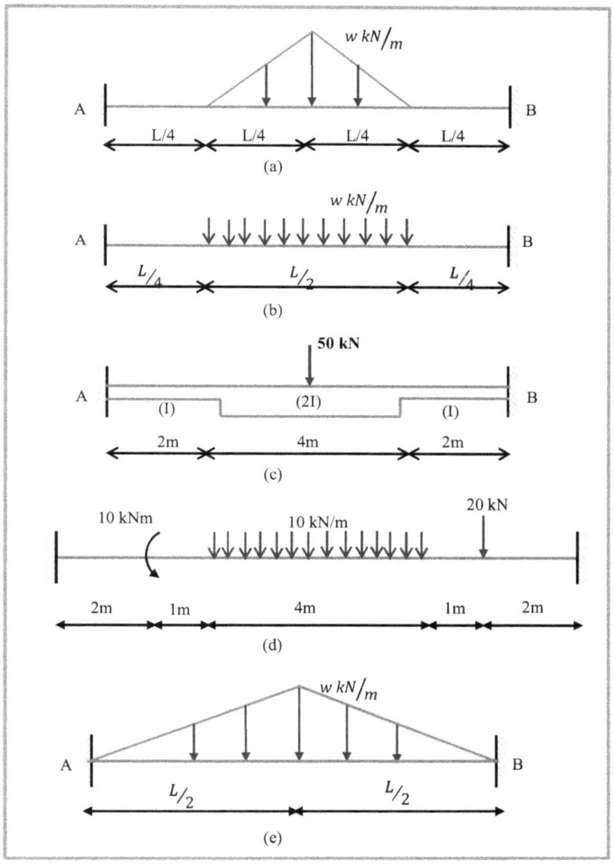

FIGURE 3.71 Indeterminate beams for analysis.

3.6.2 Solve the beam shown in Figure 3.71c, if the support 'A' rotates clockwise by 0.02 rad and settles down by 10 mm alone, without any external loads. Assume $EI = 85{,}000$ kNm^2.

3.6.3 Figure 3.72 shows the propped cantilever with an overhang loaded with UDL. Find the overhang length so that the deflection at the free end is zero. Assume EI is constant.

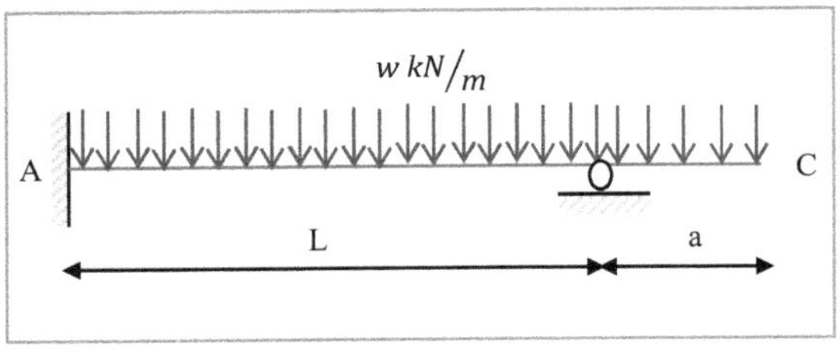

FIGURE 3.72

3.6.4 For the beam shown in Figure 3.72, find the position of the prop if the slope at the prop is zero.

3.6.5 Consider an unsymmetrical rigid-jointed plane frame shown in Figure 3.73. Compute the support reactions using the consistent deformation method. Draw the axial force, shear force, and bending moment diagrams. Also, sketch the deflected shape.

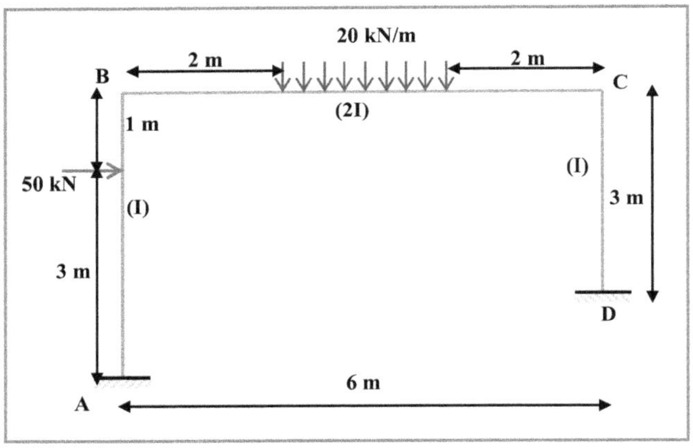

FIGURE 3.73

3.6.6 A rigid-jointed plane frame with inclined legs is shown in Figure 3.74. Analyse the frame using the consistent deformation method. Construct the axial force, shear force, and bending moment diagrams. Also, draw the elastic curve.

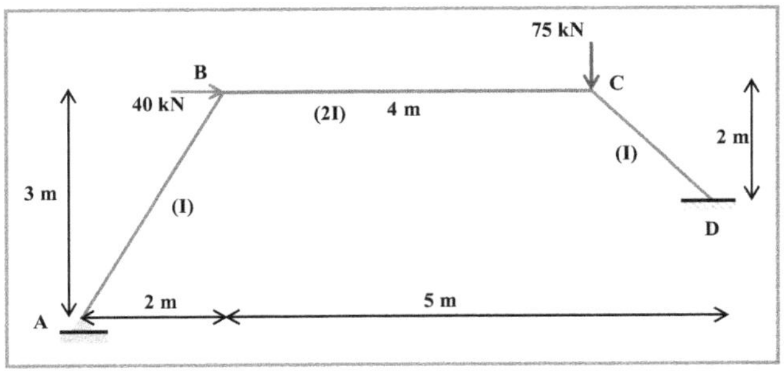

FIGURE 3.74

3.6.7 Re-analyse the rigid-jointed frame shown in Figure 3.74 using the consistent deformation method, if the support D settle down by 10 mm and rotates clockwise by 0.002 rad, in the absence of external loads. Assume $EI = 85{,}000$ kNm^2.

3.6.8 A pitched portal frame $ABCD$ is shown in Figure 3.75. Compute the support reactions using the consistent deformation method. Draw the axial force, shear force, and bending moment diagrams. Also, draw the deflected shape.

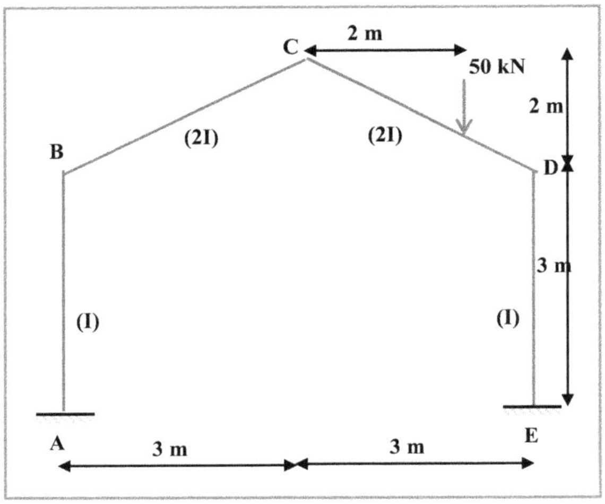

FIGURE 3.75

3.6.9 Analyse the box frame subjected to self-equilibrating loading shown in Figure 3.76, using the consistent deformation method. Construct the axial force, shear force, and bending moment diagrams. Also, sketch the elastic curve.

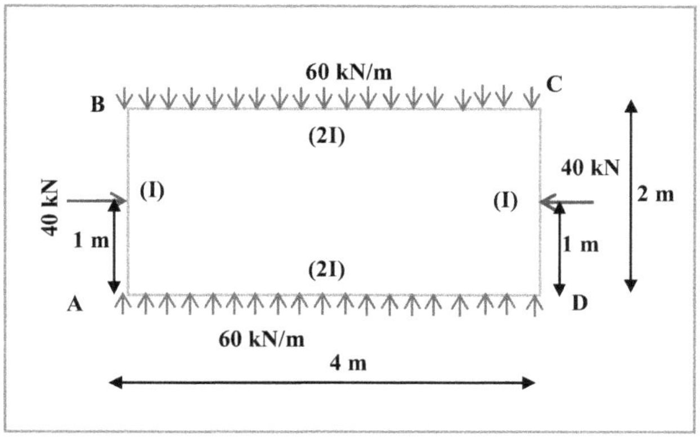

FIGURE 3.76

3.6.10 Analyse the anti-symmetric rigid-jointed plane frame shown in Figure 3.77 using the consistent deformation method. Sketch the shear force and bending moment diagrams. Also, sketch the deflected shape.

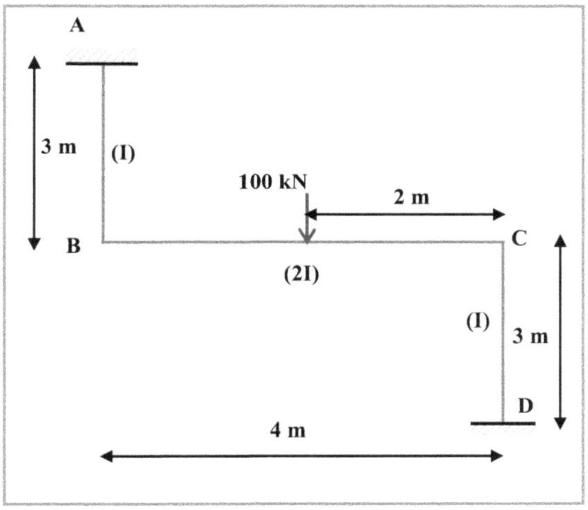

FIGURE 3.77

3.6.11 Analyse the pin-jointed plane frame shown in Figure 3.78 using the consistent deformation method. Assume all the members have the same axial rigidity.

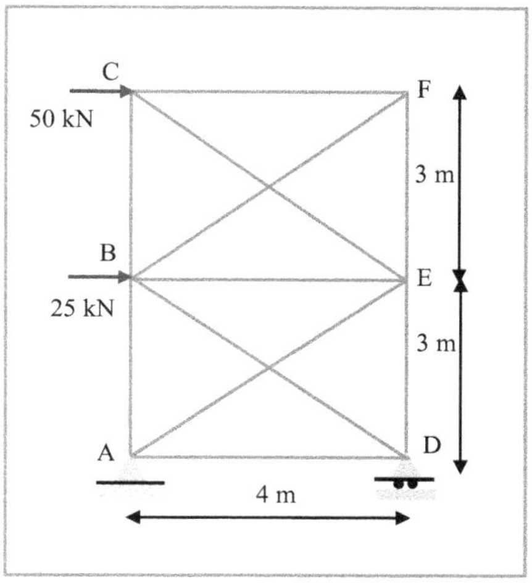

FIGURE 3.78

3.6.12 Re-analyse the truss shown in Figure 3.78, if the members *BD* and *BF* are too long by 3 mm and the members *CF* and *BE* are too short by 2 mm instead of applying external loads. Assume the axial rigidity of all diagonal members is 2.05×10^{11} kN and for all other members, it is 2.46×10^{11} kN.

3.6.13 Re-analyse the pin-jointed plane frame shown in Figure 3.78, using the consistent deformation method, if the frame is subjected to a temperature rise of 40°C instead of external loads. Assume the coefficient of thermal expansion $\alpha=11\times10^{-6}$/°C. Also, assume the axial rigidity of all the diagonal members is 2.05×10^{11} kN and for all other members is 2.46×10^{11} kN.

3.6.14 How is the method of consistent deformation different from the theorem of three moments in analysing indeterminate structures?

3.6.15 Explain briefly the step-by-step solution involved in the consistent deformation method.

3.6.16 Explain the procedure involved in the analysis of a pin-jointed plane frame when it is subjected to a temperature change.

3.6.17 A self-equilibrating rectangular box culvert is loaded symmetrically with respect to both vertical and horizontal axes. Explain with sketches how it is to be analysed using the method of consistent deformation.

3.7 ANSWER KEY

Problem No.		Answer Key at Salient Locations
	Figure 3.71a	$M_A = -M_B = -\frac{3}{100}wL^2$
	Figure 3.71b	$M_A = -M_B = -\frac{1}{192}wL^2$
	Figure 3.71c	$M_A=-M_B=-41.66$ kNm; BM@centre$=58.34$ kNm; $R_A=R_B=25$ kN
	Figure 3.71d	$M_A=-50.53$ kNm; $M_B=70.13$ kNm; $R_A=23.04$ kN; $R_B=36.96$ kN
3.6.1	Figure 3.71e	$M_A = -M_B = -\frac{5}{96}wL^2$
3.6.2	$M_A=1{,}701.85$ kNm; $M_B=779.25$ kNm; $R_A=-310.14$ kN$=R_B$	
3.6.3	$a=0.35L$	
3.6.4	$a=0.408L$	
3.6.5	$M_A=-34.17$ kNm; $M_D=48.98$ kNm; $M_B=9.24$ kNm; $M_C=44.83$ kNm	
3.6.6	$R_A=-0.58$ kN, $R_D=75.58$ kN, $M_D=-1.63$ kNm, $M_{BA}=-M_{BC}=0.94$ kNm	
3.6.7	$M_D=-23.55$ kNm; $V_A=3.00$ kN; $H_A=2.5$ kN	
3.6.8	$M_A=10.20$ kNm; $M_E=5.64$ kNm; $V_A=7.54$ kN; $V_E=42.45$ kN; $H_A=6.45$ kN; $H_E=-6.45$ kN	
3.6.9	$M_{AD}=M_{DA}=120$ kNm; $M_{AB}=-M_{BA}=-120$ kNm	
3.6.10	$M_A=-12.46$ kNm; $M_D=12.46$ kNm; $M_{BA}=12.46$ kNm; $M_{BC}=-12.46$ kNm; $M_{CB}=12.46$ kNm; $M_{CD}=-12.46$ kNm	
3.6.11	$V_A=93.75$ kN; $H_A=-75$ kN; $V_D=-93.75$ kN; $F_{AB}=62.15$ kN; $F_{AE}=52.66$ kN; $F_{BC}=23.43$ kN; $F_{CF}=18.75$ kN	
3.6.12	$V_A=0$ kN; $H_A=0$ kN; $V_D=0$ kN; $F_{AB}=29.43$ kN; $F_{AE}=49.05$ kN; $F_{BC}=43.55$ kN; $F_{CF}=58.07$ kN	
3.6.13	Since the frame is externally determinate, due to uniform temperature raise no internal forces developed in the members of the truss	

Note: The answers are provided based on the analysis using structural analysis software. The results from the exact solution may differ slightly (decimals), but these differences are negligible.

4 Strain Energy Method
Theorem of Least Work

4.1 INTRODUCTION

In Structural Analysis I, or in basic structural analysis, or in the analysis of determinate structures, we studied the determination of deflections using energy methods, specifically Castigliano's second theorem. This theorem serves as the basis for deriving the theorem of least work. Let us recall Castigliano's second theorem, which states

> In a linearly elastic, statically determinate structure, the first partial derivative of the total internal strain energy of the structure with respect to a force applied at any point in the structure gives the displacement at the point of application of that force and in the direction of it.

It should be noted that the force may be a concentrated force or a moment (couple), and the displacement may be a transverse displacement or a rotational displacement. If U represents the total internal strain energy in the structure and W is the force applied at any point in the structure, then according to Castigliano's second theorem, the displacement at that point is given by $\Delta = \dfrac{\partial U}{\partial W}$. If the force is a redundant reaction (R_i), then the displacement at the location of the redundant is $\Delta_i = \dfrac{\partial U}{\partial R_i}$ for $i = 1, 2, 3, \ldots, I_s$, where I_s is the degree of static indeterminacy, and U is the strain energy. In the case of indeterminate beams and rigid-jointed frames, the strain energy due to axial and shear forces is generally neglected, and only the strain energy due to bending is considered. Therefore, for beams and rigid-jointed frames, the displacement at the location of the redundant can be expressed as $\Delta_i = \dfrac{\partial U}{\partial R_i} = \int (M_x) \times \left(\dfrac{\partial M_x}{\partial R_i}\right) \times \dfrac{dx}{EI}$, where M_x is the bending moment due to external loads and redundants, and EI is the flexural rigidity. Similarly, in case of indeterminate pin-jointed plane frames (trusses), the above equation takes the form as $\Delta_i = \dfrac{\partial U}{\partial R_i} = \sum_j (F_j) \times \left(\dfrac{\partial F_j}{\partial R_i}\right) \times \dfrac{L_j}{A_j E}$, where F_j is the internal axial force in the jth member of the truss due to external loads and redundants, and L_j and A_j are the length and cross-sectional area of the jth member.

Now let us state the **theorem of least work**. It states:

> In a statically indeterminate and linearly elastic structure, the partial derivative of the total internal strain energy with respect to a redundant force should vanish. Because, it is the function of such redundant forces to prevent any displacement at their point of application. In other words, the redundant forces developed in such way that the total internal strain energy should be a minimum.

4.2 PROOF OF THE THEOREM OF LEAST WORK

Consider a two-span continuous beam ABC subjected to the loads $W_1, W_2, \ldots, W_n$, located at $x_1, x_2, \ldots, x_n$ from the leftmost support A. Let the displacements due to these loads, $W_1, W_2, \ldots, W_n$, are $\Delta_1, \Delta_2, \ldots, \Delta_n$, respectively (Figure 4.1a). Assume the flexural rigidity of the beam is EI. The beam is statically indeterminate to the first degree. Therefore, select the reaction at support B (R_B) as a redundant reaction (Figure 4.1b). Using the principle of superposition, Figure 4.1b can be represented as the superposition of (i) a simply supported beam AC with external loads (Figure 4.1c), and (ii) a simply supported beam AC with the unknown redundant reaction R_B (Figure 4.1d). The vertical deflection at point B in a simply supported beam AC due

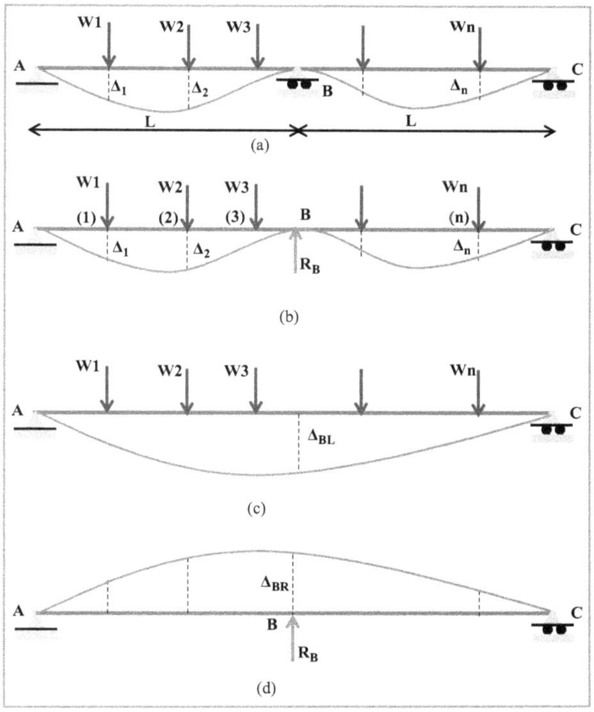

FIGURE 4.1 (a and b) Two-span continuous beams, (c) determinate beam with external loads, and (d) determinate beam with redundant R_B.

to the applied loads can be calculated using Castigliano's theorem. Since there is no external load at point B, assume a fictitious load Q acting at B. Let the deflection at B due to the applied loads be Δ_{BL}.

Then, the strain energy stored in the determinate beam is

$$U_{SL} = \frac{1}{2}W_1\Delta_1 + \frac{1}{2}W_2\Delta_2 + \cdots + \frac{1}{2}W_n\Delta_n + \frac{1}{2}Q\Delta_{BL} \quad (4.1)$$

The beam is assumed to be linearly elastic. The displacement Δ_1 under the load W_1, considering the effects of all the loads W_1, W_2, …, W_n, and the fictitious load Q, can be written as

$$\Delta_1 = W_1\delta_{11} + W_2\delta_{12} + \cdots + W_n\delta_{1n} + Q\delta_{1B} \quad (4.2a)$$

where δ_{11}, δ_{12}, δ_{1n}, δ_{1B} are the flexibility coefficients at 1 due to a unit load applied successively at 1, 2, …, n, and at B, in the determinate simply supported beam AC. Similar equations can be written for Δ_2, Δ_3, …, Δ_n, Δ_B

$$\Delta_2 = W_1\delta_{21} + W_2\delta_{22} + \cdots + W_n\delta_{2n} + Q\delta_{2B} \quad (4.2b)$$

$$\vdots$$

$$\Delta_n = W_1\delta_{n1} + W_2\delta_{n2} + \cdots + W_n\delta_{nn} + Q\delta_{nB} \quad (4.2c)$$

$$\Delta_B = W_1\delta_{B1} + W_2\delta_{B2} + \cdots + W_n\delta_{Bn} + Q\delta_{BB} \quad (4.2d)$$

Substituting equation 4.2 into equation 4.1, we can write

$$\begin{aligned} U_{SL} = &\frac{1}{2}W_1\left(W_1\delta_{11} + W_2\delta_{12} + \cdots + W_n\delta_{1n} + Q\delta_{1B}\right) \\ &+\frac{1}{2}W_2\left(W_1\delta_{21} + W_2\delta_{22} + \cdots + W_n\delta_{2n} + Q\delta_{2B}\right) \\ &\cdots+\frac{1}{2}W_n\left(W_1\delta_{n1} + W_2\delta_{n2} + \cdots + W_n\delta_{nn} + Q\delta_{nB}\right) \\ &+\frac{1}{2}Q\left(W_1\delta_{B1} + W_2\delta_{B2} + \cdots + W_n\delta_{Bn} + Q\delta_{BB}\right) \end{aligned} \quad (4.3)$$

Using Castigliano's theorem, the deflection at point B can be obtained by taking the partial derivative of the total internal strain energy with respect to Q, i.e.,

$$\begin{aligned} \Delta_{BL} = \frac{\partial U_{SL}}{\partial Q} = &\frac{1}{2}W_1\delta_{1B} + \frac{1}{2}W_2\delta_{2B} + \cdots + \frac{1}{2}W_n\delta_{nB} + Q\delta_{nB} \\ &+\frac{1}{2}\left(W_1\delta_{B1} + W_2\delta_{B2} + \cdots + W_n\delta_{Bn}\right) + \left(Q\delta_{BB}\right) \end{aligned} \quad (4.4)$$

Using Maxwell's reciprocal theorem, we can write $\delta_{ij} = \delta_{ji}$

$$\therefore \Delta_{BL} = \frac{\partial U_{SL}}{\partial Q} = W_1\delta_{1B} + W_2\delta_{2B} + \cdots + W_n\delta_{nB} + 2Q\delta_{nB} \quad (4.5)$$

Since Q is a fictitious load, substituting $Q=0$ into equation 4.5, we obtain,

$$\Delta_{BL} = \frac{\partial U_{SL}}{\partial Q} = W_1\delta_{1B} + W_2\delta_{2B} + \cdots + W_n\delta_{nB} \left(\downarrow\right) \quad (4.6)$$

The strain energy stored in the simply supported beam AC due to the redundant reaction R_B is,

$$U_{SR} = \frac{R_B^2 L^3}{3EI} \quad (4.7)$$

The deflection at B due to the redundant reaction R_B is,

$$\Delta_{BR} = \frac{\partial U_{SR}}{\partial R_B} = \frac{2R_B L^3}{3EI}\left(\uparrow\right) \quad (4.8)$$

Since the deflection at B in the original beam is zero, $\Delta_{BL} = \Delta_{BR}$

$$\begin{aligned} \therefore \frac{\partial U_{SL}}{\partial Q} &= -\Delta_{BR} \\ &= -\frac{\partial U_{SR}}{\partial R_B} \left(-\text{ve sign indicate deflection upward}\right) \end{aligned} \quad (4.9)$$

Since Q is a fictitious load, we can replace Q by R_B. Thus, equation 4.9 can be expressed as

$$\frac{\partial U_{SL}}{\partial R_B} = -\frac{\partial U_{SR}}{\partial R_B} \quad \text{(or)}$$

$$\frac{\partial}{\partial R_B}\left(U_{SL} + U_{SR}\right) = 0 \quad \text{(or)} \quad (4.10)$$

$$\frac{\partial U}{\partial R_B} = 0 \quad (4.11)$$

where $U = U_{SL} + U_{SR}$, is the total strain energy due to external loads and the redundant reaction R_B in the basic beam. Equation 4.11 represents the statement of the least work theorem.

Therefore, in general, the theorem of least work can be expressed as

$$\frac{\partial U}{\partial R_i} = 0 \quad \text{for } i = 1, 2, 3, \ldots, I_s \quad (4.11a)$$

where I_s is the degree of static indeterminacy.

The following examples illustrate the applications of the theorem of least work.

4.3 APPLICATION TO BEAMS

The procedure for applying the method of least work and Castigliano's second theorem for the analysis of indeterminate beams is outlined below.

i. As discussed in the force method in Chapter 1 and the consistent deformation method in Chapter 3, in this method we first **determine the static indeterminacy** of the given beam and then identify or select the redundant support reactions/forces.
ii. **Draw the basic determinate structure** with external loads and selected redundant reactions.
iii. **Write the compatibility equations** using the theorem of least work, as expressed by equation 4.11a as

$$\frac{\partial U}{\partial R_i} = \int (M_x) \times \left(\frac{\partial M_x}{\partial R_i}\right) \times \frac{dx}{EI} = 0 \quad \text{for } i = 1, 2, 3, \ldots, I_s \tag{4.11b}$$

iv. **Draw the appropriate free body diagrams** (FBDs) at various sections of the beam and derive the bending moment expressions due to external loads and redundant reactions using equilibrium equations. Once the moment expressions are derived in terms of loads and redundant reactions, the partial derivative terms (in the above equation) can be easily calculated.
v. **Generate the simultaneous equations** by substituting the above expressions into equation 4.11a and solve for the required unknowns.
vi. Once the redundant reactions are known, using the FBDs and the static equilibrium equations, **the force responses, i.e., the other support reactions**, and member end internal forces can be determined.
vii. After determining all the support reactions and internal forces, sketch the **shear force and bending moment diagrams**.
viii. If there is any **settlement or yielding of supports** in the beam, the theorem of least work is not applicable. However, these problems can be solved using Castigliano's second theorem. The redundant reactions should be selected so that they correspond to the location and type of yielding of supports. For instance, if support rotation occurs, the corresponding support moment should be selected as a redundant. Similarly, if settlement exists, the corresponding support reaction should be chosen as a redundant. Once the redundants are selected, compatibility equations can be written using Castigliano's second theorem. Then, follow steps (iv) to (vii) to complete the analysis of the beam.

Example 4.1

Analyse the propped cantilever beam shown in Figure 4.2 using the strain energy method. Also, draw the shear force and bending moment diagram.

Solution:

The beam is statically indeterminate to the first degree, i.e., $I_s = 3 - 2 = 1$. Select the reaction at "B," R_B (↓) as redundant. The primary beam with external loads and the redundant reaction is shown in Figure 4.3.

The compatibility condition to solve for R_B using the theorem of least work is

$$\frac{\partial U}{\partial R_B} = 0 \Rightarrow \int M_x \frac{\partial M_x}{\partial R_B} \cdot \frac{dx}{EI} = 0 \tag{4.12}$$

The expressions for M_x and $\frac{\partial M_x}{\partial R_B}$ for various segments of the beam, along with the corresponding FBDs, are conveniently shown in Table 4.1.

Substituting the expressions of M_x and $\frac{\partial M_x}{\partial R_B}$ from Table 4.1 into equation 4.12,

$$\therefore \int M_x \frac{\partial M_x}{\partial R_B} \cdot \frac{dx}{EI} = \int_0^1 (-5x^2)(0) \cdot \frac{dx}{EI}$$

$$+ \int_0^3 (-R_B x - 10x - 5)(-x) \cdot \frac{dx}{2EI}$$

$$+ \int_0^3 (40x + R_B x - 155 - 6R_B)(x - 6) \cdot \frac{dx}{2EI} = 0$$

$$= \frac{1}{2EI} \left| R_B \frac{x^3}{3} + 10\frac{x^3}{3} + 5\frac{x^2}{2} \right|_0^3$$

$$+ \frac{1}{2EI} \left| \begin{matrix} 40\frac{x^3}{3} + R_B\frac{x^3}{3} - 155\frac{x^2}{2} - 6R_B\frac{x^2}{2} \\ -240\frac{x^2}{2} - 6R_B\frac{x^2}{2} + 930x + 36R_B x \end{matrix} \right|_0^3 = 0$$

$$= (9R_B + 112.5) + 63R_B + 1{,}372.5 = 0$$

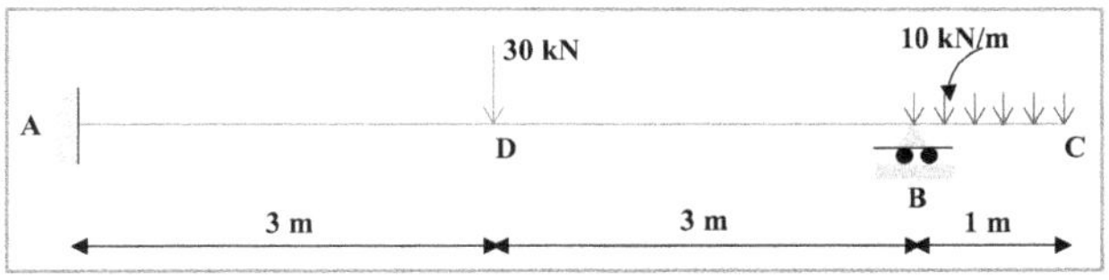

FIGURE 4.2 Propped cantilever beam.

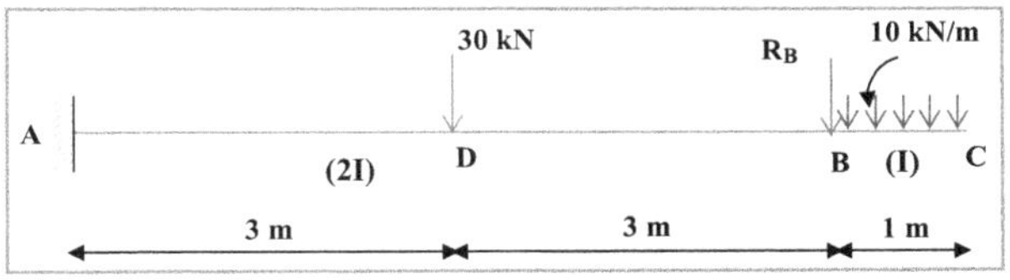

FIGURE 4.3 Primary beam with redundant R_B.

TABLE 4.1
Bending moment expressions and its derivative terms of various segments of a beam

Segment	Origin	FBD	M_x (kNm)	$\frac{\partial M_x}{\partial R_B}$	Limits (m)
BC	*C*	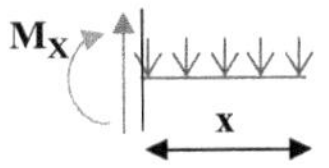	$-5x^2$	0	0–1
BD	*B*	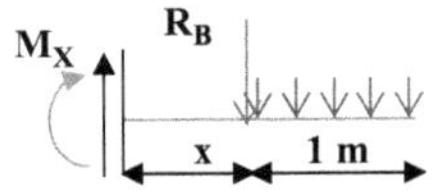	$-R_Bx-10\left(x+\frac{1}{2}\right)$	$-x$	0–3
AD	*A*	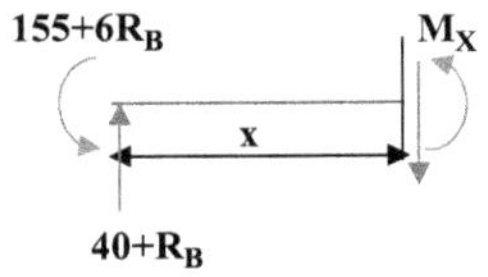	$(40+R_B)x-155-6R_B$	$(x-6)$	0–3

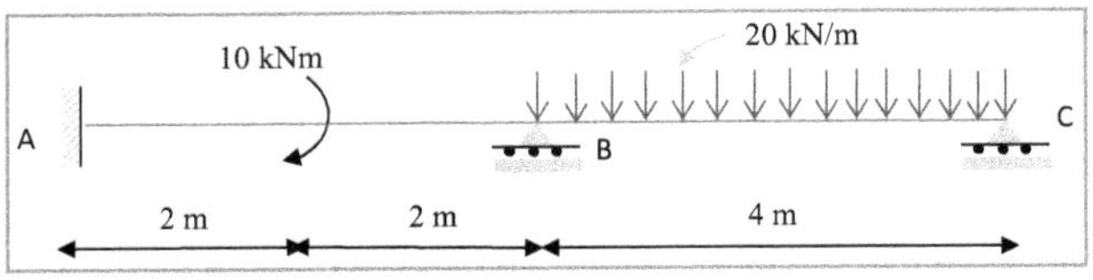

FIGURE 4.4 Two-span continuous beam.

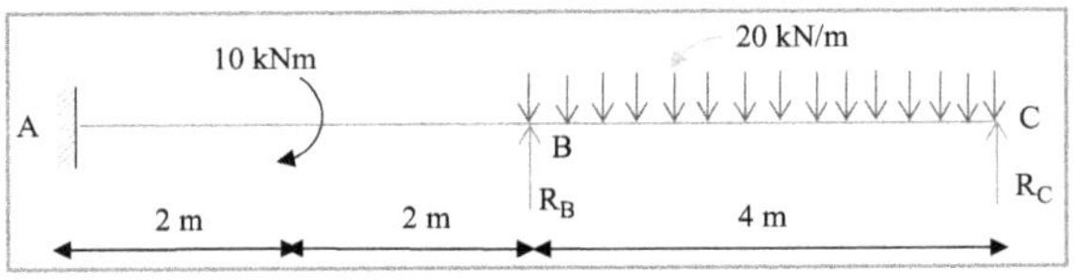

FIGURE 4.5 Primary beam with external loads and redundant reactions.

$R_B=-20.63$ kN

i.e. $R_B=20.625\ \text{kN}(\uparrow)$

$(-\text{ve sign indicates } R_B \text{ acting opposite to the assumed direction})$

This is exactly the same as the result obtained in Example 3.1, solved using the consistent deformation method. The shear force, bending moment diagrams and the deflected shape are presented in Figure 3.9.

(Note: Unlike the consistent deformation method, which involves calculating displacements due to loads and unit loads using the unit load method, the theorem of least work does not offer a specific advantage in selecting the direction of the redundant as downward. Therefore, the reaction R_B can be assumed to act upward, as it is the obvious choice.)

Example 4.2

Re-analyse Example 3.2, using the theorem of least work (Figure 4.4). Assume *EI* is constant. Plot the shear force and bending moment diagrams.

Solution:

The static indeterminacy is $I_s=4-2=2$, i.e., the beam is statically indeterminate to the second degree. Let us select the reactions at *B* (R_B) and *C* (R_C) as redundant reactions. The primary beam with the external loads and the redundant reactions, as shown in Figure 4.5.

According to the theorem of least work, the compatibility conditions can be written as

$$\frac{\partial U}{\partial R_B}=0 \Rightarrow \int M_x\frac{\partial M_x}{\partial R_B}\frac{dx}{EI}=0 \tag{4.13}$$

$$\frac{\partial U}{\partial R_C}=0 \Rightarrow \int M_x\frac{\partial M_x}{\partial R_C}\frac{dx}{EI}=0 \tag{4.14}$$

The FBDs of various beam segments, used for deriving the moment expressions, are shown in Figure 4.6. Based on Figure 4.6, the moment expressions (M_x) and their partial derivatives $\left(\frac{\partial M_x}{\partial R_B} \text{ and } \frac{\partial M_x}{\partial R_C}\right)$ are shown in Table 4.2.

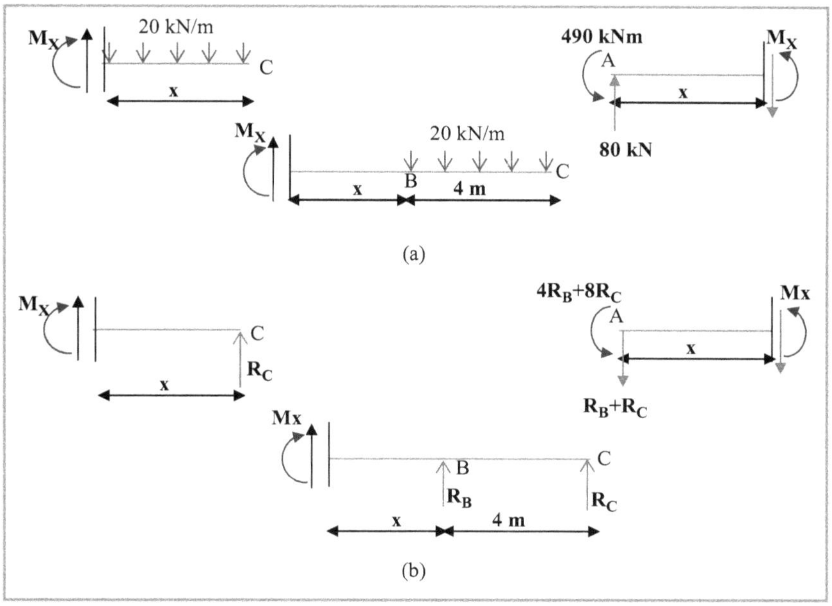

FIGURE 4.6 Free body diagrams due to (a) loads and (b) redundant reactions.

TABLE 4.2
Moment expressions due to loads and redundants and its derivatives

Segment	Origin	Moment M_x Due to: Loads	Moment M_x Due to: Redundants	$\frac{\partial M_x}{\partial R_B}$	$\frac{\partial M_x}{\partial R_C}$	Limits (m)
BC	C	$-10x^2$	$R_C x$	0	x	0–4
DB	B	$-80(x+2)=-80x-160$	$R_C(x+4)+R_B x$	x	$(x+4)$	0–2
AD	A	$80x-490$	$4R_B+8R_C-R_B x-R_C x$	$(4-x)$	$(8-x)$	0–2

Substituting the above expressions into equations 4.13 and 4.14,

$$\frac{\partial U}{\partial R_B}=\int M_x\frac{\partial M_x}{\partial R_B}\cdot\frac{dx}{EI}$$

$$=\int_0^2\left(-80x-160+R_C(x+4)+R_B x\right)(x)\cdot\frac{dx}{EI}$$

$$+\int_0^2\left(80x-490+4R_B+8R_C-R_B x-R_C x\right)(4-x)\cdot\frac{dx}{EI}=0$$

$$=\left[-80\frac{x^3}{3}-160\frac{x^2}{2}+R_C\left(\frac{x^3}{3}+\frac{4x^2}{2}\right)+R_B\frac{x^3}{3}\right]_0^2\frac{1}{EI}$$

$$+\left[\begin{matrix}80\left(\frac{4x^2}{2}-\frac{x^3}{3}\right)-490\left(4x-\frac{x^2}{2}\right)+4R_B\left(4x-\frac{x^2}{2}\right)\\+8R_C\left(4x-\frac{x^2}{2}\right)-R_B\left(\frac{4x^2}{2}-\frac{x^3}{3}\right)-R_C\left(\frac{4x^2}{2}-\frac{x^3}{3}\right)\end{matrix}\right]_0^2\frac{1}{EI}$$

$$=0$$

$$\Rightarrow -3046.67+53.33\,R_C+21.33\,R_B=0$$

$$\Rightarrow R_B+2.5\,R_C=142.83 \qquad (4.13a)$$

$$\frac{\partial U}{\partial R_C}=\int M_x\frac{\partial M_x}{\partial R_C}\cdot\frac{dx}{EI}=\int_0^4\left(-10x^2+R_C x\right)(x)\cdot\frac{dx}{EI}$$

$$+\int_0^2\left[-80(x+2)+R_B x+R_C(x+4)\right](x+4)\cdot\frac{dx}{EI}$$

$$+\int_0^2\left(80x-490+4R_B+8R_C-R_B x-R_C x\right)(8-x)\cdot\frac{dx}{EI}=0$$

$$=\frac{1}{EI}\left[\begin{matrix}-10\frac{x^3}{3}+R_C\frac{x^3}{3}-80\left(\frac{x^3}{3}+\frac{6x^2}{2}+8x\right)\\+R_B\left(\frac{x^3}{3}+\frac{4x^2}{2}\right)+R_C\left(\frac{x^3}{3}+\frac{8x^2}{2}+16x\right)\end{matrix}\right]_0^2$$

$$\Rightarrow R_B+3.2R_C=166.63 \qquad (4.14a)$$

By solving equations 4.13a and 4.14a, we obtain

$$R_C = 34 \text{ kN}$$

$$R_B = 57.83 \text{ kN}$$

The FBD, along with the support reactions, shear force, and bending moment diagrams, are shown in Figure 4.7.

Example 4.3

Analyse the two-span continuous beam shown in Figure 4.8 using the theorem of least work. Also, draw the shear force and bending moment diagrams.

Solution:

The beam is statically indeterminate to the second degree ($I_s = 4 - 2 = 2$). The vertical reactions at B (R_B) and C (R_C) are chosen as redundant reactions. The primary beam with external loads and redundant reactions is shown in Figure 4.9.

The compatibility equations according to the theorem of least work are

$$\frac{\partial U}{\partial R_B} = 0 \Rightarrow \int M_x \frac{\partial M_x}{\partial R_B} \cdot \frac{dx}{EI} = 0 \tag{4.15}$$

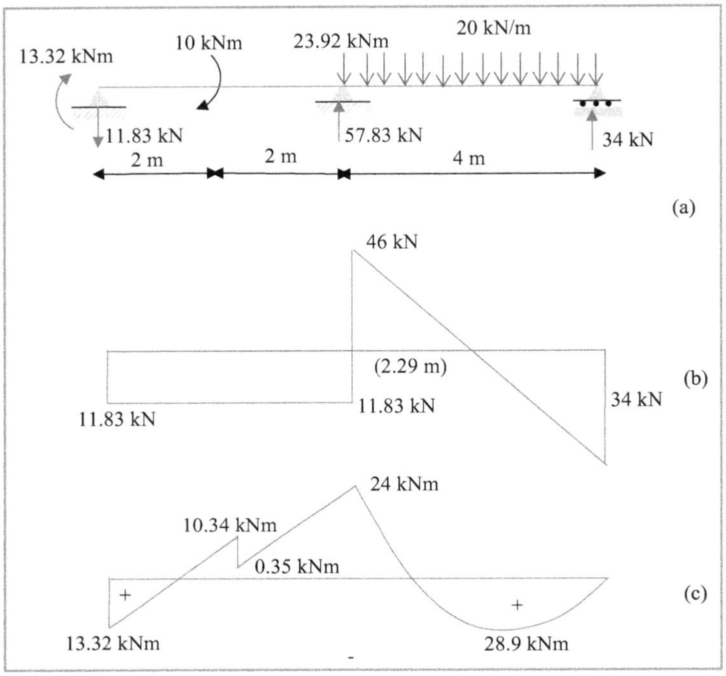

FIGURE 4.7 (a) Free body diagram, (b) shear force diagram, and (c) bending moment diagram.

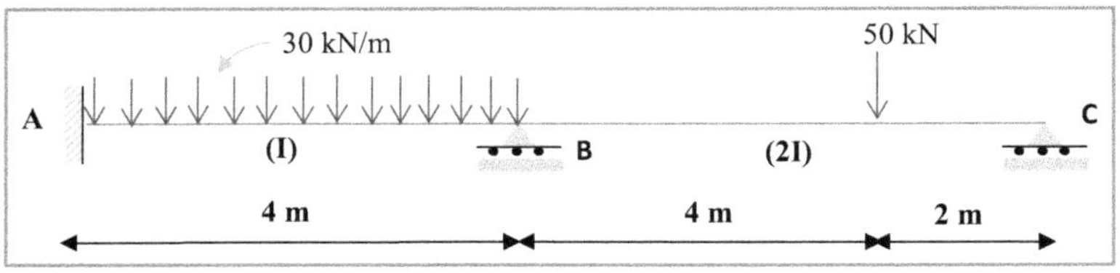

FIGURE 4.8 Two-span continuous beam.

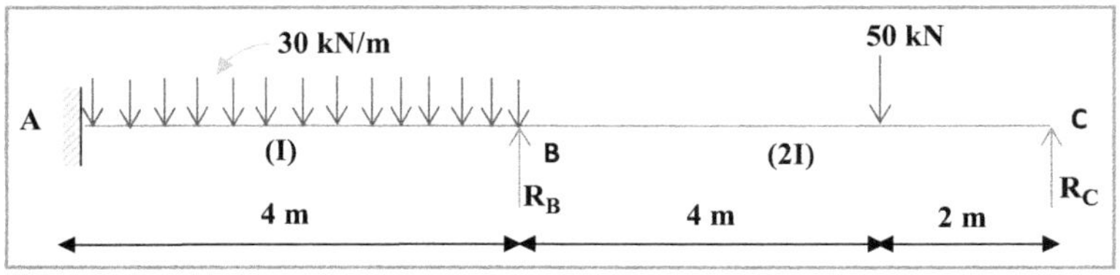

FIGURE 4.9 Primary beam with external loads and redundants.

$$\frac{\partial U}{\partial R_C} = 0 \Rightarrow \int M_x \frac{\partial M_x}{\partial R_C} \cdot \frac{dx}{EI} = 0 \qquad (4.16)$$

The FBDs of various beam segments, used for deriving the moment expressions, are shown in Figure 4.10. The moment expressions (M_x) and their partial derivatives $\left(\frac{\partial M_x}{\partial R_B} \text{ and } \frac{\partial M_x}{\partial R_C}\right)$ based on the FBDs shown in Figure 4.10 are given in Table 4.3.

$$\frac{\partial U}{\partial R_B} = \frac{1}{EI}\int_0^4 \begin{bmatrix} (170 - R_B - R_C)x \\ -(640 - 4R_B - 10R_C) - 15x^2 \end{bmatrix} (4 - x) \cdot \frac{dx}{EI} = 0$$

$$= (170 - R_B - R_C)\left[\frac{4x^2}{2} - \frac{x^3}{3}\right]_0^4 - (640 - 4R_B - 10R_C)$$

$$\left[4x - \frac{x^2}{2}\right]_0^4 - 15\left[\frac{4x^3}{3} - \frac{x^4}{4}\right]_0^4 = 0$$

$$= (170 - R_B - R_C)\left(\frac{64}{6}\right) - (640 - 4R_B - 10R_C)(8) = 320$$

$$=> 21.33R_B + 69.33R_C = 3{,}626.67$$

$$=> R_B + 3.25R_C = 170.02 \qquad (4.17)$$

$$\frac{\partial U}{\partial R_C} = \frac{1}{2EI}\int_0^2 (R_C x)(x)\,dx + \frac{1}{2EI}\int_0^2 \left[R_C(x+2) - 50x\right](x+2)dx$$

$$+\frac{1}{EI}\int_0^4 \begin{bmatrix} (170 - R_B - R_C)x \\ -(640 - 4R_B - 10R_C) - 15x^2 \end{bmatrix} (10 - x)dx = 0$$

$$= \frac{1}{2EI}\left[R_C \cdot \frac{x^3}{3}\right]_0^2 + \frac{1}{2EI}\begin{bmatrix} R_C\left(\frac{x^3}{3} + \frac{4x^2}{2} + 4x\right) \\ -50\left(\frac{x^3}{3} + \frac{2x^2}{2}\right) \end{bmatrix}_0^4$$

$$+\frac{1}{EI}\begin{bmatrix} (170 - R_B - R_C)\left(\frac{10x^2}{2} - \frac{x^3}{3}\right) \\ -(640 - 4R_B - 10R_C)\left(10x - \frac{x^2}{2}\right) \end{bmatrix}_0^4$$

$$-15\left[10\frac{x^3}{3} - \frac{x^4}{4}\right]_0^4 = 0$$

$$= \frac{1}{2EI}\left(\frac{8}{3}R_C\right) + \frac{1}{2EI}(69.33R_C - 1{,}866.67)$$

$$+\frac{1}{EI}\begin{pmatrix} 9{,}973.33 - 58.67R_B - 58.67R_C \\ -20{,}480 + 128R_B + 320R_C \end{pmatrix} = 0$$

$$=> 69.33R_B + 297.32R_C = 13{,}680$$

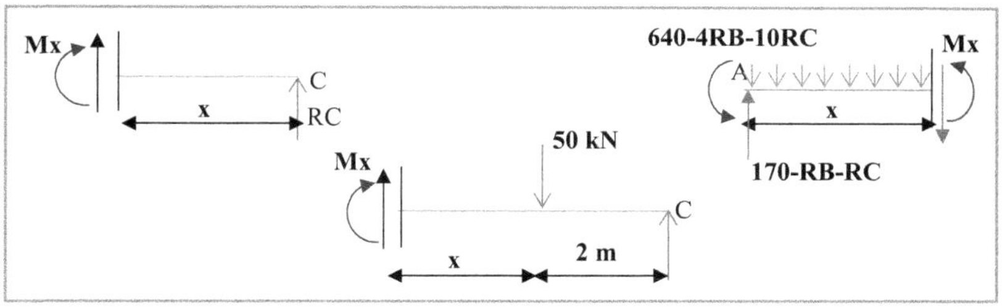

FIGURE 4.10 Free body diagrams due to loads and redundants.

TABLE 4.3
Bending moment expressions and its partial derivatives

Segment	Origin	Moment (M_x)	$\frac{\partial M_x}{\partial R_B}$	$\frac{\partial M_x}{\partial R_C}$	Limits (m)	(EI)
DC	C	$R_C x$	0	x	0–2	$2EI$
BD	D	$R_C(x+2) - 50x$	0	$(x+2)$	0–4	$2EI$
AB	A	$(170 - R_B - R_C)x$ $-(640 - 4R_B - 10R_C) - 15x^2$	$(-x+4)$	$(10 - x)$	0–4	EI

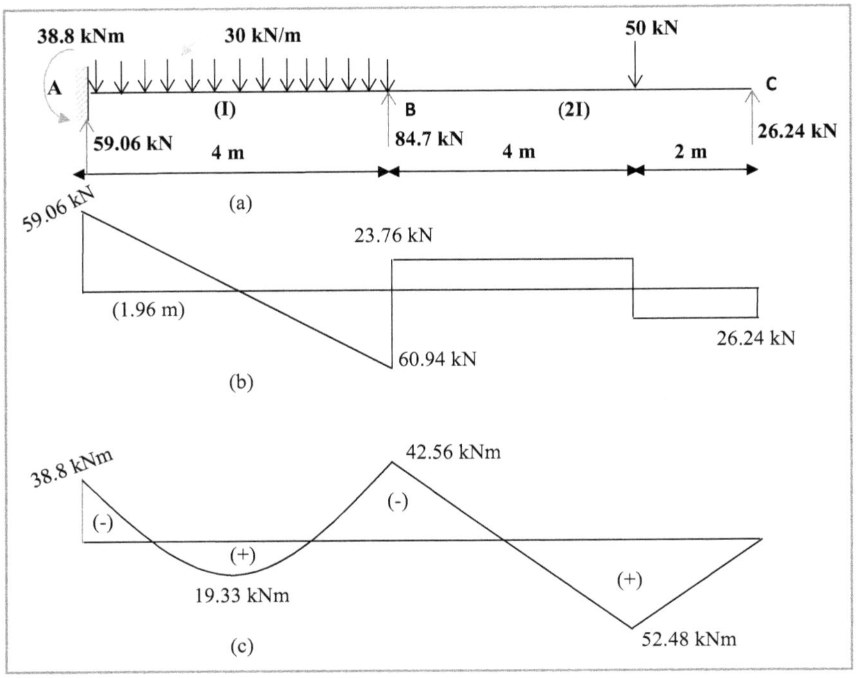

FIGURE 4.11 (a) Free body diagram, (b) shear force diagram, and (c) bending moment diagram.

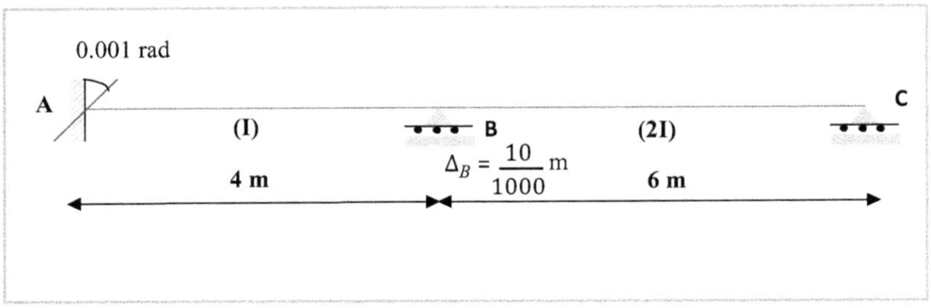

FIGURE 4.12 Two-span continuous beam with yielding of supports.

$$=> R_B + 4.29R_C = 197.31 \quad (4.18)$$

By solving equations 4.17 and 4.18, we get

$$R_C = 26.24 \text{ kN}$$

$$R_B = 84.7 \text{ kN}$$

These results are the same as those obtained in Example 3.4 using the consistent deformation method. The FBD showing the support reactions, as well as the shear force and bending moment diagrams, are shown in Figure 4.11.

Example 4.4

Re-analyse the beam shown in Example 4.3 if support A rotates by 0.001 rad clockwise and support B settles down by 10 mm in the absence of external loads (Figure 4.12). Assume $EI = 135\times10^3$ kNm^2.

Solution:

As it is mentioned in Example 4.3, static indeterminacy $I_s = 2$. Let us select the moment at B (M_B), and the vertical reaction at B (R_B) as redundants is shown in Figure 4.13.

According to the theorem of least work, the compatibility conditions can be written as,

$$\frac{\partial U}{\partial M_A} = 0.001 = \int M_x \frac{\partial M_x}{\partial M_A} \cdot \frac{dx}{EI} \quad (4.19)$$

$$\frac{\partial U}{\partial R_B} = -0.01 = \int M_x \frac{\partial M_x}{\partial R_B} \cdot \frac{dx}{EI} \quad (4.20)$$

The FBDs used for calculating the bending moment expressions for various segments of the beam are shown in Figure 4.13. The moment expressions (M_x) and their

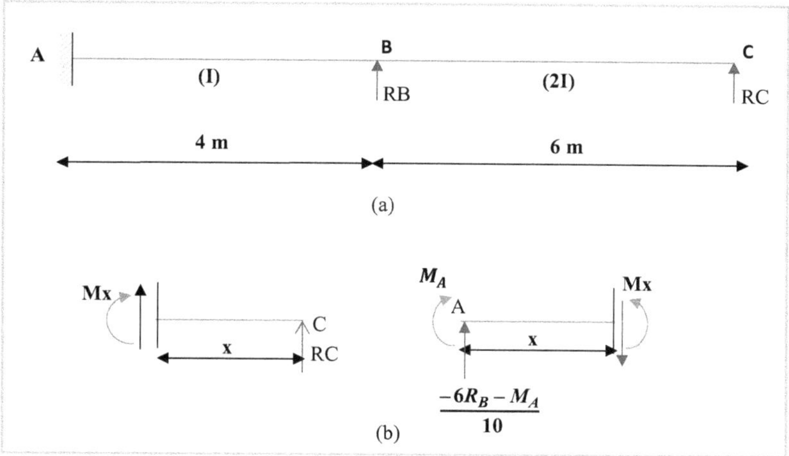

FIGURE 4.13 (a) Primary beam with redundants and (b) free body diagrams.

TABLE 4.4

Moment expressions (Mx) and their partial derivative terms $\left(\frac{\partial M_x}{\partial R_B} \text{ and } \frac{\partial M_x}{\partial R_C}\right)$

Segment	Origin	Moment M_x	$\frac{\partial M_x}{\partial M_A}$	$\frac{\partial M_x}{\partial R_B}$	Limits (m)	(*EI*)
BC	*C*	$\left(\frac{M_A - 4R_B}{10}\right)x$	$\frac{x}{10}$	$\frac{-4x}{10}$	0–6	2*EI*
AB	*A*	$M_A - \left(\frac{M_A + 6R_B}{10}\right)x$	$1 - \frac{x}{10}$	$\frac{-6x}{10}$	0–4	*EI*

partial derivative terms $\left(\frac{\partial M_x}{\partial R_B} \text{ and } \frac{\partial M_x}{\partial R_C}\right)$ are conveniently shown in Table 4.4.

Substituting the values from Table 4.4 into equations 4.19 and 4.20,

$$\frac{\partial U}{\partial M_A} = 0.001 = \frac{1}{2EI}\int_0^6 \frac{1}{10}(M_A - 4R_B)(x)\cdot\frac{x}{10}\,dx$$

$$+\frac{1}{EI}\int_0^4 \frac{1}{10}(10M_A - 6R_Bx - M_Ax)\left(\frac{10-x}{10}\right)dx$$

$$= \frac{1}{2EI}\left[\frac{1}{100}(M_A - 4R_B)\frac{x^3}{3}\right]_0^6$$

$$+\frac{1}{EI}\left[\begin{matrix}\frac{1}{100}\left(10M_A\left(10x - \frac{x^2}{2}\right)\right) \\ -6R_B\left(\frac{10x^2}{2} - \frac{x^3}{3}\right) - M_A\left(\frac{10x^2}{2} - \frac{x^3}{3}\right)\end{matrix}\right]_0^4$$

$$0.001 = \frac{1}{200EI}(72\ M_A - 288\ R_B)$$

$$+\frac{1}{100EI}(320\ M_A - 352\ R_B - 58.66\ M_A)$$

$$0.1EI = 297.34\ M_A - 496\ R_B$$

$$M_A - 1.66\ R_B = 45.40 \quad (4.19a)$$

$$\frac{\partial U}{\partial R_B} = -0.01 = \frac{1}{2EI}\int_0^6 \frac{1}{10}(M_A - 4R_B)(x)\cdot\left(\frac{-4x}{10}\right)dx$$

$$+\frac{1}{EI}\int_0^4 \frac{1}{10}(10M_A - 6R_Bx - M_Ax)\left(\frac{-6x}{10}\right)dx$$

$$= -0.01 = \frac{1}{2EI}\left[\frac{1}{100}\left(-M_A\frac{4x^3}{3} + R_B\frac{16x^3}{3}\right)_0^6\right]$$

$$+\frac{1}{EI}\left[\frac{1}{100}\left(-60M_A\frac{x^2}{2} + R_B\frac{36x^3}{3} + 6M_A\frac{x^3}{3}\right)_0^4\right]$$

$$= \frac{1}{200EI}(-288\ M_A + 1{,}152\ R_B)$$

$$+\frac{1}{100EI}(-480\ M_A + 768\ R_B + 128\ M_A)$$

$$= -0.01 \times 100EI = -496\ M_A + 1344\ R_B$$

$$-\ M_A + 2.7\ R_B = -272.17 \quad (4.20a)$$

By solving equations 4.19a and 4.20a, we get

$$R_B = -218 \text{ kN}$$

$$M_A = -316.48 \text{ kNm}$$

Using the known values of R_B and M_A, the remaining reactions can be determined by applying the static equilibrium equations.

$$\sum M_C = 0 \Rightarrow R_A = \frac{-6(-218)-(-316.48)}{10} = 162.45 \text{ kN}$$

$$\sum M_A = 0 \Rightarrow R_C = \frac{-(316.48)-4(-218)}{10} = 55.55 \text{ kN}$$

The FBD, shear force, and bending moment diagrams are shown in Figure 4.14.

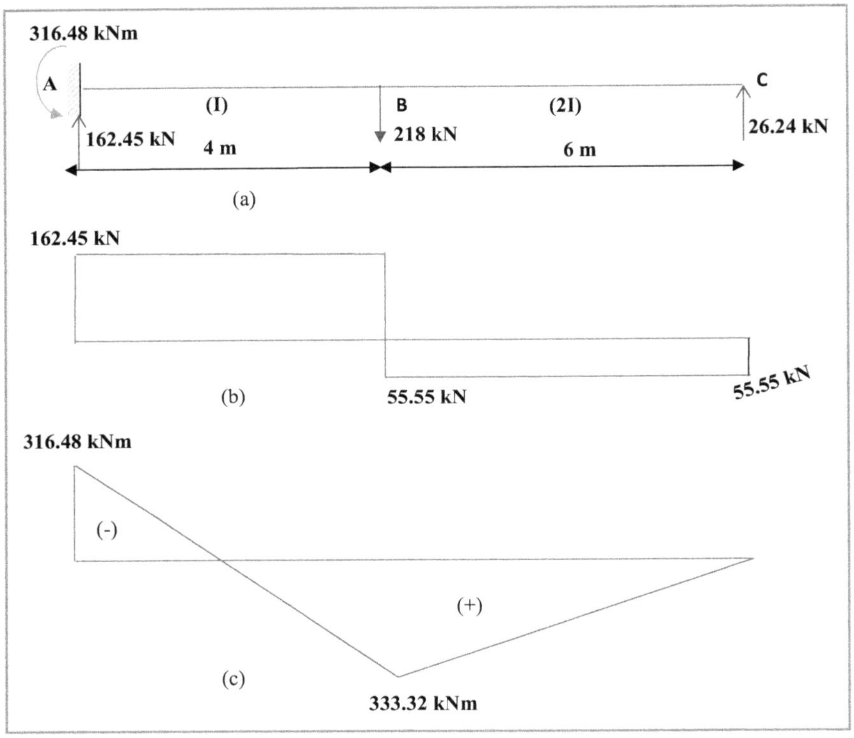

FIGURE 4.14 (a) Free body diagram, (b) shear force diagram, and (c) bending moment diagram.

Example 4.5

A symmetrical three-span continuous beam is shown in Figure 4.15. Find the support reactions and, hence, draw the shear force and bending moment diagrams.

Solution:

The actual degree of static indeterminacy of the beam is $I_s = 4 - 2 = 2$. Due to symmetry, the degree of indeterminacy is reduced to 1. The problem can be solved in two ways.

METHOD 1

Consider $R_A = R_D$ as redundant, as shown in Figure 4.16. Based on the theorem of least work, the compatibility equation can be expressed as

$$\frac{\partial U}{\partial R_A} = 2\int M_x \frac{\partial M_x}{\partial R_A} \cdot \frac{dx}{EI} = 0 \tag{4.21}$$

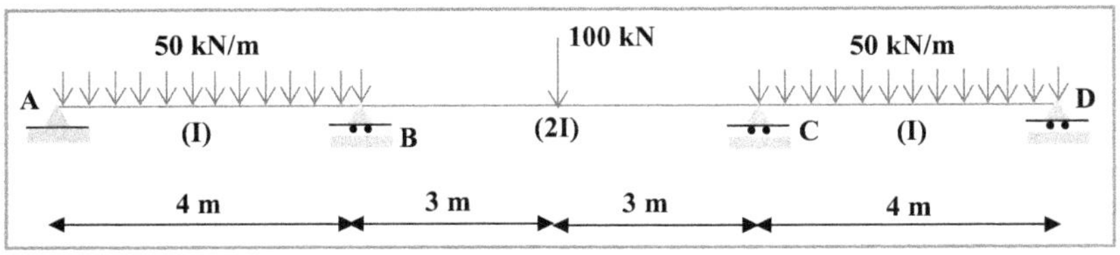

FIGURE 4.15 Symmetrical three-span continuous beam.

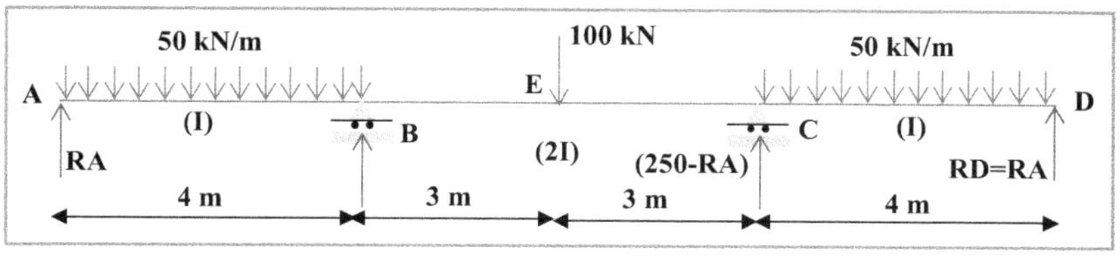

FIGURE 4.16 Primary beam with external loads and redundant.

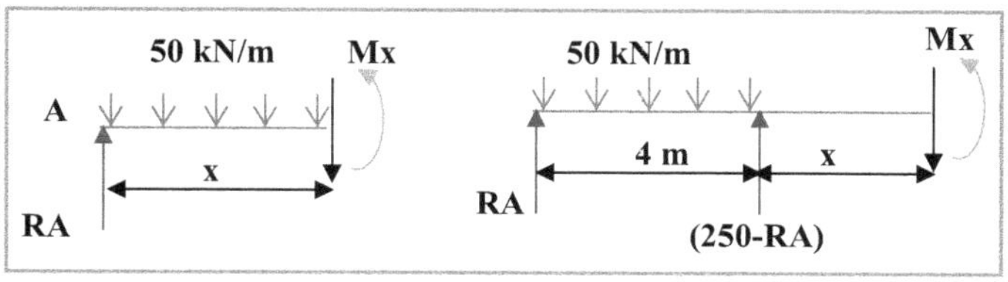

FIGURE 4.17 Free body diagrams due to loads and redundants.

TABLE 4.5
Moment expressions for different beam segments and its partial derivatives

Segment	Origin	Moment M_x	$\frac{\partial M_x}{\partial R_A}$	Limits (m)	(EI)
AB	A	$(R_A x - 25x^2)$	x	0–4	EI
BE	B	$(R_A(x+4)+(250-R_A)x-200(x+2))$	4	0–3	$2EI$

The FBDs used for writing the moment expressions in various segments of the beam are shown in Figure 4.17. The moment expressions for different beam segments and the partial derivatives of M_x are shown in Table 4.5.

Substituting the expression of M_x and $\frac{\partial M_x}{\partial R_A}$ in equation 4.21,

$$\frac{\partial U}{\partial R_A} = 2\int_0^4 (R_A x - 25x^2)(x)\frac{dx}{EI}$$

$$+2\int_0^3 [R_A(x+4)+(250-R_A)x-200(x+2)](4)\frac{dx}{2EI} = 0$$

$$\Rightarrow \frac{2}{EI}\left[R_A\frac{x^3}{3} - 25\frac{x^4}{4}\right]_0^4$$

$$+\frac{2}{2EI}\left[R_A\left(\frac{4x^2}{2}+16x\right)+(250-R_A)\frac{4x^2}{2} - 200\left(\frac{4x^2}{2}+8x\right)\right]_0^3 = 0$$

$$\Rightarrow \frac{2}{EI}(21.33\ R_A - 1{,}600) + \frac{1}{EI}(66\ R_A + 4{,}500 - 18\ R_A - 8{,}400) = 0$$

$$\Rightarrow 90.66\ R_A = 7{,}100$$

$$R_A = 78.32\ \text{kN} = R_D\ (\uparrow)$$

$\therefore$ Using $\sum F_Y = 0$, we can write

$$R_B = R_C = 250 - R_A = 171.68\ \text{kN}(\uparrow)$$

The FBD, along with the support reactions, shear force, and bending moment diagrams, is shown in Figure 4.18.

METHOD 2

Due to symmetry, only half of the beam needs to be considered for the analysis, as shown in Figure 4.19, with appropriate boundary conditions at the symmetrical line, i.e., at the centre of span BC. Since the slope is zero at the centre of span BC, the shear force is also zero. Therefore, a guided support is considered to satisfy these conditions. Select M_E as the redundant.

Therefore, the compatibility condition, as per the theorem of least work, is

$$\frac{\partial U}{\partial M_E} = \int M_x \frac{\partial M_x}{\partial M_E} \cdot \frac{dx}{EI} = 0 \qquad (4.22)$$

The rest of the procedure is left to the reader to verify the results.

Example 4.6

Analyse the non-prismatic beam shown in Figure 4.20, using the theorem of least work. Draw the shear force and bending moment diagrams.

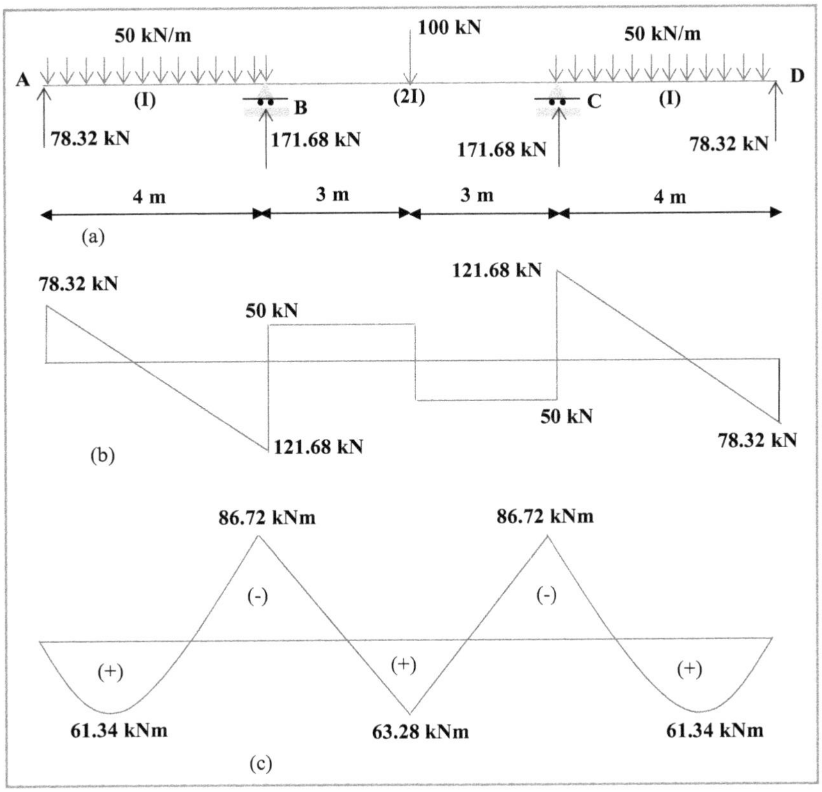

FIGURE 4.18 (a) Free body diagram, (b) shear force diagram, and (c) bending moment diagram.

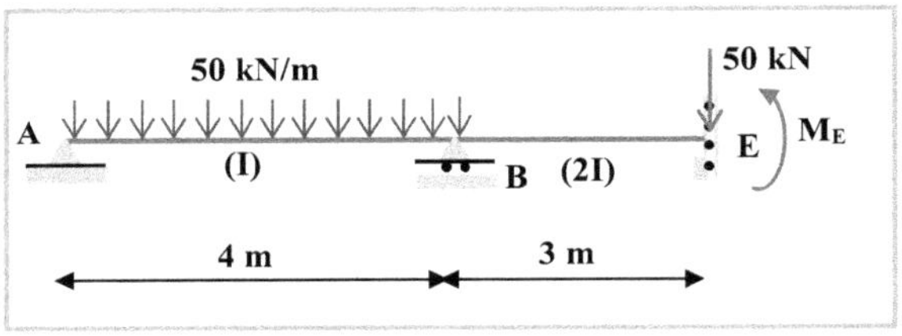

FIGURE 4.19 Symmetrical half portion of beam.

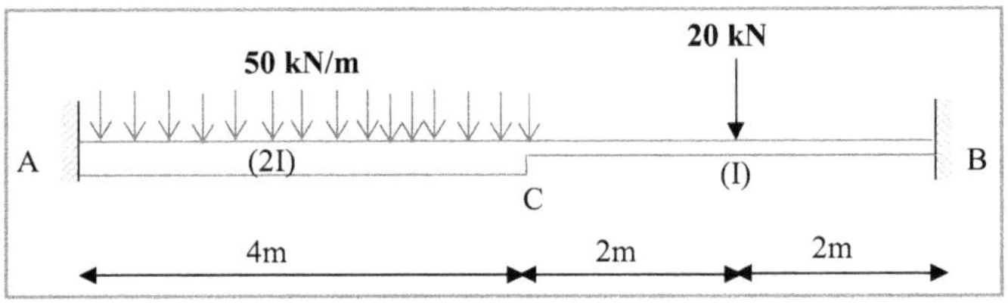

FIGURE 4.20 Non-prismatic beam *ACB*.

Solution:

The static indeterminacy, $I_s = 4 - 2 = 2$, i.e., the beam is statically indeterminate to the second degree. Let us select M_B and R_B as redundant reactions. The primary beam with external loads and the redundant reactions is shown in Figure 4.21.

Based on the theorem of least work, the compatibility conditions are expressed as

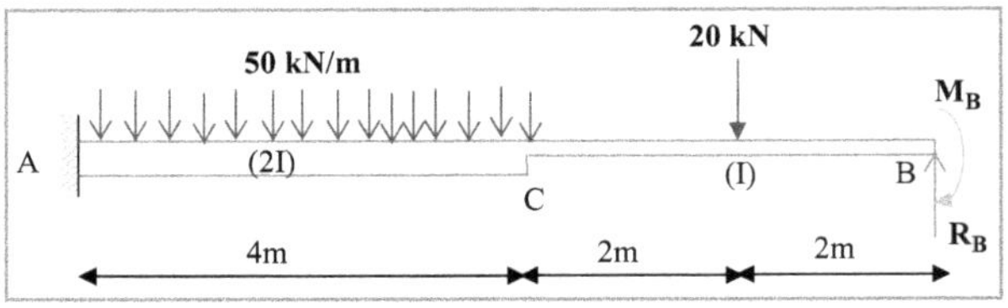

FIGURE 4.21 Primary beam with loads and redundants.

$$\frac{\partial U}{\partial M_B} = \int M_x \frac{\partial M_x}{\partial M_B} \cdot \frac{dx}{EI} = 0 \qquad (4.23)$$

$$\frac{\partial U}{\partial R_B} = \int M_x \frac{\partial M_x}{\partial R_B} \cdot \frac{dx}{EI} = 0 \qquad (4.24)$$

The FBDs used for writing the moment expressions of different beam segments are shown in Figure 4.22. The moment expressions and the partial derivatives of them are presented in Table 4.6.

On substituting the above moment expressions and their partial derivatives into equations 4.23 and 4.24

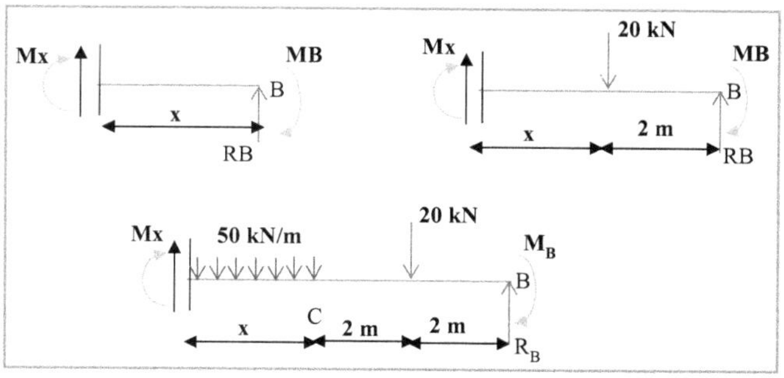

FIGURE 4.22 Free body diagrams due to loads and redundants.

TABLE 4.6
Bending moment expressions for different beam segments and its partial derivatives

Segment	Origin	Moment (M_x)	$\frac{\partial M_x}{\partial M_B}$	$\frac{\partial M_x}{\partial R_B}$	Limits (m)	(*EI*)
BD	*B*	$R_B x - M_B$	-1	x	0–2	*EI*
CD	*D*	$R_B(x+2) - 20x - M_B$	-1	$(x+2)$	0–2	*EI*
AC	*C*	$R_B(x+4) - 20(x+2) - 25x^2 - M_B$	-1	$(x+4)$	0–4	2*EI*
			-1	$(x+4)$		

$$\frac{\partial U}{\partial M_B} = \int_0^2 (R_B x - M_B)(-1)\frac{dx}{EI} + \int_0^2 \left[R_B(x+2) - 20x - M_B\right](-1)\frac{dx}{EI} + \int_0^4 \left[R_B(x+4) - 20(x+2) - 25x^2 - M_B\right](-1)\frac{dx}{2EI} = 0$$

$$= \frac{1}{EI}\left[-R_B\frac{x^2}{2} + M_B x\right]_0^2 + \frac{1}{EI}\left[-R_B\left(\frac{x^2}{2} + 2x\right) + \frac{20x^2}{2} + M_B x\right]_0^2 + \frac{1}{2EI}\left[\begin{matrix}-R_B\left(\frac{x^2}{2} + 4x\right) + 20\left(\frac{x^2}{2} + 2x\right) \\ +25\frac{x^3}{3} + M_B x\end{matrix}\right]_0^4 = 0$$

$$\Rightarrow -2R_B + 2M_B - 6R_B + 40 + 2M_B - 12R_B + 160 + 266.67 + 2M_B = 0$$

$$\Rightarrow -20R_B + 6M_B = -466.67$$

$$\Rightarrow -3.33R_B + M_B = -77.78 \qquad (4.23a)$$

$$\frac{\partial U}{\partial R_B} = \int_0^2 (R_B x - M_B)(x)\frac{dx}{EI} + \int_0^2 \left[R_B(x+2) - 20x - M_B\right](x+2)\frac{dx}{EI} + \int_0^4 \left[R_B(x+4) - 20(x+2) - 25x^2 - M_B\right](x+4)\frac{dx}{2EI} = 0$$

$$= \frac{1}{EI}\left[R_B\frac{x^3}{3} + M_B\frac{x^2}{2}\right]_0^2 + \frac{1}{EI}\left[\begin{matrix}R_B\left(\frac{x^3}{3} + \frac{4x^2}{2} + 4x\right) - 20\left(\frac{x^3}{3} + \frac{2x^2}{2}\right) \\ -M_B\left(\frac{x^2}{2} + 2x\right)\end{matrix}\right]_0^2 + \frac{1}{2EI}\left[\begin{matrix}R_B\left(\frac{x^3}{3} + \frac{8x^2}{2} + 16x\right) - 20\left(\frac{x^3}{3} + \frac{6x^2}{2} + 8x\right) \\ -25\left(\frac{x^4}{4} + \frac{4x^3}{3}\right) - M_B\left(\frac{x^2}{2} + 4x\right)\end{matrix}\right]_0^4 = 0$$

$$\Rightarrow 2.67R_B - 2M_B + 18.67R_B - 133.33 - 6M_B + 74.67R_B - 1{,}013.33 - 1{,}866.67 - 12M_B = 0$$

$$\Rightarrow 96.01R_B - 20M_B = 3{,}013.33$$

$$\Rightarrow 4.8R_B - M_B = 150.67 \quad (4.24a)$$

By solving equations 4.23a and 4.24a, we obtain

$$R_B = 49.58 \text{ kN}$$

$$M_B = 87.32 \text{ kNm}$$

The FBD indicating the support reactions, shear force and bending moment diagrams is shown in Figure 4.23.

Example 4.7

A beam *AB* of length *L* carries a uniformly distributed load of *w/m* throughout its length, as shown in Figure 4.24. The beam is fixed at end *A* and supported on a spring at end *B*. Show that the reaction at end *B* is $\frac{3wL}{8}\left[\frac{1}{1+(3EI\delta/L^3)}\right]$, where *EI* is the flexural rigidity of the beam, and δ is the compression of the spring produced by unit force.

Solution:

The static indeterminacy, $I_s = 1$, i.e., the beam is statically indeterminate to the first degree. Let us select the vertical reaction at *B* (R_B) as redundant is shown in Figure 4.25.

According to the strain energy theorem, the compatibility condition can be written as

$$\left(\frac{\partial U}{\partial R_B}\right)_{\text{beam}} + \left(\frac{\partial U}{\partial R_B}\right)_{\text{spring}} = 0 \quad (4.25)$$

$$\Rightarrow \left(\frac{\partial U}{\partial R_B}\right)_{\text{beam}} = \int M_x \frac{\partial M_x}{\partial R_B} \cdot \frac{dx}{EI}$$

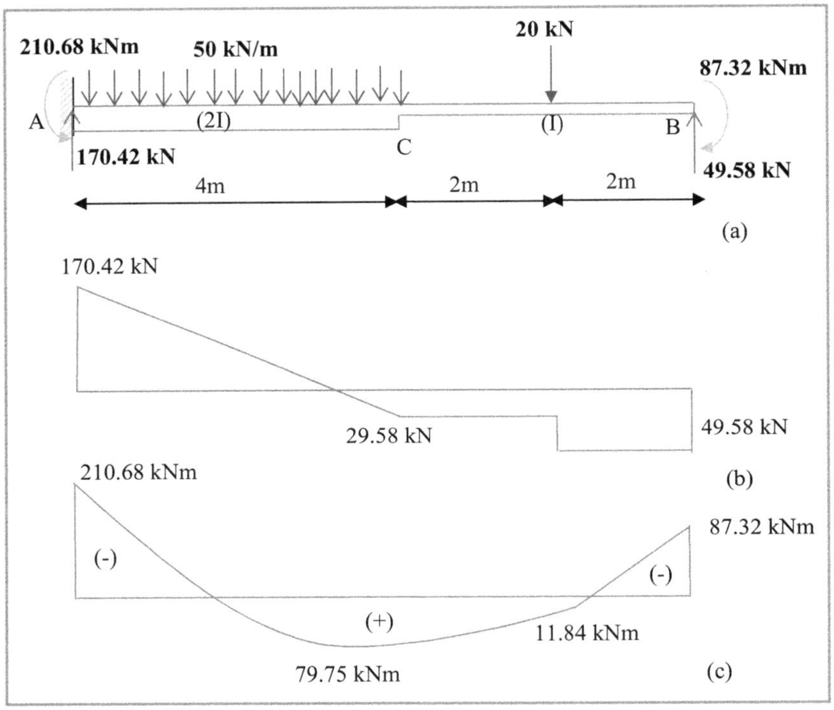

FIGURE 4.23 (a) Free body diagram, (b) shear force diagram, and (c) bending moment diagram.

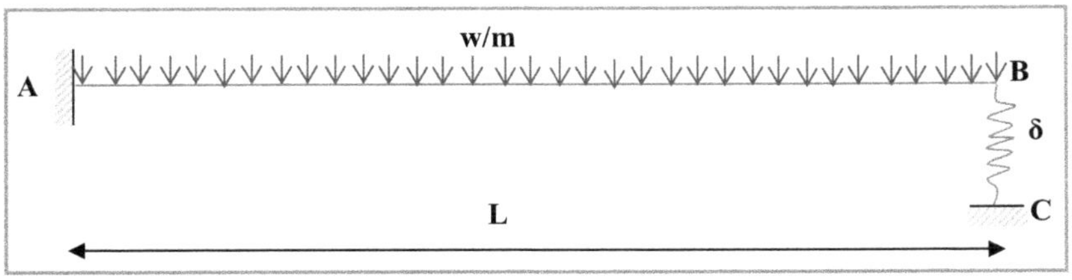

FIGURE 4.24 Beam supported with a linear spring.

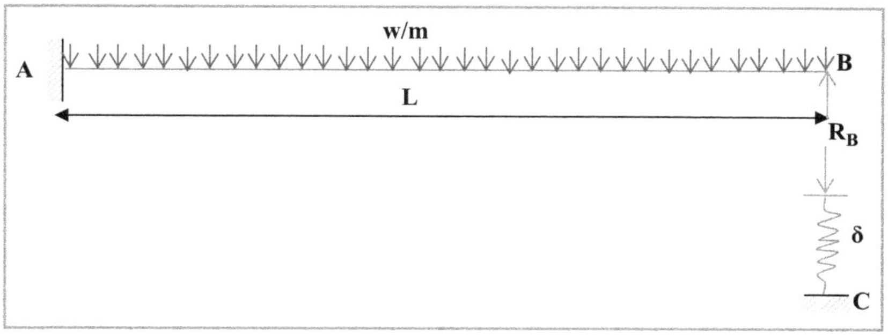

FIGURE 4.25 Free body diagram of beam and spring with loads and redundants.

$$\Rightarrow \int_0^L \left(R_B x - \frac{wx^2}{2}\right)(x)\frac{dx}{EI} = \frac{1}{EI}\left[R_B\frac{x^3}{3} - \frac{wx^4}{8}\right]_0^L$$

$$\Rightarrow \frac{1}{EI}\left[\frac{R_B L^3}{3} - \frac{wL^4}{8}\right]$$

SE in the spring

$$U_s = \frac{1}{2}K\Delta^2$$

$$= \frac{1}{2}\left(\frac{R_B}{\delta}\right)\Delta^2$$

$$= \frac{1}{2}\left(\frac{R_B}{\delta}\right)(R_B \cdot \delta)^2 \quad (\because \Delta = R_B \cdot \delta)$$

$$= \frac{1}{2}(R_B)^2 \cdot \delta$$

$$\frac{\partial U_s}{\partial R_B} = \frac{2R_B\delta}{2}$$

Substituting the above into equation 4.25, we get

$$\frac{R_B L^3}{3EI} - \frac{wL^4}{8EI} + R_B\delta = 0$$

$$\Rightarrow R_B\left(\frac{L^3}{3EI} + \delta\right) = \frac{wL^4}{8EI}$$

$$\Rightarrow R_B = \frac{wL^4}{8EI}\Big/\left(\frac{L^3}{3EI} + \delta\right)$$

$$\Rightarrow R_B = \frac{3wL^4}{8}\Big/\left(\frac{1}{L^3 + 3EI\delta}\right)$$

$$\Rightarrow R_B = \frac{3wL}{8}\Bigg/\left(\frac{1}{1 + \frac{3EI\delta}{L^3}}\right)$$

4.4 APPLICATION TO RIGID-JOINTED PLANE FRAMES

As discussed in Chapter 3, the axial force, shear force and bending moment are the primary internal resistances due to external forces in a rigid-jointed plane frame. In general, the strain energy due to axial force and shear force is neglected, and the strain energy due to only bending is considered in the rigid-jointed plane frames. Therefore, the procedure for the application of the theorem of least work for the analysis of rigid-jointed plane frames is the same as that adopted for beams in the previous section. First, we determine the static indeterminacy of the frame and identify the unknown reactions as redundant forces or reactions. Then, sketch the basic or primary structure with external loads and selected redundants. Now, using the theorem of least work, write the compatibility conditions as per equation 4.11a or 4.11b.

$$\frac{\partial U}{\partial R_i} = \int (M_x)\times\left(\frac{\partial M_x}{\partial R_i}\right)\times\frac{dx}{EI} = 0 \quad \text{for } i = 1,2,3,\ldots,I_s \tag{4.11b}$$

The FBDs are conveniently drawn with external loads and redundant forces, so that the bending moment expressions can be easily written in various segments of the frame. Now substitute these moment expressions into the compatibility equations to get the simultaneous equations, and by solving these simultaneous equations, we can obtain the desired unknowns. Once the redundants are obtained, using the FBDs and the static equilibrium equations, the member end forces, i.e., axial forces, shear forces and bending moments along with the remaining support reactions, can be easily determined. Finally, with the known values of the above, the axial force, shear force and bending moment diagrams can be drawn. As discussed in the previous section, when there is yielding of supports in the frame, the theorem of least work is not applicable; however, it can be analysed by using Castigliano's second theorem. The following example demonstrates the application of these theorems.

Example 4.8

Analyse the rigid-jointed plane frame shown in Figure 4.26, using the Theorem of least work. Draw the axial force, shear force, and bending moment diagrams. Assume EI is constant.

Solution:

The static indeterminacy $I_s = 5 - 3 = 2$, i.e., the frame is statically indeterminate to the second degree. Let us select the vertical and horizontal reactions at C (R_C and H_C) as redundant reactions. The primary structure with redundant reactions, and external loads is shown in Figure 4.27.

According to the Theorem of least work, the compatibility conditions can be expressed as,

$$\frac{\partial U}{\partial R_C} = \int M_x \frac{\partial M_x}{\partial R_C} \cdot \frac{dx}{EI} = 0 \tag{4.26}$$

$$\frac{\partial U}{\partial H_C} = \int M_x \frac{\partial M_x}{\partial H_C} \cdot \frac{dx}{EI} = 0 \tag{4.27}$$

The FBDs of different frame segments, used for deriving the bending moment expressions are shown in Figure 4.28. The calculation of M_x, $\frac{\partial M_x}{\partial R_C}$ and $\frac{\partial M_x}{\partial H_C}$ is presented in Table 4.7.

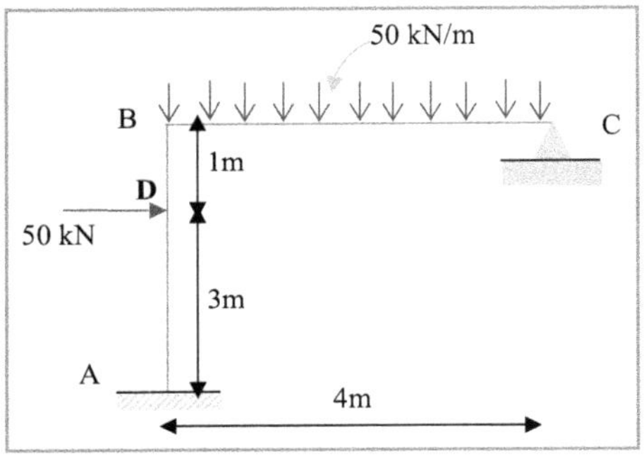

FIGURE 4.26 Rigid-jointed frame *ABC*.

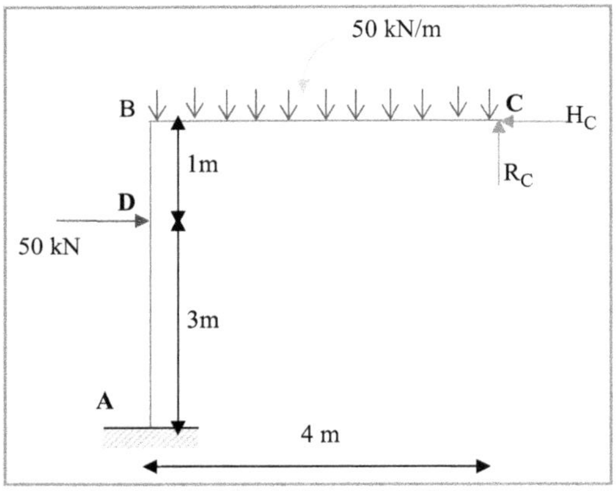

FIGURE 4.27 Primary structure with external loads and redundants.

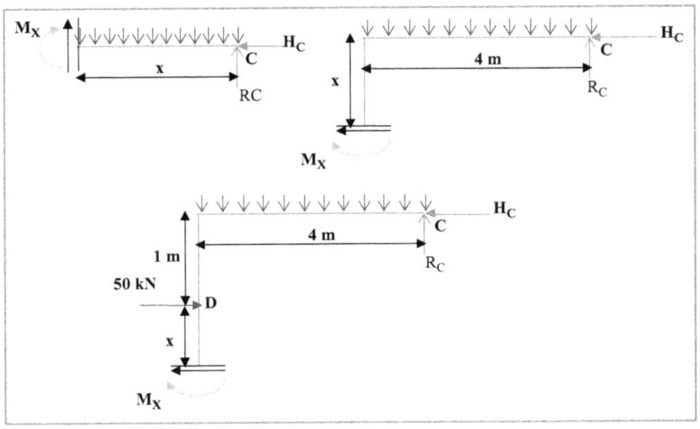

FIGURE 4.28 Free body diagrams due to loads and redundants.

$$\frac{\partial U}{\partial R_C} = \int_0^4 \left(R_C x - 25x^2\right)(x)\frac{dx}{EI} + \int_0^1 \left(4R_C + H_C x - 400\right)(4)\frac{dx}{EI}$$

$$+\int_0^1 \left(4R_C + H_C(x+1) - 400 - 50x\right)(4)\frac{dx}{EI} = 0$$

$$= \frac{1}{EI}\left[R_C\frac{x^3}{3} - 25\frac{x^4}{4}\right]_0^4 + \frac{1}{EI}\left|16R_C x + 4H_C\frac{x^2}{2} - 1{,}600x\right|_0^1$$

$$\Rightarrow 21.33R_C - 1{,}600 + 16R_C + 2H_C - 1{,}600$$

$$+48R_C + 30H_C - 4{,}800 - 900 = 0$$

$$\Rightarrow 85.33R_C + 32H_C = 8{,}900$$

$$\Rightarrow 2.67R_C + H_C = 278.13 \tag{4.26a}$$

$$\frac{\partial U}{\partial H_C} = \int_0^4 \left(R_C x - 25x^2\right)(0)\frac{dx}{EI} + \int_0^1 \left(4R_C + H_C x - 400\right)(x)\frac{dx}{EI}$$

$$+\int_0^3 \left(4R_C + H_C(x+1) - 400 - 50x\right)(x+1)\frac{dx}{EI} = 0$$

$$= \frac{1}{EI}\left|4R_C\frac{x^2}{2} + H_C\frac{x^3}{3} - 400\frac{x^2}{2}\right|_0^1$$

$$+\frac{1}{EI}\left|\begin{matrix} 4R_C\left(\frac{x^2}{2} + x\right) + H_C\frac{(x+1)^3}{3} \\ -400\left(\frac{x^2}{2} + x\right) - 50\left(\frac{x^3}{3} + \frac{x^2}{2}\right)\end{matrix}\right|_0^3$$

$$\Rightarrow 2R_C + 0.33H_C - 200 + 30R_C + 21H_C - 3{,}000 - 675 = 0$$

$$\Rightarrow 32R_C + 21.33H_C = 3{,}875$$

$$\Rightarrow 1.5R_C + H_C = 181.67 \tag{4.27a}$$

TABLE 4.7
Calculation of M_x, $\frac{\partial M_x}{\partial R_C}$ and $\frac{\partial M_x}{\partial H_C}$

Segment	Origin	Moment M_x	$\frac{\partial M_x}{\partial R_C}$	$\frac{\partial M_x}{\partial H_C}$	Limits (m)
BC	*C*	$R_C x - 25x^2$	x	0	0–4
DB	*D*	$4R_C + H_C x - 400$	4	x	0–1
AD	*D*	$4R_C + H_C(x+1) - 400 - 50x$	4	$(x+1)$	0–3

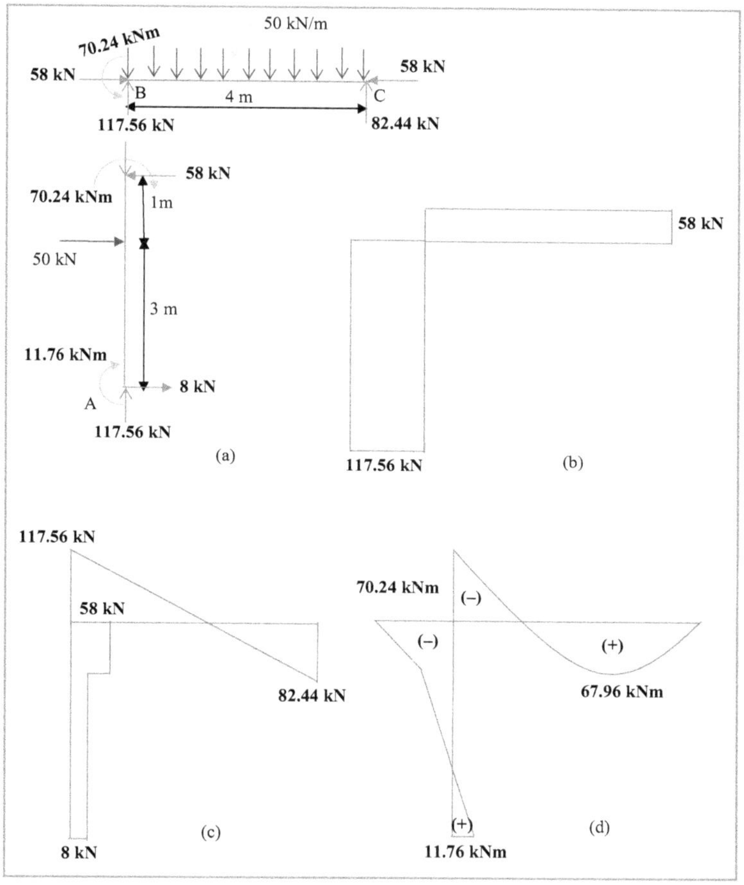

FIGURE 4.29 (a) Free body diagram, (b) axial force diagram, (c) shear force diagram, and (d) bending moment diagram.

By solving the equations 4.26a and 4.27a, we obtain

$$R_C = 82.44 \text{ kN} \quad \text{and} \quad H_C = 58 \text{ kN}$$

The FBD, axial force, shear force, and bending moment diagrams are shown in Figure 4.29.

Example 4.9

Analyse the frame *ABCD*, as shown in Figure 4.30, using the theorem of least work. Assume *EI* is constant. Draw the SF and BM diagrams.

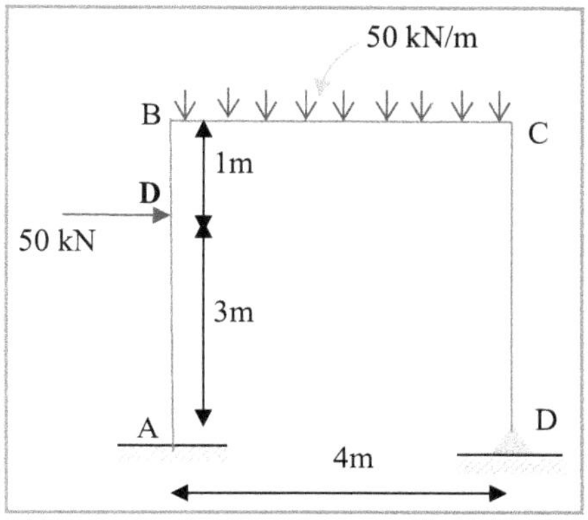

FIGURE 4.30 Rigid-jointed frame *ABCD*.

Solution:

The static indeterminacy, $I_s = 5 - 3 = 2.$; i.e., the frame is statically indeterminate to the second degree. Let us select the reactions at D (V_D and H_D) as redundant reactions. The primary frame with the redundants and the external loads is shown in Figure 4.31. The FBDs of different segments of the frame used for writing the bending moment expressions, are shown in Figure 4.32.

Based on the theorem of least work, the compatibility conditions can be written as

$$\frac{\partial U}{\partial V_D} = \int M_x \frac{\partial M_x}{\partial V_D} \cdot \frac{dx}{EI} = 0 \tag{4.28}$$

$$\frac{\partial U}{\partial H_D} = \int M_x \frac{\partial M_x}{\partial H_D} \cdot \frac{dx}{EI} = 0 \tag{4.29}$$

The calculation of M_x, $\frac{\partial M_x}{\partial V_D}$ and $\frac{\partial M_x}{\partial H_D}$ is presented in Table 4.8.

$$\frac{\partial U}{\partial V_D} = 0 + \int_0^4 \left(V_D x - 25x^2 - 4H_D\right)(x)\,\frac{dx}{EI}$$

$$+\int_0^3 \left[(50 - H_D)x + 4V_D - 550\right](+4)\frac{dx}{EI}$$

$$+\int_0^1 \left[4V_D - 400 - H_D(4 - x)\right](4)\frac{dx}{EI} = 0$$

$$= \frac{1}{EI}\left|V_D\frac{x^3}{3} - 25\frac{x^4}{4} - 4H_D\frac{x^2}{2}\right|_0^4$$

$$+\frac{1}{EI}\left|200\frac{x^2}{2} - 4H_D\frac{x^2}{2} + 16V_D x - 2{,}200x\right|_0^3$$

$$+\frac{1}{EI}\left|16V_D x - 1{,}600x - 4H_D\left(4x - \frac{x^2}{2}\right)\right|_0^1 = 0$$

$$\Rightarrow 21.33V_D - 1{,}600 - 32H_D + 900 - 18H_D$$

$$+48V_D - 6{,}600 + 16V_D - 1{,}600 - 14H_D = 0$$

$$\Rightarrow 85.33\,V_D - 64H_D = 8{,}900$$

$$\Rightarrow 1.33\,V_D - H_D = 139.06 \tag{4.28a}$$

$$\frac{\partial U}{\partial H_D} = \int_0^4 (-H_D x)(-x)\,\frac{dx}{EI} + \int_0^4 \left(V_D x - 25x^2 - 4H_D\right)(-4)\frac{dx}{EI}$$

$$+\int_0^1 \left[4V_D - 400 - H_D(4 - x)\right](x - 4)\frac{dx}{EI}$$

$$+\int_0^3 \left[(50 - H_D)x + 4V_D - 550\right](-x)\frac{dx}{EI} = 0$$

$$= \frac{1}{EI}\left|H_D\frac{x^3}{3}\right|_0^4 + \frac{1}{EI}\left|-4V_D\frac{x^2}{2} + 100\frac{x^3}{3} + 16H_D x\right|_0^4$$

$$+\frac{1}{EI}\left|4V_D\left(\frac{x^2}{2} - 4x\right) - 400\left(\frac{x^2}{2} - 4x\right) + H_D\frac{(x-4)^3}{3}\right|_0^1$$

$$+\frac{1}{EI}\left|(H_D - 50)\frac{x^3}{3} - 4V_D\frac{x^2}{2} + 550\frac{x^2}{2}\right|_0^3 = 0$$

$$\Rightarrow 21.33H_D - 32V_D + 2{,}133.33 + 64 = H_D - 14V_D + 1{,}400 - 9H_D$$

$$+21.33H_D + 9H_D - 450 - 18V_D + 2{,}475 = 0$$

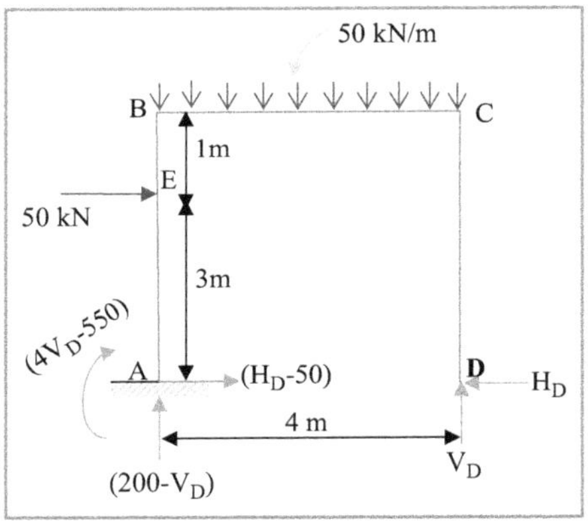

FIGURE 4.31 Rigid-jointed frame *ABC*.

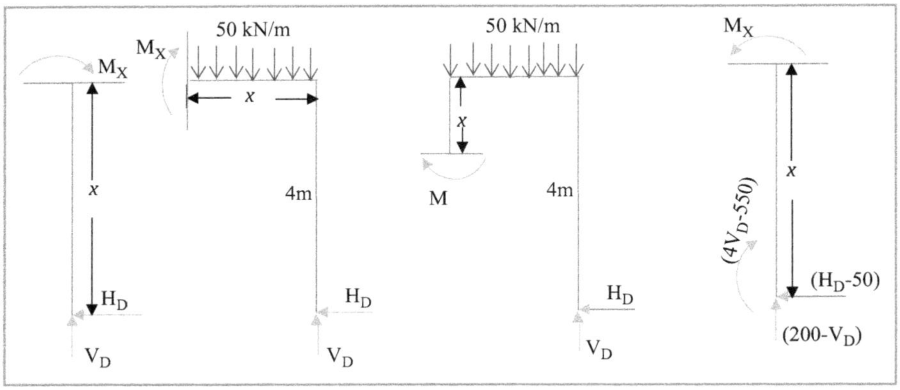

FIGURE 4.32 Free body diagrams.

TABLE 4.8
Bending moment expressions and its partial derivatives in various segments of a frame

Segment	Origin	Moment M_x	$\frac{\partial M_x}{\partial V_D}$	$\frac{\partial M_x}{\partial H_D}$	Limits (m)
CD	D	$-H_D x$	0	$-x$	0–4
BC	C	$V_D x - 25x^2 - 4H_D$	x	–4	0–4
BE	B	$4V_D - 400 - H_D(4 - x)$	4	$(x - 4)$	0–1
AE	A	$-(H_D - 50)x + 4V_D - 550$	4	$-x$	0–3

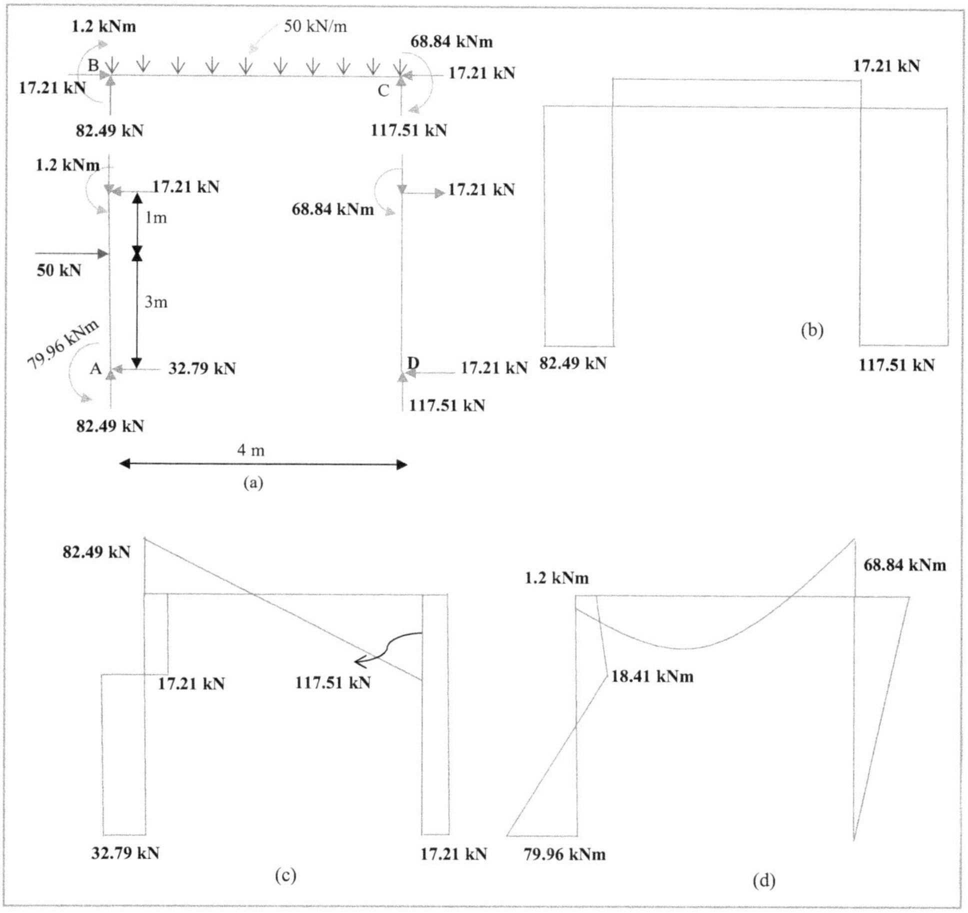

FIGURE 4.33 (a) Free body diagram, (b) axial force diagram, (c) shear force diagram, and (d) bending moment diagram.

$$\Rightarrow 106.67H_D - 64V_D = -5{,}558.33$$

$$\Rightarrow H_D - 0.59V_D = -52.11 \quad (4.29a)$$

By solving equations 4.28a and 4.29a, we get

$$V_D = 117.5 \text{ kN}$$

$$H_D = 17.21 \text{ kN}$$

The FBD, along with the internal forces, axial and shear forces, and bending moment diagrams, are shown in Figure 4.33.

Example 4.10

Analyse the rigid-jointed plane frame shown in Figure 4.34 using the theorem of least work. Draw the axial and shear force, and bending moment diagrams.

Solution:

The static indeterminacy of the frame is $I_s = 5 - 3 = 2$. Let us select H_A and V_A as redundant reactions. The primary frame with the redundant reactions and the external loads is shown in Figure 4.35.

The FBDs of different frame segments used for expressing the bending moments are shown in Figure 4.36. The moment expressions and their partial derivatives are shown in Table 4.9.

Using the theorem of least work, the compatibility conditions can be expressed as,

$$\frac{\partial U}{\partial V_A} = \int M_x \frac{\partial M_x}{\partial V_A} \cdot \frac{dx}{EI} = 0 \qquad (4.30)$$

$$\frac{\partial U}{\partial H_A} = \int M_x \frac{\partial M_x}{\partial H_A} \cdot \frac{dx}{EI} = 0 \qquad (4.31)$$

$$\frac{\partial U}{\partial V_A} = 0 + \int_0^5 \left(V_A x - 3H_A - 5\frac{x^3}{3} \right)(x)\frac{dx}{2EI}$$

$$+ \int_0^4 \left(H_A(1-x) + 5V_A - 208.33 \right)(5)\frac{dx}{EI} = 0$$

$$= \frac{1}{2EI}\left| V_A \frac{x^3}{3} - 3H_A \frac{x^2}{2} - 5\frac{x^5}{15} \right|_0^5$$

$$+ \frac{1}{EI}\left| 5H_A\left(x - \frac{x^2}{2} \right) + 25V_A x - 1{,}041.65 \right|_0^4 = 0$$

$$\Rightarrow 20.83V_A - 18.75H_A - 520.83 - 20H_A + +100V_A - 4{,}166.6 = 0$$

$$\Rightarrow 120.83V_A - 38.75H_A = 4{,}687.43$$

$$\Rightarrow 3.12V_A - H_A = 120.96 \qquad (4.30a)$$

$$\frac{\partial U}{\partial H_A} = \int_0^3 (-H_A x)(-x)\frac{dx}{EI} + \int_0^5 \left(V_A x - 3H_A - 5\frac{x^3}{3} \right)(-3)\frac{dx}{2EI}$$

$$+ \int_0^4 \left(H_A(1-x) + 5V_A - 208.33 \right)(1-x)\frac{dx}{EI} = 0$$

$$= \frac{1}{EI}\left| H_A \frac{x^3}{3} \right|_0^3 + \frac{1}{2EI}\left| -3V_A \frac{x^2}{2} + 9H_A x + 5\frac{x^4}{4} \right|_0^5$$

$$+ \frac{1}{EI}\left| H_A \frac{(x-4)^3}{3} + 5V_A\left(x - \frac{x^2}{2} \right) - 208.33\left(x - \frac{x^2}{2} \right) + \right|_0^4 = 0$$

$$\Rightarrow 9H_A + 22.5H_A - 18.75 + 390.63 + 9.33H_A - 20V_A + 833.32 = 0$$

$$\Rightarrow 40.83H_A - 38.75V_A = -1{,}223.95$$

$$\Rightarrow H_A - 0.95\ V_A = -29.97 \qquad (4.31a)$$

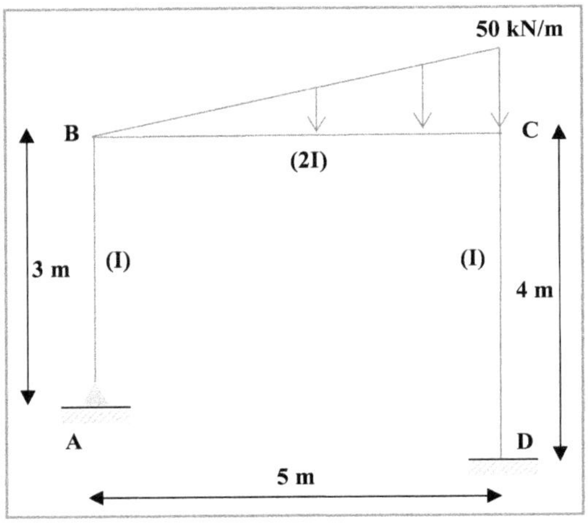

FIGURE 4.34 Unsymmetrical rigid-jointed plane frame *ABCD.*

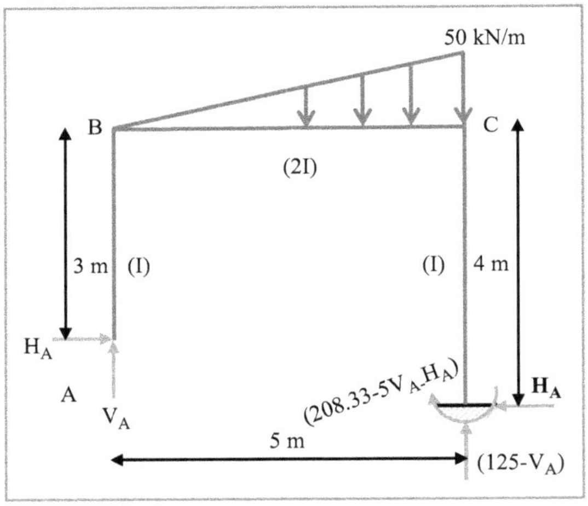

FIGURE 4.35 Primary frame with external loads and redundants.

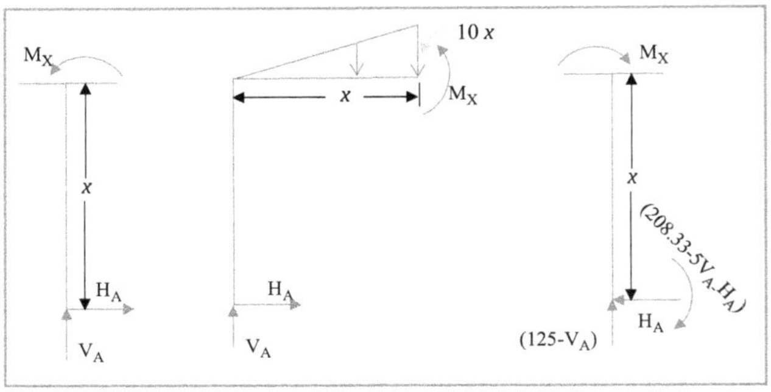

FIGURE 4.36 Free body diagrams.

TABLE 4.9
Moment expressions, and their partial derivatives

Segment	Origin	Moment (M_x)	$\frac{\partial M_x}{\partial V_A}$	$\frac{\partial M_x}{\partial H_A}$	Limits (m)	EI
AB	A	$-H_A x$	0	$-x$	0–3	EI
BC	B	$V_A x - 3H_A - 5\frac{x^3}{3}$	x	-3	0–5	2EI
CD	D	$-H_A x + H_A + 5V_A - 208.33$	5	$(1-x)$	0–4	EI

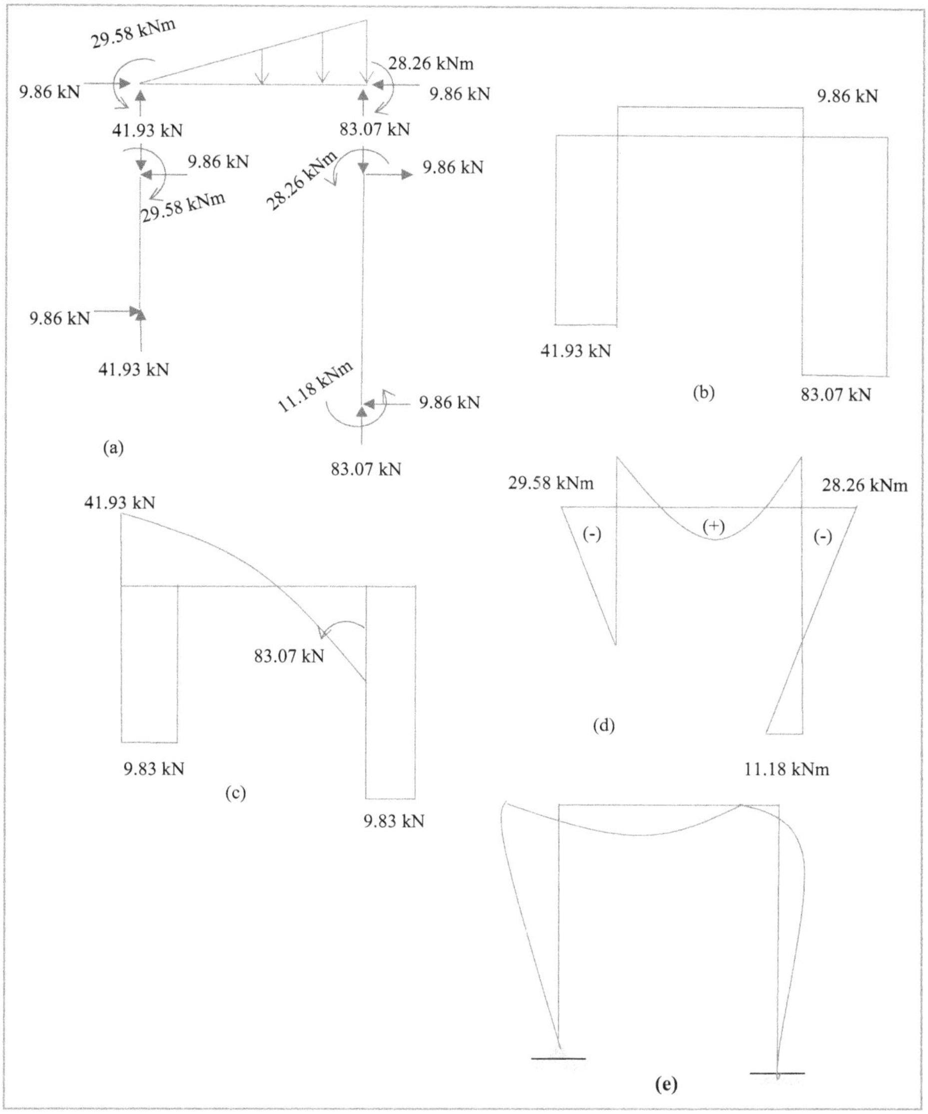

FIGURE 4.37 (a) Free body diagram, (b) axial force diagram, (c) shear force diagram, (d) bending moment diagram, and (e) probable elastic curve.

By solving equations 4.30a and 4.31a, we obtain

$$V_A = 41.93 \text{ kN}$$

$$H_A = 9.86 \text{ kN}$$

The internal forces in the FBD, axial and shear forces, bending moment diagram, and probable elastic curve are shown in Figure 4.37.

Example 4.11

Analyse the symmetrical frame shown in Figure 4.38, using the strain energy method. Draw the axial force, shear force, and bending moment diagrams.

Solution:

The frame with fixed bases in general is statically indeterminate to the third degree. The support reactions at *A* or *D* can be chosen as redundant reactions, with the cantilever serving as a basic determinate structure. However, taking advantage of symmetry can reduce the degree of static indeterminacy (Figure 4.39). By selecting the redundant reactions to act simultaneously at *A* and *D*, as shown in Figure 4.39, the symmetry of the structure can be maintained.

Let us select M_A and H_A as redundant reactions. Therefore, the compatibility conditions as per the theorem of least work are

$$\frac{\partial U}{\partial H_A} = 2\int M_x \frac{\partial M_x}{\partial H_A} \cdot \frac{dx}{EI} = 0 \tag{4.32}$$

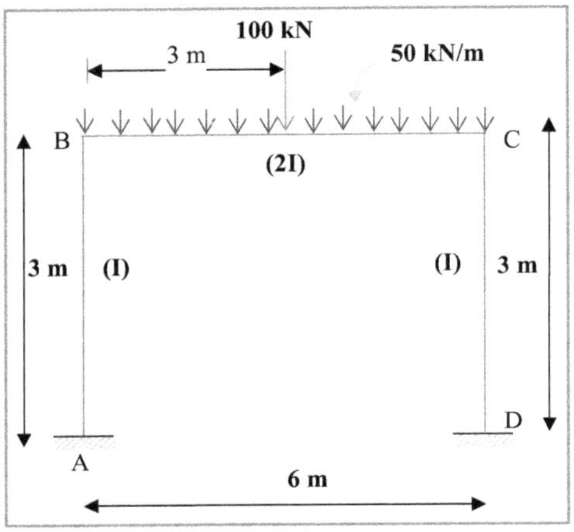

FIGURE 4.38 Symmetrical frame *ABCD*.

$$\frac{\partial U}{\partial M_A} = 2\int M_x \frac{\partial M_x}{\partial M_A} \cdot \frac{dx}{EI} = 0 \tag{4.33}$$

The FBDs of different segments of the frame used for writing the moment expressions are shown in Figure 4.40. The calculations of the M_x, $\frac{\partial M_x}{\partial H_A}$ and $\frac{\partial M_x}{\partial M_A}$ are presented in Table 4.10.

$$\frac{\partial U}{\partial H_A} = 2\left[\begin{array}{l}\int_0^3 (M_A - H_A x)(-x)\dfrac{dx}{EI} \\ +\int_0^3 \left(M_A + 200x - 3H_A - 25x^2\right)(-3)\dfrac{dx}{2EI}\end{array}\right] = 0$$

$$= 2\left[\begin{array}{l}\dfrac{1}{EI}\left|-M_A\dfrac{x^2}{2} + H_A\dfrac{x^3}{3}\right|_0^3 \\ +\dfrac{1}{2EI}\left|-3M_A x - 600\dfrac{x^2}{2} + 9H_A x + 75\dfrac{x^3}{3}\right|_0^3\end{array}\right] = 0$$

$$\Rightarrow 2\left[(-4.5M_A + 9H_A) + (-4.5M_A - 1{,}350 + 13.5H_A + 337.5)\right] = 0$$

$$\Rightarrow -9M_A + 22.5H_A = 1{,}012.5$$

$$\Rightarrow -M_A + 2.5H_A = 112.5 \tag{4.31a}$$

$$\frac{\partial U}{\partial M_A} = 2\left[\begin{array}{l}\int_0^3 (M_A - H_A x)(1)\dfrac{dx}{EI} \\ +\int_0^3 \left(M_A + 200x - 3H_A - 25x^2\right)(1)\dfrac{dx}{2EI}\end{array}\right] = 0$$

$$= \frac{1}{EI}\left|M_A x - H_A\frac{x^2}{2}\right|_0^3$$

$$+\frac{1}{2EI}\left|M_A x + 200\frac{x^2}{2} - 3H_A x - 25\frac{x^3}{3}\right|_0^3 = 0$$

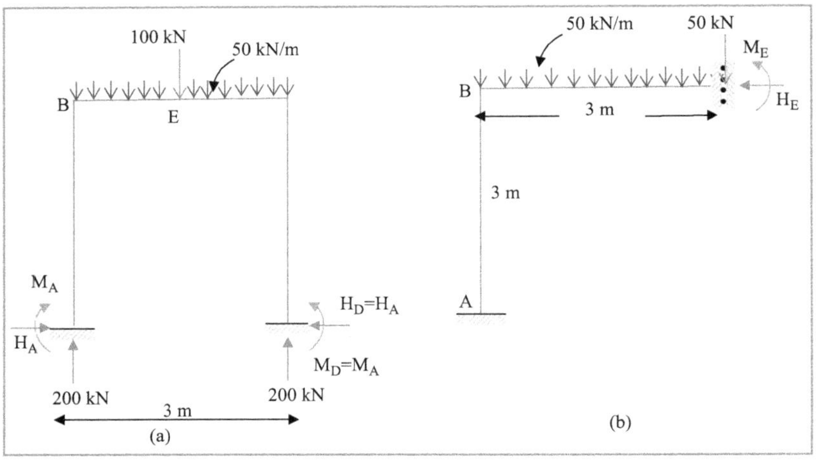

FIGURE 4.39 (a) Redundants act simultaneously at *A* and *D* and (b) half portion of frame due to symmetry.

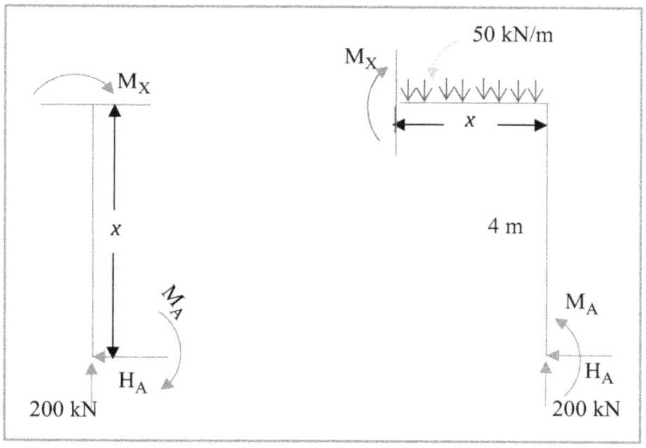

FIGURE 4.40 Free body diagrams.

TABLE 4.10

Calculation of M_x, $\frac{\partial M_x}{\partial H_A}$, and $\frac{\partial M_x}{\partial M_A}$

Segment	Origin	Moment (M_x)	$\frac{\partial M_x}{\partial H_A}$	$\frac{\partial M_x}{\partial M_A}$	Limits (m)	EI
CD	D	$M_A - H_A x$	$-x$	1	0–3	EI
BC	C	$M_A + 200x - 3H_A - 25x^2$	–3	1	0–3	2EI

$$\Rightarrow 3M_A - 4.5H_A + 1.5M_A + 450 - 4.5H_A - 112.5 = 0$$

$$\Rightarrow 4.5M_A - 9H_A = -337.50$$

$$\Rightarrow M_A - 2H_A = -75 \qquad (4.32a)$$

From equation 4.32a and 4.33a, we obtain

$$M_A = 75 \text{ kNm}$$

$$H_A = 75 \text{ kN}$$

ALTERNATE SOLUTION

Since the frame is symmetric in terms of loading, geometry, and support conditions, we can use this symmetry. The slope at the centre of *BC* is zero due to symmetry, and thus, the shear force at the centre of *BC* is also zero. To account for these conditions, we can analyse only half of the frame, as shown in Figure 4.39b, with the guided-fixed support at the cut location *E*. Select the moment M_E and the axial force H_E as redundant reactions. The compatibility conditions, based on the theorem of least work, are

$$\frac{\partial U}{\partial H_E} = \int M_x \frac{\partial M_x}{\partial H_E} \cdot \frac{dx}{EI} = 0 \qquad (4.32b)$$

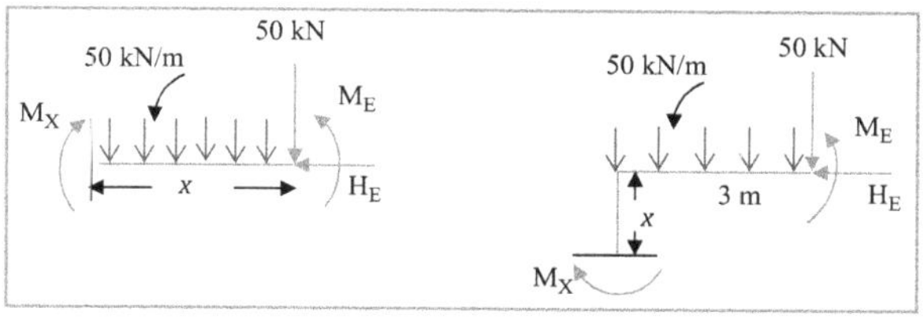

FIGURE 4.41 Free body diagrams.

TABLE 4.11

Bending moment expressions and their partial derivatives

Segment	Origin	Moment M_x	$\frac{\partial M_x}{\partial M_E}$	$\frac{\partial M_x}{\partial H_E}$	Limits (m) 0–1	EI
BE	B	$M_E - 50x - 25x^2$	1	0	0–3	2EI
AB	B	$M_E + H_E x - 375$	1	x	0–3	EI

$$\frac{\partial U}{\partial M_E} = \int M_x \frac{\partial M_x}{\partial M_E} \cdot \frac{dx}{EI} = 0 \qquad (4.33b)$$

The FBDs of different segments of a frame used for writing the moment expressions are shown in Figure 4.41. The bending moment expressions and their partial derivatives are presented in Table 4.11.

On substituting the above values into equations 4.32b and 4.33b,

$$\frac{\partial U}{\partial M_E} = \int_0^3 \left(M_E - 50x - 25x^2\right)(1)\frac{dx}{2EI}$$

$$+\int_0^3 (M_E + H_E x - 375)(1)\frac{dx}{EI} = 0$$

$$= \frac{1}{2EI}\left| M_E x - 50\frac{x^2}{2} - 25\frac{x^3}{3}\right|_0^3$$

$$+\frac{1}{EI}\left| M_E x + H_E\frac{x^2}{2} - 375x\right|_0^3 = 0$$

$$\Rightarrow 1.5M_E - 112.5 - 112.5 + 3M_E + 4.5H_E - 1{,}125 = 0$$

$$\Rightarrow 4.5M_E + 4.5H_E = 1{,}350$$

$$\Rightarrow M_E + H_E = 300 \qquad (4.32c)$$

$$\frac{\partial U}{\partial H_E} = \int_0^3 (M_E + H_E x - 375)(x)\frac{dx}{EI} = 0$$

$$= \left| M_E\frac{x^2}{2} + H_E\frac{x^3}{3} - 375\frac{x^2}{2}\right|_0^3 = 0$$

$$\Rightarrow 3M_A - 4.5H_A + 1.5M_A + 450 - 4.5H_A - 112.5 = 0$$

$$\Rightarrow 4.5M_E + 9H_E = 1{,}687.5$$

$$\Rightarrow M_E + 2H_E = 375 \qquad (4.33c)$$

From equation 4.32b and 4.33b, we get,

$$H_E = 75 \text{ kN} \quad \text{and} \quad M_E = 225 \text{ kNm}$$

The FBD along the internal forces, axial and shear force diagrams, bending moment diagram, and probable deflected shape are shown in Figure 4.42.

Example 4.12

Analyse the rigid-jointed plane frame shown in Figure 4.43, using the theorem of least work. Draw the AFD, SFD, and BMD. Also, draw the probable deflected shape. Assume *EI* is constant.

Solution:

The static indeterminacy of frame is $I_s = 5 - 3 = 2$.; i.e., the frame is statically indeterminate to the second degree. Select the reactions at *D* as redundant reactions; the primary structure with the external loads and the redundant reactions is shown in Figure 4.44.

Using the theorem of least work, the compatibility conditions can be written as,

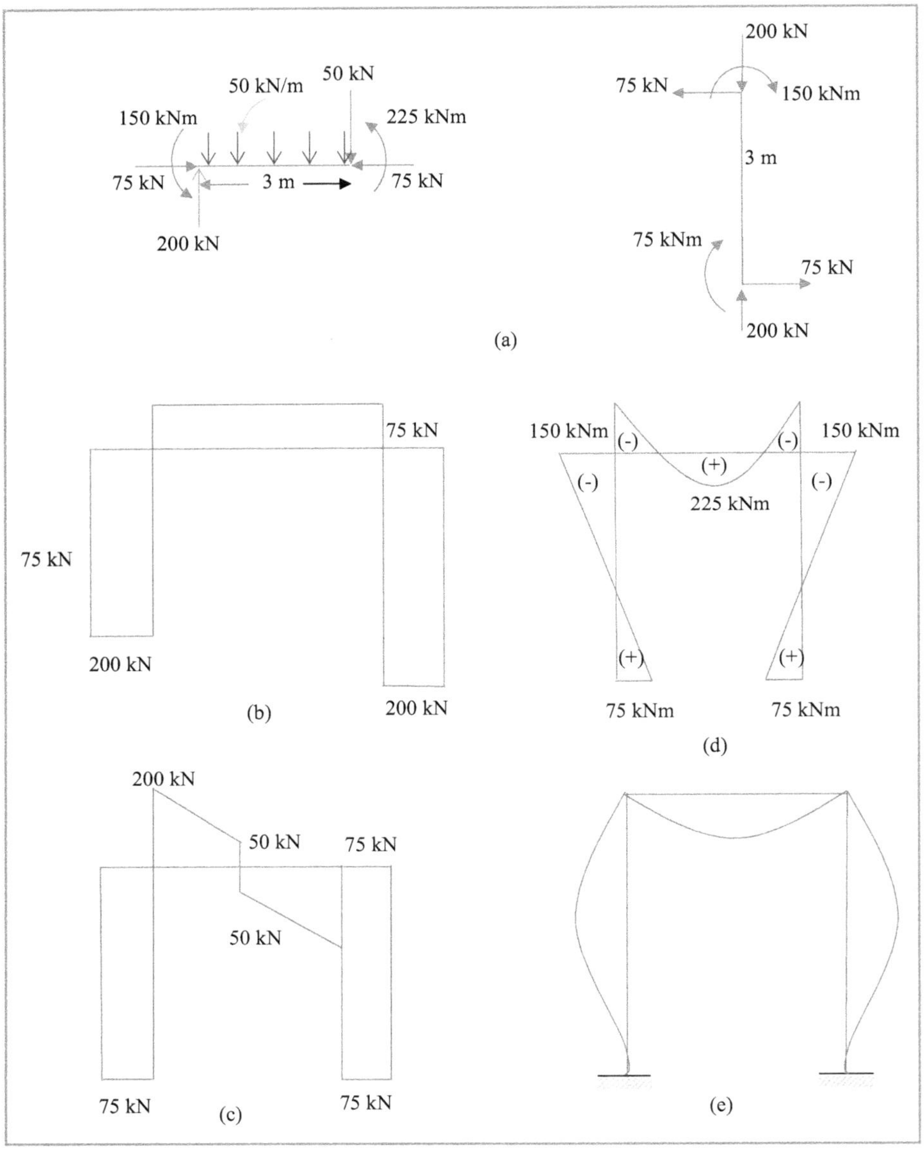

FIGURE 4.42 (a) Free body diagram, (b) axial force diagram, (c) shear force diagram, (d) bending moment diagram, and (e) probable deflected shape.

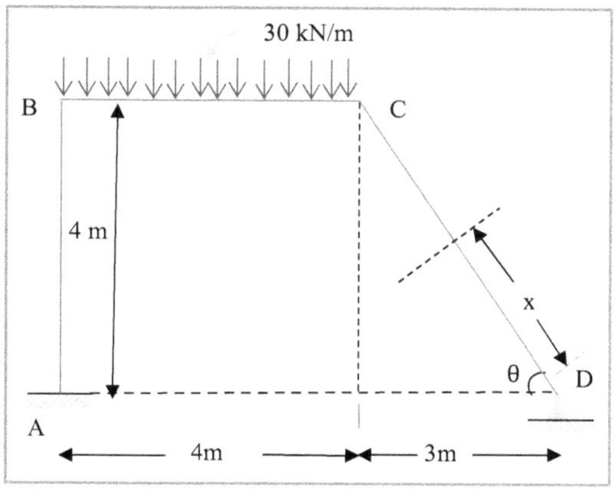

FIGURE 4.43 Rigid-jointed plane frame.

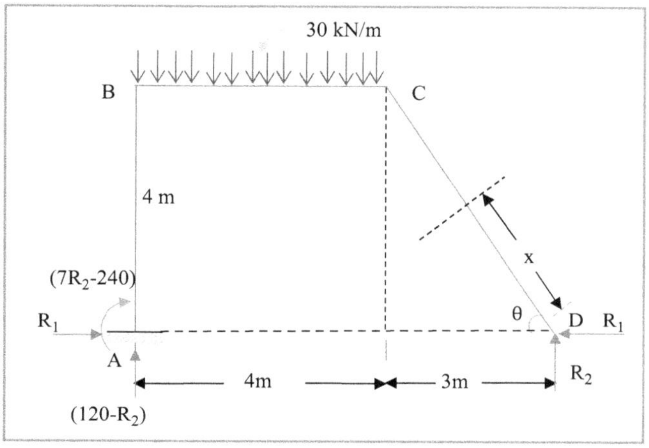

FIGURE 4.44 Primary structure with external loads and redundant reactions.

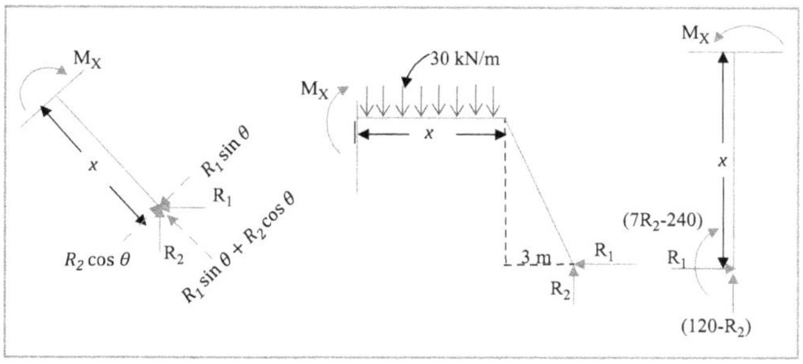

FIGURE 4.45 Free body diagram.

$$\frac{\partial U}{\partial R_1} = \int M_x \frac{\partial M_x}{\partial R_1} \cdot \frac{dx}{EI} = 0 \tag{4.34}$$

$$\frac{\partial U}{\partial R_2} = \int M_x \frac{\partial M_x}{\partial R_2} \cdot \frac{dx}{EI} = 0 \tag{4.35}$$

The FBDs of different frame segments are shown in Figure 4.45. The moment expressions and their partial derivatives are shown in Table 4.12.

TABLE 4.12
Bending moment expressions, and their partial derivatives

Segment	Origin	Moment M_x	$\frac{\partial M_x}{\partial R_1}$	$\frac{\partial M_x}{\partial R_2}$	Limits (m)
CD	D	$\left(\frac{3}{5}R_2 - \frac{4}{5}R_1\right)x$	$-\frac{4}{5}x$	$\frac{3}{5}x$	0–5
BC	C	$R_2(x+3) - 15x^2 - 4R_1$	-4	$x+3$	0–4
AB	A	$7R_2 - 240 - R_1x$	$-x$	7	0–4

$$\frac{\partial U}{\partial R_1} = \int_0^5 \left(\frac{3}{5}R_2 - \frac{4}{5}R_1\right)x\left(-\frac{4}{5}x\right)\frac{dx}{EI}$$

$$+\int_0^4 \left[R_2(x+3) - 15x^2 - 4R_1\right](-4)\frac{dx}{EI}$$

$$+\int_0^4 (7R_2 - 240 - R_1x)(-x)\frac{dx}{EI} = 0$$

$$= \left[\frac{-12}{25}R_2\frac{x^3}{3} + \frac{16}{25}R_1\frac{x^3}{3}\right]_0^5$$

$$+\left[-4R_2\left(\frac{x^2}{2} + 3x\right) + 60\frac{x^3}{3} + 16R_1x\right]_0^4$$

$$+\left[-7R_2\frac{x^2}{2} + 240\frac{x^2}{2} + R_1\frac{x^3}{3}\right]_0^4 = 0$$

$$\Rightarrow -20R_2 + 26.67R_1 - 80R_2 + 1{,}280 + 64R_1 - 56R_2 + 1{,}920 + 21.33R_1 = 0$$

$$\Rightarrow -156R_2 + 112R_1 = -3{,}200$$

$$\Rightarrow -R_2 + 0.717R_1 = -20.51 \tag{4.34a}$$

$$\frac{\partial U}{\partial R_2} = \int_0^5 \left(\frac{3}{5}R_2 - \frac{4}{5}R_1\right)x\left(\frac{3}{5}x\right)\frac{dx}{EI}$$

$$+\int_0^4 \left[R_2(x+3) - 15x^2 - 4R_1\right](x+3)\frac{dx}{EI}$$

$$+\int_0^4 (7R_2 - 240 - R_1x)(7)\frac{dx}{EI} = 0$$

$$= \left[\frac{9}{25}R_2\frac{x^3}{3} - \frac{12}{25}R_1\frac{x^3}{3}\right]_0^5$$

$$+\left[R_2\frac{(x+3)^3}{3} - 15\frac{x^4}{4} - 45\frac{x^3}{3} - 4R_1\frac{x^2}{2} - 12R_1x\right]_0^4$$

$$+\left[49R_2x - 1680x - 7R_1\frac{x^2}{2}\right]_0^4 = 0$$

$$\Rightarrow 15R_2 - 20R_1 + 105.33R_2 - 960 - 960 - 32R_1$$

$$-48R_1 + 196R_2 - 6{,}720 - 56R_1 = 0$$

$$\Rightarrow 316.33R_2 - 156R_1 = 8{,}640$$

$$\Rightarrow R_2 - 0.49R_1 = 27.31 \quad (4.35a)$$

From equation 4.34a and 4.35a, we get

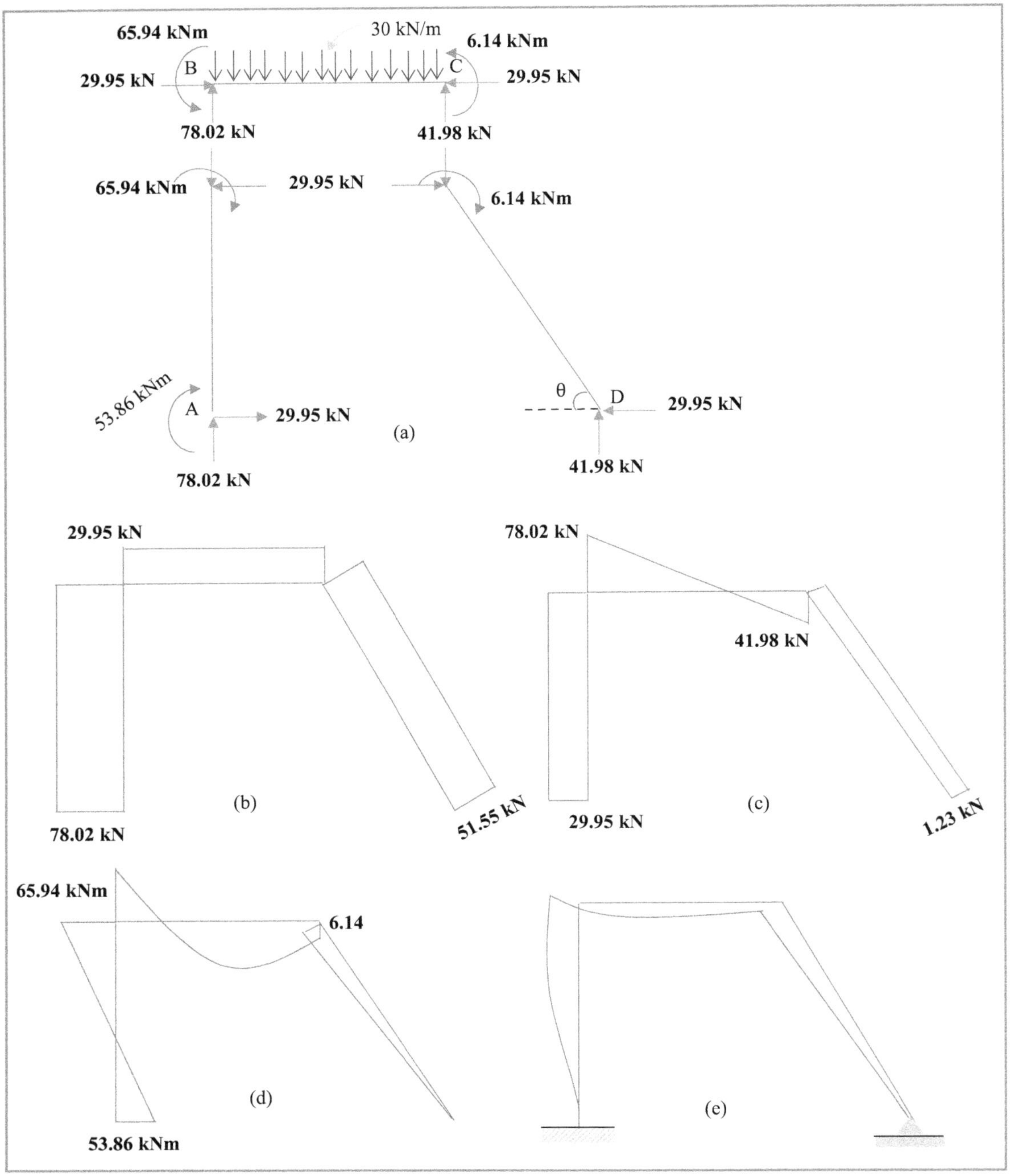

FIGURE 4.46 (a) Free body diagram, (b) axial force diagram, (c) shear force diagram, (d) bending moment diagram, and (e) probable deflected shape.

$$R_1 = 29.95 \text{ kN}$$

$$R_2 = 41.98 \text{ kN}$$

The FBD, along with internal forces, axial force and shear force diagrams, bending moment diagram, and probable deflected shape are shown in Figure 4.46.

Example 4.13

Re-analyse the Q.No. 4.11, if the support *D* is fixed and settle down by 10 mm, and rotates 0.002 rad in clockwise, in the absence of the external loads (Figure 4.47). Assume $EI = 135 \times 10^3 \text{ kNm}^2$.

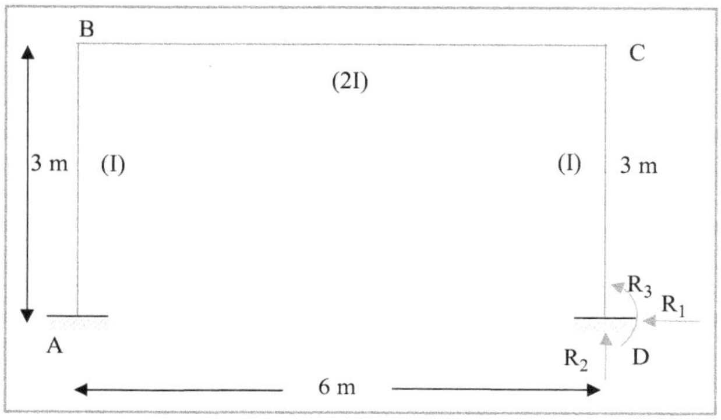

FIGURE 4.47 Symmetrical frame.

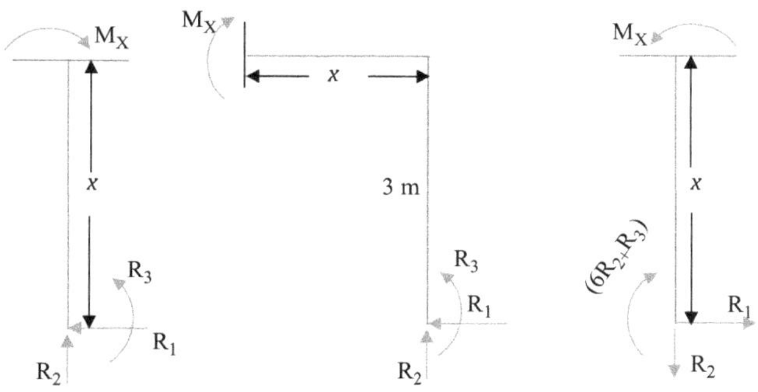

FIGURE 4.48 Free body diagrams.

Solution:

The static indeterminacy, $I_s = 3$. Select the reactions at *D*, i.e., R_1, R_2, and R_3 as redundant reactions (shown in Figure 4.47). Using Castigliano's second theorem, the compatibility equations can be expressed as

$$\frac{\partial U}{\partial R_1} = \int M_x \frac{\partial M_x}{\partial R_1} \cdot \frac{dx}{EI} = 0 \tag{4.36}$$

$$\frac{\partial U}{\partial R_2} = \int M_x \frac{\partial M_x}{\partial R_2} \cdot \frac{dx}{EI} = -0.01 \tag{4.37}$$

$$\frac{\partial U}{\partial R_3} = \int M_x \frac{\partial M_x}{\partial R_3} \cdot \frac{dx}{EI} = -0.002 \tag{4.38}$$

The FBDs of different segments of a frame, used for deriving the moment expressions, are shown in Figure 4.48.

The calculation of M_x and its derivatives is shown in Table 4.13.

$$\frac{\partial U}{\partial R_1} = \int_0^3 (R_3 - R_1 x)(-x)\frac{dx}{EI} + \int_0^6 (R_3 + R_2 x - 3R_1)(-3)\frac{dx}{2EI}$$

$$+\int_0^3 (6R_2 + R_3 - R_1 x)(-x)\frac{dx}{EI} = 0$$

$$= \frac{1}{EI}\left| -R_3\frac{x^2}{2} + R_1\frac{x^3}{3} \right|_0^3 + \frac{1}{2EI}\left| -3R_3 x - 3R_2\frac{x^2}{2} + 9R_1 x \right|_0^6$$

$$+\frac{1}{EI}\left| -6R_2\frac{x^2}{2} - R_3\frac{x^2}{2} + R_1\frac{x^3}{3} \right|_0^3 = 0$$

$$= -4.5R_3 + 9R_1 - 9R_3 - 27R_2 + 27R_1 - 27R_2$$

$$-4.5R_3 + 9R_1 = 0$$

$$\Rightarrow 45R_1 - 54R_2 - 18R_3 = 0 \tag{4.36a}$$

TABLE 4.13
Calculation of Mx and its partial derivatives

Segment	Origin	Moment M_x	$\frac{\partial M_x}{\partial R_1}$	$\frac{\partial M_x}{\partial R_2}$	$\frac{\partial M_x}{\partial R_3}$	Limits (m) 0–1	*EI*
CD	*D*	$R_3 - R_1 x$	$-x$	0	1	0–3	*EI*
BC	*C*	$R_3 + R_2 x - 3R_1$	-3	x	1	0–6	2*EI*
AB	*A*	$6R_2 + R_3 - R_1 x$	$-x$	6	1	0–3	*EI*

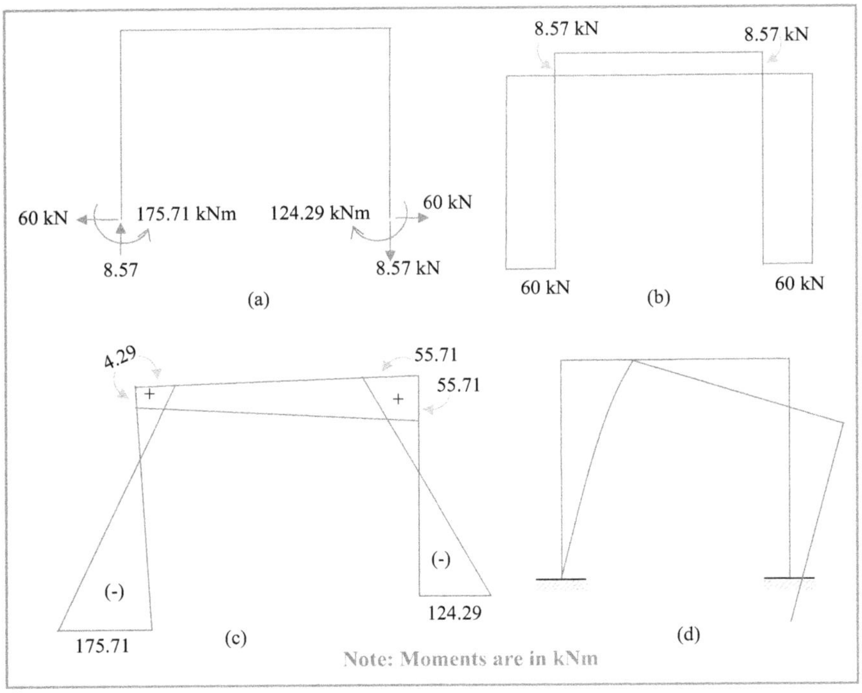

FIGURE 4.49 (a) Free body diagram, (b) shear force diagram, (c) bending moment diagram, and (d) probable deflected shape.

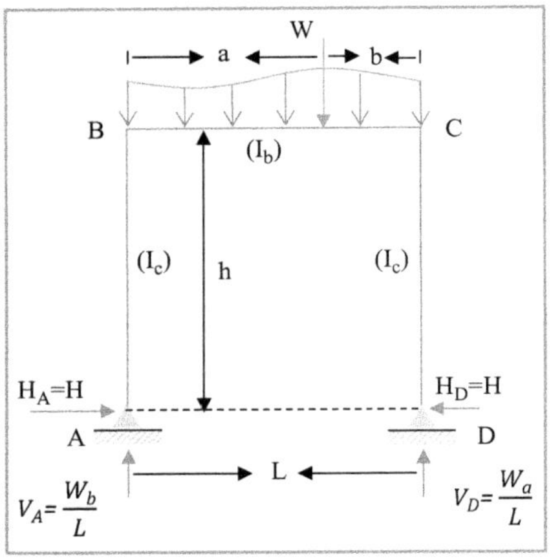

FIGURE 4.50 Two-hinged rigid-jointed frame.

$$\frac{\partial U}{\partial R_2}=\int_0^3 (R_3-R_1x)(0)\,\frac{dx}{EI}+\int_0^6 (R_3+R_2x-3R_1)(x)\frac{dx}{2EI}$$

$$+\int_0^3 (6R_2+R_3-R_1x)(6)\frac{dx}{EI}=-0.01$$

$$=\frac{1}{2EI}\left|R_3\frac{x^2}{2}+R_2\frac{x^3}{3}-3R_1\frac{x^2}{2}\right|_0^6$$

$$+\frac{1}{EI}\left|36R_2x+6R_3x-6R_1\frac{x^2}{2}\right|_0^3=-0.01$$

$$=9R_3+36R_2-27R_1+108R_2$$

$$+18\ R_3-27R_1=-0.01\times EI$$

$$\Rightarrow -54R_1+144R_2+27R_3=-1{,}350 \qquad (4.37a)$$

$$\frac{\partial U}{\partial R_3}=\int_0^3 (R_3-R_1x)(1)\,\frac{dx}{EI}+\int_0^6 (R_3+R_2x-3R_1)(1)\frac{dx}{2EI}$$

$$+\int_0^3 (6R_2+R_3-R_1x)(1)\frac{dx}{EI}=-0.002$$

$$=\frac{1}{EI}\left|R_3x-R_1\frac{x^2}{2}\right|_0^3+\frac{1}{2EI}\left|R_3x+R_2\frac{x^2}{2}-3R_1x\right|_0^6$$

$$+\left|6R_2x+R_3x-R_1\frac{x^2}{2}\right|_0^3=-0.002$$

$$=3R_3-4.5R_1+3R_3+9R_2-9.0R_1+18R_2+3R_3$$

$$-4.5R_1=-0.002\times EI$$

$$\Rightarrow -18R_1+27R_2+9R_3=-270 \qquad (4.38a)$$

Equations 4.36a – 4.38a can be written in a matrix as

$$\begin{bmatrix}45 & -54 & -18\\ -54 & 144 & 27\\ -18 & 27 & 9\end{bmatrix}\begin{bmatrix}R_1\\ R_2\\ R_3\end{bmatrix}=\begin{bmatrix}0\\ -1{,}350\\ -270\end{bmatrix}$$

By solving, we get

$R_1=-60$ kN; $R_2=-8.57$ kN; $R_3=-124.29$ kNm

FINAL END REACTIONS

The FBD, along with remaining support reactions, shear force, bending moment diagrams, and probable deflected shape, is shown in Figure 4.49.

Example 4.14

Analyse a two-hinged rigid-jointed frame shown in Figure 4.50, using the theorem of least work.

Solution:

ABCD is a two-hinged symmetrical frame subjected to gravity loading on *BC*, whose total load is, say, '*W*' acting at a distance '*a*' from '*B*'. The static indeterminacy, $I_s = 3$. Due to symmetry, the vertical reactions at *A* and *D* are statically determinate. Hence, let us select the horizontal reactions at *A* and *D* ($H_A = H_D = H$) as redundant reactions.

Using the theorem of least work, the compatibility condition can be written as,

$$\frac{\partial U}{\partial H} = \int M_x \frac{\partial M_x}{\partial H} \cdot \frac{dx}{EI} = 0 \qquad (4.39)$$

From the FBDs (Figure 4.51a and b), it can be seen that due to hinged supports at *A* and *D*, the columns are subjected to outward linear moments caused by the redundant force *H*, with its maximum value being *Hh* at joints *B* and *C*. This causes the columns to bulge outward, inducing a hanging BM in the beam *BC* (Figure 4.51c), which helps in reducing the sagging BM in the beam *BC* (Figure 4.51d). Now, we can easily find the unknown *H*, using the theorem of least work.

Let $M^0_{(x)}$ be the free bending moment in the beam *BC*. Let us calculate M_x and $\frac{\partial M_x}{\partial H}$ using the FBDs (Figure 4.52).

$$\text{For columns, } M_x = (-Hx), \quad \frac{\partial M_x}{\partial H} = -x \quad (0 \le x \le h)$$

$$\text{For beam } BC, M_x = \left(M^0_x - Hh\right), \quad \frac{\partial M_x}{\partial H} = -h \quad (0 \le x \le L)$$

$$\therefore \frac{\partial U}{\partial H} = \int_0^L \left(M^0_{(x)} - Hh\right)(-h)\frac{dx}{EI_b} + 2\int_0^h (-Hx)(-x)\frac{dx}{EI_c} = 0$$

$$= -\frac{h}{EI_b}\int_0^L M^0_{(x)} \cdot \frac{dx}{EI_b} + \frac{Hh^2L}{EI_b} + \frac{2Hh^3}{3EI_c} = 0$$

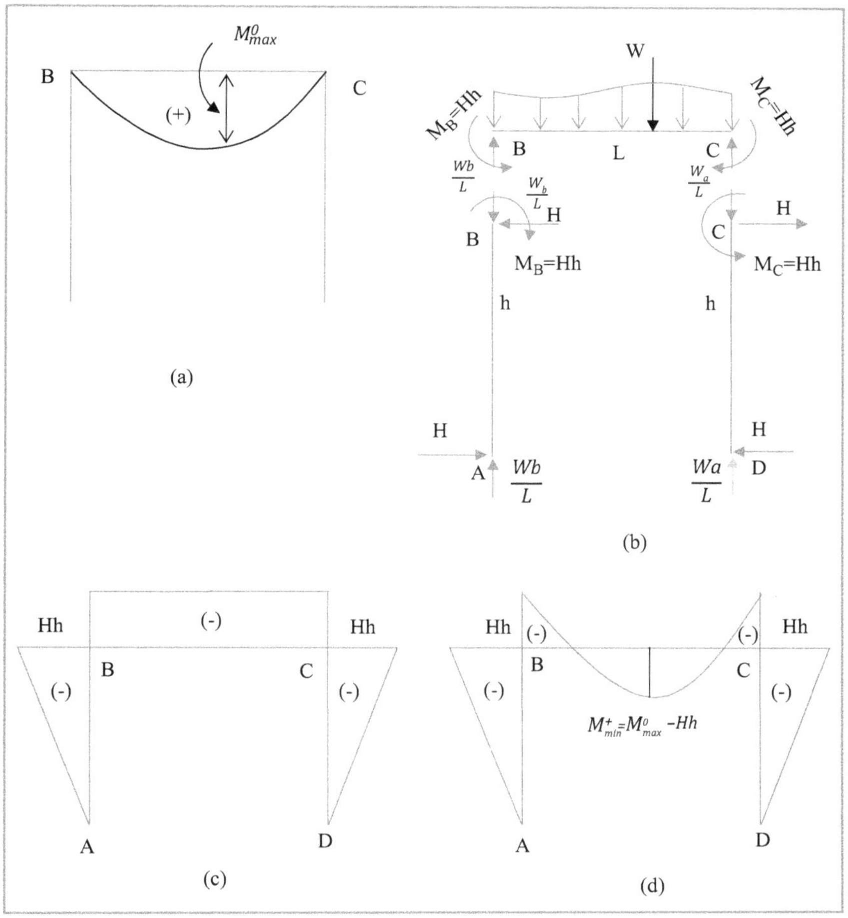

FIGURE 4.51 (a) *BMD* due to *W* on basic frame, (b) Free body diagram, (c) *BMD* due to *W* on basic frame, and (d) Resultant *BMD*.

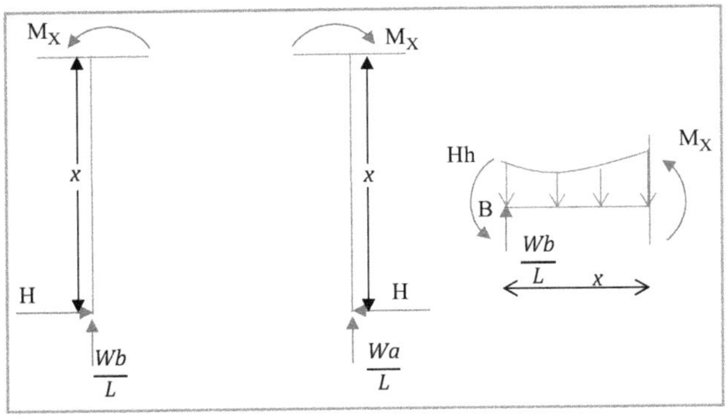

FIGURE 4.52 Free body diagrams.

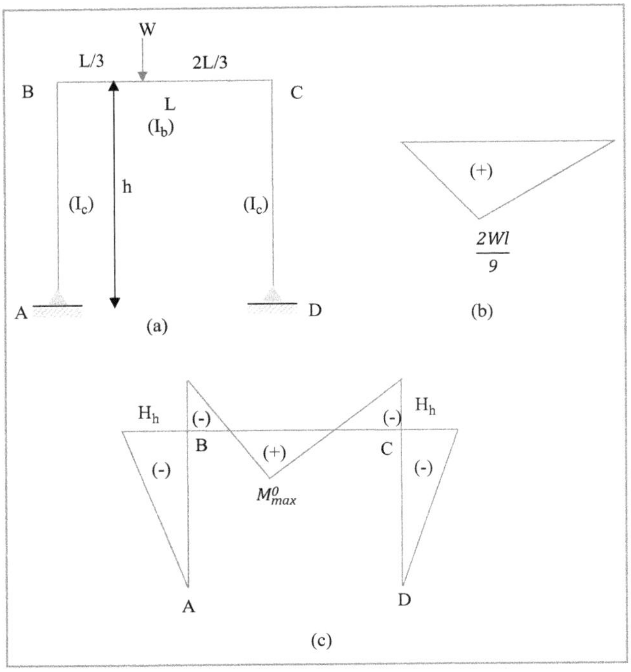

FIGURE 4.53 (a) Frame subjected to load at one-third of span *BC*, (b) free bending moment diagram, and (c) net bending moment diagram.

$$\Rightarrow \frac{h}{EI_b}\int_0^L M^0_{(x)} \cdot dx = H\left[\frac{h^2L}{EI_b}+\frac{2h^3}{3EI_c}\right]$$

$$\Rightarrow H = \frac{\dfrac{h}{EI_b}\displaystyle\int_0^L M^0_{(x)} \cdot \frac{dx}{EI}}{\left[\dfrac{h^2L}{EI_b}+\dfrac{2h^3}{3EI_c}\right]}$$

$$= \frac{A_b^0}{Lh+\dfrac{2h^3}{3I_c}\times I_b} = \frac{A_b^0}{Lh\left[1+\dfrac{2}{3}\left(\dfrac{I_b/L}{I_c/h}\right)\right]}$$

$$H = \frac{A_b^0}{Lh\left[1+\dfrac{2}{3}\cdot\propto\right]} \tag{4.40}$$

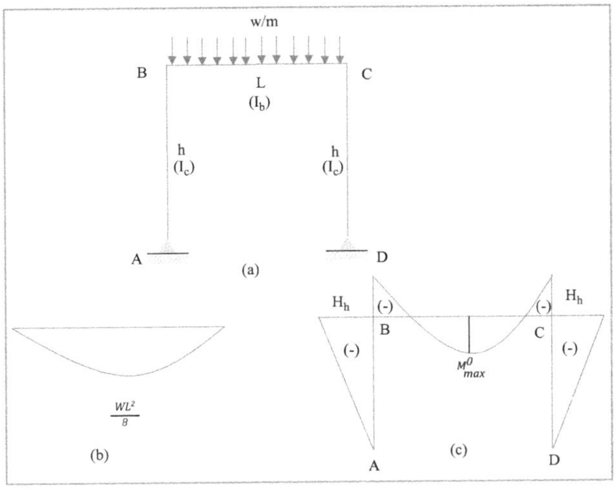

FIGURE 4.54 (a) Frame subjected to *udl BC*, (b) free bending moment diagram, and (c) net bending moment diagram.

where $\propto = \left(\dfrac{I_b/L}{I_c/h}\right)$ measures the relative stiffness of a beam to the column.

$$A_b^0 = \int_0^L M_x^0 \cdot dx = \text{Area of free bending moment diagram of beam } BC$$

Note:

1. When the beam is infinitely rigid relative to the column, α approaches infinity, indicating that the beam behaves as if it is effectively hinged at *B* and *C*. Consequently, the hogging moment developed at *B* and *C* becomes zero, resulting in a zero value of *H* as per the equation.
2. When the columns are infinitely rigid compared to the beam, the value of α approaches zero. This implies that the ends of the beam are rotationally restrained (if *W* is eccentric, the beam will translate horizontally). Therefore, the $H = \dfrac{A_b^0}{Lh}$, and the fixed end moments are equal to $\dfrac{A_b^0}{L}$, representing the average free sagging bending moment of the beam.

In practice, the α value varies between zero and infinity, and hence the horizontal reaction H lies between zero and $\dfrac{A^0}{Lh}$.

CASE I

If the load is acting at one-third of the span (Figure 4.53),

$$\therefore\ A_b^0 = \frac{1}{2}\times L\times\frac{2WL}{9} = \frac{WL^2}{9}$$

$$\therefore H = \frac{A_b^0}{Lh\left[1+\frac{2}{3}\cdot\propto\right]} = \frac{WL}{9h\left(1+\frac{2}{3}\cdot\propto\right)}$$

CASE II

If the load is uniformly distributed of *w/m* (Figure 4.54)

$$A_b^0 = \frac{2}{3}\times L\times\frac{wL^2}{8} = \frac{wL^3}{12} = \frac{WL^2}{12}$$

$$\therefore H = \frac{WL}{12h\left(1+\frac{2}{3}\cdot\propto\right)}$$

$$\therefore M_B = M_C = \frac{WL}{12h\left(1+\frac{2}{3}\cdot\propto\right)}$$

M_{max} +ve BM in the beam

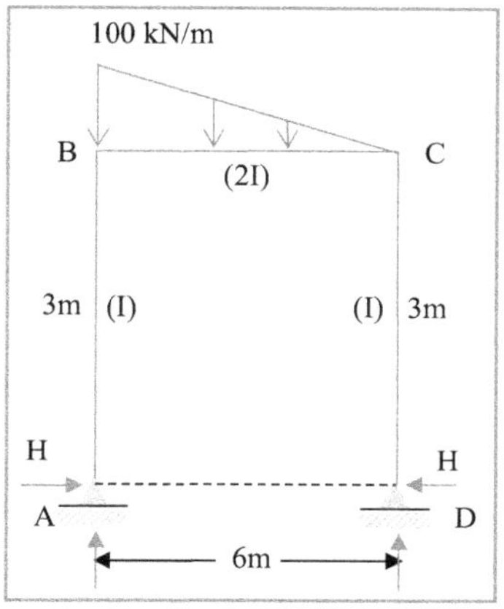

FIGURE 4.55 Two-hinged symmetrical frame subjected to unsymmetrical loading.

$$M_{max}^0 = M^0 - M_B = \frac{WL}{8} = \frac{WL}{12\left(1+\frac{2}{3}\cdot\propto\right)}$$

$$\Rightarrow M_B = M_C = Hh = \frac{WL}{9}\left(\frac{1}{1+\frac{2}{3}\cdot\propto}\right)$$

$$\Rightarrow M_{max}^0 = \frac{2WL}{9} - \frac{WL}{9}\left(\frac{1}{1+\frac{2}{3}\cdot\propto}\right)$$

Example 4.15

A two-hinged rigid-jointed frame *ABCD* is subjected to an unsymmetrical load, as shown in Figure 4.55. Analyse the frame by applying symmetry and verify the results using the expression for *H*.

Solution:

Figure 4.55 depicts a symmetric frame subjected to unsymmetrical loading. Consider its mirror image, shown in Figure 4.56b. By applying the principle of parity, it is observed that the horizontal reaction *H* is the same for both frames, i.e., in Figure 4.56a and b. Figure 4.56c represents the resultant frame, which is symmetrically loaded and obtained by applying the principle of superposition to Figure 4.56a and b. Hence, the horizontal reaction in the resultant frame is $H_1 = 2H$.

Now the resultant frame $A^1B^1C^1D^1$ is a symmetrical frame. Apply the theorem of least work

$$\frac{\partial U}{\partial H_1} = \int M_x\frac{\partial M_x}{\partial H_1}\cdot\frac{dx}{EI} = 0$$

For column (Figure 4.57)

$$M_x = -H_1x; \quad 0 \le x \le 3\text{ m}$$

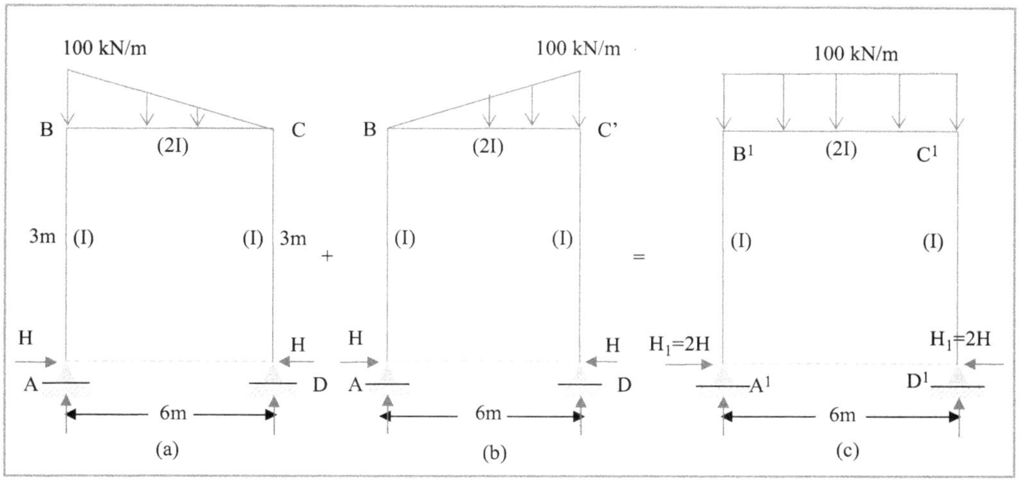

FIGURE 4.56 (a) Symmetrical frame subjected to unsymmetrical, (b) mirror image of (a), and (c) resultant frame by superposition of loading.

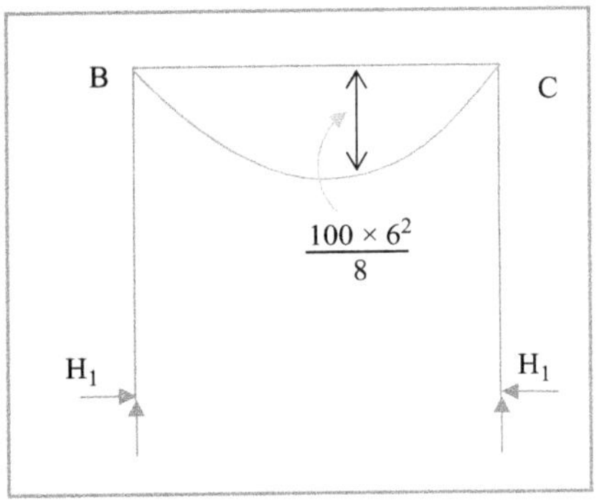

FIGURE 4.57 Free bending moment diagram.

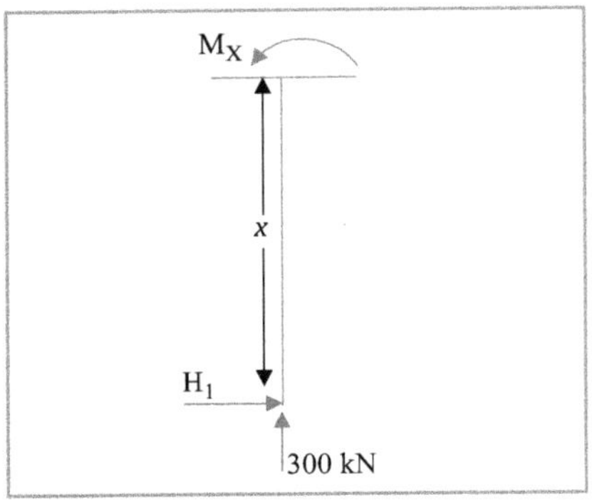

FIGURE 4.58 Free body diagram.

For beam (Figure 4.58)

$$M_x = M^0_{(x)} - H_1 h = M^0_{(x)} - 3H; \quad 0 \le x \le 6 \text{ m}$$

$$\therefore \frac{\partial U}{\partial H_1} = 2\int_0^3 (-H_1 x)(-x)\frac{dx}{EI} + \int_0^6 \left(M^0_{(x)} - 3H_1\right)(-3)\frac{dx}{2EI} = 0$$

$$= \frac{2H_1}{EI}\left(\frac{x^3}{3}\right)_0^3 - \frac{3}{2EI}\int_0^6 M^0_{(x)} \cdot dx + \frac{9H_1}{2EI} \times 6 = 0$$

$$=> 45H_1 = \frac{3}{2}\int_0^6 M^0_{(x)} \cdot dx = \frac{3}{2}\left[\frac{2}{3} \times 6 \times \left(\frac{100 \times 6^2}{8}\right)\right]$$

$$\therefore H_1 = 60 \text{ kN} \Rightarrow 2H = 60 \text{ kN} \Rightarrow H = 30 \text{ kN}$$

Check

$$H_1 = \frac{A_b^0}{Lh\left[1 + \frac{2}{3} \cdot \propto\right]}$$

$$A_b^0 = \frac{2}{3} \times 6 \times 450$$

$$= 1{,}800 \text{ kNm}^2$$

$$\propto = \left(\frac{EI_b/L}{EI_c/h}\right) = \frac{2/6}{1/3} = 1$$

$$\therefore H_1 = \frac{1{,}800}{6 \times 3\left[1 + \frac{2}{3} \times 1\right]} = 60 \text{ kN}$$

The FBD, axial and shear force diagrams, and bending moment diagram are shown in Figure 4.59.

4.5 APPLICATION TO PIN-JOINTED PLANE FRAMES (TRUSSES)

The procedure for analysing the truss using the theorem of least work is outlined below.

i. Similar to beams and rigid-jointed plane frames, the first step in analysing pin-jointed plane frames (trusses) is to determine the static indeterminacy of the truss, as explained in Chapter 1.
ii. Next, identify the redundants. If the truss is internally indeterminate, select any member force as a redundant. If the indeterminacy is external, choose the support reactions as redundants. Then, draw the basic or determinate truss with the external loads and redundants.
iii. Using the theorem of least work, write the compatibility equations using

$$\frac{\partial U}{\partial R_i} = \sum_j (F_j) \times \left(\frac{\partial F_j}{\partial R_i}\right) \times \frac{L_j}{A_j E} = 0 \quad \text{for } i = 1, 2, 3, \ldots, I_s$$

where F_j is the axial force in the members of the truss due to the external loads and redundants.
iv. Determine the forces in the truss members using either the method of joints or the tension coefficient method. The method of sections can also be employed. Once the member forces are expressed in terms of the external loads and redundants, the partial derivative terms can be easily derived.
v. Substitute the above values, along with the lengths and axial rigidities of the members, into the equation to generate a set of simultaneous equations. Solve these equations to determine the redundants.
vi. Once the redundants are obtained, the final member forces can be calculated by substituting these values into the member forces derived earlier in step (iv), expressed in terms of loads and redundants. This entire process can be conveniently summarised in a tabular format, as illustrated in the following examples.
vii. If there is any lack of fit in the truss or yielding of supports, as mentioned earlier, the theorem of least work is not applicable. However, it can be analysed by using other strain energy-based principles, such as Castigliano's second theorem. This is demonstrated in Example 4.19.

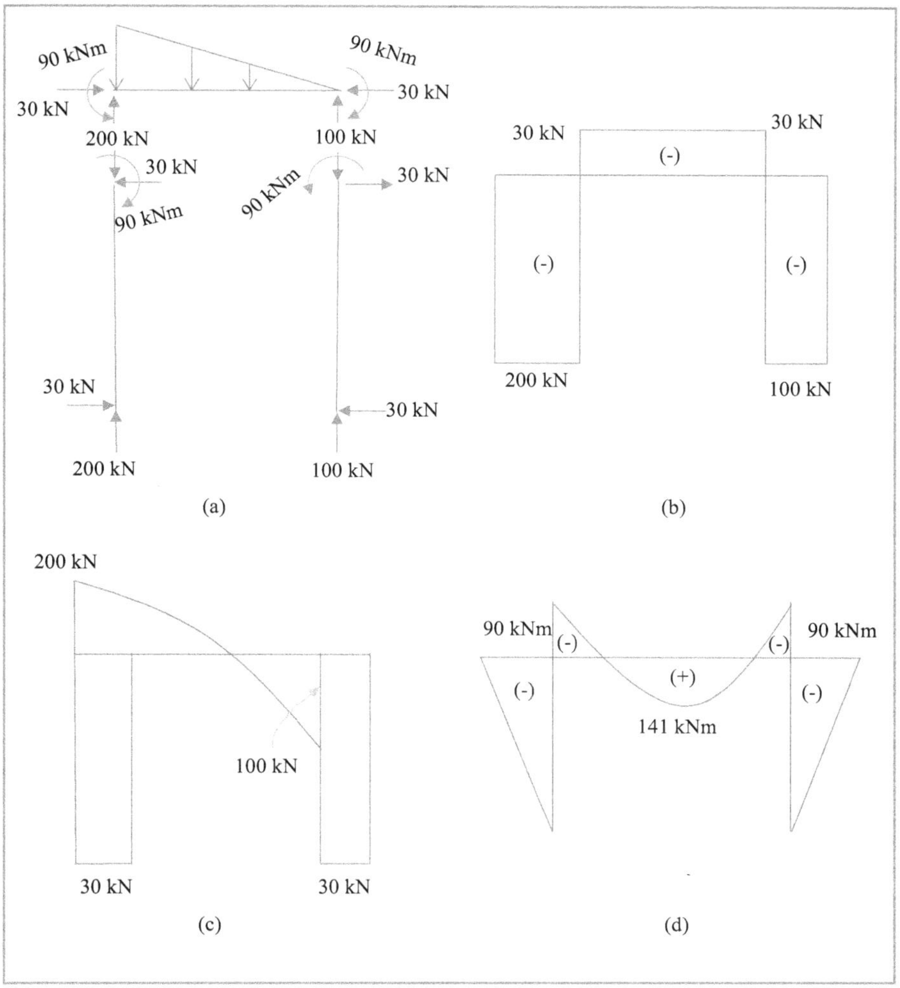

FIGURE 4.59 (a) Free body diagram, (b) axial force diagram, (c) shear force diagram, and (d) bending moment diagram.

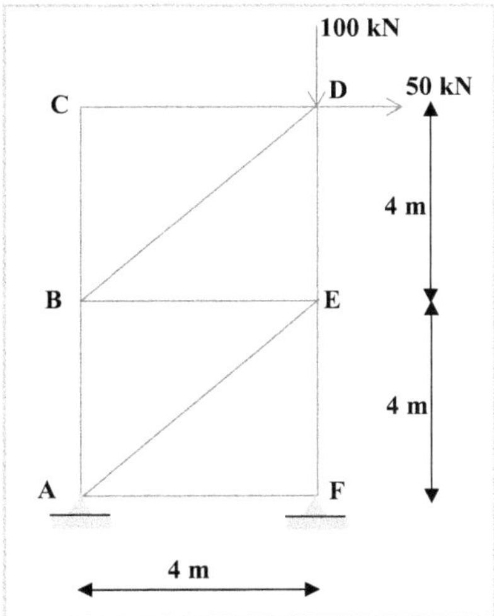

FIGURE 4.60 Indeterminate truss.

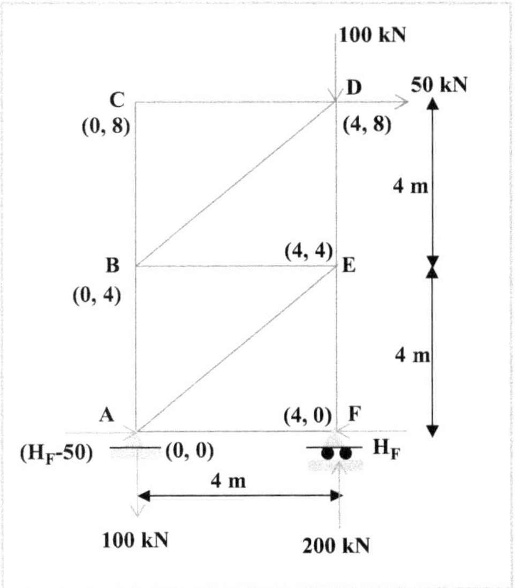

FIGURE 4.61 Basic truss with external loads and redundants.

Example 4.16

Analyse the pin-jointed plane frame shown in Figure 4.60, using the theorem of least work. Assume the axial rigidity of all members is the same.

Solution:

The static indeterminacy, $I_s = m - 2j + R = 9 - 2\times6 + 4 = 1$, i.e., the truss is statically externally indeterminate to the first degree. Let us select the horizontal reaction at *H*, i.e., H_F, as redundant. Now apply the theorem of least work

$$\frac{\partial U}{\partial H_F} = \sum_{i=1}^{n} F_i \frac{\partial F_i}{\partial H_F} \cdot \frac{L_i}{AE} = 0 \qquad (4.41)$$

One can use either the method of joints or the tension coefficient method to calculate the member forces due to the loads and redundant. Let us adopt the tension coefficient method to calculate the member forces F_i in the truss in this example (Figure 4.61).

JOINT *F*

$$\sum F_x = 0 \Rightarrow t_{FA}(0-4) + t_{FE}(4-4) - H_F = 0$$

$$t_{FA} = -\frac{H_F}{4} \Rightarrow F_{FA} = -4\times\frac{H_F}{4} = -H_F$$

$$\sum F_y = 0 \Rightarrow t_{FA}(0-0) + t_{FE}(4-0) + 200 = 0$$

$$t_{FE} = -50 \Rightarrow F_{FE} = 4\times(-50) = -200 \text{ kN}$$

JOINT *A*

$$\sum F_x = 0 \Rightarrow t_{FA}(4-0) + t_{AE}(4-0) + t_{AB}(0-0) + (H_F - 50) = 0$$

$$\Rightarrow t_{AE} = \frac{-H_F + 50 - 4(-H_F/4)}{4} = 12.5$$

$$\Rightarrow F_{AE} = +70.75 \text{ kN}$$

$$\sum F_y = 0 \Rightarrow t_{FA}(0-0) + t_{AE}(4-0) + t_{AB}(4-0) - 100 = 0$$

$$\Rightarrow t_{AB} = \frac{100 - 4(12.5)}{4} = 12.5$$

$$\Rightarrow F_{AB} = 50 \text{ kN}$$

TABLE 4.14
Member forces and their partial derivatives

Member	Length	F_i	$\frac{\partial U}{\partial H_F}$	$F_i \frac{\partial F_i}{\partial H_F} \cdot \frac{L}{AE}$
AB	4/AE	50	0	0
BC	4/AE	0	0	0
CD	4/AE	0	0	0
BD	5.66/AE	70.7	0	0
BE	4/AE	−50	0	0
DE	4/AE	−150	0	0
FA	4/AE	$-H_F$	−1	$4H_F/AE$
FE	4/AE	−200	0	0
AE	5.66/AE	70.75	0	0

JOINT *B*

$$\sum F_x = 0 \Rightarrow t_{BE}(4-0) + t_{BD}(4-0) = 0$$

$$\Rightarrow t_{BE} = -t_{BD} = -12.5$$

$$\therefore F_{BE} = -4\times12.5 = -50 \text{ kN}$$

JOINT *C*

$$\sum F_x = 0 \Rightarrow t_{CB}(0-0) + t_{CD}(4-0) = 0$$

$$t_{CD} = 0 \Rightarrow F_{CD} = 0$$

$$\sum F_y = 0 \Rightarrow t_{CB}(4-8) + t_{CD}(8-8) = 0$$

$$\Rightarrow t_{CB} = 0$$

$$\Rightarrow F_{CB} = 0$$

JOINT *D*

$$\sum F_x = 0 \Rightarrow t_{BD}(0-4) + t_{DE}(4-4) + 50 = 0$$

$$\Rightarrow t_{BD} = 12.5 \Rightarrow F_{BD} = 5.66(12.5) = 70.7 \text{ kN}$$

$$\sum F_y = 0 \Rightarrow t_{BD}(4-8) + t_{DE}(4-8) - 100 = 0$$

$$\Rightarrow t_{DE} = -37.5$$

$$\Rightarrow F_{DE} = -150 \text{ kN}$$

The member forces and their partial derivatives are presented in Table 4.14. The product of $F_i \dfrac{\partial F_i}{\partial H_F} \cdot \dfrac{L_i}{AE}$ is also shown in the table. The summation is also shown at the bottom of the table, required in the equation to solve the unknown.

From equation 4.41 and using Table 4.14,

$$\frac{\partial U}{\partial H_F} = \frac{4H_F}{AE} = 0 \Rightarrow H_F = 0$$

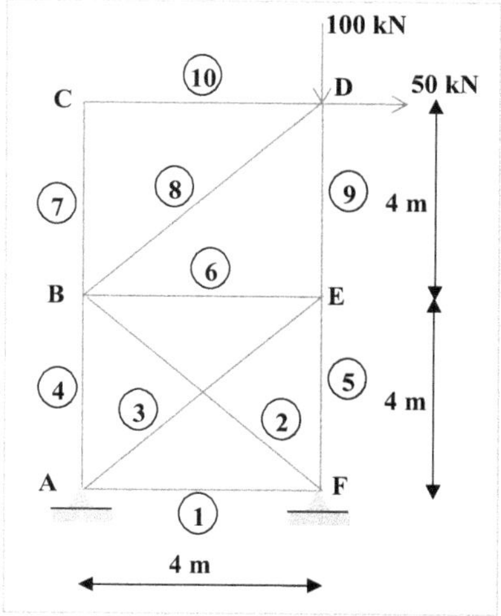

FIGURE 4.62 Indeterminate truss.

FINAL MEMBER FORCES

After obtaining H_F, the final member forces can be easily calculated by substituting the value of H_F in column 3 of Table 4.14. The final bar forces are F_{AB}=50 kN; F_{BC}=F_{CD}=0; F_{BD}=+70.7 kN; F_{BE}=−50 kN; F_{DE}=−150 kN; F_{FA}=0 kN; F_{FE}=−200 kN; F_{AE}=70.75 kN.

Example 4.17

Re-analyse the pin-jointed plane frame shown in Example 4.16, by adding one more member *BF* in the lower panel (Figure 4.62). Assume *AE* is constant.

Solution:

The total static indeterminacy, $I_s = m - 2j + R = 10 - 2\times 6 + 4 = 2$ and external indeterminacy, $I_{se} = R - 3 = 4 - 3 = 1$. Therefore, the internal indeterminacy, $I_{si} = 2 - 1 = 1$. Let us select the horizontal support reaction at *F* (H_2), and the force in member ②, i.e., F_2, as the redundants, as shown in Figure 4.63.

Based on the theorem of least work, the compatibility equations can be written as

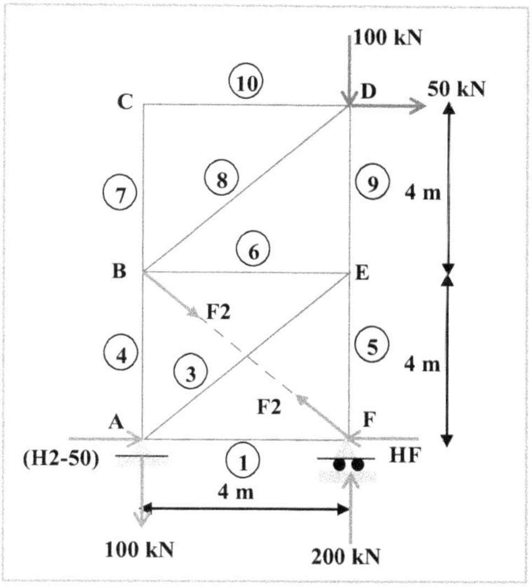

FIGURE 4.63 Basic truss with external loads and redundants.

$$\frac{\partial U}{\partial H_2} = \sum F_i \frac{\partial F_i}{\partial H_2} \cdot \left(\frac{L}{AE}\right)_i = 0 \qquad (4.42)$$

$$\frac{\partial U}{\partial F_2} = \sum F_i \frac{\partial F_i}{\partial F_2} \cdot \left(\frac{L}{AE}\right)_i = 0 \qquad (4.43)$$

Now calculate the member forces using the method of joints (Figure 4.64).

At joint 'C': $F_7 = F_{10} = 0$

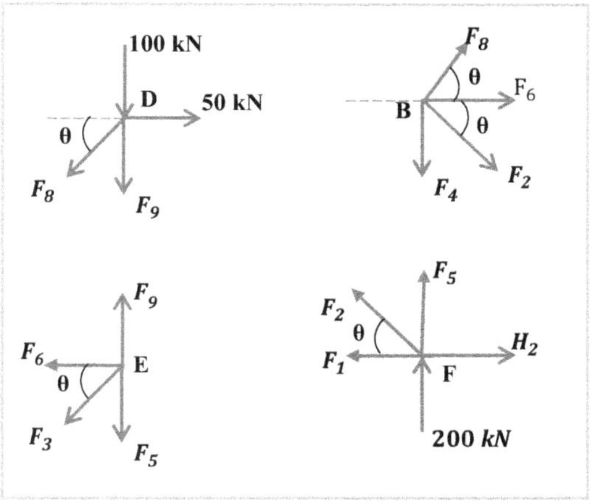

FIGURE 4.64 Free body diagrams of joints.

AT JOINT *D*

$$\sum F_x = 0 \Rightarrow F_8 = \frac{50}{\cos\theta} = 70.75 \text{ kN}$$

TABLE 4.15
Member forces, partial derivatives terms and the flexibility coefficients

Member No	$\left(\frac{L}{AE}\right)_i$	F_i	$\frac{\partial F_i}{\partial H_2}$	$\frac{\partial F_i}{\partial F_2}$	$F_i\frac{\partial F_i}{\partial H_2}\cdot\left(\frac{L}{AE}\right)_i$	$F_i\frac{\partial F_i}{\partial F_2}\cdot\left(\frac{L}{AE}\right)_i$
1	4/AE	$-H_2-0.707F_2$	−1.0	−0.7.7	$\frac{1}{AE}(4H_2+2.83F_2)$	$\frac{1}{AE}(2.83H_2+2F_2)$
2	5.66/AE	F_2	0	1	0	$5.66F_2/AE$
3	5.66/AE	$0.707+F_2$	0	1	0	$(400+5.66F_2)/AE$
4	4/AE	$50-0.707F_2$	0	−0.707	0	$(-141.4+2F_2)/AE$
5	4/AE	$-200-0.707F_2$	0	−0.707	0	$(565.6+2F_2)/AE$
6	4/AE	$-50-0.707F_2$	0	−0.707	0	$(141.4+2F_2)/AE$
7	4/AE	0	0	0	0	0
8	5.66/AE	70.75	0	0	0	0
9	4/AE	−150	0	0	0	0
10	4/AE	0	0	0	0	0

$$\sum F_y = 0 \Rightarrow F_9 = -100 - F_8\sin\theta = -150 \text{ kN}$$

AT JOINT *B*

$$\sum F_x = 0 \Rightarrow F_6 = -(F_8\cos\theta + F_2\cos\theta)$$

$$= -0.707F_8 - 0.707F_2 = -50 - 0.707F_2$$

$$\sum F_y = 0 \Rightarrow F_4 = (F_8\sin\theta - F_2\sin\theta) = 50 - 0.707F_2$$

AT JOINT *E*

$$\sum F_x = 0 \Rightarrow F_3 = \frac{-F_6}{\cos\theta} = 70.75 + F_2$$

$$\sum F_y = 0 \Rightarrow F_5 = F_9 - F_3\sin\theta$$

$$= -150 - (70.75 + F_2)\times\frac{4}{5.66} = -200 - 0.707F_2$$

AT JOINT *F*

$$\sum F_x = 0 \Rightarrow F_1 = H_2 - F_2\cos\theta = -H_2 - 0.707F_2$$

The member forces, derivative terms, and the flexibility coefficients $\left(\frac{L}{AE}\right)$, are shown in Table 4.15. The summation of $F_i\frac{\partial F_i}{\partial H_2}\cdot\left(\frac{L}{AE}\right)_i$ and $F_i\frac{\partial F_i}{\partial F_2}\cdot\left(\frac{L}{AE}\right)_i$ are shown at the bottom of the table to use in equations 4.42 and 4.43.

$$\Sigma = \frac{1}{AE}(4H_2 + 2.83F_2);\quad \Sigma = \frac{1}{AE}(2.83H_2 + 19.32F_2 + 965.6)$$

Using Table 4.15, equations 4.42 and 4.43 can be written as

$$4H_2 + 2.83F_2 = 0 \tag{4.42a}$$

$$2.83H_2 + 19.32F_2 = -965.6 \tag{4.43a}$$

By solving equations 4.42a and 4.43a, we obtain

$$H_2 = 39.44 \text{ kN};\quad F_2 = -55.75 \text{ kN}$$

FINAL MEMBER FORCES

The final member forces can be calculated by substituting the above known values of H_2 and F_2 in column 3 of the above Table 4.15. Therefore, the member forces are

$$F_1 = 0 \quad F_6 = -10.58 \text{ kN}$$

$$F_2 = -\ 55.75 \text{ kN} \quad F_7 = 0 \text{ kN}$$

$$F_3 = 15 \text{ kN} \quad F_8 = 70.75 \text{ kN}$$

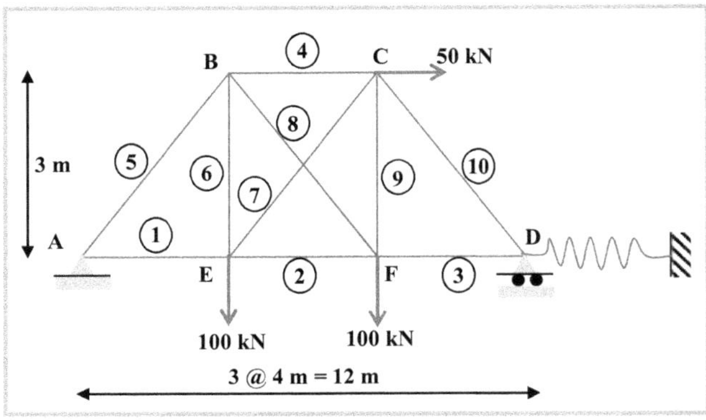

FIGURE 4.65 Indeterminate truss.

$F_4 = 89.41\text{ kN} \quad F_9 = -150\text{ kN}$

$F_5 = -160.58\text{ kN} \quad F_{10} = 0\text{ kN}$

Example 4.18

Analyse the pin jointed plane frame shown in Figure 4.65, using the theorem of least work. Assume the flexibility of all top and bottom chord members equal to 15 mm/MN, and web members equal to 10 mm/MN.

Solution:

The total static indeterminacy, I_s=10−2×6+4=2 and external indeterminacy I_{se}=R−3=4−3=1. Therefore, the internal indeterminacy, I_{si}=2−1=1. Let us select the force in the spring (R_1), and the force in member (F_8), as redundants. The primary truss with the external loads and the redundants is shown in Figure 4.66.

Using the theorem of least work, the compatibility conditions can be expressed as

$$\frac{\partial U}{\partial R_1} = \sum F_i \frac{\partial F_i}{\partial R_1}\left(\frac{L}{AE}\right)_i + \frac{\partial U_{SP}}{\partial R_1} = 0 \qquad (4.44)$$

$$\frac{\partial U}{\partial F_8} = \sum F_i \frac{\partial F_i}{\partial F_8}\left(\frac{L}{AE}\right)_i + \frac{\partial U_{SP}}{\partial F_8} = 0 \qquad (4.45)$$

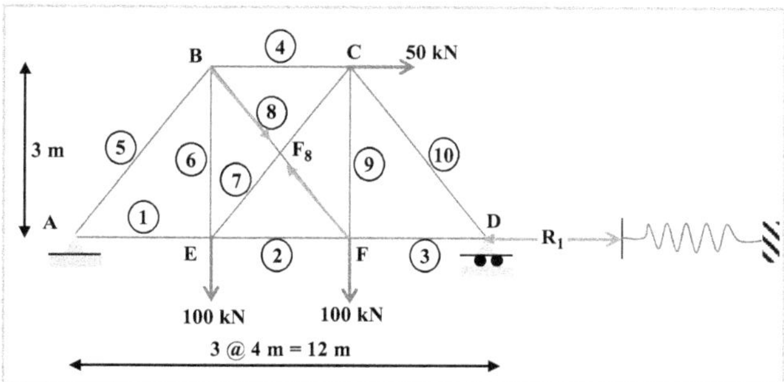

FIGURE 4.66 Basic truss with loads and redundants.

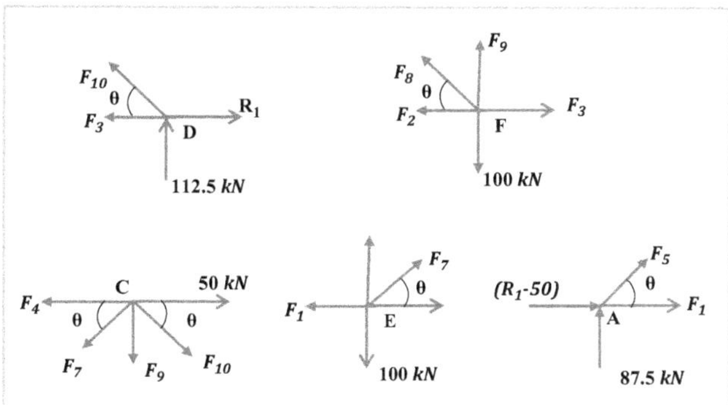

FIGURE 4.67 Free body diagrams of joints.

The FBDs of various joints for the calculation of member forces due to the external loads and the redundants using the method of joints are shown in Figure 4.67.

JOINT *D*

Assume all tensile forces are positive (+ve).

$$\sum F_y = 0 \Rightarrow F_{10} = -\frac{112.5}{\sin\theta} = -187.5\text{ kN}$$

$$\sum F_x = 0 \Rightarrow F_3 = -R_1 - F_{10}\cos\theta = -R_1 + 150$$

JOINT *F*

$$\sum F_x = 0 \Rightarrow F_2 = F_3 - F_8\cos\theta$$

$$= -R_1 + 150 - F_8 \times \frac{4}{5}$$

$$= -R_1 + 150 - 0.8F_8$$

$$\sum F_y = 0 \Rightarrow F_9 = 100 - F_8\sin\theta = 100 - 0.6F_8$$

JOINT *C*

$$\sum F_y = 0 \Rightarrow F_7 = \frac{-F_9 - F_{10}\sin\theta}{\sin\theta}$$

$$= \frac{-100 + 0.6F_8 + 112.5}{3/5} = 20.83 + F_8$$

$$\sum F_x = 0 \Rightarrow F_4 = 50 + F_{10}\cos\theta - F_7\cos\theta$$

$$= 50 - 150 - 16.66 - 0.8F_7$$

$$= -116.66 - 0.8F_8$$

JOINT *E*

$$\sum F_x = 0 \Rightarrow F_1 = F_2 + F_7\cos\theta$$

$$= -R_1 + 150 - 0.8F_8 + 16.66 + 0.8F_8$$

$$= -R_1 + 166.66$$

$$\sum F_y = 0 \Rightarrow F_6 = 100 - F_7\sin\theta$$

$$= 100 - 12.49 - 0.6F_8$$

$$= 87.51 - 0.6F_8$$

JOINT *A*

$$\sum F_y = 0 \Rightarrow F_5 = -\frac{87.5}{\sin\theta} = -145.83\text{ kN}$$

TABLE 4.16
Member forces, and their derivative terms, and the flexibility coefficients

Member No.	Flexibility $f = \left(\frac{L}{AE}\right) \times 10^{-6}$ mm/kN	Force (F_i)	$\frac{\partial F_i}{\partial R_1}$	$\frac{\partial F_i}{\partial F_8}$	$f \cdot F_i \frac{\partial F_i}{\partial R_1}$	$f \cdot F_i \frac{\partial F_i}{\partial F_8}$
1	15	$166.66 - R_1$	−1	0	$(-2{,}500 \times 15R_1) \times 10^{-6}$	0
2	15	$150 - R_1 - 0.8F_8$	−1	−0.8	$(-2{,}250 \times 15R_1 + 12F_8) \times 10^{-6}$	$-1{,}800 + 12R_1 + 9.6F_8$
3	15	$150 - R_1$	−1	0	$(-2{,}250 \times 15R_1) \times 10^{-6}$	0
4	15	$-116.66 - 0.8F_8$	0	−0.8	0	$1{,}400 + 9.6F_8$
5	10	−145.83	0	0	0	0
6	10	$87.51 - 0.6F_8$	0	−0.6	0	$-525.06 + 3.6F_8$
7	10	$20.83 + F_8$	0	1	0	$208.3 + 10F_8$
8	10	F_8	0	1	0	$10F_8$
9	10	$-100 - 0.6F_8$	0	−0.6	0	$-600 + 3.6F_8$
10	10	−187.5	0	0	0	0

The above member forces, and their derivative terms, and the flexibility coefficients are shown in Table 4.16. The summation of the product of $F_i \frac{\partial F_i}{\partial R_1}\left(\frac{L}{AE}\right)_i$ and $F_i \frac{\partial F_i}{\partial F_8} \cdot F_i$ are also shown at the bottom of the table for using these in expressions in equations 4.44 and 4.45.

From the table

$$\sum f \cdot F_i \frac{\partial F_i}{\partial R_1} = (-7{,}000 + 45R_1 + 12F_8) \times 10^{-6}$$

$$\sum f \cdot F_i \frac{\partial F_i}{\partial F_8} = (-1{,}316.76 + 12R_1 + 46.4F_8) \times 10^{-6}$$

By substituting the above in equations 4.44 and 4.45, we obtain

$$-7{,}000 + 45R_1 + 12F_8 = 0 \Rightarrow 3.75R_1 + F_8 = 583.33 \quad (4.44a)$$

The SE in the spring $U_{SP} = \frac{1}{2}Kx^2 = \frac{1}{2}K\left(\frac{P}{K}\right)^2 = \frac{1}{2}R_1^2 \cdot \frac{1}{K}$

$$\frac{\partial U}{\partial R_1} = \frac{R_1}{K} = \frac{R_1}{8 \times 10^3}$$

∴ Equation 4.45 becomes

$$(12R_1 + 46.4F_8 - 1{,}316.76) \times 10^{-6} + \frac{R_1}{8 \times 10^3} = 0$$

$$1.096R_1 + 0.371F_8 = 10.53$$

$$2.95R_1 + F_8 = 28.39 \quad (4.45a)$$

By solving equations 4.44a and 4.45a, we get

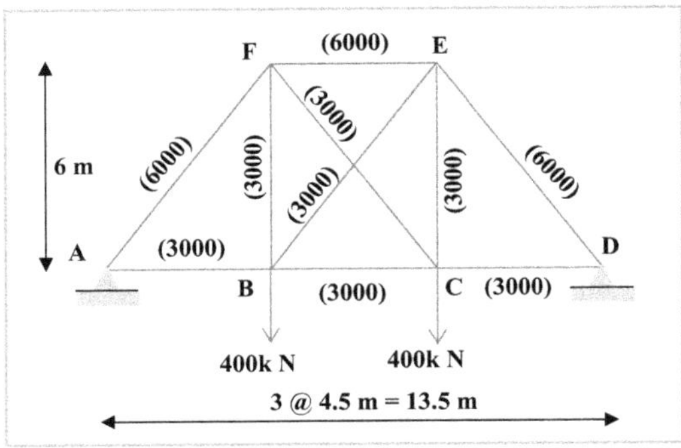

FIGURE 4.68 Indeterminate truss.

$$R_1 = 209.84 \text{ kN}$$

$$F_8 = -590.67 \text{ kN}$$

By substituting the known values R_1 and F_8 in column 3 of the above table, we can calculate the net member forces and these are shown below.

$F_1 = -43.18$ kN $F_2 = 412.69$ kN

$F_3 = -59.84$ kN $F_4 = 355.87$ kN

$F_5 = -145.83$ kN $F_6 = 441.91$ kN

$F_7 = -569.84$ kN $F_8 = -590.67$ kN

$F_9 = 250.40$ kN $F_{10} = -187.5$ kN

Example 4.19

A pin jointed plane frame is shown in Figure 4.68. The area of c/s of the members is shown in parentheses in mm². Analyse the frame in each case separately.

i. Due to external loads.
ii. If the member *BE* is too long in length by 3 mm. Assume $E=2\times10^5$ N/mm².
iii. If the member *BE* undergoes a rise of temperature of 36.4°C; Assume $\alpha=11\times10^{-6}$/°C.
iv. If the pin-jointed plane frame shown in Figure 4.68 undergoes a rise of temperature of 40°C.

Solution:

The static indeterminacy, $I_s=10-2\times6+4=2$. Let us select the horizontal reaction at *D* ($H_D=R_1$) and force in member *BE* ($F_{BE}=R_2$) as redundant forces. The primary truss with the external loads, and the redundants are shown in Figure 4.69.

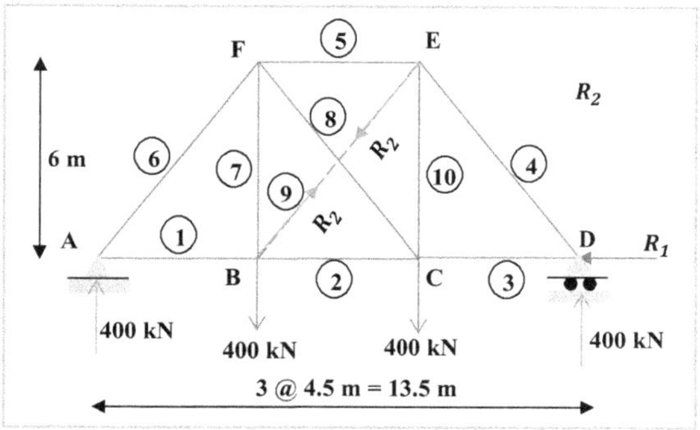

FIGURE 4.69 Primary truss with external loads and redundant forces.

$$\cos\theta=\frac{4.5}{7.5}=0.6$$

$$\sin\theta=\frac{6}{7.5}=0.8$$

∴ From the theorem of least work, we can write the compatibility conditions as

$$\frac{\partial U}{\partial R_1}=\sum_{i=1}^{n}F_i\frac{\partial F_i}{\partial R_1}\cdot\left(\frac{L}{AE}\right)_i=0 \qquad (4.46)$$

$$\frac{\partial U}{\partial R_2}=\sum_{i=1}^{n}F_i\frac{\partial F_i}{\partial R_2}\cdot\left(\frac{L}{AE}\right)_i=0 \qquad (4.47)$$

Using the method of joints, let us calculate the member forces in the basic truss due to the external loads, and the redundants, and arrange them in Table 4.17. The partial derivatives of the member forces are also shown in the table. The summation of the product of $F_i\frac{\partial F_i}{\partial R_1}\cdot\left(\frac{L}{AE}\right)$ and $F_i\frac{\partial F_i}{\partial R_2}\cdot\left(\frac{L}{AE}\right)$ are also shown at the bottom of the table.

By substituting the above summations in equations 4.46 and 4.47,

$$\frac{\partial U}{\partial R_1}=(22.5R_1+4.5R_2-6,750)\times10^{-3}=0$$

$$\Rightarrow 5R_1+R_2=1,500 \qquad (4.46a)$$

$$\frac{\partial U}{\partial R_2}=(4.5R_1+41.85R_2-7,075)\times10^{-3}=0$$

$$\Rightarrow 0.107R_1+R_2=169.00 \qquad (4.47a)$$

TABLE 4.17
Member forces due to loads and redundants, and their derivative terms, and the flexibility coefficients

Member No.	$\left(\frac{L}{AE}\right)_i$ mm/kN	F_i (kN)	$\frac{\partial F_i}{\partial R_1}$	$\frac{\partial F_i}{\partial R_2}$	$F_i\frac{\partial F_i}{\partial R_1}\cdot\left(\frac{L}{AE}\right)_i$	$F_i\frac{\partial F_i}{\partial R_2}\cdot\left(\frac{L}{AE}\right)_i$
1	7.5×10^{-3}	$300-R_1$	1	0	$(7.5R_1-2,250)10^{-3}$	0
2	7.5×10^{-3}	$300-R_1-0.6R_2$	−1	−0.6	$(7.5R_1-2,250+4.5R_2)10^{-3}$	$(-1,350+4.5R_1+2.7R_2)10^{-3}$
3	7.5×10^{-3}	$300-R_1$	−1	0	$(7.5R_1-2,250)10^{-3}$	0
4	6.25×10^{-3}	−500	0	0	0	0
5	3.75×10^{-3}	$-300-0.6R_2$	0	−0.6	0	$(675+1.35R_2)10^{-3}$
6	6.25×10^{-3}	−500	0	0	0	0
7	10×10^{-3}	$400-0.8R_2$	0	−0.8	0	$(6.4R_2-3,200)10^{-3}$
8	12.5×10^{-3}	R_2	0	1	0	$(12.5R_2)10^{-3}$
9	12.5×10^{-3}	R_2	0	1	0	$(12.5R_2)10^{-3}$
10	10×10^{-3}	$400-0.8R_2$	0	−0.8	0	$(6.4R_2-3,200)10^{-3}$
Σ					$(22.5R_1+4.5R_2-6,750)\times10^{-3}$	$(4.5R_1+41.85R_2-7,075)\times10^{-3}$

By solving equations 4.46a and 4.47a, we found $R_1 = 272$ kN and $R_2 = 139.89$ kN.

FINAL MEMBER END FORCES

The final member forces can be easily calculated by substituting the known values of R_1 and R_2 in column (3) of Table 4.17, which are

$$F_1 = F_3 = 28 \text{ kN}; \quad F_2 = -55.93 \text{ kN}; \quad F_4 = F_6 = -500 \text{ kN}$$

$$F_7 = F_{10} = 288.16 \text{ kN}; \quad F_5 = -383.88 \text{ kN}; \quad F_8 = F_9 = 139.8 \text{ kN}$$

CASE (II)

If the member *BE* is too long by 5 mm.

From Castigliano's second theorem, we can write

$$\frac{\partial U}{\partial R_1} = \sum F_i \frac{\partial F_i}{\partial R_1}\left(\frac{L}{AE}\right)_i = 0 \qquad (4.48)$$

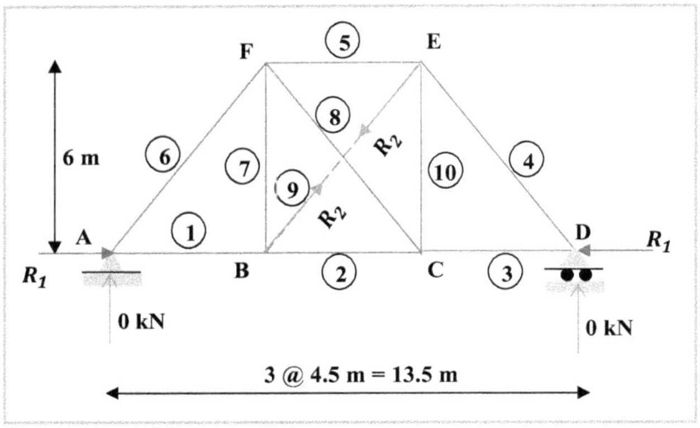

FIGURE 4.70 Primary truss with external loads and redundants.

$$\frac{\partial U}{\partial R_2} = \sum F_i \frac{\partial F_i}{\partial R_2}\left(\frac{L}{AE}\right)_i = -3 \qquad (4.49)$$

The support reactions for the calculation of member forces are shown in Figure 4.70.

The calculation of F_i and its derivatives are shown in Table 4.18.

By substituting the values from the table into equations 4.48 and 4.49, we get

$$\frac{\partial U}{\partial R_1} = (22.5R_1 + 4.5R_2) \times 10^{-3} = 0$$

$$\Rightarrow 5R_1 + R_2 = 0 \qquad (4.48a)$$

$$\frac{\partial U}{\partial R_2} = (4.5R_1 + 41.85R_2) \times 10^{-3} = -3$$

$$\Rightarrow 0.107R_1 + R_2 = -71.68 \qquad (4.49a)$$

By solving equations 4.48a and 4.49a, we obtain

$$R_1 = 14.65 \text{ kN}$$

$$R_2 = -73.24 \text{ kN}$$

FINAL MEMBER END FORCES

We can obtain the final member forces by substituting the known values of R_1 and R_2 in column (3) of Table 4.18, which are

$$F_1 = F_3 = -14.65 \text{ kN}; \quad F_2 = 29.29 \text{ kN}; \quad F_4 = F_6 = 0 \text{ kN}$$

$$F_5 = 43.94 \text{ kN}; \quad F_7 = F_{10} = 58.59 \text{ kN}; \quad F_8 = F_9 = -73.24 \text{ kN}$$

TABLE 4.18
Calculation of F_i, and its derivatives and flexibility coefficients

Member No.	$\left(\frac{L}{AE}\right) \times 10^{-3}$ mm	F_i (kN)	$\frac{\partial F_i}{\partial R_1}$	$\frac{\partial F_i}{\partial R_2}$	$F_i \frac{\partial F_i}{\partial R_1} \cdot \left(\frac{L}{AE}\right) \times 10^{-3}$ kN	$F_i \frac{\partial F_i}{\partial R_2} \cdot \left(\frac{L}{AE}\right) \times 10^{-3}$kN
1	7.5	$-R_1$	1	0	$7.5R_1$	0
2	7.5	$-R_1 - 0.6R_2$	−1	−0.6	$7.5R_1 + 4.5R_2$	$4.5R_1 + 2.7R_2$
3	7.5	$-R_1$	−1	0	$7.5R_1$	0
4	6.25	0	0	0	0	0
5	3.75	$-0.6R_2$	0	−0.6	0	$1.35R_2$
6	6.25	0	0	0	0	0
7	10	$-0.8R_2$	0	−0.8	0	$6.4R_2$
8	12.5	R_2	0	1	0	$12.5R_2$
9	12.5	R_2	0	1	0	$12.5R_2$
10	10	$-0.8R_2$	0	−0.8	0	$6.4R_2$
Σ					$(22.5R_1 + 4.5R_2) \times 10^{-3}$	$(4.5R_1 + 41.85R_2) \times 10^{-3}$

Case (iii)

If the member BE undergoes a temperature rise of 36.4°C and $\alpha = 11 \times 10^{-6}$/°C.

Free expansion of the member due to temperature rise $= L\alpha\Delta T$

$$= (7.5 \times 10^3) \times 11 \times 10^{-6} \times 36.4 = 3.003 \text{ mm}$$

∴ The compatibility conditions are

$$\frac{\partial U}{\partial R_1} = \sum F_i \frac{\partial F_i}{\partial R_1}\left(\frac{L}{AE}\right)_i = 0 \quad (4.50)$$

$$\frac{\partial U}{\partial R_2} = \sum F_i \frac{\partial F_i}{\partial R_2}\left(\frac{L}{AE}\right)_i = -3.003 \text{ mm} \quad (4.51)$$

The value of F_i, $\frac{\partial U}{\partial R_1}$, $\frac{\partial U}{\partial R_2}$, and $\frac{L}{AE}$ are the same as those obtained in case (ii), as there are no external loads, and the redundants are the same. Hence,

From equations 4.50 and 4.51, we can write

$$\therefore 5R_1 + R_2 = 0 \quad (4.50a)$$

$$(4.5R_1 + 41.85R_2) \times 10^{-3} = -3.003 \quad (4.51a)$$

$$\Rightarrow 0.107R_1 + R_2 = -71.75$$

By solving the above equations 4.50a and 4.51a, we find

$$\therefore R_1 = 14.65 \text{ kN} \quad \text{and} \quad R_2 = -73.24 \text{ kN}$$

Since the obtained values of R_1 and R_2 in this case are the same as those obtained in case (ii), the final member forces are also the same as those obtained in case (ii). Therefore,

$F_1 = F_3 = -14.65$ kN; $F_2 = 29.29$ kN; $F_4 = F_6 = 0$ kN;
$F_7 = F_{10} = 58.59$ kN; $F_5 = 43.94$ kN; $F_8 = F_9 = -73.24$ kN

Case (iv)

If the pin jointed plane frame undergoes a rise of temperature of 40°C.

A uniform change in temperature does not produce internal forces in the members of a truss, even if the truss is statically indeterminate internally. However, if the truss is externally indeterminate, it develops internal forces in its members. In the present case, since the truss is statically indeterminate externally, the uniform change in temperature will induce internal forces in the truss. The free thermal expansion of the bottom chord $AD = L\alpha\Delta T = 13{,}500 \times 40 \times 11 \times 10^{-6} = 5.94$ mm.

∴ The strain energy equations are

$$\frac{\partial U}{\partial R_1} = \sum F_i \frac{\partial F_i}{\partial R_1}\left(\frac{L}{AE}\right)_i = 5.94 \text{ mm} \quad (4.52)$$

$$\frac{\partial U}{\partial R_2} = \sum F_i \frac{\partial F_i}{\partial R_2}\left(\frac{L}{AE}\right)_i = 0 \quad (4.53)$$

Since the redundants are the same as in the previous cases, there is no change in F_i, $\frac{\partial U}{\partial R_1}$, $\frac{\partial U}{\partial R_2}$.

$$\frac{\partial U}{\partial R_1} = (22.5R_1 + 4.5R_2) \times 10^{-3} = 5.94$$

$$\Rightarrow 5R_1 + R_2 = 1{,}320 \quad (4.52a)$$

$$\frac{\partial U}{\partial R_2} = (4.5R_1 + 41.85R_2) \times 10^{-3} = 0$$

$$\Rightarrow 0.107R_1 + R_2 = 0 \quad (4.53a)$$

By solving equations 4.52a and 4.53a, we obtain

$$R_1 = 269.7 \text{ kN}$$

$$R_2 = -28.9 \text{ kN}$$

Final Member End Forces

$F_1 = F_3 = -269.7$ kN; $F_2 = -252.36$ kN; $F_4 = F_6 = 0$ kN

$F_7 = F_{10} = 23.12$ kN; $F_5 = 17.34$ kN; $F_8 = F_9 = -28.9$ kN

4.6 Problems for Practice

4.6.1 Can we apply the theorem of least work to statically indeterminate problems subjected to yielding of supports? Discuss.

4.6.2 Analyse the non-prismatic beam shown in Figure 4.71 using the theorem of least work.

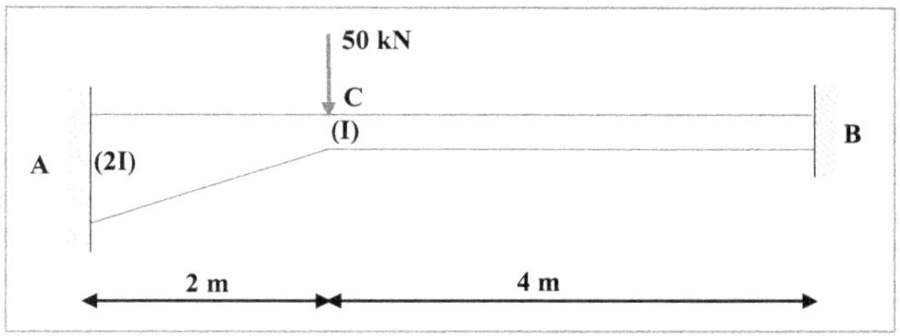

FIGURE 4.71

Analyse the continuous beam shown in Figure 4.72, using the theorem of least work. Assume EI is constant.

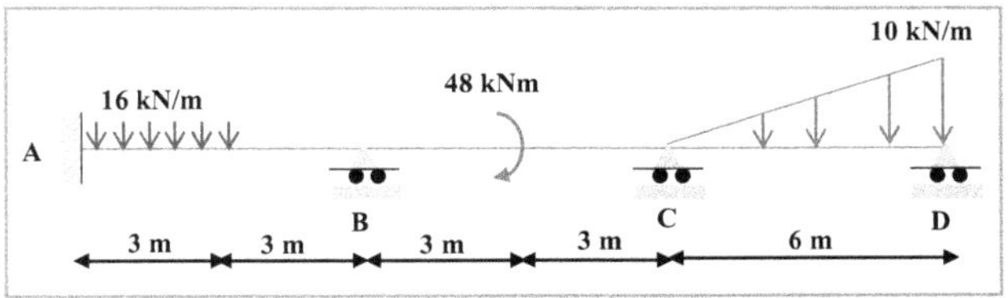

FIGURE 4.72

4.6.3 A continuous beam *ABCD*, which has an internal hinge at the middle of the span *BC*, is subjected to loading as shown in Figure 4.73. Analyse the beam using the Theorem of least work. Draw the shear force and bending moment diagrams. Also, draw the probable deflected shape.

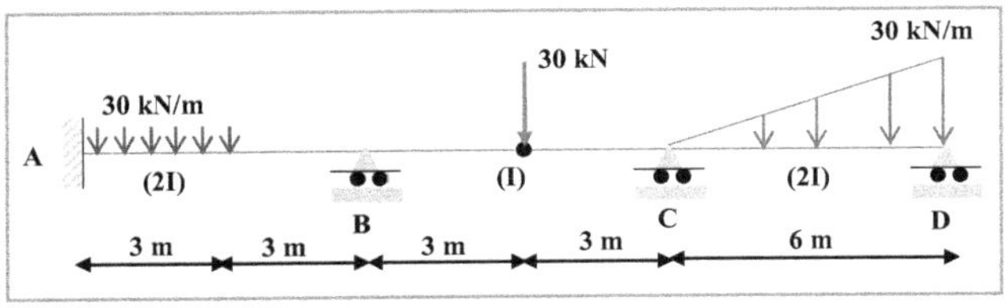

FIGURE 4.73

4.6.4 Analyse the continuous beams shown in Figure 4.74, using a suitable energy method, if supports *B* and *C* settle down by 10 and 5 mm respectively. Assume $EI = 1.35 \times 10^5$ kNm^2. Draw the shear force and bending moment diagrams.

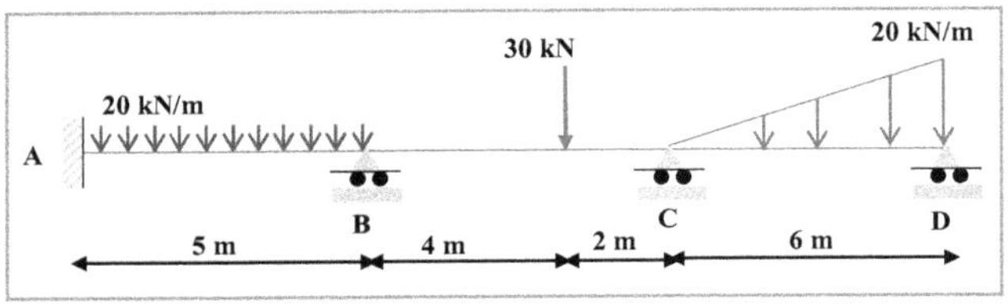

FIGURE 4.74

4.6.5 Re-analyse the beam shown in Figure 4.72, if support *A* rotates clockwise by 0.001 rad, and support *B* settle down by 5 mm instead of external loads. $EI = 1.35 \times 10^5$ kNm^2. Draw the shear force and bending moment diagrams.

4.6.6 Analyse the rigid-jointed plane frame shown in Figure 4.75, using the theorem of least work. Draw the axial force, shear force, and bending moment diagrams. Also draw the probable elastic curve.

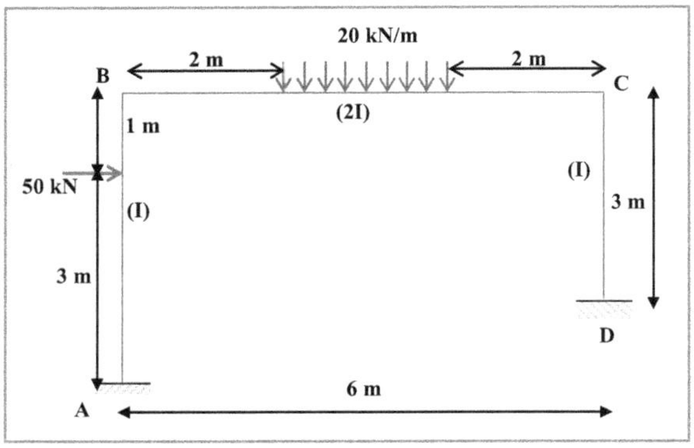

FIGURE 4.75

4.6.7 A portal frame *ABCDE* shown in Figure 4.76. Analyse it by the theorem of least work. Draw the axial force, shear force, and bending moment diagrams. Also, draw the probable elastic curve.

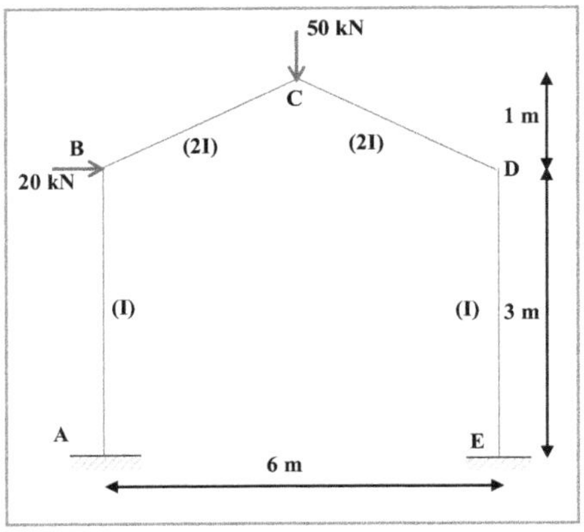

FIGURE 4.76

4.6.8 Analyse the anti-symmetric rigid-jointed frame shown in Figure 4.77. Also, construct the axial force, shear force, and bending moment diagrams. Furthermore, draw the probable elastic curve.

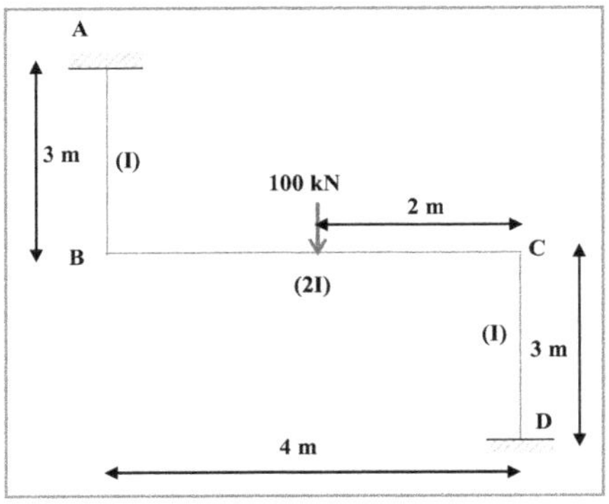

FIGURE 4.77

4.6.9 A ring of radius *R*, and of uniform cross-section is subjected to two equal and opposite vertical loads *W* at the extremities of the vertical diameter. Find the maximum +ve and −ve bending moments. Find also the decrease in length of the vertical diameter, and the corresponding increase in length of the horizontal diameter.

4.6.10 Analyse the frame shown in Figure 4.78, using the theorem of least work. Draw the bending moment diagram. Assume *EI* is constant.

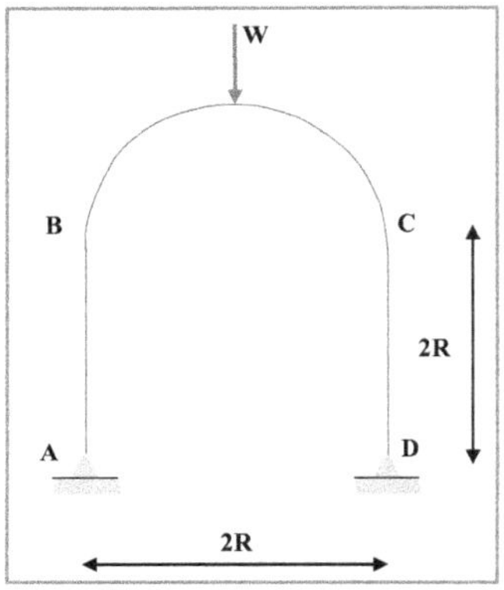

FIGURE 4.78

4.6.11 Analyse the box frame subjected to self-equilibrating loading is shown in Figure 4.79, using the theorem of least work. Construct the axial force, shear force, and bending moment diagrams. Also sketch the elastic curve.

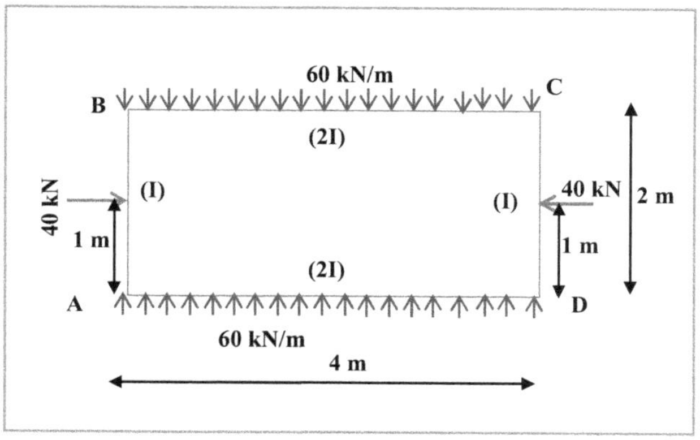

FIGURE 4.79

4.6.12 A plane braced frame is shown in Figure 4.80. The member BF in the frame initially short in length by 2.5 mm. Analyse the truss for all member forces. Assume the cross-sectional area of all members is constant equal to 800 mm², and modulus of elasticity $E=200$ kN/mm². Also, analyse the truss if member BF is subjected to a temperature rise of 20°C. $\alpha=11\times10^{-6}$/°C.

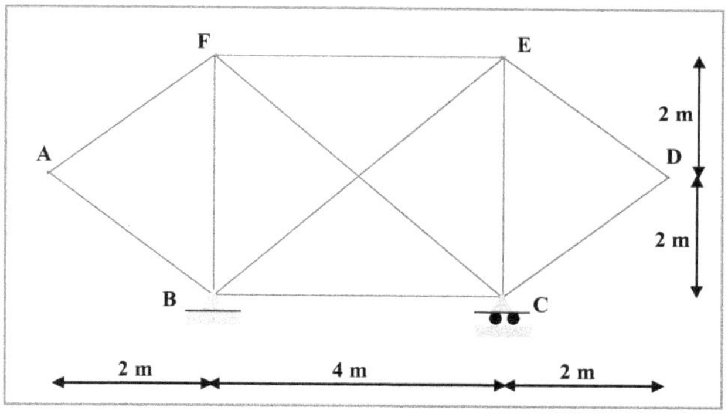

FIGURE 4.80

4.7 ANSWER KEY

Q.No.	Fig.No.	Answer at Critical Points
4.6.2	4.71	$R_B=8.39$ kN; $M_B=13.86$ kNm (hogging)
4.6.3	4.72	$R_A=42.96$ kN(↑); $M_A=40.89$ kNm (hogging); $R_B=5.39$ kN (↓); $R_C=22.71$ kN(↑)
4.6.4	4.73	$R_A=70.94$ kN(↑); $M_A=57.53$ kNm (hogging); $R_B=31.36$ kN (↑); $R_C=56.55$ kN(↑); $R_D=51.15$ kN(↑)
4.6.5	4.74	$R_A=141.3$ kN(↑); $M_A=303.69$ kNm (hogging); $R_B=69.37$ kN (↓); $R_C=90.66$ kN(↑); $R_D=27.4$ kN(↑)
4.6.6	4.72	$R_A=17.02$ kN(↑); $M_A=26.23$ kNm (hogging); $R_B=37.61$ kN (↓); $R_C=28.53$ kN(↑); $R_D=-7.94$ kN(↓)
4.6.7	4.75	$M_A=-34.17$ kNm; $M_D=48.98$ kNm; $M_B=9.24$ kNm; $M_C=44.83$ kNm
4.6.8	4.76	$V_A=20.76$ kN(↑); $M_A=3.34$ kNm (hogging); $H_A=1.48$ kN(→); $V_E=29.24$ kN (↑); $H_E=21.48$ kN(←); $M_E=31.19$ kNm (sagging)
4.6.9	4.77	$M_A=-12.46$ kNm; $M_D=12.46$ kNm; $M_{BA}=12.46$ kNm; $M_{BC}=-12.46$ kNm; $M_{CB}=12.46$ kNm; $M_{CD}=-12.46$ kNm
4.6.10		Max +ve BM $=\dfrac{WR}{\pi}$; Max – ve BM $=\dfrac{-WR}{2\pi}(\pi-2)$; decrease in vertical diameter $=\dfrac{WR^3}{4\pi EI}(\pi^2-8)$
4.6.11	4.78	$H_D=-H_A=0.0597W$
4.6.12	4.79	$M_{AD}=M_{DA}=120$ kNm; $M_{AB}=-M_{BA}=-120$ kNm
4.6.13	4.80	a. $F_{AB}=F_{AF}=F_{DE}=F_{DC}=0$; $F_{BF}=F_{BC}=F_{EF}=F_{EC}=10.36$ kN; $F_{CF}=F_{BE}=-14.65$ kN b. $F_{AB}=F_{AF}=F_{DE}=F_{DC}=0$; $F_{BF}=F_{BC}=F_{EF}=F_{EC}=-3.65$ kN; $F_{CF}=F_{BE}=5.16$ kN

Note: The answers are provided based on the analysis using structural analysis software. The results from the exact solution may differ slightly (decimals), but these differences are generally negligible.

FURTHER READING

Menon, D. (2023). *Structural Analysis*, Narosa Publishing House Pvt Ltd., New Delhi, ISBN 978-8184877342.

Reddy, C. S. (2011). *Basic Structural Analysis*, 3e, Tata McGraw-Hill Education Pvt Ltd, New Delhi, ISBN 978-0070702769.

Shah, H. J. and Junarkar, S. B. (2015). *Mechanics of Structures*, Vol. II, 24e, Charotar Publishing House, Anand, ISBN 978-9385039027.a

5 Column Analogy Method

5.1 INTRODUCTION

In the year 1932, Prof. Hardy Cross proposed the column analogy method for the analysis of fixed beams, frames, arches with fixed ends, and closed frames. It is ideally suited for calculating the stiffness factors and carryover factors for members with varying cross-sections. This method is applicable for indeterminate structures with a maximum degree of indeterminacy of three. It is based on an analogy between the moments produced by fixed ends in a statically indeterminate structure and the stresses developed in a short column due to an eccentric load.

5.2 DEVELOPMENT OF THE METHOD

Consider a fixed beam subjected to random loading, as shown in Figure 5.1.

Using the principle of superposition, the bending moment diagram is divided into two parts: (i) the bending moment diagram due to loads (M_L) on the basic determinate structure (Figure 5.1b), and (ii) the bending moment diagram due to redundants (M_R) on the basic determinate structure (Figure 5.1c). Therefore, the bending moment (M) at any point is given by

$$M = M_L - M_R \tag{5.1}$$

From the geometric requirements (using the moment area theorem)

i. The change in slope between points A and B is equal to zero, i.e., sum of $\left(\frac{M_L}{EI}\right)$, and $\left(\frac{M_R}{EI}\right)$ diagrams between A and B is zero. (1), i.e., area of $\left(\frac{M_L}{EI} - \frac{M_R}{EI}\right)$ diagrams$=0$.

   (or) Area of $\frac{M_L}{EI}$ = Area of $\frac{M_R}{EI}$

ii. The tangential deviation of point B from the tangent drawn at A is zero. This implies that the moment of area $\frac{M_L}{EI}$ and $\frac{M_R}{EI}$ diagrams between A and B, calculated about point B, is zero. In other words, the moment of area of the $\left(\frac{M_L}{EI}\right)$ diagram about B is equal to the moment of area of the $\left(\frac{M_R}{EI}\right)$ diagram about B. This means the centroids of $\left(\frac{M_L}{EI}\right)$ and $\left(\frac{M_R}{EI}\right)$ diagrams coincide at the same location. (2)

We know that when there is a short column subjected to an eccentric load P (Figure 5.2a), a non-uniform stress distribution results across the column section, as shown in Figure 5.2b. The load is assumed to act at a distance "e" from the Y-axis along the X-axis, as shown. The stress at any point, located at a perpendicular distance "y" from the Y-axis, is given by $f = \frac{P}{A} \pm \frac{Pey}{I_y}$

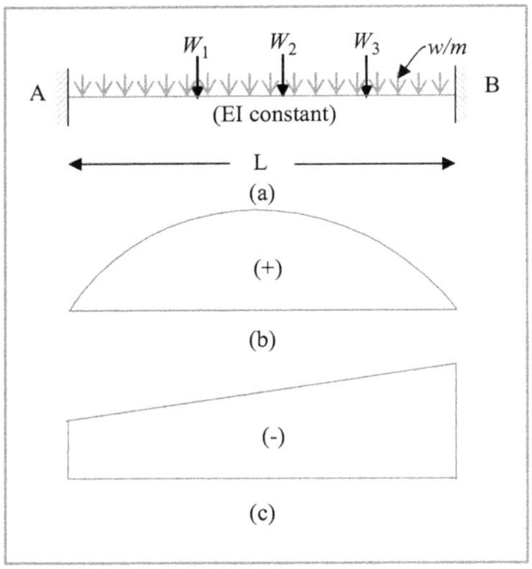

FIGURE 5.1 Effect of loading on a fixed beam.

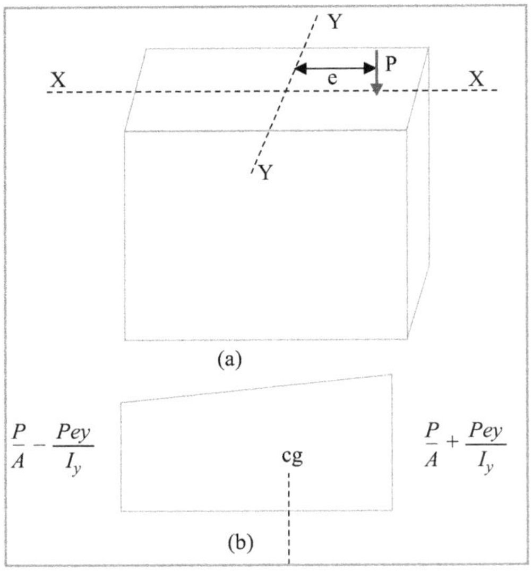

FIGURE 5.2 (a) Column subjected to eccentric load and (b) corresponding stress distribution.

DOI: 10.1201/9781003562955-5

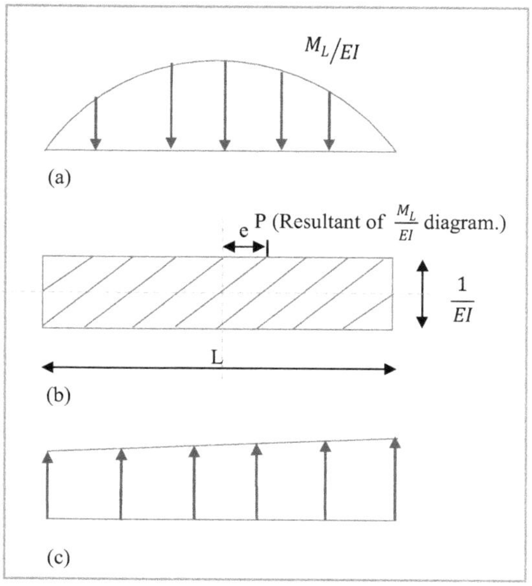

FIGURE 5.3 (a) Load on the top of an analogous column, (b) section of an analogous column, and (c) pressure at the bottom of an analogous column.

Let us now assume that the beam is replaced by a short column with a cross-section and centreline identical in shape and length to the beam. Its width at any section is equal to the $1/EI$ value of the beam at the corresponding section (Figure 5.3).

The load acting on the top of the analogous column is $\left(\frac{M_L}{EI}\right)$ diagram, as shown in Figure 5.3a. The cross-section of the analogous column is shown in Figure 5.3b, and the pressure developed at the bottom of the analogous column is shown in Figure 5.3c.

It may be observed from Figure 5.3 that for equilibrium,

i. The total load on the top of the column is equal to the pressure at the bottom, and (3)
ii. The moment of the total load about B is equal to the moment of the pressure at the bottom about B. This means that the centroid of the downward load at the top must coincide with the centroid of the upward pressure at the bottom. (4)

From the above, the conditions 1 and 2 of a fixed beam are analogous to the conditions 3 and 4 of a column. Hence, this analogy can be conveniently used for the analysis of fixed beams. Therefore, we can now analyse the analogous column subjected to loading equal to the $\left(\frac{M_L}{EI}\right)$ diagram and find the pressure developed at the bottom of the (M_R) diagram. Then, from equation 5.1, the final moment can be expressed as

$$M = M_L - M_R$$

5.3 SIGN CONVENTION

It is important to consider the sign convention regarding the nature of the applied loads $\left(\frac{M_L}{EI}\right)$ diagram and the pressure acting at the bottom of the analogous column (M_R diagram).

i. If the M_L diagram is +ve, i.e., tension at the bottom, the load acting at the top of the analogous column is downwards.
ii. Upward pressure at the base of the analogous column (M_R diagram) is considered +ve if it causes tension at the top.
iii. The moment (M) at any point is given by the equation $M = M_L - M_R$ is positive if it causes tension at the bottom.

While applying this to the beams, there is an eccentricity with respect to only the Y-axis. Hence, M_x=0.

$$\therefore M_R = \frac{P}{A} + \left(\frac{M_y}{I_{yy}}\right)x$$

where A = area of the analogous column

P = area of $\left(\frac{M_L}{EI}\right)$ diagram

I_{yy} = second moment of area of the analogous column, about the y–y axis.

e_y = eccentricity of the 'load P' with respect to the centroid of the analogous column.

$M_y = Pe_y$

The final moment $M = M_L - M_R$

5.4 APPLICATION TO BEAMS

Example 5.1

Analyse the fixed beam AB, shown in Figure 5.4a. Draw the bending moment diagram. Assume EI is constant.

Solution

Assume the fixed end moments at A and B as M_A and M_B respectively. Let us consider a simply supported beam as the basic determinate beam, as shown in Figure 5.4b. The BM diagram due to the loading on the basic beam, is shown in Figure 5.4c, and the $\left(\frac{M_L}{EI}\right)$ diagram is shown in Figure 5.4d.

The analogous column of width equal to $1/EI$ is shown in Figure 5.4e, and the stress distribution is shown in Figure 5.4f.

From the column analogy, we can write

$$M_A = \frac{P}{A} - \frac{Pey}{I_y} \tag{5.2}$$

$$M_B = \frac{P}{A} + \frac{Pey}{I_y} \tag{5.3}$$

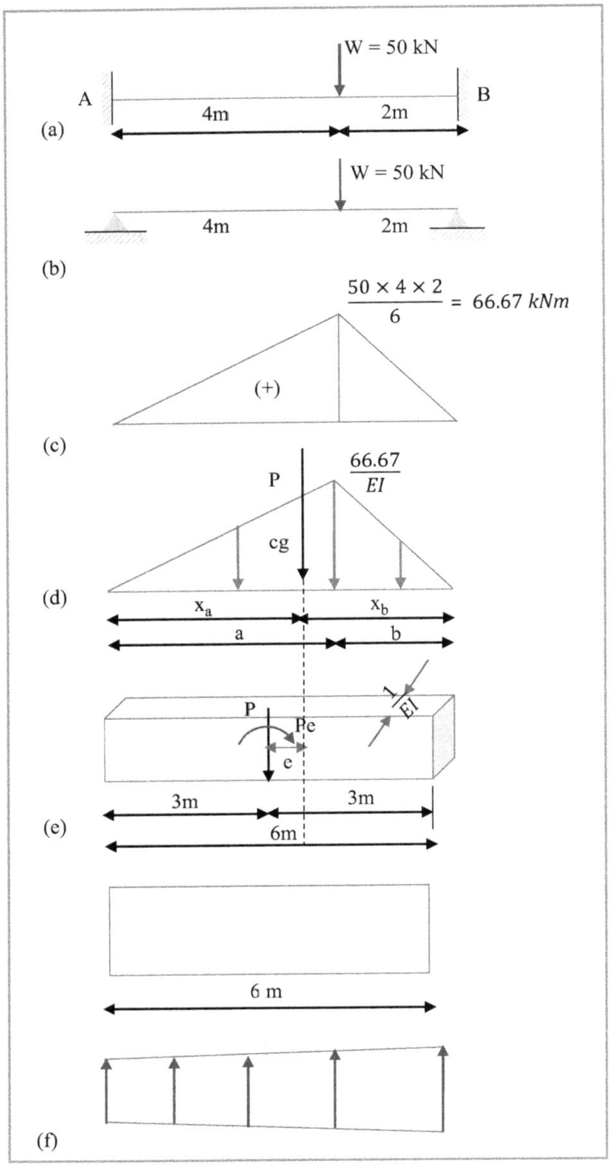

FIGURE 5.4 (a) Indeterminate fixed beam, (b) basic beam, (c) M_L diagram, (d) M_L/EI diagram, (e) section of an analogous column, and (f) pressure distribution.

where P = the axial load on the analogous column = Area of $\left(\frac{M_L}{EI}\right)$ diagram

$$= \frac{1}{2} \times 6 \times \frac{66.67}{EI} = \frac{200}{EI}$$

A = area of the analogous column

$$= 6 \times \frac{1}{EI} = \frac{6}{EI}$$

e = eccentricity with respect to the centre of the analogous column.

$$= x_a - \frac{L}{2} = \left(\frac{L+a}{3}\right) - \frac{L}{2} = \frac{6+4}{3} - 3 = 0.333 \text{ m}$$

I_y = moment of inertia of the analogous column, about the y–y axis.

$$= \frac{1}{12} \times \frac{1}{EI} \times L^3 = \frac{1}{12} \times \frac{1}{EI} \times 6^3 = \frac{18}{EI}$$

$$y = \frac{L}{2} = 3 \text{ m}$$

Substituting the above values into equations 5.2 and 5.3, we can write

$$\therefore (M_R)_A = \frac{\left(\frac{200}{EI}\right)}{\left(\frac{6}{EI}\right)} - \frac{\left(\frac{200}{EI}\right)(0.33)3}{\left(\frac{18}{EI}\right)} = 22.33 \text{ kNm}$$

$$(M_R)_B = \frac{\left(\frac{200}{EI}\right)}{\left(\frac{6}{EI}\right)} + \frac{\left(\frac{200}{EI}\right)(0.333) \times 3}{\left(\frac{18}{EI}\right)} = 44.43 \text{ kNm}$$

FINAL END MOMENTS

The final end moments can be calculated from equation 5.1 as

$$M_A = (M_L)_A - (M_R)_A = 0 - 22.23 \text{ kNm} = -22.33 \text{ kNm}$$

$$M_B = (M_L)_B - (M_R)_B = 0 - 44.43 \text{ kNm} = -44.43 \text{ kNm}$$

The –ve sign indicates the moments are hogging in nature, i.e., tension at the top.

ALTERNATIVE SOLUTION

Assume the basic determinate structure is a cantilever beam, as shown in Figure 5.5a. The M_L/EI diagram and the section of an analogous column are shown in Figure 5.5b and c, respectively.

$$P = -\frac{1}{2} \times 4 \times \frac{200}{EI} = -\frac{400}{EI}$$

$$x_a = \frac{1}{3} \times 4 = \frac{4}{3} \text{ m}$$

$$e = \frac{L}{2} - x_a = \frac{6}{2} - \frac{4}{3} = \frac{5}{3} \text{ m}$$

$$I_y = \frac{1}{12} \times \frac{1}{EI} \times 6^3 = \frac{18}{EI}$$

$$y = \frac{L}{2} = 3 \text{ m}$$

By substituting the above values into equations 5.2 and 5.3, we obtain

$$\therefore (M_R)_A = \frac{\left(-\frac{400}{EI}\right)}{\left(\frac{6}{EI}\right)} - \frac{\left(\frac{400}{EI}\right) \times \frac{5}{3} \times 3}{\left(\frac{18}{EI}\right)}$$

$$= -177.78 \text{ kNm}$$

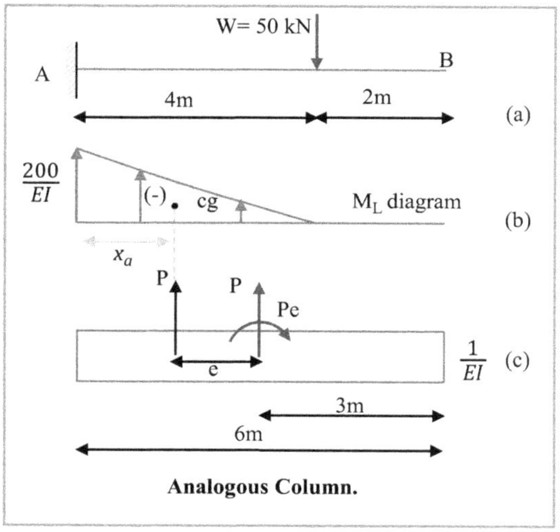

FIGURE 5.5 (a) Determinate beam, (b) M_L/EI diagram, and (c) analogous column.

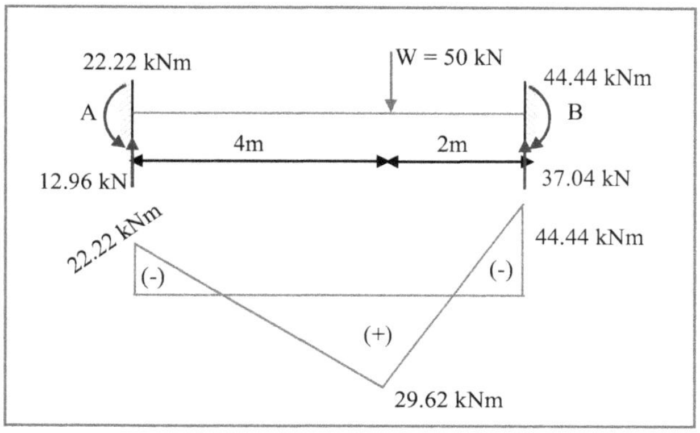

FIGURE 5.6 Bending moment diagram.

$$(M_R)_B = \frac{\left(-\frac{400}{EI}\right)}{\left(\frac{6}{EI}\right)} + \frac{\left(\frac{400}{EI}\right)\times\frac{5}{3}\times 3}{\left(\frac{18}{EI}\right)}$$

$$= 44.43 \text{ kNm}$$

Final End Moments

After determining the above values of moments, from equation 5.1, we get

$$M_A = (M_L)_A - (M_R)_A = -200 + 177.78 = -22.22 \text{ kNm}$$

$$M_B = (M_L)_B - (M_R)_B = 0 - 44.44 = -44.44 \text{ kNm}$$

These results are the same as those obtained previously. The bending moment diagram is shown in Figure 5.6.

Example 5.2

Calculate the fixed end moments for the fixed beam *AB*, shown in Figure 5.7. Assume *EI* is constant.

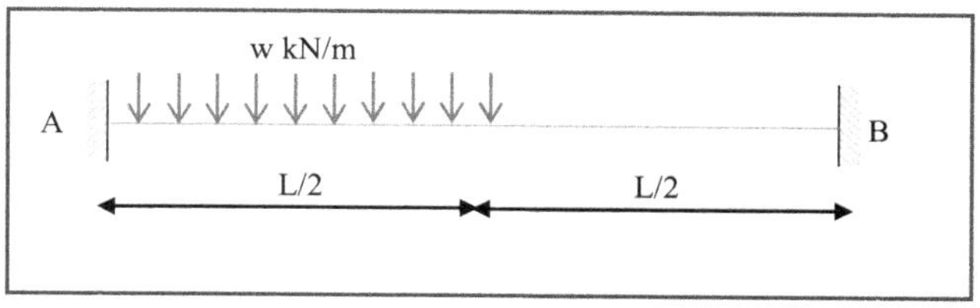

FIGURE 5.7 Indeterminate fixed beam.

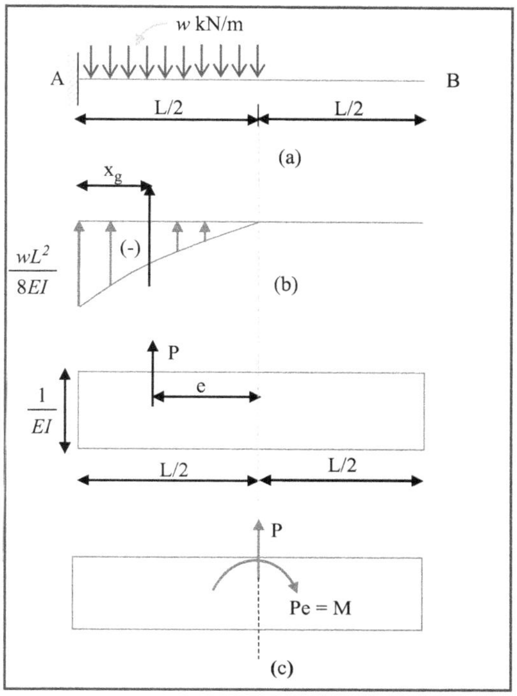

FIGURE 5.8 (a) Determinate beam, (b) M_L/EI diagram, and (c) analogous column.

Solution

Select the cantilever beam, fixed at *A*, as the basic determinate structure, as shown in Figure 5.8a. The M_L diagram is shown in Figure 5.8b. The c/s of the analogous column, along with the eccentric load and equivalent load system, is shown in Figure 5.8c.

The resultant load *P*=area of the M_L diagram.

$$P = \frac{1}{3}\times\frac{wL^2}{8EI}\times\frac{L}{2} = \frac{wL^3}{48EI}$$

c.g. of the resultant load from *A* is $x_g = \frac{1}{4}\times\frac{L}{2} = \frac{L}{8}$

$$\therefore \text{Eccentricity } (e) = \frac{L}{2} - x_g = \frac{L}{2} - \frac{L}{8} = \frac{3L}{8}$$

PROPERTIES OF THE ANALOGOUS COLUMN

$$A = L\times\frac{1}{EI} = \frac{L}{EI}$$

$$I_y = \frac{1}{12} \times \frac{1}{EI} \times L^3 = \frac{L^3}{12EI}$$

$$y = \frac{L}{2}$$

$$\therefore (M_R)_A = -\frac{P}{A} - \frac{Pe}{I_y} y \quad \left(\begin{array}{l} \because -\text{ve sign for } P \text{ is due} \\ \text{to the upward loading} \end{array} \right)$$

$$= \frac{\left(\frac{-wL^3}{48EI}\right)}{\left(\frac{L}{EI}\right)} - \frac{\left(\frac{wL^3}{48EI}\right) \times \frac{3L}{8}}{\left(\frac{L^3}{12EI}\right)} \times \frac{L}{2}$$

$$= -\frac{wL^2}{48} - \frac{3wL^2}{64} = \frac{-13wL^2}{192}$$

$$(M_R)_B = -\frac{P}{A} + \frac{Pe}{I_y} y = -\frac{-wL^2}{48} + \frac{-3wL^2}{64} = \frac{5wL^2}{192}$$

FINAL END MOMENTS

From equation 5.1, we can write the final end moments as

$$M_A = (M_L)_A - (M_R)_A = \frac{-wL^2}{48} - \left(\frac{-13wL^2}{192}\right) = \frac{-11wL^2}{192}$$

$$M_B = (M_L)_B - (M_R)_B = 0 - \frac{5wL^2}{192} = \frac{-5wL^2}{192}$$

(–ve sign indicates the moments are hogging in nature, producing tension at the top)

Example 5.3

Compute the fixed end moments of a fixed beam, shown in Figure 5.9a. Assume EI is constant.

Solution

The cantilever beam with a fixed end at B is considered as the basic determinate beam, as shown in Figure 5.9b. The M_L diagram is shown in Figure 5.9c, and the analogous column is shown in Figure 5.9d.

$$\text{The resultant load } (P) = \frac{bh}{n+1} = \frac{L \times \frac{wL^2}{6EI}}{3+1} = \frac{wL^3}{24EI}$$

$$x_g = \frac{b}{n+2} - \frac{L}{3+2} = \frac{L}{5}$$

$$\text{Eccentricity } (e) = \frac{L}{2} - \frac{L}{5} = \frac{3L}{10}$$

PROPERTIES OF THE ANALOGOUS COLUMN

$$A = L \times \frac{1}{EI} = \frac{L}{EI}$$

$$I_y = \frac{1}{12} \times \frac{1}{EI} \times L^3 = \frac{L^3}{12EI}$$

$$y = \frac{L}{2}$$

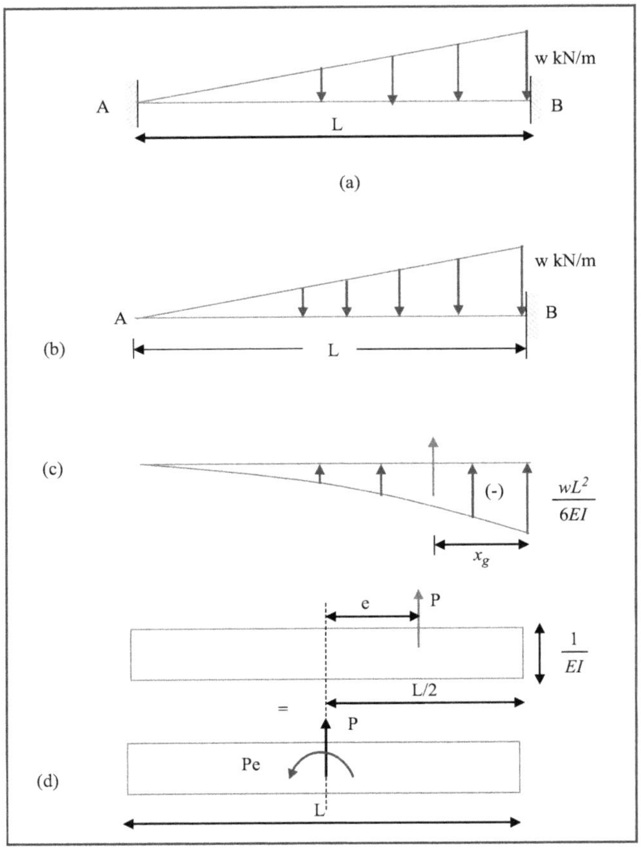

FIGURE 5.9 (a) Fixed beam subjected to linearly varying load (b) determinate beam, (c) M_L/EI diagram, and (d) analogous column.

$$\therefore (M_R)_A = -\frac{P}{A} + \frac{Pe}{I_y} y = \frac{\left(\frac{-wL^3}{24EI}\right)}{\left(\frac{L}{EI}\right)} + \frac{\left(\frac{wL^3}{24EI}\right) \times \frac{3L}{10}}{\left(\frac{L^3}{12EI}\right)} \times \frac{L}{2}$$

$$= \frac{-wL^2}{24} + \frac{3wL^2}{40} = \frac{wL^2}{30}$$

$$(M_R)_B = -\frac{P}{A} - \frac{Pe}{I_y} y = \frac{-wL^2}{24} - \frac{3wL^2}{40} = \frac{-14wL^2}{120}$$

FINAL END MOMENTS

$$M_A = (M_L)_A - (M_R)_A = 0 - \frac{wL^2}{30} = \frac{-wL^2}{30}$$

$$M_B = (M_L)_B - (M_R)_B = \frac{-wL^2}{6} + \frac{14wL^2}{120} = \frac{-wL^2}{20}$$

Example 5.4

Analyse the non-prismatic fixed beam, shown in Figure 5.10a, using the column analogy method.

Solution

Assume the beam is simply supported at *A* and *B*, forming the basic determinate beam, as shown in Figure 5.10b. The bending moment diagram is illustrated in Figure 5.10c. The M_L diagram and the cross-section of the analogous column are shown in Figure 5.10d and e, respectively.

The resultant load

$$P = \text{Area of } M_L \text{ diagram}$$

$$= 2\left(\frac{1}{2}\times\frac{50}{EI}\times 2\right)+\left(\frac{100}{EI}\times 2\right)+\left(\frac{1}{2}\times 2\times\frac{50}{EI}\right)$$

$$= \frac{350}{EI}$$

$$x_g = 3 \text{ m}$$

$$\text{Eccentricity } (e) = \frac{L}{2} - x_g = 0 \text{ i.e. c.g. of the load coincides with c.g. of the column}$$

PROPERTIES OF THE ANALOGOUS COLUMN

$$A = \frac{1}{2EI}\times 2\times 2 + \frac{1}{EI}\times 2 = \frac{4}{EI}$$

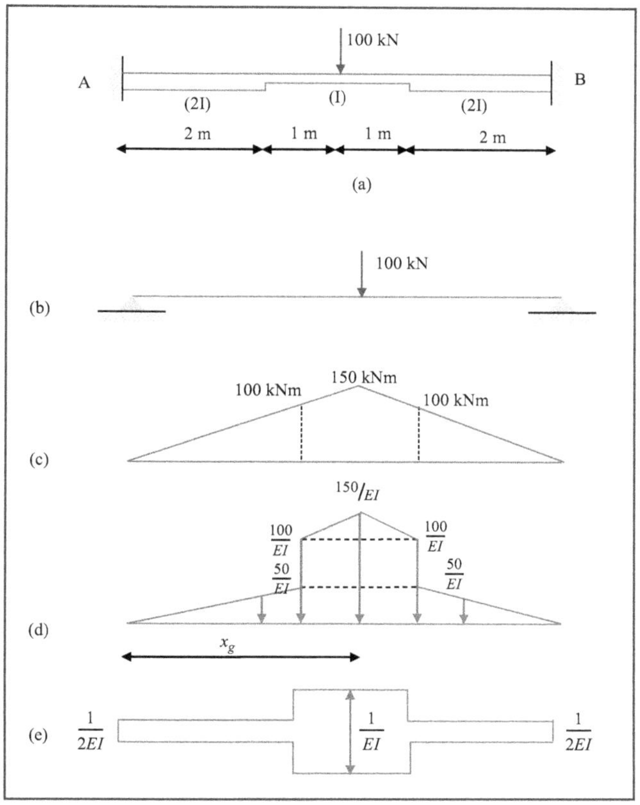

FIGURE 5.10 (a) Non-prismatic fixed beam, (b) determinate beam, (c) M_L diagram, (d) M_L/EI diagram, and (e) analogous column.

$$I_y = \frac{1}{12}\times 2^3 \times \frac{1}{2EI}\times 2 + \frac{1}{12}\times\frac{1}{EI}\times 2^3 = \frac{16}{12EI}$$

$$(M_R)_{A,B} = \frac{P}{A} \pm \frac{Pe}{I_y} y \quad (5.4)$$

Since the eccentricity *e*=0,

$$(M_R)_A = \frac{P}{A} = \frac{350/EI}{4/EI} = 87.5 = (M_R)_B$$

FINAL END MOMENTS

$$M_A = (M_L)_A - (M_R)_A = 0 - 87.5 = -87.5 \text{ kNm}$$

$$M_B = (M_L)_B - (M_R)_B = 0 - 87.5 = -87.5 \text{ kNm}$$

(–ve sign indicates the moments are hogging in nature, producing tension at the top)

Example 5.5

Analyse the non-prismatic fixed beam, shown in Figure 5.11a, using the column analogy method.

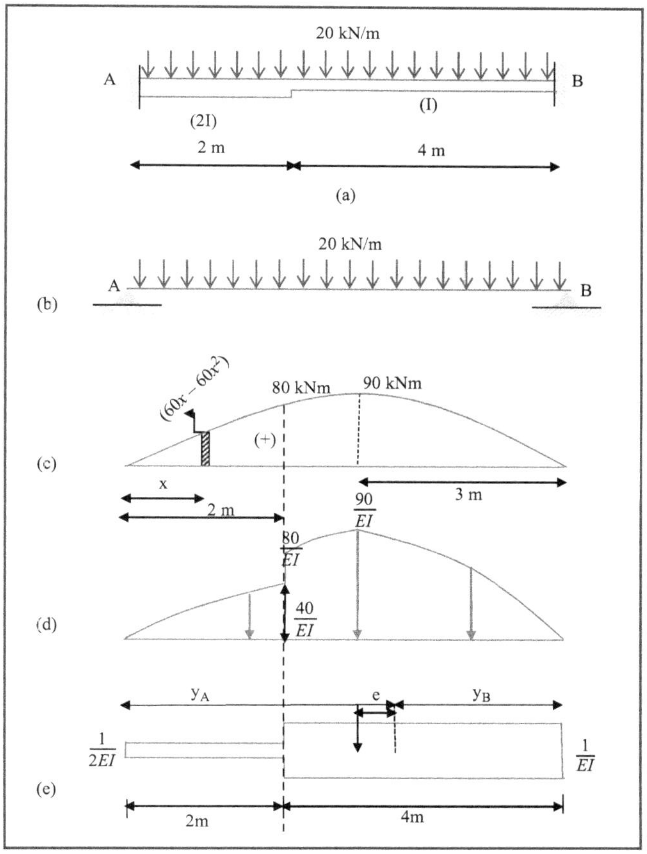

FIGURE 5.11 (a) Non-prismatic fixed beam, (b) primary beam, (c) bending moment diagram, (d) M_L/EI diagram, and (e) analogous column.

Solution

Assume the basic determinate structure is a simply supported beam, as shown in Figure 5.11b, and the corresponding bending moment diagram is shown in Figure 5.11c. The M_L diagram and the c/s of an analogous column are shown in Figure 5.11d and e, respectively.

The resultant load P = Area of M_L digram

$$P = \int M_x \cdot dx$$

$$= \int_0^2 (60x - 10x^2)\frac{dx}{2EI} + \int_2^6 (60x - 10x^2)\frac{dx}{EI}$$

$$P = \frac{1}{2EI}\left[60\frac{x^2}{2} - 10\frac{x^3}{3}\right]_0^2 + \frac{1}{EI}\left[60\frac{x^2}{2} - 10\frac{x^3}{3}\right]_2^6$$

$$= \frac{46.67}{EI} + \left(\frac{360}{EI} - \frac{93.33}{EI}\right) = \frac{313.34}{EI}$$

Location of the centroid from A (x_g)

$$x_g = \frac{\int M_x \cdot x \cdot dx}{\int M_x \cdot dx} \tag{5.5}$$

$$\int M_x \cdot dx = \int_0^2 (60x - 10x^2)x\frac{dx}{2EI} + \int_2^6 (60x - 10x^2)x\frac{dx}{EI}$$

$$= \frac{1}{2EI}\left[60\frac{x^3}{3} - 10\frac{x^4}{4}\right]_0^2 + \frac{1}{EI}\left[60\frac{x^3}{3} - 10\frac{x^4}{4}\right]_2^6$$

$$= \frac{60}{EI} + \left(\frac{1080}{EI} - \frac{120}{EI}\right) = \frac{1020}{EI}$$

Therefore, from equation 5.5,

$$x_g = \frac{1{,}020/EI}{313.34/EI} = 3.25 \text{ m}$$

PROPERTIES OF ANALOGOUS COLUMN

$$A = \frac{1}{2EI} \times 2 + \frac{1}{EI} \times 4 = \frac{5}{EI}$$

c.g. of the analogous column from A is

$$\bar{x} = \frac{\frac{1}{2EI} \times 2 \times 1 + \frac{1}{EI} \times 4 \times 4}{\frac{5}{EI}} = 3.4 \text{ m}$$

$$\therefore \text{Eccentricity } (e) = \bar{x} - x_g = 3.4 - 3.25 = 0.15 \text{ m}$$

$$I_y = \frac{1}{EI}\left[\frac{1}{12} \times \frac{1}{2} \times 2^3 + \frac{1}{2} \times 2 \times (3.4 - 1)^2\right]$$

$$+ \frac{1}{EI}\left[\frac{1}{12} \times 1 \times 4^3 + 1 \times 4 \times (4 - 3.4)^2\right]$$

$$= \frac{1}{EI}[6.093 + 6.773] = \frac{12.866}{EI}$$

MOMENTS DUE TO THE REDUNDANTS

$$(M_R)_A = \frac{P}{A} + \frac{Pe}{I_y} y_A \tag{5.6}$$

$$(M_R)_A = \frac{\left(\frac{313.34}{EI}\right)}{\left(\frac{5}{EI}\right)} + \frac{\left(\frac{313.34}{EI} \times 0.15 \times 3.4\right)}{\left(\frac{12.866}{EI}\right)} = 75.08 \text{ kNm}$$

$$(M_R)_B = \frac{P}{A} - \frac{Pe}{I_y} y_B \tag{5.7}$$

$$(M_R)_B = \frac{\left(\frac{313.34}{EI}\right)}{\left(\frac{5}{EI}\right)} - \frac{\left(\frac{313.34}{EI} \times 0.15 \times 2.6\right)}{\left(\frac{12.866}{EI}\right)} = 53.17 \text{ kNm}$$

FINAL END MOMENTS

$$M_A = (M_L)_A - (M_R)_A = 0 - 75.08 = -75.08 \text{ kNm}$$

$$M_B = (M_L)_B - (M_R)_B = 0 - 53.17 = -53.17 \text{ kNm}$$

Example 5.6

Compute the fixed end moments of a non-prismatic fixed beam shown in Figure 5.12, using the column analogy method.

Solution

Assume the simply supported beam as the basic determinate structure, as shown in Figure 5.13a. The bending moment diagram is illustrated in Figure 5.13b, the M_L diagram is shown in Figure 5.13c, and the cross-section of the analogous column is shown in Figure 5.13d.

Resultant load,

$$P = \frac{1}{2} \times 2 \times \frac{133.34}{EI} + \frac{2}{2}\left(\frac{66.67}{EI} + \frac{83.34}{EI}\right) + \frac{1}{2} \times 2 \times \frac{83.34}{EI}$$

$$= \frac{366.69}{EI}$$

c.g. of the resultant load from point A is

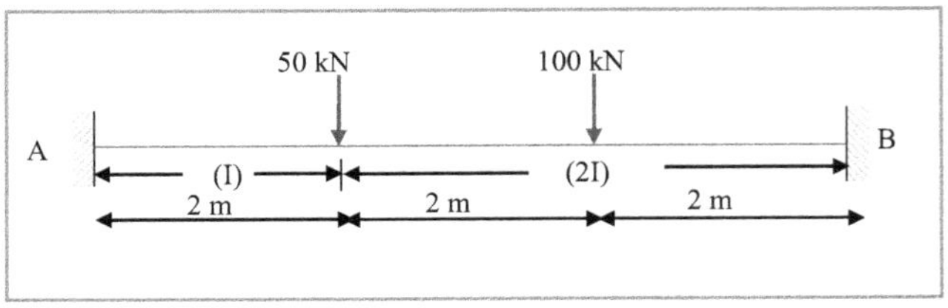

FIGURE 5.12 Indeterminate beam.

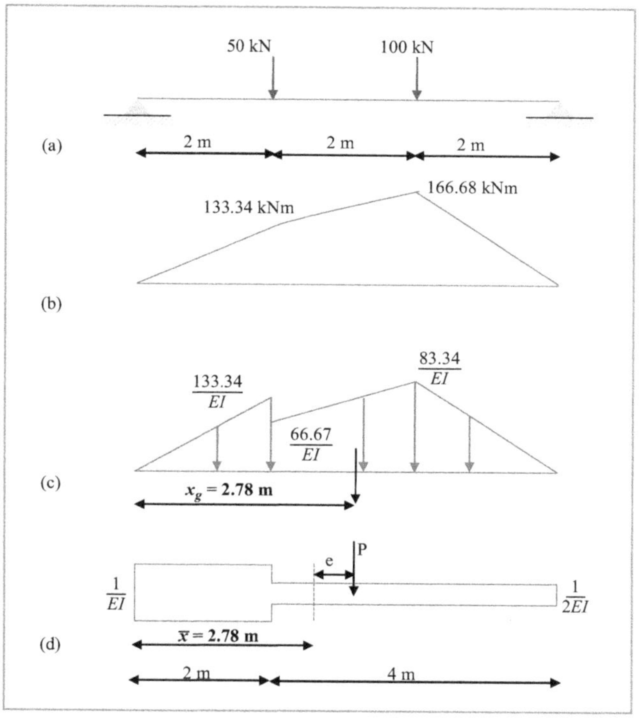

FIGURE 5.13 (a) Primary beam, (b) M_L diagram, (c) M_L/EI diagram, and (d) analogous column.

$$x_g = \frac{\left[\frac{1}{2}\times\frac{133.34}{EI}\times 2\times\frac{4}{3}\right]+\left[\frac{66.67}{EI}\times 2\times 3+\frac{1}{2}\times 2\times\frac{16.67}{EI}\times 3.33\right]+\left[\frac{1}{2}\times\frac{83.34}{EI}\times 2\times 4.67\right]}{\frac{366.69}{EI}}$$

$= 2.78$ m

PROPERTIES OF THE ANALOGOUS COLUMN

$$A = \frac{1}{EI}\times 2+\frac{1}{2EI}\times 4 = \frac{4}{EI}$$

c.g. of the analogous column from point *A* is

$$\bar{x} = \frac{\frac{1}{EI}\times 2\times 1+\frac{1}{2EI}\times 4\times 4}{\frac{4}{EI}} = 2.5 \text{ m}$$

$$\therefore \text{Eccentricity } (e) = x_g - \bar{x} = 2.78 - 2.5 = 0.28 \text{ m}$$

$$I_y = \frac{1}{EI}\left[\frac{1}{12}\times 1\times 2^3+1\times 2\times(1.5)^2+\frac{1}{12}\times\frac{1}{2}\times 4^3+\frac{1}{2}\times 4\times(1.5)^2\right]$$

$$= \frac{12.33}{EI}$$

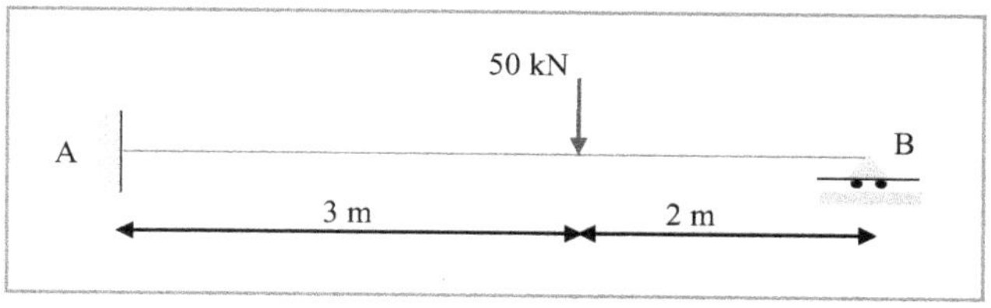

FIGURE 5.14 Propped cantilever beam.

MOMENTS DUE TO THE REDUNDANTS

$$(M_R)_A = \frac{P}{A} - \frac{Pe}{I_y} y_A \qquad (5.8)$$

$$(M_R)_A = \left(\frac{366.69}{EI}\times\frac{EI}{4}\right) - \frac{366.69}{EI}\times 0.28\times\frac{EI}{12.33}\times 2.5 = 70.85 \text{ kNm}$$

$$(M_R)_B = \frac{P}{A} + \frac{Pe}{I_y} y_B \qquad (5.9)$$

$$(M_R)_B = \left(\frac{366.69}{EI}\times\frac{EI}{4}\right) + \frac{366.69}{EI}\times 0.28\times\frac{EI}{12.33}\times 3.5 = 120.81 \text{ kNm}$$

FINAL END MOMENTS

By substituting the known values of the moments into equation 5.1, we obtain

$$M_A = (M_L)_A - (M_R)_A = 0 - 70.85 = -70.85 \text{ kNm}$$

$$M_B = (M_L)_B - (M_R)_B = 0 - 120.81 = -120.81 \text{ kNm}$$

Example 5.7

Compute the fixed end moment of the propped cantilever beam *AB* shown in Figure 5.14. Assume the *EI* is constant.

Solution

The basic determinate beam is considered by treating *AB* as a simply supported beam, as shown in Figure 5.15a. The M_L and $\frac{M_L}{EI}$ diagrams are shown in Figure 5.15b and c, respectively. The flexural rigidity *EI*=0 at the hinge support due to no resistance against rotation (stiffness). Therefore, the width of the analogous column at the hinge support is infinitely large, as shown in Figure 5.15d. Since the area of the analogous column=α, the c.g. of the analogous column lies at the hinge, i.e., at point *B*.

Resultant load (*P*)=Area of $\frac{M_L}{EI}$ diagram

$$P = \frac{1}{2}\times\frac{60}{EI}\times 5 = \frac{150}{EI}$$

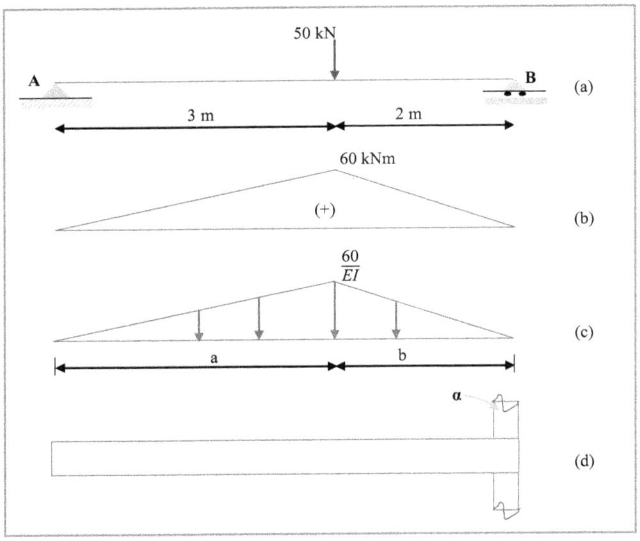

FIGURE 5.15 (a) Basic determinate beam, (b) bending moment diagram, (c) M_L/EI diagram, and (d) analogous column.

c.g. of the resultant load from point A is

$$x_g = \frac{L+a}{3} = \frac{5+3}{3} = \frac{8}{3}\text{ m}$$

$$\therefore \text{Eccentricity, } (e) = L - x_g = 5 - \frac{8}{3} = \frac{7}{3}\text{ m}$$

$$I_y = \frac{L^3}{3EI} \quad (MI \text{ about base, because c.g. lies at point } B)$$

$$I_y = \frac{5^3}{3EI} = \frac{125}{3EI}$$

Moment due to the redundant

$$(M_R)_A = \frac{P}{A} + \frac{Pe}{I_y} y_A$$

$$= \frac{\left(\frac{150}{EI}\right)}{\alpha} + \frac{\frac{150}{EI} \times \frac{7}{3}}{\frac{125}{3EI}} \times 5$$

$$= 0 + 42 = 42\text{ kNm}$$

Final moment at A is

$$M_A = (M_L)_A - (M_R)_A = 0 - 42 = -42\text{ kNm}$$

5.5 APPLICATION TO SYMMETRICAL RIGID-JOINTED PLANE FRAMES

The single-bay, single-storey rigid-jointed portal frames with a single axis of symmetry are commonly encountered in practice. The existence of this axis of symmetry in the structures confirms that it is a principal axis of the analogous column, with a perpendicular centroidal axis forming the other principal axis. Consequently, the centroid always lies on this axis of symmetry. For such frames, the column analogy can be effectively applied. The analogous column has the same overall centreline dimensions, with its thickness equal to $1/EI$. The sectional properties, viz. area A, centroidal distances, and second moments of areas with respect to the principal axes (I_{XX} and I_{YY}) can be easily determined. Thus, the stresses can also be computed using the column analogy, as demonstrated in the following examples.

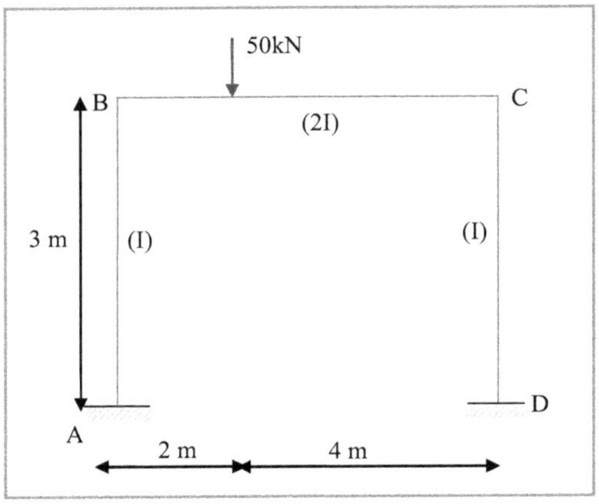

FIGURE 5.16 Indeterminate rigid-jointed frame.

Example 5.8

Analyse the symmetrical frame subjected to an unsymmetrical load shown in Figure 5.16, using the column analogy method.

Solution

The basic determinate frame with hinged support at A and roller support at D, is shown in Figure 5.17a. The M_L diagram is shown in Figure 5.17b, and $\left(\frac{M_L}{EI}\right)$ diagram and the c/s of an analogous column, are shown in Figure 5.17c and d, respectively.

$$\text{Resultant load } P = \frac{1}{2} \times 6 \times \frac{66.67}{2EI} = \frac{200}{2EI} = \frac{100}{EI}$$

$$\text{c.g. of the resultant load, } x_g = \frac{L+a}{3} = \frac{6+2}{3} = \frac{8}{3}\text{ m}$$

PROPERTIES OF THE ANALOGOUS COLUMN

$$A = 2 \times \frac{1}{EI} \times 3 + \frac{1}{2EI} \times 6 = \frac{9}{EI}$$

The frame is symmetrical about the Y–Y axis. Therefore, centroid lies on the Y–Y axis, i.e.,

$$\text{From } AB,\ x_c = \frac{6}{2} = 3\text{ m}$$

$$\text{From } BC,\ y_c = \frac{\frac{3}{EI} \times 1.5 \times 2}{\frac{9}{EI}} = 1.0\text{ m}$$

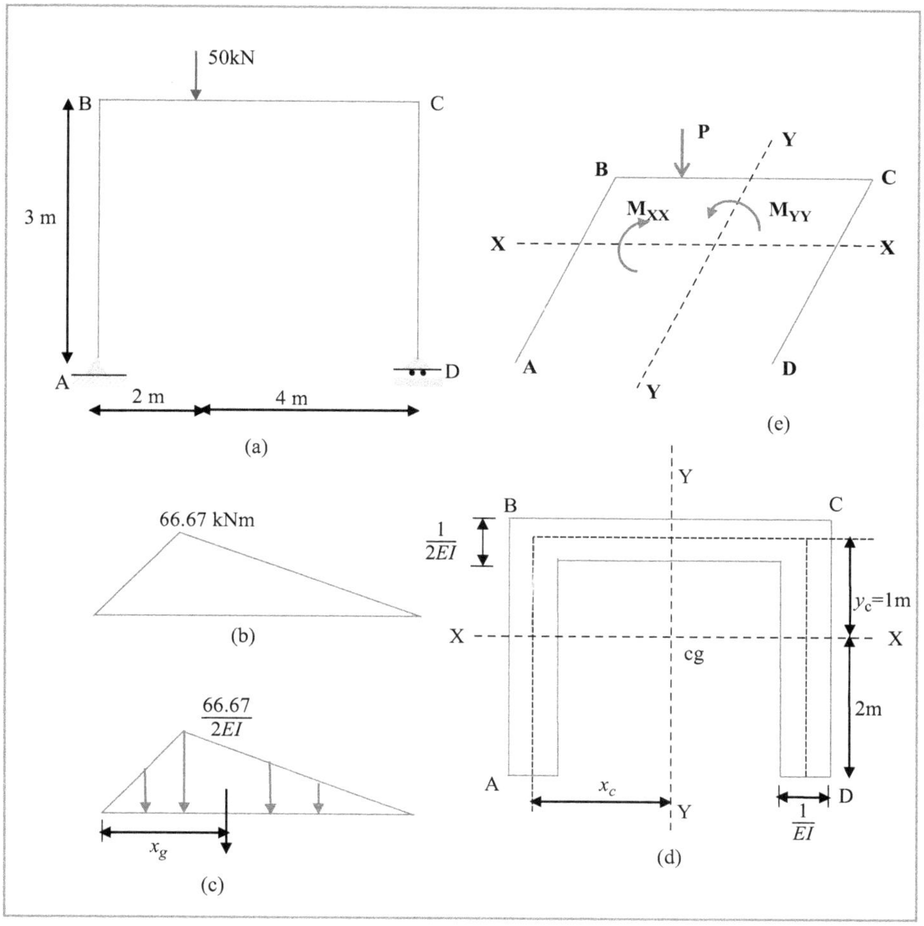

FIGURE 5.17 (a) Basic determinate frame, (b) M_L diagram, (c) M_L/EI diagram, and (d) analogous column.

$$I_{XX} = \left(\frac{1}{3}\right) \times \frac{1}{EI}\left[1^3 \times 2 + 2^3 \times 2\right] + \left[\frac{1}{2EI} \times 6 \times 1^2\right]$$

$$= \frac{9}{EI} \quad \left[\because \left(\frac{1}{2EI}\right)^3 \text{ is very small and is neglected}\right]$$

$$I_{YY} = 2\left[\frac{1}{EI} \times 3 \times 3^2\right] + \left[\frac{1}{12} \times \frac{1}{2EI} \times 6^3\right]$$

$$= \frac{63}{EI} \quad \left[\because \left(\frac{1}{EI}\right)^3 \text{ is very small and is neglected}\right]$$

Eccentricities

$$e_{XX} = 3 - \frac{8}{3} = \frac{1}{3}\text{ m}$$

$$e_{YY} = 1\text{ m}$$

$$M_{XX} = P_X e_{YY} = \frac{100}{EI} \times 1 = \frac{100}{EI}\ (\text{Clockwise})$$

$$M_{YY} = P_X e_{XX} = \frac{100}{EI} \times \frac{1}{3} = \frac{100}{3EI}\ (\text{Anticlockwise})$$

MOMENTS DUE TO THE REDUNDANTS

$$(M_R)_A = \frac{P}{A} - \frac{M_{XX}}{I_{XX}} y_A + \frac{M_{YY}}{I_{YY}} x_A$$

$$= \frac{100/EI}{9/EI} - \frac{100/EI}{9/EI} \times 2 + \frac{100/3EI}{63/EI} \times 3$$

$$= 11.11 - 22.22 + 1.58 = -9.53\text{ kNm}$$

$$(M_R)_B = \frac{P}{A} + \frac{M_{XX}}{I_{XX}} y_B + \frac{M_{YY}}{I_{YY}} x_B$$

$$= 11.11 + \frac{100}{9} \times 1 + \frac{100/3}{63} \times 3$$

$$= 23.80\text{ kNm}$$

$$(M_R)_C = \frac{P}{A} + \frac{M_{XX}}{I_{XX}} y_C - \frac{M_{YY}}{I_{YY}} x_C$$

$$= \frac{100/EI}{9/EI} + \frac{100}{EI} \times \frac{EI}{9} \times 1 - \frac{100}{3EI} \times \frac{EI}{9} \times 3$$

$$= 11.11 + 11.11 - 1.58 = 20.64\text{ kNm}$$

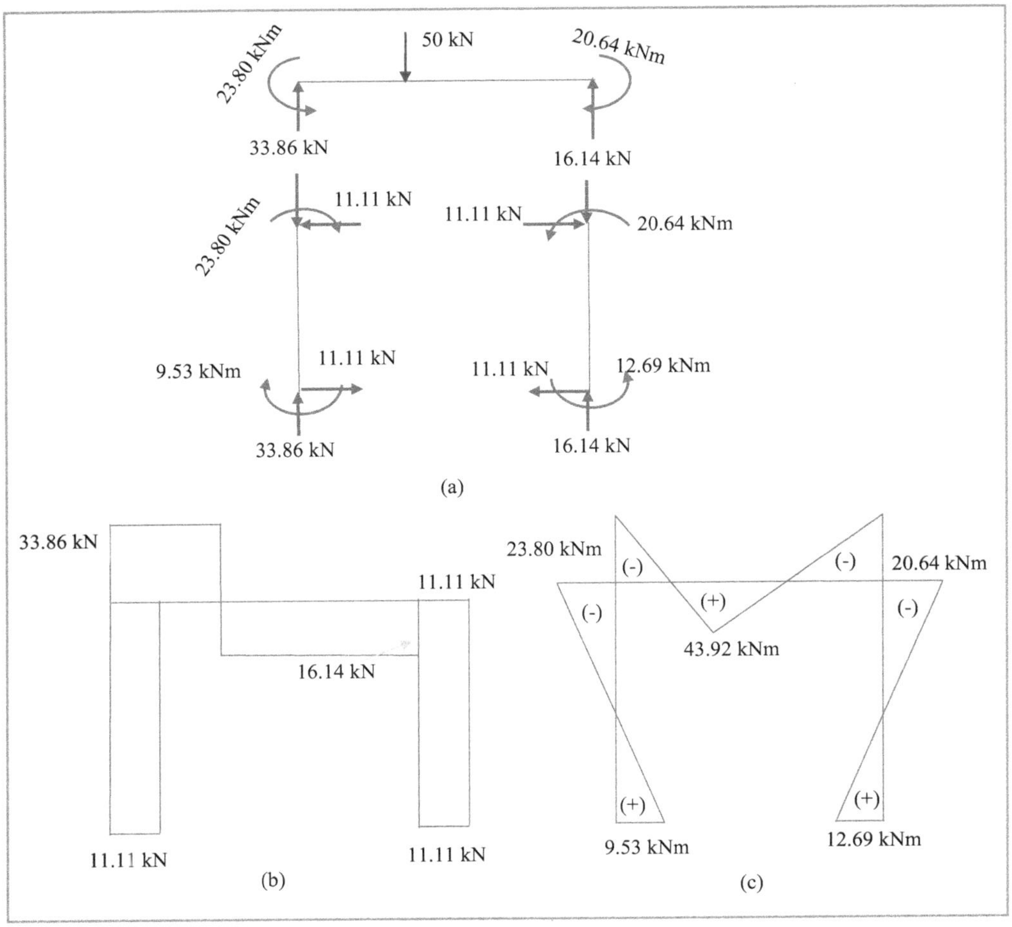

FIGURE 5.18 (a) Free body diagram, (b) shear force diagram, and (c) bending moment diagram.

$$(M_R)_D = \frac{P}{A} - \frac{M_{XX}}{I_{XX}} y_D - \frac{M_{YY}}{I_{YY}} x_D$$

$$= \frac{100/EI}{9/EI} - \frac{100}{EI} \times \frac{EI}{9} \times 2 - \frac{100}{3EI} \times \frac{EI}{63} \times 3$$

$$= 11.11 - 22.22 - 1.58 = -12.69 \text{ kNm}$$

FINAL END MOMENTS

$$M_A = (M_L - M_R)_A = 0 - (-9.53) = 9.53 \text{ kNm}$$

$$M_B = (M_L - M_R)_B = 0 - 23.80 = -23.80 \text{ kNm}$$

$$M_C = (M_L - M_R)_C = 0 - 20.64 = -20.64 \text{ kNm}$$

$$M_D = (M_L - M_R)_D = 0 - (-12.69) = 12.69 \text{ kNm}$$

Support reactions, shear force, and bending moment diagrams, are shown in Figure 5.18.

Example 5.9

Analyse the rigid-jointed plane frame shown in Figure 5.19 using the column analogy method.

Solution

Assume the basic determinate frame is a cantilever, with a fixed end at *A*, as shown in Figure 5.20a. The M_L and $\frac{M_L}{EI}$ diagrams are shown in Figure 5.20b and c, respectively. The analogous column cross-section is shown in Figure 5.20d. Since tension is developed on the outside of the frame in the M_L diagram, the loading is shown upwards (Figure 5.20e).

PROPERTIES OF LOADING DIAGRAM

$$P_1 = \frac{1}{3} \times \frac{800}{EI} \times 8 = \frac{6{,}400}{3EI}$$

$$P_2 = \frac{1{,}600}{EI} \times 4 = \frac{6{,}400}{EI}$$

$$P_3 = \frac{1}{2} \times 2 \times \frac{40}{EI} = \frac{40}{EI}$$

$$\text{Resultant load } (P) = \frac{6{,}400}{3EI} + \frac{6{,}400}{EI} + \frac{40}{EI} = \frac{25{,}720}{3EI}$$

PROPERTIES OF THE ANALOGOUS COLUMN

Due to symmetry about the Y–Y axis, $x_c = 4$ m

Let y_c be the centroid from BC.

$$y_c = \frac{\frac{1}{EI}\times 4\times 2 + 0 + \frac{1}{EI}\times 4\times 2}{\frac{1}{EI}\times 4\times 2 + \frac{1}{2EI}\times 8} = 1.33 \text{ m}$$

Area of analogous column, $A = \frac{1}{EI}\times 4\times 2 + \frac{1}{2EI}\times 8 = \frac{12}{EI}$

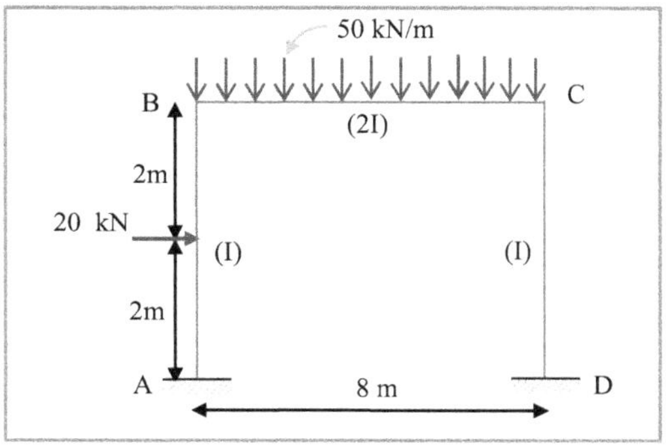

FIGURE 5.19 Rigid-jointed frame $ABCD$.

$$I_{XX} = 2\times\left[\frac{1}{EI}\times(2.67)^3\times\frac{1}{3}+\frac{1}{3}\times\frac{1}{EI}\times(1.33)^3\right] + \left[\frac{1}{2EI}\times 8\times(1.33)^2\right] = \frac{21.33}{EI}$$

$$I_{YY} = 2\times\left[\frac{1}{EI}\times 4\times(4)^2\right] + 2\left[\frac{1}{2EI}\times(4)^3\times\frac{1}{3}\right] = \frac{149.33}{EI}$$

$$M_{XX} = \frac{-6{,}400}{3EI}\times 1.33 + \frac{6{,}400}{EI}\times(2.67-2) + \frac{40}{EI}\left(2.67-\frac{2}{3}\right) = \frac{1{,}530.67}{EI}\text{ (Clockwise)}$$

[inducess compression on BC and tension at A and D]

$$M_{YY} = \frac{6{,}400}{3EI}\times(4-2) + \frac{6{,}400}{EI}\times 4 + \frac{40}{EI}\times 4 = \frac{30{,}026.67}{EI}\text{ (Clockwise)}$$

[inducess compression on CD and tension on AB]

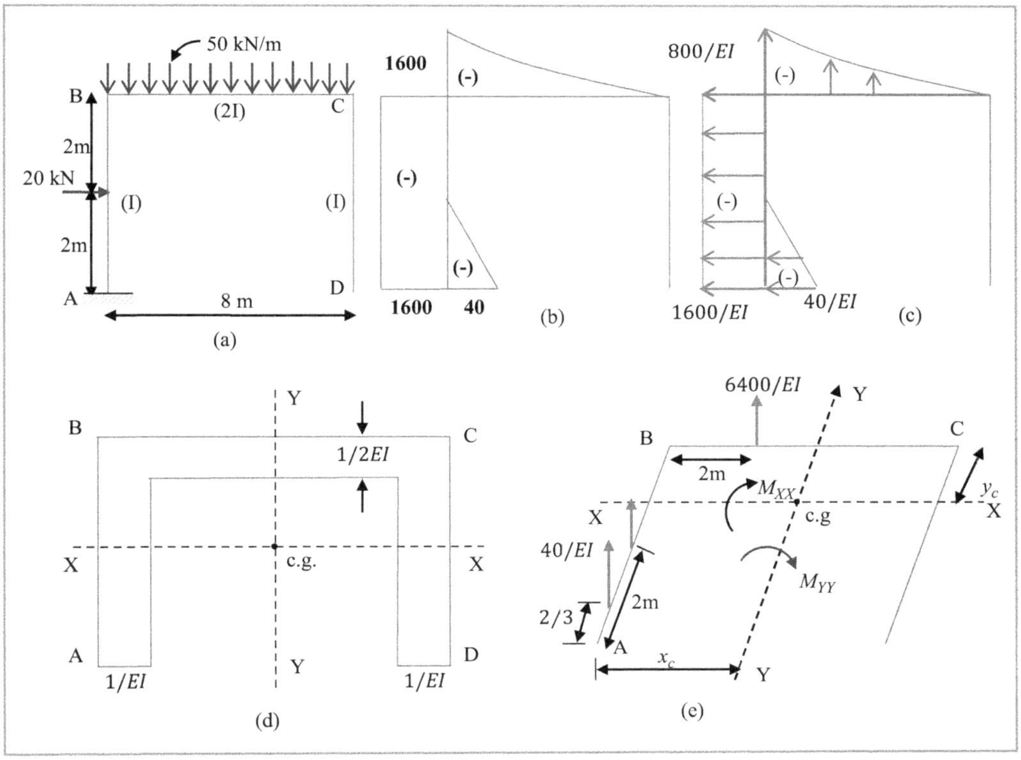

FIGURE 5.20 (a) Basic determinate frame, (b) M_L diagram, (c) $\left(\frac{M_L}{EI}\right)$ diagram, (d) analogous column, and (e) resultant forces on the analogous column.

TABLE 5.1
Moments Due to the Redundants and the Final End Moments

Location	M_L (kNm)	$\frac{P}{A}$	$\left(\frac{M_{XX}}{I_{XX}}\right)y$	$\left(\frac{M_{YY}}{I_{YY}}\right)x$	M_R (kNm)	M (kNm)
A	−1,640	−714.44	−191.60	−804.30	−1,710.34	70.34
B	−1,600	−714.44	95.44	−804.30	−1,423.30	−176.7
C	0	−714.44	95.44	804.30	185.3	−185.3
D	0	−714.44	−191.60	804.30	−101.74	101.74

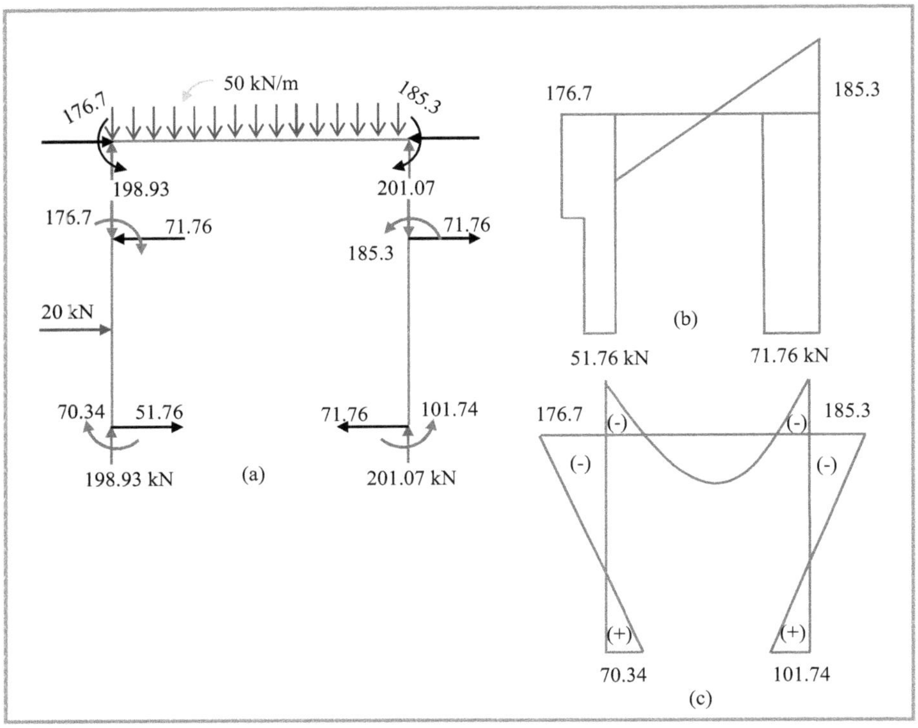

FIGURE 5.21 (a) Free body diagram, (b) shear force diagram, and (c) bending moment diagram.

MOMENTS DUE TO THE REDUNDANTS

$$(M_R)_A = \frac{-P}{A} - \frac{M_{XX}}{I_{XX}} y_A - \frac{M_{YY}}{I_{YY}} x_A$$

$$= \frac{(-25,720/3EI)}{(12/EI)} - \frac{1,530.67}{EI} \times \frac{EI}{21.33} \times 2.67$$

$$- \frac{30,026.67}{EI} \times \frac{EI}{149.33} \times 4$$

$$= -714.44 - 191.60 - 804.30 = -1,710.34 \text{ kNm}$$

The calculation of moments due to the redundants and the final end moments is shown in Table 5.1.

The axial forces, shear forces, and bending moments are shown in the free body diagram (Figure 5.21a). The shear force and bending moment diagrams are shown in Figure 5.21b and c, respectively.

Example 5.10

Analyse the rigid-jointed plane frame *ABCD* whose ends *A* and *D* are hinged, as shown in Figure 5.22a, using the column analogy method.

Solution

The primary structure is shown in Figure 5.22b, with a roller support at *D* and a hinged support at *A*. The M_L and M_L/EI diagrams are shown in Figure 5.22c and d, respectively. The flexural rigidity (*EI*) of the hinged supports at *A* and *D* is zero because these supports do not resist rotation. Consequently, the width of the analogous columns $\frac{1}{EI}$ tends to infinity at *A* and *D*, as illustrated in Figure 5.22e. This implies that the area of the analogous column becomes α. Therefore, the centroid axis of the analogous columns passes through *A* and *D*.

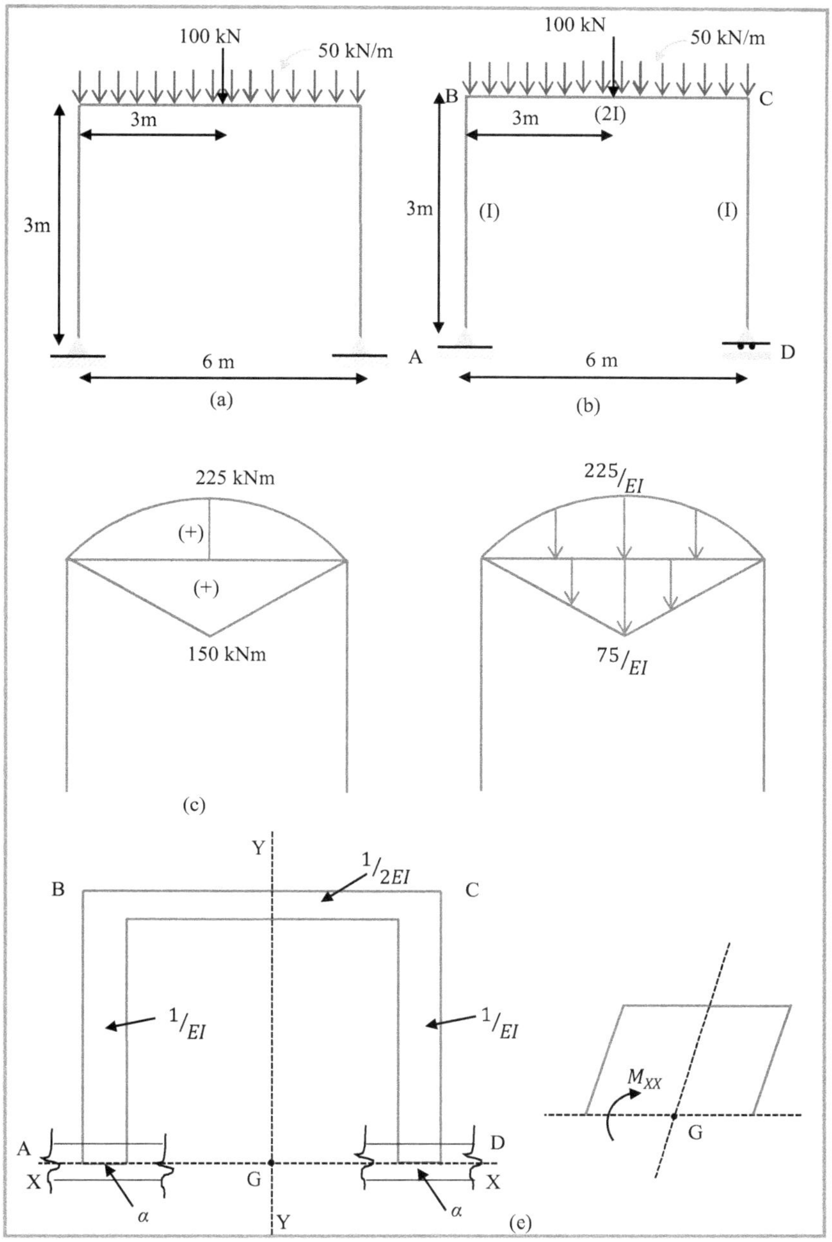

FIGURE 5.22 (a) Indeterminate rigid-jointed frame, (b) basic determinate frame, (c) M_L diagram, (d) $\left(\frac{M_L}{EI}\right)$ diagram, and (e) analogous column.

Furthermore, since the frame has vertical symmetry, the centroid lies on the symmetrical axis, as shown in Figure 5.22e.

Therefore, $x_c = 3$ m

$$y_c = 3 \text{ m} \quad \text{from } BC$$

Since, the areas at *A* and *D* are infinite, $I_{YY} \rightarrow \alpha$ with the centroid at *G*.

$$\therefore I_{YY} = 2\left[\frac{1}{EI}\times 3^3 \times \frac{1}{3}\right] + \left[\frac{1}{2EI}\times 6 \times 3^2\right] = \frac{45}{EI}$$

RESULTANT LOAD

$$P_1 = \frac{2}{3}\times 6 \times \frac{225}{2EI} = \frac{450}{EI}$$

$$P_2 = \frac{1}{2}\times 6 \times \frac{75}{EI} = \frac{225}{EI}$$

$$\text{Resultant load } P = P_1 + P_2 = \frac{675}{EI}$$

$$M_{XX} = \frac{450}{EI}\times 3 + \frac{225}{EI}\times 3 = \frac{2{,}025}{EI}$$

(anticlockwise, which produces tension at

A and D and compression in BC)

$M_{YY} = 0$ (due to the resultant load lies along $y-y$ axis.)

MOMENTS DUE TO THE REDUNDANTS

$$(M_R)_B = \frac{P}{\alpha} - \frac{M_{XX}}{I_{XX}} y_B = 0 + \frac{2{,}025}{EI}\times\frac{EI}{45}\times 3 = 135\ \text{kNm}$$

$$(M_R)_C = (M_R)_B = 135\ \text{kNm}$$

The final end moments

$$M_B = (M_L - M_R)_B = M_C = 0 - 135 = -135\ \text{kNm}$$

The internal forces and support reactions are shown in Figure 5.23a, and the shear force and bending moment diagrams are depicted in Figure 5.23b and c, respectively.

Example 5.11

Analyse the gable frame *ABCDE* shown in Figure 5.24, using the column analogy method.

Solution

The primary structure is shown in Figure 5.25a, with a hinged support at *A* and a roller support at *E*. The M_L diagram, and $\frac{M_L}{EI}$ diagrams are shown in Figure 5.25b and c, respectively. The cross-section of the analogous column is shown in Figure 5.25d.

Area of the analogous column, $A = \frac{1}{EI}\times 4\times 2 + \frac{1}{2EI}\times 5.1\times 2 = \frac{13.1}{EI}$

Since the cross-section is symmetrical about the *Y–Y* axis, the centroid lies on the *Y–Y* axis.

$$\therefore x_c = 5\ \text{m}$$

$$y_c \text{ from } A = \frac{2\times\frac{1}{EI}\times 4\times 2 + \frac{1}{2EI}\times 5.1\times 4.5\times 2}{\frac{13.1}{EI}} = 2.97\ \text{m}$$

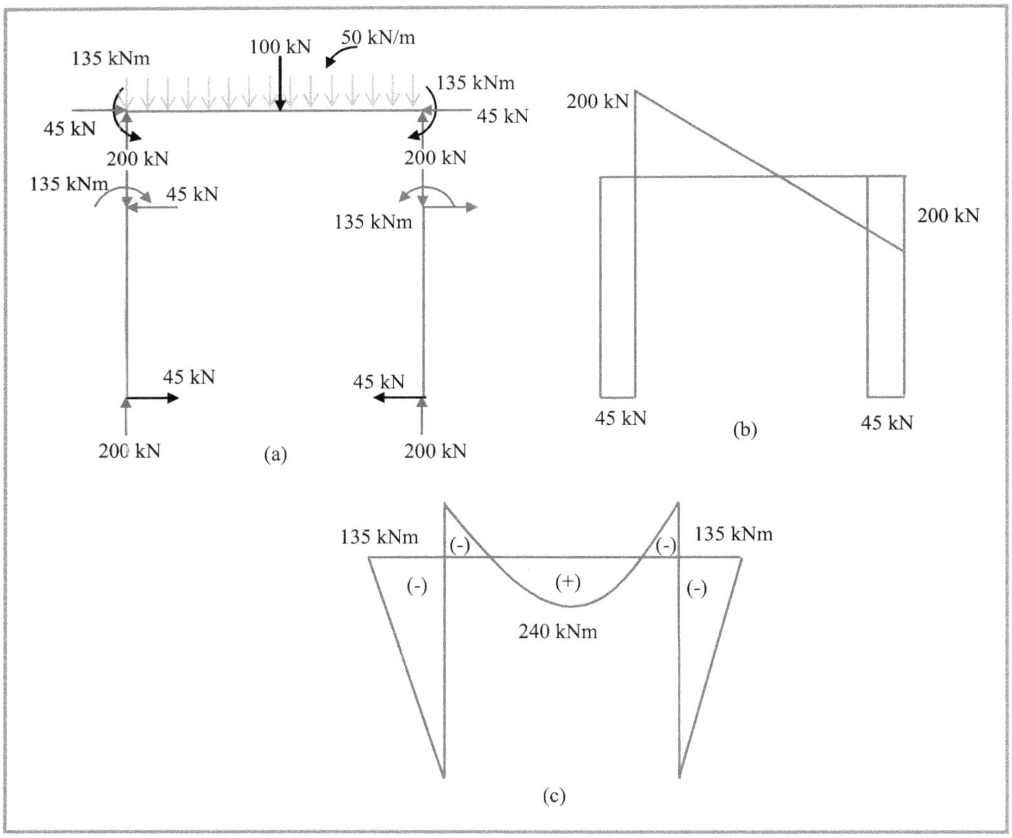

FIGURE 5.23 (a) Free body diagram, (b) shear force diagram, and (c) bending moment diagram.

$$I_{XX} = 2\left[\frac{1}{3}\times\frac{1}{EI}\times(2.97)^3 + \frac{1}{3}\times\frac{1}{EI}\times(1.03)^3\right]$$

$$+2\left[\frac{1}{12}\times\frac{1}{2EI}\sin\theta\times(5.1)^3\cdot\sin\theta + \frac{1}{2EI}\times5.1\times(1.03+0.5)^2\right]$$

$$=\frac{30.55}{EI}$$

$$I_{YY} = \frac{1}{EI}\times4\times(5)^2\times2 + \left[\begin{array}{l}\frac{1}{2EI}\cdot\cos\theta\times\frac{(5.1)^3}{12}\cdot\cos\theta\\ +\frac{1}{2EI}\times5.1\times(2.5)^2\end{array}\right]\times2$$

$$=\frac{242.49}{EI}$$

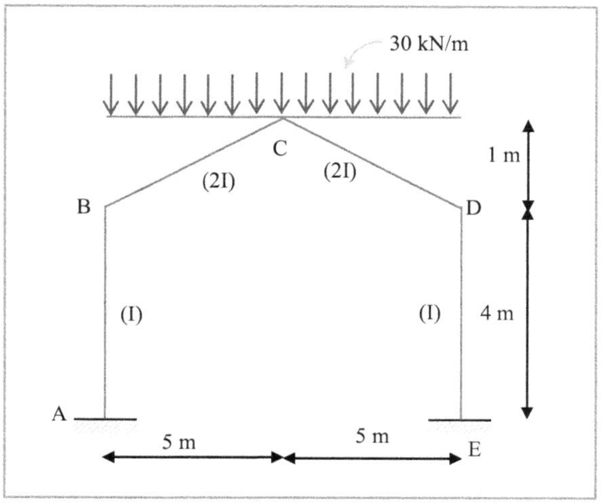

FIGURE 5.24 Indeterminate gable frame.

RESULTANT LOAD (*P*)

$$P_1 = \frac{2}{3}\times\frac{375}{2EI}\times5.1 = \frac{637.5}{EI} \quad @\ \frac{3}{8}\times5.1 = 1.91\text{ m} \quad \text{from } C$$

$$P_2 = \frac{2}{3}\times\frac{375}{2EI}\times5.1 = \frac{637.5}{EI} \quad @\ \frac{3}{8}\times5.1 = 1.91\text{ m} \quad \text{from } C$$

$$\therefore P = P_1 + P_2 = \frac{1,275}{EI}$$

$$M_{XX} = \frac{637.5}{EI}\times(1.03+3.19\cdot\sin\theta)\times2 = \frac{2,110.75}{EI}$$

(produces compression at B, C, D, and tension at A and E)

$$M_{YY} = 0 \quad \left(\text{due to symmetry of } \frac{M_L}{EI} \text{ diagram}\right)$$

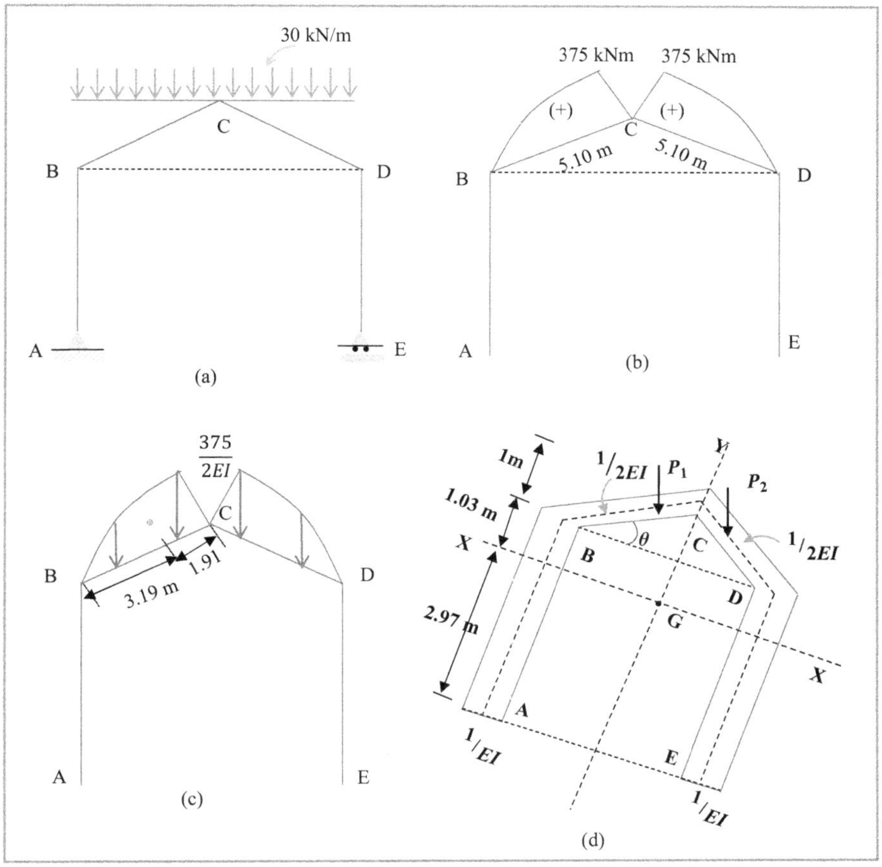

FIGURE 5.25 (a) Basic determinate frame, (b) M_L diagram, (c) $\left(\frac{M_L}{EI}\right)$ diagram, and (d) analogous column.

MOMENTS DUE TO THE REDUNDANTS

$$(M_R)_A = \frac{P}{A} - \frac{M_{XX}}{I_{XX}} y_A$$

$$= \frac{1,275}{EI} \times \frac{EI}{13.1} - \frac{2,110.75}{EI} \times \frac{EI}{30.55} \times 2.97$$

$$= 97.32 - 205.20 = -107.88 \text{ kNm} = (M_R)_E$$

$$(M_R)_B = \frac{P}{A} + \frac{M_{XX}}{I_{XX}} y_B$$

$$= 97.32 + \frac{2,110.75}{EI} \times \frac{EI}{30.55} \times 1.03$$

$$= 168.48 \text{ kNm} = (M_R)_D$$

$$(M_R)_C = \frac{P}{A} + \frac{M_{XX}}{I_{XX}} y_C$$

$$= 97.32 + \frac{2,110.75}{EI} \times \frac{EI}{30.55} \times 2.03$$

$$= 237.57 \text{ kNm}$$

FINAL END MOMENTS

$$M_A = M_E = (M_L - M_R)_A = 0 - (-107.88) = 107.88 \text{ kNm}$$

$$M_B = M_D = (M_L - M_R)_B = 0 - 168.48 = -168.48 \text{ kNm}$$

$$M_C = (M_L - M_R)_C = 375 - 237.57 = 137.43 \text{ kNm}$$

The stress resultants are shown in Figure 5.26a, and the bending moment diagram is shown in Figure 5.26b.

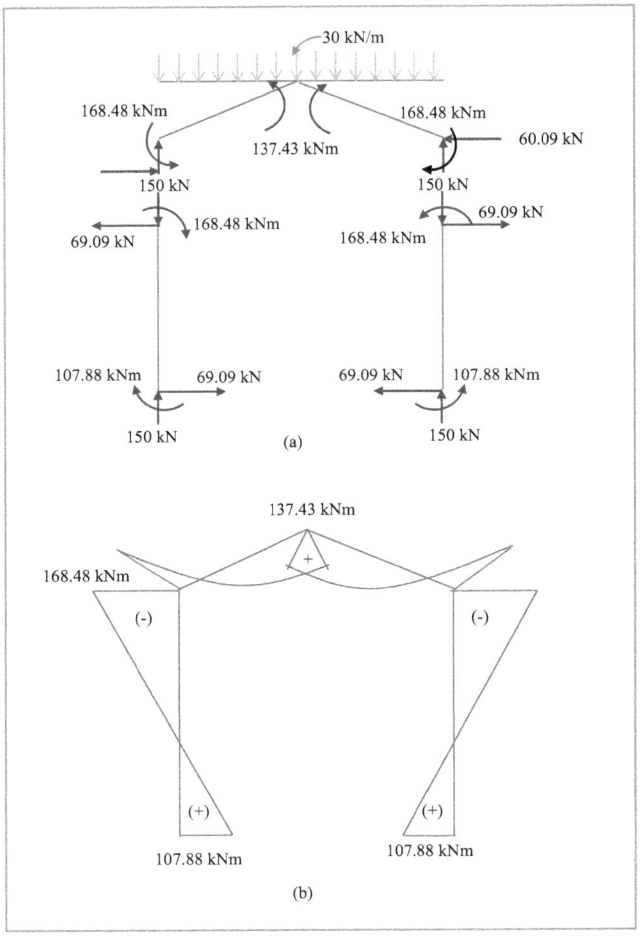

FIGURE 5.26 (a) Free body diagram and (b) bending moment diagram.

5.6 COLUMN ANALOGY FOR BOX FRAMES

Box configurations are commonly used in box culverts, box girder bridges, and tunnels. The box frames typically have axes of symmetry about both the *X–X* axis and *Y–Y* axes, ensuring that the centroid lies exactly at the centre of the box. These frames may either be supported by external supports (whose reactions are normally statically determinate) or be self-equilibrating under the action of external loads, as shown in Figure 5.27a and b. The box frame is internally indeterminate to the third degree; hence, any three internal forces can be chosen as the unknowns. To analyse the box frame, it can be cut at any suitable location (i.e., the middle of any side), as illustrated in Figure 5.27c. The bending moment diagram (M_L diagram) can then be generated by treating the cut extremes as cantilevers. The analysis then proceeds similarly to the method used for frames discussed in previous examples. The procedure is illustrated in the following examples.

Example 5.12

Analyse the rigid-jointed closed frame shown in Figure 5.28, using the column analogy method.

Solution

Support reactions

$$\sum M_A = 0 \Rightarrow V_D = \left[25 \times 2 + 50 \times 4 \times \frac{4}{2} - 50 \times 1\right]\frac{1}{4}$$

$$= 100 \text{ kN}(\uparrow)$$

$$\sum F_Y = 0 \Rightarrow V_A = 200 - 100 = 100 \text{ kN}(\uparrow)$$

$$\sum F_X = 0 \Rightarrow H_A = 50 - 25 = 25 \text{ kN } (\rightarrow)$$

The basic determinate structure with a cut at the middle of the bottom member, *AD*, is shown in Figure 5.29a. The M_L diagram and $\frac{M_L}{EI}$ diagrams are shown in Figure 5.29b and c, respectively. The section of the analogous column and the resultant loads are shown in Figure 5.29d and e, respectively.

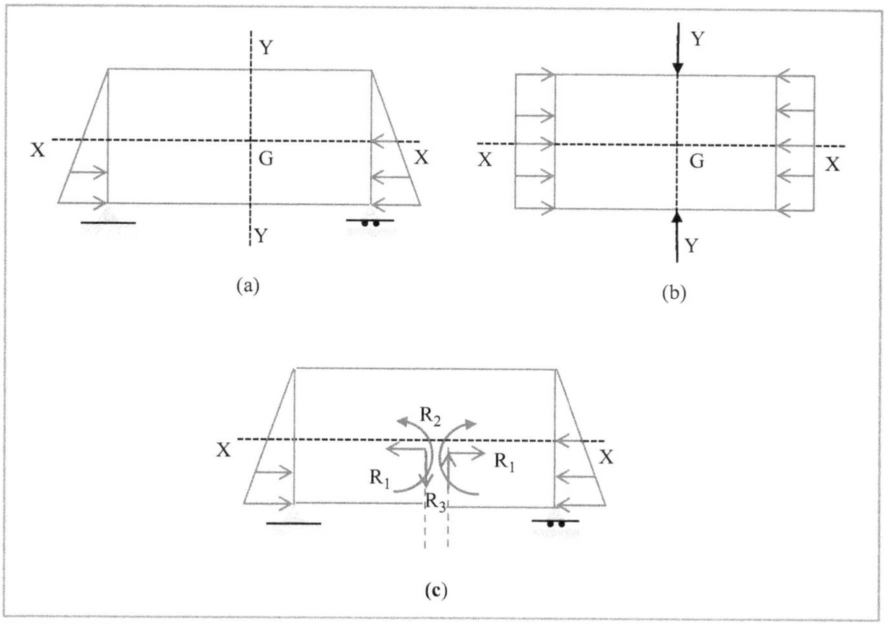

FIGURE 5.27 (a) Simply supported box frame, (b) self-equilibrating box frame, and (c) redundants on the basic frame.

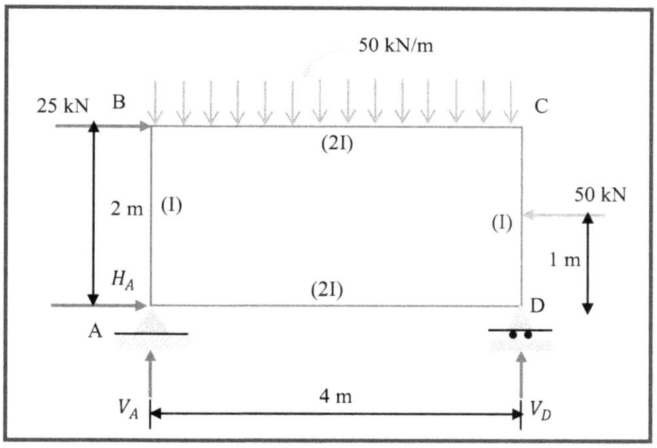

FIGURE 5.28 Box frame.

PROPERTIES OF ANALOGOUS COLUMN

$$A = 2\left[\frac{1}{2EI}\times 4+\frac{1}{EI}\times 2\right]=\frac{8}{EI};\quad P_1=\frac{1}{2}\times\frac{50}{EI}\times 1=\frac{25}{EI}\ (\uparrow)$$

$$P_2=\left[\frac{2}{3}\times 4\times\frac{100}{2EI}-50\times\frac{1}{2EI}\times 4\right]=\frac{33.33}{EI}(\downarrow);$$

$$P_3=\frac{1}{2}\times 2\times\frac{50}{EI}=\frac{50}{EI}\ (\uparrow)$$

$$\text{Resultant load } (P)=P_1+P_2+P_3=\frac{-25}{EI}+\frac{33.33}{EI}-\frac{50}{EI}$$

$$=\frac{-41.67}{EI}(\uparrow)$$

The analogous column is symmetric with respect to both the X–X axis and Y–Y axes. Therefore, the c.g. lies at the centre.

$$M_{XX}=-P_1\times 0.67+P_2\times 1-P_3\times 0.33=\frac{-25}{EI}$$

$$\times 0.67+\frac{33.33}{EI}\times 1-\frac{50}{EI}\times 0.33=\frac{0.08}{EI}$$

$$M_{YY}=(P_3-P_1)\times 2=\left(\frac{50}{EI}-\frac{25}{EI}\right)\times 2=\frac{50}{EI}\ \text{(clockwise)}$$

(produces compression on CD and tension on AB)

$$I_{XX}=\frac{1}{12}\times\frac{1}{EI}\times 2^3\times 2+\frac{1}{2EI}\times 4\times 1^2\times 2=\frac{5.33}{EI}$$

$$I_{YY}=\frac{1}{12}\times\frac{1}{2EI}\times 4^3\times 2+\frac{1}{EI}\times 2\times 2^2\times 2=\frac{21.33}{EI}$$

FINAL END MOMENTS

The final end moments are calculated and presented in Table 5.2. The bending moment diagram is shown in Figure 5.30.

Example 5.13

Analyse the self-equilibrating box frame shown in Figure 5.31, using the column analogy method.

Solution

Since both the loading and the frame are symmetric with respect to both the X–X axis and Y–Y axis, the point of application of the resultant loads coincides with the centroid of the frame. That is, $e_{XX}=e_{YY}=0$.

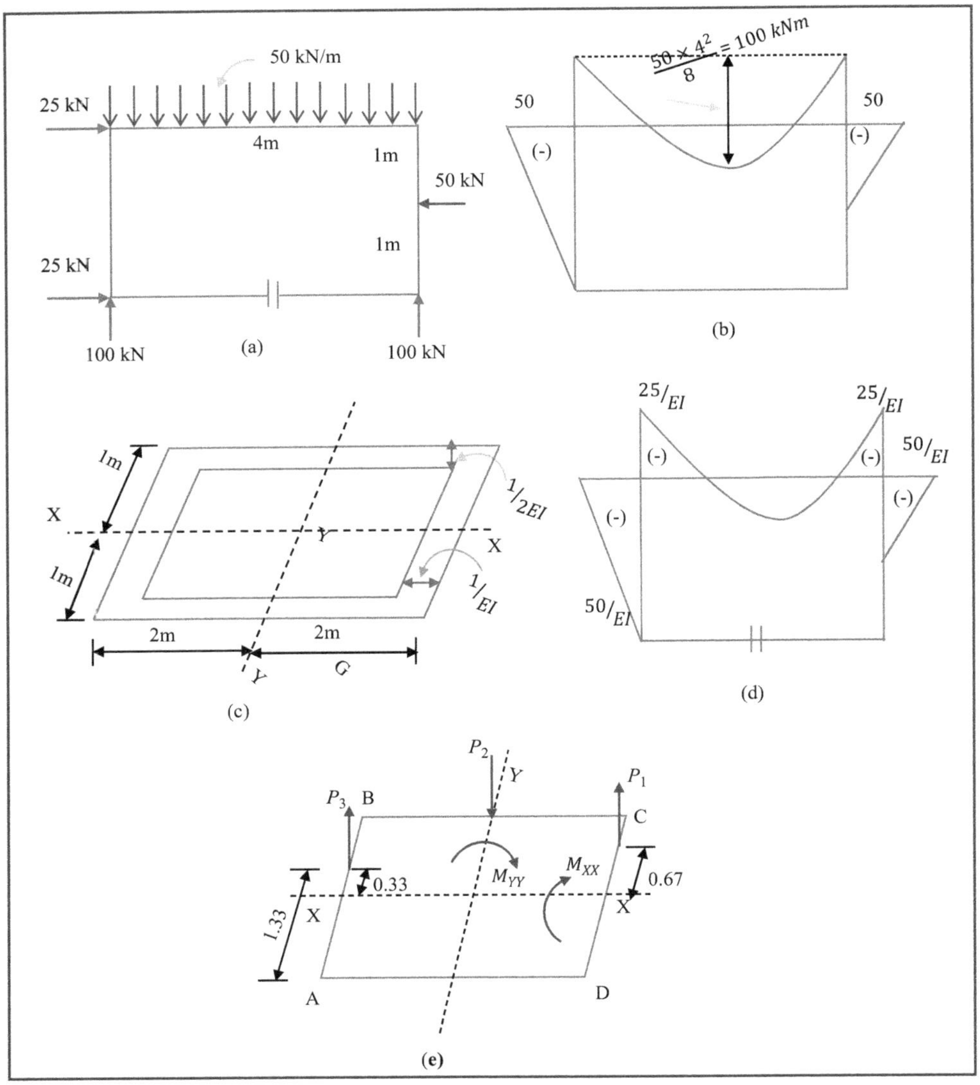

FIGURE.5.29 (a) Basic determinate frame, (b) M_L diagram, (c) $\left(\frac{M_L}{EI}\right)$ diagram, (d) analogous column, and (e) resultant loads.

TABLE 5.2
Final End Moments

Point	M_L (kNm)	$\frac{P}{A}$	$\left(\frac{M_{XX}}{I_{XX}}\right)y$	$\left(\frac{M_{YY}}{I_{YY}}\right)x$	M_R (kNm)	M (kNm)
A	0	−5.21	0	−4.68	−9.89	9.89
B	−50	−5.21	0	−4.68	−9.89	−40.11
C	−50	−5.21	0	4.68	−0.53	−49.47
D	0	−5.21	0	4.68	−0.53	0.53

Consequently, the bending moments due to the redundants, $M_R = \frac{P}{A}$.

Let us assume the basic structure has hinges at the corners so that each segment of the box frame acts as a simply supported beam. The M_L diagram and $\frac{M_L}{EI}$ diagrams are shown in Figure 5.32a and b, respectively. The resultant loads are shown in Figure 5.32c, and the cross-section of the analogous column is shown in Figure 5.32d.

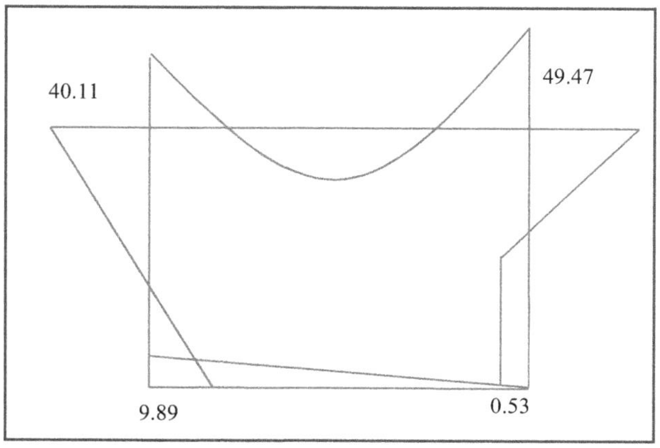

FIGURE 5.30 Bending moment diagram.

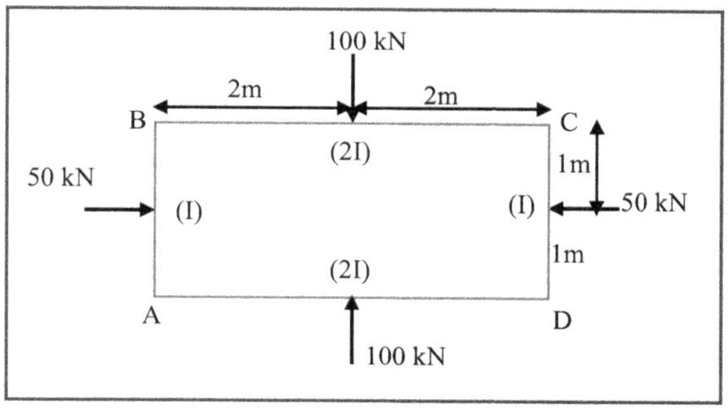

FIGURE 5.31 Self-equilibrating box frame.

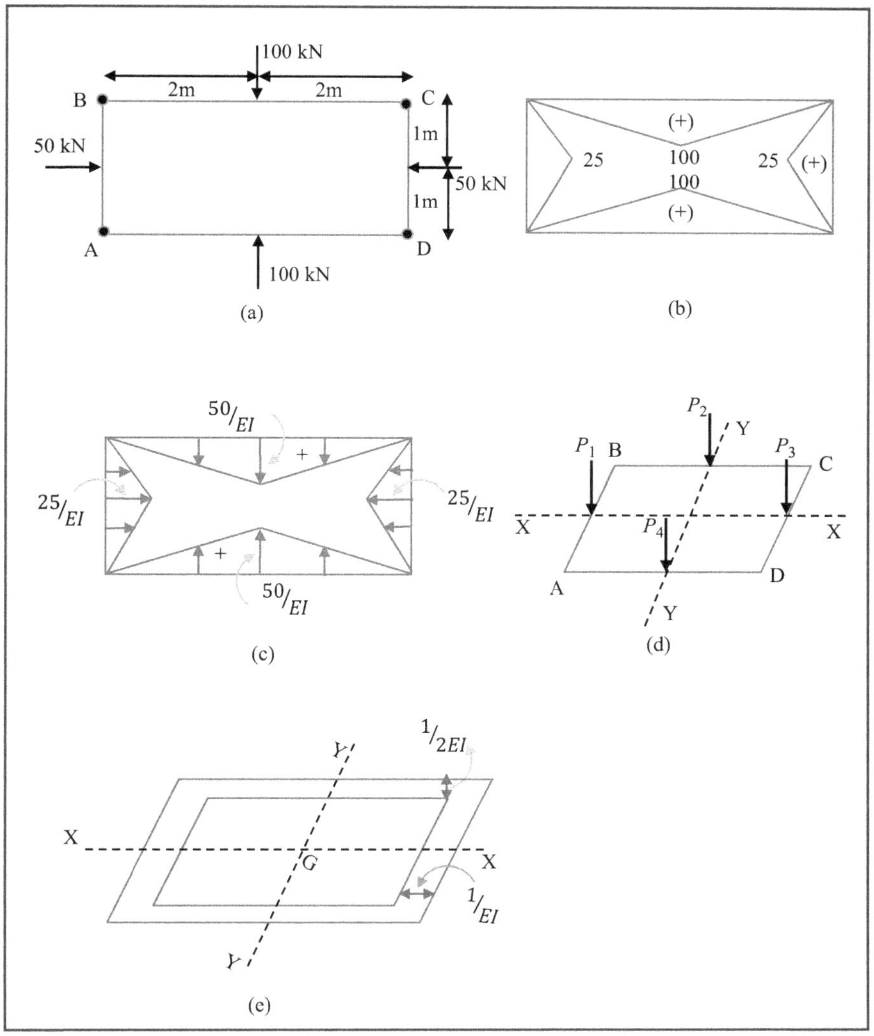

FIGURE 5.32 (a) Basic structure, (b) M_L diagram, (c) $\frac{M_L}{EI}$ diagram, (d) resultant loads, and (e) analogous column.

PROPERTIES OF THE ANALOGOUS COLUMN

$$P_1 = \frac{1}{2} \times \frac{25}{EI} \times 2 = \frac{25}{EI} = P_3$$

$$P_2 = \frac{1}{2} \times \frac{50}{EI} \times 4 = \frac{100}{EI} = P_4$$

RESULTANT LOAD (*P*)

$$P = 2(P_1 + P_2) = \frac{250}{EI}$$

$$A = 2\left[\frac{1}{2EI} \times 4 + \frac{1}{EI} \times 2\right] = \frac{8}{EI}$$

MOMENTS DUE TO THE REDUNDANTS

$$(M_R)_A = \frac{P}{A} = \frac{250}{EI} \times \frac{EI}{8} = 31.25 \text{ kNm}$$

$$= (M_R)_B = (M_R)_C = (M_R)_D.$$

Final End Moments

$$M_A = (M_L - M_R)_A = 0 - 31.25 = -31.25 \text{ kNm}$$

$$= M_B = M_C = M_D$$

ALTERNATE SOLUTION

Let us cut the frame at the middle of the beam *AD*, as shown in Figure 5.33a. The corresponding M_L diagram, M_L/EI diagram, and the resultant loads are shown in Figure 5.33b–d, respectively.

$$P_1 = 4 \times \frac{25}{EI} = \frac{100}{EI} \text{ @ 2 m from } B \text{ on side } BC \text{ (on } Y - Y \text{ axis.)}$$

$$P_2 = \frac{1}{2} \times 2 \times \frac{50}{EI} = \frac{50}{EI} \text{ @ 0.667 m from } B \text{ on side } BC$$

$$P_3 = 2 \times \frac{100}{EI} = \frac{200}{EI} \text{ @ 1 m from } A \text{ on side } AB \text{ (on } X - X \text{ axis)}$$

$$P_4 = \frac{1}{2} \times 1 \times \frac{50}{EI} = \frac{25}{EI} \text{ @ 0.667 m from } X - X \text{ axis on side } AB$$

RESULTANT LOAD

$$P = P_1 + 4P_2 + 2P_3 + 2P_4$$

$$= \frac{100}{EI} + 4 \times \frac{50}{EI} + 2 \times \frac{200}{EI} + 2 \times \frac{25}{EI}$$

$$= \frac{750}{EI}$$

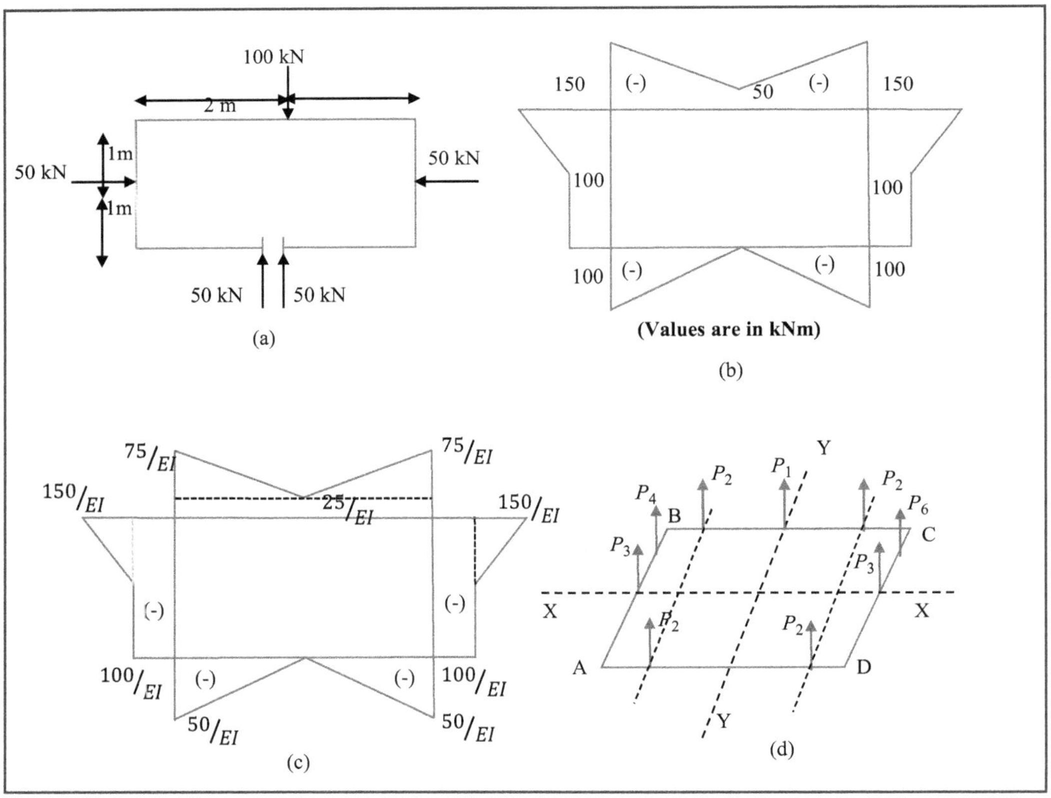

FIGURE 5.33 (a) Basic structure, (b) M_L diagram, (c) $\frac{M_L}{EI}$ diagram, and (d) resultant loads.

$$A = 2\left[\frac{1}{2EI}\times 4+\frac{1}{EI}\times 2\right]=\frac{8}{EI} \quad \text{(as calculated earlier)}$$

$$I_{XX} = 2\left[\frac{1}{12}\times\frac{1}{EI}\times 2^3+\frac{1}{2EI}\times 4\times 1^2\right]=\frac{5.33}{EI}$$

$$M_{XX} = P_1\times 1+2P_4\times 0.667$$

$$=\frac{100}{EI}\times 1+2\times\frac{25}{EI}\times 0.667 = 133.35 \text{ kNm } (\curvearrowleft)$$

(produces compression on AD and tension on BC)

$$M_{YY} = 0$$

MOMENTS DUE TO THE REDUNDANTS

$$(M_R)_A = (M_R)_D = \frac{-P}{A}+\frac{M_{XX}}{I_{XX}}y_A$$

$$=\frac{-750}{EI}\times\frac{EI}{8}+\frac{133.35}{EI}\times\frac{EI}{5.33}\times 1$$

$$= -68.73 \text{ kNm}$$

$$(M_R)_B = (M_R)_C = \frac{-P}{A}-\frac{M_{XX}}{I_{XX}}y_B$$

$$=\frac{-750}{EI}\times\frac{EI}{8}-\frac{133.35}{EI}\times\frac{EI}{5.33}\times 1$$

$$= -118.73 \text{ kNm}$$

FINAL END MOMENTS (FIGURE 5.34)

$$M_A = M_D = (M_L - M_R)_A = -100+68.73 = -31.27 \text{ kNm}$$

$$M_B = M_C = (M_L - M_R)_B = -150+118.73 = -31.27 \text{ kNm}$$

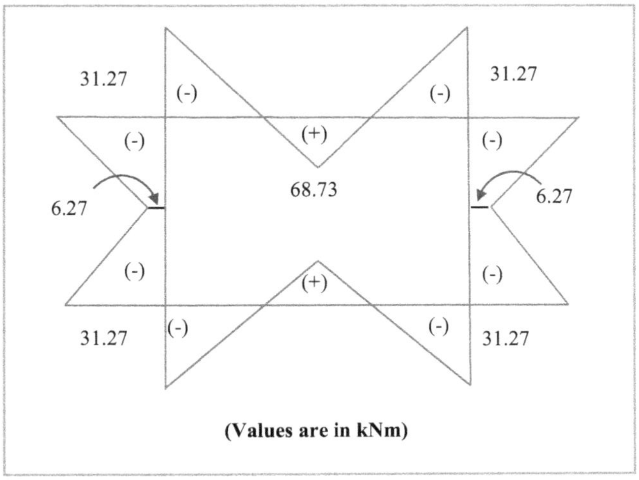

FIGURE 5.34 Bending moment diagram.

5.7 APPLICATION TO UNSYMMETRICAL RIGID-JOINTED PLANE FRAMES

In the case of frames with unsymmetrical geometry, the principle axes of the analogous column will not align with the X and Y axes. This means that when the analogous column is subjected to an eccentric load P_L with respect to the centroid (elastic centre), unsymmetrical bending occurs. As a result, the simple elastic stress equation does not apply in the column analogy. In such cases, the following modified equation 5.10 is used.

$$M_R = \frac{P_L}{A}\pm\frac{M'_{XX}}{I'_{XX}}y\pm\frac{M'_{yy}}{I'_{yy}}x \quad (5.10)$$

where

$$M'_{XX} = M_{XX}-\frac{I_{XY}}{I_{YY}}M_{YY};\quad M'_{YY} = M_{YY}-\frac{I_{XY}}{I_{XX}}M_{XX}$$

$$I'_{XX} = I_{XX}\left[1-\frac{I_{XY}^2}{I_{XX}\cdot I_{YY}}\right];\quad I'_{YY} = I_{YY}\left[1-\frac{I_{XY}^2}{I_{XX}\cdot I_{YY}}\right]$$

Example 5.14

Analyse the unsymmetrical rigid-jointed plane frame shown in Figure 5.35, using the column analogy method.

Solution

The basic determinate frame as a cantilever with fixed end at A is shown in Figure 5.36a. The corresponding M_L and $\frac{M_L}{EI}$ diagrams are shown in Figure 5.36b and c, respectively. The analogous column and the resultant loads are shown in Figure 5.37.

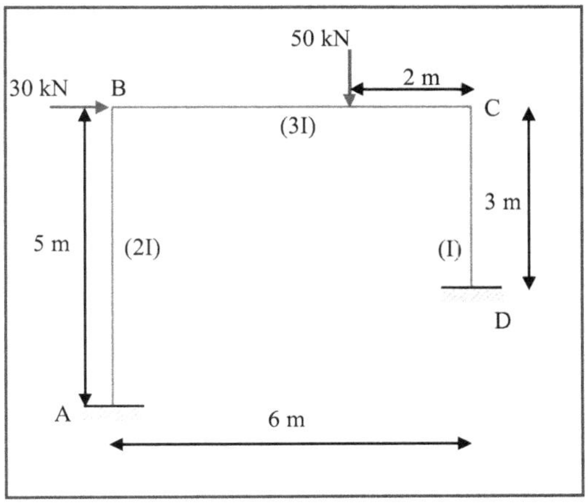

FIGURE 5.35 Unsymmetrical rigid-jointed frame.

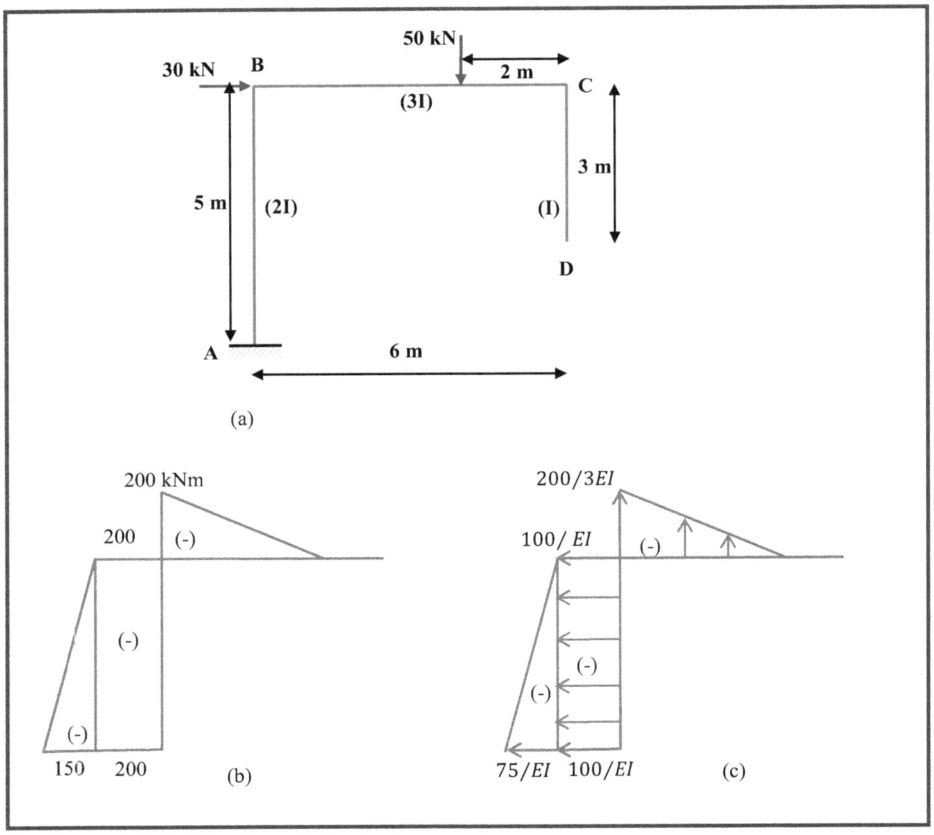

FIGURE 5.36 (a) Basic determinate frame, (b) M_L diagram, and (c) $\left(\frac{M_L}{EI}\right)$ diagram.

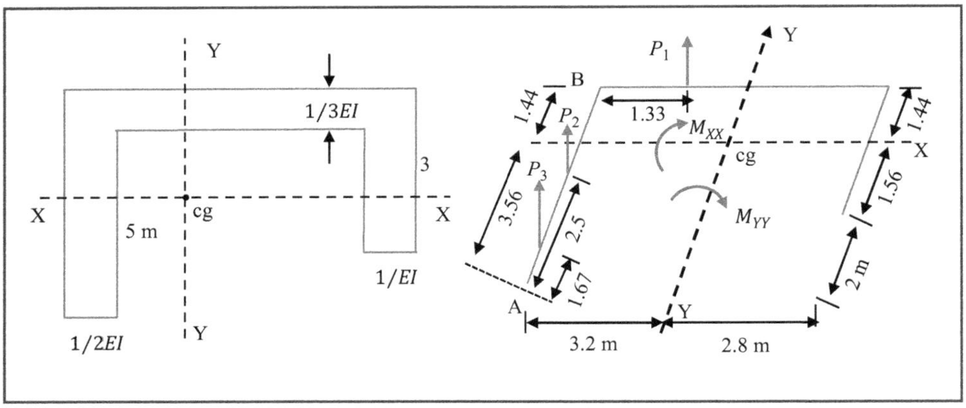

FIGURE 5.37 Analogous column and the resultant loads.

Properties of the analogous column

$$A = \left(3 \times \frac{1}{EI}\right) + \left(6 \times \frac{1}{3EI}\right) + \left(5 \times \frac{1}{2EI}\right) = \frac{7.5}{EI}$$

Location of the centroid (x_g, y_g) with reference to *AB* (Figure 5.37)

$$y_g = \frac{\left(\frac{1}{2EI} \times 5 \times \frac{5}{2}\right) + \left(\frac{1}{3EI} \times 6 \times 5\right) + \left(\frac{1}{EI} \times 3\right)\left(2 + \frac{3}{2}\right)}{\frac{7.5}{EI}}$$

$$= 3.56 \text{ m}$$

$$x_g = \frac{\left(\frac{1}{2EI}\times 5\times 0\right)+\left(\frac{1}{3EI}\times 6\times 3\right)+\left(\frac{1}{EI}\times 3\times 6\right)}{\frac{7.5}{EI}} = 3.2 \text{ m}$$

$$I_{XX} = \left[\frac{1}{3}\times\frac{1}{2EI}\times(3.56)^3+\frac{1}{3}\times\frac{1}{2EI}\times(1.44)^3\right]$$

$$+\left[\frac{1}{3EI}\times 6\times(1.44)^2\right]+\left[\frac{1}{3}\times\frac{1}{EI}\times\left[(1.56)^3+(1.44)^3\right]\right]$$

$$=\frac{8.017}{EI}+\frac{4.147}{EI}+\frac{2.261}{EI}=\frac{14.425}{EI}$$

$$I_{YY} = \left[\frac{1}{2EI}\times 5\times(3.2)^2\right]+\left[\frac{1}{3}\times\frac{1}{3EI}\times\left[(3.2)^3+(2.8)^3\right]\right]$$

$$+\left[\frac{1}{EI}\times 3\times(2.8)^2\right]=\frac{55.2}{EI}$$

$$I_{XY} = \left[\frac{1}{2EI}\times 5\times(-3.2)(2.5-3.56)\right]$$

$$+\left[\frac{1}{3EI}\times 6\times(3-3.2)(5-3.56)\right]$$

$$+\left[\frac{1}{EI}\times 3\times(6-3.2)(3.5-3.56)\right]$$

$$=\frac{8.48}{EI}-\frac{0.576}{EI}-\frac{0.504}{EI}=\frac{7.4}{EI}$$

With respect to the principal axis

$$I'_{XX} = I_{XX}\left[1-\frac{I_{XY}^2}{I_{XX}\cdot I_{YY}}\right]=\frac{14.425}{EI}\left[1-\frac{\left(\frac{7.4}{EI}\right)^2}{\left(\frac{14.425}{EI}\times\frac{55.2}{EI}\right)}\right]$$

$$=\frac{13.43}{EI}$$

$$I'_{YY} = I_{YY}\left[1-\frac{I_{XY}^2}{I_{XX}\cdot I_{YY}}\right]=\frac{55.2}{EI}\left[1-\frac{\left(\frac{7.4}{EI}\right)^2}{\left(\frac{14.425}{EI}\times\frac{55.2}{EI}\right)}\right]=\frac{51.39}{EI}$$

Loads on the analogous column

$$P_1=\frac{1}{2}\times\frac{200}{3EI}\times 4=\frac{400}{3EI} \quad \text{acting at } \frac{4}{3}=1.33 \text{ m from } B$$

$$P_2=\frac{100}{EI}\times 5=\frac{500}{EI} \quad \text{acting at } 2.5 \text{ m from } A$$

$$P_3=\frac{1}{2}\times\frac{75}{EI}\times 5=\frac{187.5}{EI} \quad \textit{acting at } \frac{5}{3}=1.67 \textit{ m from } A$$

$$\text{Resultant load } (P)=P_1+P_2+P_3=\frac{820.83}{EI}$$

$$M_{XX}=\frac{-400}{3EI}\times 1.44+\frac{500}{EI}(3.56-2.5)+\frac{187.5}{EI}(3.56-1.67)$$

$$=\frac{-192}{EI}+\frac{530}{EI}+\frac{354.37}{EI}=\frac{692.37}{EI}\ (\curvearrowright)$$

(Produces compression on *BC* and tension at *A* and *D*)

$$M_{YY}=\frac{400}{3EI}(3.2-1.33)+\frac{500}{EI}\times 3.2+\frac{187.5}{EI}\times 3.2$$

$$=\frac{2{,}449.33}{EI}\ (\curvearrowright)$$

(Produces compression on *CD* and tension on *AB*)

$$M'_{XX}=M_{XX}-\frac{I_{XY}}{I_{YY}}M_{YY}=\frac{692.37}{EI}-\frac{7.4/EI}{55.2/EI}\times 2{,}449.33$$

$$=\frac{364.02}{EI}$$

$$M'_{YY}=M_{YY}-\frac{I_{XY}}{I_{XX}}.M_{XX}=\frac{2{,}449.33}{EI}-\frac{7.4/EI}{14.425/EI}\times\frac{692.37}{EI}$$

$$=\frac{2{,}094.14}{EI}$$

MOMENTS DUE TO THE REDUNDANTS

$$M_R=\frac{-P}{A}\pm\frac{M'_{YY}}{I'_{YY}}.x_i\pm\frac{M'_{XX}}{I'_{XX}}y_i$$

$$(M_R)_A=\frac{-820.83}{7.5}-\frac{2{,}094.14}{51.39}\times 3.2-\frac{364.02}{13.43}\times 3.57$$

$$=-109.44-130.39-96.76=-336.59 \text{ kNm}$$

$$(M_R)_B=\frac{-820.83}{7.5}-\frac{2{,}094.14}{51.39}\times 3.2+\frac{364.02}{13.43}\times 1.44$$

$$=-109.44-130.39+39.03=-200.8 \text{ kNm}$$

$$(M_R)_C=\frac{-820.83}{7.5}+\frac{2{,}094.14}{51.39}\times 2.8+\frac{364.02}{13.43}\times 1.44$$

$$=-109.44+114.09+39.03=43.68 \text{ kNm}$$

$$(M_R)_D=\frac{-820.83}{7.5}+\frac{2{,}094.14}{51.39}\times 2.8-\frac{364.02}{13.43}\times 1.57$$

$$=-109.44+114.09-42.55=-37.9 \text{ kNm}$$

FINAL BENDING MOMENTS

$$M_A=(M_L-M_R)_A=-350+336.59=-13.41 \text{ kNm}$$

$$M_B=(M_L-M_R)_B=-200+200.8=0.8 \text{ kNm}$$

$$M_C=(M_L-M_R)_C=0-43.68=-43.68 \text{ kNm}$$

$$M_D=(M_L-M_R)_D=0+37.9=37.9 \text{ kNm}$$

5.8 STIFFNESS FACTORS OF NON-PRISMATIC MEMBERS

Prismatic members:

Stiffness (rotational): It is defined as the moment required to produce a unit rotation (1 rad) at one end of the beam when the other end is fixed (Figure 5.38).

Cary over factor: The carryover factor from A to B for a beam AB is equal to the ratio of the moment developed at B to the moment applied at A. Let us consider the fixed beam, as shown in Figure 5.39. Apply the moment M_A at A to produce a unit rotation at A. As a result a moment will develop at point B, denoted as M_B.

Steps to find the carryover factors using the column analogy method

1. Draw the analogous column with depth=length of the beam and width equal $\frac{1}{EI}$ (it will vary if EI varies).
2. Apply a unit downward load at A on the c/s of the analogous column corresponding to a unit rotation at A.
3. Calculate the moment required to produce a unit rotation at A, which is the pressure developed in the analogous column at edge A. Also, calculate the moment developed at end B, which is the pressure developed in the analogous column at edge B.
4. Calculate the carryover factor from A to $B = -\frac{M_B}{M_A}$

Example 5.15

Determine the stiffness factor and carryover factor from A to B for a fixed beam of length L as shown in Figure 5.40, using the column analogy method. Assume EI is constant.

Solution

First, draw the analogous column (Figure 5.41).

i. Apply a unit load at A corresponding to the unit rotation at A (Figure 5.41).
ii. Calculate the pressures developed at A and B using the analogous column.

$$p = \frac{P}{A} \pm \frac{M}{I}(x_A)$$

where

P = Total resultant load on the analogous column = 1

$A = c/s$ area of the analogous column $= L \times \frac{1}{EI} = \frac{L}{EI}$

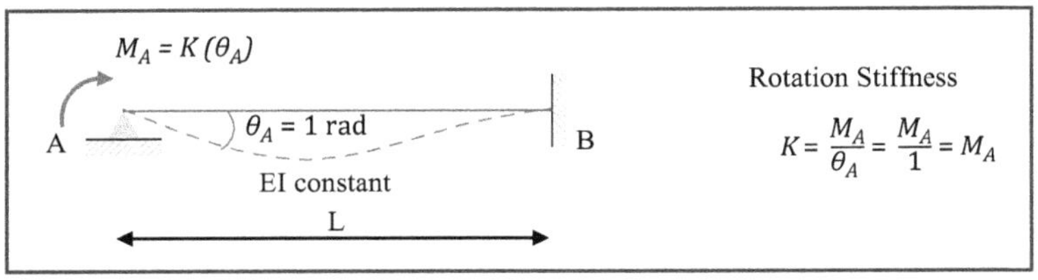

FIGURE 5.38 Beam with one end hinged and the other end fixed.

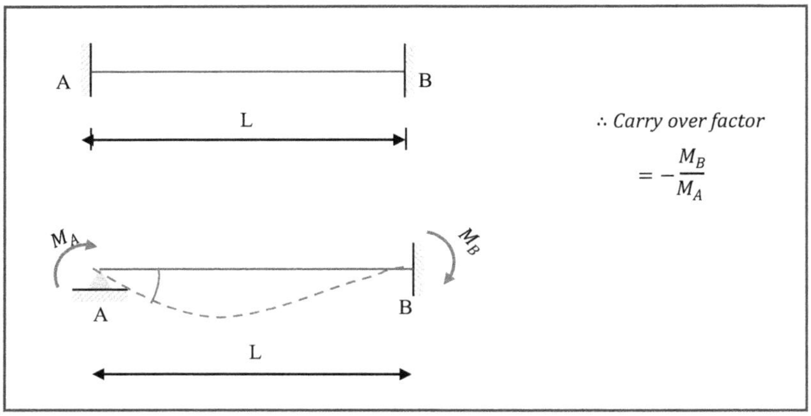

FIGURE 5.39 Carryover factor.

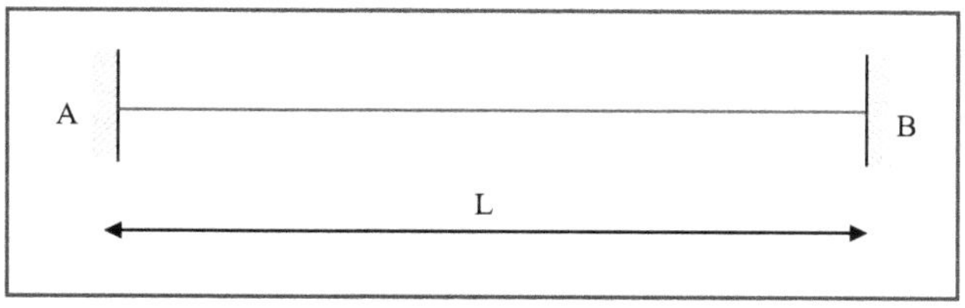

FIGURE 5.40 Fixed beam *AB*.

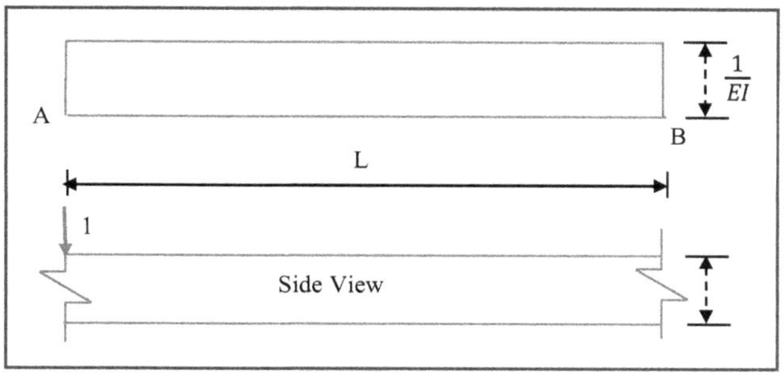

FIGURE 5.41 Analogous column.

$$\text{Eccenticity } (e) = \frac{L}{2}$$

$$\therefore \text{Moment } (M) = P \times e = 1 \times \frac{L}{2} = \frac{L}{2}$$

$$x_A = \frac{L}{2}; \quad I_{YY} = \frac{1}{12} \times L^3 \times \frac{1}{EI} = \frac{L^3}{12EI}$$

$$M_A = \text{Pressure at } A = p_A = \frac{1}{(L/EI)} + \frac{1 \times (L/2)}{L^3/12EI} \times \frac{L}{2}$$

$$= \frac{EI}{L} + \frac{3EI}{L} = \frac{4EI}{L}$$

$$\therefore \text{The stiffness factor, } K_A = \left(\frac{M_A}{1}\right) = \frac{4EI}{L} = \frac{M_A}{\theta_A}$$

$$\therefore \text{Moment at } B = \text{Pressure at } B = \frac{P}{A} - \frac{M}{I_{YY}} \times x_B$$

$$M_B = p_B = \frac{1}{(L/EI)} - \frac{1 \times (L/2)}{\left(L^3/12EI\right)} \times \frac{L}{2}$$

$$= \frac{EI}{L} - \frac{3EI}{L} = \frac{-2EI}{L}$$

$$\therefore \text{Carry over factor from } A \text{ to } B = -\frac{p_B}{p_A} = -\frac{M_B}{M_A}$$

$$= -\left(\frac{-2EI/L}{4EI/L}\right) = \frac{1}{2}$$

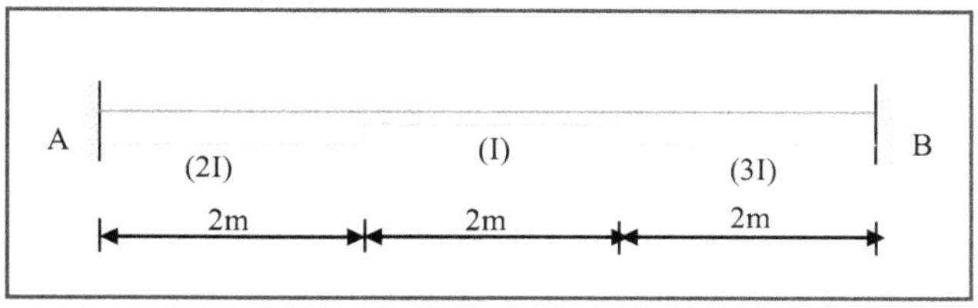

FIGURE 5.42 Non-prismatic fixed beam.

Example 5.16

Using the column analogy method, find the stiffness factor at *A*, and carryover factor from *A* to *B* for the non-prismatic fixed beam shown in Figure 5.42.

Solution

The *c/s* of the analogous column is shown in Figure 5.43. To calculate the stiffness factor at *A*, apply a unit load at *A* on the analogous column corresponding to the unit rotation at *A* (Figure 5.43).

The pressures at *A* and *B* are

$$p_{A,B} = \frac{P}{A} \pm \frac{M}{I} \times x$$

where

$$A = \frac{1}{2EI} \times 2 + \frac{1}{EI} \times 2 + \frac{1}{3EI} \times 2 = \frac{22}{6EI}$$

$$P = 1 \text{ unit}$$

c.g. of the analogous column *c/s* is x_g from point *A* is

$$x_g = \frac{\sum A_i x_i}{\sum A_i} = \frac{\left(\frac{1}{2EI} \times 2 \times 1\right) + \left(\frac{1}{EI} \times 2 \times 3\right) + \left(\frac{1}{3EI} \times 2 \times 4\right)}{\frac{22}{6EI}}$$

$$= \frac{29}{11EI} = 2.44 \text{ m}$$

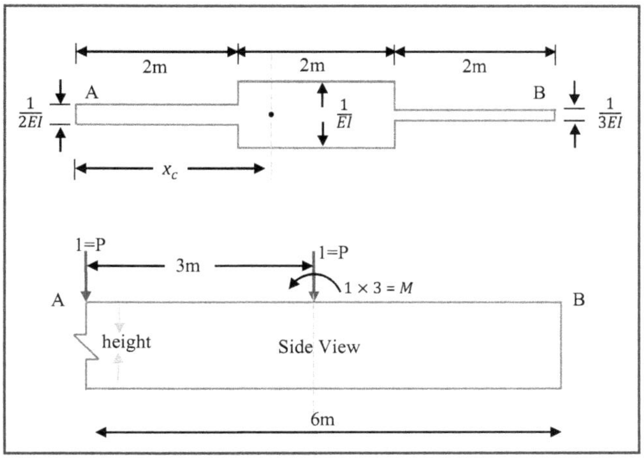

FIGURE 5.43 Cross-section of the analogous column.

$$I_{YY} = \left[\frac{1}{12}\times\frac{1}{2EI}\times 2^3 + \frac{1}{2EI}\times 2\times(2.44-1)^2\right]$$

$$+\left[\frac{1}{12}\times\frac{1}{EI}\times 2^3 + \frac{1}{EI}\times 2\times(3-2.44)^2\right]$$

$$+\left[\frac{1}{12}\times\frac{1}{3EI}\times 2^3 + \frac{1}{3EI}\times 2\times(5-2.44)^2\right]$$

$$= \frac{8.29}{EI}$$

$$x_A = 2.4 \text{ m}, \quad x_B = 6-2.44 = 3.56 \text{ m}$$

$$\therefore\ M_A = \text{Pressure at } A = p_A = \frac{P}{A} + \frac{M}{I}x_A$$

$$= \frac{1}{(22/6EI)} + \frac{1\times 2.44}{8.29/EI}\times 2.44$$

$$= 0.272\ EI + 0.718\ EI = 0.99\ EI$$

$$\text{Pressure at } B = \frac{P}{A} - \frac{M}{I}\times x_B = \frac{1}{(22/6EI)} - \frac{1\times 2.44}{8.29/EI}\times 3.56$$

$$= 0.272\ EI - 1.047\ EI = -0.775\ EI$$

$$\text{Stiffness factor } K_A = \frac{M_A}{\theta_A} = \frac{0.99\ EI}{1} = 0.99\ EI$$

$$\therefore \text{Carry over factor from } A \text{ to } B = -\frac{M_B}{M_A} = -\left(\frac{-0.775\ EI}{0.99\ EI}\right)$$

$$= +0.782$$

5.9 PROBLEMS FOR PRACTICE

5.9.1 Explain what changes need to be made when hinged supports occur in the beams and frames while applying the column analogy method.

5.9.2 How do you account for the settlement of supports that occurs in the fixed beam during the application of the column analogy method?

5.9.3 Determine the end moments of a fixed beam with a span of 6 m subjected to an eccentric clockwise couple of 20 kNm at 2 m from the left support using the column analogy method. Assume the flexural rigidity is constant. Draw the shear force and bending moment diagrams.

5.9.4 Analyse the fixed beam shown in Figure 5.44 using the column analogy method. Draw the bending moment diagram.

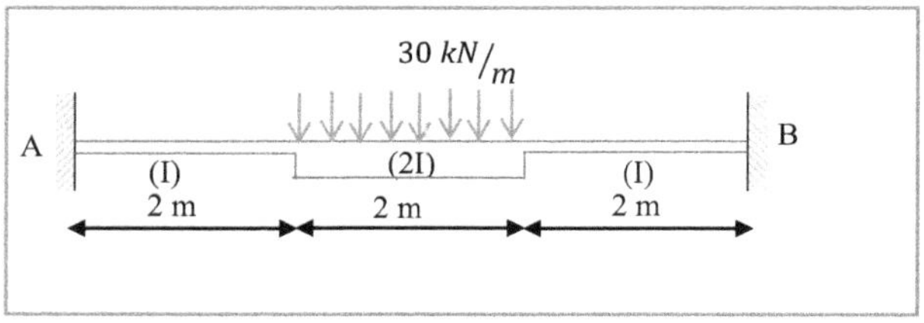

FIGURE 5.44

5.9.5 A propped cantilever beam is shown in Figure 5.45. Find the fixed end moment at A using the column analogy method. Draw the bending moment diagram.

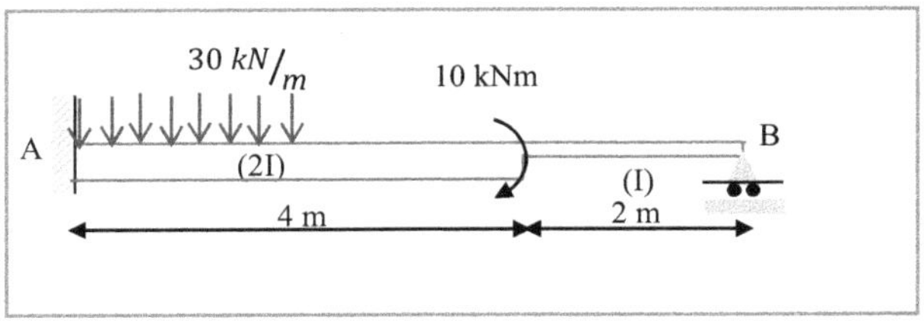

FIGURE 5.45

5.9.6 If the fixed support A of the propped cantilever beam shown in Figure 5.45 rotates by 0.001 rad in the absence of external loads, find the fixed end moment using the column analogy method. Draw the bending moment diagram. Assume $EI = 138{,}375$ kNm^2.

5.9.7 In the absence of external loads on the fixed beam shown in Figure 5.44 (Q.No. 5.9.6), if support B settles by 5 mm, analyse the beam using the column analogy method. Draw the bending moment diagram. Assume $EI = 138{,}375$ kNm^2.

5.9.8 Analyse the symmetrical rigid-jointed frame shown in Figure 5. 46, using the column analogy method. Draw the shear force and bending moment diagrams.

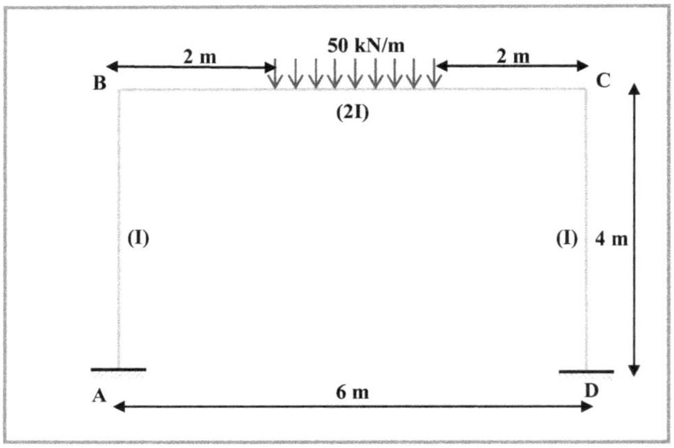

FIGURE 5.46

5.9.9 A rigid-jointed portal frame is shown in Figure 5.47. Analyse it using the column analogy method. Draw the shear force and bending moment diagrams.

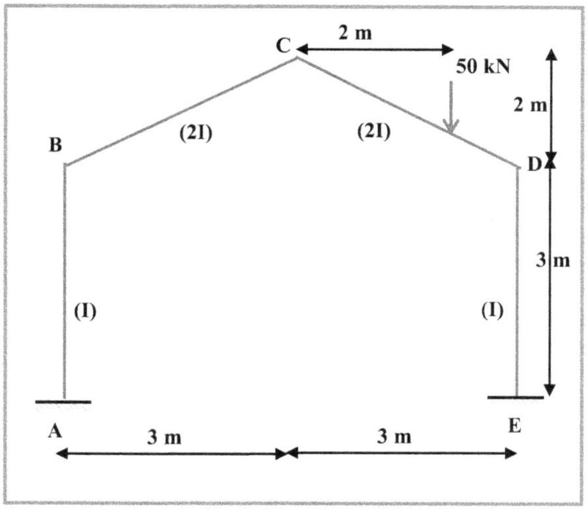

FIGURE 5.47

5.9.10 Analyse the anti-symmetric frame shown in Figure 5.48 using the column analogy method. Draw the bending moment diagram.

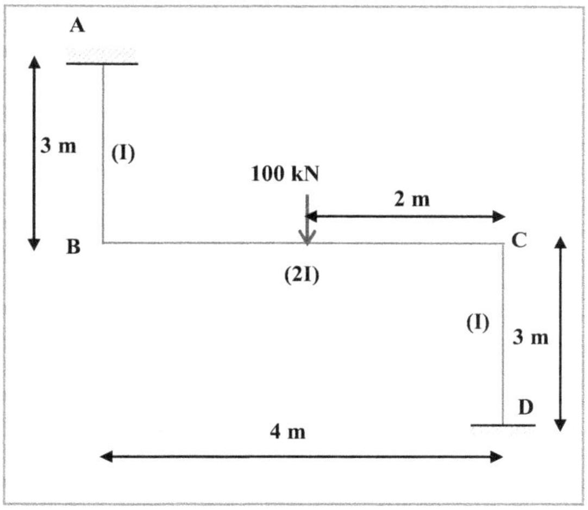

FIGURE 5.48

5.9.11 Analyse the box frame subjected to symmetrical loading shown in Figure 5.49 using the column analogy method. Draw the bending moment diagram.

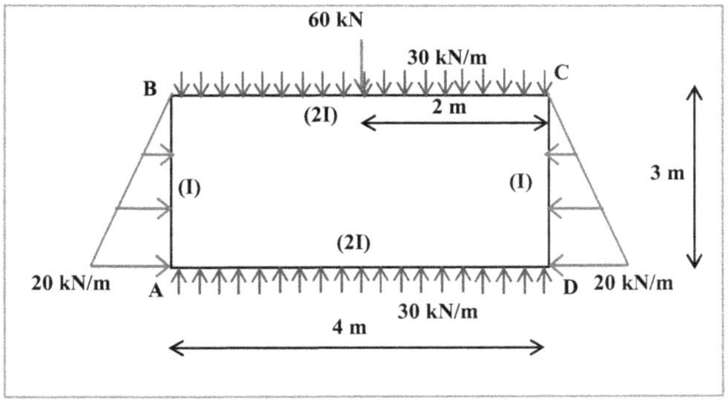

FIGURE 5.49

5.9.12 Analyse the unsymmetrical rigid-jointed frame shown in Figure 5.50, using the column analogy method. Draw the shear force and bending moment diagrams.

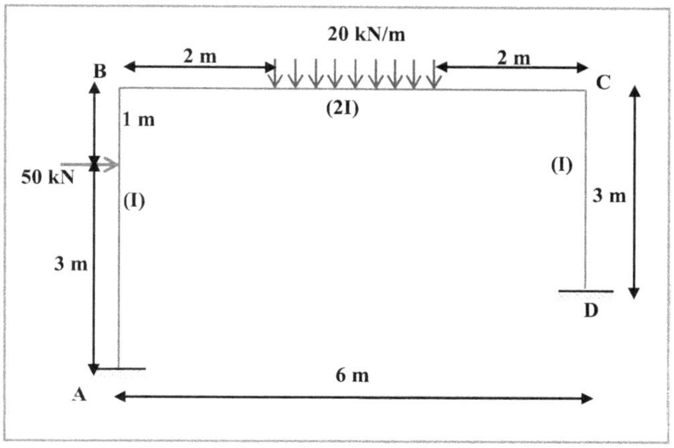

FIGURE 5.50

5.10 ANSWER KEY

S.No.	Problem No.	Answers at Key Points
1	5.9.4	$M_A=0$, $M_B=-6.67$ kNm
2	5.9.5	$M_A=-M_B=-38$ kNm
3	5.9.6	$M_A=-39.44$ kNm, $M_B=0$
4	5.9.7	$M_A=133.04$ kNm
5	5.9.8	$M_A=M_B=-116.67$ kNm
6	5.9.9	$M_A=21.21$ kNm$=M_D$; $M_B=-43.07$ kNm$=M_C$
7	5.9.10	$M_A=10.20$ kNm; $M_E=5.64$ kNm; $V_A=7.54$ kN; $V_E=42.45$ kN; $H_A=6.45$ kN; $H_E=-6.45$ kN
8	5.9.11	$M_A=-12.46$ kNm; $M_D=12.46$ kNm; $M_{BA}=12.46$ kNm; $M_{BC}=-12.46$ kNm; $M_{CB}=12.46$ kNm; $M_{CD}=-12.46$ kNm
9	5.9.12	$M_A=-17.08$ kNm, $M_B=-35.86$ kNm, $M_C=-35.86$ kNm, $M_D=-17.08$ kNm
10	5.9.13	$M_A=-34.17$ kNm; $M_D=48.98$ kNm; $M_B=9.24$ kNm; $M_C=44.83$ kNm

Note: The answers are provided based on analysis using structural analysis software. The results from the exact solution may differ slightly (in decimals), but these differences are generally negligible.

FURTHER READING

Bhavikatti, S. S. (2011). Structural *Analysis-II*, Vikas Publishing House, Noida, ISBN 978-9325968806, 978-9325968800.

Menon, D. (2023). *Structural Analysis*, Narosa Publishing House Pvt Ltd, New Delhi, ISBN 978-8184877342.

Reddy, C. S. (2011). *Basic Structural Analysis*, 3e, Tata McGraw-Hill Education Pvt Ltd, New Delhi, ISBN 978-0070702769.

Shah, H. J. and Junarkar, S. B. (2015). *Mechanics of Structures*, Vol. II, (24 E) Charotar Publishing House, Anand, ISBN 978-9385039027.

Wang, C. K. (1953). *Statically Indeterminate Structures*, McGraw Hill Higher Education, ISBN 978-0070681309.

Williams, A. (2009). *Structural Analysis in Theory and Practices*, Elsevier, Burlington, MA.

6 Slope–Deflection Method

6.1 INTRODUCTION

The slope–deflection (SD) method, introduced by George Maney in 1915, is used for analysing rigid-jointed structures. The basic principle of this method is similar to that of the displacement method discussed in Chapter 1. It is suitable for analysing both statically determinate and indeterminate rigid-jointed structures, such as beams and frames. The kinematic indeterminacy – i.e., the unknown joint or support degrees of freedom (DoFs) of the structure – must first be determined in terms of the unknown rotations and linear displacements. In a three-dimensional rigid-jointed structure, each node or joint has three rotational and three translational DoF. In contrast, for two-dimensional structures, each joint has one rotational and two translational DoF. These unknown rotations and displacements are then related to the unknown member end moments and the known fixed-end moments for each member using the force–displacement relationship, expressed through the SD equations. The unknown slopes and deflections are then determined by applying the appropriate equilibrium equations for each unknown displacement. The end moments for each member are calculated by substituting these displacements into the force–displacement equations (SD equations). Using the known end moments, the given loading, and the free-body diagram, the remaining support reactions can be determined using simple static equilibrium equations.

6.2 FORMULATION OF SD EQUATIONS

To develop the general form of the SD equations, consider a typical span *CD* of a continuous beam subjected to random loading, as shown in Figure 6.1. Assume the flexural rigidity *EI* is constant. The following sign convention is used in the derivation of the SD equations:

(i) Moments acting at the ends of members are positive if they act in a clockwise direction, and (ii) the rotations, i.e., the slope of the tangent to the elastic curve with respect to its original position, are positive if they are clockwise. Similarly, the clockwise chord rotation $\varphi_{CD} = \frac{\Delta_D - \Delta_C}{L}$, due to the settlement of supports Δ_C and Δ_D at *C* and *D*, respectively, is positive.

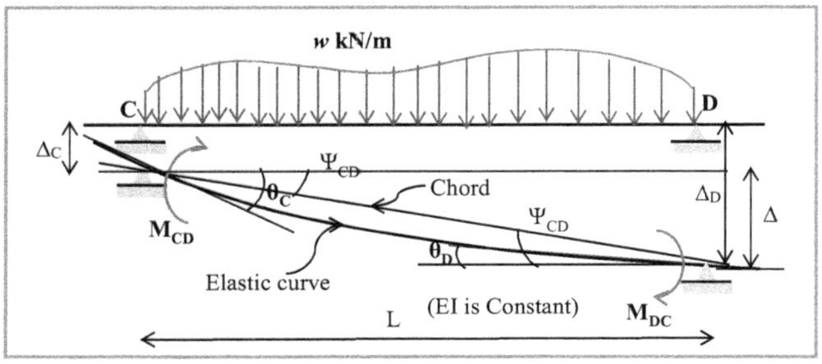

FIGURE 6.1 Typical span *CD* of a continuous beam.

Let M_{CD} and M_{DC} be the internal end moments at the supports *C* and *D*, respectively, in the +ve direction.

Let θ_C and θ_D be the angular rotations (slopes) at *C* and *D*, respectively.

Let Δ be the relative settlement between the supports.

Let φ_{CD} be the chord rotation in the +ve direction.

In structural analysis, the structures are generally assumed to be linearly elastic. We can use the principle of superposition to calculate the end moments resulting from angular rotations, settlement of supports, and loads.

6.2.1 End Moments Due to the Rotational Displacement at C (θ_C)

Consider the member *CD*, which is kinematically free, i.e., fixed at both ends, and the end *C* slips by θ_C, as shown in Figure 6.2a. To determine the moment required to cause the rotation θ_C, the principle of superposition is applied, followed by the conjugate beam method, as illustrated in Figure 6.2b and c, respectively.

From Figure 6.2b, the rotation at *C*, i.e., θ_{C1}, due to the moment M_{CD} in the real beam, is equal to the shear force at C_1 in the conjugate beam C_1D_1. To determine this, let us apply the moment equilibrium equation about D_1.

$$\sum M_{D1} = 0 \Rightarrow V_{C1} = \frac{1}{L}\left[\frac{1}{2} \times L \times \frac{M_{CD1}}{EI} \times \frac{2L}{3}\right]$$

$$= \frac{M_{CD1}L}{3EI} = \theta_{C1}\,(\text{clockwise rotation})$$

Similarly, the rotation at *D*, i.e., θ_{D1}, due to the moment M_{CD1} in real beam, is equal to the shear force at D_1 in the conjugate beam C_1D_1. Let us apply the vertical force equilibrium equation to find this rotation.

$$\sum F_Y = 0 \Rightarrow V_{C1} + V_{D1} = \left[\frac{1}{2} \times L \times \frac{M_{CD1}}{EI}\right] \Rightarrow V_{D1}$$

$$= \frac{M_{CD1}L}{6EI}\,(\text{anticlockwise rotation}) = \theta_{D1}$$

Similarly, from Figure 6.2c, the rotations at *C* (θ_{C2}), and *D* (θ_{D2}) due to the moment M_{DC1} at *D* in the real beam can be determined using the conjugate beam C_2D_2, as

$$\theta_{C2} = \frac{M_{DC1}L}{6EI}\ (\text{anticlockwise rotation})$$

DOI: 10.1201/9781003562955-6

and

$$\theta_{D2} = \frac{M_{DC1}L}{3EI}\,(\text{clockwise rotation})$$

From Figure 6.2a, since the end D is fixed, the rotation at D is zero.

$$\therefore \theta_D = \theta_{D1} + \theta_{D2} = 0 \Rightarrow -\frac{M_{CD1}L}{6EI} + \frac{M_{DC1}L}{3EI} = 0 \Rightarrow M_{CD1}$$

$$= 2M_{DC1}$$

Using the principle of superposition, the rotation at C in Figure 6.2a is equal to the summation of the rotations at C in Figure 6.2b and c, resulting from the moments M_{CD1} and M_{DC1}, respectively.

$$\therefore \theta_C = \theta_{C1} + \theta_{C2} \Rightarrow \frac{M_{CD1}L}{3EI} - \frac{M_{DC1}L}{6EI} = \frac{M_{CD1}L}{3EI}$$

$$-\frac{M_{CD1}L}{2\times 6EI} = \frac{M_{CD1}L}{4EI}$$

(or)

$$M_{CD1} = \frac{4EI}{L}\theta_C \quad \text{and} \quad M_{DC1} = \frac{2EI}{L}\theta_C \text{ in clockwise} \qquad (6.0a)$$

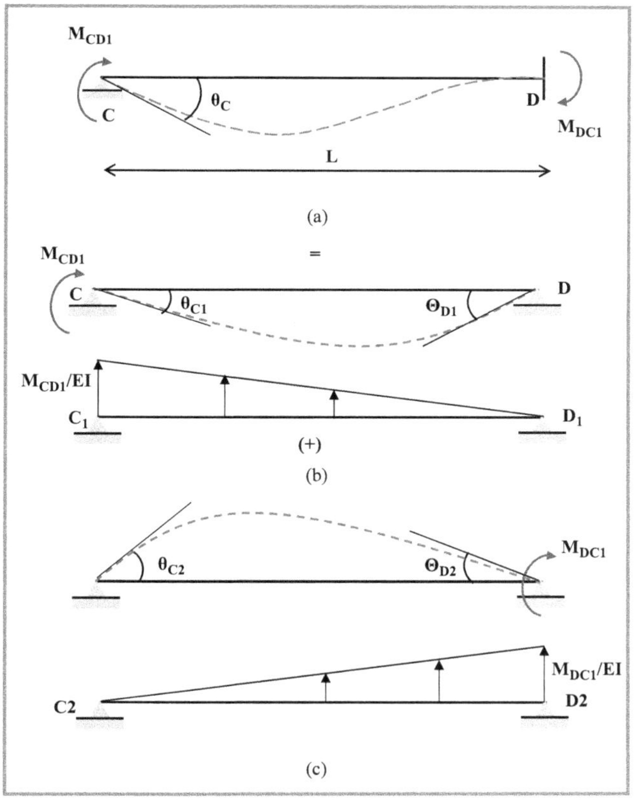

FIGURE 6.2 (a) Far end fixed and near end rotated by θ_C, (b) conjugate beam due to end moment M_{CD} at C, (c) conjugate beam due to end moment M_{DC} at D.

6.2.2 End Moments Due to the Rotational Displacement at D (θ_D)

Figure 6.3 shows the fixed beam CD subjected to a rotational slip θ_D at D.

By adopting a similar procedure as above, the end moments M_{CD2} and M_{DC2} can be obtained, respectively, at C and D as

$$M_{CD2} = \frac{2EI}{L}\theta_D \quad \text{and} \quad M_{DC2} = \frac{4EI}{L}\theta_D \qquad (6.0b)$$

6.2.3 End Moments Due to the Relative Settlement of Supports between C and D (φ_{CD})

Let Δ_C and Δ_D be the settlements of supports at C and D, respectively, and let Δ be the relative settlement of D with respect to C. This results in the chord rotating clockwise by $\varphi_{CD} = \frac{\Delta_D - \Delta_C}{L} = \frac{\Delta}{L}$, as shown in Figure 6.4a. This chord rotation induces an anticlockwise moment M and opposite shear forces at both ends C and D, as shown in the figure The moment M can be related to the displacement Δ using the conjugate beam method, as discussed earlier. The conjugate beam is shown in Figure 6.4b with free ends, because the ends in the original/real beam are fixed supports. Furthermore, Δ is shown at the end D^1 in the conjugate beam due to the presence of settlement/displacement at D in the original beam, as shown in Figure 6.4b.

Applying the moment equilibrium equation with respect to D^1, we can write

$$\sum M_{D^1} = \Delta + \frac{1}{2}\times L \times \frac{M}{EI}\times\frac{L}{3} - \frac{1}{2}\times L\times\frac{M}{EI}\times\frac{2L}{3} = 0$$

$$\Delta M = \frac{6EI}{L^2}\Delta\,(\text{Anticlcokwise})$$

$$\therefore M_{CD3} = M_{DC3} = \frac{6EI}{L^2}\Delta\,(\text{Anticlcokwise}) \qquad (6.0c)$$

6.2.4 Fixed End Moments Due to the External Loads

In the previous cases, the force–displacement relationships were established by relating angular and linear displacements to the corresponding end moments at C and D. To formulate the SD equations, the effect of external loads acting on the member must be converted into an equivalent moment at

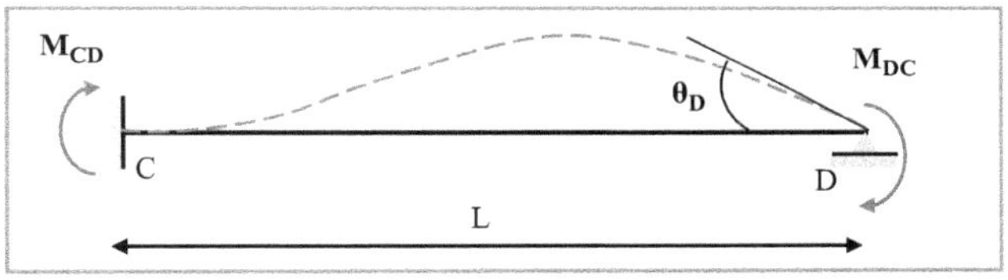

FIGURE 6.3 Fixed beam with the far end rotated by θ_D.

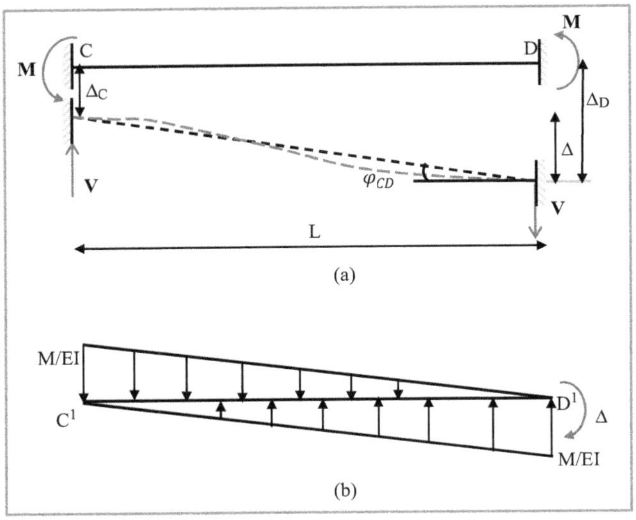

FIGURE 6.4 Member end moments due to the relative settlement of supports. (a) Original/real beam with relative settlement Δ at D. (b) Conjugate beam.

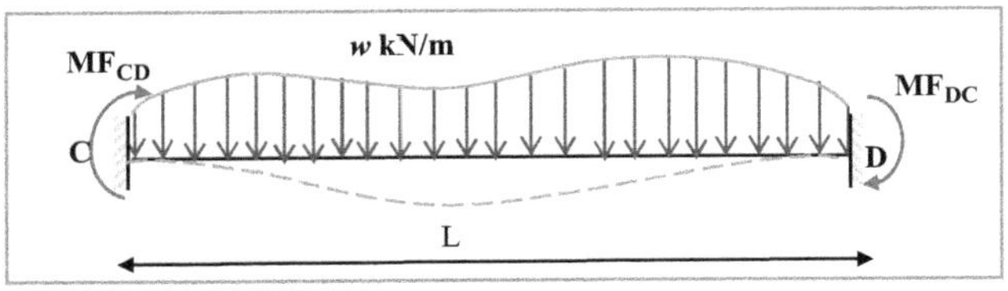

FIGURE 6.5 Beam with the fixed end moments due to external loads.

the supports. These equivalent moments are known as fixed end moments. Let us consider a kinematically free beam, i.e., a fixed beam CD, with zero rotations and settlements, subjected to arbitrary loading as shown in Figure 6.5. The fixed end moments induced by the loads are MF_{CD} and MF_{DC}, respectively, at C and D, as shown in the figure.

6.2.5 SD Equations

As mentioned earlier, assuming the structure is linearly elastic, the principle of superposition can be used to calculate the resulting end moments due to rotational displacements (equation 6.0a–b), settlement of supports (equation 6.0c), and applied loads. Therefore, we can write

$$M_{CD} = MF_{CD} + M_{CD1} + M_{CD2} + M_{CD3}$$

$$M_{CD} = MF_{CD} + \frac{4EI}{L}\theta_C + \frac{2EI}{L}\theta_D - \frac{6EI}{L}\varphi_{CD} \quad (6.1)$$

$$M_{CD} = MF_{CD} + \frac{2EI}{L}\left[2\theta_C + \theta_D - 3\varphi_{CD}\right] \quad (6.2)$$

$$M_{DC} = MF_{DC} + \frac{2EI}{L}\theta_C + \frac{4EI}{L}\theta_D - \frac{6EI}{L}\varphi_{CD} \quad (6.3)$$

$$M_{DC} = MF_{DC} + \frac{2EI}{L}\left[\theta_C + 2\theta_D - 3\varphi_{CD}\right] \quad (6.4)$$

Equations 6.2 and 6.4 are similar. Therefore, we can express these two equations in a single generalised equation by referring to the near end as N, the far end as F, member stiffness as $k = I/L$, and the span's chord rotation as $\varphi = \Delta/L$, as

$$M_N = MF_N + 2Ek\left[2\theta_N + \theta_F - 3\varphi\right] \quad (6.4a)$$

where

M_N is an internal moment (member end moment) at the near end of the span. This moment is *positive*, if it acts in a clockwise direction.

MF_N *is* a fixed end moment at the near end support due to external loads, and it is positive if it is acting in the clockwise direction. The fixed end moments for various loading conditions can be obtained from Table 6.1.

θ_N and θ_F *represents* the near end and far end slopes or rotations (in radians) at the supports, and they are considered positive when measured in the clockwise direction.

$\varphi = \Delta/L$ *represents* the chord rotation (in radians) due to the linear displacement in the span, and it is considered positive when measured in the clockwise direction.

E is the elastic modulus of the material.

equation 6.4a is a generalised SD equation, which is applied twice for each span (CD), i.e., from C to D and from D to C.

6.3 MODIFIED SD EQUATIONS

The SD equations 6.2 and 6.4 account for all possible displacements at the ends; hence, these equations are referred to as the general SD equations, and they can be applied to solve any type of prismatic beam or rigid-jointed plane frame problems. However, in certain cases, the degree of kinematic indeterminacy can be reduced by neglecting the end rotations or translations (unknown) in the structure. Such cases typically arise when the far end (extreme end) of a beam or frame element in a structure is free to rotate or translate, i.e., when the extreme end is supported by a hinge, roller, or guided/sliding support. These cases are demonstrated in the following sections.

6.3.1 Extreme Support is Hinge or Roller

Let the extreme end support of a beam or frame be a roller or hinge support, as shown in Figure 6.6a. Since the moment at the hinge or roller support is zero, the rotation at this end can be ignored. Therefore, the general SD equation 6.2 can be modified as shown below and applied only once to the span CD, rather than twice.

Substituting M_{DC}=0 and MF_{DC}=0 (due to loading) into equation 6.4, we obtain

$$0 = 0 + \frac{2EI}{L}\left[\theta_C + 2\theta_D - 3\varphi_{CD}\right] \Rightarrow \theta_D = \frac{1}{2}\left(3\varphi_{CD} - \theta_C\right)$$

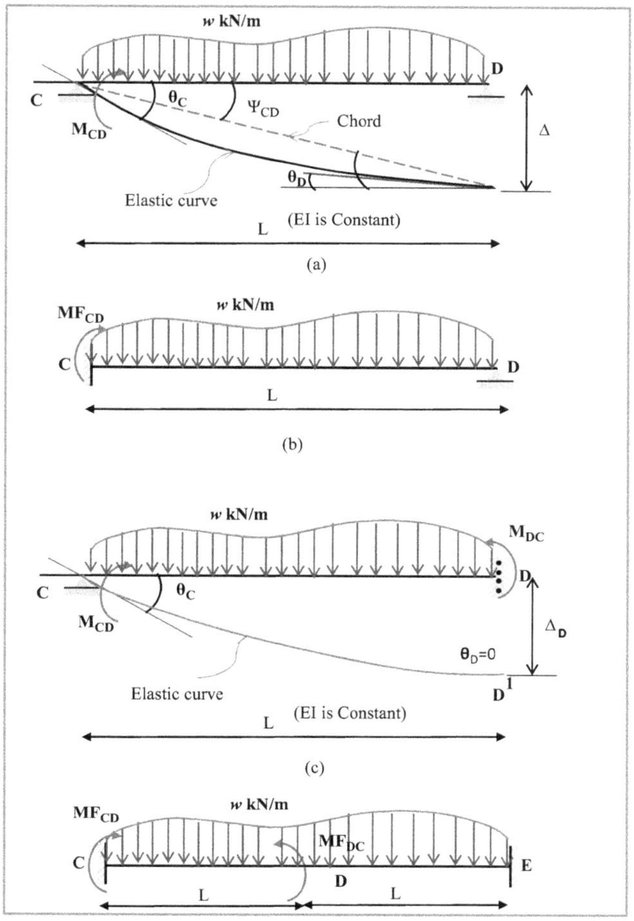

FIGURE 6.6 (a) Beam with the far end hinged, (b) beam with modified fixed end moments due to external loads, (c) beam with a far end guided support, and (d) symmetrical beam *CDE* with modified fixed end moments due to external loads.

On substituting the above θ_D into equation 6.2, equation 6.2 can be modified as

$$M_{CD} = MF_{CD} + \frac{2EI}{L}\left[2\theta_C + \frac{1}{2}(3\varphi_{CD} - \theta_C) - 3\varphi_{CD}\right]$$

$$= MF_{CD} + \frac{3EI}{L}[\theta_C - \varphi_{CD}] \tag{6.5}$$

Equation 6.5 is a modified SD equation that can be used when the far or extreme end is a roller or hinge support. It incorporates the modified stiffness and modified fixed end moment for the propped cantilever, as shown in Figure 6.6b. Equation 6.5 simplifies the solution, as the general SD equations 6.2 and 6.4 require application for span *CD* which involves an extra unknown, i.e., the far end rotation θ_D.

The modified stiffness for span *CD* is $K = \dfrac{3EI}{L}$

6.3.2 Guided Support at Far End *D*

Let the far end of a beam be supported by a guided support, as shown in Figure 6.6c. Since the rotation at the guided support θ_D is zero, the only unknown displacement in the beam at support *C* is the rotation θ_C. By taking the modified stiffness of a prismatic beam element exhibiting cantilever action, the unknown translational displacement at support *D*, i.e., Δ_D can be neglected. Therefore, considering the modified stiffness ($K=EI/L$), and $\theta_D=0$, the general SD equations 6.2 and 6.4 can be modified as follows

$$M_{CD} = MF_{CD} + \frac{EI}{L}\theta_C \tag{6.6}$$

$$M_{DC} = MF_{DC} - \frac{EI}{L}\theta_C \tag{6.7}$$

The fixed end moments MF_{CD} and MF_{DC} must be modified accordingly, by simply visualising beam *CD* as the left half of a symmetrical beam *CDE*, as shown in Figure 6.6d. Using similar logic, if the near end (End *C*) is a guided support ($\theta_C=0$), the general SD equations 6.2 and 6.4 can take the form as

$$M_{CD} = MF_{CD} - \frac{EI}{L}\theta_D \tag{6.8}$$

$$M_{DC} = MF_{DC} + \frac{EI}{L}\theta_D \tag{6.9}$$

6.4 ANALYSIS PROCEDURE

i. **Determine the degree of kinematic indeterminacy**: Determine the kinematic indeterminacy (unknown joint DoF) of the structure and identify them. Assume these DoF are in the positive direction, with rotations and chord rotations considered positive in the clockwise direction.
ii. **Calculate fixed end moments**: Calculate the fixed end moments for each span due to external loading using the standard expressions for various loading conditions provided in Table 6.1. Use modified fixed end moments wherever possible.
iii. **Apply SD equations**: Normally, the SD equation (equation 6.4a) must be applied twice for each span of a beam or frame. However,
    a. If the far end of a beam or frame is pin or roller supported, then the modified SD (equation 6.5) can be applied only to the restrained end.
    b. Similarly, if the far or near end has a guided or sliding support, use equations 6.6–6.9 to generate the SD equations.
iv. **Formulate moment equilibrium equations**: Write the moment equilibrium equations corresponding to each unknown joint DoF. Each of these equations should be expressed in terms of the unknown internal moments, as mentioned in the SD equations. In the case of frames subjected

TABLE 6.1
Fixed End Moments for Various Loading Conditions

Beam	Moment at *A*	Moment at *B*
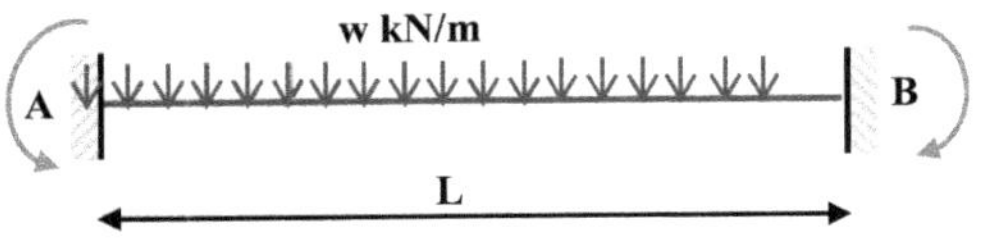	$\frac{wL^2}{12}$	$\frac{wL^2}{12}$
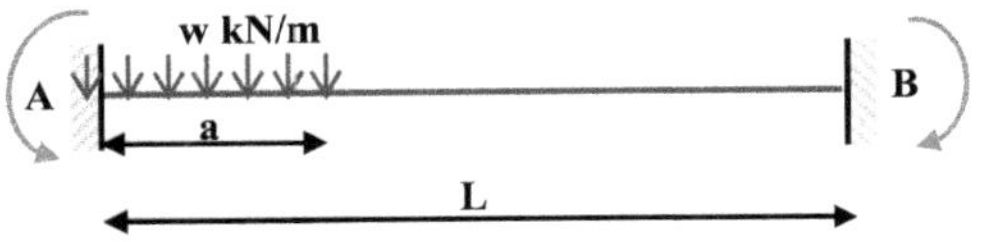	$\frac{wa^2\left(3a^2-8aL+6L^2\right)}{12L^2}$	$\frac{wa^3(4L-3a)}{12L^2}$
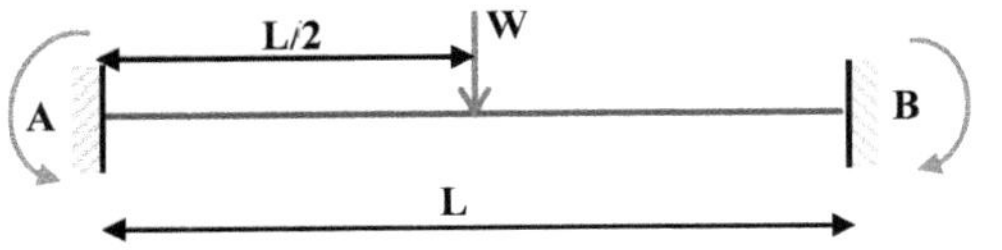	$\frac{WL}{8}$	$\frac{WL}{8}$
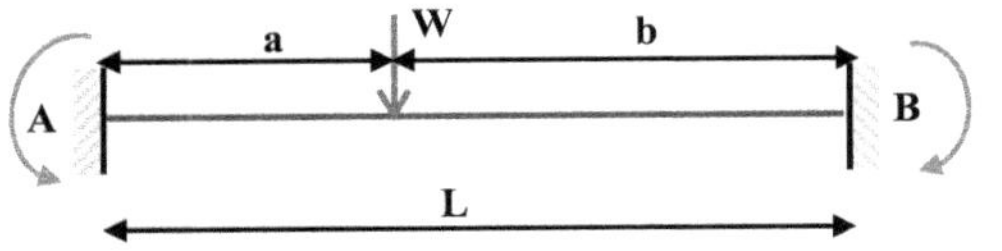	$\frac{Wab^2}{L^2}$	$\frac{Wa^2b}{L^2}$
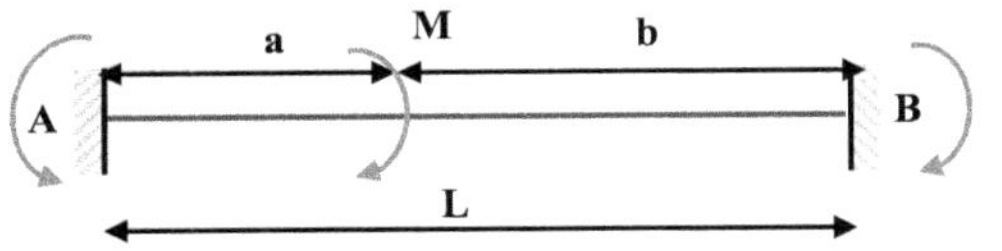	$\frac{Mb(b-2a)}{L^2}$	$\frac{Ma(2b-a)}{L^2}$
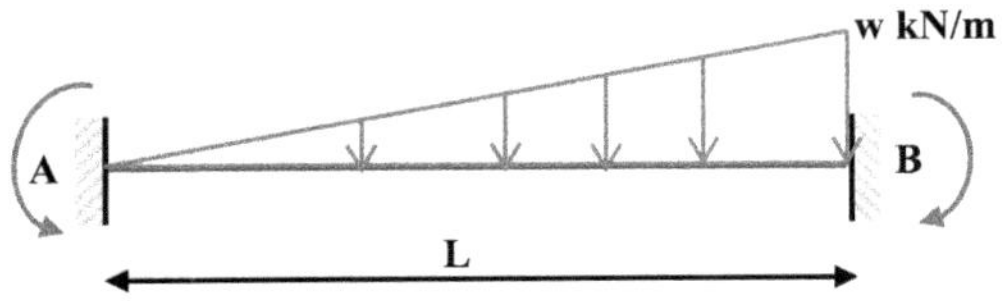	$\frac{wL^2}{30}$	$\frac{wL^2}{20}$
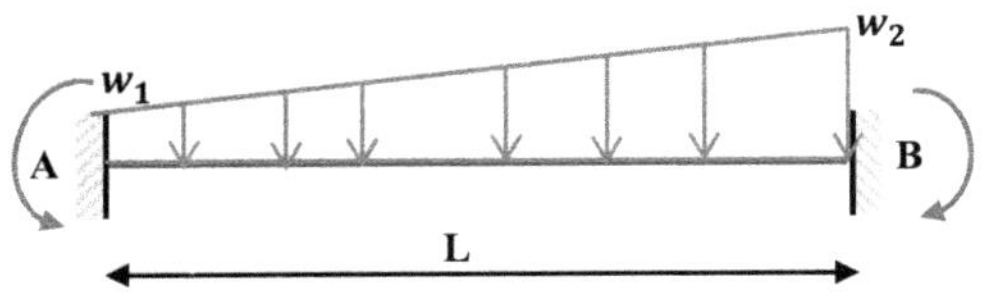	$\frac{L^2}{10}\left(\frac{w_1}{2}+\frac{w_2}{3}\right)$	$\frac{L^2}{10}\left(\frac{w_1}{3}+\frac{w_2}{2}\right)$
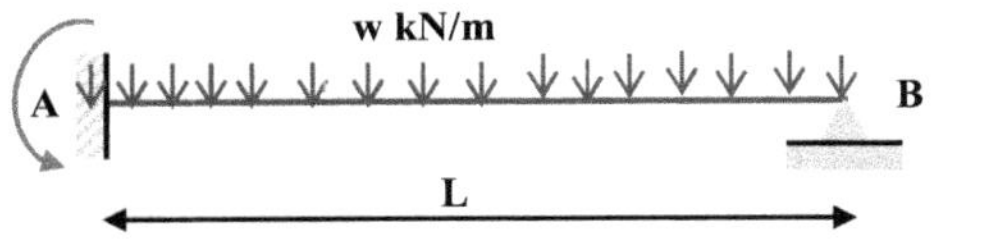	$\frac{WL^2}{8}$	
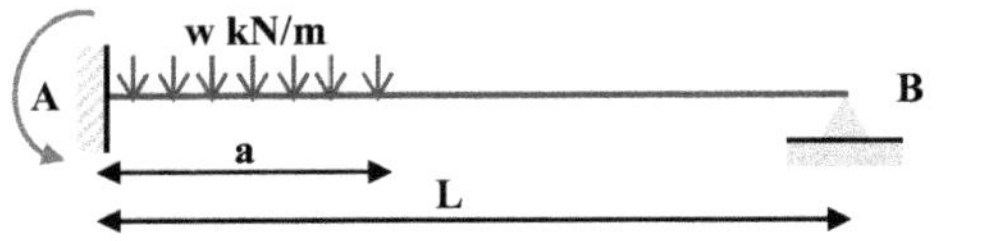	$\frac{wa^2\left(a^2-4aL+4L^2\right)}{8L^2}$	
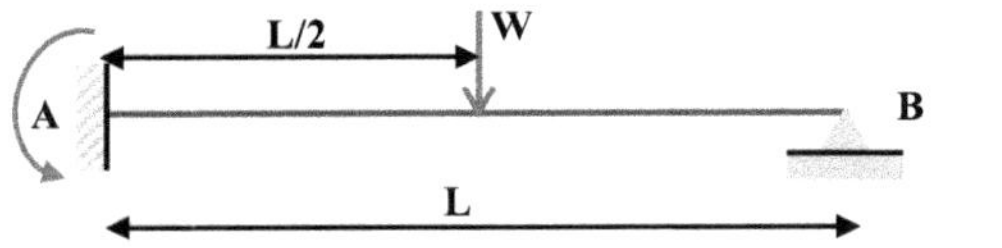	$\frac{3WL}{16}$	
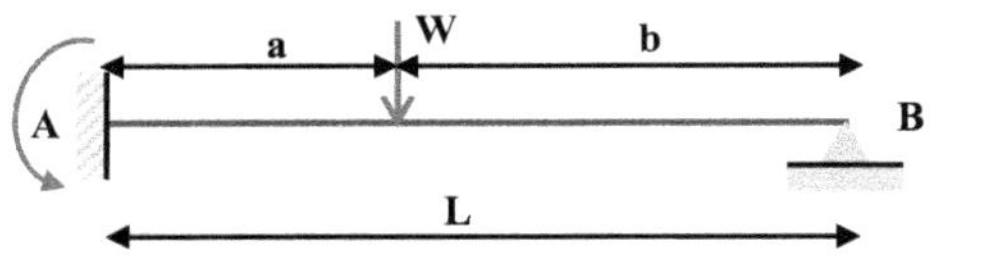	$\frac{Wab(L+b)}{2L^2}$	
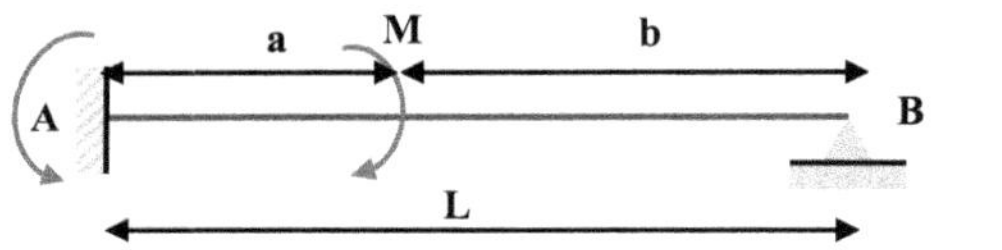	$M\left(1-\frac{3a}{L}+\frac{3a^2}{2L^2}\right)$	

(Continued)

TABLE 6.1 (*Continued*)
Fixed End Moments for Various Loading Conditions

Beam	Moment at *A*	Moment at *B*
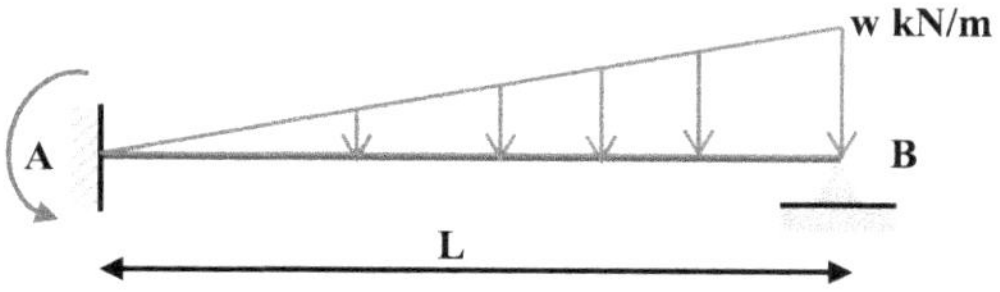	$\frac{7wL^2}{120}$	
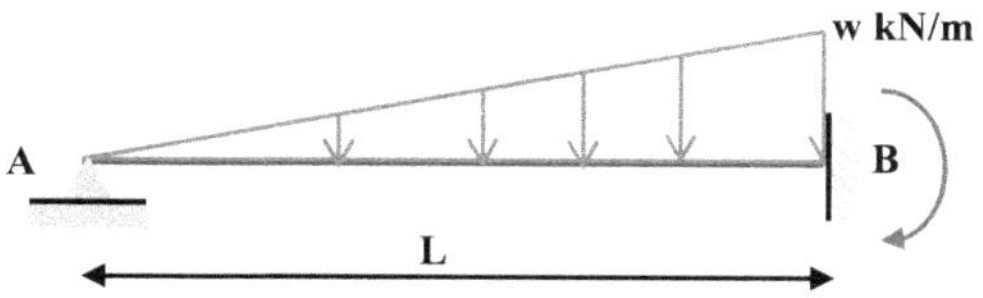		$\frac{wL^2}{15}$
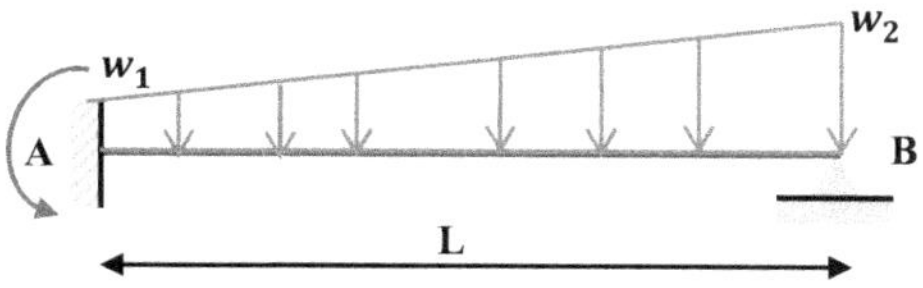	$L^2\left(\frac{w_1}{15}+\frac{7w_2}{120}\right)$	

to lateral sway, additional column shear equilibrium equations must be written corresponding to each sway displacement. Column shears should be expressed in terms of the internal unknown column moments as specified in SD equations.

v. **Solve for unknowns**: Substitute the SD equations into the equilibrium equations obtained in step iv to find the unknown DoF or joint displacements. Member end moments can then be calculated by substituting the known joint displacements back into the SD equations.

Note: Negative results indicate that the rotations and moments act in the anti-clockwise direction.

vi. **Shear force and bending moment diagrams**: Using free body diagrams and static equilibrium equations, calculate the member end shears. Construct the shear force and bending moment diagrams. Finally, draw the probable elastic curve using the known results of the DoFs.

6.5 NUMERICAL PROBLEMS

6.5.1 Applications to the Beams Without Yielding of Supports

Example 6.1

Analyse the two-span continuous beam as shown in Figure 6.7 using SD method. Assume *EI* is constant.

Solution:

Neglect axial deformations. The kinematic indeterminacy (unknown DoF), $I_k = 2j - R = 2\times3-5=1$. Since supports *A* and *C* are fixed supports, $\theta_A = \theta_C = 0$. Therefore, the unknown DoF is the rotational displacement θ_B.

FIXED END MOMENTS

The fixed end moments due to the external loads on each span (Figure 6.8) can be calculated using the expressions given in Table 6.1, as shown below:

$$MF_{AB} = -\frac{Wab^2}{L^2} = -\frac{30\times2\times3^2}{5^2} = -21.66 \text{ kNm}$$

$$MF_{BA} = \frac{Wba^2}{L^2} = \frac{30\times3\times2^2}{5^2} = 14.40 \text{ kNm}$$

$$MF_{BC} = -\frac{wL^2}{12} = -\frac{15\times5^2}{12} = -31.25 \text{ kNm}$$

$$MF_{CB} = \frac{wL^2}{12} = \frac{15\times5^2}{12} = 31.25 \text{ kNm}$$

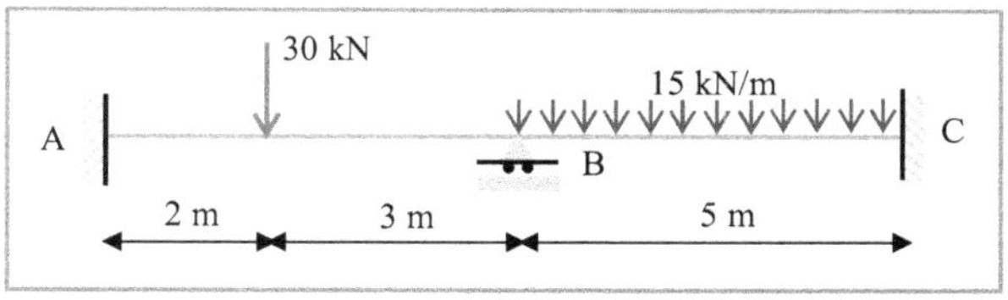

FIGURE 6.7 Two-span continuous beam.

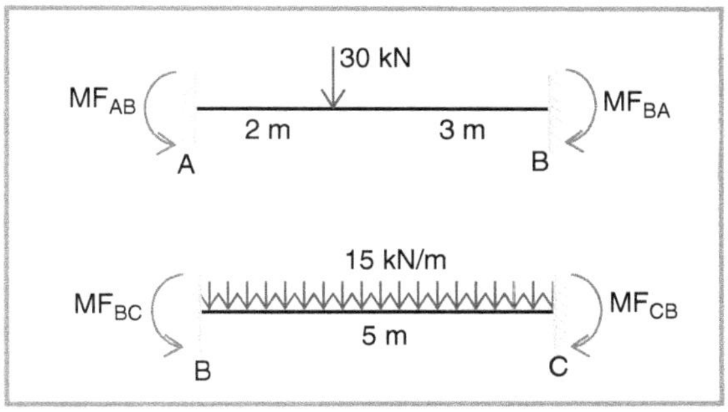

FIGURE 6.8 Fixed end moments.

SD Equations

Span AB: For span AB, $\theta_A = 0,\ \theta_B = ?, \varphi_{AB} = 0, L = 5$ m, $I = I$ Therefore, using equation 6.4a, we can write the SD equations for each span as shown below

$$M_N = MF_N + \frac{2EI}{L}(2\theta_N + \theta_F - 3\varphi)$$

$$M_{AB} = MF_{AB} + \frac{2EI}{L}(2\theta_A + \theta_B - 3\varphi_{AB})$$

$$M_{AB} = -21.66 + \frac{2EI}{5}(2(0) + \theta_B - 3(0)) = -21.66 + \frac{2EI\theta_B}{5} \quad (6.10)$$

$$M_{BA} = MF_{BA} + \frac{2EI}{L}(2\theta_B + \theta_A - 3\varphi_{AB})$$

$$M_{BA} = 14.40 + \frac{2EI}{5}(2(\theta_B) + 0 - 3(0)) = 14.40 + \frac{4EI\theta_B}{5} \quad (6.11)$$

For span BC: ($\theta_B = ?, \theta_C = 0, \varphi_{BC} = 0, L = 5$ m, $I = I$)

$$M_N = MF_N + \frac{2EI}{L}(2\theta_N + \theta_F - 3\varphi)$$

$$M_{BC} = MF_{BC} + \frac{2EI}{L}(2\theta_B + \theta_C - 3\varphi_{BC})$$

$$M_{BC} = -31.25 + \frac{2EI}{5}(2\theta_B + 0 - 3(0)) = -31.25 + \frac{4EI\theta_B}{5} \quad (6.12)$$

$$M_{CB} = MF_{CB} + \frac{2EI}{L}(\theta_B + 2\theta_C - 3\varphi_{BC})$$

$$M_{CB} = 31.25 + \frac{2EI}{5}(\theta_B + 2(0) - 3(0)) = 31.25 + \frac{2EI\theta_B}{5} \quad (6.13)$$

For convenience, the above equations can be written in a matrix as follows:

$$\begin{bmatrix} M_{AB} \\ M_{BA} \\ M_{BC} \\ M_{CB} \end{bmatrix} = \begin{bmatrix} -21.66 \\ 14.40 \\ -31.25 \\ 31.25 \end{bmatrix} + \begin{bmatrix} 2/5 \\ 4/5 \\ 4/5 \\ 2/5 \end{bmatrix} [EI\theta_B] \quad (6.14)$$

Joint Equilibrium Equations

The above four equations contain five unknowns. The necessary fifth equation can be obtained from the moment equilibrium equation at joint B (corresponding to the unknown DoF θ_B) using the free body diagram shown in Figure 6.9.

$$M_{BA} + M_{BC} = 0 \quad (6.15)$$

(Since segment (support B) is of defferential length, moment due to beam shears is neglected)

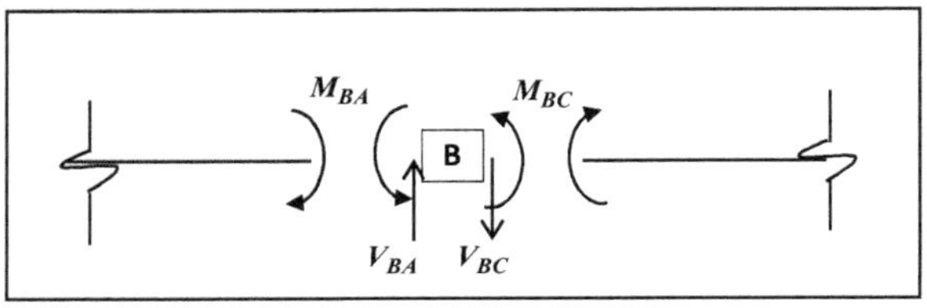

FIGURE 6.9 Free body diagram of joint B.

Substituting equations 6.11 and 6.12 into equation 6.15, we obtain

$$14.40 + \frac{4EI}{5}\theta_B - 31.25 + \frac{4EI}{5}\theta_B = 0$$

$$\frac{8EI}{5}\theta_B = 16.85$$

$$\therefore \theta_B = \frac{10.531}{EI} \text{ (clockwise)}$$

Final End Moments

Substituting the above result into equation 6.14, we get the final end moments as shown below.

$$\begin{bmatrix} M_{AB} \\ M_{BA} \\ M_{BC} \\ M_{CB} \end{bmatrix} = \begin{bmatrix} -21.66 \\ 14.40 \\ -31.25 \\ 31.25 \end{bmatrix} + \begin{bmatrix} 2/5 \\ 4/5 \\ 4/5 \\ 2/5 \end{bmatrix} [EI\theta_B = 10.531]$$

$$= \begin{bmatrix} -17.448 \\ 22.825 \\ -22.825 \\ 35.462 \end{bmatrix} \text{kNm}$$

Using the known values of end moments, the shear forces at the end of each span can be calculated using the static equilibrium equations, as shown in Figure 6.10a. The shear force and bending moment diagrams are shown in Figure 6.10b and c, respectively. The probable elastic curve is shown in Figure 6.10d.

Position of Zero Shear in Span BC

From similar triangles, we can write

$$\frac{x}{40.02} = \frac{5-x}{34.973} \Rightarrow x = 2.67 \text{ m}$$

$$\therefore \text{Maximum +ve BM} = 40.02 \times 2.67 - 35.462 - 15 \times 2.67 \times \frac{2.67}{2} = 17.925 \text{ kNm}$$

Maximum +ve BM in Span AB

$$M_{max} = 16.925 \times 2 - 17.448 = 16.402 \text{ kNm}$$

Position of Contraflexure in Span BC

To get the position of the contraflexure point, equate the moment to zero in span BC.

$$\therefore M_x = 40.02x - 15\frac{x^2}{2} - 35.462 = 0 \Rightarrow -80.04x + 15x^2 + 70.924 = 0$$

$$\therefore x = 1.12 \text{ m and } 4.21 \text{ m}$$

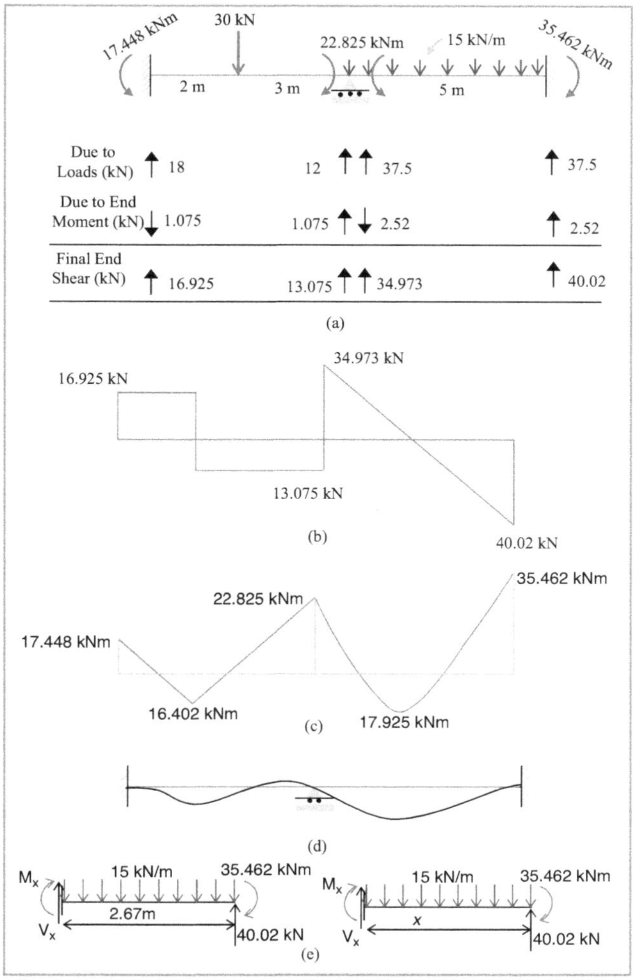

FIGURE 6.10 (a) Free body diagram, (b) shear force diagram, (c) bending moment diagram, and (d) probable deflected shape.

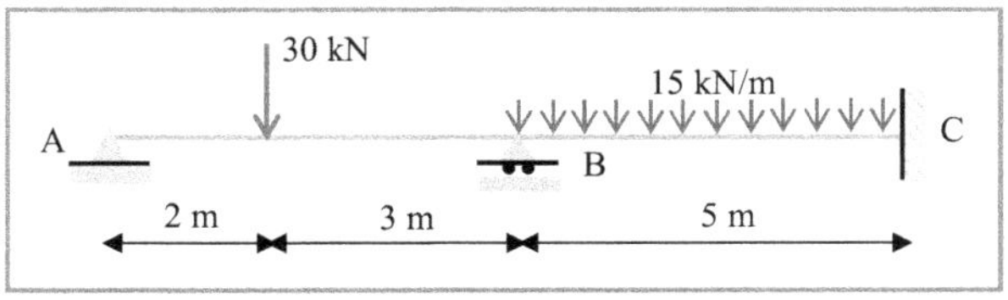

FIGURE 6.11 Two-span continuous beam with a hinge support at *A*.

Example 6.2

Re-analyse Problem No. 6.1 by treating end *A* as a hinged support instead of a fixed support (Figure 6.11).

Solution:

In general, the kinematic indeterminacy of the beam is 2, i.e., θ_A and θ_B. However, taking advantage of the fact that the moment at *A* is zero due to the simply supported end, the rotation at *A* can be neglected by applying the modified stiffness, fixed end moments, and the modified SD equation for *AB*. Therefore, the degree of kinematic indeterminacy is reduced to only one, i.e., θ_B.

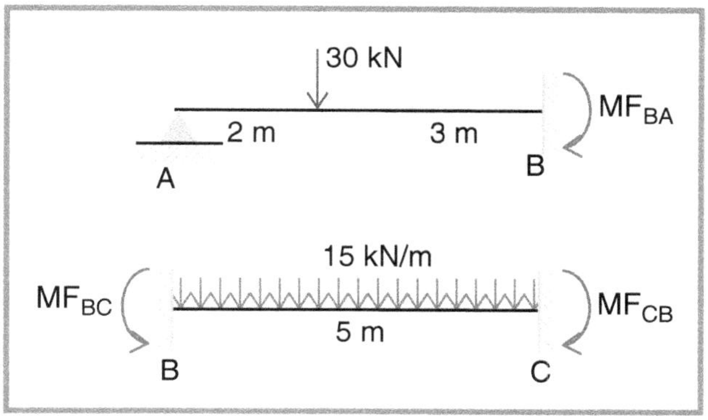

FIGURE 6.12 Fixed end moments.

FIXED END MOMENTS

The fixed moments are shown in Figure 6.12 and using the expressions provided in Table 6.1, we have

$$MF_{BA} = Wab(L+a)/2L^2 = 25.20 \text{ kNm}$$

$$MF_{BC} = -\frac{wL^2}{12} = -\frac{15\times 5^2}{12} = -31.25 \text{ kNm}$$

$$MF_{CB} = \frac{wL^2}{12} = \frac{15\times 5^2}{12} = 31.25 \text{ kNm}$$

SD EQUATIONS

For span *AB*: (θ_A = negelcted, $\theta_B = ?$, $\varphi_{AB} = 0$, $L = 5\text{m}$, $I = I$), Using modified SD equation, the SD equation for span *AB* can be written as below

$$M_{BA} = MF_{BA} + \frac{3EI}{L}(\theta_B - \varphi_{AB})$$

$$M_{BA} = 25.20 + \frac{3EI}{5}(\theta_B - 3(0)) = 25.20 + \frac{3EI\theta_B}{5} \quad (6.16)$$

For span *BC*: ($\theta_B = ?$, $\theta_C = 0$, $\varphi_{BC} = 0$, $L = 5$ m, $I = I$)

$$M_N = MF_N + \frac{2EI}{L}(2\theta_N + \theta_F - 3\varphi)$$

$$M_{BC} = MF_{BC} + \frac{2EI}{L}(2\theta_B + \theta_C - 3\varphi_{BC})$$

$$M_{BC} = -31.25 + \frac{2EI}{5}(2\theta_B + 0 - 3(0)) = -31.25 + \frac{4EI\theta_B}{5} \quad (6.17)$$

$$M_{CB} = MF_{CB} + \frac{2EI}{L}(\theta_B + 2\theta_C - 3\varphi_{BC})$$

$$M_{CB} = 31.25 + \frac{2EI}{5}(\theta_B + 2(0) - 3(0)) = 31.25 + \frac{2EI\theta_B}{5} \quad (6.18)$$

For convenience, the above SD equations can be written in matrix form as shown below.

$$\begin{bmatrix} M_{BA} \\ M_{BC} \\ M_{CM} \end{bmatrix} = \begin{bmatrix} 25.20 \\ -31.25 \\ 31.25 \end{bmatrix} + \begin{bmatrix} 3/5 \\ 4/5 \\ 2/5 \end{bmatrix} [EI\theta_B] \tag{6.19}$$

JOINT EQUILIBRIUM EQUATIONS

The above three equations contain four unknowns. The necessary fourth equation can be obtained from the moment equilibrium equation at joint *B*. Using the free body diagrams of joint *B* shown in Figure 6.13, the moment equilibrium equation can be written, as shown below.

$$M_{BA} + M_{BC} = 0 \tag{6.20}$$

(Moment due to shear is neglected as segment is of differential length)

Substituting equations 6.17 and 6.18 into equation 6.20, we have

$$25.20 + \frac{3EI}{5}\theta_B - 31.25 + \frac{4EI}{5}\theta_B = 0$$

$$\frac{7EI}{5}\theta_B = 6.05 \Rightarrow EI\theta_B = 4.32 \text{ kNm (or) } \theta_B = \frac{4.32}{EI} \text{(clockwise)}$$

FINAL END MOMENTS

To get the final end moments for each span in the beam, substitute the θ_B value into equation 6.19.

$$\therefore \begin{bmatrix} M_{BA} \\ M_{BC} \\ M_{CB} \end{bmatrix} = \begin{bmatrix} 25.20 \\ -31.25 \\ 31.25 \end{bmatrix} + \begin{bmatrix} 3/5 \\ 4/5 \\ 2/5 \end{bmatrix} [EI\theta_B = 4.32]$$

$$= \begin{bmatrix} 27.792 \\ -27.794 \\ 32.978 \end{bmatrix} \text{kNm}$$

With the known values of final end moments, and using the static equilibrium equations, we can find out the end shears (Reactions), as shown in Figure 6.14a. The shear force and bending moment diagrams are shown in Figure 6.14b and c, respectively. The probable deflected shape is shown in Figure 6.14d.

Example 6.3

Analyse the continuous beam *ABCD*, shown in Figure 6.15, using the SD method. Assume *EI* is constant.

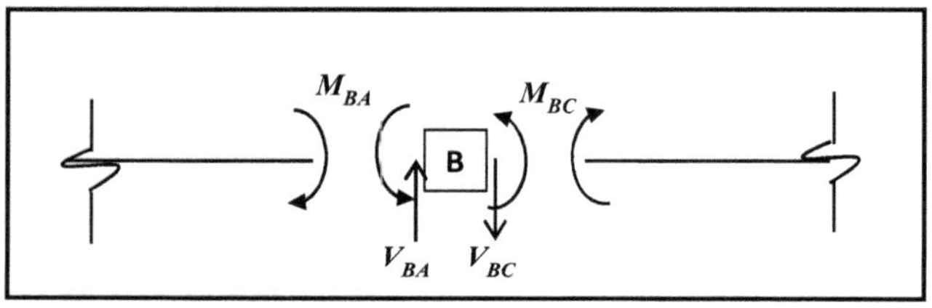

FIGURE 6.13 Free body diagram of joint *B*.

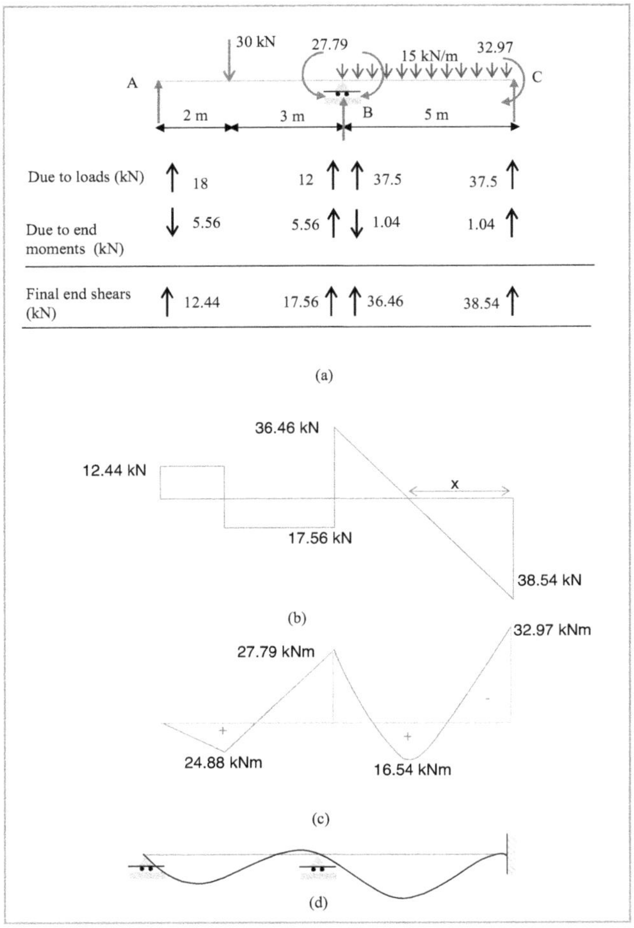

FIGURE 6.14 (a) Free body diagram, (b) shear force diagram, (c) bending moment diagram, and (d) probable deflected shape.

Solution:

The kinematic indeterminacy, $I_k=4$, i.e., $\theta_A, \theta_B, \theta_C$ and Δ_A. Since overhang portion is a determinate, its effect can be converted into a known concentrated moment at *B*. Further, as the end moment at *B* is a known value, θ_B can be ignored. Therefore, the DoF is limited to only one, i.e., θ_C. However, appropriate modifications needs to done while calculating the fixed end moments, stiffness and SD equation for span *BC*.

FIXED END MOMENTS

The fixed end moments are shown in Figure 6.16.

Using the expressions provided in Table 6.1, we have

$$MF_{CB} = \frac{3}{2}\frac{wL^2}{12} - \frac{M_0}{2} = \frac{3 \times 20 \times 5^2}{2 \times 12} - \frac{10}{2} = 57.5 \text{ kNm}$$

(modified FEM)

$$MF_{CD} = -\left(\frac{20 \times 1 \times 4^2}{5^2} + \frac{20 \times 4 \times 1^2}{5^2}\right) = -16 \text{ kNm}$$

$$MF_{DC} = \left(\frac{20 \times 1^2 \times 4}{5^2} + \frac{20 \times 1 \times 4^2}{5^2}\right) = 16 \text{ kNm}$$

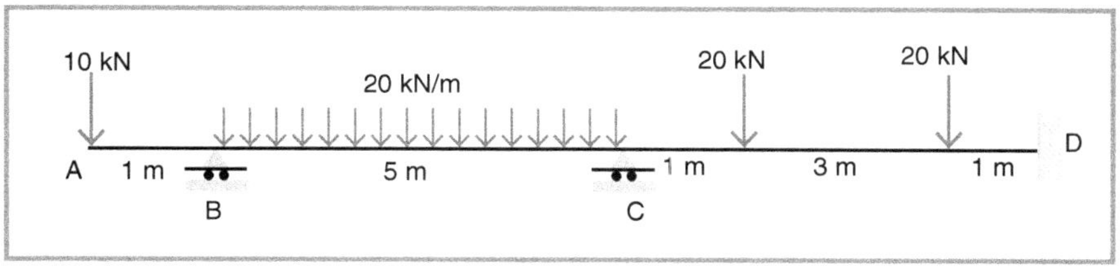

FIGURE 6.15 Two-span continuous beam with one side overhang.

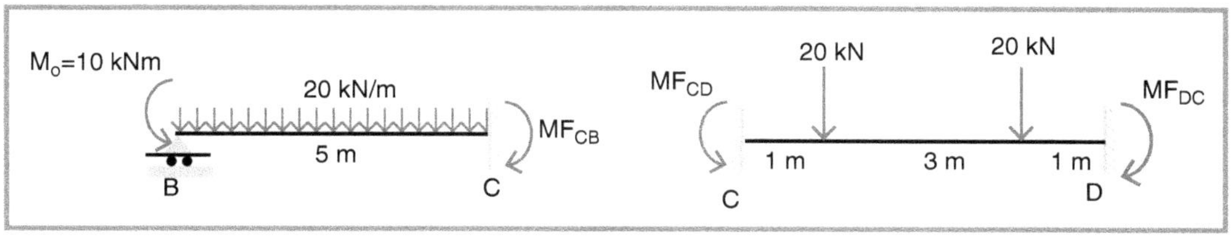

FIGURE 6.16 Fixed end moments.

SD EQUATIONS

For span *BC*: Use modified SD equation. θ_B = negelcted, $\theta_C = ?, \varphi_{BC} = 0, L = 5$ m, $I = I, M_{BC} = -10$ kNm)

$$M_{CB} = MF_{CB} + \frac{3EI}{L}(\theta_C - \varphi_{BC})$$

$$M_{CB} = 57.5 + \frac{3EI}{5}(\theta_C - 0) = 57.5 + \frac{3EI\theta_C}{5} \quad (6.21)$$

For span *CD*: ($\theta_C = ?, \theta_D = 0, \varphi_{CD} = 0, L = 5$ m, $I = I$)

$$M_N = MF_N + \frac{2EI}{L}(2\theta_N + \theta_F - 3\varphi)$$

$$M_{CD} = MF_{CD} + \frac{2EI}{L}(2\theta_C + \theta_D - 3\varphi_{CD})$$

$$M_{CD} = -16 + \frac{2EI}{5}(2\theta_C + 0 - 3(0)) = -16 + \frac{4EI\theta_C}{5} \quad (6.22)$$

$$M_{DC} = MF_{DC} + \frac{2EI}{L}(\theta_C + 2\theta_D - 3\varphi_{CD})$$

$$M_{DC} = 16 + \frac{2EI}{5}(\theta_C + 0 - 3(0)) = 16 + \frac{2EI\theta_C}{5} \quad (6.23)$$

The above SD equations can be conveniently written in matrix form as shown below:

$$\begin{bmatrix} M_{CB} \\ M_{CD} \\ M_{DC} \end{bmatrix} = \begin{bmatrix} 57.50 \\ -16 \\ 16 \end{bmatrix} + \begin{bmatrix} 3/5 \\ 4/5 \\ 2/5 \end{bmatrix} [EI\theta_C] \quad (6.24)$$

JOINT EQUILIBRIUM EQUATIONS

The above three equations contain four unknowns. The necessary fourth equation can be obtained from the moment equilibrium equation at joint *C* corresponding to unknown DoF θ_C, as shown in Figure 6.17.

Applying the moment equilibrium equation at joint *C*, we have

$$M_{CB} + M_{CD} = 0 \quad (6.25)$$

(Moment due to shear is neglected as segment is of differential length)

Substituting equations 6.21 and 6.22 into equation 6.25, we obtain

$$57.5 + \frac{3EI\theta_C}{5} - 16 + \frac{4EI\theta_C}{5} = 0 \Rightarrow EI\theta_C = -29.64 \quad \text{or}$$

$$\theta_C = \frac{-29.64}{EI} \text{ radians}$$

The final end moments can be calculated by back substituting the above known value of θ_C into equation 6.24. Therefore,

$$\begin{bmatrix} M_{CB} \\ M_{CD} \\ M_{DC} \end{bmatrix} = \begin{bmatrix} 57.50 \\ -16 \\ 16 \end{bmatrix} + \begin{bmatrix} 3/5 \\ 4/5 \\ 2/5 \end{bmatrix} [EI\theta_C = -29.64]$$

$$= \begin{bmatrix} 39.716 \\ -39.716 \\ 4.144 \end{bmatrix} \text{kNm}$$

With the know values of the final end moments, using the free body diagram shown in Figure 6.18a, and the static equilibrium equations, the end shears (reactions) can be determined. The shear force and bending moment diagrams, and probable elastic curve are shown in Figure 6.18b–d, respectively.

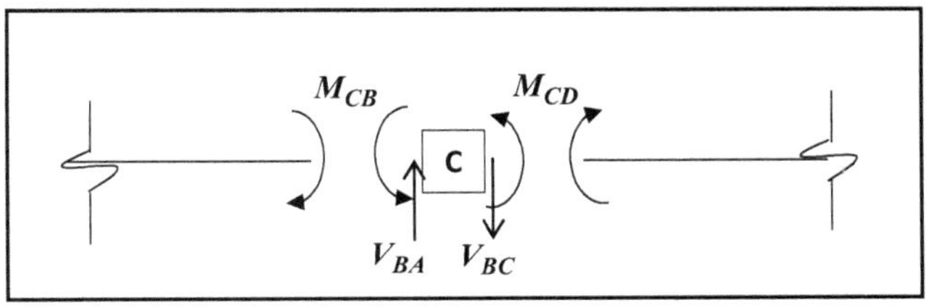

FIGURE 6.17 Free body diagram of joint *CD*.

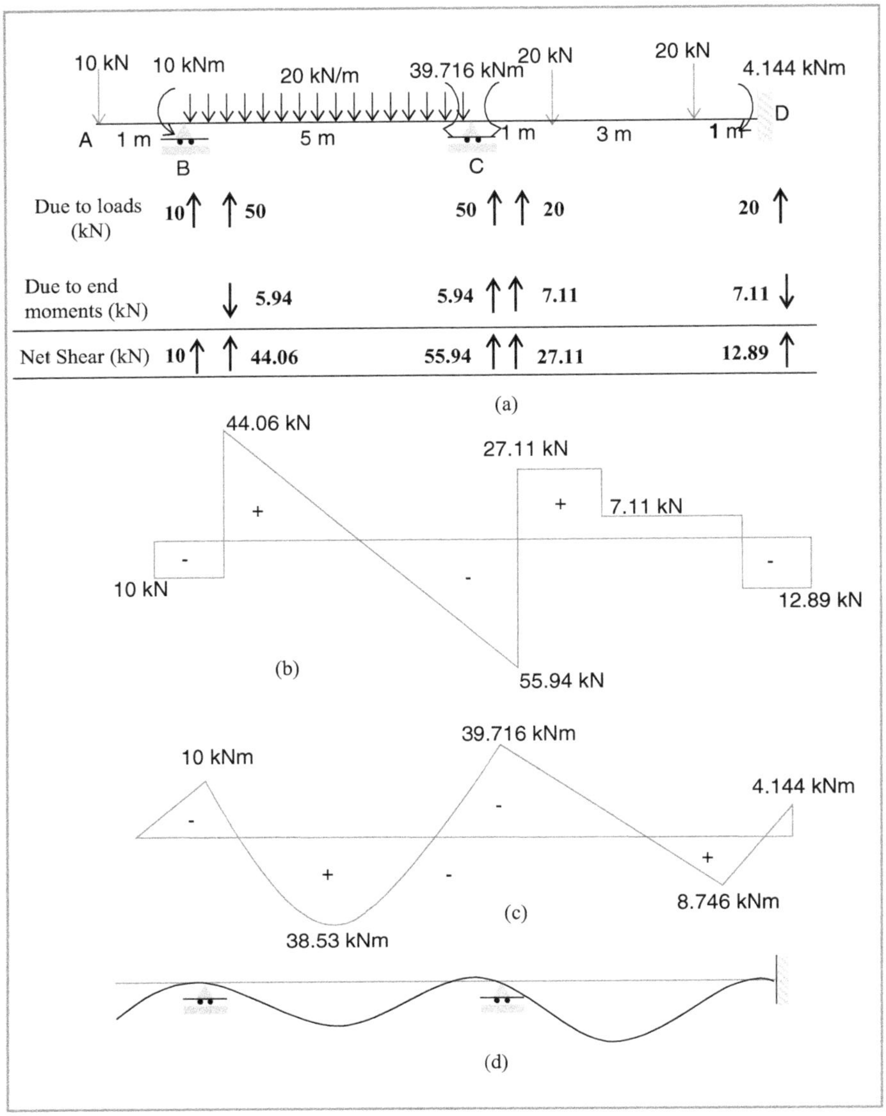

FIGURE 6.18 (a) Free body diagram, (b) shear force diagram, (c) bending moment diagram, and (d) probable deflected shape.

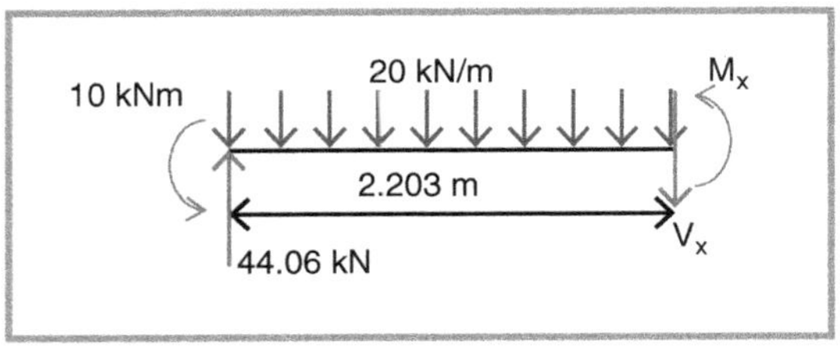

FIGURE 6.19 Free body diagram.

Maximum +ve BM in *BC*: Using free body diagrams shown in Figure 6.19, the maximum bending moment can be determined. From Figure 6.18b,

$$\frac{x}{44.06} = \frac{5-x}{55.94} \Rightarrow x = 2.203 \text{ m}$$

$$\therefore \text{Maximum +ve BM} = 44.06 \times 2.203 - 10 - 20 \times 2.203 \times \frac{2.203}{2} = 38.53 \text{ kNm}$$

Example 6.4

Analyse the three-span continuous beam *ABCD* as shown in Figure 6.20 using SD method. Assume EI is constant.

Solution:

The moment at end *A* is zero due to simply supported condition, and the rotation at *A* can therefore be neglected. Hence, the kinematic indeterminacy is reduced to 2, i.e., θ_B and θ_C. Consequently, the modified fixed end moments and modified SD equations for span *AB* (propped cantilever) can be applied.

FIXED END MOMENTS

The fixed end moments are shown in Figure 6.21.

Using the expressions given in Table 6.1, we have the fixed end moments

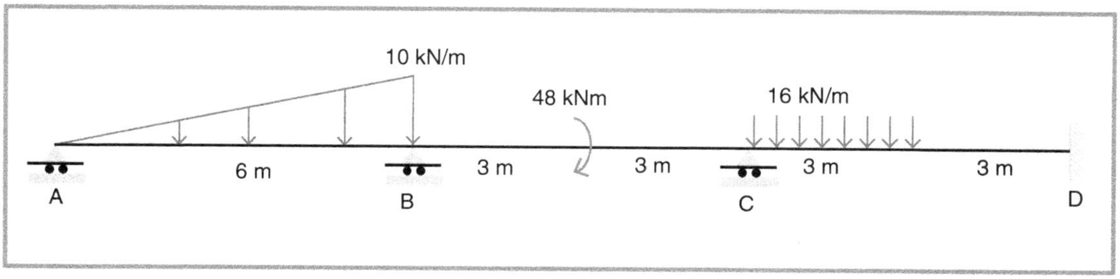

FIGURE 6.20 A three-span continuous beam.

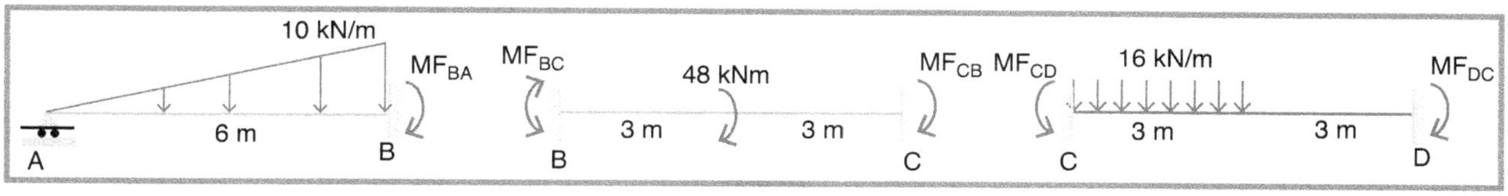

FIGURE 6.21 Fixed end moments.

$$MF_{BA} = \frac{wL^2}{20} + \frac{1}{2}\frac{wL^2}{30} = \frac{wL^2}{15} = \frac{10\times 6^2}{15} = 24 \text{ kNm}$$

$$MF_{BC} = \frac{M}{4} = \frac{48}{4} = 12 \text{ kNm} = MF_{CB}$$

$$MF_{CD} = -\frac{11}{192}wL^2 = -\frac{11\times 16\times 6^2}{192} = -33 \text{ kNm}$$

$$MF_{DC} = \frac{5}{192}wL^2 = \frac{5\times 16\times 6^2}{192} = 15 \text{ kNm}$$

SD EQUATIONS

For span *AB*: (Use modified SD equation. θ_A = negelcted, $\theta_B = ?, \varphi_{AB} = 0, L = 6$ m, $I = I$)

$$M_{BA} = MF_{BA} + \frac{3EI}{L}(\theta_B - \varphi_{AB})$$

$$M_{BA} = 24 + \frac{3EI}{6}(\theta_B - 0) = 24 + \frac{EI\theta_B}{2} \quad (6.26)$$

For span *BC*: ($\theta_C = ?, \theta_D = 0, \varphi_{CD} = 0, L = 5$ m, $I = I$), we have the SD equations

$$M_N = MF_N + \frac{2EI}{L}(2\theta_N + \theta_F - 3\varphi)$$

$$M_{BC} = MF_{BC} + \frac{2EI}{L}(2\theta_B + \theta_C - 3\varphi_{BC})$$

$$M_{BC} = 12 + \frac{2EI}{6}(2\theta_B + \theta_C - 3(0)) = 12 + \frac{2EI\theta_B}{3} + \frac{EI\theta_C}{3} \quad (6.27)$$

$$M_{CB} = MF_{CB} + \frac{2EI}{L}(\theta_B + 2\theta_C - 3\varphi_{BC})$$

$$M_{CB} = 12 + \frac{2EI}{6}(\theta_B + 2(\theta_C) - 3(0)) = 12 + \frac{EI\theta_B}{3} + \frac{2EI\theta_C}{3} \quad (6.28)$$

For span *CD*: ($\theta_C = ?, \theta_D = 0, \varphi_{CD} = 0, L = 6$ m, $I = I$)

$$M_N = MF_N + \frac{2EI}{L}(2\theta_N + \theta_F - 3\varphi)$$

$$M_{CD} = MF_{CD} + \frac{2EI}{L}(2\theta_C + \theta_D - 3\varphi_{CD})$$

$$M_{CD} = -33 + \frac{2EI}{6}(2\theta_C + 0 - 3(0)) = -33 + \frac{2EI\theta_C}{3} \quad (6.29)$$

$$M_{DC} = MF_{DC} + \frac{2EI}{L}(\theta_C + 2\theta_D - 3\varphi_{CD})$$

$$M_{DC} = 15 + \frac{2EI}{6}(\theta_C + 0 - 3(0)) = 15 + \frac{EI\theta_C}{3} \quad (6.30)$$

The above SD equations can be conveniently written in matrix form as shown below.

$$\begin{bmatrix} M_{BA} \\ M_{BC} \\ M_{CB} \\ M_{CD} \\ M_{DC} \end{bmatrix} = \begin{bmatrix} 24 \\ 12 \\ 12 \\ -33 \\ 15 \end{bmatrix} + \begin{bmatrix} 1/2 & 0 \\ 2/3 & 1/3 \\ 1/3 & 2/3 \\ 0 & 2/3 \\ 0 & 1/3 \end{bmatrix} \begin{bmatrix} EI\theta_B \\ EI\theta_C \end{bmatrix} \quad (6.31)$$

JOINT EQUILIBRIUM EQUATIONS

The free body diagram of joints *B* and *C* corresponding to the unknown DoF, i.e., θ_B and θ_C, is shown in Figure 6.22. Using the free body diagram, the moment equilibrium equations at joints *B* and *C* can be expressed as shown below.

At joint *B*

$$\sum M_B = M_{BA} + M_{BC} = 0 \Rightarrow 24 + \frac{EI\theta_B}{2} + 12 + \frac{2EI\theta_B}{3} + \frac{EI\theta_C}{3} = 0$$

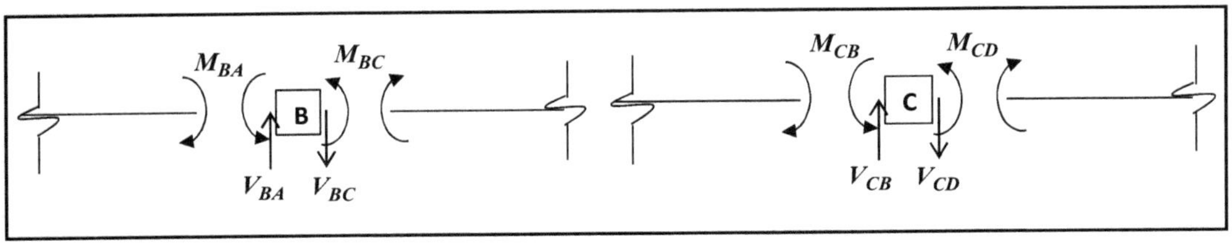

FIGURE 6.22 Free body diagrams of joints B and C.

$$\frac{7EI\theta_B}{6}+\frac{EI\theta_C}{3}=-36 \Rightarrow 7EI\theta_B+2EI\theta_C=-216 \qquad (6.32)$$

At joint C

$$M_{CB}+M_{CD}=0$$

$$\Rightarrow 12+\frac{EI\theta_B}{3}+\frac{2EI\theta_C}{3}+-33+\frac{2EI\theta_C}{3}=0$$

$$\Rightarrow EI\theta_B+4EI\theta_C=63 \qquad (6.33)$$

By solving equations 6.32 and 6.33, we get

$$EI\theta_B=-38.07\text{ kNm} \quad \text{and} \quad EI\theta_C=25.26\text{ kNm}$$

FINAL END MOMENTS

The final end moments can be determined by back substituting the above values of $EI\theta_B$ and $EI\theta_C$ into equation 6.31, as shown below.

$$\begin{bmatrix} M_{BA} \\ M_{BC} \\ M_{CB} \\ M_{CD} \\ M_{DC} \end{bmatrix}=\begin{bmatrix} 24 \\ 12 \\ 12 \\ -33 \\ 15 \end{bmatrix}+\begin{bmatrix} 1/2 & 0 \\ 2/3 & 1/3 \\ 1/3 & 2/3 \\ 0 & 2/3 \\ 0 & 1/3 \end{bmatrix}\begin{bmatrix} EI\theta_B=-38.07 \\ EI\theta_C=\ \ 25.26 \end{bmatrix}$$

$$=\begin{bmatrix} 4.96 \\ -4.96 \\ 16.16 \\ -16.16 \\ 23.42 \end{bmatrix}\text{kNm}$$

END SHEARS

Using the static equilibrium equations and with the help of the free body diagram shown in Figure 6.23a, the member end shears can be calculated. The shear force and bending moment diagrams are presented in Figure 6.23b and c, respectively. The probable deflected shape is shown in Figure 6.23d.

ZERO SHEAR IN SPAN *CD*

From Figure 6.23b

$$\frac{x}{34.79}=\frac{3-x}{13.21}\Rightarrow x=2.17\text{ m}$$

$$\therefore \text{Maximum +ve BM}=34.79\times 2.17-16.16-16\times 2.17 \times\frac{2.17}{2}=21.66\text{ kNm}$$

Zero shear in span AB, from Figure 6.23a

$$V_x=8.35-\frac{1}{2}\times x\times\frac{10x}{6}=0\Rightarrow x=3.17\text{ m}$$

$$\therefore \text{Maximum +ve BM}=8.35\times 3.17-0.5\times\frac{3.17\times 10}{6}\times 3.17\times\frac{3.17}{3}=17.62\text{ kNm}$$

Example 6.5

Analyse the two-span continuous beam whose one end is fixed and the other end is guided support, as shown in Figure 6.24, using the SD method. Assume EI is constant.

Solution:

The kinematic indeterminacy of the given beam is 2, i.e., θ_B & Δ_C. This would be considered as a sway type problem. However, for span BC, the modified SD equation and modified fixed end moments can be used to ignore Δ_C. To use the modified SD equation, the beam $ABCB^!A^!$ shall be visualised as symmetrical with respect to point C, with ABC representing the left half of $ABCB^!A^!$, as shown Figure 6.25. The moment at C is nothing but the mid span moment in the beam $BCB^!$. Consequently, the kinematic indeterminacy is assumed to be 1, i.e., θ_B.

FIXED END MOMENTS

Using the expressions provided in Table 6.1, and the free body diagrams shown in Figure 6.26, the fixed moments can be calculated as shown below.

$$MF_{AB}=-\frac{WL}{8}=-\frac{25\times 4}{8}=-12.5\text{ kNm}=-MF_{BA}$$

$$MF_{BC}=-\frac{wL^2}{12}=-\frac{15\times 4^2}{12}=-20\text{ kNm}$$

$$MF_{CB}=-\left(\frac{wL^2}{8}-\frac{wL^2}{12}\right)=-\left(\frac{wL^2}{24}\right)=-\frac{15\times 4^2}{24}=-10\text{ kNm}$$

SD EQUATIONS

For span *AB*: ($\theta_A=0, \theta_B=?, \varphi_{AB}=0, L=4\text{ m}, I=I$)

$$M_{AB}=MF_{AB}+\frac{2EI}{L}(2\theta_A+\theta_B-3\varphi_{AB})$$

$$=-12.5+\frac{2EI}{4}(2(0)+\theta_B-3(0))$$

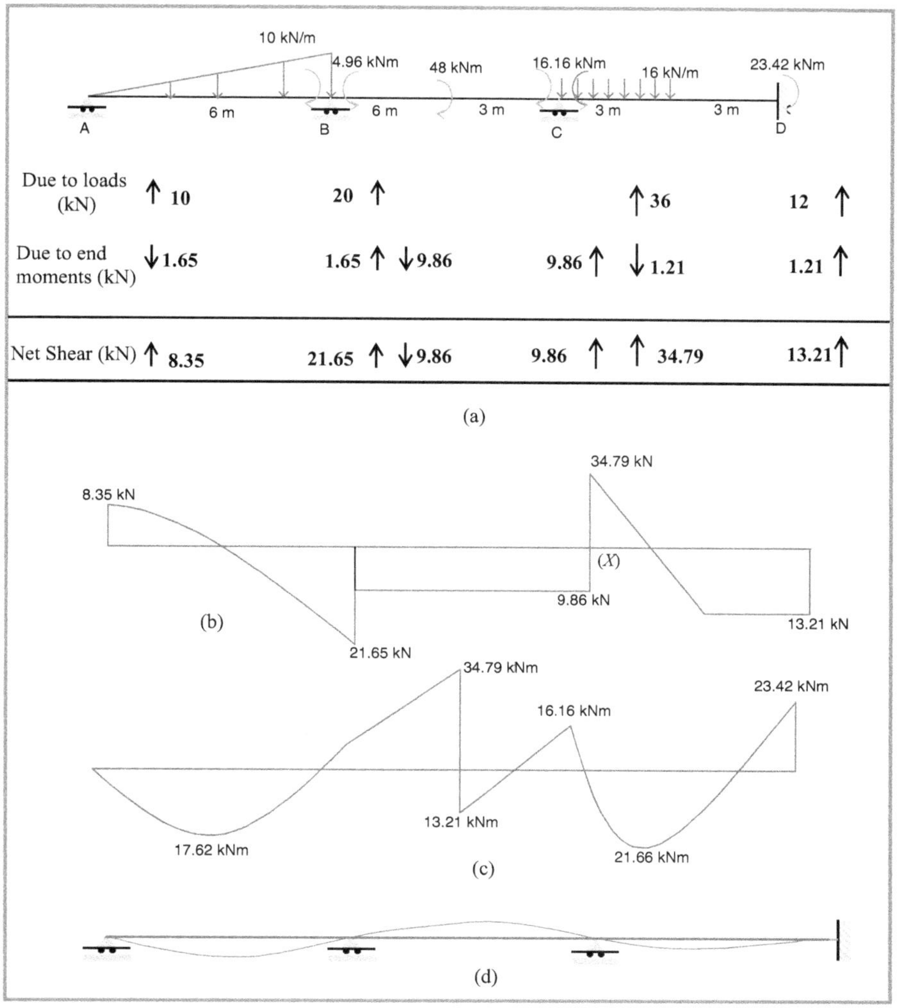

FIGURE 6.23 (a) Free body diagram, (b) shear force diagram, (c) bending moment diagram, and (d) probable elastic curve.

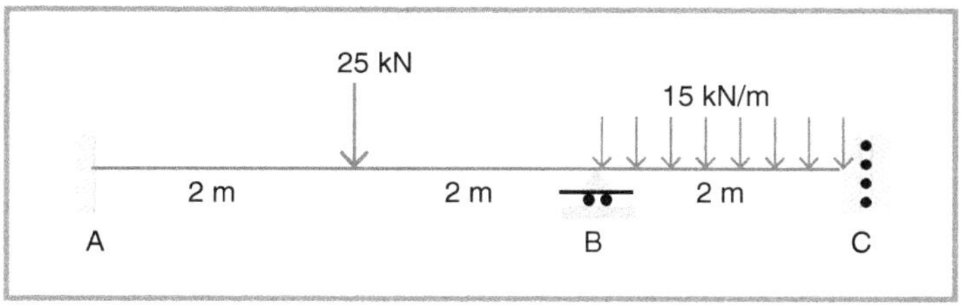

FIGURE 6.24 Two-span continuous beam with guided support.

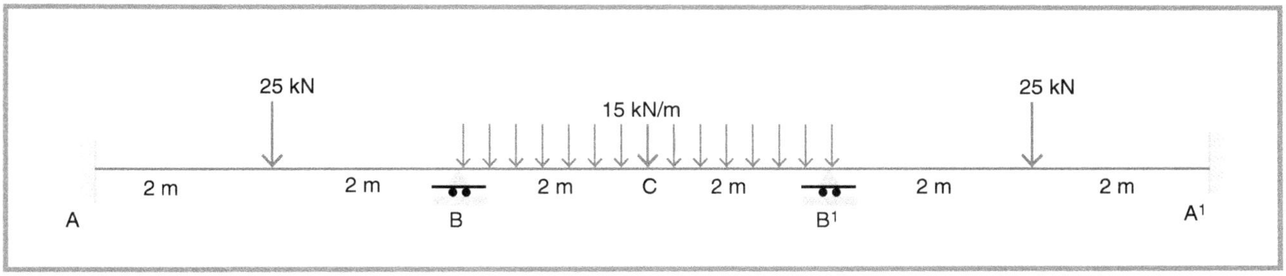

FIGURE 6.25 Visualised symmetrical beam $ABCB^1A^1$.

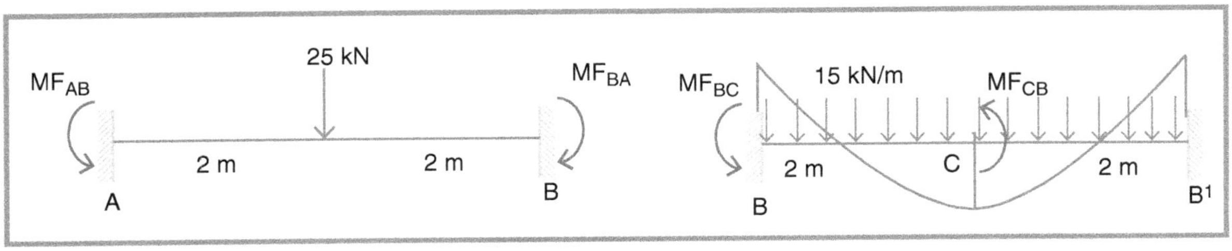

FIGURE 6.26 Fixed end moments.

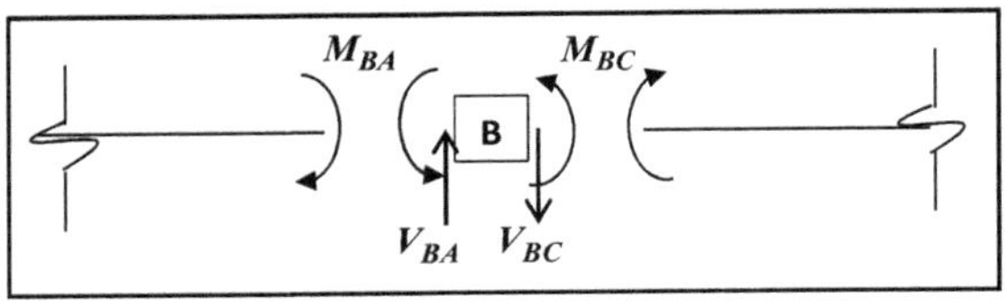

FIGURE 6.27 Free body diagram of joint *B*.

$$M_{AB} = -12.5 + \frac{EI\theta_B}{2} \quad (6.34)$$

$$M_{BA} = MF_{BA} + \frac{2EI}{L}(\theta_A + 2\theta_B - 3\varphi_{AB})$$
$$= 12.5 + \frac{2EI}{4}(0 + 2\theta_B - 3(0))$$

$$M_{BA} = 12.5 + EI\theta_B \quad (6.35)$$

For span *BC*: ($\theta_B = ?, \theta_C = 0, \varphi_{BC} =$ neglected, $L = 2$ m, $I = I$)

$$M_{BC} = MF_{BC} + \frac{EI}{L}(\theta_B) = -20 + \frac{EI}{2}(\theta_B) \quad (6.36)$$

$$M_{CB} = MF_{CB} - \frac{EI}{L}(\theta_B) = -10 - \frac{EI}{2}(\theta_B) \quad (6.37)$$

For convenience, the above SD equations can be written in matrix form as shown below.

$$\begin{bmatrix} M_{AB} \\ M_{BA} \\ M_{BC} \\ M_{CB} \end{bmatrix} = \begin{bmatrix} -12.5 \\ 12.5 \\ -20 \\ -10 \end{bmatrix} + \begin{bmatrix} 1/2 \\ 1 \\ 1/2 \\ -1/2 \end{bmatrix} \times [EI\theta_B] \quad (6.38)$$

JOINT EQUILIBRIUM EQUATIONS

The moment equilibrium equation at joint *B* corresponding to the unknown joint rotation θ_B using the free body diagram shown in Figure 6.27 can be expressed as follows.

$$M_{BA} + M_{BC} = 0$$

$$12.5 + EI\theta_B - 20 + \frac{EI}{2}(\theta_B) = 0 \Rightarrow EI\theta_B = 5 \text{ kNm}$$

FINAL END MOMENTS

By back substituting the above known value of $EI\theta_B$ in equation 6.38, the final end moments can be calculated as shown below.

$$\begin{bmatrix} M_{AB} \\ M_{BA} \\ M_{BC} \\ M_{CB} \end{bmatrix} = \begin{bmatrix} -12.5 \\ 12.5 \\ -20 \\ -10 \end{bmatrix} + \begin{bmatrix} 1/2 \\ 1 \\ 1/2 \\ -1/2 \end{bmatrix} \times [5] = \begin{bmatrix} -10 \\ 17.5 \\ -17.5 \\ -12.5 \end{bmatrix} \text{kNm}$$

SHEAR FORCE AND BENDING MOMENT DIAGRAMS

Using the known values of these end moment, the member end shears are shown in the free body diagram in Figure 6.28a. The shear force and bending moment diagrams are presented in Figure 6.28b and c, respectively. The probable deflected shape is shown in Figure 6.28d.

6.5.2 Application to the Beams with Yielding of Supports

Example 6.6

A two-span continuous beam is shown in Figure 6.29. Assume that the support *A* subjected to a clockwise rotational slip of 0.002 rad and *EI* is constant. Analyse the beam using SD method. Take EI=75,000 kNm^2.

Solution:

Since the supports *A* and *C* are fixed, the kinematic indeterminacy of the beam is 1, i.e., θ_B.

FIXED END MOMENTS

Using the formulae given in Table 6.1, the fixed end moments are calculated, as shown below.

DUE TO LOADS

$$MF_{AB} = MF_{BA} = 0 \text{ (since no laods on span } AB)$$

$$MF_{BC} = -MF_{CB} = -\frac{wL^2}{12} = -\frac{20 \times 4^2}{12} = -26.67 \text{ kNm}$$

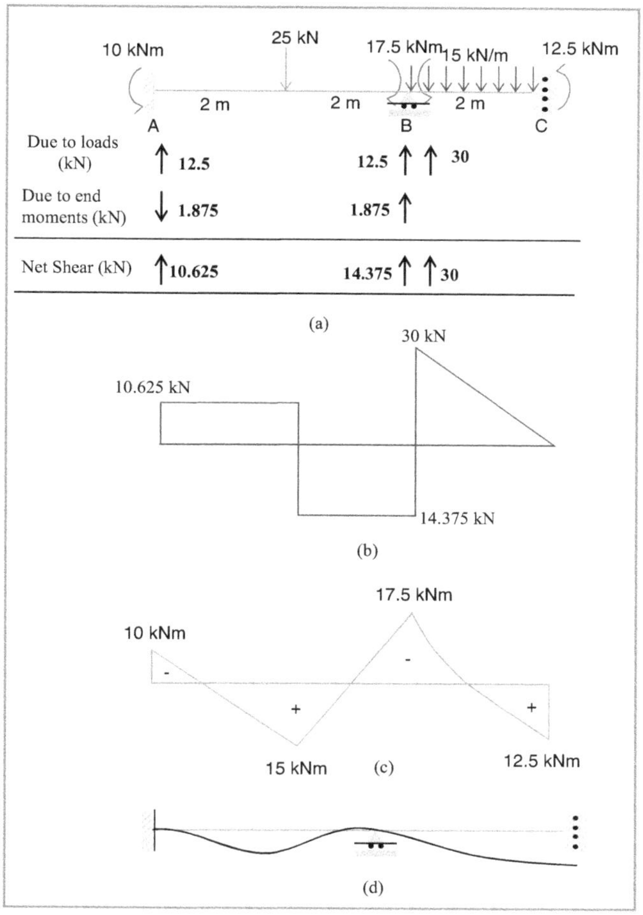

FIGURE 6.28 (a) Free body diagram, (b) shear force diagram, (c) bending moment diagram, and (d) probable deflected shape.

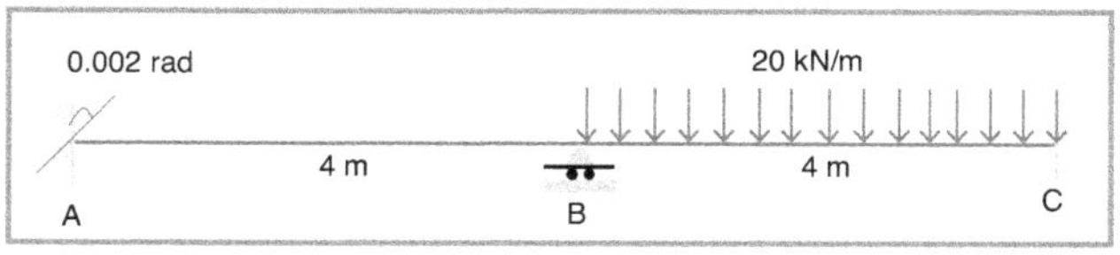

FIGURE 6.29 Two-span continuous beam.

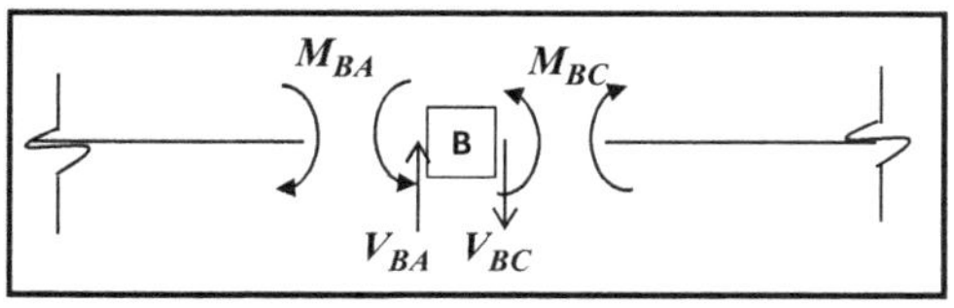

FIGURE 6.30 Free body diagram of joint *B*.

SD EQUATIONS

For span *AB*: The support *A* is subjected to rotational slip, $\theta_A = 0.002$ rad, $\theta_B = ?, \varphi_{AB} = 0, L = 4$ m, $I = I$, we have the SD equations:

$$M_{AB} = MF_{AB} + \frac{2EI}{L}(2\theta_A + \theta_B - 3\varphi_{AB})$$

$$= 0 + \frac{2EI}{4}(2(0.002) + \theta_B - 3(0))$$

$$M_{AB} = 0.002EI + \frac{EI\theta_B}{2} \tag{6.39}$$

$$M_{BA} = MF_{BA} + \frac{2EI}{L}(\theta_A + 2\theta_B - 3\varphi_{AB})$$

$$= 0 + \frac{2EI}{4}(0.002 + 2\theta_B - 3(0))$$

$$M_{BA} = 0.001EI + EI\theta_B \tag{6.40}$$

For span *BC*: $(\theta_B = ?, \theta_C = 0, \varphi_{BC} = 0, L = 4 \text{ m}, I = I)$

$$M_{BC} = MF_{BC} + \frac{2EI}{L}(2\theta_B + \theta_C - 3\varphi_{BC})$$

$$= -26.67 + \frac{2EI}{4}(2\theta_B) = -26.67 + EI\theta_B \tag{6.41}$$

$$M_{CB} = MF_{CB} + \frac{2EI}{L}(\theta_B + 2\theta_C - 3\varphi_{BC}) = 26.67 + \frac{2EI}{4}(\theta_B)$$

$$= 26.67 + 0.5EI\theta_B \tag{6.42}$$

JOINT EQUILIBRIUM EQUATIONS

Using the free body diagram shown in Figure 6.30, applying the moment equilibrium equation at joint *B* corresponding to the unknown joint rotation θ_B, we can write

$$M_B = M_{BA} + M_{BC} = 0 \tag{6.43}$$

$$\Rightarrow 0.001EI + EI\theta_B - 26.67 + EI\theta_B = 0 \Rightarrow EI\theta_B = -24.165 \text{ kNm}$$

FINAL END MOMENTS

By substituting the above value of $EI\theta_B$ into SD equations 6.39–6.42, we can get the final end moments as shown below.

$$\begin{bmatrix} M_{AB} \\ M_{BA} \\ M_{BC} \\ M_{CB} \end{bmatrix} = \begin{bmatrix} 150 \\ 75 \\ -26.67 \\ -26.67 \end{bmatrix} + \begin{bmatrix} 1/2 \\ 1 \\ 1 \\ 1/2 \end{bmatrix} \times [-24.165]$$

$$= \begin{bmatrix} 137.91 \\ 50.835 \\ -50.835 \\ 14.587 \end{bmatrix} \text{kNm}$$

END SHEARS

The member end shears are shown in the free body diagram (Figure 6.31a). The shear force and bending

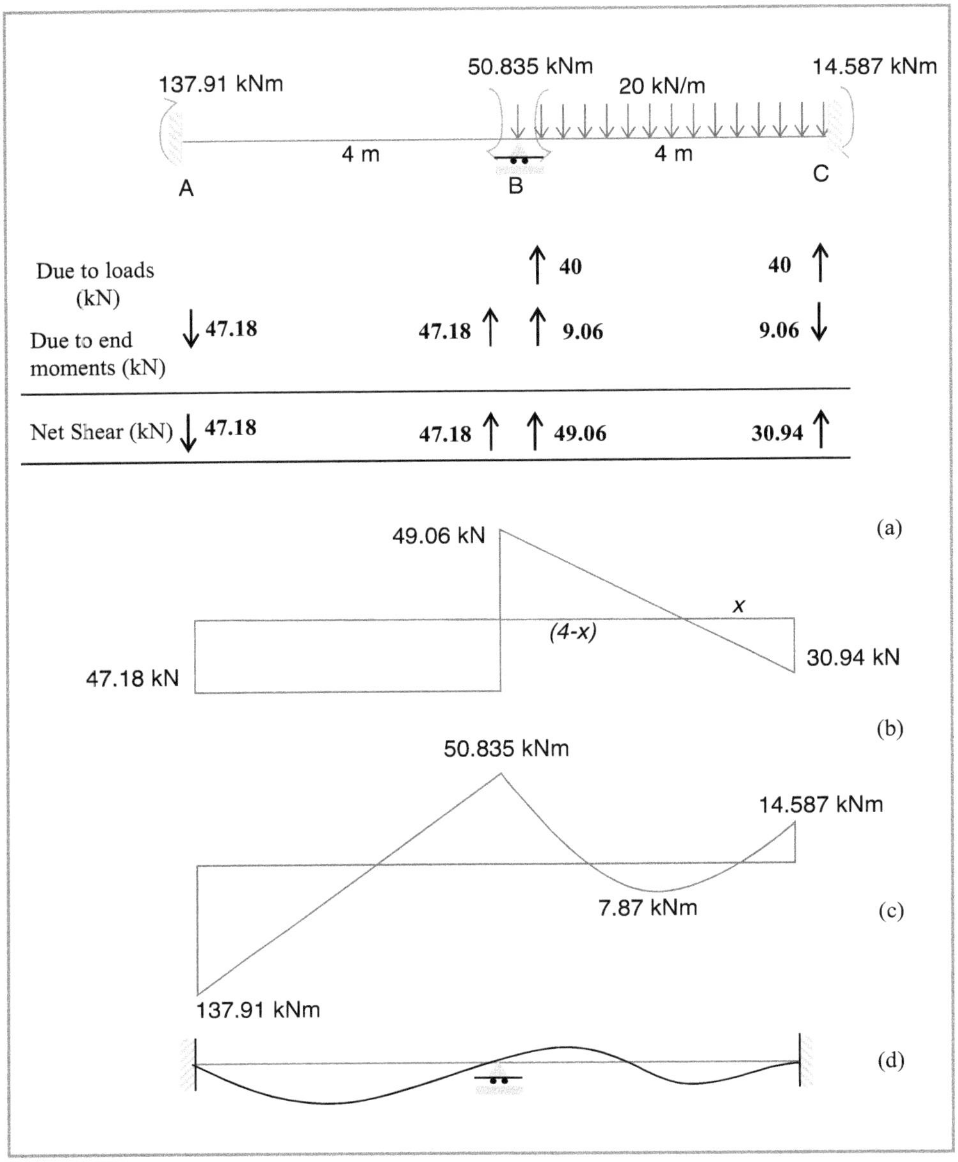

FIGURE 6.31 (a) Free body diagram, (b) shear force diagram, (c) bending moment diagram, and (d) probable deflected shape.

moment diagrams are shown in Figure 6.31b and c, respectively. The probable deflected shape is shown in Figure 6.31d.

Maximum +ve BM in span *BC*

From Figure 6.31b

$$\frac{x}{30.94} = \frac{4-x}{49.06} \Rightarrow x = 1.55 \text{ m}$$

$$\text{Max. +ve BM} = 30.94 \times 1.55 - 14.587 - 20 \times 1.55^2 \times 0.5$$

$$= 9.35 \text{ kNm}$$

Example 6.7

A three-span continuous beam with an internal hinge at *B* is shown in Figure 6.32. Analyse the beam using the SD method.

Solution:

Since at point *B* there is an internal hinge, for spans *BA* and *BC*, the modified SD equation and modified fixed end moments can be applied; therefore, θ_{BA} and θ_{BC} can be ignored. Consequently, the kinematic indeterminacy is limited to only 2, i.e., θ_C and Δ_B.

FIXED END MOMENTS

Using Table 6.1 and Figure 6.33a, the fixed end moments can be expressed as

$$MF_{AB} = MF_{CB} = 0 \left(\text{since no loads on spans } AB \text{ and } BC\right)$$

$$MF_{CD} = MF_{DC} = \frac{M}{4} = \frac{10}{4} = 2.5 \text{ kNm}$$

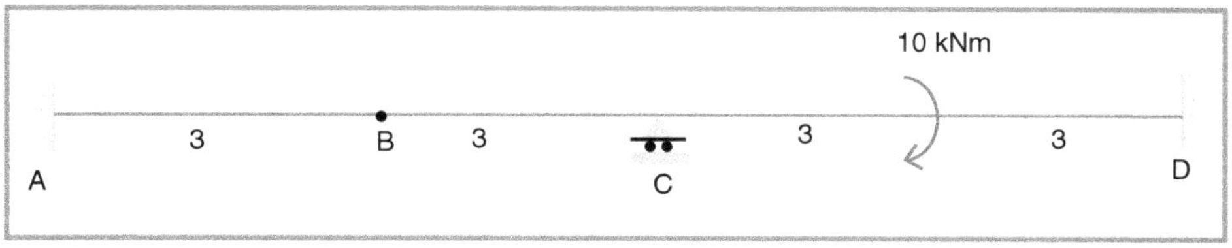

FIGURE 6.32 A three-span continuous beam with an internal hinge.

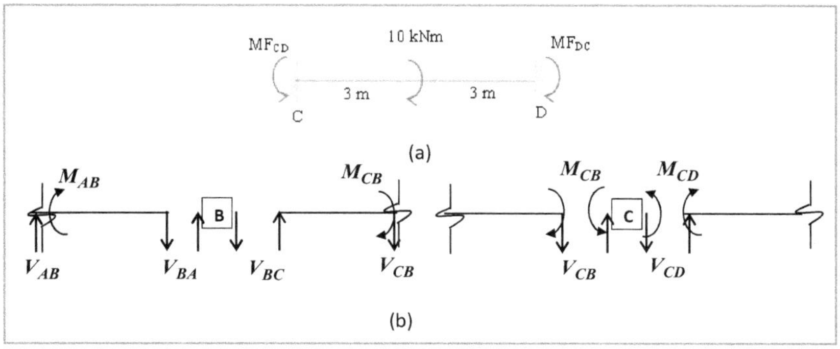

FIGURE 6.33 (a) Fixed end moments and (b) free body diagrams of joints C and B.

SD EQUATIONS

Due to the internal hinge present at point B, the modified stiffness equation can be applied to spans BA and BC; thus, we have

For span *AB*: $(\theta_A = 0, \theta_{BA} : \text{ignored?}, \varphi_{AB} = \frac{\Delta_B}{3}, L = 3 \text{ m}, I = I)$

$$M_{AB} = MF_{AB} + \frac{3EI}{L}(\theta_A - \varphi_{AB}) = 0 + \frac{3EI}{3}\left(0 - \frac{\Delta_B}{3}\right) = -\frac{EI\Delta_B}{3} \tag{6.44}$$

For span *BC*: $(\theta_C = ?, \theta_{BC} : \text{ignored?}, \varphi_{BC} = \frac{-\Delta_B}{3}, L = 3 \text{ m}, I = I)$

$$M_{CB} = MF_{CB} + \frac{3EI}{L}(\theta_C - \varphi_{AB}) = 0 + \frac{3EI}{3}\left(\theta_C - \frac{-\Delta_B}{3}\right)$$

$$= EI\theta_C + \frac{EI\Delta_B}{3} \tag{6.45}$$

For span *CD*: $(\theta_C = ?, \theta_D = 0, \varphi_{CD} = 0, L = 6 \text{ m}, I = I)$

$$M_{CD} = MF_{CD} + \frac{2EI}{L}(2\theta_C + \theta_D - 3\varphi_{CD})$$

$$= 2.5 + \frac{2EI}{6}(2\theta_C) = 2.5 + 0.667EI\theta_C \tag{6.46}$$

$$M_{DC} = MF_{DC} + \frac{2EI}{L}(\theta_C + 2\theta_D - 3\varphi_{CD})$$

$$= 2.5 + \frac{2EI}{6}(\theta_C) = 2.5 + 0.333EI\theta_C \tag{6.47}$$

The above SD equations can be expressed in matrix form as

$$\begin{bmatrix} M_{AB} \\ M_{CB} \\ M_{CD} \\ M_{DC} \end{bmatrix} = \begin{bmatrix} 0 \\ 0 \\ 2.5 \\ 2.5 \end{bmatrix} + \begin{bmatrix} 0 & -0.333 \\ 1 & 0.333 \\ 0.667 & 0 \\ 0.333 & 0 \end{bmatrix} \times \begin{bmatrix} EI\theta_C \\ EI\Delta_B \end{bmatrix} \tag{6.48}$$

JOINT EQUILIBRIUM EQUATIONS

Using the free body diagram at joint C corresponding to the unknown joint rotation θ_C, as shown in Figure 6.33b, the moment equilibrium equation can be expressed as

$$M_C = M_{CB} + M_{CD} = 0$$

$$EI\theta_C + \frac{EI\Delta_B}{3} + 2.5 + 0.667EI\theta_C = 0 \Rightarrow 1.667EI\theta_C$$
$$+ 0.333EI\Delta_B = -2.5 \tag{6.49}$$

And the shear equilibrium equation at joint B can be written as

$$V_B = V_{BA} - V_{BC} = 0$$

$$V_{BA} = -\frac{M_{AB}}{3} \quad \text{and} \quad V_{BC} = -\frac{M_{CB}}{3}$$

$$-\frac{M_{AB}}{3}+\frac{M_{CB}}{3}=0 \quad (\text{or}) \quad -M_{AB}+M_{CB}=0$$

$$\Rightarrow \frac{EI\Delta_B}{3}+EI\theta_C+\frac{EI\Delta_B}{3}=0 \Rightarrow 0.667EI\Delta_B+EI\theta_C=0 \tag{6.50}$$

By solving equations 6.49 and 6.50, we get

$$EI\Delta_B = 3.21 \text{ kNm}; \quad EI\theta_C = -2.14 \text{ kNm}$$

FINAL END MOMENTS

By substituting the above known values of $EI\theta_C$ and $EI\Delta_B$ into SD equations 6.44–6.47, we can get the final end moments as

$$\begin{bmatrix} M_{AB} \\ M_{CB} \\ M_{CD} \\ M_{DC} \end{bmatrix} = \begin{bmatrix} 0 \\ 0 \\ 2.5 \\ 2.5 \end{bmatrix} + \begin{bmatrix} 0 & -1/3 \\ 1 & 1/3 \\ 2/3 & 0 \\ 1/3 & 0 \end{bmatrix} \times \begin{bmatrix} -2.14 \\ 3.21 \end{bmatrix}$$

$$= \begin{bmatrix} -1.07 \\ -1.07 \\ 1.07 \\ 1.78 \end{bmatrix} \text{kNm}$$

The member end forces are shown in the free body diagram (Figure 6.34a). The shear force and bending moment diagrams are shown in Figure 6.34b and c, respectively. The probable deflected shape is presented in Figure 6.34d.

Example 6.8

Analyse the continuous beam *ABC* shown in Figure 6.35 using the SD method. The stiffness of the spring *k* is $\frac{12EI}{L^3}$

Solution

The spring support allows the beam to rotate as well as translate at *C*, and the beam is supported by a hinge at *A* and continuous at *B*; therefore, the degree of kinematic indeterminacy is 4, i.e., $\theta_A, \theta_B, \theta_C$ and Δ_C. Since the moments at *A* and *C* are zero, we can use the modified SD equations with appropriate modifications in the fixed end moments and stiffness for spans *AB* and *BC*, θ_A and θ_C can be neglected. Thus, the kinematic indeterminacy is limited to only two, i.e., θ_B and Δ_C

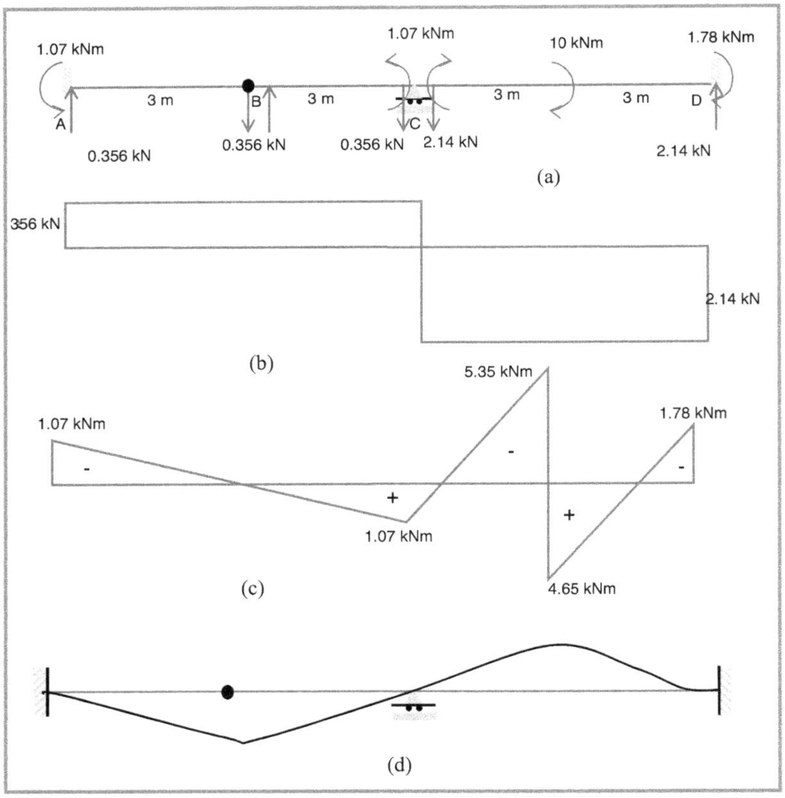

FIGURE 6.34 (a) Free body diagram, (b) shear force diagram, (c) bending moment diagram, and (d) probable elastic curve.

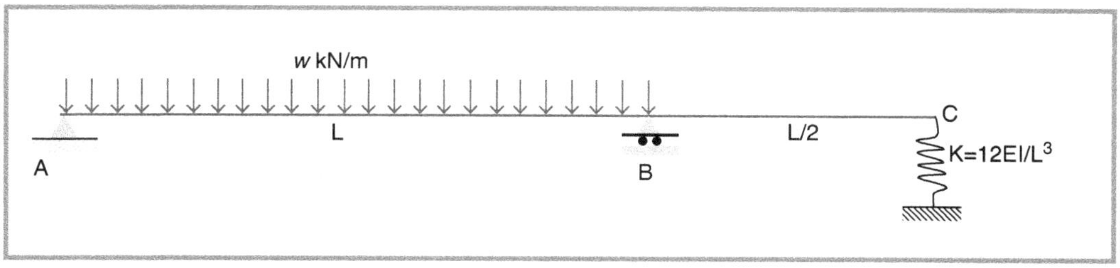

FIGURE 6.35 Beam with spring at C.

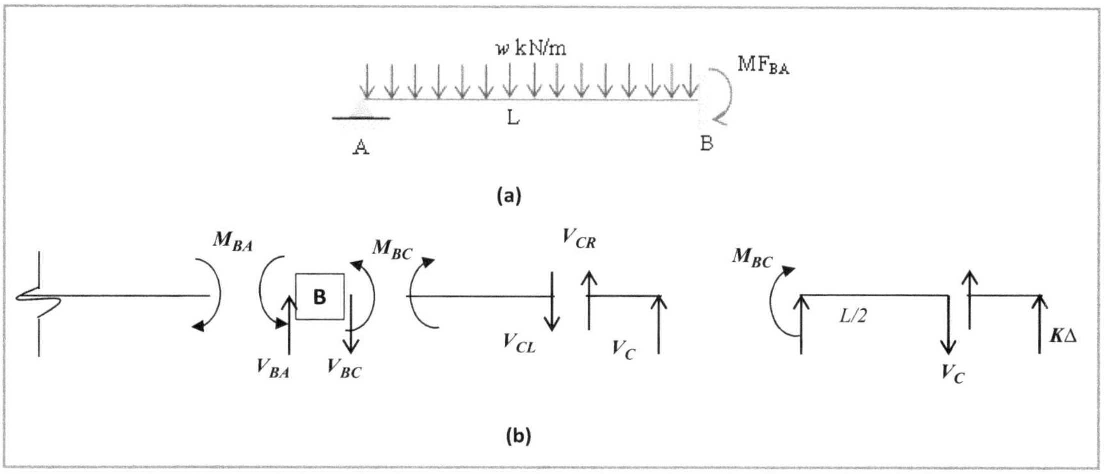

FIGURE 6.36 (a) Fixed end moments and (b) free body diagram of joints B and C.

FIXED END MOMENTS

Using Table 6.1 and Figure 6.36a, the fixed end moments are

$$MF_{BA} = \frac{wL^2}{12} + \frac{wL^2}{24} = \frac{wL^2}{8} \text{ kNm}$$

$$MF_{BC} = 0 \,(\text{since no loads on } BC)$$

SD EQUATIONS

As mentioned above, the modified SD equations for

Span *AB*: $(\theta_A : \text{ignored}, \theta_B = ?, \varphi_{AB} = 0, L = L, I = I)$

$$M_{BA} = MF_{BA} + \frac{3EI}{L}(\theta_B - \varphi_{AB}) = \frac{WL^2}{8} + \frac{3EI}{L}(\theta_B - 0)$$

$$= \frac{WL^2}{8} + \frac{3EI}{L}(\theta_B) \tag{6.51}$$

Span *BC*: $(\theta_C : \text{ignored}, \theta_B = ?, \varphi_{BC} = \Delta_C/(L/2) = 2\Delta_C/L, L = L/2, I = I)$

$$M_{BC} = MF_{BC} + \frac{3EI}{L}(\theta_B - \varphi_{BC}) = 0 + \frac{3EI}{L/2}\left(\theta_B - \frac{2\Delta_C}{L}\right)$$

$$= \frac{6EI}{L}\left(\theta_B - \frac{2\Delta_C}{L}\right) \tag{6.52}$$

EQUILIBRIUM EQUATIONS

At joint B, apply the moment equilibrium equation corresponding to the unknown joint rotation θ_B, as shown in Figure 6.36b.

$$M_B = M_{BA} + M_{BC} = 0$$

$$\frac{WL^2}{8} + \frac{3EI}{L}(\theta_B) + \frac{6EI}{L}\left(\theta_B - \frac{2\Delta_C}{L}\right) = 0 \tag{6.53}$$

At joint *C*: The shear equilibrium equation corresponding to Δ_C, can be expressed as

$$V_C = \frac{-M_{BC}}{L/2} = \frac{-2M_{BC}}{L} = -k\Delta_C$$

$$\frac{-12EI}{L^2}\left(\theta_B - \frac{2\Delta_C}{L}\right) = \frac{-12EI\Delta_C}{L^3} \Rightarrow \theta_B - \frac{2\Delta_C}{L}$$

$$= \frac{\Delta_C}{L} => \theta_B = \frac{3\Delta_C}{L} \qquad (6.54)$$

By solving equations 6.53 and 6.54, we obtain

$$\frac{WL^2}{8} + \frac{3EI}{L}\left(\frac{3\Delta_C}{L}\right) + \frac{6EI}{L}\left(\frac{3\Delta_C}{L} - \frac{2\Delta_C}{L}\right) = 0 \Rightarrow \Delta_C$$

$$= \frac{-wL^4}{120EI} \quad \text{and} \quad \theta_B = \frac{-wL^3}{40EI}$$

FINAL END MOMENTS

By back substituting the above known values of θ_B and Δ_C into equations 6.51 and 6.52, we get

$$M_{BA} = \frac{WL^2}{8} + \frac{3EI}{L}\left(\frac{-wL^3}{40EI}\right) = \frac{wL^2}{20}$$

$$M_{BC} = \frac{6EI}{L}\left(\frac{-wL^3}{40EI} - \frac{2\frac{-wL^4}{120EI}}{L}\right) = -\frac{wL^2}{20}$$

Maximum +ve BM in span AB

From Figure 6.37b

$$\frac{x}{9wL/20} = \frac{L-x}{11wL/20} \Rightarrow x = \frac{9L}{20}$$

$$\text{Max +ve BM in } AB = \frac{9wL}{20} \times \frac{9L}{20} - w \times \frac{9L}{20} \times \frac{9L}{40} = \frac{81wL^2}{800}$$

The free body diagram representing the member end forces, along with the shear force and bending moment diagrams, is shown in Figure 6.37a–c, respectively.

Example 6.9

Analyse the non-prismatic fixed beam *ABC* shown in Figure 6.38, using the SD method. Draw the shear force and bending moment diagrams.

Solution:

Since both ends of the beam are fixed, the beam is kinematically constrained. Therefore, to solve the problem, we need to consider a node at point *B*, so that the DoFs at point *B* can be taken as I_k=2, i.e., θ_C and Δ_B. Now, the SDs can be applied for spans *BA* and *BC*.

FIXED END MOMENTS

Since the loads are acting at joint *B*, and there are no loads on spans *AB* and *BC*, the fixed end moments are $MF_{AB} = MF_{BA} = MF_{BC} = MF_{CB} = 0$

SD EQUATIONS

For span *AB*: $(\theta_A = 0, \theta_B = ?, \varphi_{AB} = \frac{\Delta_B}{8}, L = 8\text{ m}, I = 2I)$

$$M_{AB} = MF_{AB} + \frac{2EI}{L}(2\theta_A + \theta_B - 3\varphi_{AB})$$

$$M_{AB} = 0 + \frac{2E(2I)}{8}\left(2(0) + \theta_B - 3\frac{\Delta_B}{8}\right) = \frac{EI}{2}\left(\theta_B - 3\frac{\Delta_B}{8}\right) \qquad (6.55)$$

$$M_{BA} = 0 + \frac{2E(2I)}{8}\left(0 + 2\theta_B - 3\frac{\Delta_B}{8}\right) = \left[\frac{EI}{2}\left(2\theta_B - 3\frac{\Delta_B}{8}\right)\right] \qquad (6.56)$$

For span *BC*: $(\theta_C = 0, \theta_B = ?, \varphi_{BC} = \frac{-\Delta_B}{4}, L = 4\text{ m}, I = I)$

$$M_{BC} = MF_{BC} + \frac{2EI}{L}(2\theta_B + \theta_C - 3\varphi_{BC})$$

$$M_{BC} = 0 + \frac{2EI}{4}\left(2\theta_B + 0 - 3\left(\frac{-\Delta_B}{4}\right)\right) = \frac{EI}{2}\left(2\theta_B + 3\left(\frac{\Delta_B}{4}\right)\right) \qquad (6.57)$$

$$M_{CB} = 0 + \frac{2EI}{4}\left(\theta_B + 2(0) - 3\left(\frac{-\Delta_B}{4}\right)\right) = \frac{EI}{2}\left(\theta_B + 3\left(\frac{\Delta_B}{4}\right)\right) \qquad (6.58)$$

We can express the above equations 6.55 – 6.58 in a matrix form as

$$\begin{bmatrix} M_{AB} \\ M_{BA} \\ M_{BC} \\ M_{CB} \end{bmatrix} = \begin{bmatrix} 0 \\ 0 \\ 0 \\ 0 \end{bmatrix} + \begin{bmatrix} 1/2 & -3/16 \\ 1 & -3/16 \\ 1 & 3/8 \\ 1/2 & 3/8 \end{bmatrix} \times \begin{bmatrix} EI\theta_B \\ EI\Delta_B \end{bmatrix} \qquad (6.59)$$

EQUILIBRIUM EQUATIONS

The free body diagrams of the joints are shown in Figure 6.39.

At joint *B*: Using the free body diagram shown in Figure 6.39, the moment equilibrium equation corresponding to θ_B at joint B can be expressed as

$$M_B = M_{BA} + M_{BC} = -40$$

$$\frac{EI}{2}\left(2\theta_B - 3\frac{\Delta_B}{8}\right) + \frac{EI}{2}\left(2\theta_B + 3\left(\frac{\Delta_B}{4}\right)\right) = -40$$

$$32EI\theta_B + 3EI\Delta_B = -640 \qquad (6.60)$$

SHEAR EQUILIBRIUM EQUATION

Applying the shear equilibrium equation at joint *C* for the free body diagram shown in Figure 6.39, we have

$$V_B = V_{BA} - V_{BC} = 10$$

$$V_{BA} = -\frac{M_{BA} + M_{AB}}{8} \quad \text{and} \quad V_{BC} = -\frac{M_{BC} + M_{CB}}{4}$$

Therefore,

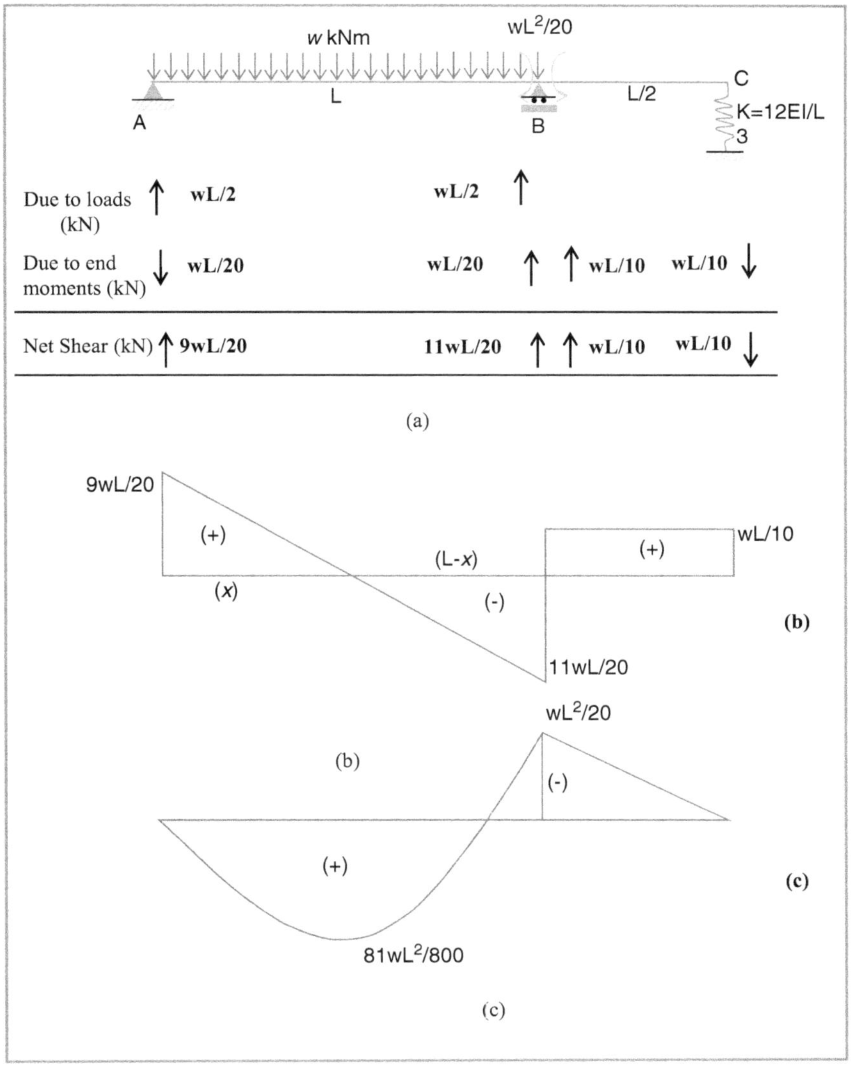

FIGURE 6.37 (a) Free body diagram, (b) shear force diagram, and (c) bending moment diagram.

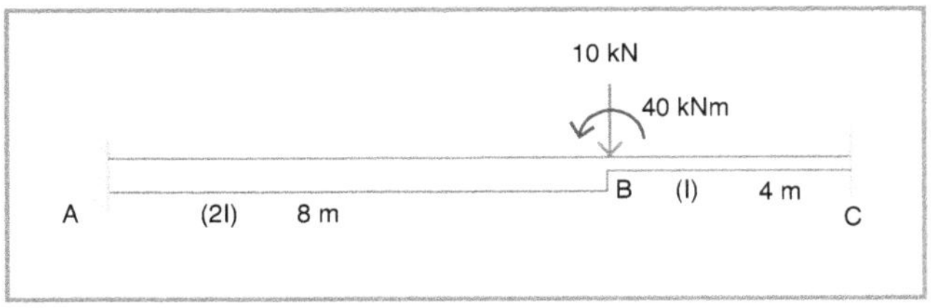

FIGURE 6.38 Non-prismatic beam.

$$V_B = -\frac{M_{BA}+M_{AB}}{8} - \left(-\frac{M_{BC}+M_{CB}}{4}\right) = 10$$

$$= -\frac{1}{8}\left[\frac{EI}{2}\left(2\theta_B - 3\frac{\Delta_B}{8}\right) + \frac{EI}{2}\left(\theta_B - 3\frac{\Delta_B}{8}\right)\right]$$

$$+\frac{1}{4}\left[\frac{EI}{2}\left(2\theta_B + 3\left(\frac{\Delta_B}{4}\right)\right) + \frac{EI}{2}\left(\theta_B + 3\left(\frac{\Delta_B}{4}\right)\right)\right] = 10$$

$$= -\frac{EI}{16}\left(3\theta_B - \frac{3\Delta_B}{4}\right) + \frac{EI}{8}\left(3\theta_B + \frac{3\Delta_B}{2}\right) = 10$$

$$3EI\theta_B + \frac{15EI\Delta_B}{4} = 160 \tag{6.61}$$

By solving equations 6.60 and 6.61, we obtain $EI\theta_B = -\frac{2{,}880}{111}$ and $EI\Delta_B = \frac{7{,}040}{111}$

FINAL END MOMENTS

By back substituting these values of $EI\theta_B$ and $EI\Delta_B$ into equation 6.59, we can determine the final end moments as

$$\begin{bmatrix} M_{AB} \\ M_{BA} \\ M_{BC} \\ M_{CB} \end{bmatrix} = \begin{bmatrix} 0 \\ 0 \\ 0 \\ 0 \end{bmatrix} + \begin{bmatrix} 1/2 & -3/16 \\ 1 & -3/16 \\ 1 & 3/8 \\ 1/2 & 3/8 \end{bmatrix} \times \begin{bmatrix} -2880/111 \\ 7040/111 \end{bmatrix}$$

$$= \begin{bmatrix} -24.86 \\ -37.84 \\ -2.16 \\ 10.81 \end{bmatrix} \text{kNm}$$

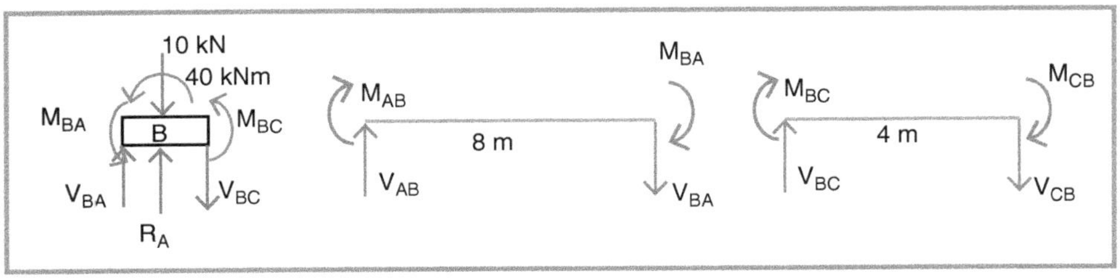

FIGURE 6.39 Free body diagram of joints.

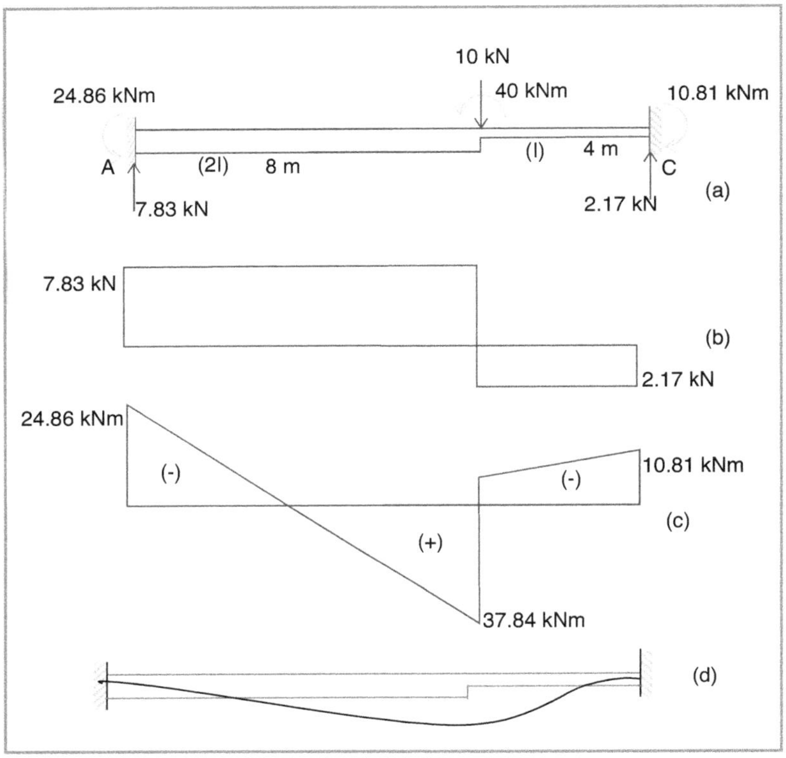

FIGURE 6.40 (a) Free body diagram, (b) shear force diagram and (c) bending moment diagram.

The member end forces are shown in the free body diagram in Figure 6.40a. The shear force and bending moment diagrams are shown in Figure 6.40b and c, respectively. The deflected shape is presented in Figure 6.40d.

6.5.3 Application to the Rigid-Jointed Plane Frames (without Side Sway)

Normally, the axial deformations in rigid-jointed plane frames are neglected, as their effect is smaller compared to the effect due to bending. When frames are properly restrained or exhibit symmetry with respect to loading, geometry, and support conditions, they are referred to as non-sway frames or frames without side sway. The following are examples of frames without side sway. In such frames, the chord rotation (φ) is equal to zero in the SD equations. Therefore, the analysis of these frames follows the same procedure as that used for continuous beams without supports yielding. However, the key difference is that, in beams, only two members can meet at a joint, whereas in frames, more than two members may meet at a joint. This is illustrated in Figure 6.41a and d. At joint C in Figure 6.41a and d, and at joint D in Figure 6.41d, three members are connected. The FBD of joint C is shown in Figure 6.41e.

Thus, the moment equilibrium equation at joint C can be expressed as

$$\sum M_C = 0 \Rightarrow M_{CB} + M_{CD} + M_{CE} = 0$$

Each joint is a rigid joint, and all members meeting at these joints rotate by the same amount. Therefore, each joint has only one unknown rotation.

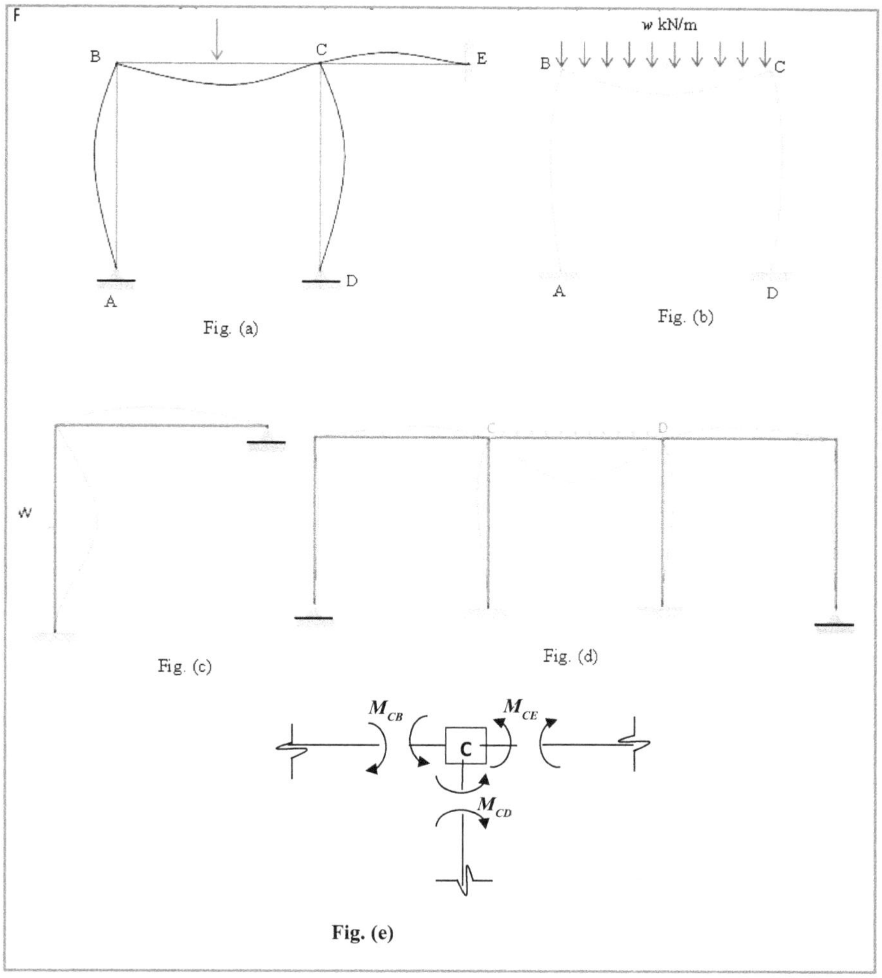

FIGURE 6.41 (a–d) Frames without side-sway and (e) free body diagram of joint C.

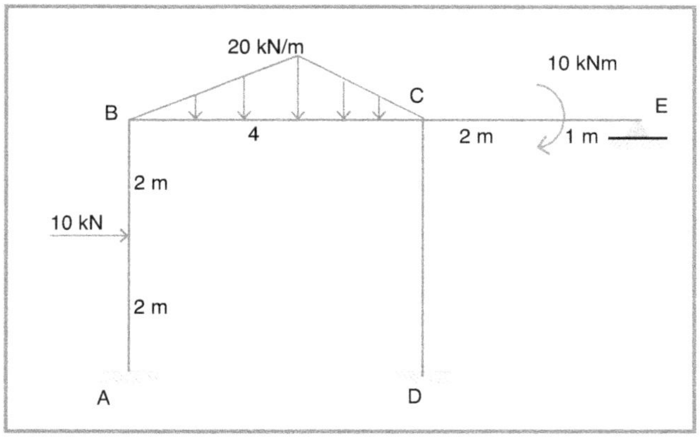

FIGURE 6.42 Rigid-jointed plane frame.

Example 6.10

A rigid-jointed plane frame is shown in Figure 6.42. Compute the support reactions using the SD method, and draw the shear force and bending moment diagrams. Also draw the probable deflected shape. Assume EI is constant.

Solution:

Usually, the degree of kinematic indeterminacy is 3. However, due to the hinge support at E, the moment is zero. Hence, the rotation at E can be ignored, provided the modified stiffness, fixed moment, and modified SD are used for span CE. Therefore, the kinematic indeterminacy is limited to 2, i.e., θ_B and θ_C.

Fixed end moments: The fixed end moments are shown in the free body diagram in Figure 6.43. These can be determined using the expressions provided in Table 6.1 as shown below.

$$MF_{AB} = -MF_{BA} = -\frac{WL}{8} = -\frac{10\times 4}{8} = -5 \text{ kNm}$$

$$MF_{BC} = -MF_{CB} = -\frac{5}{96}wL^2 = -\frac{5}{96}\times 20\times 4^2 = -16.67 \text{ kNm}$$

$$MF_{CE} = \frac{Mb}{L^2}(2a-b) - \frac{1}{2}\frac{Ma}{L^2}(2b-a)$$

$$= \frac{10\times 1}{3^2}(2\times 2-1) - \frac{1}{2}\frac{10\times 2}{3^2}(2\times 1-2)$$

$$= 3.33 \text{ kNm}$$

$$MF_{CD} = MF_{DC} = 0$$

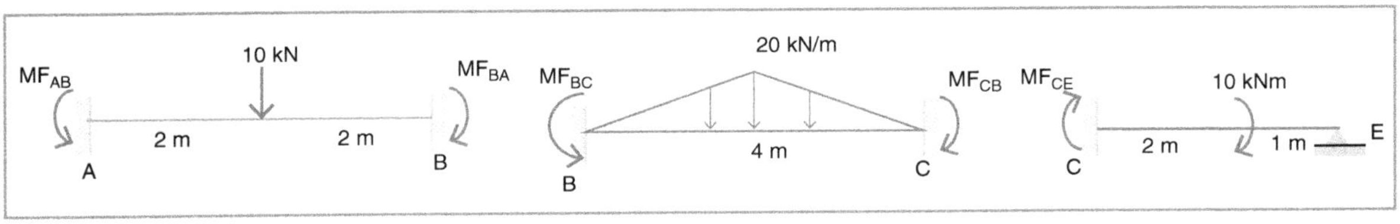

FIGURE 6.43 Free body diagram for fixed end moments.

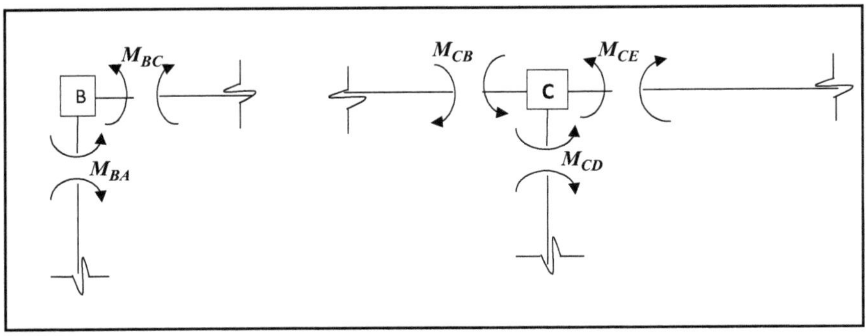

FIGURE 6.44 Free body diagram of joints *B* and *C*.

SD EQUATIONS

For span *AB*: $(\theta_A = 0, \theta_B = ?, \varphi_{AB} = 0, L = 4\text{ m}, I = I)$

$$M_{AB} = MF_{AB} + \frac{2EI}{L}(2\theta_A + \theta_B - 3\varphi_{AB})$$

$$M_{AB} = -5 + \frac{2E(I)}{4}(2(0) + \theta_B - 3(0)) = -5 + \frac{EI}{2}(\theta_B) \tag{6.62}$$

$$M_{BA} = 5 + \frac{2E(I)}{4}(0 + 2\theta_B - 3(0)) = 5 + EI\theta_B \tag{6.63}$$

Span *BC*: $(\theta_B = ?, \theta_C = ?, \varphi_{BC} = 0, L = 4\text{ m}, I = I)$

$$M_{BC} = MF_{BC} + \frac{2EI}{L}(2\theta_B + \theta_C - 3\varphi_{BC})$$

$$M_{BC} = -16.67 + \frac{2EI}{4}(2\theta_B + \theta_C - 3(0))$$

$$= -16.67 + EI\theta_B + \frac{EI\theta_C}{2} \tag{6.64}$$

$$M_{CB} = 16.67 + \frac{2EI}{4}(\theta_B + 2\theta_C - 3(0)) = 16.67 + \frac{EI\theta_B}{2} + EI\theta_C \tag{6.65}$$

Span *CD*: $(\theta_D = ?, \theta_C = ?, \varphi_{CD} = 0, L = 4\text{ m}, I = I)$

$$M_{CD} = MF_{CD} + \frac{2EI}{L}(2\theta_C + \theta_D - 3\varphi_{CD})$$

$$M_{CD} = 0 + \frac{2EI}{4}(2\theta_C + 0 - 3(0)) = EI\theta_C \tag{6.66}$$

$$M_{DC} = 0 + \frac{2EI}{4}(\theta_C + 2(0) - 3(0)) = \frac{EI\theta_C}{2} \tag{6.67}$$

Span *CE*: $(\theta_C = ?, \theta_E : \text{ignored}, \varphi_{CE} = 0, L = 3\text{ m}, I = I)$

$$M_{CE} = MF_{CE} + \frac{3EI}{L}(\theta_C - \varphi_{CE}) = 3.33 + \frac{3EI}{3}(\theta_C - 0)$$
$$= 3.33 + EI\theta_C \tag{6.68}$$

The above equations can be written in matrix form for convenience, as presented below.

$$\begin{bmatrix} M_{AB} \\ M_{BA} \\ M_{BC} \\ M_{CB} \\ M_{CD} \\ M_{DC} \\ M_{CE} \end{bmatrix} = \begin{bmatrix} -5 \\ 5 \\ -16.67 \\ 16.67 \\ 0 \\ 0 \\ 3.33 \end{bmatrix} + \begin{bmatrix} 1/2 & 0 \\ 1 & 0 \\ 1 & 1/2 \\ 1/2 & 1 \\ 0 & 1 \\ 0 & 1/2 \\ 0 & 1 \end{bmatrix} \times \begin{bmatrix} EI\theta_B \\ EI\theta_C \end{bmatrix} \tag{6.69}$$

JOINT EQUILIBRIUM EQUATIONS

To solve for the unknown rotations, write the joint equilibrium equations corresponding to the rotations θ_B and θ_C at joints *B* and *C*, respectively, as shown in Figure 6.44.

Using the free body diagrams shown in Figure 6.44, the joint moment equilibrium equations can be expressed as follows.

Joint *B*:

$$\sum M_B = 0 \Rightarrow M_{BA} + M_{BC} = 0$$

$$5 + EI\theta_B - 16.67 + EI\theta_B + \frac{EI\theta_C}{2} = 0 \Rightarrow 4EI\theta_B + EI\theta_C = 23.34 \tag{6.70}$$

Joint *C*:

$$\sum M_C = 0 \Rightarrow M_{CB} + M_{CD} + M_{CE} = 0$$

$$16.67+\frac{EI\theta_B}{2}+EI\theta_C+EI\theta_C+3.33+EI\theta_C=0$$

$$\Rightarrow EI\theta_B+6EI\theta_C=-40 \qquad (6.71)$$

From equations 6.70 and 6.71, we get $EI\theta_B = 7.827$ kNm and $EI\theta_C = -7.968$ kNm. By back substituting these values into equation 6.69, we obtain the final end moments as shown below.

FINAL END MOMENTS

$$\begin{bmatrix} M_{AB} \\ M_{BA} \\ M_{BC} \\ M_{CB} \\ M_{CD} \\ M_{DC} \\ M_{CE} \end{bmatrix} = \begin{bmatrix} -5 \\ 5 \\ -16.67 \\ 16.67 \\ 0 \\ 0 \\ 3.33 \end{bmatrix} + \begin{bmatrix} 1/2 & 0 \\ 1 & 0 \\ 1 & 1/2 \\ 1/2 & 1 \\ 0 & 1 \\ 0 & 1/2 \\ 0 & 1 \end{bmatrix} \times \begin{bmatrix} 7.827 \\ -7.968 \end{bmatrix}$$

$$= \begin{bmatrix} -1.086 \\ 12.827 \\ -12.827 \\ 12.616 \\ -7.968 \\ -3.984 \\ -4.638 \end{bmatrix} \text{kNm}$$

SHEAR FORCE AND BENDING MOMENT DIAGRAM

The member end forces are shown in free body diagrams (Figure 6.45a). The shear force, bending moment diagrams, and probable deflected shape are shown in Figure 6.45b–d, respectively.

Zero shear in span *BC*

$$0.5\times 10x\times x-19.895=0\Rightarrow x=1.99\text{ m}$$

Maximum +ve BM in span *BC*

$$\text{Max +ve BM} = 19.895\times 1.99-12.616-0.5\times 1.99\times 19.9\times\frac{1.99}{3}=13.84\text{ kNm}$$

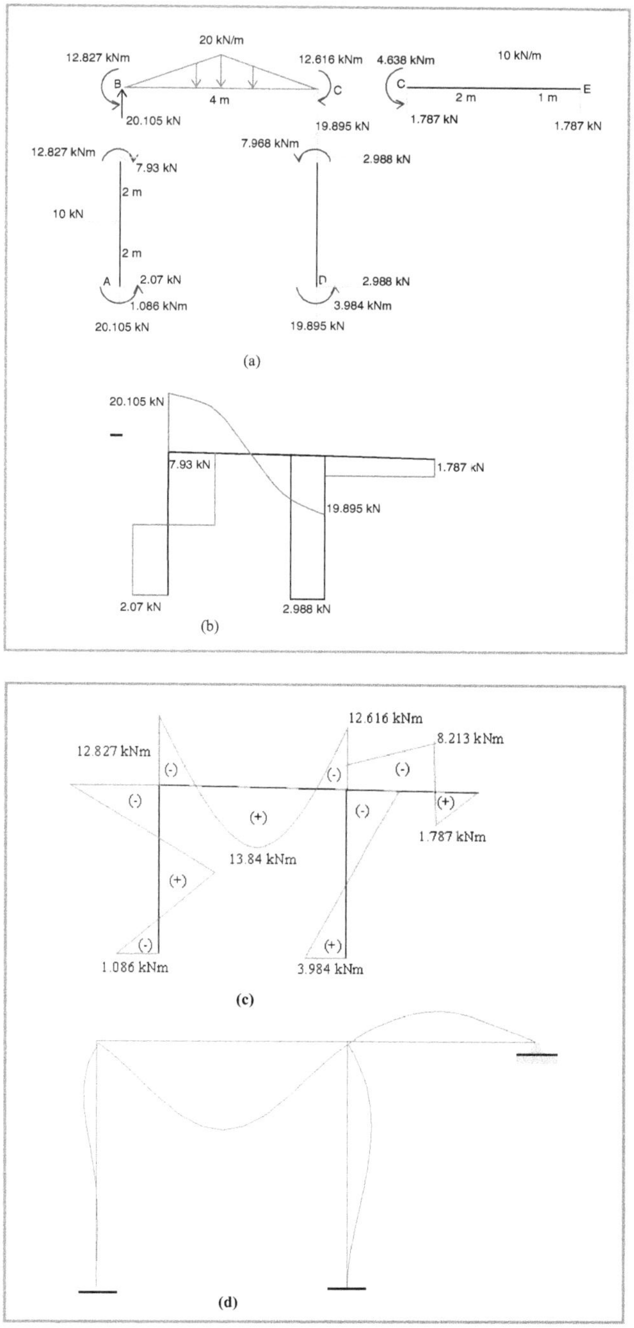

FIGURE 6.45 (a) Free body diagram, (b) shear force diagram, (c) bending moment diagram, and (d) probable elastic curve.

Example 6.11

Analyse the symmetric frame shown in Figure 6.46, using the SD method. Take advantage of symmetry.

Solution:

Due to symmetry, the frame is not subjected to side sway. Only half of the frame is sufficient for the analysis, as shown in Figure 6.47. At section *E*, symmetry ensures that the slope is zero, but there is a vertical displacement. Hence, a guided support is appropriate to apply at *E*. As a result, the kinematic indeterminacy is reduced to 1, i.e., θ_B, provided that modified SD equations are used for span *AB*.

FIXED END MOMENTS

The free body diagrams are shown in Figure 6.48a and using the expressions given in Table 6.1, the fixed end moments can be calculated as

$$MF_{AB}=MF_{BA}=0$$

$$MF_{BE}=-\frac{WL}{8}=-\frac{100\times 4}{8}=-50\text{ kNm}$$

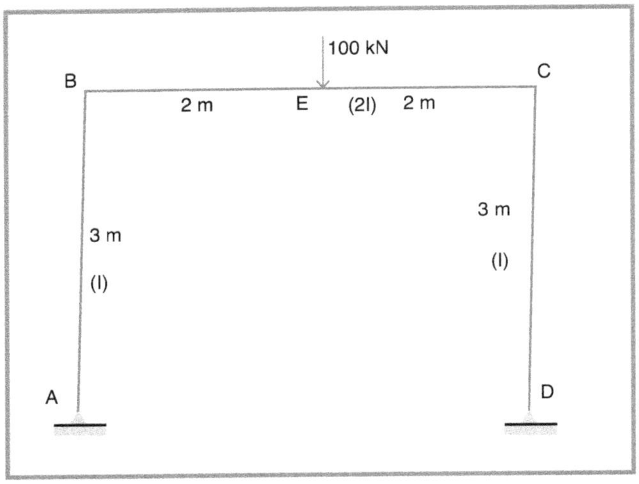

FIGURE 6.46 Rigid-jointed frame.

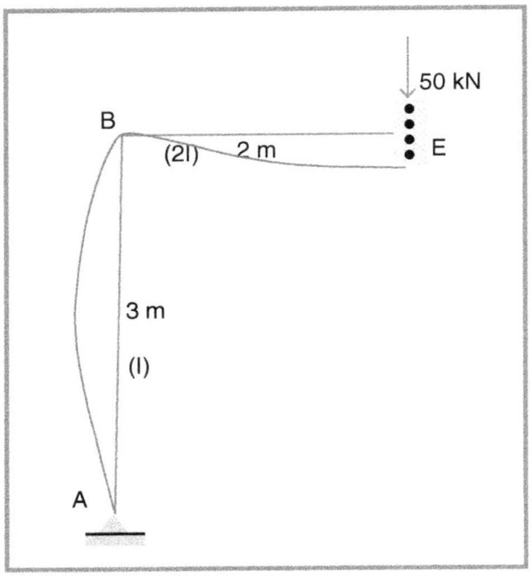

FIGURE 6.47 Visualised as half of symmetrical frame.

$$MF_{EB} = -\left(\frac{WL}{4} - \frac{WL}{8}\right) = -\frac{WL}{8} = -\frac{100 \times 4}{8}$$

$$= -50 \text{ kNm (sagging moment at mid span)}$$

SD EQUATIONS

Span AB: ($\theta_B = ?, \theta_A$: ignored, $\varphi_{AB} = 0, L = 3$ m, $I = I$)

$$M_{BA} = MF_{BA} + \frac{3EI}{L}(\theta_B - \varphi_{AB}) = 0 + \frac{3EI}{3}(\theta_B - 0) = EI\theta_B \tag{6.72}$$

Span BE: ($\theta_B = ?, \theta_E = 0, \varphi_{BE}$: *ignored*, $L = 2$ *m*, $I = 2I$)

$$M_{BE} = MF_{BE} + \frac{EI}{L}(\theta_B) = -50 + \frac{2EI}{2}(\theta_B) = -50 + EI\theta_B \tag{6.73}$$

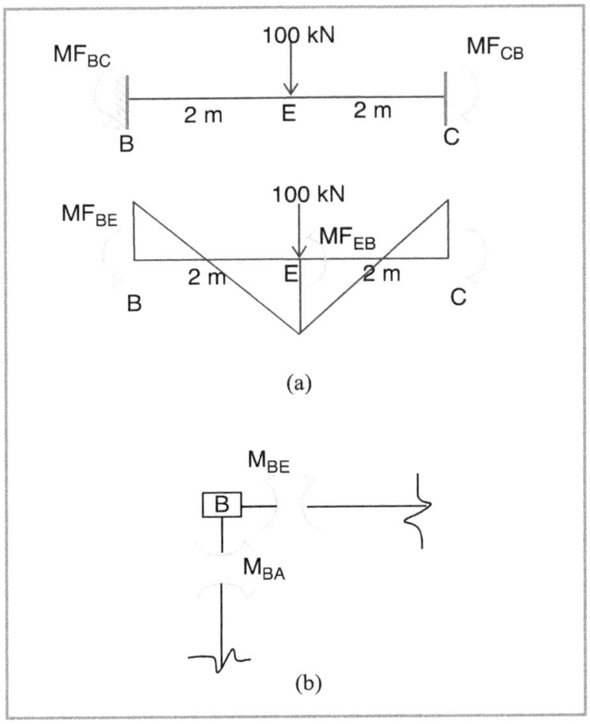

FIGURE 6.48 (a) Free body diagram for fixed end moments and (b) free body diagram of joint.

$$M_{EB} = MF_{EB} - \frac{EI}{L}(\theta_B) = -50 - \frac{2EI}{2}(\theta_B) = -50 - EI\theta_B \tag{6.74}$$

JOINT EQUILIBRIUM EQUATIONS

To solve for the unknown rotation, apply the joint equilibrium equation corresponding to θ_B at joints *B* (Figure 6.48b)

Joint *B*:

$$\sum M_B = 0 \Rightarrow M_{BA} + M_{BE} = 0$$

$$\Rightarrow EI\theta_B - 50 + EI\theta_B = 0 \Rightarrow EI\theta_B = 25 \text{ kNm}$$

FINAL END MOMENTS

By substituting the above value of $EI\theta_B$ into SD equations, the final end moments can be calculated as shown below.

$$M_{BA} = 25 \text{ kNm}; \quad M_{BE} = -50 + 25 = -25 \text{ kNm};$$

$$M_{EB} = -50 - 25 = -75 \text{ kNm}$$

The member end forces are shown in the free body diagram in Figure 6.49a. The shear force, bending moment diagrams, and the probable deflected shape are shown in Figure 6.49b–d, respectively.

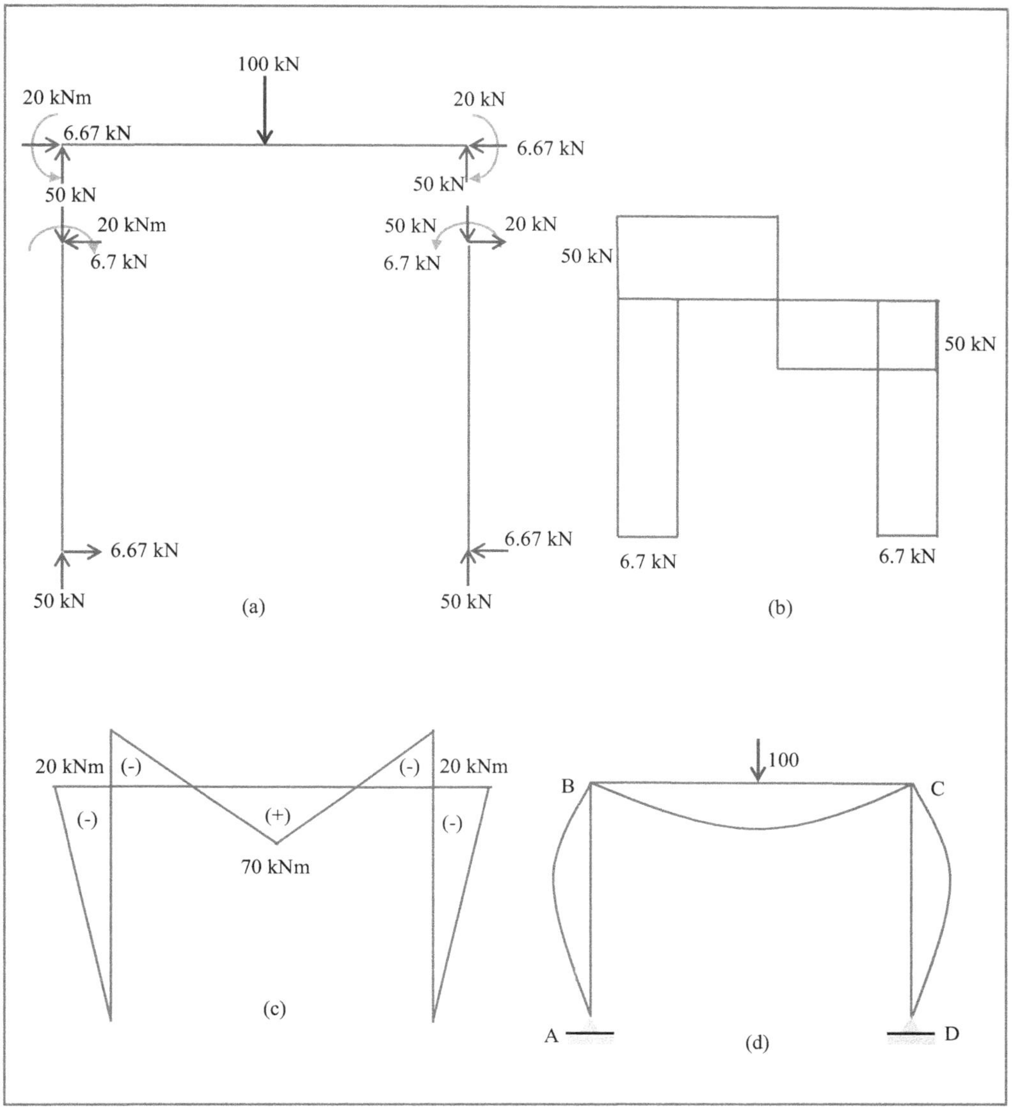

FIGURE 6.49 (a) Free body diagram, (b) probable deflected shape, (c) shear force diagram, and (d) bending moment diagram.

Example 6.12

Analyse the frame shown in Figure 6.50 using the SD method. Draw the SF and BM diagrams. Also, draw the probable elastic curve.

Solution:

Since the supports at points A and D are hinged, the moments at A and D are zero. By using the modified SD equations for spans AB and CD, the rotations at A and D can be ignored. Therefore, the kinematic indeterminacy is limited to 2, i.e., θ_B and θ_C. Furthermore, the frame is not subjected to lateral sway.

Fixed end moments: Using the expressions given in Table 6.1, the fixed end moments can be calculated.

$$MF_{BA} = \frac{wL^2}{12} + \frac{wL^2}{24} = \frac{wL^2}{8} = \frac{20\times 4^2}{8} = 40 \text{ kNm}$$

$$MF_{BC} = MF_{CB} = \frac{M}{4} = \frac{10}{4} = 2.5 \text{ kNm}$$

$$MF_{CD} = -\frac{wL^2}{20} - \frac{1}{2}\frac{wL^2}{30} = -\frac{wL^2}{15} = -\frac{20\times 4^2}{15} = -21.33 \text{ kNm}$$

$$MF_{EB} = -\frac{wab^2}{L^2} = -\frac{50\times 2\times 1^2}{3^2} = -11.11 \text{ kNm}$$

$$MF_{BE} = \frac{wba^2}{L^2} = -\frac{50\times 1\times 2^2}{3^2} = 22.22 \text{ kNm}$$

$$MF_{FC} = MF_{CF} = 0$$

SD EQUATIONS

For span AB: ($\theta_B = ?, \theta_A$: ignored, $\varphi_{AB} = 0, L = 4$ m, $I = 2I$)

$$M_{BA} = MF_{BA} + \frac{3EI}{L}(\theta_B - \varphi_{AB}) = 40 + \frac{3E(2I)}{4}(\theta_B - 0)$$

$$= 40 + 1.5EI\theta_B \tag{6.75}$$

Span BC: ($\theta_B = ?, \theta_C = ?, \varphi_{BC} = 0, L = 4$ m, $I = 2I$)

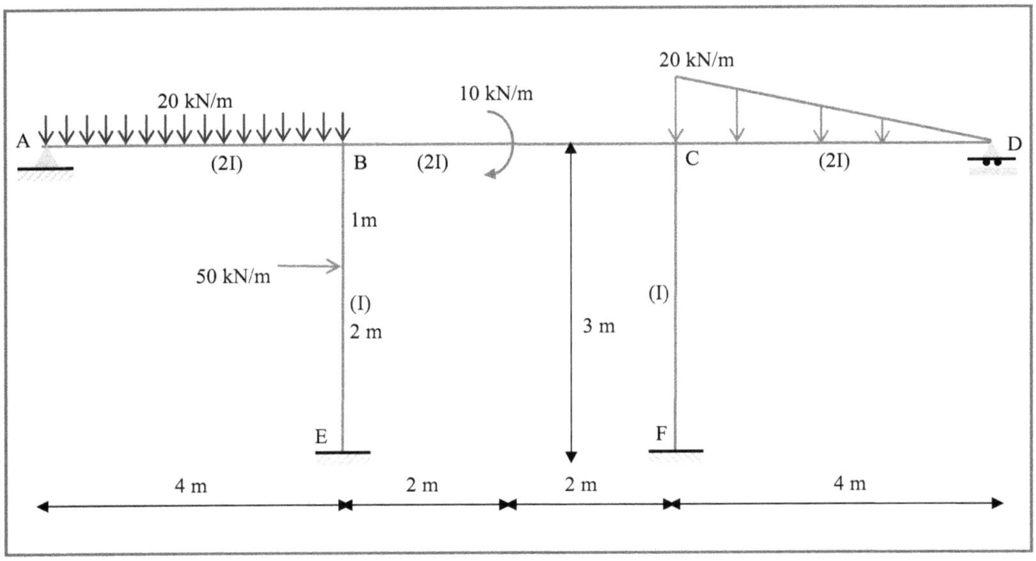

FIGURE 6.50 Rigid-jointed frame.

$$M_{BC} = MF_{BC} + \frac{2EI}{L}(2\theta_B + \theta_C - 3\varphi_{BC})$$

$$= 2.5 + \frac{2E(2I)}{4}(2\theta_B + \theta_C - 3(0))$$

$$M_{BC} = 2.5 + 2EI\theta_B + EI\theta_C \tag{6.76}$$

$$M_{CB} = MF_{CB} + \frac{2EI}{L}(\theta_B + 2\theta_C - 3\varphi_{BC})$$

$$= 2.5 + \frac{2E(2I)}{4}(\theta_B + 2\theta_C - 3(0))$$

$$M_{CB} = 2.5 + EI\theta_B + 2EI\theta_C \tag{6.77}$$

Span *CD*: ($\theta_C = ?, \theta_D$: ignored, $\varphi_{CD} = 0, L = 4$ m, $I = 2I$)

$$M_{CD} = MF_{CD} + \frac{3EI}{L}(\theta_C - \varphi_{CD}) = -21.33 + \frac{3E(2I)}{4}(\theta_C - 0)$$

$$M_{CD} = -21.33 + 1.5EI\theta_C \tag{6.78}$$

Span *BE*: ($\theta_B = ?, \theta_E = 0, \varphi_{BE} = 0, L = 3$ m, $I = I$)

$$M_{BE} = MF_{BE} + \frac{2EI}{L}(2\theta_B + \theta_E - 3\varphi_{BE})$$

$$= 22.22 + \frac{2EI}{3}(2\theta_B + 0 - 3(0))$$

$$M_{BE} = 22.22 + 1.333EI\theta_B \tag{6.79}$$

$$M_{EB} = MF_{EB} + \frac{2EI}{L}(\theta_B + 2\theta_E - 3\varphi_{BE})$$

$$= -11.11 + \frac{2EI}{3}(\theta_B + 2(0) - 3(0))$$

$$M_{EB} = -11.11 + 0.667EI\theta_B \tag{6.80}$$

Span *CF*: ($\theta_C = ?, \theta_F = 0, \varphi_{CF} = 0, L = 3$ m, $I = I$)

$$M_{CF} = MF_{CF} + \frac{2EI}{L}(2\theta_C + \theta_F - 3\varphi_{CF})$$

$$= 0 + \frac{2EI}{3}(2\theta_C + 0 - 3(0))$$

$$M_{CF} = 1.333EI\theta_C \tag{6.81}$$

$$M_{FC} = MF_{FC} + \frac{2EI}{L}(\theta_C + 2\theta_F - 3\varphi_{CF})$$

$$= 0 + \frac{2EI}{3}(\theta_C + 2(0) - 3(0))$$

$$M_{FC} = 0.667EI\theta_C \tag{6.82}$$

Conveniently, the above SD equations can be expressed in matrix form as shown below.

$$\begin{bmatrix} M_{BA} \\ M_{BC} \\ M_{CB} \\ M_{CD} \\ M_{BE} \\ M_{EB} \\ M_{CF} \\ M_{FC} \end{bmatrix} = \begin{bmatrix} 40 \\ 2.5 \\ 2.5 \\ -21.33 \\ 22.22 \\ -11.11 \\ 0 \\ 0 \end{bmatrix} + \begin{bmatrix} 1.5 & 0 \\ 2 & 1 \\ 1 & 2 \\ 0 & 1.5 \\ 1.333 & 0 \\ 0.667 & 0 \\ 0 & 1.333 \\ 0 & 0.667 \end{bmatrix} \times \begin{bmatrix} EI\theta_B \\ EI\theta_C \end{bmatrix} \tag{6.83}$$

JOINT EQUATIONS OF EQUILIBRIUM

To solve for the unknown rotations, apply the joint equilibrium equations corresponding to θ_B and θ_C at joints B and C, respectively, as shown in Figure 6.51.

Joint *B*:

$$\sum M_B = 0 \Rightarrow M_{BA} + M_{BC} + M_{BE} = 0$$

$$40 + 1.5EI\theta_B + 2.5 + 2EI\theta_B + EI\theta_C + 22.22 + 1.333EI\theta_B = 0$$

$$4.833EI\theta_B + EI\theta_C = -64.72 \tag{6.84}$$

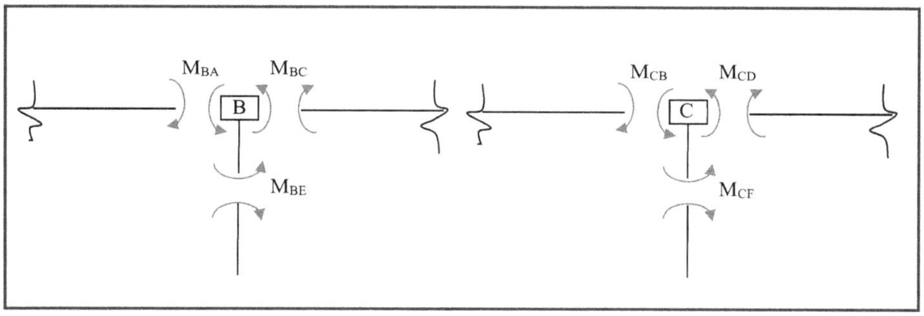

FIGURE 6.51 Free body diagram of joints *B* and *C*.

Joint C:

$$\sum M_C = 0 \Rightarrow M_{CB} + M_{CD} + M_{CF} = 0$$

$$2.5 + EI\theta_B + 2EI\theta_C - 21.33 + 1.5EI\theta_C + 1.333EI\theta_C = 0$$

$$EI\theta_B + 4.833EI\theta_C = 18.83 \tag{6.85}$$

By solving equations 6.84 and 6.85, we obtain

$$EI\theta_B = -14.832 \text{ kNm} \quad \text{and} \quad EI\theta_C = 6.964 \text{ kNm}.$$

FINAL END MOMENTS

By substituting the above values into equation 6.83, we can get the final end moments, as shown below.

$$\begin{bmatrix} M_{BA} \\ M_{BC} \\ M_{CB} \\ M_{CD} \\ M_{BE} \\ M_{EB} \\ M_{CF} \\ M_{FC} \end{bmatrix} = \begin{bmatrix} 40 \\ 2.5 \\ 2.5 \\ -21.33 \\ 22.22 \\ -11.11 \\ 0 \\ 0 \end{bmatrix} + \begin{bmatrix} 1.5 & 0 \\ 2 & 1 \\ 1 & 2 \\ 0 & 1.5 \\ 1.333 & 0 \\ 0.667 & 0 \\ 0 & 1.333 \\ 0 & 0.667 \end{bmatrix}$$

$$\times \begin{bmatrix} EI\theta_B = -14.832 \\ EI\theta_C = 6.964 \end{bmatrix} = \begin{bmatrix} 17.75 \\ -20.2 \\ 1.6 \\ -10.88 \\ 2.45 \\ -21 \\ 9.28 \\ 4.64 \end{bmatrix} \text{kNm}$$

With the known values of end moments, using the static equilibrium equations and the free body diagram shown in Figure 6.52a, the member end shear and axial forces can be calculated. The probable elastic curve and shear force diagrams are shown in Figure 6.52b and c, respectively. The bending moment diagram is shown in Figure 6.52d.

6.5.4 Application to the Rigid-Jointed Plane Frames with Lateral Sway

In the previous section, we studied rigid-jointed plane frames not subjected to sway. This occurs when the frames are restrained against sway or are symmetrical in terms of loading, geometry, and end conditions. However, in practice, loading is rarely symmetrical, and the geometry of the columns and beams in a frame may vary. Therefore, frames are often subjected to sway, which cannot be avoided. Some examples of frames subjected to lateral sway are shown in Figure 6.53a–e below.

In Figure 6.53a, the frame is symmetrical, but the load is asymmetric. Due to the unsymmetrical loading on the beam, the end moments M_{BC} and M_{CB} are different. If $a>b$, then $M_{BC} > M_{CB}$, causing joints *B* and *C* to displace to the right with an unknown displacement Δ (sway). Therefore, the total unknown DoFs, i.e., unknown displacements in the frame, are θ_B, θ_C and Δ. Corresponding to the rotational displacements, we consider the moment equilibrium equations at these joints. Similarly, for the unknown translational displacement, we consider the horizontal force equilibrium equation.

While writing the SD equations, we account for the chord rotation $\varphi = \Delta/h$ as an unknown. The SD equations for the columns are the same as those for beams subjected to support settlement, as one end (*B*) of the column undergoes a linear displacement of Δ relative to the other end (A) of the column. However, in this case, Δ is unknown.

The chord rotations for Figure 6.53b and c are shown as follows.

$$\varphi_{AB} = \varphi_{DC} = \frac{\Delta}{h}$$

$$\varphi_{AB} = \frac{\Delta}{h_1}; \quad \varphi_{DC} = \frac{\Delta}{h_2}$$

The chord rotations of the frame shown in Figure 6.53d are shown in Figure 6.54.

$$\varphi_{AB} = \frac{\Delta_1}{h_1}; \quad \text{But } \Delta_1 = \frac{\Delta}{\cos\alpha}$$

$$\therefore \varphi_{AB} = \frac{\Delta}{h_1 \cos\alpha}$$

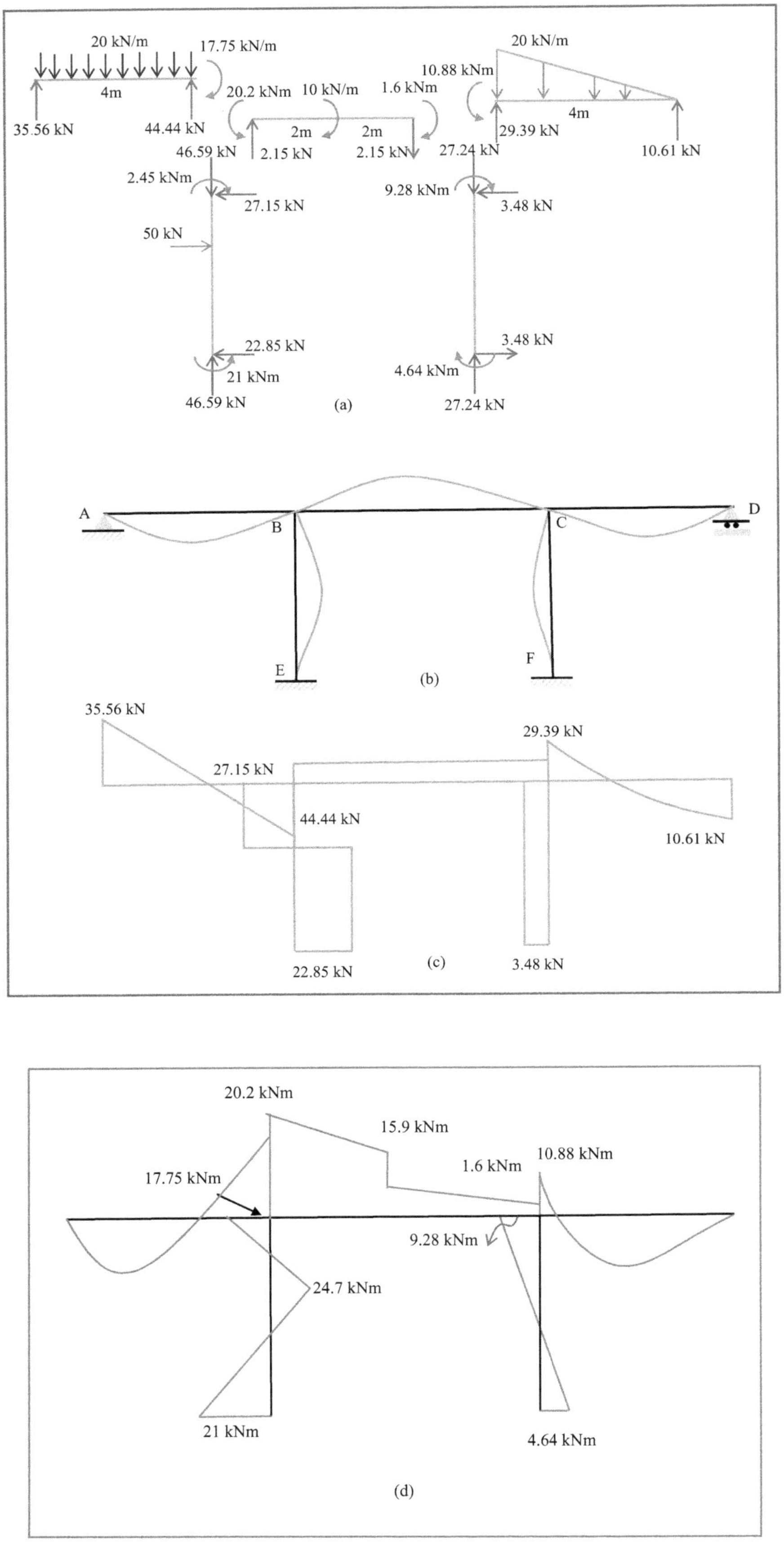

FIGURE 6.52 (a) Free body diagram, (b) probable deflected shape, (c) shear force diagram, and (d) bending moment diagram.

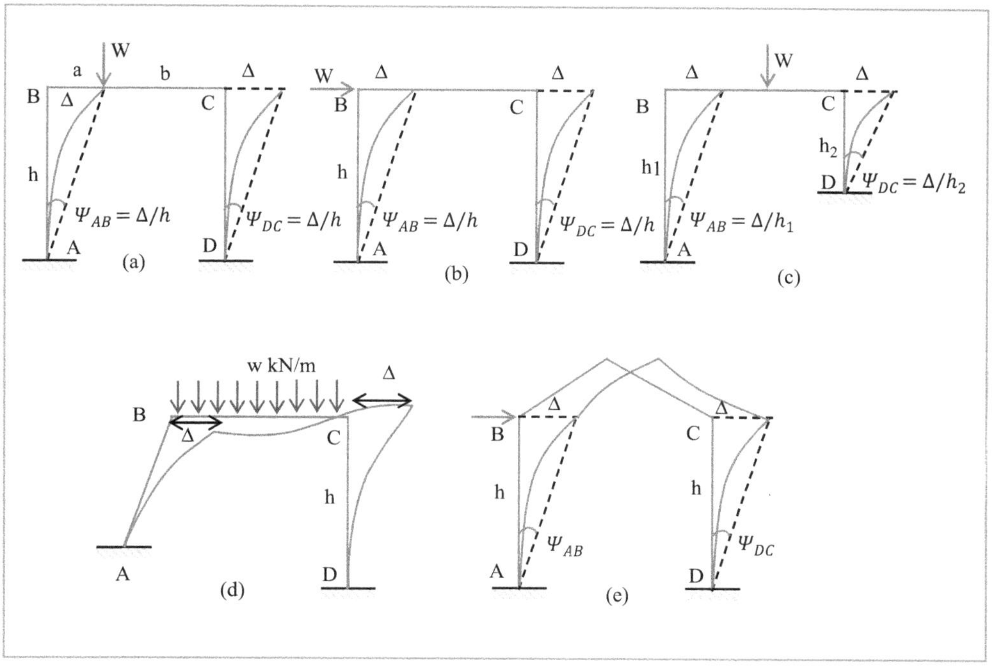

FIGURE 6.53 (a) Frame is symmetrical but loaded unsymmetrically, (b) frame is symmetrical but subjected to horizontal load, (c) frame is symmetrical but geometry of columns is not the same, (d) unsymmetrical frame, and (e) unsymmetrical frame.

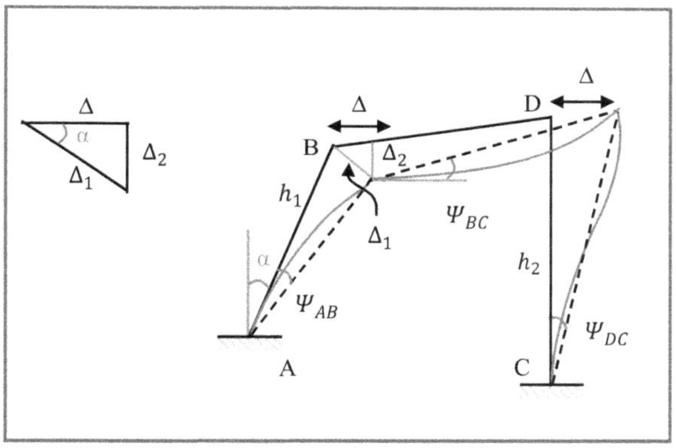

FIGURE 6.54 Chord rotations.

$$\varphi_{CD} = \frac{\Delta}{h_2} = \frac{\Delta_2}{L_{BC}} = \frac{\Delta \cdot \tan\alpha}{L_{BC}}$$

The chord rotations for the frame whose both columns are inclined, as shown in Figure 6.55a, can be expressed as follows.

$$\varphi_{AB} = \frac{\Delta_1}{h_1};\quad \text{But } \cos\alpha = \frac{\Delta}{\Delta_1} \Rightarrow \Delta_1 = \frac{\Delta}{\cos\alpha}$$

$$\varphi_{AB} = \frac{\Delta}{h_1 \cos\alpha} \approx \frac{\Delta}{h_1}$$

$$\varphi_{CD} = \frac{CC^1}{h_2} = \frac{\Delta/\cos\beta}{h_2} \approx \frac{\Delta}{h_2}$$

$$\varphi_{BC} = \frac{BB^1 + C^1C^{11}}{L_{BC}} = \frac{\Delta\tan\alpha + \Delta\tan\beta}{L_{BC}} = (\tan\alpha + \tan\beta)\frac{\Delta}{L_{BC}}$$

where α and β are the inclinations of columns AB and CD with verticals, respectively.

The chord rotations for the gable frame are shown in Figure 6.55b.

$$\text{No of unknown translations} = \text{No. of joints} \times 2 - \text{No. of members}$$
$$= 3 \times 2 - 4 = 2 \quad \text{i.e. } \Delta_B \text{ and } \Delta_D$$

$$\varphi_{AB} = \frac{\Delta_B}{h} \quad \text{and} \quad \varphi_{DE} = \frac{\Delta_D}{h}$$

$$\varphi_{BC} = -\frac{\Delta_B - \Delta_D}{2d\sin\alpha} \quad \text{and} \quad \varphi_{CD} = \frac{\Delta_B - \Delta_D}{2d\sin\alpha}$$

In the case of multi-storey frames, each floor is associated with one side-sway along with rotational DoFs at each joint.

Consider a three-bay, three-storey frame as shown in Figure 6.56. The frame has twelve joints and three storeys; each joint has one rotational DoF, and each storey has one translational DoF. Therefore, the frame has a total of 12 rotational DoFs and 3 translational DoFs.

6.5.4.1 Equilibrium Equations

To solve for the unknown rotations and horizontal translations (sway), we need to write the moment and shear equilibrium equations. As discussed earlier, for each unknown rotational displacement, one joint moment equilibrium equation must be written. Similarly, for each horizontal displacement (sway) at a floor level, one horizontal force equilibrium equation must be written. For example, in Figure 6.57, corresponding to Δ_1, the horizontal force equilibrium equation at the bottom of the floor level is written, as shown below.

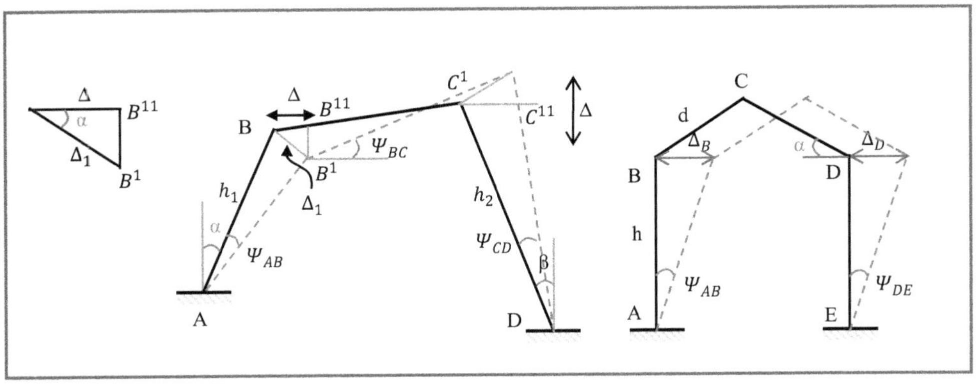

FIGURE 6.55 Chord rotations of (a) frame with inclined columns and (b) gabled frame.

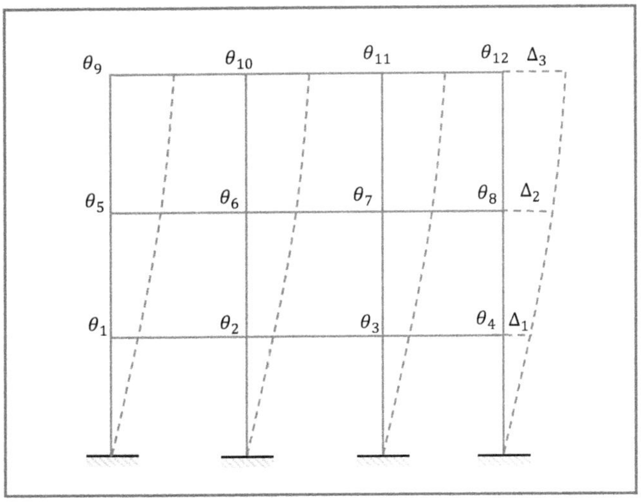

FIGURE 6.56 Degree of freedom in a three-bay, three-storey frame.

$$\sum F_X = 0 \Rightarrow V_1 + V_2 = W_1 + W_2$$

Similarly, corresponding to Δ_2, $\sum F_X = 0 \Rightarrow V_3 + V_4 = W_2$

These member end shears can be expressed in terms of the end moments of the corresponding members in SD equations as

$$V_1 = -\frac{(M_{BA} + M_{AB})}{h_1} \quad \text{and} \quad V_2 = -\frac{(M_{EF} + M_{FE})}{h_1}$$

Analysis of sway problems is demonstrated in the following examples.

Example 6.13

Analyse the simple frame with unsymmetrical loading as shown in Figure 6.58, using the SD method.

Solution:

The given frame is symmetrical, but the loading on *BC* is unsymmetrical. Hence, the frame is subjected to lateral sway, Δ, assumed to occur towards the right (neglecting axial deformations in the members). The other unknown displacements are the rotations at *B* and *C*, i.e., θ_B and θ_C. Therefore, the total kinematic indeterminacy is 3, i.e., θ_B , θ_C and Δ.

FIXED END MOMENTS

The fixed end moments can be calculated using the expressions given in Table 6.1, as shown below.

$$MF_{AB} = MF_{BA} = MF_{CD} = MF_{DC} = 0$$

$$MF_{BC} = -\frac{wL^2}{20} = -\frac{20 \times 6^2}{20} = -36 \text{ kNm}$$

$$MF_{CB} = \frac{wL^2}{30} = +\frac{20 \times 6^2}{30} = 24 \text{ kNm}$$

SD EQUATIONS

Span *AB*: $(\theta_B = ?, \theta_A = 0, \varphi_{AB} = \frac{\Delta}{4}, L = 4 \text{ m}, I = I)$

(Chords of elastic curve of $AB^1$and DC^1 rotates in the clockwise direction)

$$M_{AB} = MF_{AB} + \frac{2EI}{L}(2\theta_A + \theta_B - 3\varphi_{AB})$$

$$= 0 + \frac{2EI}{4}\left(0 + \theta_B - 3\frac{\Delta}{4}\right) = 0.5EI\theta_B - \frac{3EI\Delta}{8} \tag{6.86}$$

$$M_{BA} = MF_{BA} + \frac{2EI}{L}(\theta_A + 2\theta_B - 3\varphi_{AB})$$

$$= 0 + \frac{2EI}{4}\left(0 + 2\theta_B - 3\frac{\Delta}{4}\right) = EI\theta_B - \frac{3EI\Delta}{8} \tag{6.87}$$

Span *BC*: $(\theta_B = ?, \theta_C = ?, \varphi_{BC} = 0, L = 6 \text{ m}, I = 2I)$

$$M_{BC} = MF_{BC} + \frac{2EI}{L}(2\theta_B + \theta_C - 3\varphi_{BC})$$

$$= +24 + \frac{4EI}{6}(2\theta_B + \theta_C - 0)$$

$$M_{BC} = -36 + \frac{4EI\theta_B}{3} + \frac{2EI\theta_C}{3} \tag{6.88}$$

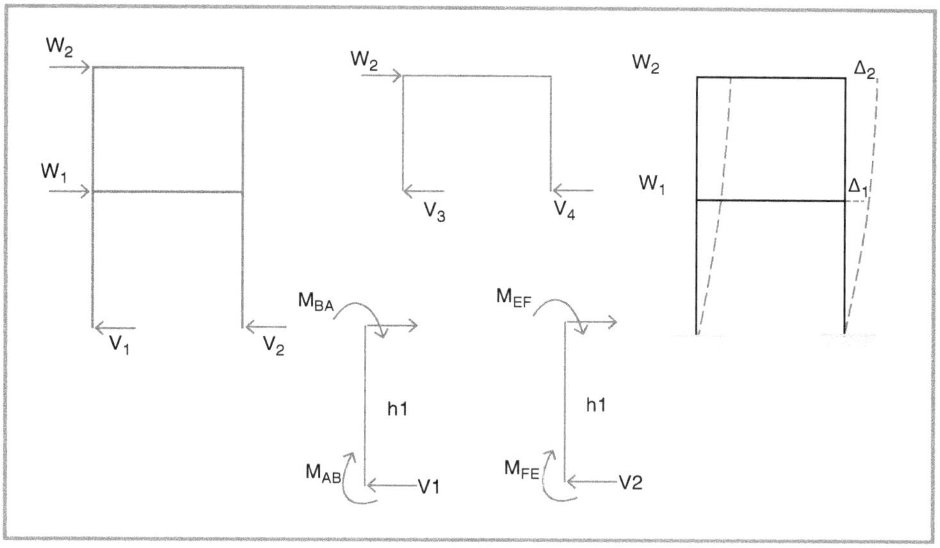

FIGURE 6.57 Free body diagrams.

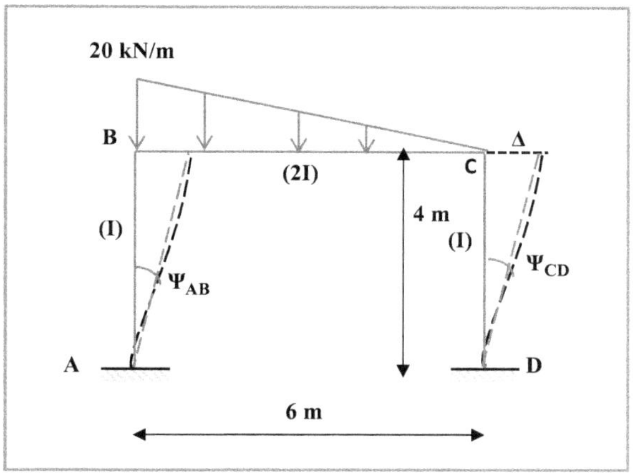

FIGURE 6.58 Symmetrical rigid-jointed frame with unsymmetrical loading.

$$M_{CB} = MF_{CB} + \frac{2EI}{L}(\theta_B + 2\theta_C - 3\varphi_{BC})$$

$$= +24 + \frac{4EI}{6}(\theta_B + 2\theta_C - 0)$$

$$M_{CB} = 24 + \frac{2EI\theta_B}{3} + \frac{4EI\theta_C}{3} \quad (6.89)$$

Span *CD*: $\left(\theta_C = ?, \theta_D = 0, \varphi_{DC} = \frac{\Delta}{4}, L = 4 \text{ m}, I = I\right)$

$$M_{CD} = MF_{CD} + \frac{2EI}{L}(2\theta_C + \theta_D - 3\varphi_{CD})$$

$$= 0 + \frac{2EI}{4}\left(2\theta_C + 0 - 3\frac{\Delta}{4}\right) = EI\theta_C - \frac{3EI\Delta}{8} \quad (6.90)$$

$$M_{DC} = MF_{DC} + \frac{2EI}{L}(\theta_C + 2\theta_D - 3\varphi_{CD})$$

$$= 0 + \frac{2EI}{4}\left(\theta_C + 0 - 3\frac{\Delta}{4}\right) = 0.5EI\theta_C - \frac{3EI\Delta}{8} \quad (6.91)$$

The above SD equations can be written in matrix form, as shown below.

$$\begin{bmatrix} M_{AB} \\ M_{BA} \\ M_{BC} \\ M_{CB} \\ M_{CD} \\ M_{DC} \end{bmatrix} = \begin{bmatrix} 0 \\ 0 \\ -36 \\ 24 \\ 0 \\ 0 \end{bmatrix} + \begin{bmatrix} 1/2 & 0 & -3/8 \\ 1 & 0 & -3/8 \\ 4/3 & 2/3 & 0 \\ 2/3 & 4/3 & 0 \\ 0 & 1 & -3/8 \\ 0 & 1/2 & -3/8 \end{bmatrix} \begin{bmatrix} EI\theta_B \\ EI\theta_C \\ EI\Delta \end{bmatrix} \quad (6.92)$$

JOINT EQUILIBRIUM EQUATIONS

Corresponding to θ_B and θ_C, apply the moment equilibrium equations at joints *B* and *C*, and corresponding to Δ, apply the horizontal force equilibrium equation (shear equilibrium equation) using the free body diagrams shown in Figure 6.59.

Joint *B*: $\sum M_B = 0 \Rightarrow M_{BA} + M_{BC} = 0$

$$EI\theta_B - \frac{3EI\Delta}{8} - 36 + \frac{4EI\theta_B}{3} + \frac{2EI\theta_C}{3} = 0$$

$$56EI\theta_B + 16EI\theta_C - 9EI\Delta = 864 \quad (6.93)$$

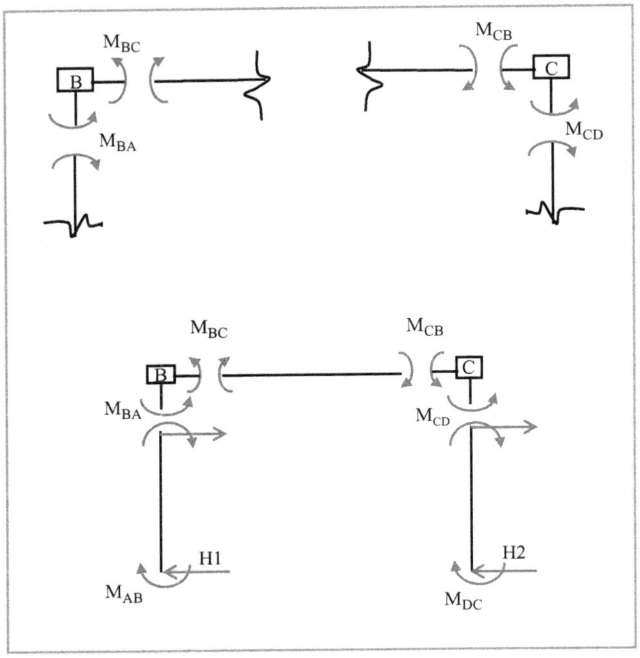

FIGURE 6.59 Free body diagrams.

At Joint *C*

$$\sum M_C = 0 \Rightarrow M_{CB} + M_{CD} = 0$$

$$24 + \frac{2EI\theta_B}{3} + \frac{4EI\theta_C}{3} + EI\theta_C - \frac{3EI\Delta}{8} = 0$$

$$16EI\theta_B + 56EI\theta_C - 9EI\Delta = -576 \qquad (6.94)$$

SHEAR EQUILIBRIUM EQUATION

$$\sum F_X = 0 \Rightarrow H_1 + H_2 = 0$$

$$-\frac{(M_{AB} + M_{BA})}{4} - \frac{(M_{DC} + M_{CD})}{4} = 0$$

$$M_{AB} + M_{BA} + M_{DC} + M_{CD} = 0$$

$$0.5EI\theta_B - \frac{3EI\Delta}{8} + EI\theta_B - \frac{3EI\Delta}{8} + EI\theta_C - \frac{3EI\Delta}{8}$$
$$+0.5EI\theta_C - \frac{3EI\Delta}{8} = 0$$

$$\Rightarrow EI\theta_B + EI\theta_C = EI\Delta \qquad (6.95)$$

Equations 6.93–6.95 can be written in matrix form as

$$\begin{bmatrix} 56 & 16 & -9 \\ 16 & 56 & -9 \\ 1 & 1 & -1 \end{bmatrix} \begin{bmatrix} EI\theta_B \\ EI\theta_C \\ EI\Delta \end{bmatrix} = \begin{bmatrix} 864 \\ -576 \\ 0 \end{bmatrix}$$

$$\Rightarrow \begin{bmatrix} EI\theta_B \\ EI\theta_C \\ EI\Delta \end{bmatrix} = \begin{bmatrix} 20.67 \\ -15.33 \\ 5.33 \end{bmatrix} \text{kNm}$$

FINAL END MOMENTS

Substituting the above values of displacements into equation 6.92, we can obtain the final end moments. The member end forces are shown in Figure 6.60a. The probable deflected shape is shown in Figure 6.60b. The shear force and bending moment diagrams are shown in Figure 6.60c and d, respectively.

$$\begin{bmatrix} M_{AB} \\ M_{BA} \\ M_{BC} \\ M_{CB} \\ M_{CD} \\ M_{DC} \end{bmatrix} = \begin{bmatrix} 0 \\ 0 \\ -36 \\ 24 \\ 0 \\ 0 \end{bmatrix} + \begin{bmatrix} 1/2 & 0 & -3/8 \\ 1 & 0 & -3/8 \\ 4/3 & 2/3 & 0 \\ 2/3 & 4/3 & 0 \\ 0 & 1 & -3/8 \\ 0 & 1/2 & -3/8 \end{bmatrix}$$

$$\begin{bmatrix} EI\theta_B = 20.67 \\ EI\theta_C = -15.33 \\ EI\Delta = 5.33 \end{bmatrix} = \begin{bmatrix} 8.33 \\ 18.67 \\ -18.67 \\ 17.33 \\ -17.33 \\ -9.67 \end{bmatrix} \text{kNm}$$

Example 6.14

Analyse the frame shown in Figure 6.61, using SD method. Draw the SF and BM diagrams.

Solution:

The kinematic indeterminacy of the frame, $I_k = 3$ i.e. θ_B, θ_C and Δ.

FIXED END MOMENTS

The fixed end moments can be calculated using Table 6.1, as shown below.

$$MF_{AB} = \frac{-Wab^2}{L^2} = \frac{-30 \times 3 \times 1^2}{4^2} = -5.625 \text{ kNm}$$

$$MF_{BA} = \frac{+Wba^2}{L^2} = \frac{+30 \times 1 \times 3^2}{4^2} = 16.875 \text{ kNm}$$

$$MF_{BC} = -MF_{CB} = -\frac{wL^2}{12} = -\frac{50 \times 4^2}{12} = -66.67 \text{ kNm}$$

$$MF_{CD} = MF_{DC} = 0 \text{ (since no external load on span } CD)$$

SD EQUATIONS

Span *AB*: $(\theta_B = ?, \theta_A = 0, \varphi_{AB} = \frac{\Delta}{4}, L = 4 \text{ m}, I = I)$

(Chords of elastic curve of $AB^1$and DC^1 rotates in the clockwise direction)

$$M_{AB} = MF_{AB} + \frac{2EI}{L}(2\theta_A + \theta_B - 3\varphi_{AB})$$

$$= -5.625 + \frac{2EI}{4}\left(0 + \theta_B - 3\frac{\Delta}{4}\right)$$

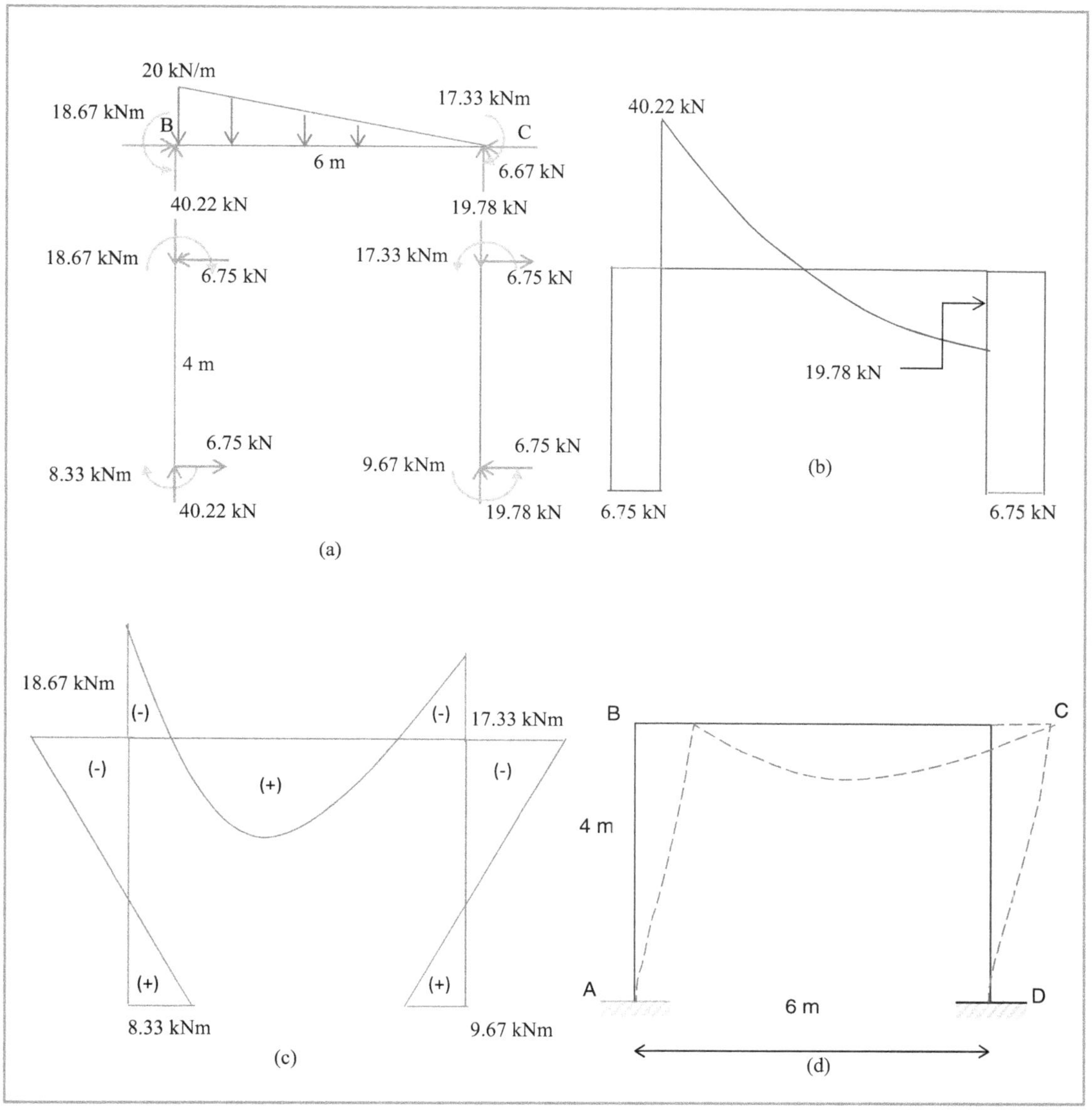

FIGURE 6.60 (a) Free body diagram, (b) shear force diagram, (c) bending moment diagram, and (d) probable deflected shape.

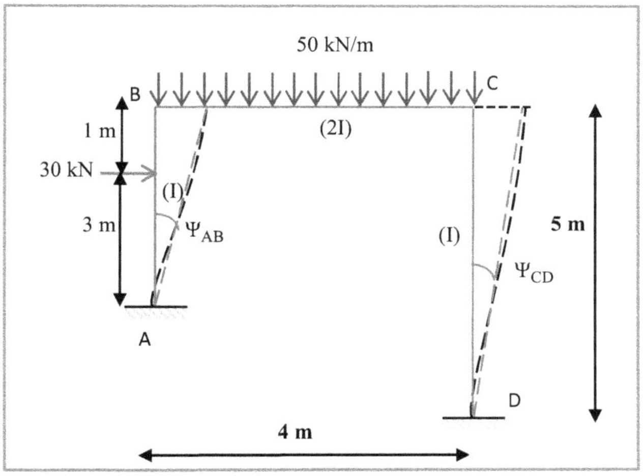

FIGURE 6.61 Rigid-jointed frame with unequal height of columns.

$$M_{AB} = -5.625 + 0.5EI\theta_B - \frac{3EI\Delta}{8} \tag{6.96}$$

$$M_{BA} = MF_{BA} + \frac{2EI}{L}(\theta_A + 2\theta_B - 3\varphi_{AB}) = 16.875$$

$$+\frac{2EI}{4}\left(0 + 2\theta_B - 3\frac{\Delta}{4}\right)$$

$$M_{BA} = 16.875 + EI\theta_B - \frac{3EI\Delta}{8} \tag{6.97}$$

Span *BC*: $(\theta_B = ?, \theta_C = ?, \varphi_{BC} = 0, L = 4\text{ m}, I = 2I)$

$$M_{BC} = MF_{BC} + \frac{2EI}{L}(2\theta_B + \theta_C - 3\varphi_{BC})$$

$$= -66.67 + \frac{4EI}{4}(2\theta_B + \theta_C - 0)$$

$$M_{BC} = -66.67 + 2EI\theta_B + EI\theta_C \tag{6.98}$$

$$M_{CB} = MF_{CB} + \frac{2EI}{L}(\theta_B + 2\theta_C - 3\varphi_{BC})$$

$$= 66.67 + \frac{4EI}{4}(\theta_B + 2\theta_C - 0)$$

$$M_{CB} = 66.67 + EI\theta_B + 2EI\theta_C \tag{6.99}$$

Span *CD*: $(\theta_C = ?, \theta_D = 0, \varphi_{DC} = \frac{\Delta}{5}, L = 5\text{ m}, I = I)$

$$M_{CD} = MF_{CD} + \frac{2EI}{L}(2\theta_C + \theta_D - 3\varphi_{CD})$$

$$= 0 + \frac{2EI}{5}\left(2\theta_C + 0 - 3\frac{\Delta}{5}\right)$$

$$M_{CD} = 0.8EI\theta_C - \frac{6EI\Delta}{25} \tag{6.100}$$

$$M_{DC} = MF_{DC} + \frac{2EI}{L}(\theta_C + 2\theta_D - 3\varphi_{CD})$$

$$= 0 + \frac{2EI}{5}\left(\theta_C + 0 - 3\frac{\Delta}{5}\right)$$

$$M_{DC} = 0.4EI\theta_C - \frac{6EI\Delta}{25} \tag{6.101}$$

The above equations 6.96–6.101 can be conveniently written in matrix form, as shown below.

$$\begin{bmatrix} M_{AB} \\ M_{BA} \\ M_{BC} \\ M_{CB} \\ M_{CD} \\ M_{DC} \end{bmatrix} = \begin{bmatrix} -5.625 \\ 16.875 \\ -66.67 \\ 66.67 \\ 0 \\ 0 \end{bmatrix} + \begin{bmatrix} 0.5 & 0 & -3/8 \\ 1 & 0 & -3/8 \\ 2 & 1 & 0 \\ 1 & 2 & 0 \\ 0 & 0.8 & -6/25 \\ 0 & 0.4 & -6/25 \end{bmatrix} \begin{bmatrix} EI\theta_B \\ EI\theta_C \\ EI\Delta \end{bmatrix} \tag{6.102}$$

JOINT EQUILIBRIUM EQUATIONS

Using the free body diagram shown in Figure 6.62, let us write the moment equilibrium equations at B and C, corresponding to θ_B and θ_C, and the horizontal force equilibrium equation (shear equilibrium equation) corresponding to Δ.

Joint *B*: $\sum M_B = 0 \Rightarrow M_{BA} + M_{BC} = 0$

$$16.875 + EI\theta_B - \frac{3EI\Delta}{8} - 66.67 + 2EI\theta_B + EI\theta_C = 0$$

$$\Rightarrow 3EI\theta_B - \frac{3EI\Delta}{8} + EI\theta_C = 49.795 \tag{6.103}$$

At joint *C*: $\sum M_C = 0 \Rightarrow M_{CB} + M_{CD} = 0$

$$66.67 + EI\theta_B + 2EI\theta_C + 0.8EI\theta_C - \frac{6EI\Delta}{25} = 0$$

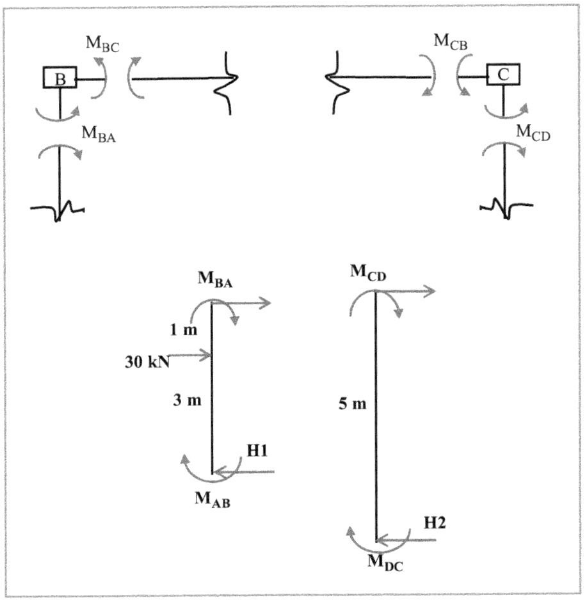

FIGURE 6.62 Free body diagrams of joints B and C and columns AB and CD.

$$\Rightarrow EI\theta_B + 2.8EI\theta_C - \frac{6EI\Delta}{25} = -66.67 \tag{6.104}$$

Shear equilibrium equation, i.e., corresponding to Δ at joint C

$$\sum F_X = 0 \Rightarrow H_1 + H_2 = 0$$

$$-\frac{(M_{AB} + M_{BA} - 30 \times 1)}{4} - \frac{(M_{DC} + M_{CD})}{5} = 30$$

$$\Rightarrow 5M_{AB} + 5M_{BA} + 4M_{DC} + 4M_{CD} = -450$$

$$\Rightarrow 5\left(-5.625 + 0.5EI\theta_B - \frac{3EI\Delta}{8}\right) + 5\left(16.875 + EI\theta_B - \frac{3EI\Delta}{8}\right)$$

$$+4\left(0.4EI\theta_C - \frac{6EI\Delta}{25}\right) + 4\left(0.8EI\theta_C - \frac{6EI\Delta}{25}\right) = -450$$

$$\Rightarrow 7.5EI\theta_B + 4.8EI\theta_C - 5.67EI\Delta = -506.25 \tag{6.105}$$

We can express equations 6.103–6.105 in matrix form as

$$\begin{bmatrix} 3 & 1 & -3/8 \\ 1 & 2.8 & -6/25 \\ 7.5 & 4.8 & -5.67 \end{bmatrix} \begin{bmatrix} EI\theta_B \\ EI\theta_C \\ EI\Delta \end{bmatrix} = \begin{bmatrix} 49.795 \\ -66.67 \\ -506.25 \end{bmatrix}$$

By solving, we obtain as

$$\begin{bmatrix} EI\theta_B \\ EI\theta_C \\ EI\Delta \end{bmatrix} = \begin{bmatrix} 40.9401 \\ -28.1824 \\ 119.5812 \end{bmatrix} \text{kNm} \tag{6.106}$$

FINAL END MOMENTS

By substituting equation 6.106 into equation 6.102, we can get the final end moments as follows:

$$\begin{bmatrix} M_{AB} \\ M_{BA} \\ M_{BC} \\ M_{CB} \\ M_{CD} \\ M_{DC} \end{bmatrix} = \begin{bmatrix} -5.625 \\ 16.875 \\ -66.67 \\ 66.67 \\ 0 \\ 0 \end{bmatrix} + \begin{bmatrix} 0.5 & 0 & -3/8 \\ 1 & 0 & -3/8 \\ 2 & 1 & 0 \\ 1 & 2 & 0 \\ 0 & 0.8 & -6/25 \\ 0 & 0.4 & -6/25 \end{bmatrix}$$

$$\begin{bmatrix} 40.9401 \\ -28.1824 \\ 119.5812 \end{bmatrix} = \begin{bmatrix} -30 \\ 12.97 \\ -12.97 \\ 51.25 \\ -51.25 \\ -39.97 \end{bmatrix} \text{kNm}$$

SHEAR FORCE AND BENDING MOMENT DIAGRAMS

The member end forces are shown in Figure 6.63a. The shear force and bending moment diagrams are shown in Figure 6.63b and c, respectively, and the probable deflected shape is shown in Figure 6.63d.

Example 6.15

Analyse the frame with inclined legs as shown in Figure 6.64, using the SD method.

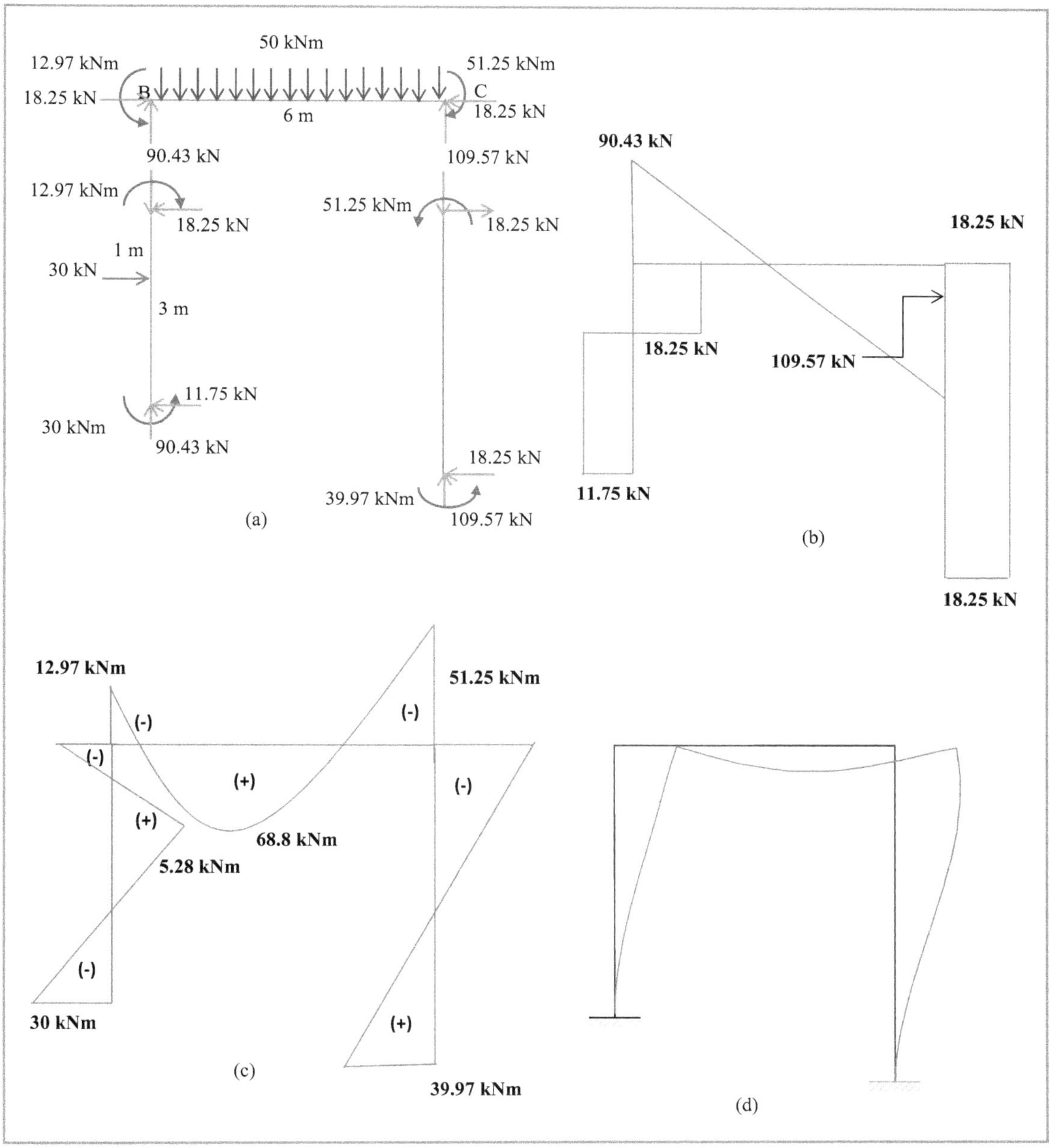

FIGURE 6.63 (a) Free body diagram, (b) shear force diagram, (c) bending moment diagram, and (d) probable elastic curve.

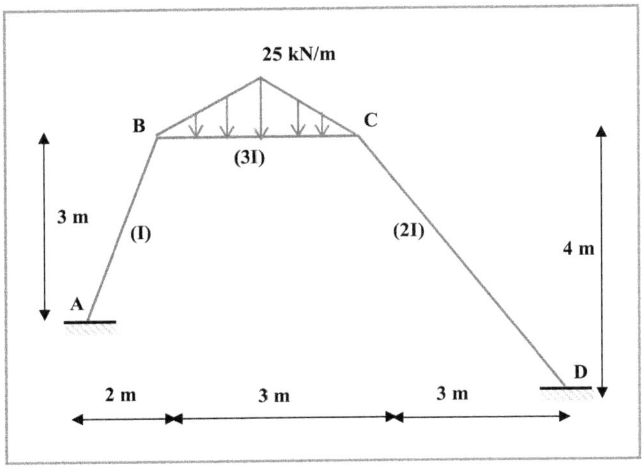

FIGURE 6.64 Rigid-jointed frame with inclined legs.

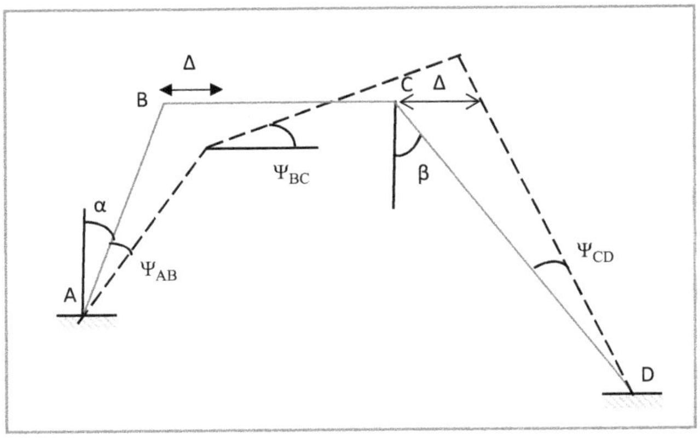

FIGURE 6.65 Chord rotations.

Solution:

The kinematic indeterminacy of the given frame is 3, i.e., θ_B, θ_C, and Δ. Assume the clockwise chord rotations to be +ve. The rotations at *B* and *C*, i.e., θ_B and θ_C are assumed to be +ve (clockwise), and translational displacement is considered positive when directed towards the right.

FIXED END MOMENTS

Using Table 6.1, the fixed end moments can be calculated as follows:

$$MF_{AB} = MF_{BA} = MF_{CD} = MF_{DC} = 0$$

$$MF_{BC} = \frac{-5wL^2}{96} = \frac{-5\times 25\times 3^2}{96} = -11.719 \text{ kNm} = -MF_{CB}$$

SD EQUATIONS

The chord rotations are shown in Figure 6.65.

Span *AB*: $(\theta_B = ?, \theta_A = 0, \varphi_{AB} = \frac{\Delta}{3}, L = 3 \text{ m}, I = I)$

(Chords of elastic curve of $AB^1$and DC^1 rotates in the clockwise direction)

$$M_{AB} = MF_{AB} + \frac{2EI}{L}(2\theta_A + \theta_B - 3\varphi_{AB})$$

$$= 0 + \frac{2EI}{3}\left(0 + \theta_B - 3\frac{\Delta}{3}\right)$$

$$M_{AB} = 0.667EI\theta_B - 0.667EI\Delta \quad (6.107)$$

$$M_{BA} = MF_{BA} + \frac{2EI}{L}(\theta_A + 2\theta_B - 3\varphi_{AB})$$

$$= 0 + \frac{2EI}{3}\left(0 + 2\theta_B - 3\frac{\Delta}{3}\right)$$

$$M_{BA} = 1.333EI\theta_B - 0.667EI\Delta \quad (6.108)$$

Span *BC*:

$$(\theta_B = ?, \theta_C = ?, \varphi_{BC} = \frac{-\Delta(\tan\alpha + \tan\beta)}{L_{BC}} = \frac{-\Delta((2/3)+(3/4))}{3}$$

$$= -0.472\Delta, L = 3 \text{ m}, I = 3I)$$

$$M_{BC} = MF_{BC} + \frac{2EI}{L}(2\theta_B + \theta_C - 3\varphi_{BC}) = -11.719$$

$$+\frac{6EI}{3}(2\theta_B + \theta_C - 3(-0.472\Delta))$$

$$M_{BC} = -11.79 + 4EI\theta_B + 2EI\theta_C + 2.832EI\Delta \quad (6.109)$$

$$M_{CB} = MF_{CB} + \frac{2EI}{L}(\theta_B + 2\theta_C - 3\varphi_{BC})$$

$$= 11.79 + \frac{6EI}{3}(\theta_B + 2\theta_C - 3(-0.472\Delta))$$

$$M_{CB} = 11.79 + 2EI\theta_B + 4EI\theta_C + 2.832EI\Delta \quad (6.110)$$

Span *CD*: $(\theta_C = ?, \theta_D = 0, \varphi_{DC} = \frac{\Delta}{4}, L = 4 \text{ m}, I = 2I)$

$$M_{CD} = MF_{CD} + \frac{2EI}{L}(2\theta_C + \theta_D - 3\varphi_{CD})$$

$$= 0 + \frac{4EI}{4}\left(2\theta_C + 0 - 3\frac{\Delta}{4}\right)$$

$$M_{CD} = 2EI\theta_C - 0.75EI\Delta \quad (6.111)$$

$$M_{DC} = MF_{DC} + \frac{2EI}{L}(\theta_C + 2\theta_D - 3\varphi_{CD})$$

$$= 0 + \frac{4EI}{4}\left(\theta_C + 0 - 3\frac{\Delta}{4}\right)$$

$$M_{DC} = EI\theta_C - 0.75EI\Delta \quad (6.112)$$

The above SD equations 6.107–6.112 can be expressed in matrix form as

$$\begin{bmatrix} M_{AB} \\ M_{BA} \\ M_{BC} \\ M_{CB} \\ M_{CD} \\ M_{DC} \end{bmatrix} = \begin{bmatrix} 0 \\ 0 \\ -11.719 \\ 11.719 \\ 0 \\ 0 \end{bmatrix} + \begin{bmatrix} 0.667 & 0 & -0.667 \\ 1.333 & 0 & -0.667 \\ 4 & 2 & 2.832 \\ 2 & 4 & 2.832 \\ 0 & 2 & -0.75 \\ 0 & 1 & -0.75 \end{bmatrix} \begin{bmatrix} EI\theta_B \\ EI\theta_C \\ EI\Delta \end{bmatrix} \quad (6.113)$$

JOINT EQUILIBRIUM EQUATIONS

Using the free body diagrams shown in Figure 6.66, corresponding to θ_B and θ_C, we write the moment equilibrium equations at *B* and *C*. Similarly, we apply the horizontal force equilibrium equation (shear equilibrium equation) corresponding to Δ, as follows.

Joint *B*: $\sum M_B = 0 \Rightarrow M_{BA} + M_{BC} = 0$

$$1.333EI\theta_B - 0.667EI\Delta - 11.79 + 4EI\theta_B + 2EI\theta_C + 2.832EI\Delta = 0$$

$$\Rightarrow 5.333EI\theta_B + 2EI\theta_C + 2.165EI\Delta = 11.79 \quad (6.114)$$

At joint *C*: $\sum M_C = 0 \Rightarrow M_{CB} + M_{CD} = 0$

$$11.79 + 2EI\theta_B + 4EI\theta_C + 2.832EI\Delta + 2EI\theta_C - 0.75EI\Delta = 0$$

$$\Rightarrow 2EI\theta_B + 6EI\theta_C + 2.082EI\Delta = -11.79 \quad (6.115)$$

Shear equilibrium equation, i.e., corresponding to Δ at joint *C* as follows:

$$\sum F_X = 0 \Rightarrow H_A + H_D = 0$$

$$-\frac{(M_{AB} + M_{BA} + 2V_{BC})}{3} - \frac{(M_{DC} + M_{CD} + 3V_{CB})}{4} = 0$$

$$V_{BC} = \frac{(0.5 \times 25 \times 3 \times 1.5) - M_{BC} - M_{CB}}{3} = \frac{56.25 - M_{BC} - M_{CB}}{3}$$

$$56.25 + 11.79 - 4EI\theta_B - 2EI\theta_C - 2.832EI\Delta$$

$$= \frac{-11.79 - 2EI\theta_B - 4EI\theta_C - 2.832EI\Delta}{3}$$

$$= -2EI\theta_B - 2EI\theta_C - 1.88EI\Delta + 18.75$$

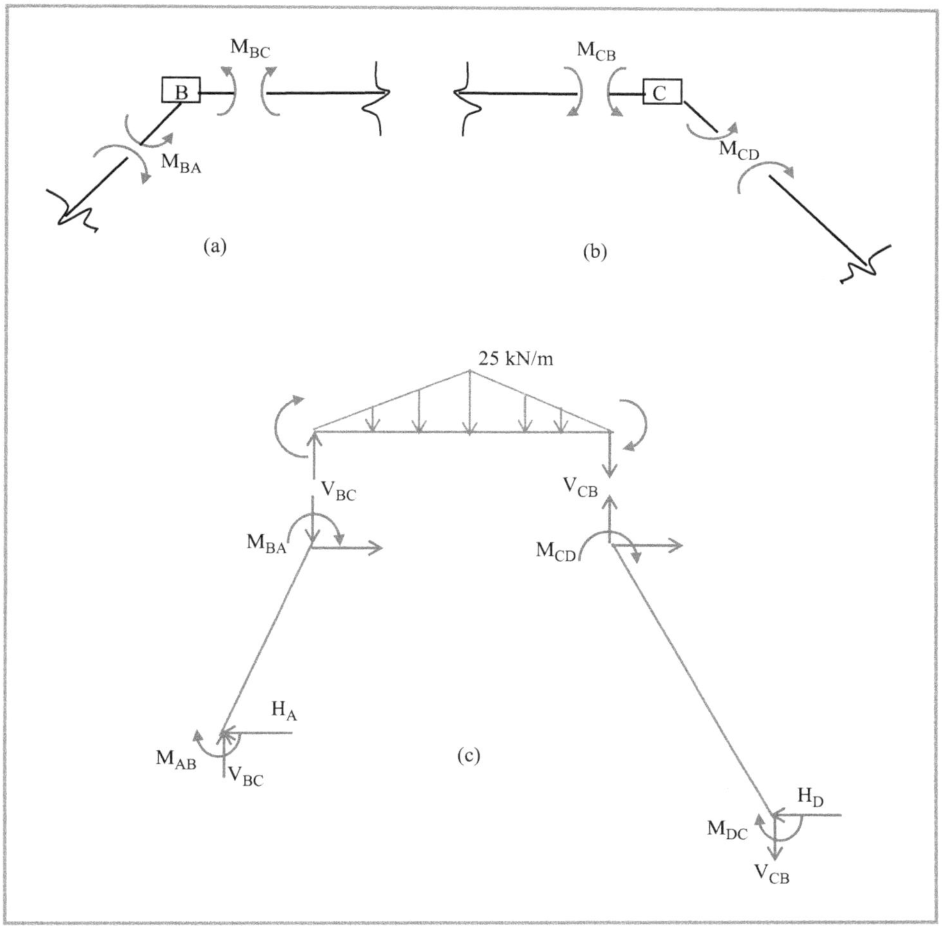

FIGURE 6.66 Free body diagrams of (a) joint *A*, (b) joint *B* and (c) columns of *AB* and *CD*.

$$V_{CB} = \frac{-56.25 - M_{BC} - M_{CB}}{3}$$

$$= \frac{-56.25 + 11.79 - 4EI\theta_B - 2EI\theta_C - 2.832EI\Delta}{-11.79 - 2EI\theta_B - 4EI\theta_C - 2.832EI\Delta} \over 3$$

$$= -2EI\theta_B - 2EI\theta_C - 1.88EI\Delta - 18.75$$

$$H_A + H_D = \frac{1}{3}\begin{pmatrix} 0.667EI\theta_B - 0.667EI\Delta + 1.333EI\theta_B - 0.667EI\Delta \\ -4EI\theta_B - 4EI\theta_C - 3.76EI\Delta + 37.5 \end{pmatrix}$$

$$+\frac{1}{4}\begin{pmatrix} EI\theta_C - 0.75EI\Delta + 2EI\theta_C - 0.75EI\Delta \\ -6EI\theta_B - 6EI\theta_C - 5.64EI\Delta - 56.25 \end{pmatrix} = 0$$

$$H_A + H_D = -2.16EI\theta_B - 2.083EI\theta_C - 3.48EI\Delta = 1.56 \quad (6.116)$$

Equations 6.101–6.103 can be expressed in matrix form as

$$\begin{bmatrix} 5.33 & 2 & 2.167 \\ 2 & 6 & 2.08 \\ -2.16 & -2.083 & -3.48 \end{bmatrix}\begin{bmatrix} EI\theta_B \\ EI\theta_C \\ EI\Delta \end{bmatrix} = \begin{bmatrix} 11.79 \\ -11.79 \\ 1.56 \end{bmatrix}$$

By solving the above, we obtain

$$\begin{bmatrix} EI\theta_B \\ EI\theta_C \\ EI\Delta \end{bmatrix} = \begin{bmatrix} 3.702 \\ -2.835 \\ -1.049 \end{bmatrix} \text{kNm} \quad (6.117)$$

FINAL END MOMENTS

By substituting equation 6.117 into equation 6.113, the final end moments can be obtained as follows:

$$\begin{bmatrix} M_{AB} \\ M_{BA} \\ M_{BC} \\ M_{CB} \\ M_{CD} \\ M_{DC} \end{bmatrix} = \begin{bmatrix} -3.17 \\ 5.64 \\ -5.64 \\ 4.81 \\ -4.81 \\ -2.05 \end{bmatrix} \text{kNm}$$

The free body diagram in Figure 6.67a shows the member end forces. The probable deflected shape is illustrated in Figure 6.67b. The shear force and bending moment diagrams are shown in Figure 6.67c and d, respectively.

Example 6.16

Analyse the anti-symmetric rigid-jointed frame shown in Figure 6.68 using the SD method.

Solution:

The given frame is anti-symmetric; hence, it is subjected to lateral sway (Δ), assumed to act toward the right. Additionally, the frame is also subjected to rotations at B and C, i.e., θ_B and θ_C, which are assumed to be +ve (clockwise).

FIXED END MOMENTS

Using Table 6.1, the fixed end moments can be calculated as follows:

$$MF_{AB} = MF_{BA} = MF_{CD} = MF_{DC} = 0$$

$$MF_{BC} = -MF_{CB} = -\frac{100 \times 4}{8} = -50 \text{ kNm}$$

SD EQUATIONS

Span *AB*: $(\theta_B = ?, \theta_A = 0, \varphi_{AB} = \frac{-\Delta}{3}, L = 3 \text{ m}, I = I)$

(The chords of the elastic curve of $AB^1$and DC^1 rotates in the clockwise direction)

$$M_{AB} = MF_{AB} + \frac{2EI}{L}(2\theta_A + \theta_B - 3\varphi_{AB})$$

$$= 0 + \frac{2EI}{3}\left(0 + \theta_B - 3 \times \frac{-\Delta}{3}\right)$$

$$M_{AB} = 0.667EI\theta_B + 0.667EI\Delta \quad (6.118)$$

$$M_{BA} = MF_{BA} + \frac{2EI}{L}(\theta_A + 2\theta_B - 3\varphi_{AB})$$

$$= 0 + \frac{2EI}{3}\left(0 + 2\theta_B - 3 \times \frac{-\Delta}{3}\right)$$

$$M_{BA} = 1.333EI\theta_B + 0.667EI\Delta \quad (6.119)$$

Span *BC*: $(\theta_B = ?, \theta_C = ?, \varphi_{BC} = 0, L = 4 \text{ m}, I = 2I)$

$$M_{BC} = MF_{BC} + \frac{2EI}{L}(2\theta_B + \theta_C - 3\varphi_{BC})$$

$$= -50 + \frac{4EI}{4}(2\theta_B + \theta_C - 3(0))$$

$$M_{BC} = -50 + 2EI\theta_B + EI\theta_C \quad (6.120)$$

$$M_{CB} = MF_{CB} + \frac{2EI}{L}(\theta_B + 2\theta_C - 3\varphi_{BC})$$

$$= 50 + \frac{4EI}{4}(\theta_B + 2\theta_C - 3(0))$$

$$M_{CB} = 50 + EI\theta_B + 2EI\theta_C \quad (6.121)$$

Span *CD*: $(\theta_C = ?, \theta_D = 0, \varphi_{DC} = \frac{\Delta}{3}, L = 3 \text{ m}, I = I)$

$$M_{CD} = MF_{CD} + \frac{2EI}{L}(2\theta_C + \theta_D - 3\varphi_{CD})$$

$$= 0 + \frac{2EI}{3}\left(2\theta_C + 0 - 3\frac{\Delta}{3}\right)$$

$$M_{CD} = 1.333EI\theta_C - 0.667EI\Delta \quad (6.122)$$

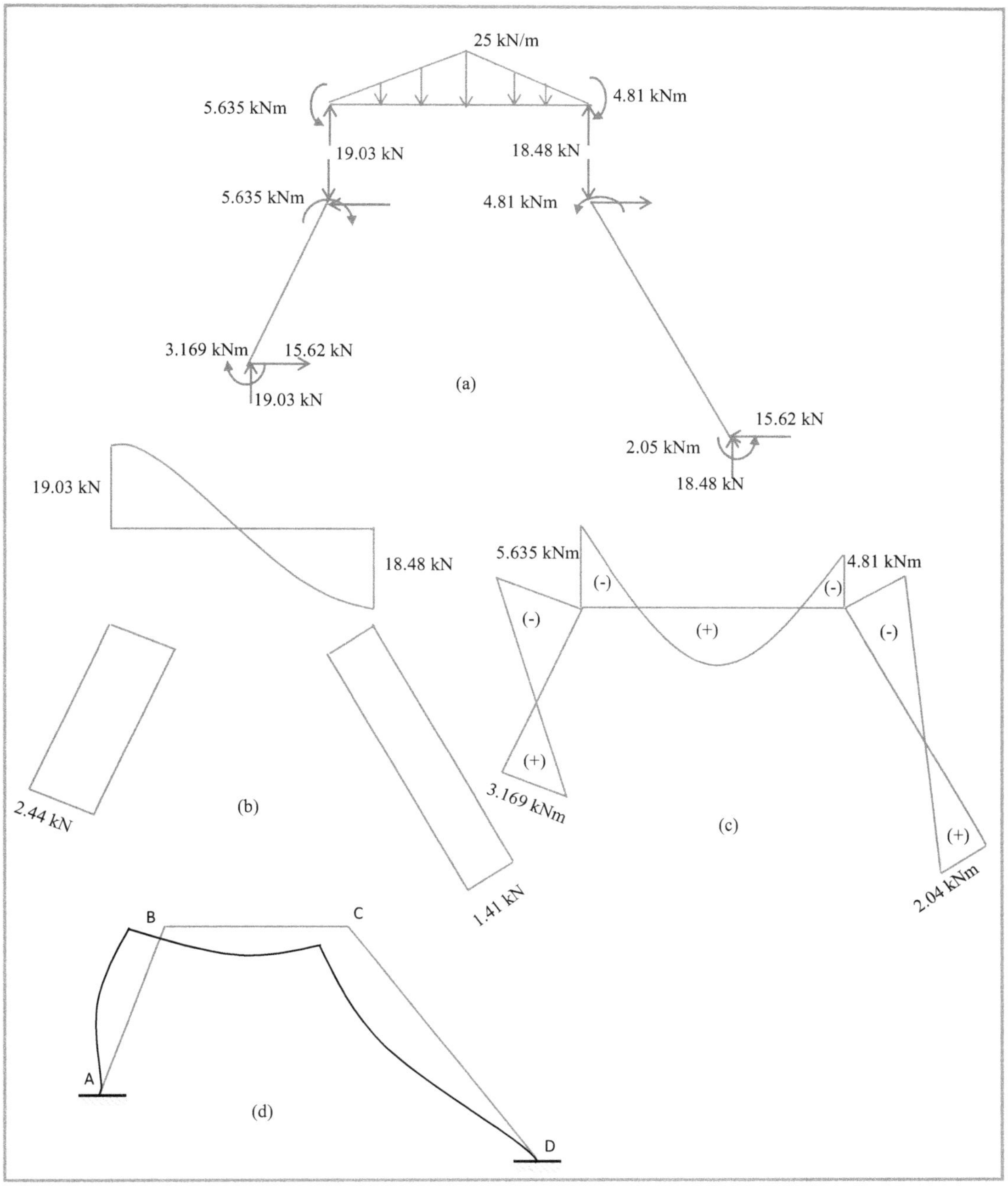

FIGURE 6.67 (a) Free body diagram (b) shear force diagram (c) bending moment diagram, and (d) probable deflected shape.

$$M_{DC} = MF_{DC} + \frac{2EI}{L}(\theta_C + 2\theta_D - 3\varphi_{CD})$$

$$= 0 + \frac{2EI}{3}\left(\theta_C + 0 - 3\frac{\Delta}{3}\right)$$

$$M_{DC} = 0.667EI\theta_C - 0.667EI\Delta \tag{6.123}$$

The above equations 6.118–6.123 can be written conveniently in matrix form, as shown below.

$$\begin{bmatrix} M_{AB} \\ M_{BA} \\ M_{BC} \\ M_{CB} \\ M_{CD} \\ M_{DC} \end{bmatrix} = \begin{bmatrix} 0 \\ 0 \\ -50 \\ 50 \\ 0 \\ 0 \end{bmatrix} + \begin{bmatrix} 0.667 & 0 & -0.667 \\ 1.333 & 0 & -0.667 \\ 2 & 1 & 0 \\ 1 & 2 & 0 \\ 0 & 1.333 & -0.667 \\ 0 & 0.667 & -0.667 \end{bmatrix} \begin{bmatrix} EI\theta_B \\ EI\theta_C \\ EI\Delta \end{bmatrix} \tag{6.124}$$

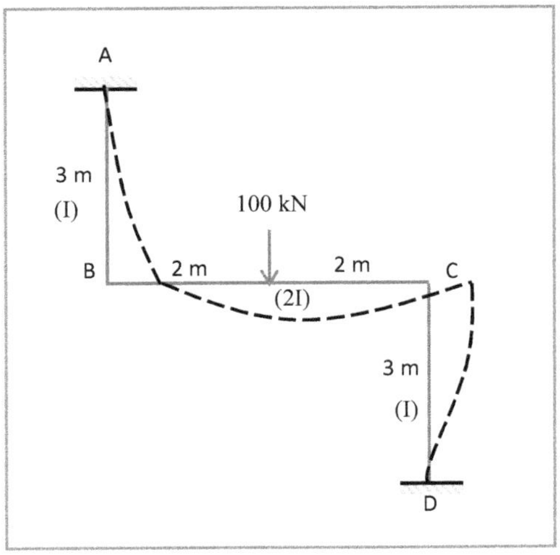

FIGURE 6.68 Anti-symmetric rigid-jointed frame.

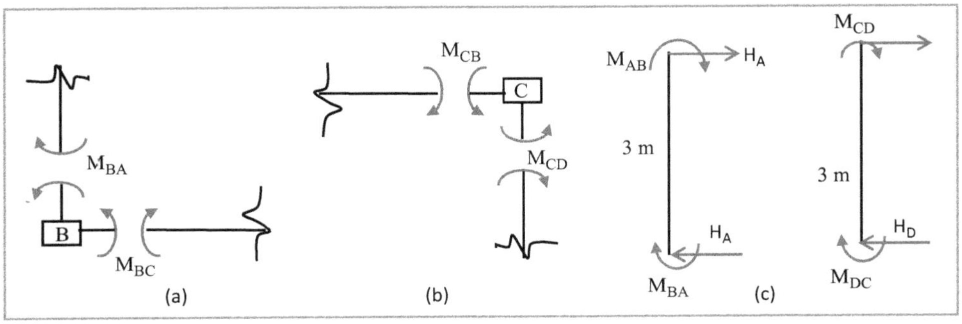

FIGURE 6.69 Free body diagrams of (a) joint B, (b) joint C, and (c) columns AB and CD.

JOINT EQUILIBRIUM EQUATIONS

Using the free body diagrams shown in Figure 6.69, apply the moment equilibrium equations at B and C, corresponding to θ_B and θ_C. Similarly, corresponding to Δ, write the horizontal force equilibrium equation (shear equilibrium equation) as follows:

Joint B: $\sum M_B = 0 \Rightarrow M_{BA} + M_{BC} = 0$

$$1.333EI\theta_B + 0.667EI\Delta - 50 + 2EI\theta_B + EI\theta_C = 0$$

$$\Rightarrow 3.333EI\theta_B + EI\theta_C + 0.667EI\Delta = 50 \qquad (6.125)$$

At joint C: $\sum M_C = 0 \Rightarrow M_{CB} + M_{CD} = 0$

$$50 + EI\theta_B + 2EI\theta_C + 1.333EI\theta_C - 0.667EI\Delta = 0$$

$$\Rightarrow EI\theta_B + 3.333EI\theta_C - 0.667EI\Delta = -50 \qquad (6.126)$$

Shear equilibrium equation, i.e., corresponding to Δ at level BC (Figure 6.69c)

$$\sum F_X = 0 \Rightarrow H_A - H_D = 0 \Rightarrow -\frac{(M_{AB} + M_{BA})}{3} - \frac{-(M_{DC} + M_{CD})}{3} = 0$$

$$-M_{AB} - M_{BA} + M_{DC} + M_{CD} = 0$$

$$-0.667EI\theta_B - 0.667EI\Delta - 1.333EI\theta_B - 0.667EI\Delta + 1.333EI\theta_C$$
$$-0.667EI\Delta + 0.667EI\theta_C - 0.667EI\Delta = 0$$

$$\Rightarrow -2EI\theta_B + 2EI\theta_C + 2.668EI\Delta = 0 \qquad (6.127)$$

Equations 6.125–6.127 can be conveniently expressed in matrix form as follows:

$$\begin{bmatrix} 3.33 & 1 & 0.667 \\ 1 & 3.333 & -0.667 \\ -2 & 2 & -2.668 \end{bmatrix} \begin{bmatrix} EI\theta_B \\ EI\theta_C \\ EI\Delta \end{bmatrix} = \begin{bmatrix} 50 \\ -50 \\ 0 \end{bmatrix}$$

By solving the above, we obtain

$$\begin{bmatrix} EI\theta_B \\ EI\theta_C \\ EI\Delta \end{bmatrix} = \begin{bmatrix} 37.594 \\ -37.594 \\ -56.110 \end{bmatrix} \text{kNm} \quad (6.128)$$

–ve sign indicates the sway is acting toward the left.

FINAL END MOMENTS

By substituting equation 6.128 into equation 6.124, we can get the final end moments as

$$\begin{bmatrix} M_{AB} \\ M_{BA} \\ M_{BC} \\ M_{CB} \\ M_{CD} \\ M_{DC} \end{bmatrix} = \begin{bmatrix} 0 \\ 0 \\ -50 \\ 50 \\ 0 \\ 0 \end{bmatrix} + \begin{bmatrix} 0.667 & 0 & -0.667 \\ 1.333 & 0 & -0.667 \\ 2 & 1 & 0 \\ 1 & 2 & 0 \\ 0 & 1.333 & -0.667 \\ 0 & 0.667 & -0.667 \end{bmatrix}$$

$$\begin{bmatrix} 37.594 \\ -37.594 \\ -56.110 \end{bmatrix} = \begin{bmatrix} -12.406 \\ 12.406 \\ -12.406 \\ 12.406 \\ -12.406 \\ 12.406 \end{bmatrix} \text{kNm}$$

FINAL END SHEARS

With the help of the free body diagram shown in Figure 6.70a and using the static equilibrium equations, the member end forces are calculated. The shear force and bending moment diagrams are shown in Figure 6.70b and c, respectively. The probable elastic curve is shown in Figure 6.70d.

Example 6.17

Analyse the frame shown in Figure 6.71 using the SD method.

Solution:

The kinematic indeterminacy of the given frame is 3, i.e., rotations at B and C (θ_B and θ_C) are assumed in the +ve direction (clockwise), and the horizontal sway (Δ) is assumed to act toward the right.

FIXED END MOMENTS

$$MF_{AB} = MF_{BA} = MF_{BC} = MF_{CB} = MF_{CD} = MF_{DC} = 0$$

SD EQUATIONS

Span *AB*: $\left(\theta_B = ?, \theta_A = 0, \varphi_{AB} = \frac{\Delta}{3}, L = 3\text{ m}, I = I\right)$

(The chords of the elastic curve of AB^1 and DC^1 rotates in the clockwise direction)

$$M_{AB} = MF_{AB} + \frac{2EI}{L}(2\theta_A + \theta_B - 3\varphi_{AB})$$

$$= 0 + \frac{2EI}{3}\left(0 + \theta_B - 3 \times \frac{\Delta}{3}\right)$$

$$M_{AB} = 0.667\theta_B - 0.667\Delta \quad (6.129)$$

$$M_{BA} = MF_{BA} + \frac{2EI}{L}(\theta_A + 2\theta_B - 3\varphi_{AB})$$

$$= 0 + \frac{2EI}{3}\left(0 + 2\theta_B - 3 \times \frac{\Delta}{3}\right)$$

$$M_{BA} = 1.333EI\theta_B - 0.667EI\Delta \quad (6.130)$$

Span *BC*: $(\theta_B = ?, \theta_C = ?, \varphi_{BC} = 0, L = 6\text{ m}, I = 2I)$

$$M_{BC} = MF_{BC} + \frac{2EI}{L}(2\theta_B + \theta_C - 3\varphi_{BC})$$

$$= 0 + \frac{4EI}{6}(2\theta_B + \theta_C - 3(0))$$

$$M_{BC} = 1.333EI\theta_B + 0.667EI\theta_C \quad (6.131)$$

$$M_{CB} = MF_{CB} + \frac{2EI}{L}(\theta_B + 2\theta_C - 3\varphi_{BC})$$

$$= 0 + \frac{4EI}{6}(\theta_B + 2\theta_C - 3(0))$$

$$M_{CB} = 0.667EI\theta_B + 1.333EI\theta_C \quad (6.132)$$

Span *CD*: $\left(\theta_C = ?, \theta_D = 0, \varphi_{DC} = \frac{\Delta}{3}, L = 3\text{ m}, I = I\right)$

$$M_{CD} = MF_{CD} + \frac{2EI}{L}(2\theta_C + \theta_D - 3\varphi_{CD})$$

$$= 0 + \frac{2EI}{3}\left(2\theta_C + 0 - 3\frac{\Delta}{3}\right)$$

$$M_{CD} = 1.333EI\theta_C - 0.667EI\Delta \quad (6.133)$$

$$M_{DC} = MF_{DC} + \frac{2EI}{L}(\theta_C + 2\theta_D - 3\varphi_{CD})$$

$$= 0 + \frac{2EI}{3}\left(\theta_C + 0 - 3\frac{\Delta}{3}\right)$$

$$M_{DC} = 0.667EI\theta_C - 0.667EI\Delta \quad (6.134)$$

The above equations 6.129–6.134 can be written conveniently in matrix form as

$$\begin{bmatrix} M_{AB} \\ M_{BA} \\ M_{BC} \\ M_{CB} \\ M_{CD} \\ M_{DC} \end{bmatrix} = \begin{bmatrix} 0 \\ 0 \\ 0 \\ 0 \\ 0 \\ 0 \end{bmatrix} + \begin{bmatrix} 0.667 & 0 & -0.667 \\ 1.333 & 0 & -0.667 \\ 1.333 & 0.667 & 0 \\ 0.667 & 1.333 & 0 \\ 0 & 1.333 & -0.667 \\ 0 & 0.667 & -0.667 \end{bmatrix}$$

$$\begin{bmatrix} EI\theta_B \\ EI\theta_C \\ EI\Delta \end{bmatrix} \quad (6.135)$$

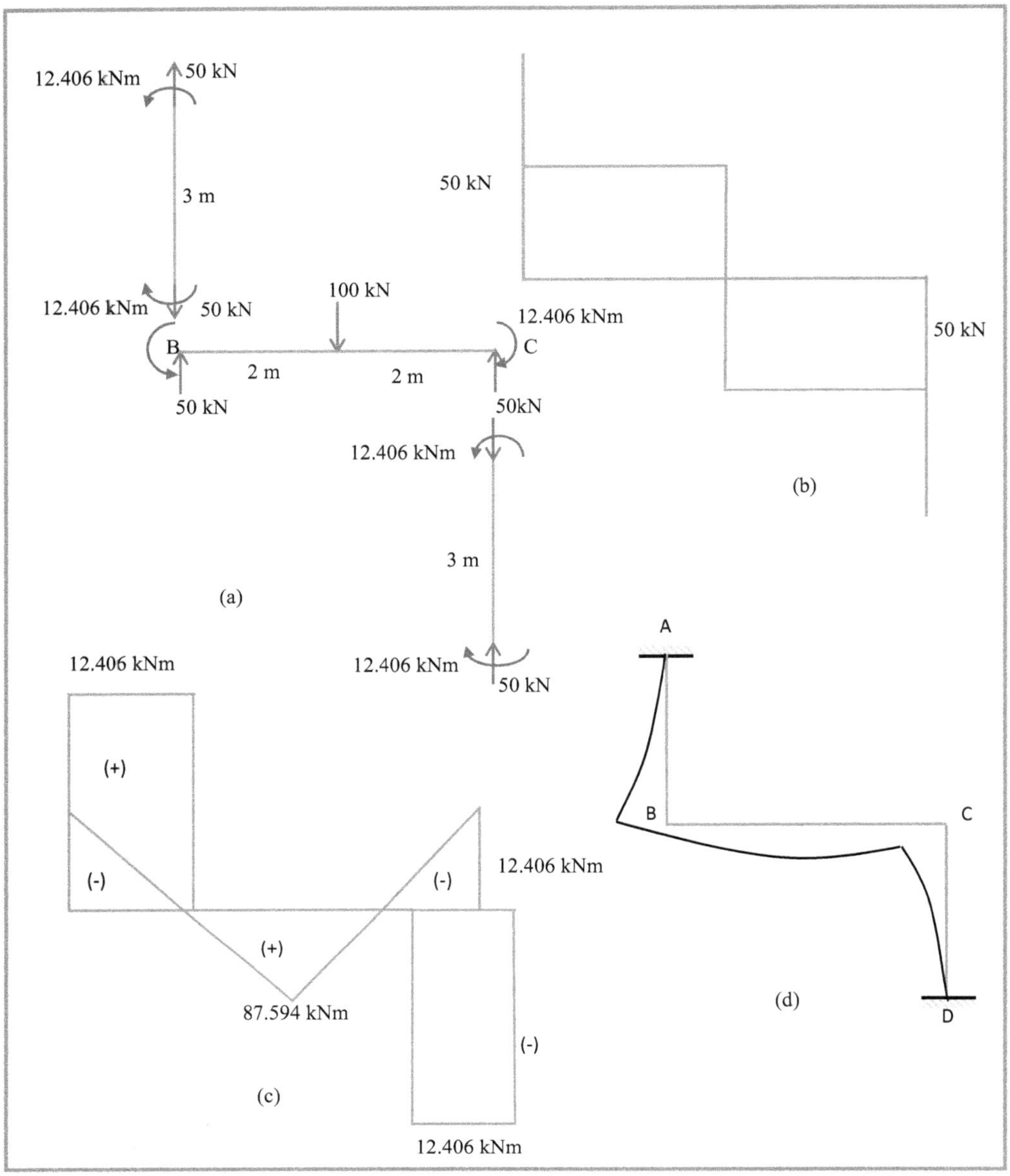

FIGURE 6.70 (a) Free body diagram, (b) shear force diagram, (c) bending moment diagram, and (d) probable elastic curve.

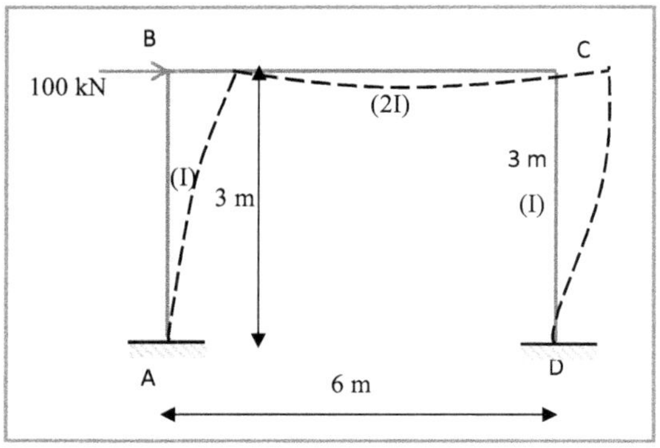

FIGURE 6.71 Rigid frame subjected to a horizontal load at joint *B*.

JOINT EQUILIBRIUM EQUATIONS

Using the free body diagrams shown in Figure 6.72, write the moment equilibrium equations corresponding to θ_B and θ_C, at *B* and *C*. Similarly, write the horizontal force equilibrium equation (shear equilibrium equation) corresponding to Δ.

Joint *B*:

$$\sum M_B = 0 \Rightarrow M_{BA} + M_{BC} = 0$$

$$\Rightarrow 1.333EI\theta_B - 0.667EI\Delta + 1.333EI\theta_B + 0.667EI\theta_C = 0$$

$$\Rightarrow 2.666EI\theta_B + 0.667EI\theta_C - 0.667EI\Delta = 0 \quad (6.136)$$

$$\sum M_C = 0 \Rightarrow M_{CB} + M_{CD} = 0$$

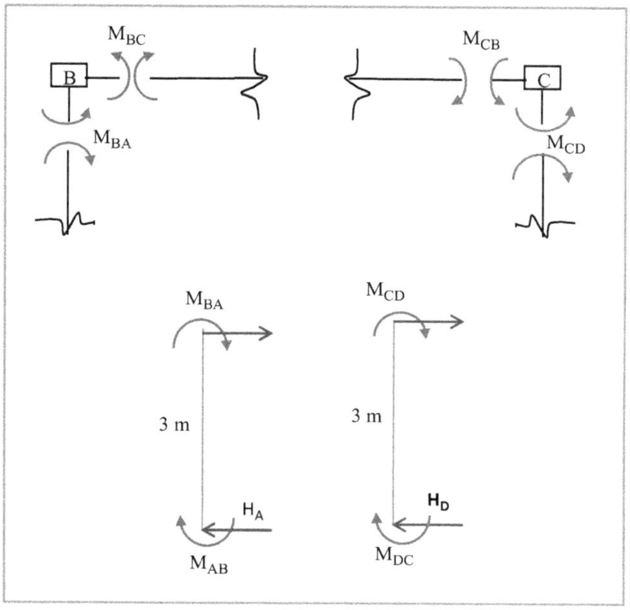

FIGURE 6.72 Free body diagrams of joints *B* and *C* and columns *AB* and *CD*.

$$\Rightarrow 0.667EI\theta_B + 1.333EI\theta_C + 1.333EI\theta_C - 0.667EI\Delta = 0$$

$$\Rightarrow 0.667EI\theta_B + 2.666EI\theta_C - 0.667EI\Delta = 0 \quad (6.137)$$

Shear equilibrium equation, i.e., corresponding to Δ at joint *C*

$$\sum F_X = 0 \Rightarrow H_A + H_D = 100 \Rightarrow -\frac{(M_{AB} + M_{BA})}{3} + \frac{-(M_{DC} + M_{CD})}{3} = 100$$

$$\Rightarrow M_{AB} + M_{BA} + M_{DC} + M_{CD} = -300$$

$$\Rightarrow 0.667\theta_B - 0.667\Delta + 1.333EI\theta_B - 0.667EI\Delta + 1.333EI\theta_C - 0.667EI\Delta + 0.667EI\theta_C - 0.667EI\Delta = -300$$

$$\Rightarrow 2EI\theta_B + 2EI\theta_C - 2.668EI\Delta = -300 \quad (6.138)$$

The above equations 6.136–6.138 can be expressed in matrix form as

$$\begin{bmatrix} 2.666 & 0.667 & -0.667 \\ 0.667 & 2.666 & -0.667 \\ 2 & 2 & -2.668 \end{bmatrix} \begin{bmatrix} EI\theta_B \\ EI\theta_C \\ EI\Delta \end{bmatrix} = \begin{bmatrix} 0 \\ 0 \\ -300 \end{bmatrix}$$

By solving the above, we get

$$\begin{bmatrix} EI\theta_B \\ EI\theta_C \\ EI\Delta \end{bmatrix} = \begin{bmatrix} 32.147 \\ 32.147 \\ 160.64 \end{bmatrix} \text{kNm}$$

By substituting the above rotational and translational displacements into equation 6.135, the final end moments can be obtained as

$$\begin{bmatrix} M_{AB} \\ M_{BA} \\ M_{BC} \\ M_{CB} \\ M_{CD} \\ M_{DC} \end{bmatrix} = \begin{bmatrix} 0 \\ 0 \\ 0 \\ 0 \\ 0 \\ 0 \end{bmatrix} + \begin{bmatrix} 0.667 & 0 & -0.667 \\ 1.333 & 0 & -0.667 \\ 1.333 & 0.667 & 0 \\ 0.667 & 1.333 & 0 \\ 0 & 1.333 & -0.667 \\ 0 & 0.667 & -0.667 \end{bmatrix}$$

$$\begin{bmatrix} 32.147 \\ 32.147 \\ 160.64 \end{bmatrix} = \begin{bmatrix} -85.705 \\ -64.295 \\ 64.295 \\ 64.295 \\ -64.295 \\ -85.705 \end{bmatrix} \text{kNm}$$

The member end forces, probable elastic curve, and the shear force and bending moment diagrams are shown in Figure 6.73.

The deflected shape and bending moment diagrams are shown in the figure. It is clear that the beam is subjected to an anti-symmetric bending moment diagram with a point of inflection at the centre of span *BC*. The deflection diagram is also anti-symmetric. The points of inflection of the columns occur approximately at mid-height (0.57*h*) of each column, due to the flexibility of the beam–column joint.

6.5.4.2 Avoiding Lateral Sway

It is also possible to avoid lateral sway (horizontal translation) in frames during calculations in certain special cases. This has already been demonstrated in examples such as symmetrically loaded frames and guided support conditions, where we neglected translation at guided fixed support by using the modified stiffness of the beam element (which behaves like a cantilever) and modified fixed end moments. The same condition was observed in the case of symmetrical frames. Similar short-cuts can also be applied to frames where the loading can be resolved into symmetric components that produce only axial forces in the beams at different floor levels, and an anti-symmetric component, in which we can find the contra-flexure points at the midpoints of the beams, as shown in Figures 6.74 and 6.75.

In these cases, it is sufficient to consider only half of such frames, with a roller support at the midpoint of various beams, as shown above. Therefore, it is sufficient to consider the rotational DoFs and neglect the translational DoF. However, the fixed end moments must be calculated carefully. These are explained in the following examples.

Example 6.18

Solve Example 6.17 (Figure 6.76) by appropriately making the loading anti-symmetric and neglecting the horizontal translation using the SD method.

Solution:

Using the principle of superposition, the frame (Figure 6.76) can be divided into two parts, as shown

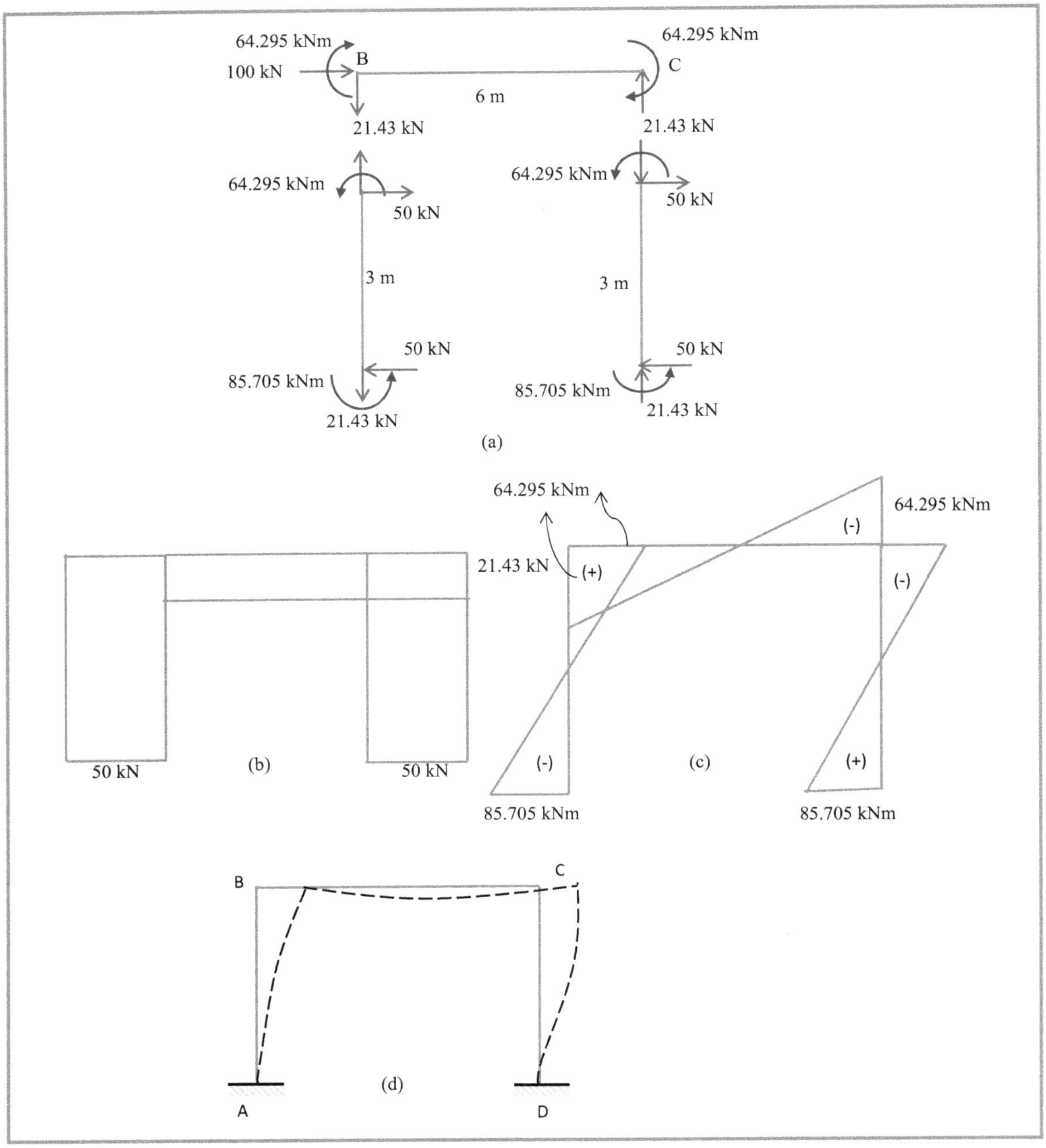

FIGURE 6.73 (a) Free body diagram, (b) shear force diagram, (c) bending moment diagram, and (b) probable elastic curve.

in Figure 6.77. By taking advantage of anti-symmetry, it is sufficient to consider only half of the frame (ABE) for the analysis (Figure 6.77), where E is the midpoint of *BC* and is provided with a roller support. Now, the kinematic indeterminacy is limited to 1, i.e., θ_B (neglecting the translational displacement), provided appropriate modifications are made in the FEMs and stiffness calculations.

For the element *AB*, the end conditions are shown in Figure 6.78.

$$MF_{AB} = MF_{BA} = \frac{-Wh}{2} = -\frac{50\times 3}{2} = -75 \text{ kNm}$$

$$MF_{BE} = 0 \left(\text{since no external load on span } BC\right)$$

SD EQUATIONS

Span *AB*: $(\theta_B = ?, \theta_A = 0, \varphi_{AB} : \text{neglected}, L = 3 \text{ m}, I = I)$

$$M_{AB} = MF_{AB} - \frac{EI}{L}(\theta_B) = -75 - \frac{EI}{3}(\theta_B) = -75 - 0.333EI\theta_B \tag{6.139}$$

$$M_{BA} = MF_{BA} + \frac{EI}{L}(\theta_B) = -75 + \frac{EI}{3}(\theta_B) = -75 + 0.333EI\theta_B \tag{6.140}$$

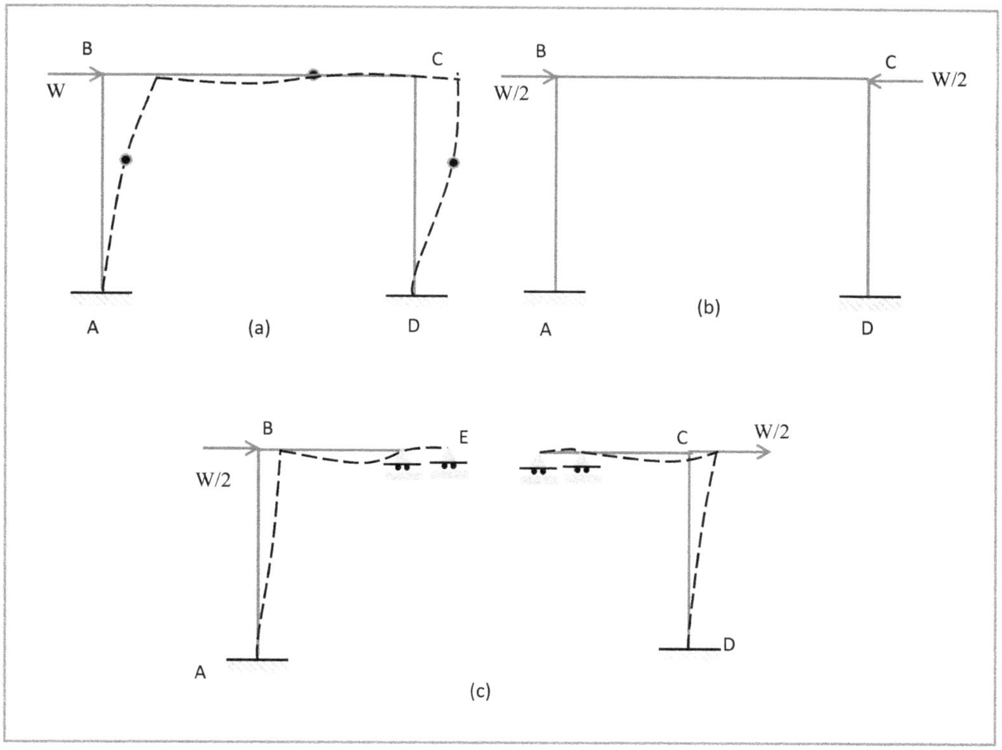

FIGURE 6.74 Single bay single storey frame (a) loading diagram with deflected shape, (b) symmetric loading part, and (c) anti-symmetric part.

Span *BE*: $(\theta_B = ?, \theta_E = 0, \varphi_{BE} : \text{neglected}, L = 3 \text{ m}, I = 2I)$

$$M_{BE} = MF_{BE} + \frac{3EI}{L}(\theta_B - \varphi_{BE}) = 0 + \frac{6EI}{3}(\theta_B - 0) = 2EI\theta_B \quad (6.141)$$

JOINT EQUILIBRIUM EQUATIONS

Joint *B*: Using Figure 6.79,

$$\sum M_B = 0 \Rightarrow M_{BA} + M_{BE} = 0$$

$$= -75 + 0.333EI\theta_B + 2EI\theta_B = 0$$

$$\Rightarrow EI\theta_B = 32.147 \text{ kNm}$$

FINAL END MOMENTS

By back substituting the known value of $EI\theta_B$ into equations 6.139–6.141, we can obtain the final end moments as follows:

$$M_{AB} = -75 - 0.333 \times 32.147 = -85.704 \text{ kNm}$$

$$M_{BA} = -75 + 0.333 \times 32.147 = -64.29 \text{ kNm}$$

$$M_{BE} = 2 \times 32.147 = 64.29 \text{ kNm}$$

The above results are exactly the same as those obtained in the solution of Example 6.19. Therefore, it is clearly demonstrated that the alternate solution (method) provides a simplified solution, as the kinematic indeterminacy is reduced from 3 to 1. The BMD and deflection diagrams are the same as those of Example 6.17.

Example 6.19

Analyse the frame shown in Figure 6.80 using the SD method. Draw the shear force and bending moment diagrams.

Solution:

The kinematic indeterminacy can be limited to 1, i.e., θ_B. The moment at *C*, i.e., M_C=0 (due to the roller support), allows θ_C to be neglected by using the modified stiffness and FEM. Furthermore, due to the roller support at *C*, the transverse displacement Δ_{BC} can be neglected by providing the guided support at *B* and using the modified stiffness and FEM for column *AB*, as shown in Figure 6.81. This can be visualised as a fixed beam (AB^1) whose length=2×*AB* and is subjected to 30 kN at 2 m from both ends *A* and B^1.

Therefore, the fixed end moments for column *AB* can be expressed as

$$MF_{AB} = \frac{-Wab^2}{L^2} + \frac{-Wba^2}{L^2} = -\left(\frac{30 \times 2 \times 6^2}{8^2} + \frac{30 \times 6 \times 2^2}{8^2}\right)$$

$$= -45 \text{ kNm}$$

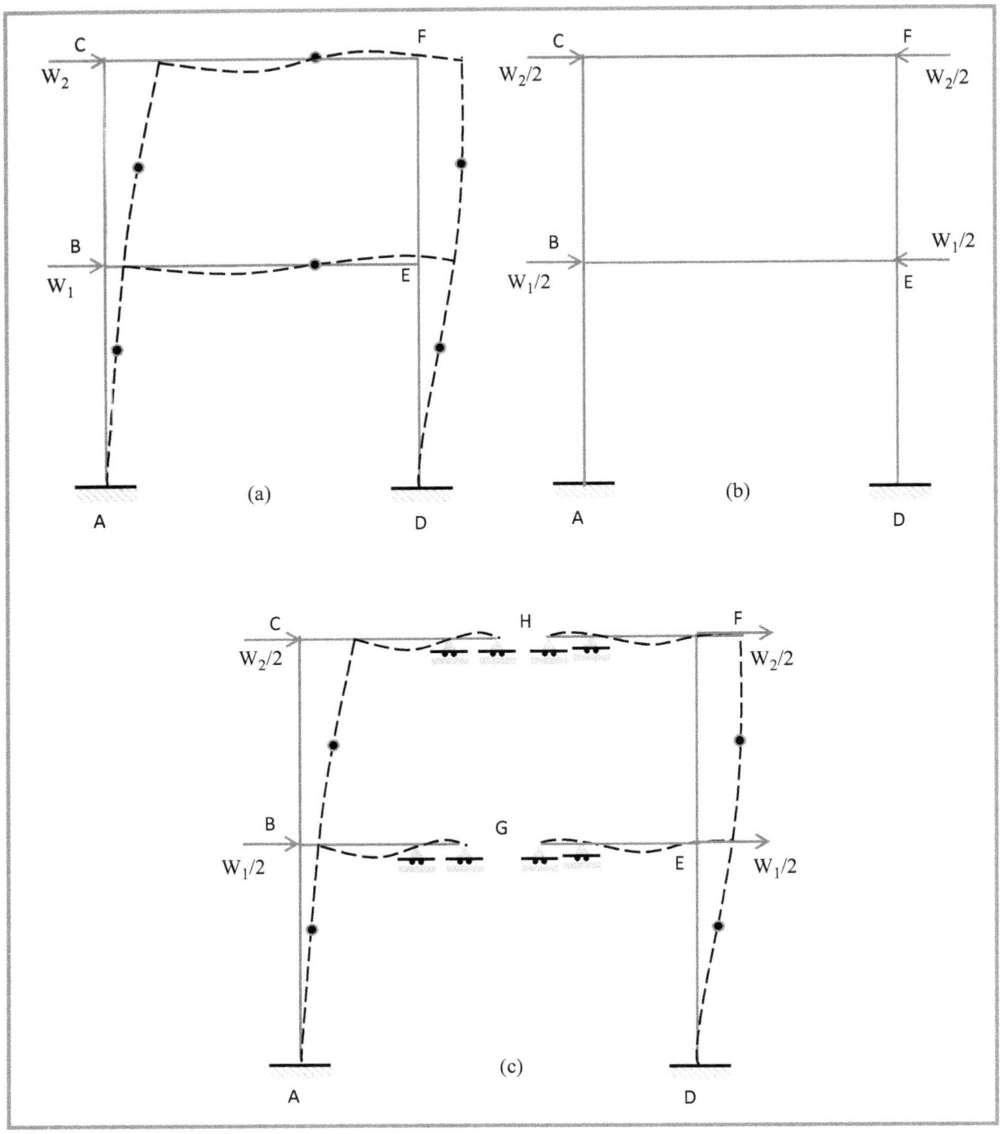

FIGURE 6.75 Single-bay two-storey frame: (a) loading diagram with deflected shape, (b) symmetric loading part, and (c) anti-symmetric part.

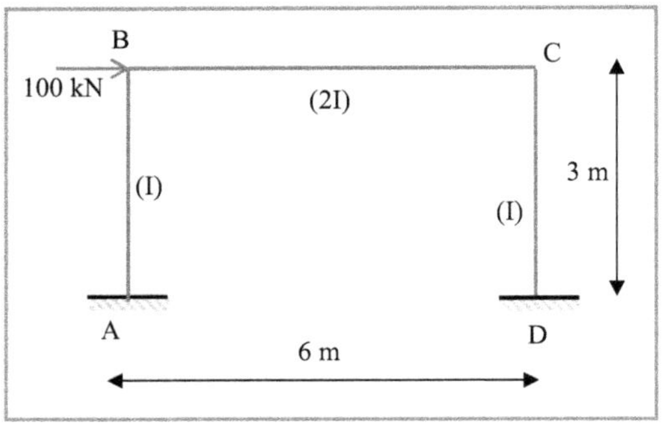

FIGURE 6.76 Frame with lateral loading.

$$MF_{BA} = -(30\times 4 - 45 - 30\times 2) = -15\ \text{kNm}$$
$$(\text{sagging BM at mid span } ABB^1)$$

$$MF_{BC} = -\frac{3}{2}\frac{wL^2}{12} = -\frac{3\times 20\times 3^2}{24} = -22.5\ \text{kNm}$$

SD Equations:

Span AB: ($\theta_B = ?, \theta_A = 0, \varphi_{AB}$: neglected, $L = 4$ m, $I = I$)

$$M_{AB} = MF_{AB} - \frac{EI}{L}(\theta_B) = -45 - \frac{EI}{4}(\theta_B) = -45 - 0.25EI\theta_B \tag{6.142}$$

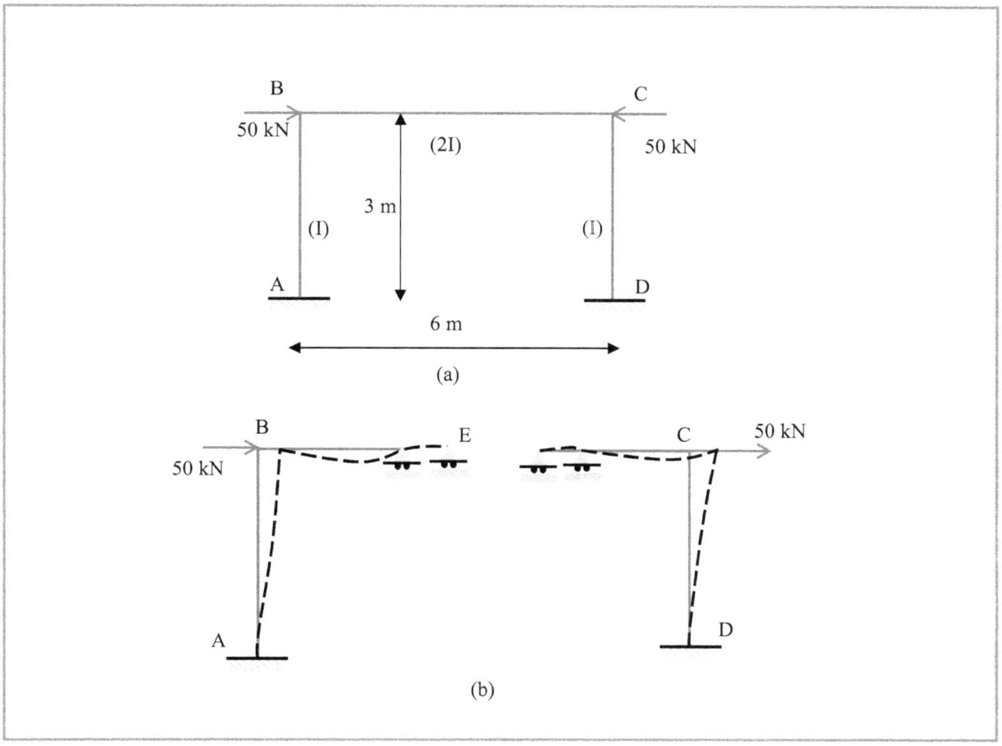

FIGURE 6.77 (a) Symmetric loading (only axial compression in *BC* and (b) anti-symmetric loading (using the principle of superposition).

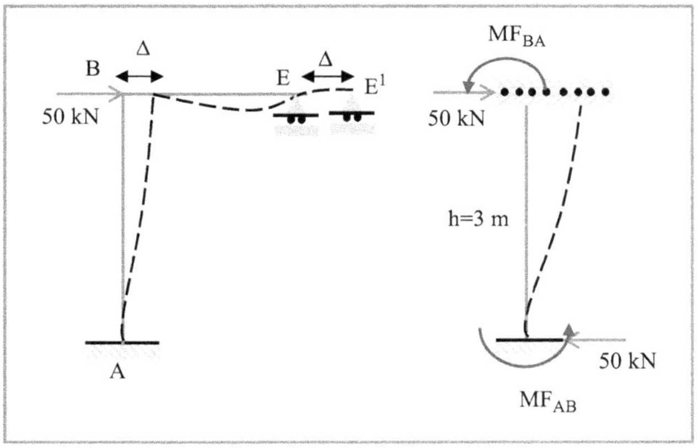

FIGURE 6.78 Half of the frame and fixed end moments.

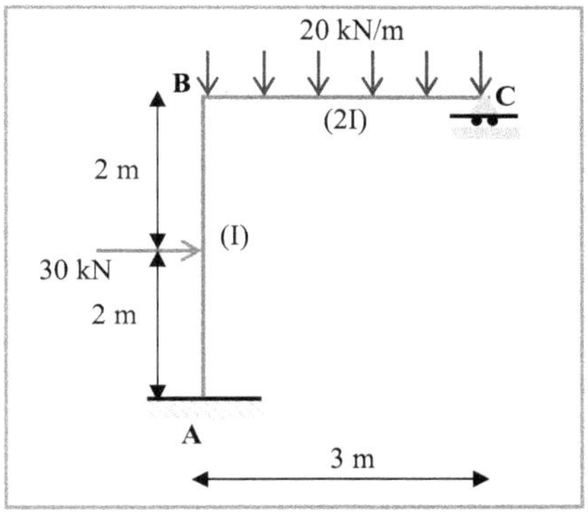

FIGURE 6.80 Rigid-jointed frame.

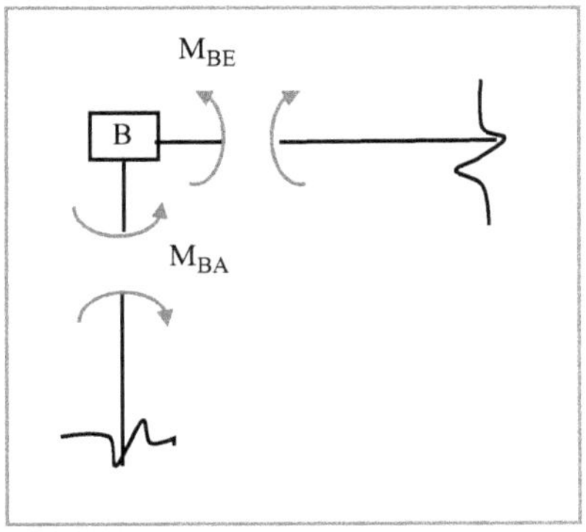

FIGURE 6.79 Free body diagram of joint *B*.

$$M_{BA} = MF_{BA} + \frac{EI}{L}(\theta_B) = -15 + \frac{EI}{4}(\theta_B) = -15 + 0.25EI\theta_B \tag{6.143}$$

Span *BC*: $(\theta_B = ?, \theta_C = 0, \varphi_{BC} = 0, L = 3\text{ m}, I = 2I)$

$$M_{BC} = MF_{BC} + \frac{3EI}{L}(\theta_B - \varphi_{BC}) = -22.5 + \frac{6EI}{3}(\theta_B - 0)$$
$$= -22.5 + 2EI\theta_B \tag{6.144}$$

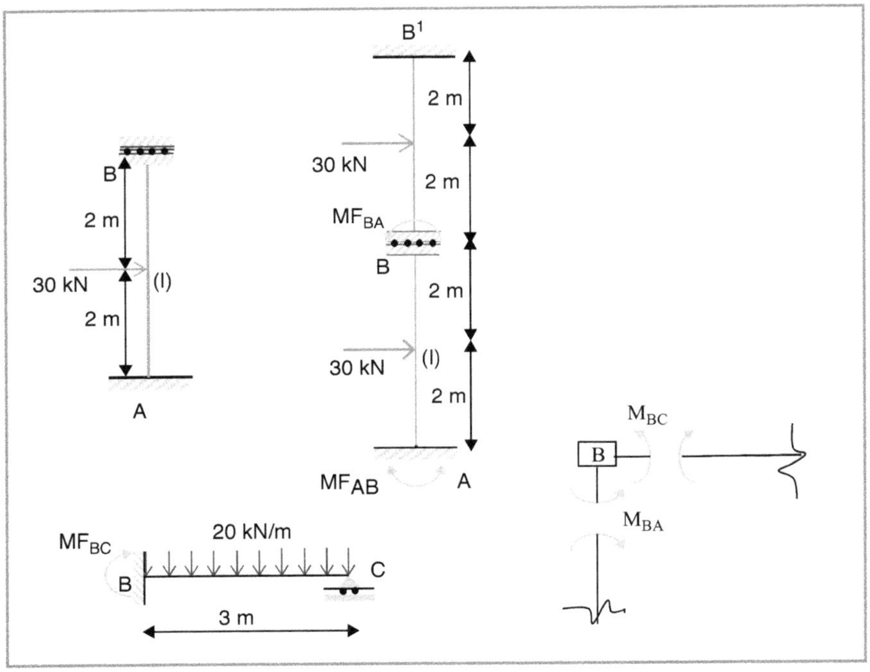

FIGURE 6.81 Fixed end moments and free body diagram of joint *B*.

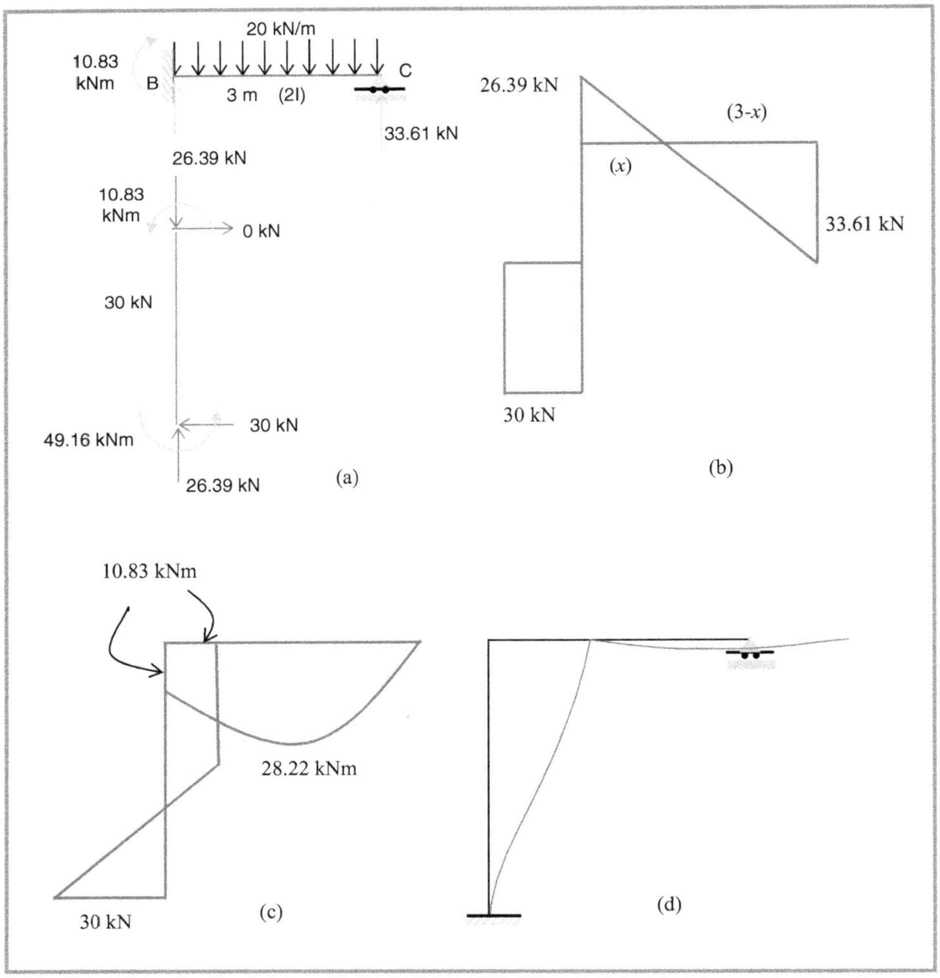

FIGURE 6.82 (a) Free body diagram, (b) shear force diagram, (c) bending moment diagram, and (d) probable elastic curve.

JOINT EQUILIBRIUM EQUATIONS

Joint *B*: From Figure 6.81, we can write

$$\sum M_B = 0 \Rightarrow M_{BA} + M_{BC} = 0$$

$$= -15 + 0.25EI\theta_B - 22.5 + 2EI\theta_B = 0$$

$$\Rightarrow EI\theta_B = 16.667 \text{ kNm}$$

FINAL END MOMENTS

By substituting the above value of $EI\theta_B$ into equations 6.142–6.144, we can obtain the final end moments as

$$M_{AB} = -45 - 0.25 \times 16.667 = -49.16 \text{ kNm}$$

$$M_{BA} = -15 + 0.25 \times 16.667 = -10.83 \text{ kNm}$$

$$M_{BC} = -22.5 + 2 \times 16.667 = 10.83 \text{ kNm}$$

The free body diagram shows the member end forces in Figure 6.82a. The shear force and bending moment diagrams are illustrated in Figure 6.82b and c, respectively. The probable deflected shape is shown in Figure 6.82d.

6.6 PROBLEMS FOR PRACTICE

6.6.1 Write the general SD equations for a beam element subjected to rotational DoFs as well as sway DoFs.

6.6.2 How do you account for the external yielding of supports in the SD method?

6.6.3 How do you identify the direction of lateral sway in a rectangular frame subjected to a concentrated gravity load acting to the right of the centre line of the frame?

6.6.4 What difficulties arise when analysing portal frames with inclined legs, and how can these difficulties be resolved?

6.6.5 Explain the procedure involved in the analysis of non-prismatic fixed beam using the SD method?

6.6.6 Analyse the two-span continuous beam *ABC*, as shown in Figure 6.83, using the SD method. Draw the shear force and bending moment diagrams. Assume *EI* is constant.

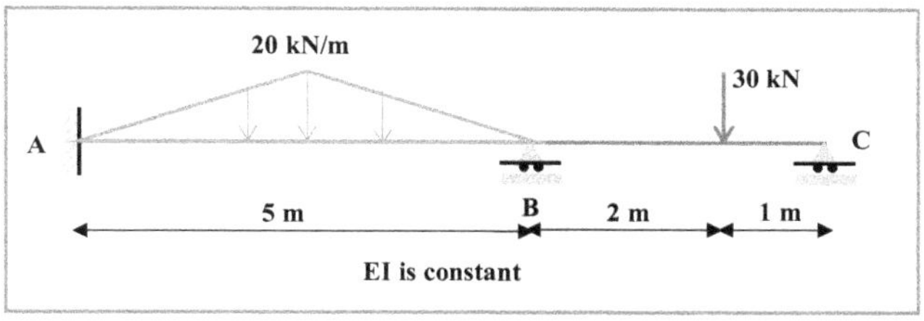

FIGURE 6.83

6.6.7 Evaluate the support reactions of the continuous beam shown in Figure 6.84 using the SD method. Assume flexural rigidity is constant.

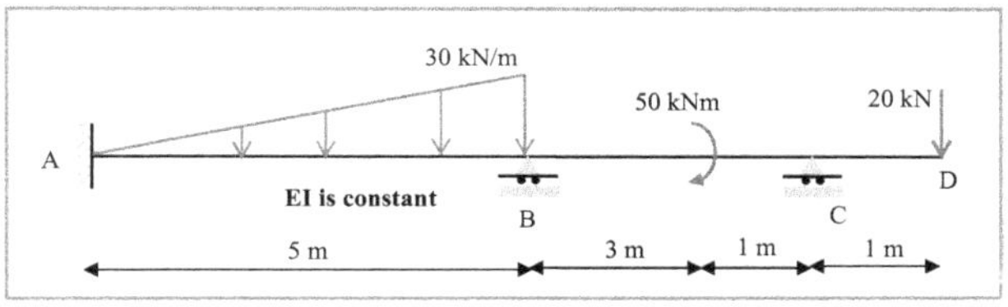

FIGURE 6.84

6.6.8 Analyse the three-span continuous beam with overhang *ABCDE*, as shown in Figure 6.85 using the SD method. Construct the SFD and BMD. Assume *EI* is constant.

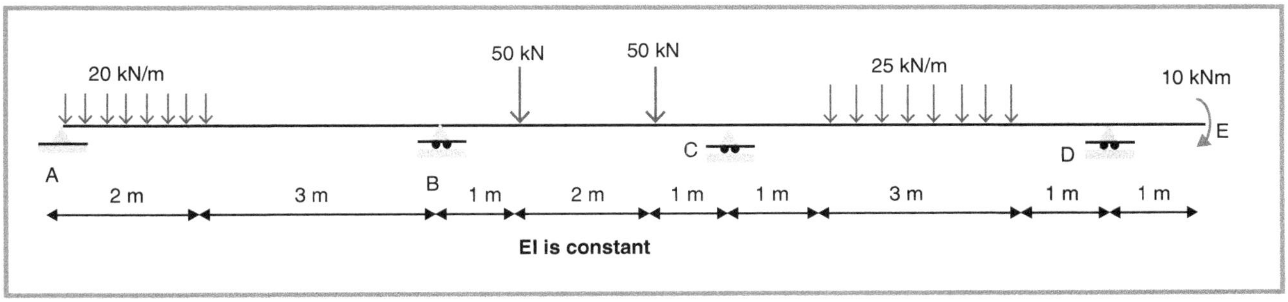

FIGURE 6.85

6.6.9 Analyse the symmetrical three-span continuous beam shown in Figure 6.86, using the SD method. Draw the shear force and bending moment diagrams.

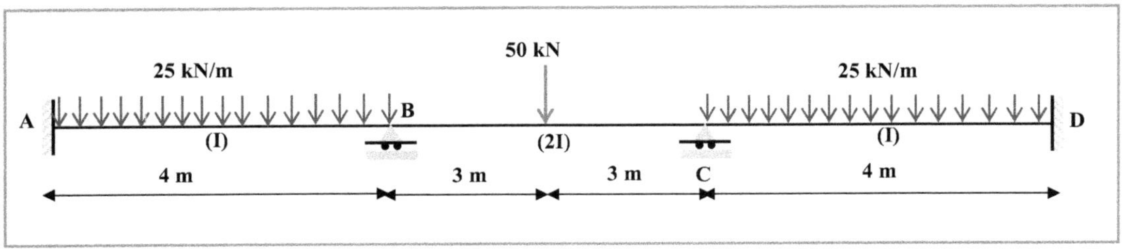

FIGURE 6.86

6.6.10 Analyse the non-prismatic fixed beam as shown in Figure 6.87, using the SD method. Construct the SFD and BMD.

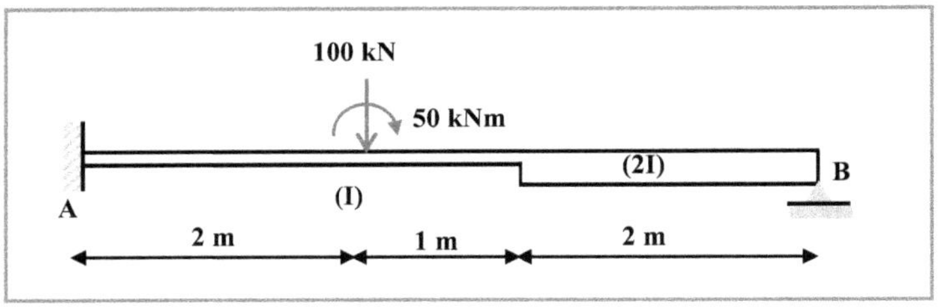

FIGURE 6.87

6.6.11 A two-span continuous beam *ABD* with an internal hinge at *C* as shown in Figure 6.88, using the SD method. Draw the shear force and bending moment diagrams.

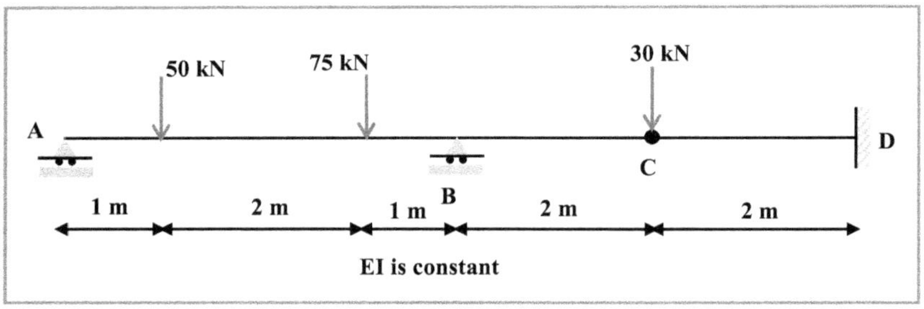

FIGURE 6.88

6.6.12 Re-analyse the beam shown in Figure 6.83, if support *A* is a guided support, instead of a fixed support. Construct the shear force and bending moment diagrams.

6.6.13 Re-analyse the beam shown in Figure 6.83 using the SD method, if support *A* rotates by 0.002 rad in the clockwise direction. Assume $EI = 138375$ kNm^2. Construct the shear force and bending moment diagrams.

6.6.14 Analyse the continuous beam shown in Figure 6.89 using the SD method, if supports *B* and *C* settle down by 10 and 5 mm, respectively. Assume $EI = 14660$ kNm^2.

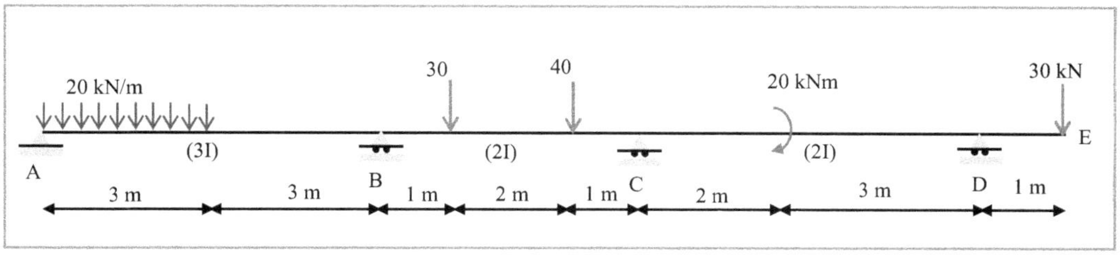

FIGURE 6.89

6.6.15 Analyse the symmetrical rigid-jointed frame shown in Figure 6.90, using the SD method. Draw the SFD and BMD.

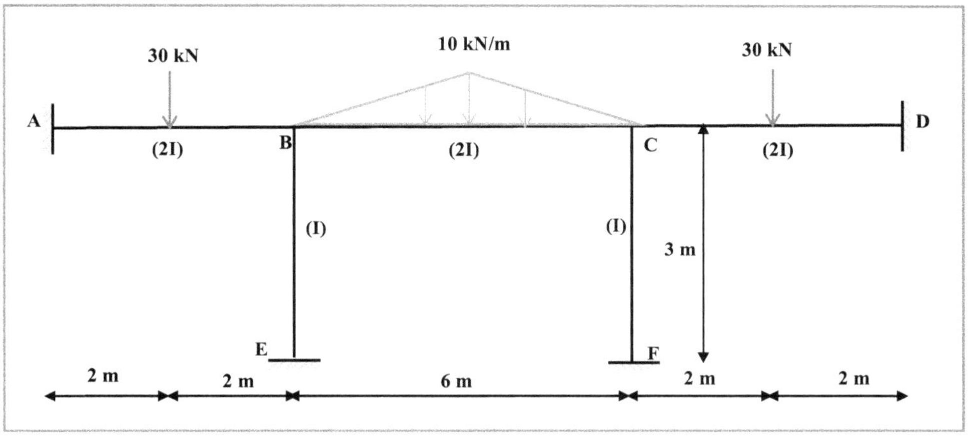

FIGURE 6.90

6.6.16 Re-analyse the frame shown in Figure 6.90, using the SD method by considering all the supports as pinned instead of fixed. Construct the shear force and bending moment diagrams.

6.6.17 A rigid-jointed frame ABCD is subjected to horizontal and gravity loading as shown in Figure 6.91. Analyse the frame using the SD method, and draw the SFD and BMD.

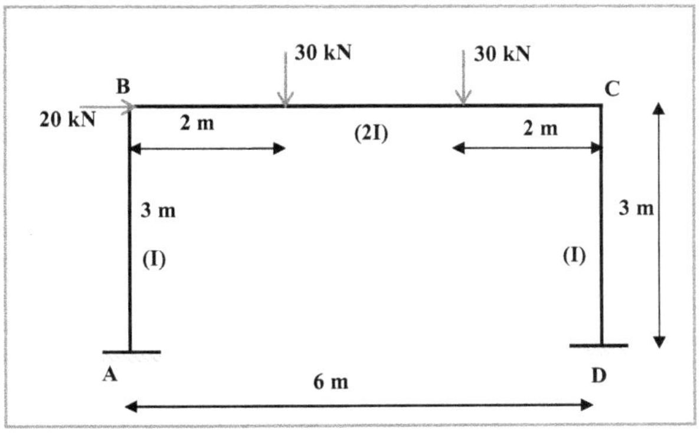

FIGURE 6.91

6.6.18 Analyse the frame shown in Figure 6.91, using the SD method, if support *D* settles down by 10 mm instead of under external loading. Assume $EI = 138375$ kNm^2.

6.6.19 An unsymmetrical rigid-jointed frame *ABCD* is shown in Figure 6.92. Analyse it using the SD method.

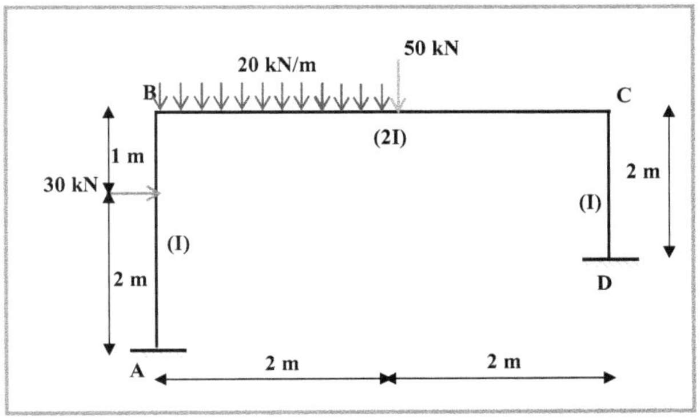

FIGURE 6.92

6.6.20 Analyse the box frame *ABCD* shown in Figure 6.93 using the SD method. Take advantage of symmetry. Draw the SFD and BMD.

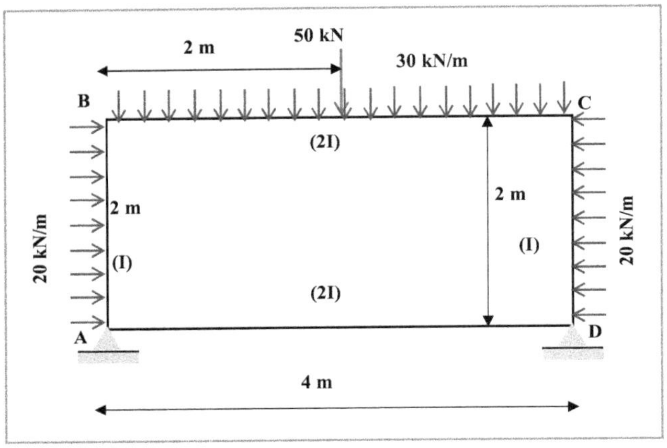

FIGURE 6.93

6.6.21 A frame *ABCD* with an inclined column *CD* is shown in Figure 6.94. Analyse the frame using the SD method and construct the SFD and BMD. Also, plot the probable deflected shape.

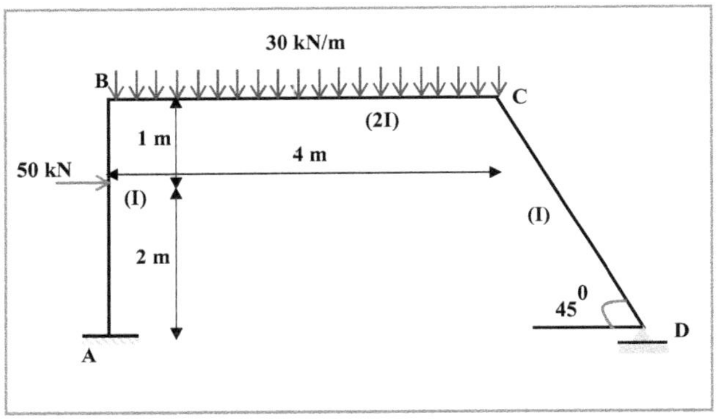

FIGURE 6.94

6.6.22 A frame *ABCD* with inclined columns *AB* and *CD* is shown in Figure 6.95. Analyse the frame using the SD method, and construct the SFD and BMD. Also, plot the probable deflected shape.

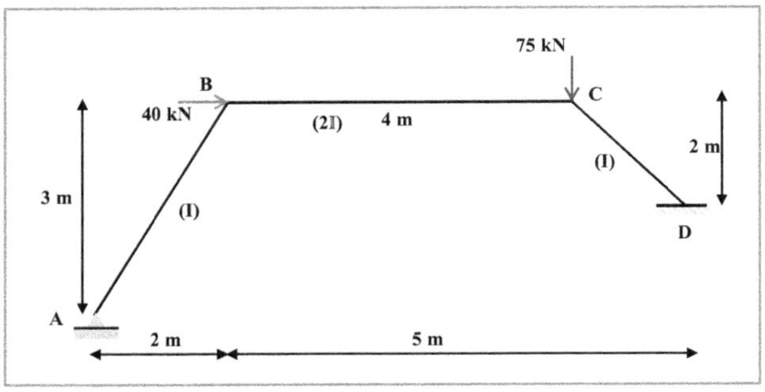

FIGURE 6.95

6.6.23 A single bay two-storey frame is shown in Figure 6.96. Analyse the frame using the SD method. Draw the SFD and BMD.

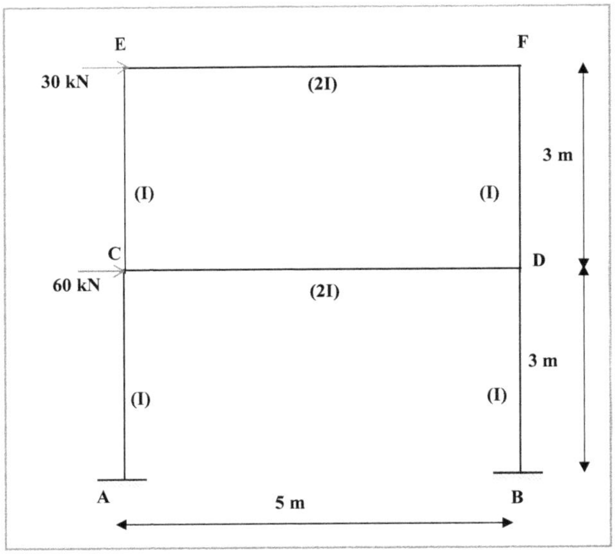

FIGURE 6.96

6.6.24 For Figure 6.96, if the supports are pinned instead of fixed, analyse the same using the SD method. Draw the SFD and BMD.

6.6.25 A frame is shown in Figure 6.97 and analysed using the SD method. Draw the SFD and BMD.

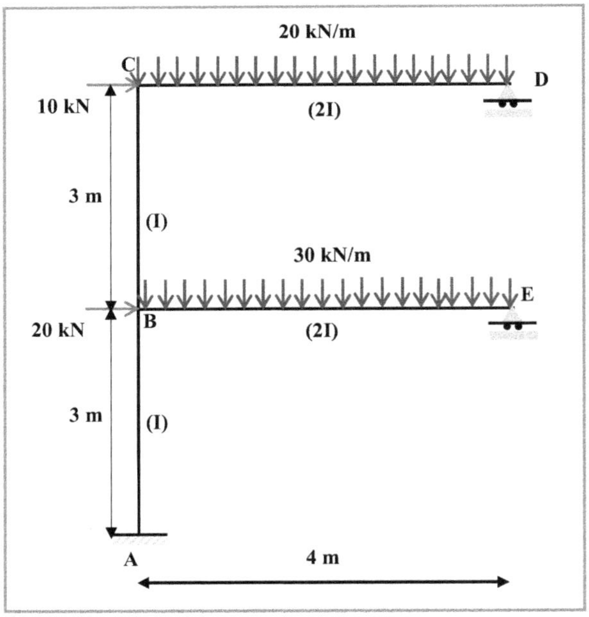

FIGURE 6.97

6.6.26 If support *A* is hinged instead of fixed for the frame shown in Figure 6.97, analyse it using the SD method. Construct the SFD and BMD.

6.6.27 Analyse the gable frame *ABCDE* shown in Figure 6.98 using the SD method. Draw the probable elastic curve, SFD, and BMD.

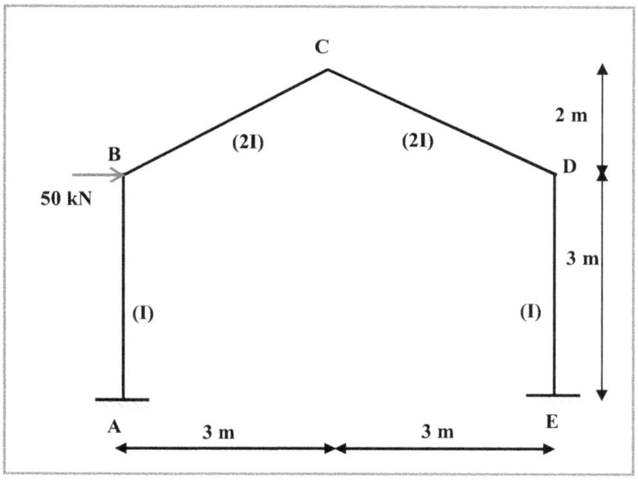

FIGURE 6.98

6.7 ANSWER KEY

Problem No.	Answer Key at Salient Locations
6.6.6	Ans: M_A=−28.87 kNm, M_B=−20.33 kNm
6.6.7	M_A=−42.30 kNm, M_B=−2.65 kNm, M_C=−20 kNm, R_A=32.93 kN
6.6.8	M_B=−23.84 kNm, M_C=−51.20 kNm, R_A=27.23 kN, R_B=55.93 kN, R_C=102.58 kN, R_D=29.26 kN
6.6.9	M_A=−32.01 kNm, M_B=−35.83 kNm, R_A=49.06 kN, R_B=75.93 kN
6.6.10	M_A=−92.58 kNm, R_B=31.48 kN, R_A=68.52 kN
6.6.11	M_{BA}=51.84 kNm, M_{BC}=−51.84, R_B=107.06 kN, R_A=43.48 kN, R_C=4.46 kN
6.6.12	M_A=−49.09 kNm, M_B=−75.91 kNm, R_A=0 kN, R_B=85.3 kN, R_C=5.34 kN
6.6.13	M_A=166.72 kNm, M_{BA}=−M_{BA}=80.84 kNm, R_A=−24.51 kN, R_B=111.46 kN, R_C=−6.95 kN
6.6.14	M_A=−97.95 kNm, M_{BA}=−M_{BC}=−21.79 kNm, R_A=64.96 kN, R_B=17.78 kN, R_D=36.55 kN
6.6.15	M_A=−14.46 kNm, M_E=0.69 kNm, R_A=14.46 kN, R_E=30.54 kN
6.6.16	R_A=9.87 kN, R_E=35.12 kN, M_{BA}=20.50 kNm, M_{BC}=−19.46 kNm, M_{BE}=−1.04 kNm
6.6.17	R_A=25.72 kN, R_D= 34.29 kN, M_A=−4.37 kNm, M_D=−29.96 kNm
6.6.18	M_A=−70.17 kNm, M_{BA}=−M_{BA}=79.29 kNm, M_D=−95.69 kNm, R_A=47.64 kN, R_B=12.36 kN
6.6.19	R_A=53.38 kN, R_D=36.61 kN, M_A =−5.0 kNm, M_D=−22.18 kNm

(*Continued*)

(*Continued*)

Problem No.	Answer Key at Salient Locations
6.6.20	M_A=0 kNm, M_{BA}=−M_{BA}=M_{BC}=−M_{CD}=42.54 kNm, M_D=0 kNm, R_A=85 kN=R_B
6.6.21	R_A=70.42 kN, R_D=49.58 kN, M_A=7.06 kNm, M_{BA}=−M_{BC}=47.72 kNm
6.6.22	R_A=−0.58 kN, R_D=75.58 kN, M_D=−1.63 kNm, M_{BA}=−M_{BC}=0.94 kNm
6.6.23	R_A=−40.83 kN, R_B=40.83 kN, M_A=−78.15 kNm, M_B=−77.70 kNm, M_{EC}=−M_{EF}=−28.67 kNm
6.6.24	R_A=−72 kN, R_B=72 kN, M_A=0 kNm, M_{EC}=−M_{EF}=−37.22 kNm
6.6.25	R_A=86.35 kN, R_E=68.17 kN, R_D=45.47 kN, M_A=−65.42 kNm, M_{CD}=−M_{CB}=37.22 kNm
6.6.26	R_A=70.00 kN, R_E=82.00 kN, R_D=47.99 kN, M_{CD}=−M_{CB}=31.98 kNm
6.6.27	M_A=−55.46 kNm, M_D=−32.15 kNm, R_A=−10.39 kN, R_E=10.39 kN

Note: The answers are provided based on the analysis using structural analysis software. The results from the exact solution may differ slightly (decimals), but these differences are generally negligible.

FURTHER READING

Bhavikatti, S. S. (2013). *Structural Analysis-II*, Vikas Publishing House, New Delhi, ISBN 978-9325968806.

Derucher, K., Putcha, C., and Kim, U. (2013). *Indeterminate Structural Analysis*, The Edwin Mellen Press, New York, ISBN 978-0773444706.

Hibbeler, R. C. (2022). *Structural Analysis*, Pearson Education (Singapore) Pte. Ltd., Delhi, ISBN 978-9354497841.

Khalfallah, S. (2018). *Structural Analysis 2: Statically Indeterminate Structures*, John Wiley & Sons, Inc, Hoboken, NJ, ISBN 978-1786303394.

Leet, K., Uang, C.-M., and Lanning, J. (2018). *Fundamentals of Structural Analysis* (6 E), McGraw Hill Education Pvt Ltd, ISBN 978-1260477245.

Menon, D. (2023). *Structural Analysis*, Narosa Publishing House Pvt Ltd., New Delhi, ISBN: 978-8184877342.

Reddy, C. S. (2011). *Basic Structural Analysis* (3 E), Tata McGraw-Hill Education Pvt Ltd, New Delhi, ISBN 978-0070702769.

Sahoo, D. R. (2023). *Indeterminate Structural Analysis*, PHI Learning Pvt Ltd, Delhi, ISBN 978-9354432958.

Shah, H. J. and Junarkar, S. B. (2015). *Mechanics of Structures*, Vol. II (24 E), Charotar Publishing House, Anand, ISBN 978-9385039027.

Thandavmoorthy, T. S. (2011). *Structural Analysis*, Oxford University Press, ISBN 978-0198069188, 978-0198069189.

Vijayanand, M., Muthu, K. U., Narendra, H., and Janardhana, M. (2019). *Indeterminate Structural Analysis*, Dreamtech Press Publisher, New Delhi, ISBN 978-9389307177.

Wang, C. K. (1953). *Statically Indeterminate Structures*, McGraw Hill Higher Education, New York, ISBN 978-0070681309.

7 Moment Distribution Method

7.1 INTRODUCTION

As discussed in the previous chapter, the slope–deflection method was the first displacement method used to analyse indeterminate beams and rigid joint plane frames. In this method, the unknown rotational and translational degrees of freedom of the structure are determined using equilibrium equations. Subsequently, the member end moments are calculated by substituting these degrees of freedom into the slope–deflection equations. This method is exact and can be efficiently programmed into an electronic computer by arranging the slope–deflection equations in matrix form, enabling the rapid and cost-effective solution of a large number of simultaneous equations.

The moment distribution method is another displacement method for analysing indeterminate beams and rigid joint frames. This method was first proposed in 1930 by Prof. Hardy Cross. Unlike other methods, it allows for the direct determination of end moments through an iterative distribution procedure without explicitly calculating the unknown joint displacements. However, once the final member end moments are determined, the unknown slopes and deflections can also be calculated, as demonstrated in some of the examples that follow.

In this method, the unknown joint rotations and translational displacements are initially restrained by treating the joints as fixed. These joints are then released sequentially, one at a time. The process of locking these joints yields locking moments accumulated at various joints and therefore needs to be balanced or released. During each release, it can be assessed how the balanced moment gets distributed to the various connecting elements based on their stiffness, and how some of the end moments are carried over to the remote ends of various connecting elements. Some of these carried-over moments may cause further imbalances at the joints and therefore need to be released again. The iterative process of locking, releasing, and distributing moments continues until the unbalanced moments dissipate or become negligible at all joints. This systematic balancing procedure gives the method its name: **moment distribution.**

The problems analysed using this method may involve one or more rotational degrees of freedom. For problems with a single rotational degree of freedom, an exact solution can be obtained in a single step. However, for problems with multiple rotational degrees of freedom, several iterative steps are required to determine the final end moments. These steps can be conveniently organised in a tabular format known as the moment distribution table.

Like the slope–deflection method, the moment distribution method is also not convenient for solving problems involving lateral sway. For sway problems, multiple moment distribution tables must be constructed, and equilibrium equations need to be explicitly solved. Unlike the slope–deflection method, the moment distribution method is not well suited for computer programming and was more widely used before the advent of digital computers. Nevertheless, this method is simple and can be easily applied to problems with one or two rotational degrees of freedom. Furthermore, this method can be applied quickly to check the moments approximately in the frames. However, for problems with multiple degrees of freedom, the iterative process of locking joints, balancing, and distributing moments may not yield exact results.

The analysis of problems involving single degrees of freedom, multiple degrees of freedom, and lateral sway is discussed in the subsequent sections.

7.2 SIGN CONVENTION

The same sign convention is used in the moment distribution method as that considered in the slope–deflection method: clockwise moments are considered positive, while counterclockwise moments are considered negative.

7.3 FLEXURAL (ROTATIONAL) STIFFNESS OF A BEAM ELEMENT

The flexural stiffness or stiffness factor of a member is defined as the moment required to produce a unit rotation at one end. In this section, the flexural stiffness of a beam element with different end conditions is derived.

7.3.1 Case 1: When the Far End Is Fixed and the Near End Is Subjected to a Clockwise Moment, M_O

Let us consider the prismatic beam OA, where the far end A is fixed, and the near end O is subjected to a clockwise moment M_O, as shown in Figure 7.1a. This moment causes the support O to rotate by an angle θ_O in the clockwise direction. Let the moment at A be M_A. The corresponding conjugate beam is shown in Figure 7.1b.

Using the conjugate beam, the rotation 'θ_O' can be calculated by determining the shear force in the conjugate beam shown in Figure 7.1b. Alternatively, by applying the principle of superposition, the scenario in Figure 7.1a can be split into two simpler cases, as illustrated in Figure 7.1c and d. The compatibility condition at A can then be written as:

$$\theta_A = 0 \Rightarrow \theta_{A1} - \theta_{A2} = 0$$

DOI: 10.1201/9781003562955-7

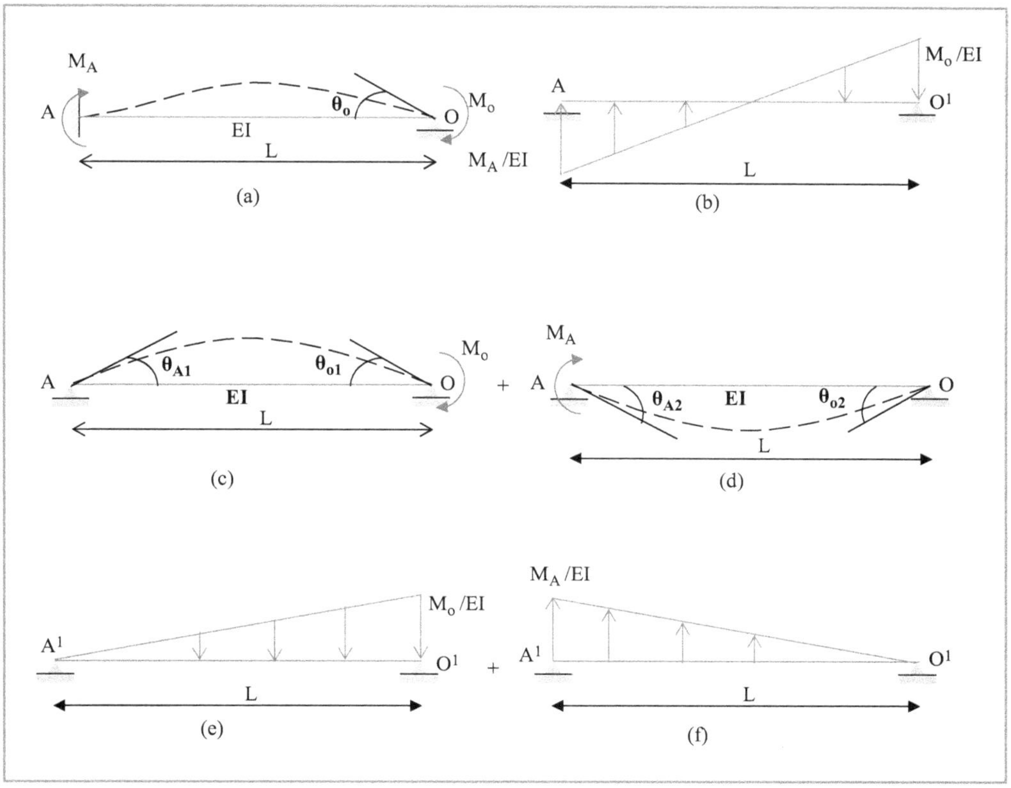

FIGURE 7.1 Analysis of beam when the far end is fixed and the near end is subjected to moment M_O. (a) Beam with deflected shape, (b) conjugate beam, (c) Beam with end moment M_O at O, (d) Beam with end moment MA at A, (e) conjugate beam for (c), and (f) conjugate beam for (d).

Using the conjugate beam, $\theta_A = V_A^1$

$$\theta_A = \frac{\frac{1}{2}\times\frac{M_O}{EI}\times L\times\frac{1}{3}L}{L} - \frac{\frac{1}{2}\times\frac{M_A}{EI}\times L\times\frac{2}{3}L}{L} \qquad (7.1)$$

$$= 0 \Rightarrow M_A = \frac{M_O}{2}$$

Now calculate θ_O using the conjugate beam (Figure 7.1e and f),

$$\theta_O = V_{O^1} = \frac{\frac{1}{2}\times\frac{M_O}{EI}\times L\times\frac{2}{3}L}{L} - \frac{\frac{1}{2}\times\frac{M_A}{EI}\times L\times\frac{1}{3}L}{L}$$

$$= \frac{2M_OL}{6EI} - \frac{M_AL}{6EI}$$

$$\Rightarrow \frac{2M_OL}{6EI} - \frac{M_OL}{12EI} = \frac{M_OL}{4EI}$$

$$(\text{or})\ \frac{M_O}{\theta_O} = \frac{4EI}{L}$$

Therefore, according to the definition, the flexural stiffness or stiffness factor

$$k = \frac{4EI}{L} \qquad (7.2)$$

From equation 7.1 and Figure 7.1b, it is evident that the contraflexure point occurs exactly at one-third of the distance from A.

Furthermore, from equation 7.1, the moment at the far end A is given by $M_A = M_O/2$. This is referred to as the carry-over moment (**COM**), and the ratio of the COM at the far end A to the applied moment at the near end O is called the **carry-over factor** (**COF**). In this case, the COF=+1/2.

7.3.2 Case 2: When Far End is Pinned (Simply Supported) and the Near End is Subjected to a Moment, M_O

Consider a prismatic beam with constant flexural rigidity 'EI', as shown in Figure 7.2, where the far end A is pinned and the near end O is subjected to a clockwise moment 'M_O'. This moment causes the support to rotate 'θ_O' in the clockwise direction, as illustrated in the figure. We will now relate 'M_O' and 'θ_O' using the conjugate beam method.

The rotation 'θ_O' at O in the real beam is equal to the shear force at O^1 in the conjugate beam, i.e.,

$$\theta_O = V_{O^1} = \frac{\frac{1}{2}\times\frac{M_O}{EI}\times L\times\frac{2}{3}L}{L} = \frac{M_OL}{3EI} \Rightarrow \frac{M_O}{\theta_O} = \frac{3EI}{L}$$

Therefore, the flexural stiffness or stiffness factor,

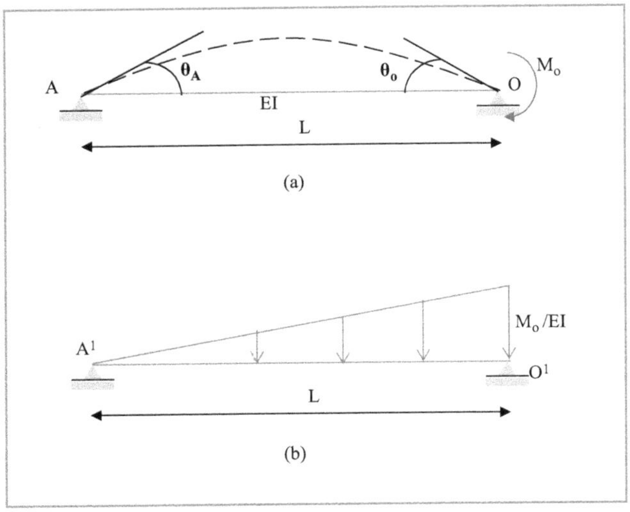

FIGURE 7.2 Far end is hinged.

$$k = \frac{M_O}{\theta_O} = \frac{3EI}{L} \quad (7.3)$$

Similarly, the rotation at A in the real beam is equal to the shear force at A^1 in the conjugate beam, i.e.,

$$\theta_A = V_{A^1} = \frac{\frac{1}{2} \times \frac{M_O}{EI} \times L \times \frac{1}{3} L}{L} = \frac{M_O L}{6EI} = \frac{\theta_O}{2}$$

Since the far end is simply supported, there is no COM. Hence, the COF is zero in this case.

From equations 7.2 and 7.3, it is evident that when the far end is fixed, the rotation produced at O is equal to ¾ of the rotation caused by a simply supported beam subjected to the same moment M_O, at O. Consequently, the flexural stiffness of this element is higher by a factor of 4/3.

7.3.3 Case 3: When the Far End is Sliding/Guided Support

Let us consider a prismatic beam OA, whose far end 'A' is a guided support, and the near end O is subjected to a clockwise moment M_O, as shown in Figure 7.3a. This moment causes the support O to rotate by an angle θ_O in the clockwise direction. The deflected shape of the beam clearly suggests behaviour resembling that of cantilever action, as shown in Figure 7.3c.

The key difference between the two beams lies in the fact that, in the given beam, the fixed end A is free to translate vertically while the other end O is restrained against translation. In contrast, the reverse is true for the cantilever beam, as illustrated.

From the moment equilibrium equation about O, $M_A = M_O$.

Using the conjugate beam, we can calculate θ_O, as shown below.

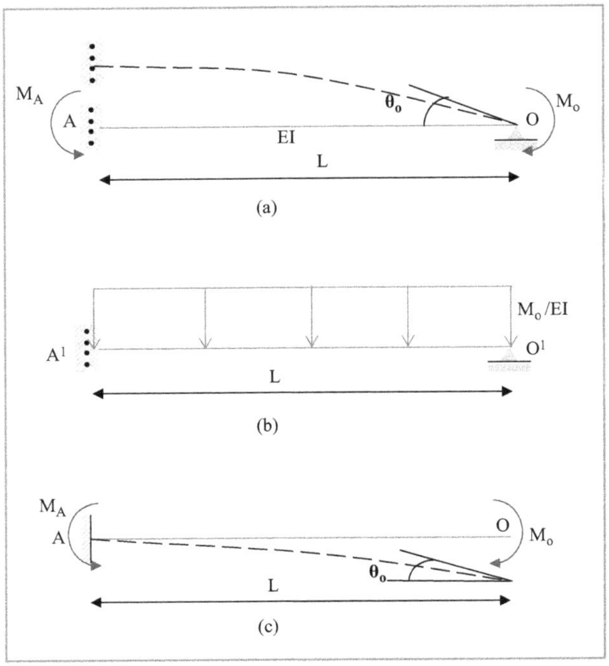

FIGURE 7.3 (a) Beam with deflected shape, (b) conjugate beam, and (c) cantilever beam with deflected shape.

$$\theta_O = V_{O^1} = \frac{M_O}{EI} \times L = \frac{M_O L}{EI} \Rightarrow \frac{M_O}{\theta_O} = \frac{EI}{L}$$

Therefore, flexural stiffness,

$$k = \frac{M_O}{\theta_O} = \frac{EI}{L} \quad (7.4)$$

By comparing equations 7.2 and 7.4, it can be observed that the flexural stiffness of the beam with the far end fixed is four times greater than that of the beam with the far end as a guided support when subjected to the same moment M_O at O. Furthermore, the COM $M_A = M_O$, but it acts in the opposite direction. Therefore, the COF in this case is −1.

7.4 JOINT STIFFNESS FACTOR AND DISTRIBUTION FACTOR

Let us consider a concentrated moment M_O acting at a joint O, as shown in Figure 7.4. Assume that the members OA, OB, OC and OD meet at joint O, with stiffness values K_{OA}, K_{OB}, K_{OC}, and K_{OD}, respectively. According to the principle of superposition, the total stiffness of joint O is the sum of the stiffnesses of the members meeting at joint O.

This total stiffness is given by $K_O = \sum K = K_{OA} + K_{OB} + K_{OC} + K_{OD}$, and it is referred to as the **joint stiffness** or **joint stiffness factor**. This

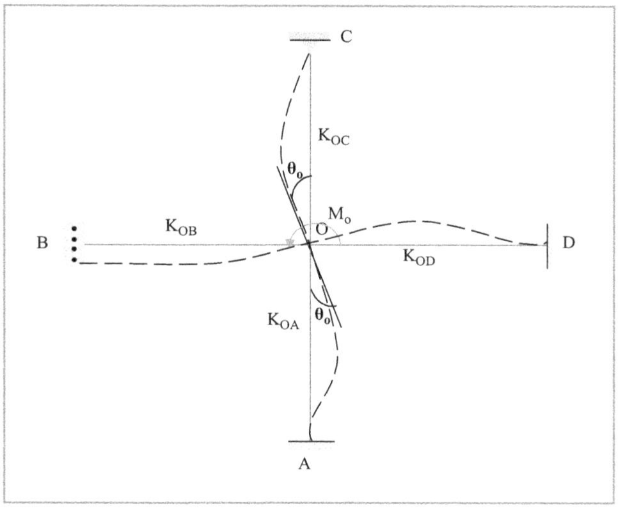

FIGURE 7.4 Frame subjected to moment at O and its deflected shape.

represents the amount of moment required to rotate joint O through a unit angle of rotation. As shown in the figure, the moment M_O causes joint O to rotate by angle θ_O in the direction of the applied moment. The moment M_O is distributed among the connecting members (OA, OB, OC and OD) in proportion to their relative stiffnesses.

The moment M_O distributed to the connecting members OA, OB, OC, and OD is given by

$$M_{OA} = K_{OA} \times \theta_O; \quad M_{OB} = K_{OB} \times \theta_O;$$

$$M_{OC} = K_{OC} \times \theta_O; \quad \text{and} \quad M_{OD} = K_{OD} \times \theta_O$$

$$\text{For equilibrium,} \quad M_O = M_{OA} + M_{OB} + M_{OC} + M_{OD}$$
$$= K_{OA} \times \theta_O + K_{OB} \times \theta_O + K_{OC} \times \theta_O + K_{OD} \times \theta_O$$

In general, if there are 'n' members meeting at joint O, then we can write

$$M_O = M_{O1} + M_{O2} + \cdots + M_{Oi} + \cdots + M_{On}$$
$$= K_{O1} \times \theta_O + K_{O2} \times \theta_O + \cdots + K_{Oi} \times \theta_O + \cdots + K_{On} \times \theta_O$$
$$= \theta_O \times \sum_{i=1}^{n} K_{Oi}$$

Therefore, in general, the moment distributed to the ith member meeting at joint O, is given by

$$M_{Oi} = \frac{K_{Oi}}{\sum_{i=1}^{n} K_{Oi}} \times M_O \Rightarrow \frac{M_{Oi}}{M_O} = \frac{K_{Oi}}{\sum_{i=1}^{n} K_{Oi}} = d_{Oi}, \text{ where}$$

d_{Oi} is the measure of relative stiffness, called the **distribution factor.** It is the fraction of the total unbalanced moment at the joint that is distributed to the member ends meeting at that joint.

$$\therefore \text{The distribution factor } (d_{Oi}) = \frac{K_{Oi}}{\sum_{i=1}^{n} K_{Oi}} \tag{7.5}$$

where K_{Oi} is the stiffness factor of ith member and $\sum_{i=1}^{n} K_{Oi}$ is the summation of stiffness factors of members meeting at joint O.

7.5 CHECK ON THE MOMENT DISTRIBUTION

Based on the relative stiffness and fixed-end moments, a check can be performed to ensure that the correct final end moments have been obtained in the moment distribution table. As a primary check, one can verify whether the end moments at the interior supports have been balanced and whether the end moments at the exterior hinged or roller supports have been reduced to zero. This verification is based on the condition of statics. Additionally, another check is based on the condition of geometry, which involves determining the relative values of rotations at each joint. We can recall that the slope–deflection equations (without sway) for end moments can be expressed in terms of fixed-end moments and end rotations as follows.

$$M_{AB} = MF_{AB} + \frac{2EI}{L}(2\theta_A + \theta_B) = MF_{AB} + K(2\theta_A + \theta_B) \tag{7.6}$$

$$M_{BA} = MF_{BA} + \frac{2EI}{L}(2\theta_B + \theta_A) = MF_{BA} + K(2\theta_B + \theta_A) \tag{7.7}$$

where K=relative stiffness=$2EI/L$

From equation 7.6, $\theta_B = \frac{(M_{AB} - MF_{AB})}{K} - 2\theta_A$

substitute θ_B into equation 7.7,

$$M_{BA} = MF_{BA} + K\left[\frac{2(M_{AB} - MF_{AB})}{K} - 4\theta_A + \theta_A\right]$$

From the above equations, the end rotations can be expressed in terms of the end moments and the fixed-end moments as

$$(\theta_A)_{rel} = \frac{(M_{AB} - MF_{AB}) - \frac{1}{2}(M_{BA} - MF_{BA})}{1.5K} \tag{7.8}$$

$$(\theta_B)_{rel} = \frac{(M_{BA} - MF_{BA}) - \frac{1}{2}(M_{AB} - MF_{AB})}{1.5K} \tag{7.9}$$

Since the above equations express the relative values of the rotations θ_A and θ_B, the pure number 1.5 in the denominator can be neglected. Therefore,

$$(\theta_A)_{rel} = \frac{(M_{AB} - MF_{AB}) - \frac{1}{2}(M_{BA} - MF_{BA})}{K} \quad (7.10)$$

$$(\theta_B)_{rel} = \frac{(M_{BA} - MF_{BA}) - \frac{1}{2}(M_{AB} - MF_{AB})}{K} \quad (7.11)$$

The terms $(M_{AB} - MF_{AB})$ and $(M_{BA} - MF_{BA})$ in the above equations can be referred to as the change in the moments from the fixed-end moment to the final end moments. Thus, the relative rotation of any one end of a member is simply the change in the moment at the near end minus half of the change in the moment at the far end divided by K. That is

$$\theta_{\text{Near end}} = \frac{(\text{Change in moment})_{\text{Near end}} + \frac{-1}{2}(\text{Change in moment})_{\text{Far end}}}{K} \quad (7.12)$$

Using the above equation, the relative values of the rotations at each end of a member can be determined. The check on the condition of geometry is that the rotations at the ends of all the members meeting at a joint should be equal, and the rotation at a fixed support should be zero.

It is important to note that this geometry check is not related to verifying the correctness of the stiffness factors and fixed-end moments used at the beginning of the moment distribution table. In fact, the geometry condition may still be satisfied even if there are errors in the stiffness factors or fixed-end moments.

The best way to understand the successive locking and unlocking of joints, along with the subsequent balancing and distribution of moments and the carry-over of moments to the far ends in the moment distribution method, is by solving examples. In the first example, the process of distributing the moment at a joint to its connecting members is explained. In the second example, a more detailed explanation is provided, covering the successive locking and unlocking of joints, balancing and distributing moments, and accounting for COMs.

7.6 NUMERICAL PROBLEMS

Example 7.1

Analyse the simple frame subjected to 100 kNm in an anti-clockwise direction at a joint O, as shown in Figure 7.5, using the moment distribution method. Assume L and EI are constant for all the members.

Solution:

The behaviour of the individual members is shown in Figure 7.6.

Before starting, let us first calculate the stiffness factors for each span (member) based on their end conditions, using equations 7.1–7.4, and determine the distribution factors using equation 7.5. It is important to note that the sum of the distribution factors at each joint must be equal to unity, except at a fixed support, where the distribution factor is zero because it requires an infinite moment to produce a unit rotation.

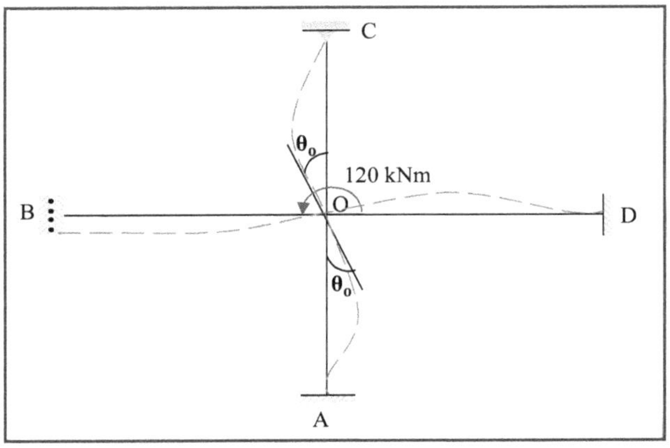

FIGURE 7.5 Frame subjected to moment at O and its deflected shape.

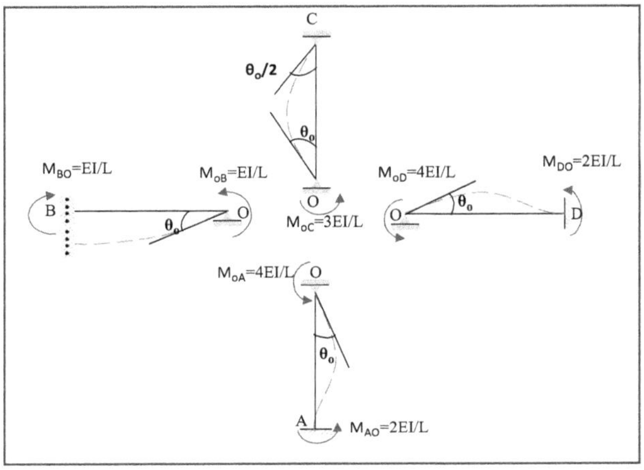

FIGURE 7.6 Individual element behaviour.

STIFFNESS FACTORS

$$K_{OA} = \frac{4EI}{L} = K_{OD}; \quad K_{OB} = \frac{EI}{L}; \quad \text{and} \quad K_{OC} = \frac{3EI}{L}$$

$$\sum_{1}^{4} K_{Oi} = \frac{4EI}{L} + \frac{EI}{L} + \frac{3EI}{L} + \frac{4EI}{L} = \frac{12EI}{L}$$

DISTRIBUTION FACTORS

$$(DF)_{OA} = (DF)_{OD} = \frac{K_{OA}}{\sum_{1}^{4} K_{Oi}} = \frac{4}{12}; \quad (DF)_{OB} = \frac{K_{OB}}{\sum_{1}^{4} K_{Oi}} = \frac{1}{12};$$

$$(DF)_{OC} = \frac{K_{OC}}{\sum_{1}^{4} K_{Oi}} = \frac{3}{12}$$

(Check: At O, sum of DFs = 4/12 + 4/12 + 1/12 + 3/12 = 1)

FIXED-END MOMENTS

In the second step, the fixed-end moments for each span are calculated. In this case, since no loads are acting on the members, the fixed-end moments (FEMs) are zero. However, there is a moment of 120 kNm acting at joint O in the counter-clockwise direction, which must be distributed among the connected members according to their distribution factors.

MOMENT DISTRIBUTION FOR EACH MEMBER

$$M_{OA} = (DF)_{OA} \times M_O = \frac{4}{12} \times (-120) = -40 \text{ kNm}$$

$$M_{OB} = (DF)_{OB} \times M_O = \frac{1}{12} \times (-120) = -10 \text{ kNm} = -M_{BO}$$

$$M_{OC} = (DF)_{OC} \times M_O = \frac{3}{12} \times (-120) = -30 \text{ kNm}$$

$$M_{OD} = (DF)_{OD} \times M_O = \frac{4}{12} \times (-120) = -40 \text{ kNm}$$

$$M_{DO} = \frac{M_{OD}}{2} = -20 \text{ kNm}; \quad M_{AO} = \frac{M_{OA}}{2} = -20 \text{ kNm}$$

The bending moment diagram is shown in Figure 7.7.

7.6.1 Application to Beams

7.6.1.1 General Procedure

The analysis of indeterminate beams using the moment distribution method is outlined below.

i. **Fix all the joints** and calculate the fixed-end moments due to the external loads for each span using the expressions given in Table 6.1. If the far end is a hinged or roller support, calculate the modified fixed-end moment, as the moment at the hinge is zero.

ii. **Determine the stiffness factors** using the appropriate equations (equations 7.2–7.4) for each span. If the far end is fixed, use $K = 4EI/L$, if the far end is hinged, use $K = 3EI/L$, if the far end is guided support, use $K = EI/L$, and use $K = EI/(L/2) = 2EI/L$ when the span is symmetrical.

iii. After calculating the stiffness factors, find the **distribution factors** for each joint, using equation 7.5. Note that the distribution factor for a hinge or roller support is 1, while for a fixed support, it is zero.

iv. Note the **COFs for each joint**: For fixed supports and interior supports, the COF is (1/2); for guided support, it is (−1); and for hinged support, it is zero.

v. **Calculate the unbalanced moment** at each joint and distribute the balanced moment (−ve sign of unbalanced moment) to the connecting members according to their distribution factors. Then apply the COMs to the far ends of the connecting members using the COFs noted in step (iv).

vi. **Repeat step (v)** until the COMs become negligible. Once this occurs, stop the iterative process and calculate the final member end moments at all the joints by summing all the locking and unlocking moments from each cycle.

vii. **Apply the geometry check** using equation 7.12.

viii. Once the final member end moments are obtained, determine the other internal forces and support reactions using free body diagrams and the static equilibrium equations.

ix. Finally, sketch the shear force and bending moment diagrams based on the calculated results.

The above procedure is demonstrated in the following examples in a more convenient table format.

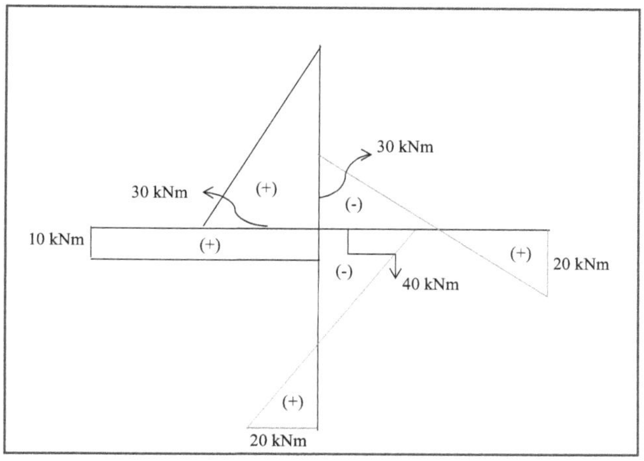

FIGURE 7.7 Bending moment diagram.

Example 7.2

Analyse the two-span continuous beam shown in Figure 7.8, using the moment distribution method. Assume constant flexural rigidity *EI*. (The same problem has been solved in Example 6.1 using the slope–deflection method.)

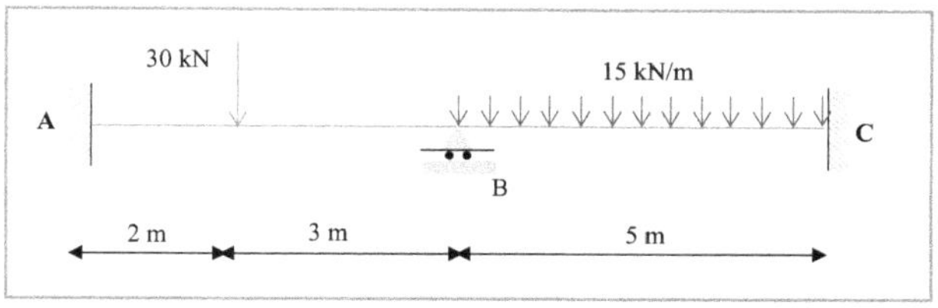

FIGURE 7.8 Two-span continuous beam.

Solution:

FIXED-END MOMENTS

Let us first calculate the locking (fixed-end) moments by treating all the joints as fixed using the expressions given in Table 6.1, as shown in Figure 7.9. Assume the clockwise moments are +ve.

$$MF_{AB} = -\frac{Wab^2}{L^2} = -\frac{30 \times 2 \times 3^2}{5^2} = -21.66 \text{ kNm}$$

$$MF_{BA} = \frac{Wba^2}{L^2} = \frac{30 \times 3 \times 2^2}{5^2} = 14.40 \text{ kNm}$$

$$MF_{BC} = -\frac{wL^2}{12} = -\frac{15 \times 5^2}{12} = -31.25 \text{ kNm}$$

$$MF_{CB} = \frac{wL^2}{12} = \frac{15 \times 5^2}{12} = 31.25 \text{ kNm}$$

STIFFNESS FACTORS

In the second step, calculate the stiffness factors for each span. In this case, the far ends of the members *BA* and *BC* are fixed. Therefore, the stiffness factors using equation 7.2 are

$$K_{BA} = \frac{4EI}{L} = \frac{4EI}{5}; \quad K_{BC} = \frac{4EI}{L} = \frac{4EI}{5}$$

DISTRIBUTION FACTORS

Using the stiffness factors of the members calculated above, determine the distribution factors of each joint using equation 7.5.

At joints *A* and *C*

Since *A* and *C* are fixed ends, they would theoretically require an infinite moment to rotate by unit radian. Hence, the stiffness of the fixed end can be considered infinite.

$$(DF)_{AB} = \frac{K_{AB}}{\infty + K_{AB}} = \frac{\frac{4EI}{L}}{\infty + \frac{4EI}{L}} = 0 \quad \text{and}$$

$$(DF)_{CB} = \frac{K_{BC}}{K_{BC} + \infty} = \frac{\frac{4EI}{L}}{\infty + \frac{4EI}{L}} = 0$$

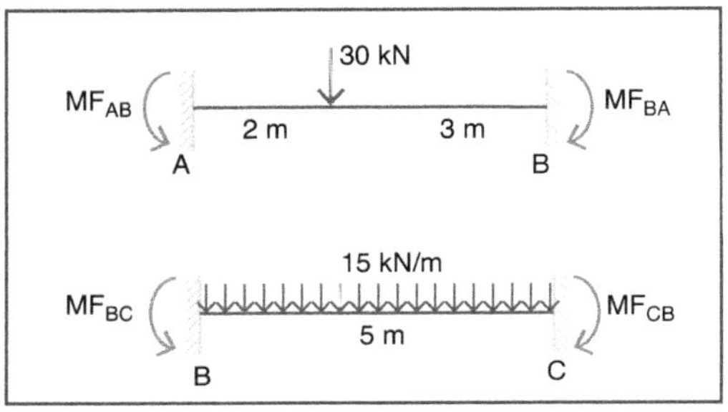

FIGURE 7.9 Fixed-end moments.

In other words, a fixed end can resist any amount of moment assigned to it and, therefore does not need to be released. In such cases, distribution factors do not apply.

At joint *B*

$$(DF)_{BA} = \frac{K_{BA}}{K_{BA} + K_{BC}} = \frac{\frac{4EI}{5}}{\frac{4EI}{5} + \frac{4EI}{5}} = 0.5$$

$$(DF)_{BC} = \frac{K_{BC}}{K_{BA} + K_{BC}} = \frac{\frac{4EI}{5}}{\frac{4EI}{5} + \frac{4EI}{5}} = 0.5$$

$$(DF)_{BA} + (DF)_{BC} = 0.5 + 0.5 = 1.0$$

Note: The sum of the distribution factors of various members meeting at a joint is always unity.

MOMENT DISTRIBUTION TABLE

After calculating the fixed-end moments and distribution factors, the iterative procedure of successive locking and unlocking of each joint, as well as the distribution and COMs, can be systematically presented in a tabular form called the moment distribution table. First, the members corresponding to each joint, along with their respective distribution factors and COFs, are entered into the moment distribution table. In Cycle 1, the locking moments or unbalanced moments are the fixed-end moments (FEMs), which are entered into the table. Since joints *A* and *C* are fixed, no balancing is required, so the balancing moment at these joints is zero. Next, joint *B* is released and balanced. At joint *B*, there is a clockwise moment of 14.40 kNm and an anti-clockwise moment of 31.250 kNm, resulting in an unbalanced moment of −16.85 kNm (anti-clockwise). Therefore, the balancing moment is +16.85 kNm (clockwise), which is distributed to the members *BA* and *BC* according to their distribution factors. The balancing moment for *BA* is calculated as 0.5×16.85=8.425 kNm, and for *BC*, it is also 0.5×16.85=8.425 kNm. Note that in the moment distribution table, the sum of the balancing moments of all the members meeting at a joint in each cycle must equal the unbalanced moment at that joint. This ensures that the sum of the end moments of all members meeting at a joint is zero. Next, the COMs are calculated as (+1/2) times the balancing moment at the far end of the respective members from the preceding row of the table. These COMs become the new unbalanced moments and are balanced in a similar way as described above. However, in this case, the COMs are at joints *A* and *C*, which are fixed. As a result, no further balancing is required at these joints.

Also, there is no unbalanced moment at joint *B*. Hence, no further moment distribution is required. This process is referred to as a single cycle moment distribution. The final end moments are then obtained by summing the moments in the respective columns (FEM+Distributed moment+COM). The entire process is illustrated in Table 7.1. The final end moments

TABLE 7.1
Moment Distribution Table

Cycle	Joint	*A*	*B*		*C*
	Member	*AB*	*BA*	*BC*	*CB*
	DF	0.000	0.500	0.500	0.000
	COF	0.500	0.000	0.000	0.500
	FEMs	−21.600	14.400	−31.250	31.250
1	Balance and distribution	0.000	8.425	8.425	0.000
	COM	4.213			4.213
	Final end moments (kNm)	−17.388	22.825	−22.825	35.463
	Check:	4.212	8.425	8.425	4.213
	Change				
	−1/2 (change)	−4.212	−2.106	−2.206	−4.212
	Sum	0.000	6.319	6.319	0.000
	$\theta_{rel} = \frac{\text{Sum}}{(-K)}$	0.000	10.531/*EI*	10.531/*EI*	0.000
		Check θ_A	Check θ_B		Check θ_C

obtained from Table 7.1 using the moment distribution method fairly match the results obtained from the slope–deflection method in Example 6.1.

The check on moment distribution is also carried out using equations7.8 and 7.9 or 7.12 in the table. It was found that the joint equilibrium is satisfied. Furthermore, the geometry check, i.e., the rotations obtained at fixed ends are zero, and the rotation values at joint *B* are also satisfied.

FINAL END SHEARS

After obtaining the final member end moments from Table 7.1, the member end shears are calculated using the static equilibrium equations, as shown in the free body diagram in Figure 7.10a. The shear force and bending moment diagrams are shown in Figure 7.10b and c, respectively.

Position of zero shear in span *BC* (Figure 7.10b):

$$\frac{x}{40.02} = \frac{5-x}{34.973} \Rightarrow x = 2.67 \text{ m}$$

$$\therefore \text{Maximum +ve BM} = 40.02 \times 2.67 - 35.462 - 15 \times 2.67 \times \frac{2.67}{2}$$

$$= 17.925 \text{ kNm}$$

Maximum+ve BM in span *AB*

$$M_{\max} = 16.925 \times 2 - 17.448 = 16.402 \text{ kNm}$$

Position of contraflexure point in span *BC*: To get the position of the contraflexure point, equate the moment to zero in span *BC*. From Figure 7.10d

$$M_x = 40.02x - 15\frac{x^2}{2} - 35.462$$

$$= 0 \Rightarrow -80.04x + 15x^2 + 70.924 = 0$$

$$\therefore x = 1.12 \text{ m and } 4.21 \text{ m}$$

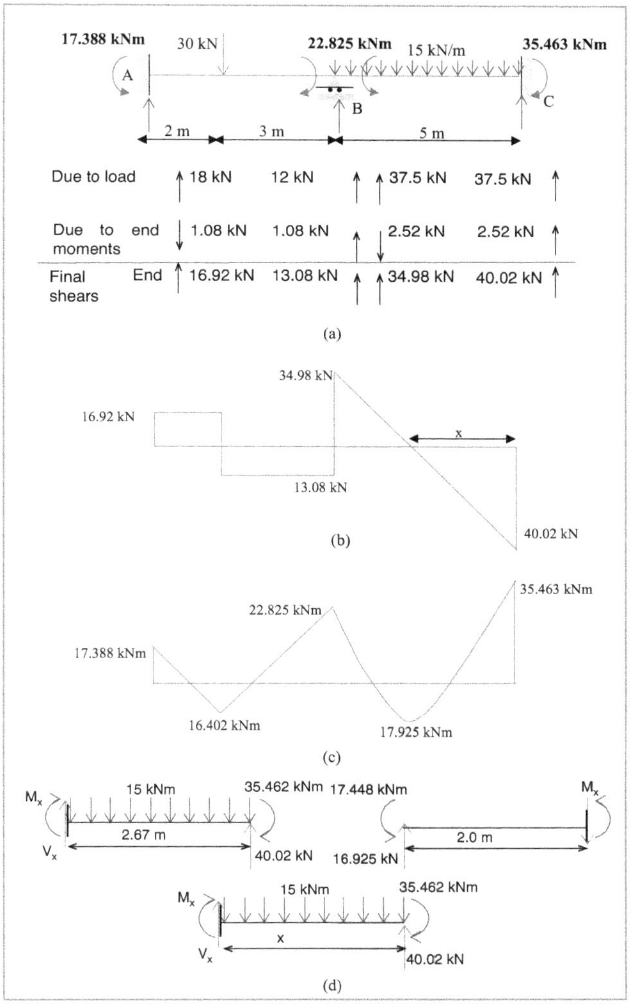

FIGURE 7.10 (a) Free body diagram, (b) shear force diagram, (c) bending moment diagram, and (d) free body diagrams of beam segments.

Example 7.3

Re-analyse Example 7.2 by treating end A as a hinged support instead of a fixed support, as shown in Figure 7.11. The rest of the data remains unaltered. (The same is solved in Example 6.2 using the slope–deflection method.)

Solution:

The problem is analysed without taking advantage of the known moment at hinged support A ($M_A=0$), i.e., without modifying the stiffness factor and fixed-end moments for span AB (A is pinned).

FIXED-END MOMENTS

Using Table 6.1, the fixed moments can be calculated as follows:

$$MF_{AB} = -\frac{Wab^2}{L^2} = -\frac{30\times2\times3^2}{5^2} = -21.66 \text{ kNm}$$

$$MF_{BA} = \frac{Wba^2}{L^2} = \frac{30\times3\times2^2}{5^2} = 14.40 \text{ kNm}$$

$$MF_{BC} = -\frac{wL^2}{12} = -\frac{15\times5^2}{12} = -31.25 \text{ kNm}$$

$$MF_{CB} = \frac{wL^2}{12} = \frac{15\times5^2}{12} = 31.25 \text{ kNm}$$

STIFFNESS FACTORS

Using equation7.2, the stiffness factors of the members BA and BC are calculated as shown below:

$$K_{BA} = \frac{4EI}{L} = \frac{4EI}{5};\quad K_{BC} = \frac{4EI}{L} = \frac{4EI}{5}$$

DISTRIBUTION FACTORS

The distribution factors for all the joints can be calculated using equation7.5.

At joint *A*: In this case, only one member meets at joint A, which is a hinged support. The distribution factor for AB is

$$(DF)_{AB} = \frac{\frac{4EI}{L}}{\frac{4EI}{L}} = 1$$

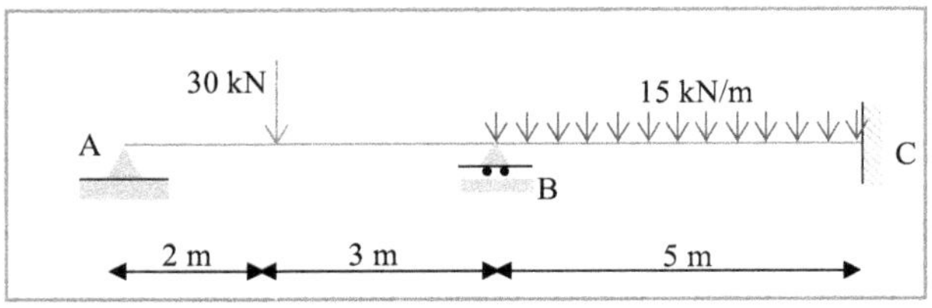

FIGURE 7.11 Two-span continuous beam.

The distribution factors for the other members are the same as those used in the previous Example 7.2.

At joint B

$$(DF)_{BA} = \frac{K_{BA}}{K_{BA}+K_{BC}} = \frac{\frac{4EI}{5}}{\frac{4EI}{5}+\frac{4EI}{5}} = 0.5$$

$$(DF)_{BC} = \frac{K_{BC}}{K_{BA}+K_{BC}} = \frac{\frac{4EI}{5}}{\frac{4EI}{5}+\frac{4EI}{5}} = 0.5$$

$$(DF)_{BA} + (DF)_{BC} = 0.5+0.5 = 1.0$$

MOMENT DISTRIBUTION CYCLE

The iterative procedure of successive locking and unlocking of each joint, along with the distribution and COMs, is presented in Table 7.2. First, the members corresponding to each joint are entered into the moment distribution table, followed by the corresponding distribution factors (DF) and CoFs in the subsequent lines. In cycle 1, the locking moments or unbalanced moments are the fixed-end moments (FEMs), which are then entered in the table below the carry-over factors (CoF). Since joint C is fixed, no balancing is required, so the balancing moment at joint C is zero. Next, joint A is released and balanced. At joint A, an anti-clockwise moment of −21.60 kNm acts as the unbalanced moment, which needs to be balanced. Therefore, the balancing moment is +21.60 kNm (clockwise). At joint B, the unbalanced moment is calculated as +14.40−31.25=− 16.85 kNm (anti-clockwise). The balancing moments for the members connected to joint B are distributed as follows:

For BA: 0.5×(+16.85)=+8.425 kNm
For BC: 0.5×(+16.85)=+ 8.425 kNm.

The COMs are then calculated as (+1/2) times the balancing moment at the far end of the respective members from the preceding line. These COMs become the new unbalanced moments, which are balanced in the same way as described in the previous cycle. This process is repeated iteratively. In this problem, it is noted that by the sixth cycle, the COMs become negligible. At this point, the balancing process can be stopped, and no further moment distribution cycles are required. The final end moments for the members are obtained by summing the moments in the respective columns (FEM+Bal and Distributed moments+COM). The entire procedure is detailed in Table 7.2. A geometry check is also performed in the table, verifying and satisfying the rotation values at the joints and the fixed end.

ALTERNATIVE SOLUTION USING THE MODIFIED STIFFNESS FACTOR

Since end A is pinned, the moment at end A is zero. Therefore, the above problem can be simplified by applying the modified stiffness and modified fixed-end moment for span AB, as demonstrated below.

TABLE 7.2
Moment Distribution Table

Cycle	Joint	A	B		C
	Member	*AB*	*BA*	*BC*	*CB*
	DF	1.000	0.500	0.500	0.000
	CoF	0.500	0.500	0.500	0.500
1	FEM	−21.600	14.400	−31.250	31.250
	Balance and distribution	21.600	8.425	8.425	
2	COM	4.213	10.800		4.213
	Balance and distribution	−4.213	−5.400	−5.400	
3	COM	−2.700	−2.106		−2.700
	Balance and distribution	2.700	1.053	1.053	
4	COM	0.527	1.350		0.527
	Balance and distribution	−0.527	−0.675	−0.675	
5	COM	−0.338	−0.263		−0.338
	Balance and distribution	0.338	0.132	0.132	
6	COM	0.066	0.169		0.066
	Balance and distribution	−0.066	−0.084	−0.084	
	Final end moments (kNm)	0.000	27.801	−27.80	33.018
	Check: Change	21.600	13.401	3.449	1.768
	−1/2 (change)	−6.700	−10.800	−0.884	−1.725
	Sum	14.900	2.601	2.565	0.043
	$\theta_{rel} = \frac{\text{Sum}}{(-K)}$	37.25/*EI*	6.502/*EI*	6.412/*EI*	0.107/*EI*
		Check θ_A	Check (θ_B)		Check θ_C

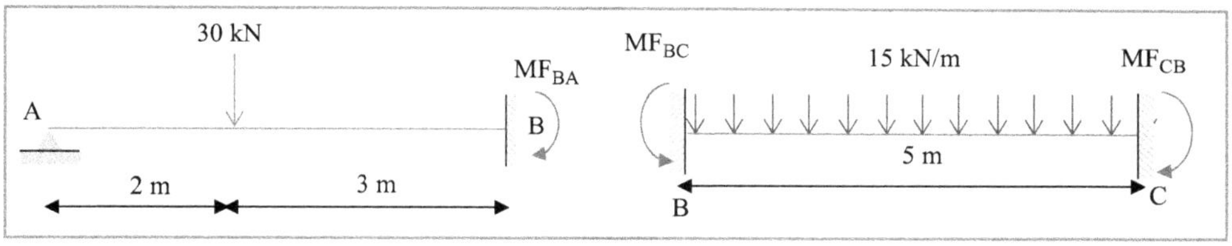

FIGURE 7.12 Fixed-end moments.

FIXED-END MOMENTS

Using Table 6.1 and the free body diagrams shown in Figure 7.12, the fixed-end moments are calculated as shown below.

$$MF_{BA} = \frac{30}{5^2}\left(2^2 \times 3 + 2 \times \frac{3^2}{2}\right) = 25.2 \text{ kNm}$$

$$MF_{BC} = -\frac{wL^2}{12} = -\frac{15 \times 5^2}{12} = -31.25 \text{ kNm} = -MF_{CB}$$

Modified stiffness factor for *AB*: Since the far end *A* is hinged,

$$K_{AB} = K_{BA} = \frac{3EI}{L} = \frac{3EI}{5}$$

Stiffness Factor for *BC*

$$K_{BC} = \frac{4EI}{L} = \frac{4EI}{5}$$

Distribution Factors

At joint *A*: Since *A* is a hinged support, the $(DF)_{AB}=1$

At joint B:

$$(DF)_{BA} = \frac{K_{BA}}{K_{BA}+K_{BC}} = \frac{\frac{3EI}{5}}{\frac{3EI}{5}+\frac{4EI}{5}} = 0.429$$

$$(DF)_{BC} = \frac{K_{BC}}{K_{BA}+K_{BC}} = \frac{\frac{4EI}{5}}{\frac{3EI}{5}+\frac{4EI}{5}} = 0.571$$

MOMENT DISTRIBUTION CYCLE

The procedure of successive locking and unlocking of each joint, along with the distribution and COMs, is presented systematically in Table 7.3.

FINAL END SHEARS

After knowing the final end moments from Table 7.3, the final end shears are calculated using the static equilibrium equations with the help of the free body diagram, as shown in Figure 7.13a. The shear force and bending moment diagrams are shown in Figure 7.13b and c, respectively.

Position of zero shear in span *BC* (Figure 7.13b):

$$\frac{x}{38.54} = \frac{5-x}{36.46} \Rightarrow x = 2.569 \text{ m}$$

$$\therefore \text{Maximum +ve BM} = 38.54 \times 2.569 - 32.97$$

$$-15 \times 2.569 \times \frac{2.569}{2} = 16.54 \text{ kNm}$$

Maximum+ve BM in span *AB*

$$M_{max} = 12.11 \times 2 = 24.22 \text{ kNm}$$

TABLE 7.3
Moment Distribution Table

Joint	A	B		C
Member	*AB*	*BA*	*BC*	*CB*
DF	1.000	0.429	0.571	0.000
CoF	0.000	0.000	0.000	0.500
FEM	0.000	25.200	–31.250	31.250
Balance and distribution		2.593	3.457	
COM				1.729
Final end moments (kNm)	0.000	27.79	–27.79	32.979

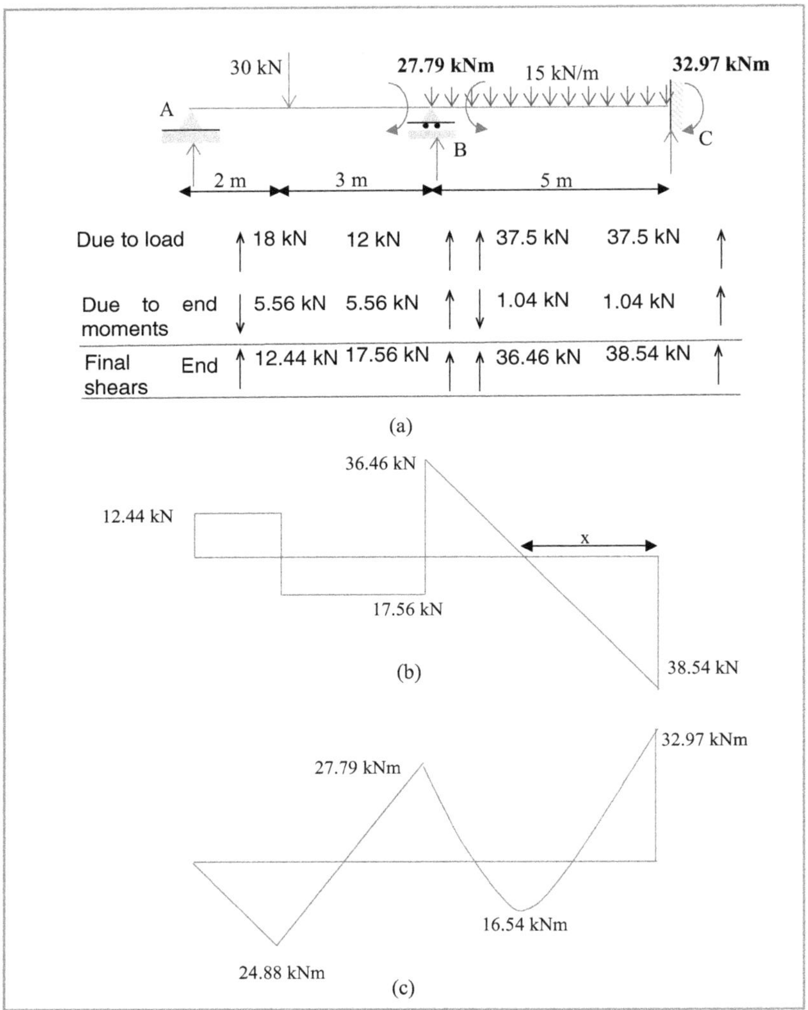

FIGURE 7.13 (a) Free body diagram, (b) shear force diagram, and (c) bending moment diagram.

Example 7.4

Analyse the continuous beam *ABCD*, as shown in Figure 7.14, using the moment distribution method. Assume *EI* is constant. Draw the shear force and bending moment diagrams. (The same problem has been solved in Example 6.3 using the slope–deflection method.)

Solution:

The cantilever portion *AB* in the figure is statically determinate; hence, its effect can be represented by applying a counter-clockwise moment of 10 kNm at *B*. Furthermore, taking advantage of the known moment at *B* (M_{BA}=−10 kNm), the rotation at support *B* can be neglected by applying the modified stiffness factor and FEM for member *BC*. Additionally, the rotation at *D* is zero because end *D* is fixed. The only unknown rotation is at *C*, denoted as θ_C.

FIXED-END MOMENTS

Using Table 6.1, the fixed-end moments can be calculated as shown below in Figure 7.15.

$$MF_{CB} = \frac{3}{2}\frac{wL^2}{12} - \frac{M_0}{2} = \frac{3\times 20\times 5^2}{2\times 12} - \frac{10}{2} = 57.5 \text{ kNm}$$

$$MF_{CD} = -\left(\frac{20\times 1\times 4^2}{5^2} + \frac{20\times 4\times 1^2}{5^2}\right) = -16 \text{ kNm}$$

$$MF_{DC} = \left(\frac{20\times 1^2\times 4}{5^2} + \frac{20\times 1\times 4^2}{5^2}\right) = 16 \text{ kNm}$$

STIFFNESS FACTORS

Modified stiffness factor for *BC*

$$K_{BC} = K_{CB} = \frac{3EI}{L} = \frac{3EI}{5}$$

Stiffness Factor for *CD*

$$K_{CD} = K_{DC} = \frac{4EI}{L} = \frac{4EI}{5}$$

DISTRIBUTION FACTORS

At joint *C*:

$$(DF)_{CB} = \frac{K_{CB}}{K_{CB}+K_{CD}} = \frac{\frac{3EI}{5}}{\frac{3EI}{5}+\frac{4EI}{5}} = 0.429$$

$$(DF)_{CD} = \frac{K_{CD}}{K_{CB}+K_{CD}} = \frac{\frac{4EI}{5}}{\frac{3EI}{5}+\frac{4EI}{5}} = 0.571$$

MOMENT DISTRIBUTION CYCLE

The successive locking and unlocking of each joint, along with the balancing moments, distributed moments, and COMs, are carried out as discussed in the previous problem and are presented in moment distribution Table 7.4. It is important to note that no distribution occurs at joint *B*.

Maximum +ve BM in *BC*

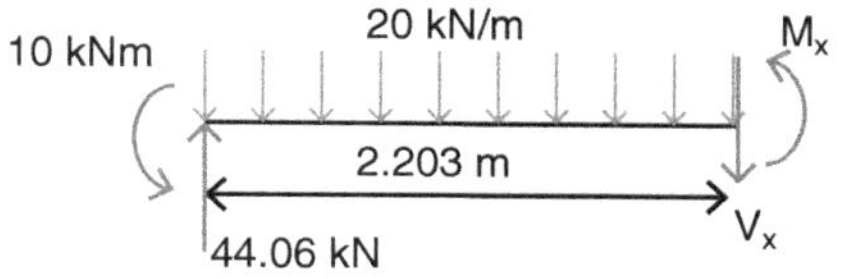

$$\frac{x}{44.06} = \frac{5-x}{55.94} \Rightarrow x = 2.203 \text{ m}$$

$$\therefore \text{Maximum +ve BM} = 44.06 \times 2.203 - 10 - 20 \times 2.203 \times \frac{2.203}{2} = 38.53 \text{ kNm}$$

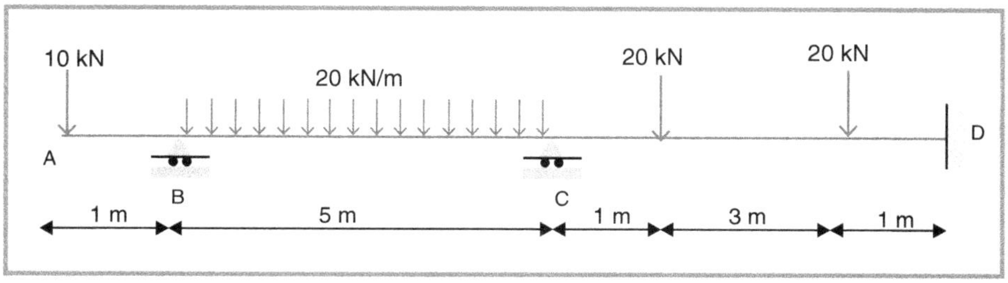

FIGURE 7.14 Two-span continuous beam with one side overhang.

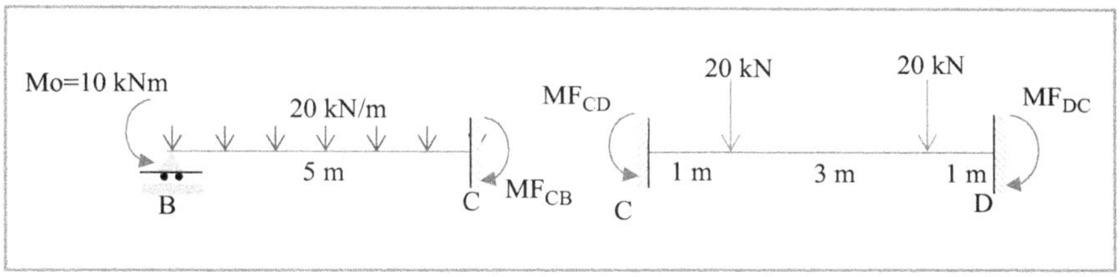

FIGURE 7.15 Fixed-end moments.

TABLE 7.4
Moment Distribution Table

Joint	B	C		D
Member	BA	CB	CD	DC
DF	1.000	0.429	0.571	0.000
COF	0.000	0.000	0.000	0.500
FEM	−10.000	57.500	−16.000	16.000
Balance and distribution		−17.786	−23.714	
COM				−11.857
Final end moments (kNm)	−10.000	39.714	−39.714	4.143

The support reactions, shear force, and bending moment diagrams are shown in Figure 7.16a–c, respectively.

Example 7.5

Analyse the beam shown in Figure 7.17, using the MD method without taking advantage of the known moment at support *A* (hinged support at *A*). (The same was solved in Example 6.4 using the SD method.)

Solution:

Assuming the moment at end A is unknown, the kinematic indeterminacy or degrees of freedom are the rotations at A, B, and C, i.e., θ_A, θ_B, and θ_C respectively.

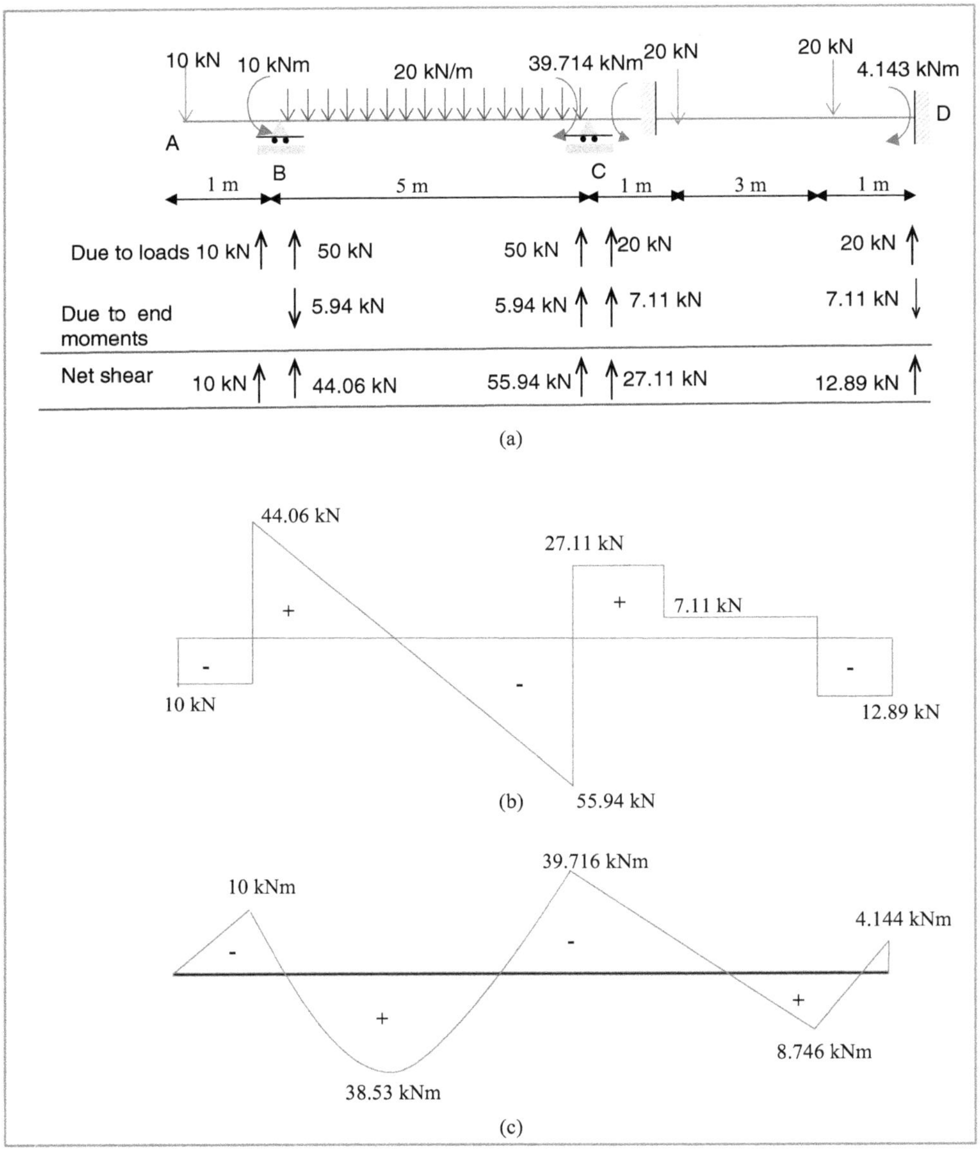

FIGURE 7.16 (a) Free body diagram, (b) shear force diagram, and (c) bending moment diagram.

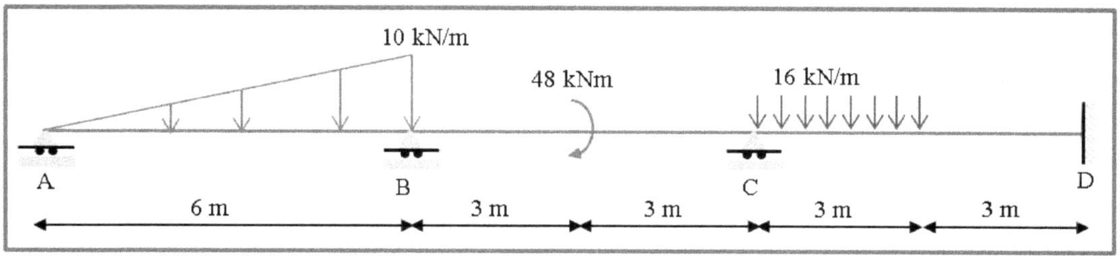

FIGURE 7.17 Three-span continuous beam *ABCD*.

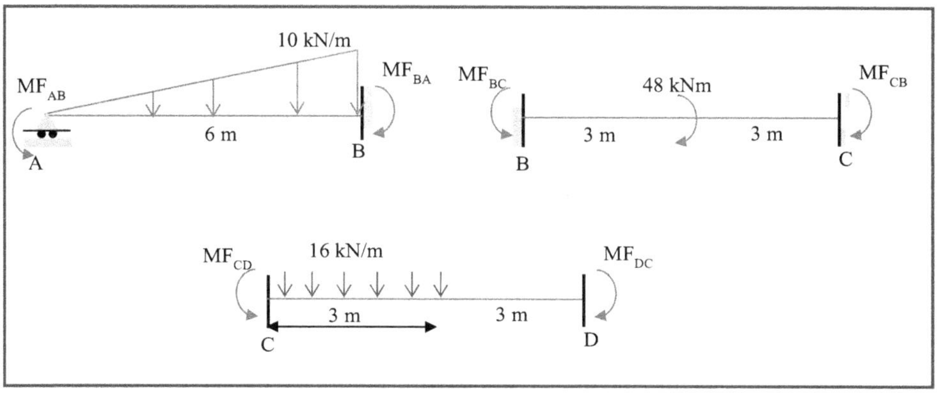

FIGURE 7.18 Fixed-end moments.

Fixed-end moments: With the help of the expressions given in Table 6.1 and using the free body diagrams shown in Figure 7.18, the fixed moments can be calculated as follows.

$$MF_{AB} = -\frac{wL^2}{30} = -\frac{10 \times 6^2}{30} = -12 \text{ kNm}$$

$$MF_{BA} = \frac{wL^2}{20} = \frac{10 \times 6^2}{20} = 18 \text{ kNm}$$

$$MF_{BC} = \frac{M}{4} = \frac{48}{4} = 12 \text{ kNm} = MF_{CB}$$

$$MF_{CD} = -\frac{11}{192} wL^2 = -\frac{11 \times 16 \times 6^2}{192} = -33 \text{ kNm}$$

$$MF_{DC} = \frac{5}{192} wL^2 = \frac{5 \times 16 \times 6^2}{192} = 15 \text{ kNm}$$

STIFFNESS FACTORS

Span AB: $K_{AB} = K_{BA} = \frac{4EI}{L} = \frac{4EI}{6}$

Span BC: $K_{BC} = K_{CB} = \frac{4EI}{L} = \frac{4EI}{6}$

Span CD: $K_{CD} = K_{DC} = \frac{4EI}{L} = \frac{4EI}{6}$

DISTRIBUTION FACTORS

At joint A: Since joint A is a pinned support, the distribution factor is unity. $(DF)_{AB} = 1.0$ (Hinged support)

At joint *B*:

$$(DF)_{BA} = \frac{K_{BA}}{K_{BA} + K_{BC}} = \frac{\frac{4EI}{6}}{\frac{4EI}{6} + \frac{4EI}{6}} = 0.5$$

$$(DF)_{BC} = \frac{K_{BC}}{K_{BA} + K_{BC}} = \frac{\frac{4EI}{6}}{\frac{4EI}{6} + \frac{4EI}{6}} = 0.5$$

At joint *C*:

$$(DF)_{CB} = \frac{K_{CB}}{K_{CB} + K_{CD}} = \frac{\frac{4EI}{5}}{\frac{4EI}{5} + \frac{4EI}{5}} = 0.5 = (DF)_{CD}$$

MOMENT DISTRIBUTION CYCLE

The procedure of successive locking and unlocking of each joint, along with the balancing moments, distribution moments, and the COMs, is presented in Table 7.5. The check on moment distribution is also shown in the table.

TABLE 7.5
Moment Distribution Table

Cycle	Joint	*A*	*B*		*C*		*D*
	Member	*AB*	*BA*	*BC*	*CB*	*CD*	*DC*
	DF	1.000	0.500	0.500	0.500	0.500	
	COF	0.500	0.500	0.500	0.500	0.000	0.500
1	FEM	−12.000	18.000	12.000	12.000	−33.000	15.000
	Balance and distribution	12.000	−15.000	−15.000	10.500	10.500	
2	COM	−7.500	6.000	5.250	−7.500	0.000	5.250
	Balance and distribution	7.500	−5.625	−5.625	3.750	3.750	
3	COM	−2.813	3.750	1.875	−2.813	0.000	1.875
	Balance and distribution	2.813	−2.813	−2.813	1.406	1.406	
4	COM	−1.406	1.406	0.703	−1.406	0.000	0.703
	Balance and distribution	1.406	−1.055	−1.055	0.703	0.703	
5	COM	−0.527	0.703	0.352	−0.527	0.000	0.352
	Balance and distribution	0.527	−0.527	−0.527	0.264	0.264	
6	COM	−0.264	0.264	0.132	−0.264		0.132
	Balance and distribution	0.264	−0.198	−0.198	0.132	0.132	
7	COM	−0.099	0.132	0.066	−0.099		0.066
	Balance and distribution	0.099	−0.099	−0.099	0.049	0.049	
8	COM	−0.049	0.049	0.025	−0.049		0.025
	Balance and distribution	0.049	−0.037	−0.037	0.025	0.025	
9	COM	−0.019	0.025	0.012	−0.019		0.012
	Balance and distribution	−0.009	−0.019	−0.019	0.009	0.009	0.005
	Final end moments (kNm)	−0.028	4.957	−4.957	16.162	−16.162	23.419
	Check for Moment Distribution						
	Change in near end moments	11.972	−13.043	−16.957	4.162	16.838	8.419
	−1/2 (change in far end moments)	6.521	−5.986	−2.081	8.479	−4.210	−8.419
	Sum	18.494	−19.029	−19.038	12.640	12.629	0.000
	$\theta_{rel} = \dfrac{\text{Sum}}{K}$	27.740	−28.543	−28.557	18.960	18.943	0.000
			Check (θ_B)		Check (θ_C)		(θ_D)

Example 7.6

Re-analyse the three-span continuous beam *ABCD* shown in Figure 7.17, taking advantage of the hinged support at *A*. Also, draw the shear force and bending moment diagrams. Assume constant flexural rigidity. (The same problem was solved in Example 6.4 using the slope–deflection method.)

Solution:

Since support *A* is a hinged support, the moment at end *A* is zero. As a result, the rotation at end *A* is neglected. By taking advantage of the hinged support at *A*, the problem can be simplified by applying the modified fixed-end moments and modified stiffness factor for span *AB* (propped cantilever).

FIXED-END MOMENTS

The fixed-end moments are calculated using Table 6.1 and are shown in Figure 7.19.

$$MF_{BA} = \frac{wL^2}{20} + \frac{1}{2}\frac{wL^2}{30} = \frac{wL^2}{15} = \frac{10 \times 6^2}{15} = 24 \text{ kNm}$$

$$MF_{BC} = \frac{M}{4} = \frac{48}{4} = 12 \text{ kNm} = MF_{CB}$$

$$MF_{CD} = -\frac{11}{192}wL^2 = -\frac{11 \times 16 \times 6^2}{192} = -33 \text{ kNm}$$

$$MF_{DC} = \frac{5}{192}wL^2 = \frac{5 \times 16 \times 6^2}{192} = 15 \text{ kNm}$$

STIFFNESS FACTORS

Span *AB*: Since end *A* is simply supported, adopt the modified stiffness factor.

$$K_{AB} = K_{BA} = \frac{3EI}{L} = \frac{3EI}{6}$$

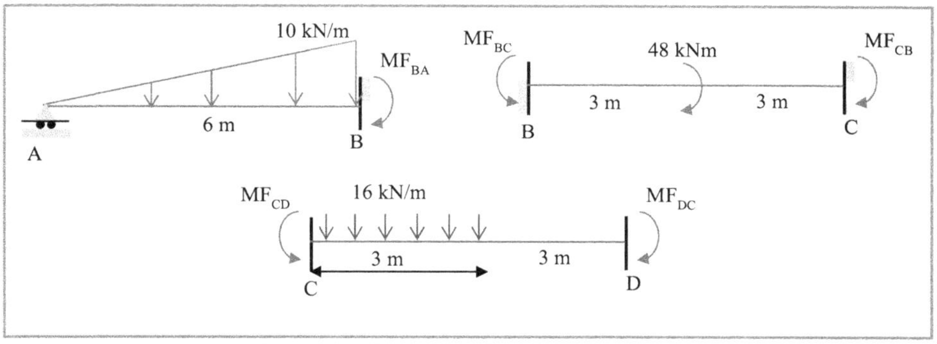

FIGURE 7.19 Fixed-end moments.

TABLE 7.6
Moment Distribution Table

Cycle	Joint	A	B		C		D
	Member	*AB*	*BA*	*BC*	*CB*	*CD*	*DC*
	DF	1.000	0.429	0.571	0.500	0.500	0.000
	CoF	0.000	0.000	0.500	0.500	0.000	0.500
1	FEM	0.000	24.000	12.000	12.000	−33.000	15.000
	Balance and distribution		−15.429	−20.571	10.500	10.500	
2	COM			5.250	−10.286		5.250
	Balance and distribution		−2.250	−3.000	5.143	5.143	
3	COM			2.571	−1.500		2.571
	Balance and distribution		−1.102	−1.469	0.750	0.750	
4	COM			0.375	−0.735		0.375
	Balance and distribution		−0.161	−0.214	0.367	0.367	
5	COM			0.184	−0.107		0.184
	Balance and distribution		−0.079	−0.105	0.054	0.054	
6	COM			0.027	−0.052		0.027
	Balance and distribution		−0.011	−0.015	0.026	0.026	0.013
	Final end moments (kNm)	0.000	4.96	−4.96	16.160	−16.160	23.41

Stiffness factor for *BC*:

$$K_{BC} = K_{CB} = \frac{4EI}{L} = \frac{4EI}{6}$$

Stiffness factor for *CD*:

$$K_{CD} = K_{DC} = \frac{4EI}{L} = \frac{4EI}{6}$$

DISTRIBUTION FACTORS

At joint *B*:

$$(DF)_{BA} = \frac{K_{BA}}{K_{BA} + K_{BC}} = \frac{\frac{3EI}{6}}{\frac{3EI}{6} + \frac{4EI}{6}} = 0.429$$

$$(DF)_{BC} = \frac{K_{BC}}{K_{BA} + K_{BC}} = \frac{\frac{4EI}{6}}{\frac{3EI}{6} + \frac{4EI}{6}} = 0.571$$

At joint *C*:

$$(DF)_{CB} = \frac{K_{CB}}{K_{CB} + K_{CD}} = \frac{\frac{4EI}{6}}{\frac{4EI}{6} + \frac{4EI}{6}} = 0.5 = (DF)_{CD}$$

MOMENT DISTRIBUTION CYCLE

The procedure of successive locking and unlocking of each joint, along with the balancing moments, distribution moments, and the COMs, is presented in Table 7.6. It is observed that by the sixth cycle, the COM is negligible at all the joints; hence, there is no need to continue with further cycles. Furthermore, it is observed that the final end moments obtained by the MD method in Table 7.6 are well matched with the results obtained in Problem No. 6.4 using the slope–deflection method. The check on moment distribution method can also be carried out, as explained in the earlier problems.

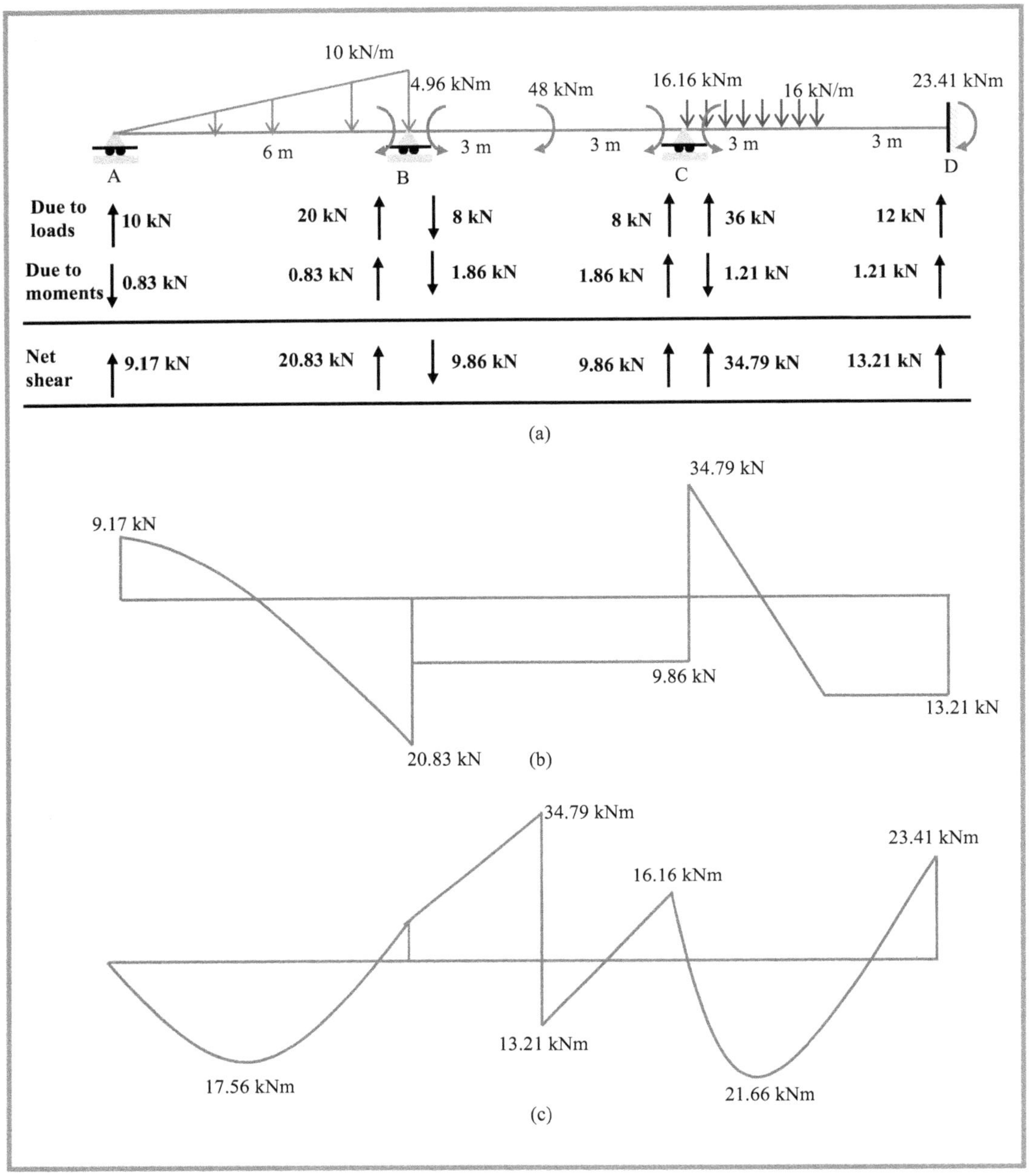

FIGURE 7.20 (a) Free body diagram, (b) shear force diagram, and (c) bending moment diagram.

SHEAR FORCE AND BENDING MOMENT DIAGRAMS

The free body diagram (Figure 7.20a) shows the member end shears. The shear force and bending moment diagrams are shown in Figure 7.20b and c, respectively.

Zero shear in span *CD* (Figure 7.20b)

$$\frac{x}{34.79} = \frac{3-x}{13.21} \Rightarrow x = 2.17 \text{ m}$$

$$\therefore \text{Maximum +ve BM} = 34.79 \times 2.17 - 16.16 - 16 \times 2.17 \times \frac{2.17}{2}$$

$$= 21.66 \text{ kNm}$$

Zero shear in span *AB*

$$V_x = 9.17 - \frac{1}{2} \times x \times \frac{10x}{6} = 0 \Rightarrow x = 3.31 \text{ m}$$

$$\therefore \text{Maximum +ve BM} = 8.35 \times 3.31 - 0.5 \times \frac{3.31 \times 10}{6} \times 3.31 \times \frac{3.31}{3} = 17.56 \text{ kNm}$$

Example 7.7

Analyse the two-span continuous beam, one end of which is fixed and the other end is a guided support, as shown in Figure 7.21, using the moment distribution method. Assume *EI* is constant. (The same problem has been solved in Example 6.5 using the SD method.)

Solution:

The kinematic indeterminacy of the given beam is 2, i.e., θ_B and Δ_C. This would be considered as a sway-type problem. However, for span *BC*, modified stiffness and modified fixed-end moments can be used to neglect Δ_C. To apply the modified stiffness, the beam $ABCB^1A^1$ should be visualised as symmetrical about point *C*, with *ABC* representing the left half of $ABCB^1A^1$, as shown in Figure 7.22. The moment at *C* is nothing but the mid-span moment of the symmetrical beam BCB^1.

FIXED-END MOMENTS

Using Table 6.1 and the free body diagram shown in Figure 7.23, the fixed-end moments can be calculated as shown below.

$$MF_{AB} = -\frac{WL}{8} = -\frac{25\times 4}{8} = -12.5\text{ kNm} = -MF_{BA}$$

$$MF_{BC} = -\frac{WL^2}{12} = -\frac{15\times 4^2}{12} = -20\text{ kNm}$$

$$MF_{CB} = -\left(\frac{WL^2}{8} - \frac{WL^2}{12}\right) = -\left(\frac{WL^2}{24}\right) = -\frac{15\times 4^2}{24} = -10\text{ kNm}$$

STIFFNESS FACTORS

Span AB: $K_{AB} = K_{BA} = \frac{4EI}{L} = \frac{4EI}{4} = EI$

Stiffness factor for *BC*: Since support *C* is a fixed guided support, use the modified stiffness factor.

$$K_{BC} = K_{CB} = \frac{EI}{L} = \frac{EI}{2}$$

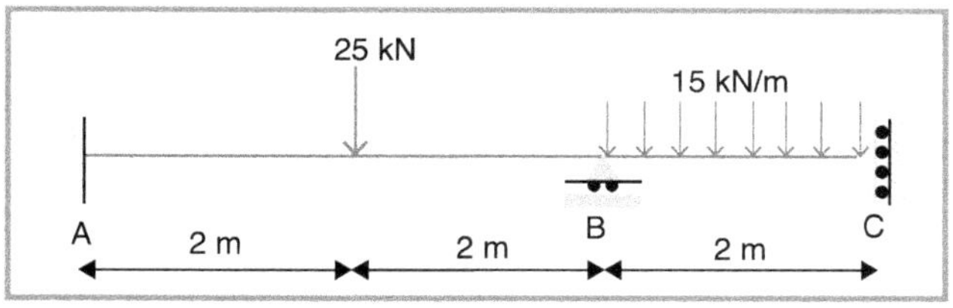

FIGURE 7.21 Two-span continuous beam *ABC* with guided support at *C*.

DISTRIBUTION FACTORS

At joint *B*:

$$(DF)_{BA} = \frac{K_{BA}}{K_{BA}+K_{BC}} = \frac{\frac{4EI}{4}}{\frac{4EI}{4}+\frac{EI}{2}} = 0.667$$

$$(DF)_{BC} = \frac{K_{BC}}{K_{BA}+K_{BC}} = \frac{\frac{EI}{2}}{\frac{4EI}{4}+\frac{EI}{2}} = 0.333$$

MOMENT DISTRIBUTION TABLE

The iterative procedure of successive locking and unlocking of each joint, along with the distribution and COMs, is presented in Table 7.7. It should be noted that due to the fixed guided support at *C*, the COF for *CB* is −1.

FINAL END SHEARS

Using the final end moments obtained in Table 7.7 and the static equilibrium equations, the member end shears can be calculated and are shown in Figure 7.24a. The shear force and bending moment diagrams are shown in Figure 7.24b and c, respectively.

Example 7.8

Analyse the three-span continuous beam shown in Figure 7.25 by taking advantage of symmetry using the moment distribution method. Assume constant flexural rigidity.

Solution:

The given beam *ABCD* is symmetrical with respect to the centre of span *BC*. Therefore, the slope of the elastic curve at the centre of span BC is zero. To maintain this condition, a fixed guided support can be provided at the centre of span *BC*, allowing the advantage of symmetry to be utilised. As a result, only half of the beam, as shown in Figure 7.26a, needs to be considered for the analysis, provided that the modified stiffness factor and modified fixed-end moments are used for span *BC*. The fixed-end moments are calculated using Figure 7.26b and Table 6.1, as shown below.

$$MF_{AB} = -\frac{wL^2}{30} = -\frac{20\times 25}{30} = -16.667\text{ kNm}$$

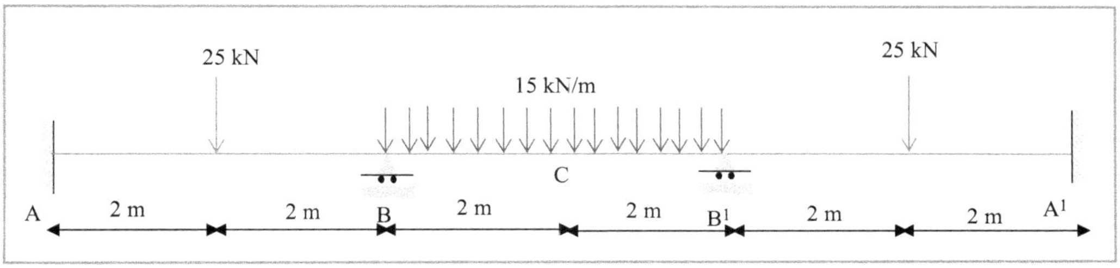

FIGURE 7.22 Visualised as symmetrical beam $ABCB^1A^1$ at *C*.

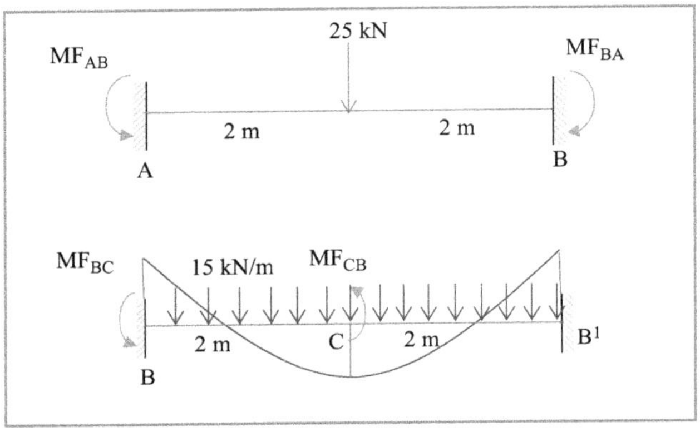

FIGURE 7.23 Fixed-end moments.

TABLE 7.7
Moment Distribution Table

Joint	A	B		C
Member	*AB*	*BA*	*BC*	*CB*
DF		0.667	0.333	
COF	0.500	0.000	0.000	−1.000
FEM	−12.500	12.500	−20.000	−10.000
Balance and distribution		5.000	2.500	
COM	2.500			−2.500
Final end moments (kNm)	−10.000	17.500	−17.500	−12.500

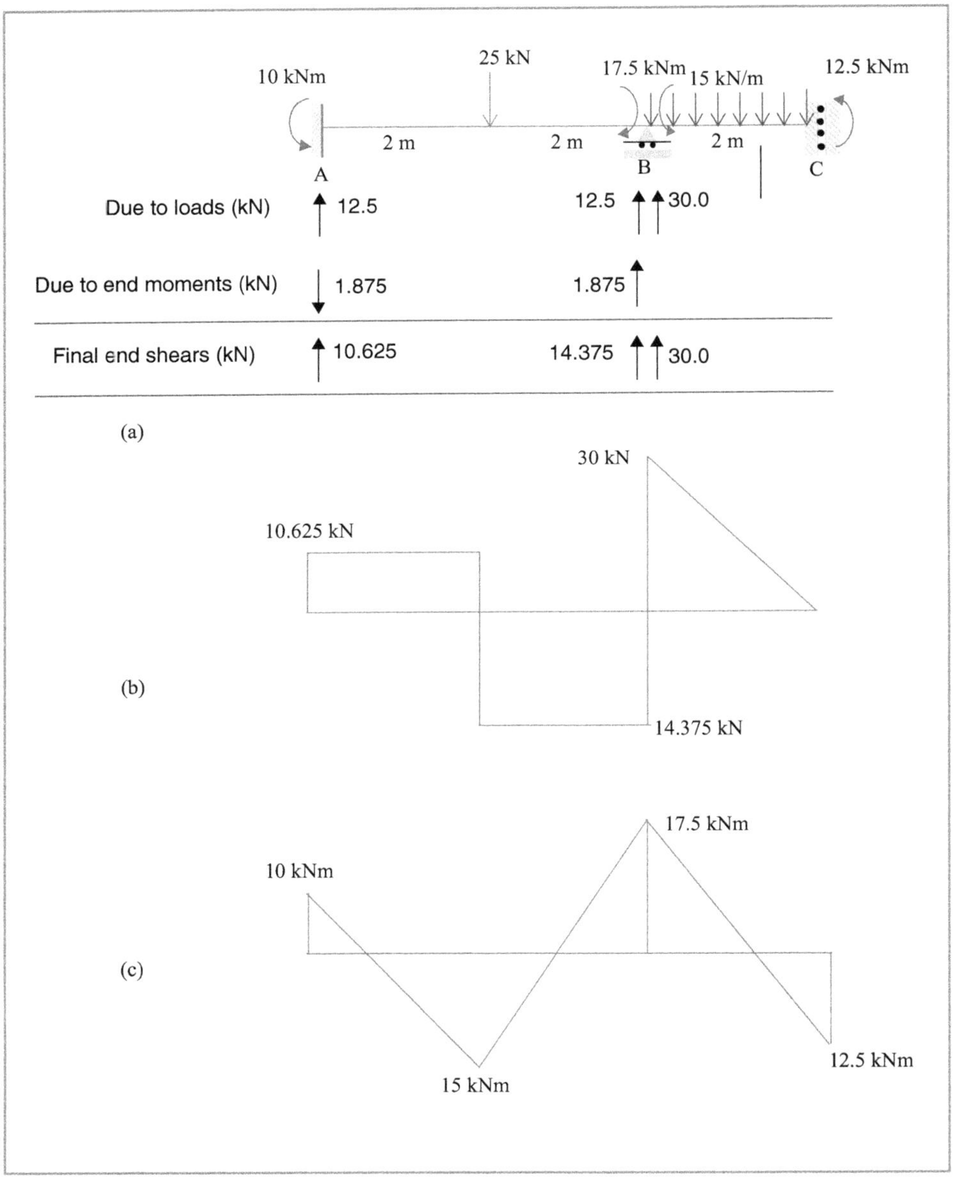

FIGURE 7.24 (a) Free body diagram, (b) shear force diagram, and (c) bending moment diagram.

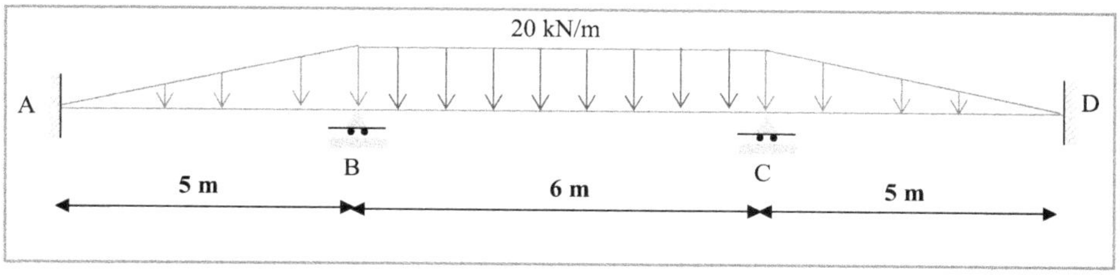

FIGURE 7.25 Symmetrical three-span continuous beam *ABCD*.

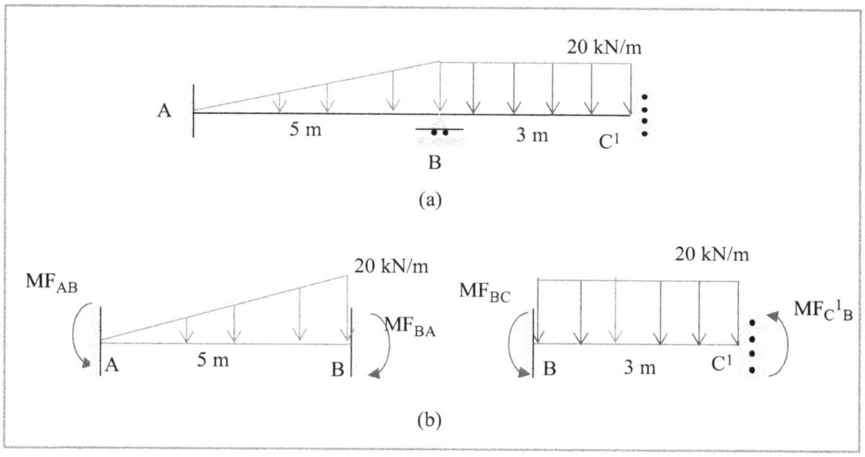

FIGURE 7.26 (a) Half portion of beam ABC^1 and (b) fixed-end moments.

$$MF_{BA} = \frac{wL^2}{20} = \frac{20 \times 25}{20} = 25 \text{ kNm}$$

$$MF_{BC} = -\frac{wL^2}{12} = -\frac{20 \times 36}{12} = -60 \text{ kNm}$$

$$MF_{C^1B} = -(60 \times 3 - 20 \times 3 \times 1.5 - 60) = -30 \text{ kNm}$$

STIFFNESS FACTORS

Span AB: $K_{AB} = K_{BA} = \frac{4EI}{L} = \frac{4EI}{5} = 0.8EI$

Stiffness factor for *BC*: (Since a guided support is provided at point *C*, use the modified stiffness factor)

$$K_{BC} = K_{CB} = \frac{2EI}{L} = \frac{2EI}{6} = 0.333EI$$

DISTRIBUTION FACTORS

At joint *B*:

$$(DF)_{BA} = \frac{K_{BA}}{K_{BA} + K_{BC}} = \frac{0.8EI}{0.8EI + 0.333EI} = 0.706$$

$$(DF)_{BC} = \frac{K_{BC}}{K_{BA} + K_{BC}} = \frac{0.333EI}{0.8EI + 0.333EI} = 0.294$$

MOMENT DISTRIBUTION TABLE

The iterative procedure of successive locking and unlocking of each joint, along with the balance and distribution moments, and COMs, are presented in Table 7.8.

TABLE 7.8
Moment Distribution Table

Joint	A	B		C
Member	AB	BA	BC	CB
DF	—	0.706	0.294	
COF	0.500	0.000	0.000	−1
FEM	−16.667	25.000	−60.000	−30.00
Balance and distribution		24.706	10.294	
COM	12.353			−10.294
Final end moments (kNm)	−4.314	49.706	−49.706	−40.294

FINAL END SHEARS

After obtaining the final member end moments from Table 7.8, using the static equilibrium equations, the member end shears are calculated, as shown in Figure 7.27a. The shear force and bending moment diagrams are shown in Figure 7.27b and c, respectively.

Zero shear in spans *AB* and *CD*: Using the free body diagram shown below

$$\sum F_y = 0 \Rightarrow 7.59 - 0.5 \times x \times 4x = 0$$

$$x = 1.95 \text{ m}$$

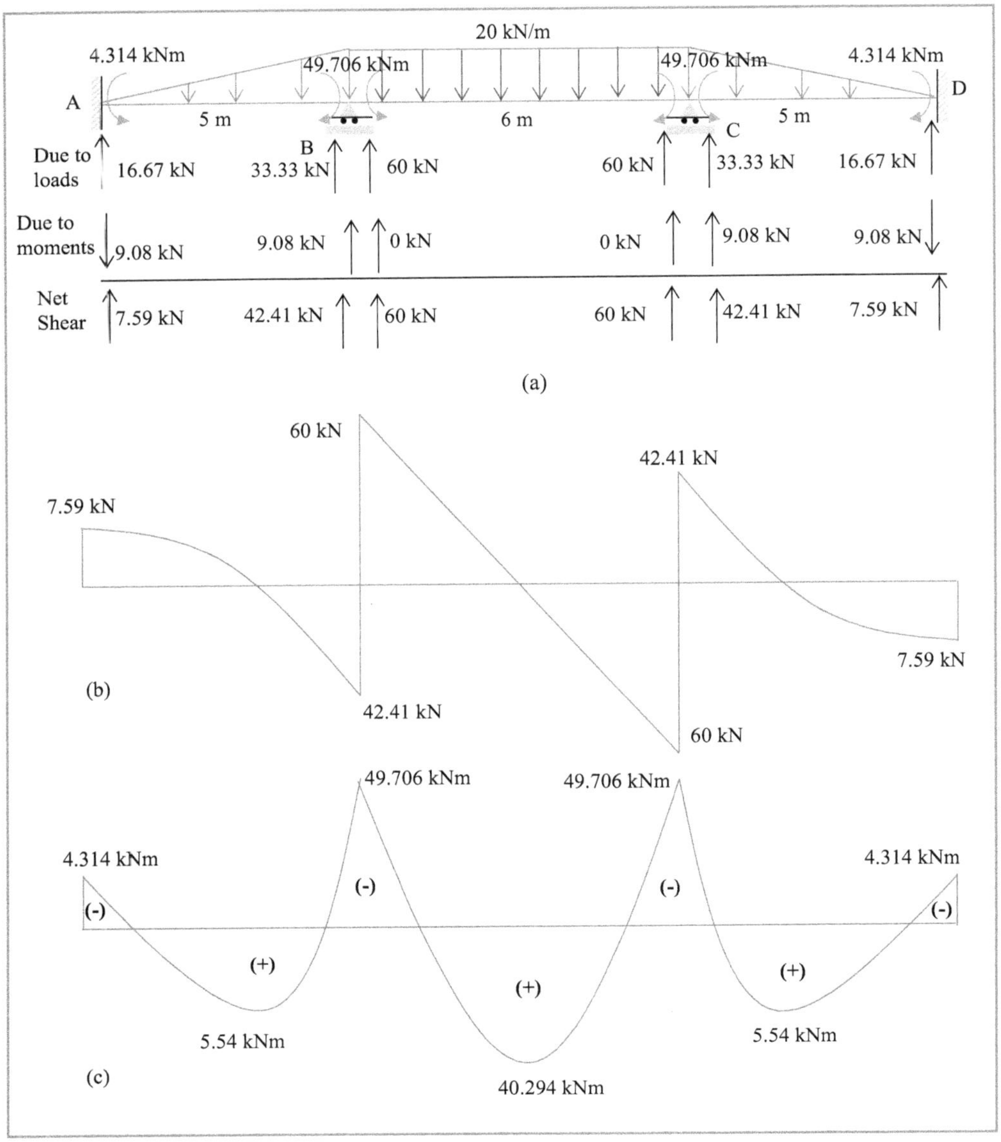

FIGURE 7.27 (a) Free body diagram, (b) shear force diagram, and (c) bending moment diagram.

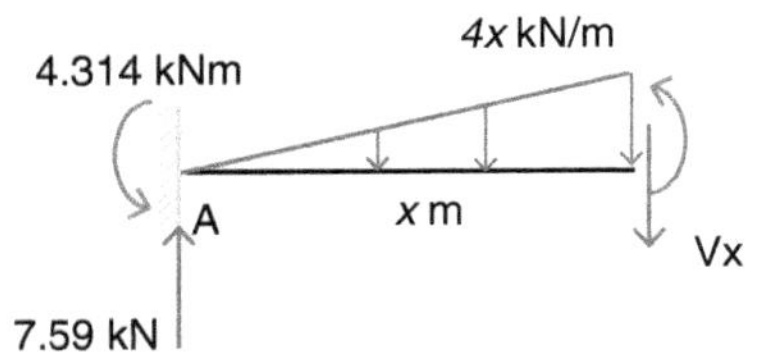

Maximum +ve bending moment in spans *AB* and *CD*

$$M_x = 7.59\times1.95 - 4.314 - \left(0.5\times1.95\times4\times1.95\times\frac{1.95}{3}\right)$$

$$= 5.54 \text{ kNm}$$

Maximum bending moment in span *BC*

$$M_x = 60\times3 - 49.706 - (0.5\times3\times20\times1.5) = 40.294 \text{ kNm}$$

EFFECT OF YIELDING OF SUPPORTS

It is apparent now that to apply the moment distribution method, we should start with the fixed-end moments. In the previous examples, the fixed-end moments due to various loadings have been considered alone in the process of moment distribution. As discussed in previous chapters, the yielding of supports induces additional moments in the structure. To account for these effects in the moment distribution method, the additional fixed end moments due to the yielding (rotational slip or settlement) of supports must be considered. Therefore, now the fixed-end moments consist of two components: (i) due to the external loading, and (ii) due to the yielding of supports. The fixed-end moments due to external loading can be calculated using Table 6.1, as seen in the previous examples, while those due to the yielding of supports can be calculated using the standard formulae derived in Section 6.2. The net fixed end moments, resulting from both external loads and the yielding of supports, are then appropriately entered in the moment distribution table, and the moment distribution is carried out as earlier. This process is explained in the following example.

Example 7.9

A two-span continuous beam is shown in Figure 7.28, in which the support *A* is subjected to a clockwise rotational slip of 0.002 rad. Analyse the beam using the moment distribution method. Assume EI=75,000 kNm². (The same problem was solved in Example 6.6 using the slope–deflection method.)

Solution:

FIXED-END MOMENTS

Due to external loads: Span *AB* is subjected to no external load, and span *BC* is subjected to 20 kNm, as shown in Figure 7.29. Therefore, using Table 6.1, the fixed-end moments are as follows:

$$(MF_{AB})_L = 0 = (MF_{BA})_L;$$

$$(MF_{BC})_L = -\frac{wL^2}{12} = -\frac{20\times 4^2}{12} = -26.67 \text{ kNm} = (-MF_{CB})_L$$

Due to the yielding of support (rotational slip of support *A*): The support A rotates clockwise by 0.002 rad, which develops the fixed-end moments at *A* and *B* only, as shown in Figure 7.29.

Using Section 6.2, the fixed-end moments due to the rotational slip are calculated as shown below.

$$(MF_{AB})_Y = \frac{4EI\theta}{L} = \frac{4\times 75{,}000\times 0.002}{4} = 150 \text{ kNm}$$

$$(MF_{BA})_Y = \frac{2EI\theta}{L} = \frac{2\times 75{,}000\times 0.002}{4} = 75 \text{ kNm};$$

$$(MF_{BC})_Y = (MF_{CB})_Y = 0$$

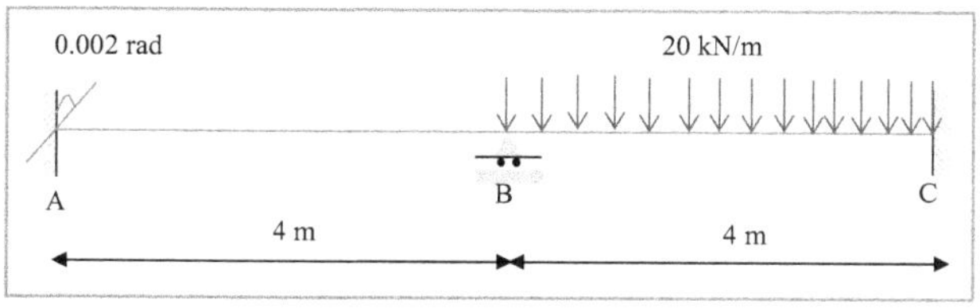

FIGURE 7.28 Two-span continuous beam *ABC*.

NET FIXED-END MOMENTS

$$MF_{AB} = 0 + 150 = 150 \text{ kNm}; \quad MF_{BA} = 0 + 75 \text{ kNm}$$

$$MF_{BC} = -26.67 + 0 = -26.67 \text{ kNm} = -MF_{CB}$$

STIFFNESS FACTORS

$$K_{BA} = K_{BC} = \frac{4EI}{L} = \frac{4EI}{4} = EI$$

DISTRIBUTION FACTORS

At joint B:

$$(DF)_{BA} = (DF)_{BC} = \frac{EI}{EI + EI} = 0.5$$

MOMENT DISTRIBUTION TABLE

The iterative procedure of locking and unlocking joints, along with the balance and distribution of moments, is shown in Table 7.9. The final end moments in the last row of the table are fairly matched with the results obtained in Example 6.6 by the slope–deflection method.

FINAL END SHEARS

After obtaining the final member end moments, the member end shears can be calculated using the static equilibrium equations with the help of the free body diagram, as shown in Figure 7.30a. The shear force and bending moment diagrams are shown in Figure 7.30b and c, respectively.

Maximum +ve BM in span *BC*

$$\frac{x}{30.94} = \frac{4-x}{49.06} \Rightarrow x = 1.55 \text{ m}$$

$$\text{Max. +ve BM} = 30.94\times 1.55 - 14.587 - 20\times 1.55^2 \times 0.5$$
$$= 9.34 \text{ kNm}$$

Example 7.10

Analyse the beam shown in Figure 7.31, if the support *B* settles down by 30 mm, using the moment distribution method. Assume EI=104,335.6 kNm².

Solution:

The beam is kinematically indeterminate to the second degree $(\theta_B \text{ and } \theta_C)$. Since the beam is subjected to both external loads and the settlement of support *B*, the

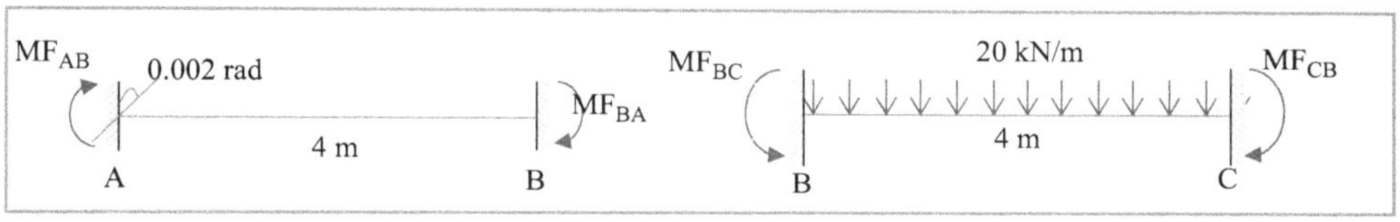

FIGURE 7.29 Fixed-end moments.

TABLE 7.9
Moment Distribution Table

Joint	A	B		C
Member	AB	BA	BC	CB
DF	—	0.500	0.500	—
COF	0.500	0.000	0.000	0.500
FEM	150.000	75.000	−26.670	26.670
Balance and distribution		−24.165	−24.165	
COM	−12.083		0.000	−12.083
Final end moments (kNm)	137.918	50.835	−50.835	14.588

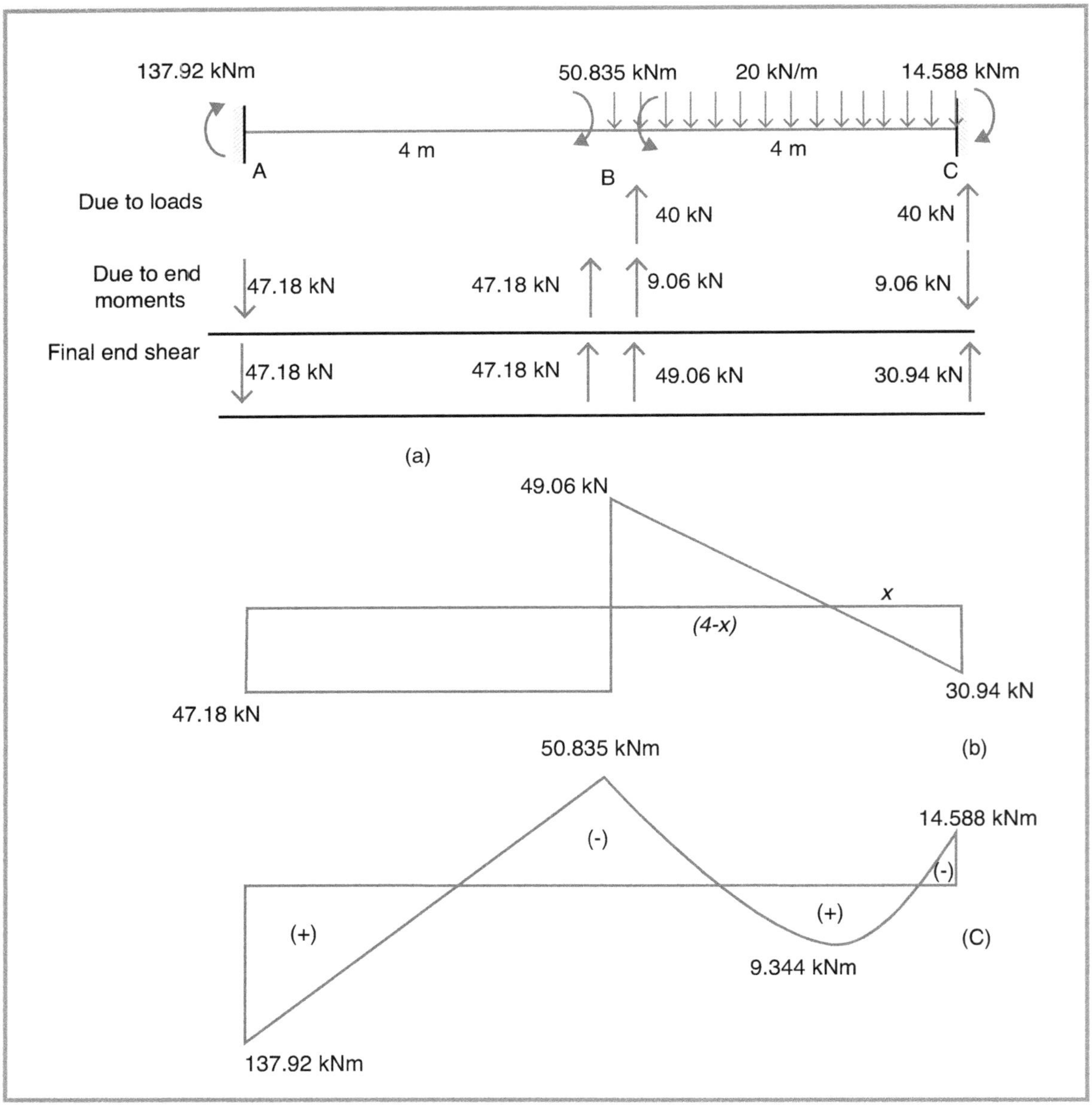

FIGURE 7.30 (a) Free body diagram, (b) shear force diagram, and (c) bending moment diagram.

fixed-end moments are calculated separately and then added to get the net fixed-end moments for each span.

Fixed-end moments due to external loads (Figure 7.32)

$$(MF_{AB})_L = -\frac{22.5\times 7.2^2}{12} = -97.2\ \text{kNm} = (-MF_{BA})_L$$

$$(MF_{BC})_L = (MF_{CB})_L = (MF_{CD})_L = (MF_{DC})_L = 0$$

Fixed-end moments due to the support B settlement (Figure 7.32)

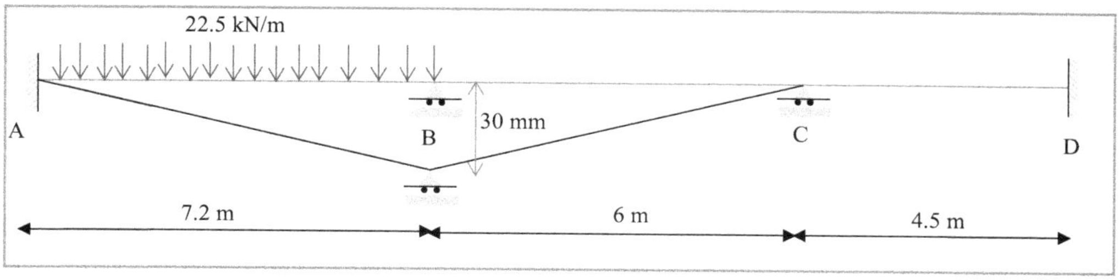

FIGURE 7.31 Three-span continuous beam with support *B* settle down by 30 mm.

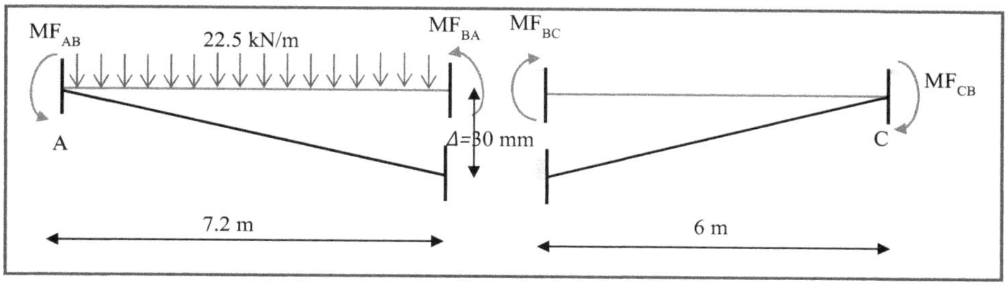

FIGURE 7.32 Fixed-end moments.

$$(MF_{AB})_s = (MF_{BA})_s = -\frac{6EI\Delta}{L^2} = -\frac{6\times 104,335.6\times 0.03}{7.2^2}$$

$$= -362.276 \text{ kNm}$$

$$(MF_{BC})_s = (MF_{CB})_s = \frac{6EI\Delta}{L^2} = \frac{6\times 104,335.6\times 0.03}{6^2}$$

$$= 521.678 \text{ kNm}$$

$$(MF_{CD})_s = (MF_{DC})_s = 0$$

Net Fixed-End Moments

$$MF_{AB} = -97.2 - 362.276 = -459.479 \text{ kNm}$$

$$MF_{BA} = 97.2 - 362.276 = -265.076 \text{ kNm}$$

$$MF_{BC} = MF_{CB} = 521.678 \text{ kNm}$$

$$MF_{CD} = MF_{DC} = 0$$

STIFFNESS FACTORS

$$K_{BA} = \frac{4EI}{L} = \frac{4EI}{7.2} = 0.556EI;$$

$$K_{BC} = K_{CB} = \frac{4EI}{L} = \frac{4EI}{6} = 0.667EI$$

$$K_{CD} = \frac{4EI}{L} = \frac{4EI}{4.5} = 0.889EI$$

DISTRIBUTION FACTORS

At joint *B*:

$$(DF)_{BA} = \frac{0.556EI}{0.556EI + 0.667EI} = 0.455$$

$$(DF)_{BC} = \frac{0.667EI}{0.556EI + 0.667EI} = 0.545$$

At joint *C*:

$$(DF)_{CB} = \frac{0.667EI}{0.667EI + 0.889EI} = 0.429$$

$$(DF)_{CD} = \frac{0.889EI}{0.667EI + 0.889EI} = 0.571$$

MOMENT DISTRIBUTION

The successive locking and unlocking of joints, along with the balancing and distribution of moments and COMs in iterative procedure, is shown in moment distribution Table 7.10. It is observed that by the end of ninth cycle, the moments have converged, and hence, the moment distribution cycles are stopped. The final end moments are shown in the table. The geometry check is also carried out in the table, and it is verified that the relative slope at *A* and *D* are zero, as they are fixed supports, and the relative slopes of the members corresponding to the joints *B* and *C* are the same. Hence, the geometry check is satisfied. The reactions are shown in the free body diagram (Figure 7.33a), and the shear force and bending moment diagrams are shown in Figure 7.33b and c, respectively.

TABLE 7.10
Moment Distribution Table

Cycle	Joint	A	B		C		D
	Member	*AB*	*BA*	*BC*	*CB*	*CD*	*DC*
	DF		0.455	0.545	0.429	0.571	
	COF	0.500	0.000	0.500	0.500	0.000	0.500
1	FEM	−459.476	−265.076	521.678	521.678	0.000	0.000
	Balance and distribution		−116.637	−139.965	−223.576	−298.102	
2	COM	−58.319		−111.788	−69.982		−149.051
	Balance and distribution		50.813	60.975	29.992	39.990	
3	COM	25.406		14.996	30.488		19.995
	Balance and distribution		−6.816	−8.180	−13.066	−17.422	
4	COM	−3.408		−6.533	−4.090		−8.711
	Balance and distribution		2.970	3.563	1.753	2.337	
5	COM	1.485		0.876	1.782		1.169
	Balance and distribution		−0.398	−0.478	−0.764	−1.018	
6	COM	−0.199		−0.382	−0.239		−0.509
	Balance and distribution		0.174	0.208	0.102	0.137	
7	COM	0.087		0.051	0.104		0.068
	Balance and distribution		−0.023	−0.028	−0.045	−0.060	
8	COM	−0.012		−0.022	−0.014		−0.030
	Balance and distribution		0.010	0.012	0.006	0.008	
9	COM	0.005		0.003	0.006		0.004
	Balance and distribution	−0.001	−0.001	−0.002	−0.003	−0.003	−0.002
	Final end moments (kNm)	−494.431	−334.987	334.987	274.133	−274.133	−137.067
	Check						
	Change in moment at near end	−34.96	−69.91	−186.69	−241.52	−274.13	−137.06
	−1/2 (change in moment at far end)	+34.96	+17.48	+120.76	+93.36	+68.53	+137.06
	Rotation = Sum/(−*K*) (rad)	0.00	0.0018	0.0018	0.0044	0.0044	0.00
		Check θ_A	Check θ_B		Check θ_C		Check θ_D

Example 7.11

Analyse a three-span continuous beam shown in Figure 7.34, subjected to external loading, with supports *B* and *C* settling down by 5 and 10 mm, respectively, using the moment distribution method. Assume the flexural rigidity $EI = 80{,}000$ kNm². Draw the shear force and bending moment diagrams.

Solution:

Since the overhang portion *DE* is statically determinate, the moment at *D* is known. The only unknown rotations to be considered are at *B* and *C*. The effect of the overhang portion can be accounted for by applying a 50 kNm clockwise moment at support *D*. Since the moment at *D* is known, the modified fixed-end moments and stiffness factors for span *CD* can be calculated to further simplify the problem. The fixed moments due to the external loads and settlement of supports are calculated separately using Table 6.1 and Section 6.2, respectively.

FIXED-END MOMENTS

Due to loads (using Table 6.1)

$$(MF_{AB})_L = -\frac{30 \times 8^2}{12} = -160 \text{ kNm} = -(MF_{BA})_L$$

$$(MF_{BC})_L = -\frac{100 \times 6}{8} - \frac{5}{192} \times 20 \times 6^2 = -93.75 \ kNm$$

$$(MF_{CB})_L = \frac{100 \times 6}{8} + \frac{11}{192} \times 20 \times 6^2 = 116.25 \text{ kNm}$$

$$(MF_{CD})_L = -\frac{150 \times 2 \times 4^2}{6^2} - \frac{1}{2} \times \frac{150 \times 2^2 \times 4}{6^2} + \frac{50}{2} = -141.67 \text{ kN}$$

$$M_{DC} = 50 \text{ kNm}$$

Due to settlement of supports (using Section 6.2)

$$(MF_{AB})_s = (MF_{BA})_s = -\frac{6EI\Delta_B}{L^2} = -\frac{6 \times 4 \times 80{,}000}{8^2} \times \frac{5}{1{,}000} = -15$$

$$(MF_{BC})_s = (MF_{CB})_s = -\frac{6EI(\Delta_C - \Delta_B)}{L^2}$$

$$= \frac{6 \times 3 \times 80{,}000}{6^2} \times \frac{(10-5)}{1{,}000} = -200 \text{ kNm}$$

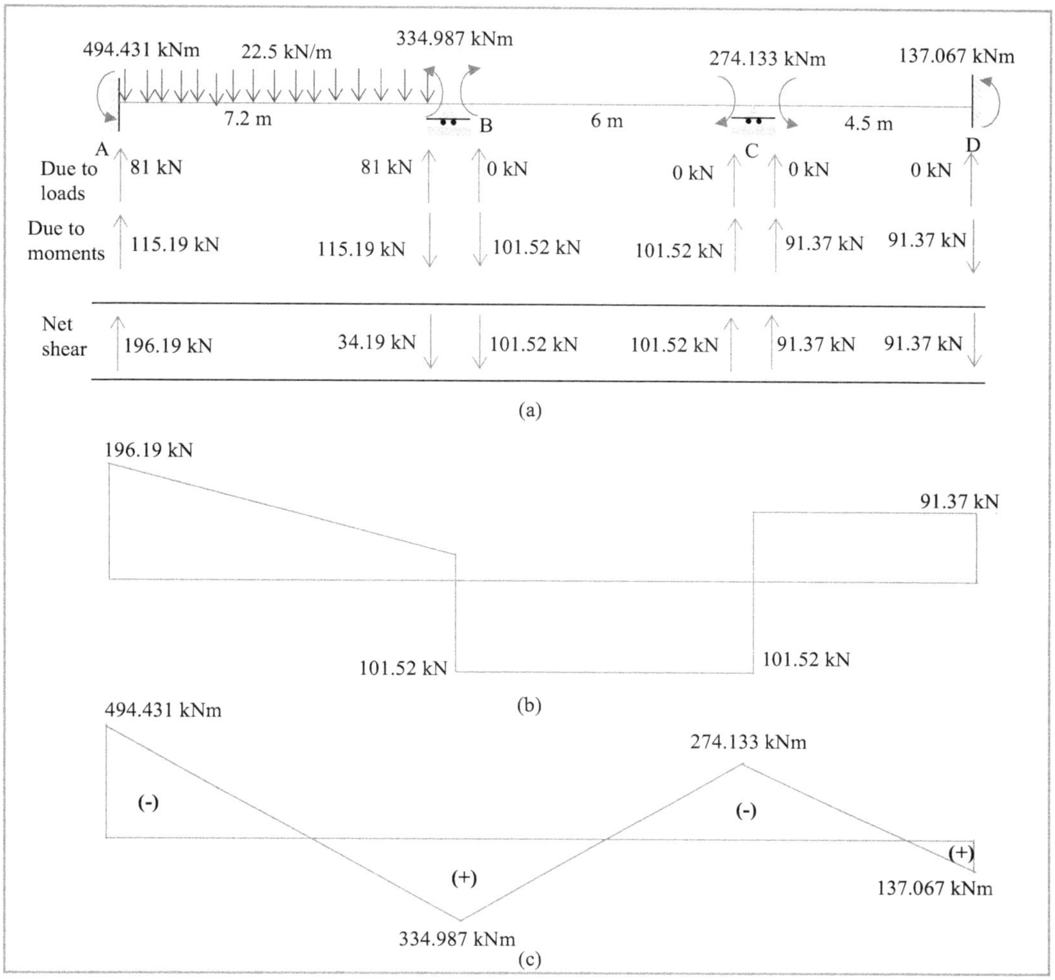

FIGURE 7.33 (a) Free body diagram, (b) shear force diagram, and (c) bending moment diagram.

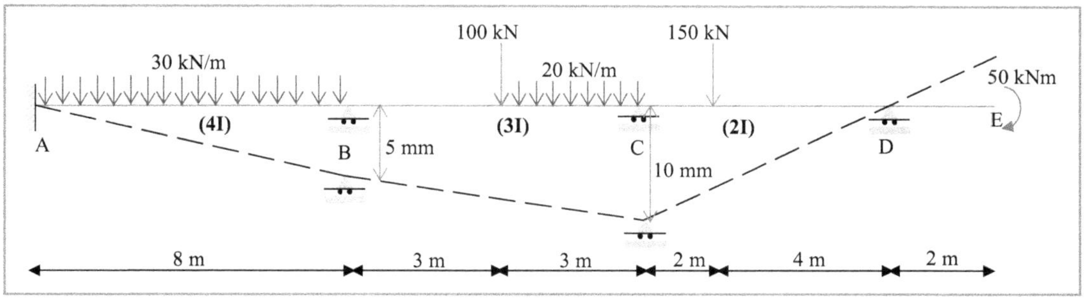

FIGURE 7.34 Three-span continuous beam with support supports settlement.

$$(MF_{CD})_s = \frac{3EI\Delta_C}{L^2} = \frac{3\times 2\times 80{,}000}{6^2}\times\frac{10}{1{,}000} = 133.33 \text{ kNm}$$

Net Fixed-End Moments

$$MF_{AB} = -160 - 150 = -310 \text{ kNm}$$

$$MF_{BA} = 160 - 150 = 10 \text{ kNm}$$

$$MF_{BC} = -93.75 - 200 = -293.75 \text{ kNm}$$

$$MF_{CB} = 116.25 - 200 = -83.75 \text{ kNm}$$

$$MF_{CD} = -141.67 + 133.33 = -8.33 \text{ kNm}$$

STIFFNESS FACTORS

Joint *B*:

$$K_{BA} = \frac{4EI}{L} = \frac{4E(4I)}{8} = 2EI$$

$$K_{BC} = \frac{4EI}{L} = \frac{4E(3I)}{6} = 2EI$$

Joint *C*:

$$K_{CB} = \frac{4EI}{L} = \frac{4E(3I)}{6} = 2EI$$

$$K_{CD} = \frac{3EI}{L} = \frac{3E(2I)}{6} = EI$$

DISTRIBUTION FACTORS

At joint *B*

$$(DF)_{BA} = \frac{2EI}{2EI + 2EI} = 0.5$$

$$(DF)_{BC} = \frac{2EI}{2EI + 2EI} = 0.5$$

At joint *C*

$$(DF)_{CB} = \frac{2EI}{2EI + EI} = 0.667$$

$$(DF)_{CD} = \frac{EI}{2EI + EI} = 0.333$$

MOMENT DISTRIBUTION TABLE

Table 7.11 shows the distribution of moments and COMs carried out in an iterative procedure. Since the modified stiffness and modified FEM are used for span *CD*, no further distribution of moments is applicable at joint *D*.

FINAL END SHEARS

The shear force and bending moment diagrams are shown in Figure 7.35.

7.6.2 Application to Rigid Joint Plane Frames (Non-Sway)

The application of the moment distribution method to the analysis of statically indeterminate rigid joint frames without lateral sway (side sway) is very similar to its application to indeterminate beams, as discussed in the previous section. However, in the case of frames, more than two members may meet at a joint. In such cases, the balanced moment at any joint must be distributed to the ends of the several members meeting at that joint according to their distribution factors. These distributions can be conveniently arranged in a tabular form while solving the problem. The analysis of rigid joint frames without sway is demonstrated in the following examples.

TABLE 7.11
Moment Distribution Table

Cycle	Joint	A	B		C		D	
	Member	*AB*	*BA*	*BC*	*CB*	*CD*	*DC*	*DE*
	DF		0.500	0.500	0.667	0.333		
	COF	0.500	0.000	0.500	0.500	0.000	0.000	
1	FEM	−310.000	10.000	−293.750	−83.750	−8.330	50.000	−50.00
	Balance and distribution		141.875	141.875	61.387	30.693		
2	COM	70.938		30.693	70.938			
	Balance and distribution		−15.347	−15.347	−47.292	−23.646		
3	COM	−7.673		−23.646	−7.673			
	Balance and distribution		11.823	11.823	5.116	2.558		
4	COM	5.911		2.558	5.911			
	Balance and distribution		−1.279	−1.279	−3.941	−1.970		
5	COM	−0.639		−1.970	−0.639			
	Balance and distribution		0.985	0.985	0.426	0.213		
6	COM	0.493		0.213	0.493			
	Balance and distribution		−0.107	−0.107	−0.328	−0.164		
7	COM	−0.053		−0.164	−0.053			
	Balance and distribution		0.082	0.082	0.036	0.018		
8	COM	0.041		0.018	0.041			
	Balance and distribution		−0.009	−0.009	−0.027	−0.014		
9	COM	−0.004		−0.014	−0.004			
	Balance and distribution		0.007	0.007	0.003	0.001		
	Final end moments (kNm)	−240.98	148.03	−148.03	0.64	−0.64	50.000	−50.00

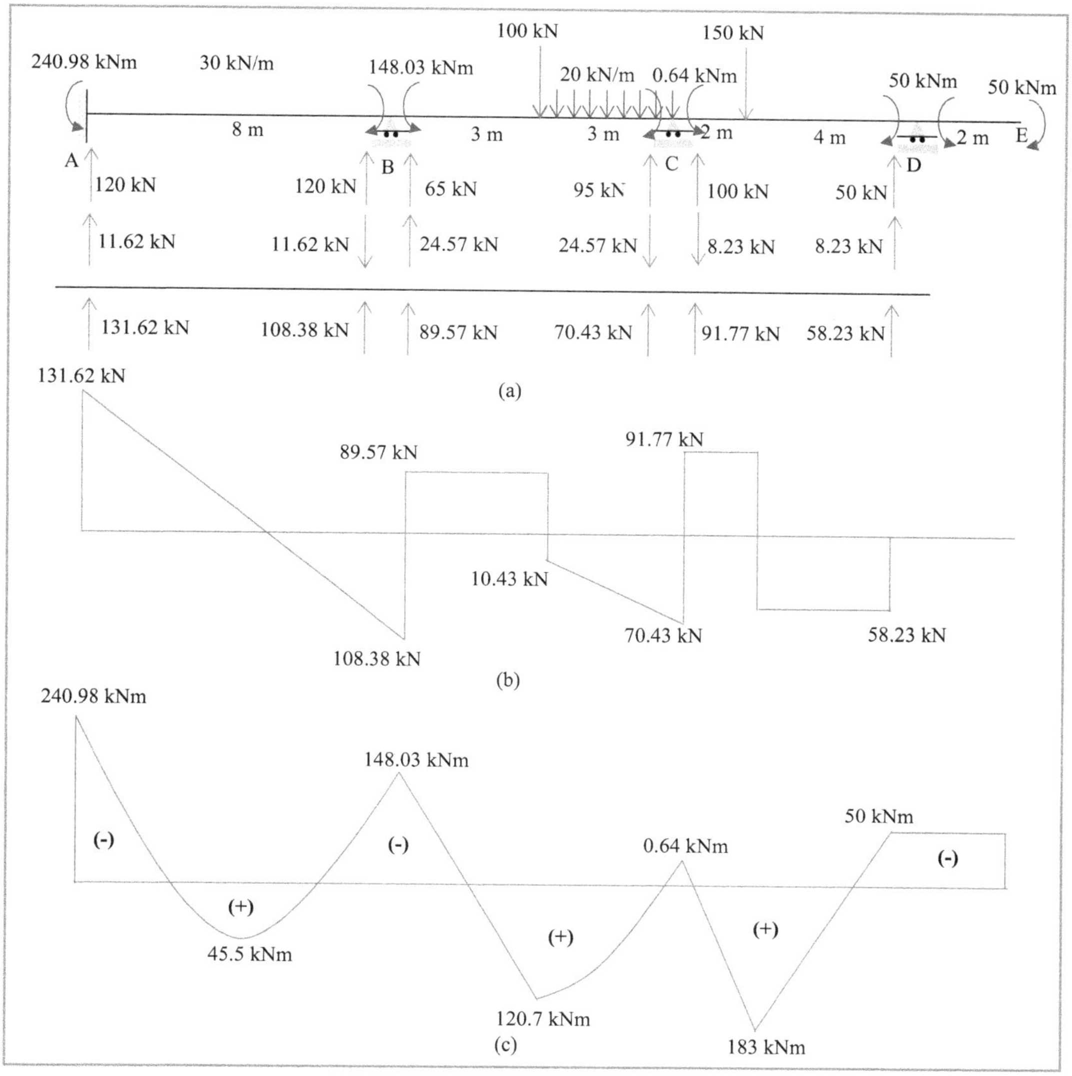

FIGURE 7.35 (a) Free body diagram, (b) shear force diagram, and (c) bending moment diagram.

Example 7.12

Analyse the frame shown in Figure 7.36 using the moment distribution method. Also, construct the shear force and bending moment diagrams.

Solution:

As explained in Example 6.19, the moment at *C* is zero; hence, the rotation at *C* can be neglected by using the modified stiffness and fixed-end moments for span *BC*. Further, the roller support at *C* allows for transverse displacement (Δ_{BC}), which can be eliminated by providing a guided support at *B* and using the modified stiffness and FEM for column *AB*, as shown in Figure 7.37.

This can be visualised as a fixed beam (AB^1) whose length = 2 × *AB*, and subjected to 30 kN at 2 m from both ends *A* and B^1. Therefore, the fixed-end moments for column *AB* are

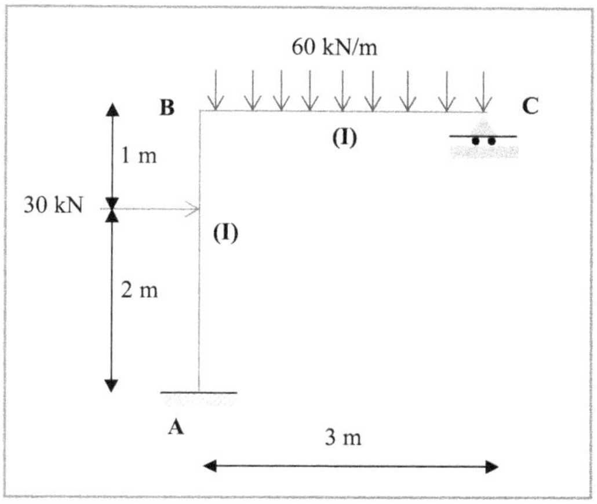

FIGURE 7.36 A simple frame *ABC*.

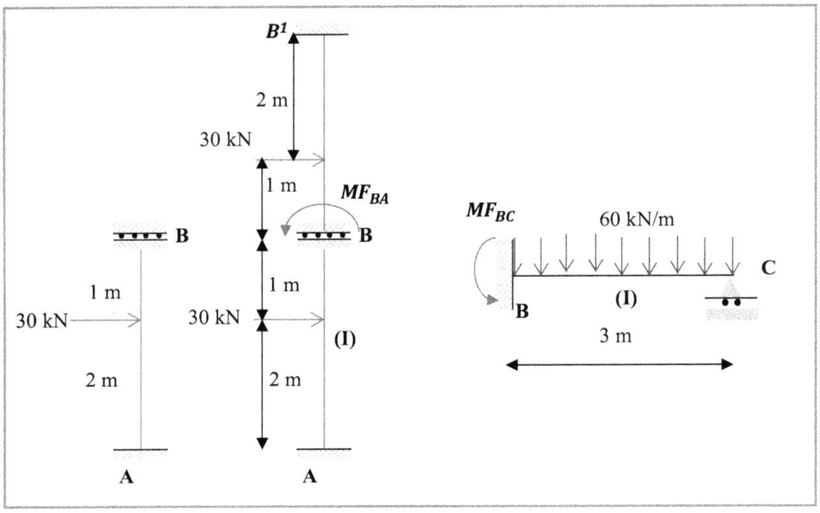

FIGURE 7.37 Fixed-end moments.

$$MF_{AB} = -\frac{Wab}{L^2}(a+b) = -\frac{30\times 2\times 4(2+4)}{6^2} = -40 \text{ kNm}$$

$MF_{BA} = -(30\times 3 - 40 - 30\times 1) = -20$ kNm (Sagging bending moment at mid-span of visualised span ABB^1)

$$MF_{BC} = -\frac{wL^2}{8} = -\frac{60\times 3^2}{8} = -67.5 \text{ kNm}$$

STIFFNESS FACTORS

Joint *B*

$$K_{BA} = \frac{EI}{L} = \frac{EI}{3} = 0.333EI$$

$$K_{BC} = \frac{3EI}{L} = \frac{3EI}{3} = EI$$

DISTRIBUTION FACTORS

The moment distribution shall be carried out at joint *B*.
Joint *B*

$$(DF)_{BA} = \frac{0.333EI}{0.333EI + EI} = 0.25$$

$$(DF)_{BC} = \frac{EI}{0.333EI + EI} = 0.75$$

MOMENT DISTRIBUTION TABLE

The distributed moments and COMs are shown in Table 7.12. The COM occurs in *AB* from *B* to *A* with a COF equal to –1. The member end shears are shown in the free body diagram, and shear force and bending moment diagrams are shown in Figure 7.38.

TABLE 7.12
Moment Distribution Table

Joint	A	B	
Member	*AB*	*BA*	*BC*
DF	—	0.250	0.750
COF	–1.000	0.000	0.000
FEM	–40.000	–20.000	–67.500
Balance and distribution		21.875	65.625
COM	–21.875		
Final end moments (kNm)	–61.875	1.875	–1.875

Example 7.13

Analyse the symmetrically loaded frame (without taking advantage of symmetry), as shown in Figure 7.39, using the moment distribution method.

Solution:

The frame is analysed by taking the kinematic indeterminacy as two, i.e., the rotations at *B* and *C* (without taking advantage of symmetry).

FIXED-END MOMENTS

Using Table 6.1, the fixed-end moments can be calculated as follows:

$$MF_{AB} = -\frac{wL^2}{30} = -\frac{20\times 3^2}{30} = -6 \text{ kNm} = -MF_{DC}$$

$$MF_{BA} = \frac{wL^2}{20} = \frac{20\times 3^2}{20} = 9 \text{ kNm} = -MF_{CD}$$

$$MF_{BC} = -MF_{CB} = -\frac{WL}{8} = -\frac{50\times 4}{8} = -25 \text{ kNm}$$

STIFFNESS FACTORS

Joints *B* and *C*:

$$K_{BA} = K_{CD} = \frac{4EI}{L} = \frac{4EI}{3} = 1.333EI$$

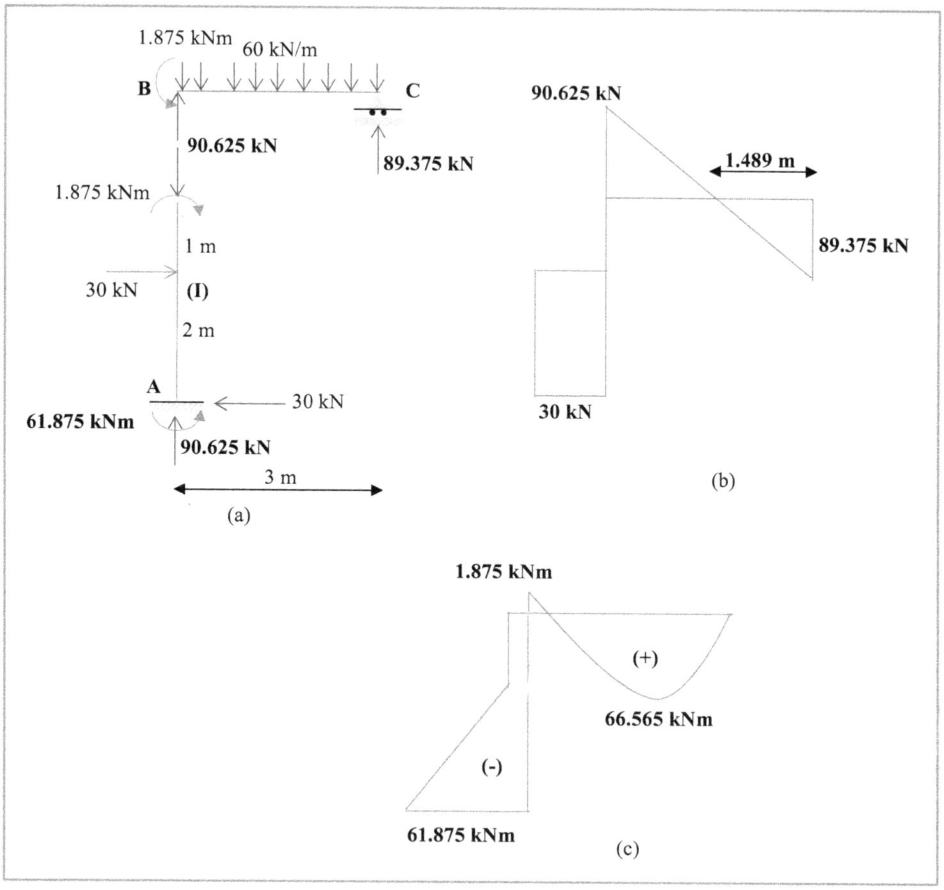

FIGURE 7.38 (a) Free body diagram, (b) shear force diagram, and (c) bending moment diagram.

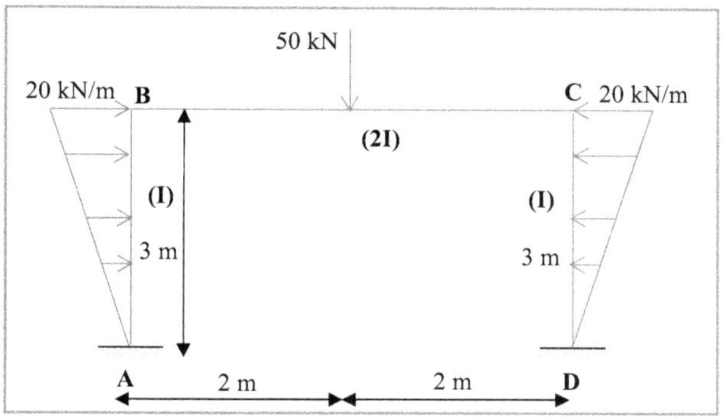

FIGURE 7.39 Symmetrical frame *ABCD*.

$$K_{BC} = K_{CB} = \frac{4EI}{L} = \frac{4E(2I)}{4} = 2EI$$

DISTRIBUTION FACTORS

Joint *B*:

$$(DF)_{BA} = (DF)_{DC} = \frac{1.333EI}{1.333EI + 2EI} = 0.4$$

$$(DF)_{BC} = (DF)_{CB} = \frac{2EI}{1.333EI + 2EI} = 0.6$$

MOMENT DISTRIBUTION TABLE

The iterative procedure of the moment distribution and COMs in each cycle are presented conveniently in a moment distribution Table 7.13.

The free body diagram shows the member end forces in Figure 7.40a. The shear force and bending moment diagrams are shown in Figure 7.40b and c, respectively.

Example 7.14

Re-analyse the Example 7.13 by taking advantage of symmetry.

Solution:

The frame is symmetrical with respect to the vertical centroid axis, i.e., with respect to the centre of span *BC*. Due to this symmetry, the slope of the elastic curve at the centre of span *BC*, i.e., at *E*, is zero. This implies that the shear force at *E* is also zero. By taking advantage of this symmetry, the frame can be truncated at *E* by providing a guided support, as shown in Figure 7.41a. This allows us to limit the kinematic indeterminacy to one, namely the rotation at *B* (θ_B) (neglecting the other degree of freedom Δ_E), by applying the modified stiffness and fixed-end moments for the spans *AB* and *BE*, as shown in Figure 7.41b, thus significantly simplifying the problem.

TABLE 7.13
Moment Distribution Table

Cycle	Joint	A	B		C		D
	Member	*AB*	*BA*	*BC*	*CB*	*CD*	*DC*
	DF		0.400	0.600	0.600	0.400	
	COF	0.500	0.000	0.500	0.500	0.000	0.500
	FEM	−6.000	9.000	−25.000	25.000	−9.000	6.000
1	Balance and distribution		6.400	9.600	−9.600	−6.400	
	COM	3.200		−4.800	4.800		−3.200
2	Balance and distribution		1.920	2.880	−2.880	−1.920	
	COM	0.960		−1.440	1.440	0.000	−0.960
3	Balance and distribution		0.576	0.864	−0.864	−0.576	
	COM	0.288		−0.432	0.432		−0.288
4	Balance and distribution		0.173	0.259	−0.259	−0.173	
	COM	0.086		−0.130	0.130		−0.086
5	Balance and distribution		0.052	0.078	−0.078	−0.052	
	COM	0.026		−0.039	0.039		−0.026
6	Balance and distribution		0.016	0.023	−0.023	−0.016	
	COM	0.008		−0.012	0.012		−0.008
7	Balance and distribution		0.005	0.007	−0.007	−0.005	
	COM	0.002		−0.003	0.003		−0.0025
8	Balance and distribution		0.001	0.002	−0.002	−0.001	
	COM	0.001		−0.001	0.001		−0.0005
	Balance and distribution		0.0004	0.0006	−0.0006	0.0004	
	Final end moments (kNm)	−1.429	18.143	−18.143	18.143	−18.143	1.429
	Check						
	Change	4.571	9.143	6.857	−6.857	−9.143	−4.573
	−1/2 change	−4.571	−2.285	3.429	−3.429	2.287	4.571
	Sum	−0.001	6.857	10.286	−10.286	−6.856	−0.002
	Sum/(*K*)=(Slopes)	0.00	5.143	5.143	−5.142	−5.142	0.00
		Check	Check		Check		Check

FIXED-END MOMENTS

Using Table 6.1,

$$MF_{AB} = -\frac{wL^2}{30} = -\frac{20\times 3^2}{30} = -6 \text{ kNm}$$

$$MF_{BA} = \frac{wL^2}{20} = \frac{20\times 3^2}{20} = 9 \text{ kNm}$$

$$MF_{BE} = -\frac{WL}{8} = -\frac{50\times 4}{8} = -25 \text{ kNm}$$

$$MF_{EB} = -\frac{WL}{8} = -\frac{50\times 4}{8} = -25 \text{ kNm (Mid span moment)}$$

STIFFNESS FACTORS

Joints *B*:

$$K_{BA} = \frac{4EI}{L} = \frac{4EI}{3} = 1.333EI$$

$$K_{BE} = \frac{2EI}{L} = \frac{2E(2I)}{4} = EI$$

DISTRIBUTION FACTORS

Joint *B*:

$$(DF)_{BA} = \frac{1.333EI}{1.333EI + EI} = 0.571$$

$$(DF)_{BE} = \frac{EI}{1.333EI + EI} = 0.429$$

MOMENT DISTRIBUTION TABLE

The locking and unlocking of joint *B*, along with the distribution and COMs is shown in Table 7.14. The COF for *EB* is −1 and for *AB* is +1/2.

The above results are the same as those obtained in Example 7.15.

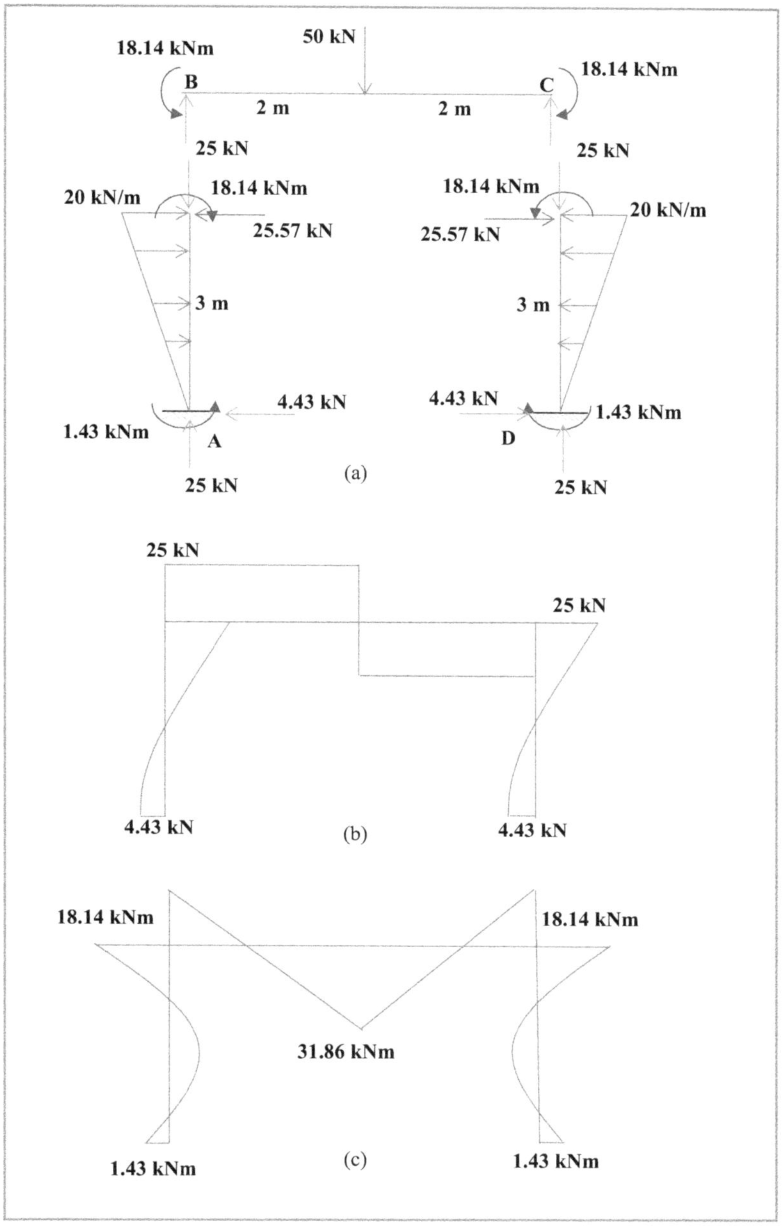

FIGURE 7.40 (a) Free body diagram, (b) shear force diagram, and (c) bending moment diagram.

TABLE 7.14
Moment Distribution Table

Joint	A	B		E
Member	*AB*	*BA*	*BE*	*EB*
DF	—	0.571	0.429	
COF	0.500	0.000	0.000	−1.000
FEM	−6.000	9.000	−25.000	−25.000
Balance and distribution		9.136	6.864	
COM	4.568			−6.864
Final end moments (kNm)	−1.432	18.136	−18.136	−31.864

Example 7.15

Analyse the self-equilibrating box frame shown in Figure 7.42, using the moment distribution method. Draw the shear force and bending moment diagrams.

Solution:

The frame is symmetrical about both vertical and horizontal centroidal axes, i.e., *AB* and *BC*. Therefore, by leveraging this symmetry, it is sufficient to analyse only one-fourth of the frame, with appropriate modifications to the stiffness factors and fixed-end moments for spans *AB* and *BC*, as shown in Figure 7.43.

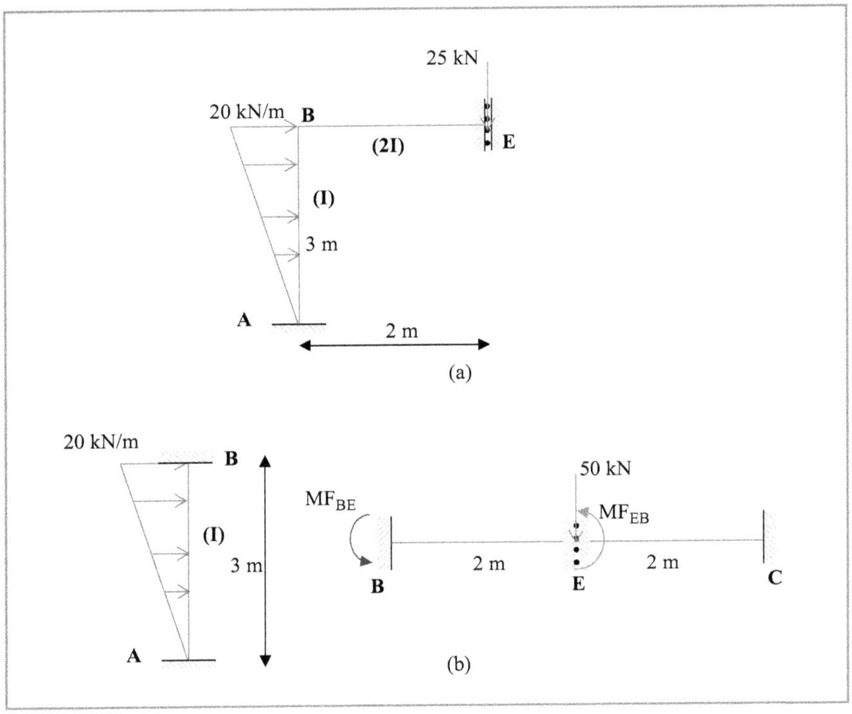

FIGURE 7.41 (a) Truncated half portion of frame *ABE* and (b) fixed-end moments.

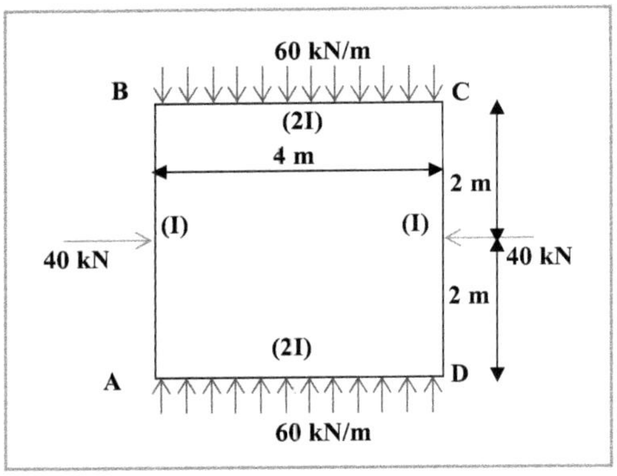

FIGURE 7.42 Symmetrical box frame *ABCD*.

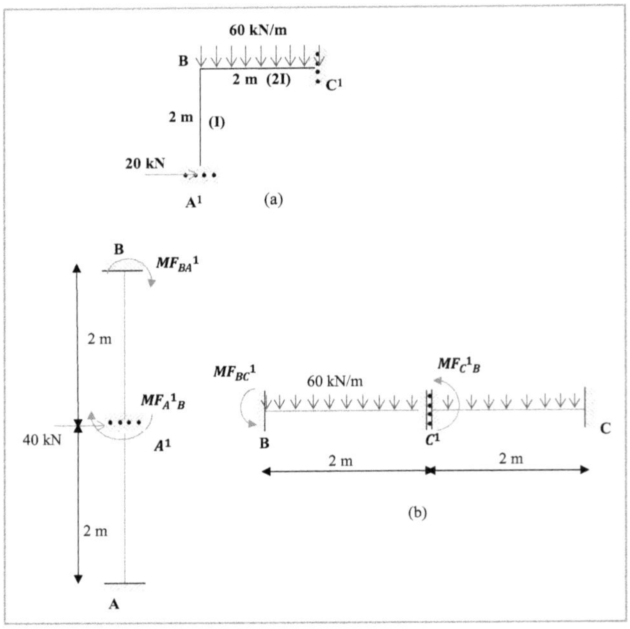

FIGURE 7.43 (a) Truncated half portion of frame A^1BC^1 and (b) fixed-end moments.

FIXED-END MOMENTS

The fixed-end moments, as shown in Figure 7.43a, are calculated using the expressions given in Table 6.1.

$$MF_{BA^1} = \frac{WL}{8} = -\frac{40 \times 4}{8} = 20 \text{ kNm}$$

$$MF_{BC^1} = -\frac{wL^2}{12} = -\frac{60 \times 4^2}{12} = -80 \text{ kNm}$$

$$MF_{C^1B} = -\left(\frac{wL^2}{8} - \frac{wL^2}{12}\right) = -\frac{wL^2}{24} = -\frac{60 \times 4^2}{24} = -40 \text{ kNm}$$

$$MF_{A^1B} = \left(\frac{WL}{4} - \frac{WL}{8}\right) = \frac{WL}{8} = \frac{40 \times 4}{8} = 20 \text{ kNm}$$

STIFFNESS FACTORS

The modified stiffness factors for A^1B and BC^1 are used due to the guided support provided at the centre of AB and BC.

Joint B:

$$K_{BA^1} = \frac{2EI}{L} = \frac{2EI}{4} = 0.5EI$$

$$K_{BC^1} = \frac{2EI}{L} = \frac{2E(2I)}{4} = EI$$

DISTRIBUTION FACTORS

Joint B:

$$(DF)_{BA^1} = \frac{0.5EI}{0.5EI + EI} = 0.333$$

$$(DF)_{BC^1} = \frac{EI}{0.5EI + EI} = 0.667$$

MOMENT DISTRIBUTION TABLE

The balancing and distribution of moments and COM are shown in Table 7.15. It is to be noted that the COF for C^1B and A^1B is −1.

FINAL END SHEARS

The member end shears are calculated using the static equilibrium equations after obtaining the final member end moments from the table, as shown in Figure 7.44a.

TABLE 7.15
Moment Distribution Table

Joint	A	B		C¹
Member	AB^1	B^1A	BC^1	C^1B
DF	—	0.333	0.667	—
COF	−1.000	0.000	0.000	−1.000
FEM	20.000	20.000	−80.000	−40.000
Balance and distribution		19.98	40.02	
COM	−19.98			−40.02
Final end moments (kNm)	0.00	39.98	−39.98	−80.02

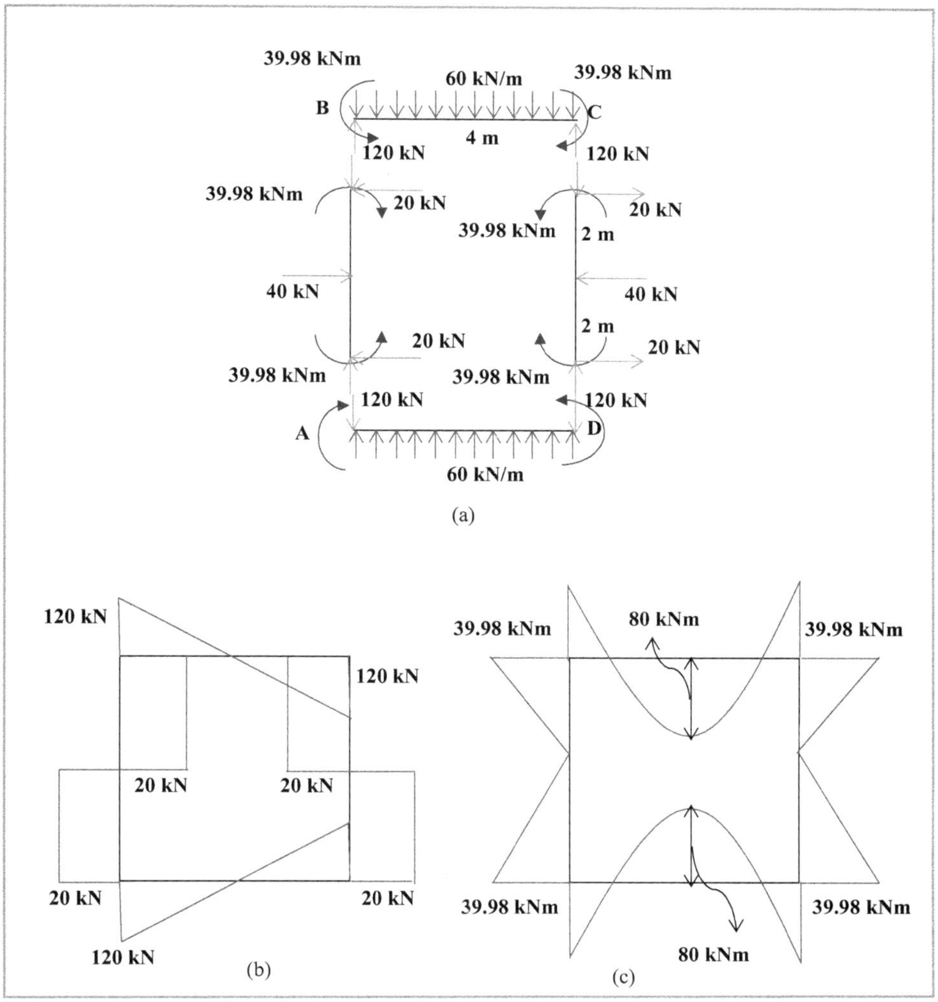

FIGURE 7.44 (a) Free body diagram, (b) shear force diagram, and (c) bending moment diagram.

The shear force and bending moment diagrams are shown in Figure 7.44b and c, respectively.

Example 7.16

Analyse the symmetrical frame with a horizontal nodal load of 100 kN shown in Figure 7.45, using the moment distribution method. Assume EI=108,000 kNm^2. (The same problem was solved in Example 6.18 using the SD method.)

Solution:

The symmetric frame subjected to a nodal load is shown in Figure 7.45. This load induces anti-symmetry in the deflected shape, as illustrated in the figure. Specifically, the deflection at the centre of span BC is zero. Using the principle of superposition, the given frame can now be split into two parts by introducing a roller support at the centre of span BC, i.e., at point E, as shown in Figure 7.46a. Thus, the analysis can be simplified by leveraging the anti-symmetry of the response and analysing only the half-frame ABE, where point E (the midpoint of BC) is provided with a roller support. Further, by taking advantage of the roller support and using the modified stiffness for BE, the rotation at E can be neglected. Consequently, the kinematic indeterminacy is limited to the rotation at point B, denoted as θ_B. When this rotation is restrained, the column AB behaves as a fixed beam subjected to chord rotation, which induces equal fixed-end moments in column AB and zero moments in beam BE, as a roller support is provided at E. The fixed-end moments are illustrated in Figure 7.46b. The procedure for solving this problem is the same as that used in Example 7.12.

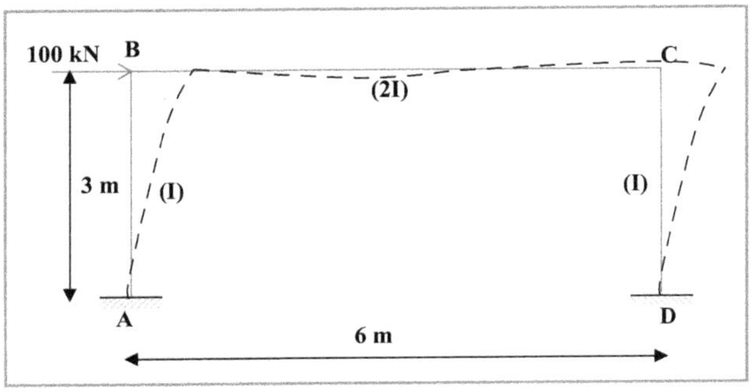

FIGURE 7.45 Symmetrical frame with anti-symmetric loading.

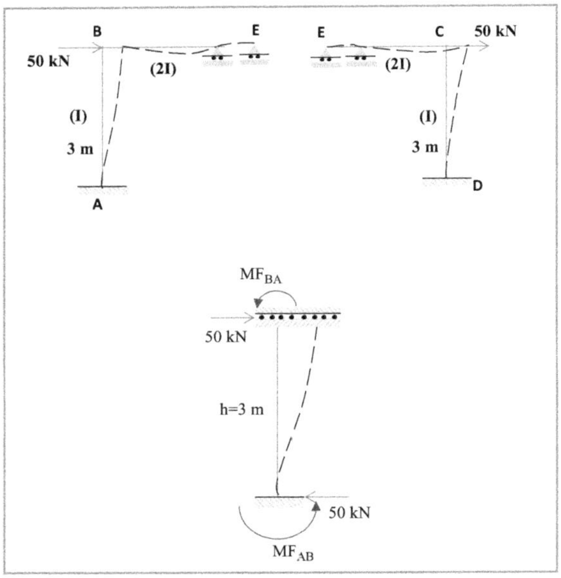

FIGURE 7.46 (a) Anti-symmetric loading and response and (b) fixed-end moments.

FIXED-END MOMENTS

The chord rotation induces an equal amount of the fixed-end moments in AB (or refer to the calculation of fixed moments in Example 7.12 in Figure 7.37).

$$MF_{AB} = MF_{BA} = -\frac{W \times h}{2} = -\frac{50 \times 3}{2} = -75 \text{ kNm}$$

$$MF_{BE} = 0 \quad (\text{Since roller support at } B)$$

STIFFNESS FACTORS

Use modified stiffness factors for AB and BE, as shown below.

$$K_{BA} = \frac{EI}{L} = \frac{EI}{3} = 0.333EI$$

$$K_{BE} = \frac{3EI}{L} = \frac{3E(2I)}{3} = 2EI$$

DISTRIBUTION FACTORS

$$(DF)_{BA} = \frac{0.333EI}{0.333EI + 2EI} = 0.143$$

$$(DF)_{BE} = \frac{2EI}{0.333EI + 2EI} = 0.857$$

MOMENT DISTRIBUTION TABLE

The distribution of balancing, distribution, and COMs is presented in Table 7.16. The COF for AB from B to A is –1 due to the guided support at B. The member end forces, shear force, and bending moment diagrams are shown in Figure 7.47.

Example 7.17

ABCDE is a rigid joint plane frame shown in Figure 7.48, using the moment distribution method. Assume the flexural rigidity is constant.

Solution:

The kinematic indeterminacy of the given frame is 2, i.e., rotations at B and C.

TABLE 7.16
Moment Distribution Table

Joint	*A*	*B*	*E*
Member	*AB*	*BA*	*BE*
DF	1.000	0.143	0.857
COF	−1.000	0.000	0.000
FEM	−75.000	−75.000	0.000
Balance and distribution		10.714	64.286
COM	−10.714		
Final end moments (kNm)	−85.714	−64.286	64.286

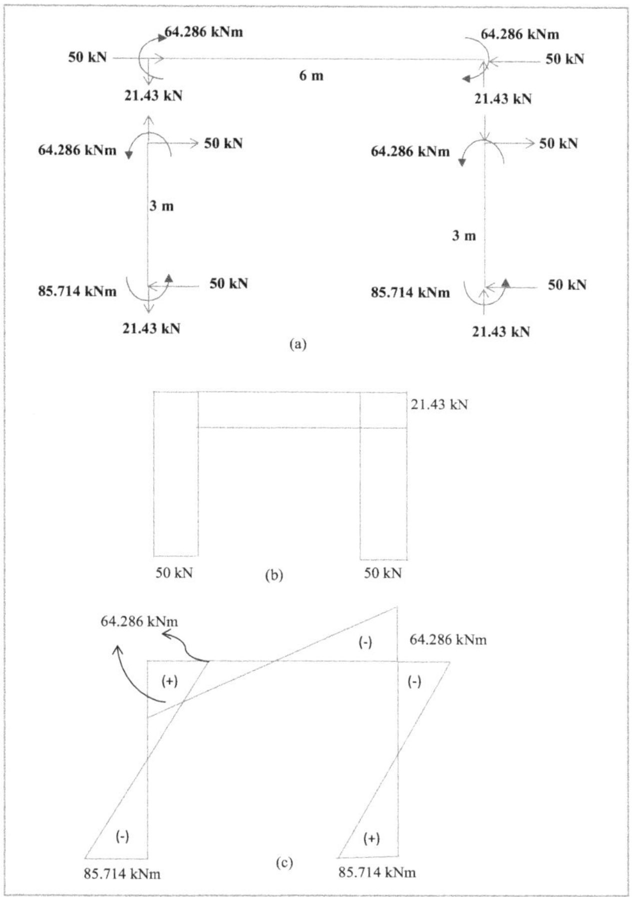

FIGURE 7.47 (a) Free body diagram, (b) shear force diagram, and (c) bending moment diagram.

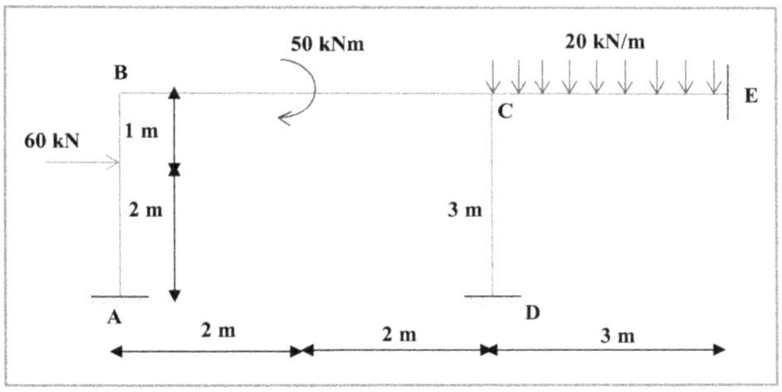

FIGURE 7.48 Rigid joint frame.

FIXED-END MOMENTS

The fixed-end moments can be calculated using Table 6.1, as shown below.

$$MF_{AB} = -\frac{60 \times 2 \times 1^2}{3^2} = -\ 13.333 \text{ kNm}$$

$$MF_{BA} = \frac{60 \times 1 \times 2^2}{3^2} = 26.667 \text{ kNm}$$

$$MF_{BC} = MF_{CB} = \frac{50}{4} = 12.5 \text{ kNm}$$

$$MF_{CD} = MF_{DC} = 0\,(\text{since no load})$$

$$MF_{CE} = -MF_{CE} = -\frac{20 \times 3^2}{12} = -15 \text{ kNm}$$

STIFFNESS FACTORS

Joint *B*

$$K_{BA} = \frac{4EI}{L} = \frac{4EI}{3} = 1.333EI$$

$$K_{BC} = \frac{4EI}{L} = \frac{4EI}{4} = EI$$

Joint *C*

$$K_{CB} = \frac{4EI}{L} = \frac{4EI}{4} = EI$$

$$K_{CD} = K_{CE} = \frac{4EI}{L} = \frac{4EI}{3} = 1.333EI$$

DISTRIBUTION FACTORS

Joint *B*

$$(DF)_{BA} = \frac{1.333EI}{1.333EI + EI} = 0.571$$

$$(DF)_{BC} = \frac{EI}{1.333EI + EI} = 0.429$$

Joint *C*

$$(DF)_{CB} = \frac{EI}{2 \times 1.333EI + EI} = 0.273$$

$$(DF)_{CD} = \frac{1.333EI}{2 \times 1.333EI + EI} = 0.364$$

$$(DF)_{CE} = \frac{1.333EI}{2 \times 1.333EI + EI} = 0.364$$

MOMENT DISTRIBUTION TABLE

At joint *C*, three members meet and are conveniently arranged for moment distribution analysis. The successive operations of moment distribution and COMs

TABLE 7.17
Moment Distribution Table

Cycle	Joint	A	B		C		D		E
	Member	*AB*	*BA*	*BC*	*CB*	*CD*	*DC*	*CE*	*EC*
	DF	—	0.571	0.429	0.273	0.364		0.364	—
	COF	0.500	0.000	0.500	0.500	0.000	0.500	0.000	0.5
1	FEM	−13.333	26.667	12.500	12.500	0.000	0.000	−15.000	15
	Balance and distribution		−22.381	−16.786	0.682	0.909		0.909	
2	COM	−11.191		0.341	−8.393		0.455		0.4545
	Balance and distribution		−0.195	−0.146	2.289	3.052		3.052	
3	COM	−0.097		1.144	−0.073		1.526		1.5260
	Balance and distribution		−0.654	−0.490	0.020	0.027		0.027	
4	COM	−0.327		0.010	−0.245		0.013		0.0133
	Balance and distribution		−0.006	−0.004	0.067	0.089		0.089	
5	COM	−0.003		0.033	−0.002		0.045		0.0446
	Balance and distribution		−0.019	−0.014	0.001	0.001		0.001	
6	COM	−0.010		0.0005	−0.007		0.0005		0.0004
	Balance and distribution		0.0002	0.0003	0.002	0.0025		0.0025	
	Final end moments (kNm)	−24.961	3.412	−3.412	6.839	4.080	2.038	−10.920	17.0401
	Check for rotations								
	Change	−11.628	−23.255	−15.912	−5.661	4.080	2.038	4.080	2.040
	−1/2 change	11.628	5.814	2.830	7.956	−1.019	−2.040	−1.019	−2.040
	Sum	0.000	−17.441	−13.081	2.295	3.061	−0.002	3.061	0.000
	Sum/(*K*)	0.000	13.081	13.081	−2.295	−2.296	0.002	−2.296	0.000
		Check (θ_A)		Check (θ_B)		Check (θ_C)	Check (θ_D)	Check (θ_C)	Check (θ_E)

in each cycle are presented in the moment distribution Table 7.17. The changes in COMs at all joints become negligible after the sixth cycle. Hence, no further cycles are required. The final member end moments are summarised in the table, where a moment distribution check is also performed. It is observed that the rotations at the fixed ends are zero, and the rotations at joints *B* and *C* are also satisfied.

The shear force and bending moment diagrams are shown in Figure 7.49.

7.6.3 Application to Rigid Joint Plane Frames (Sway)

In the previous sections, continuous beams and rigid joint plane frames subjected to rotational degrees of freedom (non-sway) were discussed and demonstrated with numerous examples. In these cases, moment distribution could be conveniently carried out in a single moment distribution table. Additionally, some examples included beams and frames subjected to support yielding, where chord rotations were calculated based on known values of support movement. The additional locking moments induced by these effects were incorporated into the moment distribution table.

However, when a frame is subjected to both rotational degrees of freedom and translational degrees of freedom (lateral sway), the chord rotations depend on the unknown lateral sway, making it impossible to directly calculate the additional locking moments due to sway. Therefore, the procedure used in previous examples is not directly applicable to such problems.

To address these cases, moment distribution must be carried out in two stages:

Stage 1: Perform moment distribution as usual, assuming the frame is subjected only to rotational degrees of freedom. This is referred to as *non-sway analysis.*

Stage 2: Determine the lateral force causing the lateral sway by applying the horizontal force equilibrium equation using the end moments obtained in Stage 1. Then, perform moment distribution considering the frame subjected only to lateral sway.

The desired end moments are obtained by superimposing the end moments from Stage 1 and Stage 2. This approach is demonstrated in detail in the following examples.

Example 7.18

Analyse the rigid joint frame *ABCD* shown in Figure 7.50 using the moment distribution method. Draw the shear force and bending moment diagrams.

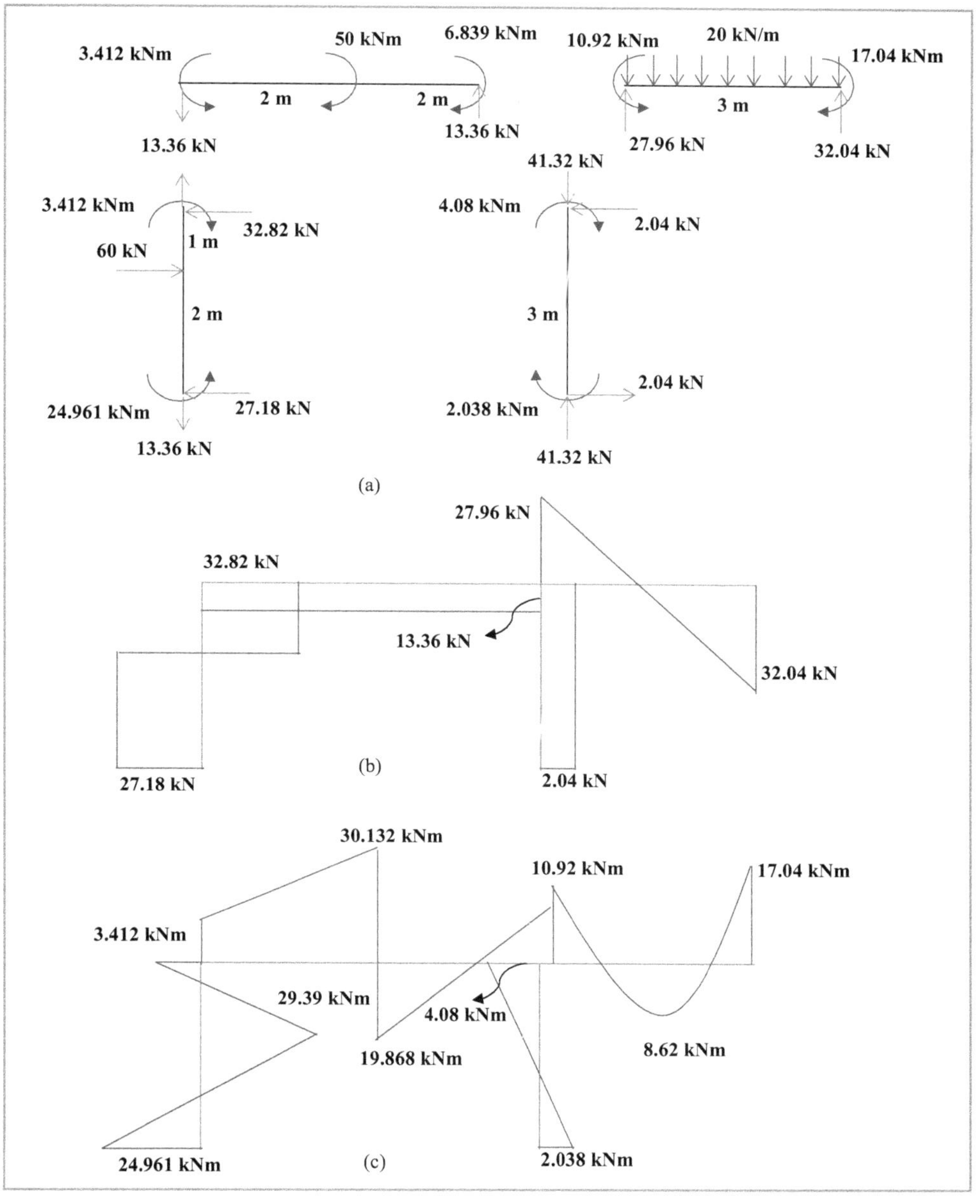

FIGURE 7.49 (a) Free body diagram, (b) shear force diagram, and (c) bending moment diagram.

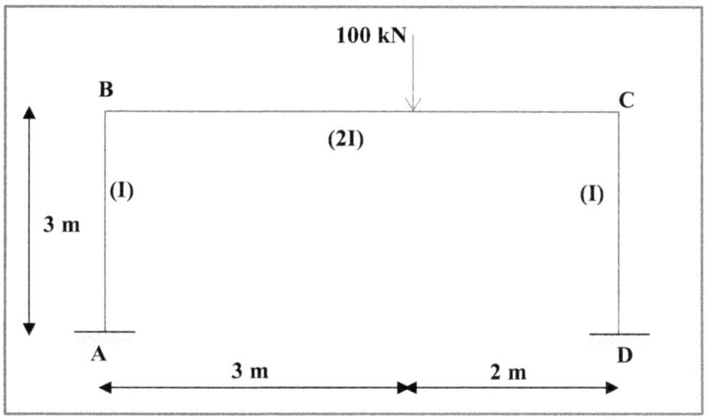

FIGURE 7.50 Rigid joint frame *ABCD* with unsymmetrical loading.

Solution:

Although the given frame is symmetrical, the beam *BC* is subjected to an eccentric point load, causing it to experience side sway. The kinematic indeterminacy of the frame is 3, consisting of the rotations at points *B* and *C* and the lateral sway of *BC* (neglecting axial deformations). Problems of this nature are analysed in two stages:

1. Frame without side sway
2. Frame subjected only to side sway (without loading)

WITHOUT SIDE SWAY

The analysis procedure of this stage is the same as that adopted in the previous examples. Neglect the side sway and consider the rotations at B and C as unknown rotations in the frame. Calculate the fixed-end moments accordingly.

FIXED-END MOMENTS

Using Table 6.1, the fixed-end moments can be calculated as

$$MF_{BC} = -\frac{100\times3\times2^2}{5^2} = -\ 48\ \text{kNm}$$

$$MF_{CB} = \frac{100\times2\times3^2}{5^2} = 72\ \text{kNm}$$

$$MF_{AB} = MF_{BA} = MF_{CD} = MF_{DC} = 0\,(\text{since no load})$$

STIFFNESS FACTORS

Joint *B*:

$$K_{BA} = \frac{4EI}{L} = \frac{4EI}{3} = 1.333EI$$

$$K_{BC} = \frac{4EI}{L} = \frac{4E(2I)}{5} = 1.6EI$$

Joint *C*:

$$K_{CB} = \frac{4EI}{L} = \frac{4E(2I)}{5} = 1.6EI$$

$$K_{CD} = \frac{4EI}{L} = \frac{4EI}{3} = 1.333EI$$

DISTRIBUTION FACTORS

Joint *B*:

$$(DF)_{BA} = \frac{1.333EI}{1.333EI+1.6EI} = 0.455$$

$$(DF)_{BC} = \frac{1.6EI}{1.333EI+1.6EI} = 0.545$$

Joint *C*:

$$(DF)_{CB} = 0.545$$

$$(DF)_{CD} = 0.455$$

MOMENT DISTRIBUTION TABLE

The moment distribution and COMs in each cycle are presented in the moment distribution Table 7.18.

TABLE 7.18
Moment Distribution Table (Without Side Sway)

Joint	A	B		C		D
Member	*AB*	*BA*	*BC*	*CB*	*CD*	*DC*
DF		0.455	0.545	0.545	0.455	
COF	0.500	0.000	0.500	0.500	0.000	0.500
FEM	0.000	0.000	−48.000	72.000	0.000	0.000
Balance and distribution		21.818	26.182	−39.273	−32.727	
COM	10.909		−19.636	13.091		−16.364
Balance and distribution		8.926	10.711	−7.140	−5.950	
COM	4.463		−3.570	5.355		−2.975
Balance and distribution		1.623	1.947	−2.921	−2.434	
COM	0.811		−1.461	0.974		−1.217
Balance and distribution		0.664	0.797	−0.531	−0.443	
COM	0.332		−0.266	0.398		−0.221
Balance and distribution		0.121	0.145	−0.217	−0.181	
COM	0.060		−0.109	0.072		−0.091
Balance and distribution		0.049	0.059	−0.040	−0.033	
COM	0.025		−0.020	0.030		−0.016
Balance and distribution		0.009	0.011	−0.016	−0.013	
COM	0.004		−0.008	0.005		−0.007
Balance and distribution		0.004	0.004	−0.003	−0.002	
COM	0.002		−0.001	0.002		−0.001
Balance and distribution	−0.001	0.0004	0.0006	−0.0009	−0.0011	
Final end moments (kNm)	16.606	33.214	−33.214	41.785	−41.785	−20.894
Check						
Change	16.606	33.214	14.786	−30.215	−41.785	−20.894
−1/2 change	−16.607	−8.303	15.107	−7.393	10.447	20.893
Sum	−0.001	24.911	29.893	−37.608	−31.338	−0.001
Sum/(−*K*)	0.001	−18.683	−18.683	23.505	23.504	0.001

FRAME WITH SIDE SWAY ANALYSIS

The free body diagrams of the columns *AB* and *CD* are shown in Figure 7.51. Apply the horizontal force equilibrium equation as shown below.

$$\sum F_X = 0 \Rightarrow H_A + H_D = 0$$

$$\Rightarrow -\left(\frac{M_{AB}+M_{BA}}{h}\right)-\left(\frac{M_{CD}+M_{DC}}{h}\right)=0$$

$$\Rightarrow -\left(\frac{16.606+33.214}{3}\right)-\left(\frac{-41.785-20.894}{3}\right)=0$$

$$\Rightarrow 4.317 \text{ kN} \neq 0$$

Thus, $\sum F_X = 0$ is not satisfied by the solution obtained in previously. In the above analysis, the frame was analysed by preventing side sway. This implies that the lateral deflection of beam *BC* was restrained by some form of lateral support applied at either point *B* or *C*, as shown in Figure 7.52. To satisfy the equilibrium condition $\sum F_X = 0$, the frame must be analysed separately for a pure sway loading case, as shown in figure, where the frame is subjected to a lateral load of 4.317 kN (←) acting towards the left. To obtain the final results, the outcomes of the non-sway analysis and the sway analysis need to be superimposed.

FIXED-END MOMENTS

Using Figure 7.52, the fixed-end moments can be calculated as

$$MF_{AB} = MF_{BA} = MF_{CD} = MF_{DC} = \frac{6EI\delta}{L^2} = \frac{6EI\delta}{9} = 0.667EI\delta$$

$$MF_{BC} = MF_{CB} = 0$$

Now we shall carry out the moment distribution with the above fixed-end moments, as carried out earlier. For convenience, let us take $MF_{AB} = MF_{BA} = MF_{CD} = MF_{DC} = 6{,}670$ kNm and then apply a multiplication factor=$0.0001EI\delta$ to the final results.

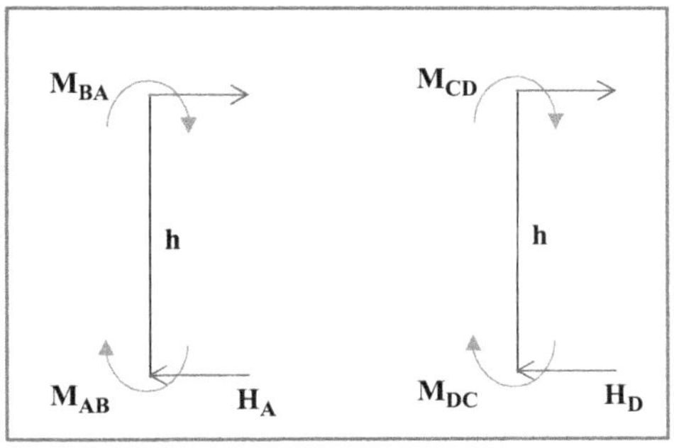

FIGURE 7.51 Free body diagram of the columns.

MOMENT DISTRIBUTION

The distribution factors calculated earlier can be used. The moment distribution and COMs in each cycle are carried out conveniently in a moment distribution Table 7.19, as shown.

To get the final δ, we use $\sum F_X = 0$.

$$\sum F_X = H_A + H_D + P = 0$$

$$\Rightarrow -\left(\frac{M_{AB}+M_{BA}}{h}\right)-\left(\frac{M_{CD}+M_{DC}}{h}\right)+P=0$$

$$\Rightarrow \frac{5{,}478.965+4{,}287.837}{3}\times 2\times 0.0001EI\delta = P = 4.317$$

$$\therefore EI\delta = 6.603 \text{ kNm}^3$$

FINAL END MOMENTS DUE TO SWAY

To get the final end moments, multiply the end moments obtained in Table 7.19 with $EI\delta$ obtained as above.

$$M_{AB} = M_{DC} = 5{,}478.965\times(6.603\times 0.0001) = 3.617 \text{ kNm}$$

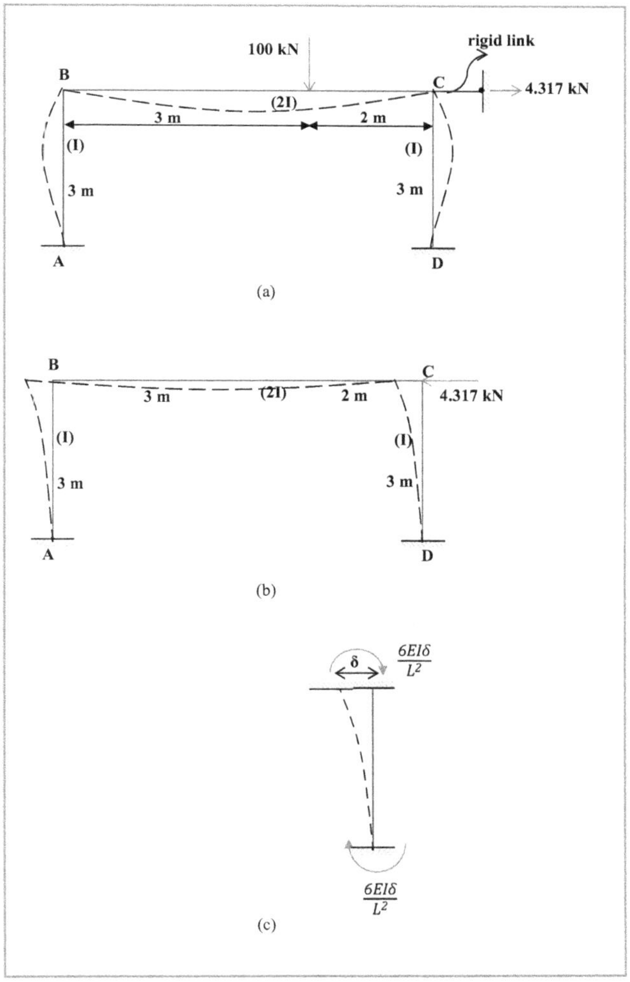

FIGURE 7.52 Rigid frame subjected to non-sway and pure sway loading.

TABLE 7.19
Moment Distribution Table (with Side-Sway Only)

Joint	*A*	*B*		*C*		*D*
Member	*AB*	*BA*	*BC*	*CB*	*CD*	*DC*
DF		0.455	0.545	0.545	0.455	
COF	0.500	0.000	0.500	0.500	0.000	0.500
FEM	6,670.000	6,670.000	0.000	0.000	6,670.000	6,670.000
Balance and distribution		−3,031.818	−3,638.182	−3,638.182	−3,031.818	
COM	−1,515.909		−1,819.091	−1,819.091		−1,515.909
Balance and distribution		826.860	992.231	992.231	826.860	
COM	413.430		496.116	496.116		413.430
Balance and distribution		−225.507	−270.609	−270.609	−225.507	
COM	−112.754		−135.304	−135.304		−112.754
Balance and distribution		61.502	73.802	73.802	61.502	
COM	30.751		36.901	36.901		30.751
Balance and distribution		−16.773	−20.128	−20.128	−16.773	
COM	−8.387		−10.064	−10.064		−8.387
Balance and distribution		4.575	5.489	5.489	4.575	
COM	2.287		2.745	2.745		2.287
Balance and distribution		−1.248	−1.497	−1.497	−1.248	
COM	−0.624		−0.749	−0.749		−0.624
Balance and distribution		0.340	0.408	0.408	0.340	
COM	0.170		0.204	0.204		0.170
Balance and distribution	−0.001	−0.093	−0.111	−0.111	−0.093	−0.002
Final end moments (kNm)	5,478.965	4,287.837	−4,287.837	−4,287.837	4,287.837	5,478.963
Change	−1,191.035	−2,382.163	−4,287.837	−4,287.837	−2,382.163	−1,191.037
−1/2 change	1,191.081	595.518	2,143.919	2,143.919	595.518	1,191.081
Sum	0.046	−1,786.645	−2,143.919	−2,143.919	−1,786.644	0.044
Sum/(−*K*)	−0.034	1,339.984	1,339.949	1,339.949	1,339.983	−0.033

$$M_{BA} = M_{CD} = 4,287.837 \times (6.603 \times 0.0001) = 2.831 \text{ kNm}$$

$$M_{BC} = M_{CB} = -4,287.837 \times (6.603 \times 0.0001) = -2.831 \text{ kNm}$$

RESULTANT MOMENTS

The resultant end moments can be obtained by summing the moments obtained from the analysis without sway and those from pure sway analysis, as presented below.

$$M_{AB} = 16.606 + 3.617 = 20.223 \text{ kNm}$$

$$M_{BA} = 33.214 + 2.831 = 36.045 \text{ kNm}$$

$$M_{BC} = -33.214 - 2.831 = -36.045 \text{ kNm}$$

$$M_{CB} = 41.785 - 2.831 = 38.954 \text{ kNm}$$

$$M_{CD} = -41.785 + 2.831 = -38.954 \text{ kNm}$$

$$M_{DC} = -20.894 + 3.617 = -17.277 \text{ kNm}$$

The free body diagram shows the reactions, and the shear force and bending moment diagrams are shown in Figure 7.53.

ALTERNATE SOLUTION FOR SWAY ANALYSIS

By leveraging the anti-symmetry of the frame, the problem can be solved more easily. As demonstrated in Example 7.16, the fixed-end moments, stiffness factors, and distribution factors are calculated as follows. The frame with anti-symmetric loading is shown in Figure 7.54.

FIXED-END MOMENTS

$$MF_{AB} = MF_{BA} = MF_{CD} = MF_{DC} = \frac{4.317 \times 3}{2 \times 2} = 3.237 \text{ kNm}$$

$$MF_{CE} = MF_{EC} = 0$$

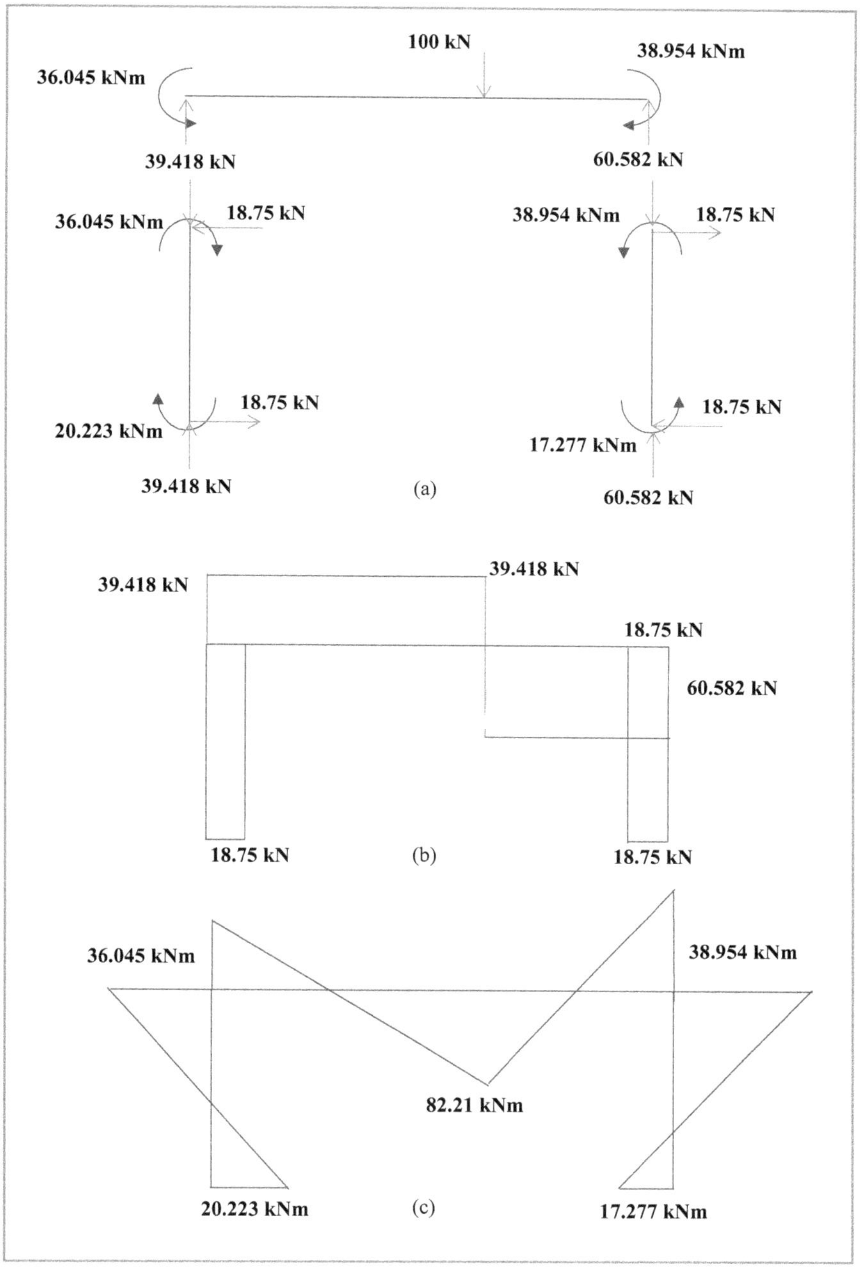

FIGURE 7.53 (a) Free body diagram, (b) shear force diagram, and (c) bending moment diagram.

STIFFNESS FACTORS

$$K_{CD} = \frac{EI}{L} = \frac{EI}{3}$$

$$K_{CE} = \frac{3EI}{L} = \frac{3E(2I)}{2.5} = 2.4EI$$

DISTRIBUTION FACTORS

$$(DF)_{CD} = \frac{\frac{1}{3}EI}{\frac{1}{3}EI + 2.4EI} = 0.122$$

$$(DF)_{CE} = \frac{2.4EI}{\frac{1}{3}EI + 2.4EI} = 0.878$$

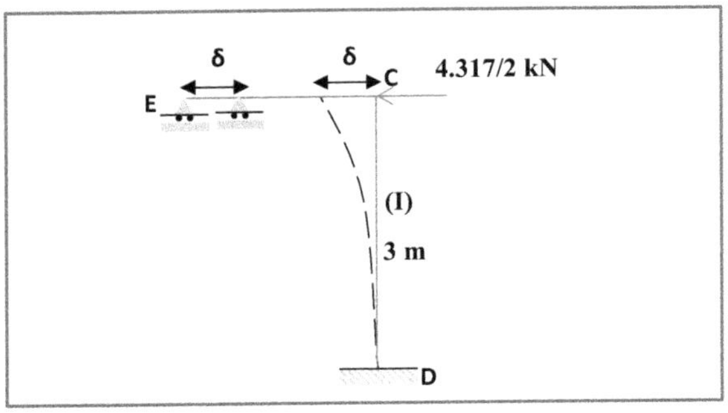

FIGURE 7.54 Anti-symmetric loading and response.

TABLE 7.20
Moment Distribution Table

Joint	D	C	
Member	*DC*	*CD*	*CE*
DF	—	0.122	0.878
COF	−1.0	0.000	0.000
FEM	3.237	3.237	0.000
Balance and distribution		−0.394	−2.843
COM	0.394		
Balance and distribution	3.631	2.843	−2.843

The moments distribution is shown in Table 7.20.

The above results are the same as those obtained previously in the sway analysis.

RESULTANT MOMENTS

The net moments can be obtained by summing the moments from the analysis without sway and those from pure sway analysis, i.e., the moments obtained in Table 7.20.

$$M_{AB} = 16.606 + 3.631 = 20.237 \text{ kNm}$$

$$M_{BA} = 33.214 + 2.843 = 36.057 \text{ kNm}$$

$$M_{BC} = -33.214 - 2.843 = -36.057 \text{ kNm}$$

$$M_{CB} = 41.785 - 2.843 = 38.9424 \text{ kNm}$$

$$M_{CD} = -41.785 + 2.843 = -38.942 \text{ kNm}$$

$$M_{DC} = -20.894 + 3.631 = -17.263 \text{ kNm}$$

Example 7.19

Analyse the unsymmetrical frame *ABCD* as shown in Figure 7.55 using the moment distribution method.

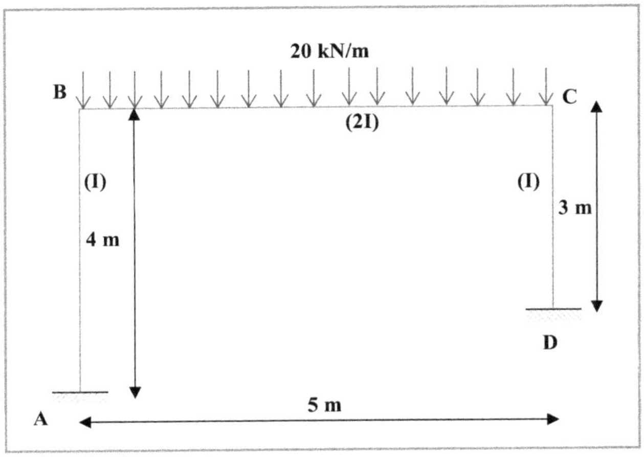

FIGURE 7.55 Rigid frame *ABCD* with unequal column heights.

Solution:

WITHOUT SWAY ANALYSIS

Neglect the side sway and consider the rotations at *B* and *C* as unknown rotations in the frame and calculate the fixed-end moments.

FIXED-END MOMENTS

Using Table 6.1, the fixed moments can be calculated as follows:

$$MF_{AB} = MF_{BA} = MF_{CD} = MF_{DC} = 0$$

$$MF_{BC} = -MF_{CB} = -\frac{20 \times 5^2}{12} = -41.667 \text{ kNm}$$

STIFFNESS FACTORS

Joint *B*

$$K_{BA} = \frac{4EI}{L} = \frac{4EI}{4} = EI$$

$$K_{BC} = \frac{4EI}{L} = \frac{4E(2I)}{5} = 1.6EI$$

Joint *C*

$$K_{CB} = \frac{4EI}{L} = \frac{4E(2I)}{5} = 1.6EI$$

$$K_{CD} = \frac{4EI}{L} = \frac{4EI}{3} = 1.333EI$$

DISTRIBUTION FACTORS

Joint *B*

$$(DF)_{BA} = \frac{EI}{EI + 1.6EI} = 0.385$$

$$(DF)_{BC} = \frac{1.6EI}{EI + 1.6EI} = 0.615$$

Joint *C*

$$(DF)_{CB} = \frac{1.6EI}{1.6EI + 1.333EI} = 0.545$$

$$(DF)_{CD} = \frac{1.333EI}{1.6EI + 1.333EI} = 0.455$$

MOMENT DISTRIBUTION TABLE

The moment distribution and COMs in each cycle is presented in the moment distribution Table 7.21.

FRAME WITH SIDE SWAY ANALYSIS

The free body diagrams of the columns *AB* and *CD* are shown in Figure 7.56. Apply the horizontal force equilibrium equation.

$$\sum F_X = 0 \Rightarrow H_A + H_D = 0$$

$$\Rightarrow -\left(\frac{M_{AB} + M_{BA}}{h}\right) - \left(\frac{M_{CD} + M_{DC}}{h}\right) = 0$$

$$\Rightarrow \left(\frac{11.131 + 22.264}{4}\right) - \left(\frac{-27.035 - 13.519}{3}\right) = 0$$

$$\Rightarrow 5.169 \text{ kN} \neq 0$$

Thus, $\sum F_X = 0$ is not satisfied by the solution obtained previously. In the above analysis, the frame was analysed by preventing the side-sway. This implies that the lateral deflection in the beam *BC* was arrested by some form of lateral support applied at either *B* or *C*, as shown in Figure 7.57. To satisfy the equilibrium condition $\sum F_X = 0$, the frame must be analysed for a pure sway loading case, as shown in the figure, where the frame is subjected to a lateral load of 5.169 kN (←) acting towards the left.

FIXED-END MOMENTS

$$MF_{AB} = MF_{BA} = \frac{6EI\delta}{L^2} = \frac{6EI\delta}{16} = 0.375EI\delta$$

$$MF_{CD} = MF_{DC} = \frac{6EI\delta}{L^2} = \frac{6EI\delta}{9} = 0.6666EI\delta$$

$$MF_{BC} = MF_{CB} = 0$$

Now we should carry out the moment distribution with the above fixed-end moments as carried out earlier. For convenience, let us take $MF_{AB} = MF_{BA} = 3{,}750$ kNm and $MF_{CD} = MF_{DC} = 6{,}666$ kNm,

TABLE 7.21
Moment Distribution Table

Joint	A	B		C		D
Member	*AB*	*BA*	*BC*	*CB*	*CD*	*DC*
DF	—	0.385	0.615	0.545	0.455	—
COF	0.500	0.000	0.500	0.500	0.000	0.500
FEM	0.000	0.000	−41.667	41.667	0.000	0.000
Balance and distribution		16.026	25.641	−22.727	−18.940	
COM	8.013		−11.364	12.821		−9.470
Balance and distribution		4.371	6.993	−6.993	−5.828	
COM	2.185		−3.497	3.497		−2.914
Balance and distribution		1.345	2.152	−1.907	−1.589	
COM	0.672		−0.954	1.076		−0.795
Balance and distribution		0.367	0.587	−0.587	−0.489	
COM	0.183		−0.293	0.293		−0.245
Balance and distribution		0.113	0.181	−0.160	−0.133	
COM	0.056		−0.080	0.090		−0.067
Balance and distribution		0.031	0.049	−0.049	−0.041	
COM	0.015		−0.025	0.025		−0.021
Balance and distribution		0.009	0.015	−0.013	−0.011	
COM	0.005		−0.007	0.008		−0.006
Balance and distribution		0.003	0.004	−0.004	−0.003	
COM	0.001		−0.002	0.002		−0.002
Balance and distribution	−0.001	0.0007	0.0013	−0.0011	−0.0009	
Final end moments (kNm)	11.131	22.264	−22.264	27.035	−27.035	−13.519

and then apply a multiplication factor=0.0001$EI\delta$ to the final results.

MOMENT DISTRIBUTION

The distribution factors calculated earlier can be used. The moment distribution and COMs in each cycle are conveniently carried out in a moment distribution (Table 7.22).

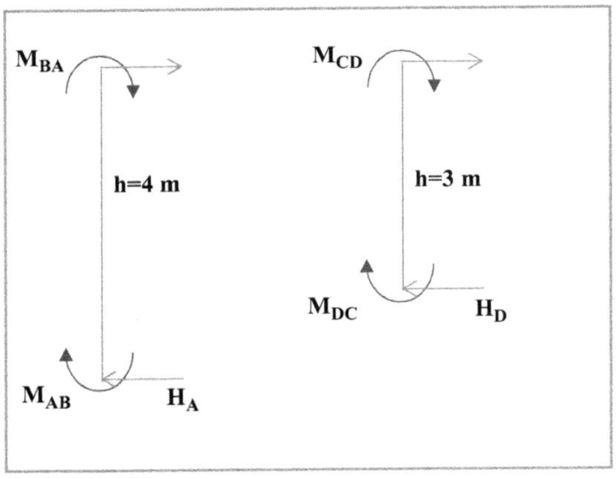

FIGURE 7.56 Free body diagram of the columns.

To get final δ, let us apply $\sum F_X = 0$.

$$\sum F_X = H_A + H_D + P = 0$$

$$\Rightarrow -\left(\frac{M_{AB} + M_{BA}}{h}\right) - \left(\frac{M_{CD} + M_{DC}}{h}\right) + P = 0$$

$$\Rightarrow \left(\frac{3{,}344.447 + 2{,}938.824}{4} + \frac{3{,}930.947 + 5{,}298.547}{3}\right)$$

$$\times 0.0001EI\delta = P = 5.169$$

$$\therefore EI\delta = 11.122 \text{ kNm}^3$$

FINAL END MOMENTS DUE TO SWAY

The final end moments can be obtained by multiplying the above known value of $EI\delta$ with the final end moments obtained in Table 7.22. Therefore, the moments due to pure sway analysis are as follows:

$$M_{AB} = 3{,}344.447 \times (11.122 \times 0.0001) = 3.72 \text{ kNm}$$

$$M_{BA} = 2{,}938.824 \times (11.122 \times 0.0001) = 3.27 \text{ kNm}$$

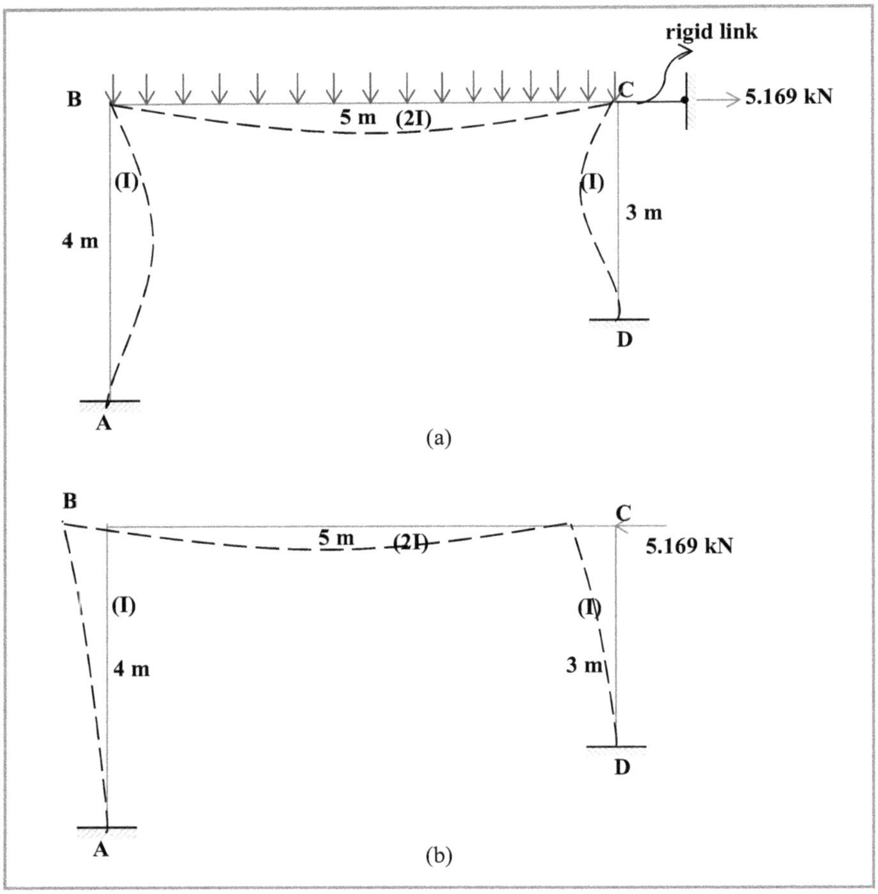

FIGURE 7.57 Rigid frame subjected to non-sway and pure sway loading.

TABLE 7.22
Moment Distribution Table (with Side Sway)

Joint	A	B		C		D
Member	AB	BA	BC	CB	CD	DC
DF	1.000	0.385	0.615	0.545	0.455	
COF	0.500	0.000	0.500	0.500	0.000	0.500
FEM	3,750.000	3,750.000	0.000	0.000	6,666	6,666.000
Balance and distribution		−1,442.308	−2,307.692	−3,636	−3,030	
COM	−721.154		−1,818.000	−1,153.846		−1,515.000
Balance and distribution		699.231	1,118.769	629.371	524.476	
COM	349.615		314.685	559.385		262.238
Balance and distribution		−121.033	−193.653	−305.119	−254.266	
COM	−60.516		−152.559	−96.826		−127.133
Balance and distribution		58.677	93.883	52.814	44.012	
COM	29.338		26.407	46.941		22.006
Balance and distribution		−10.157	−16.251	−25.604	−21.337	
COM	−5.078		−12.802	−8.125		−10.668
Balance and distribution		4.924	7.878	4.432	3.693	
COM	2.462		2.216	3.939		1.847
Balance and distribution		−0.852	−1.364	−2.149	−1.791	
COM	−0.426		−1.074	−0.682		−0.895
Balance and distribution		0.413	0.661	0.372	0.310	
COM	0.207		0.186	0.331		0.155
Balance and distribution	−0.036	−0.072	−0.114	−0.180	−0.151	−0.0075
Final end moments (kNm)	3,344.447	2,938.824	−2,938.824	−3,930.947	3,930.947	5,298.547

$$M_{BC} = -2,938.824 \times (11.122 \times 0.0001) = -3.27 \text{ kNm}$$

$$M_{CB} = -3,930.947 \times (11.122 \times 0.0001) = -4.37 \text{ kNm}$$

$$M_{CD} = 3,930.947 \times (11.122 \times 0.0001) = 4.37 \text{ kNm}$$

$$M_{DC} = 5,298.547 \times (11.122 \times 0.0001) = 5.89 \text{ kNm}$$

RESULTANT MOMENTS

The net member end moments can be obtained by summing the end moments obtained in Table 7.21 and the end moments obtained above in the pure sway analysis, as shown below.

$$M_{AB} = 11.13 + 3.72 = 14.85 \text{ kNm}$$

$$M_{BA} = 22.26 + 3.27 = 25.53 \text{ kNm}$$

$$M_{BC} = -22.26 - 3.27 = -25.53 \text{ kNm}$$

$$M_{CB} = 27.03 - 4.37 = 22.66 \text{ kNm}$$

$$M_{CD} = -27.03 + 4.37 = -22.66 \text{ kNm}$$

$$M_{DC} = -13.52 + 5.89 = -7.63 \text{ kNm}$$

SHEAR FORCE AND BENDING MOMENT DIAGRAMS

With the known values of the net moments of the members and using the static equilibrium equations, the other member end forces can be calculated, as shown in the free body diagram (Figure 7.58a). The shear force and bending moment diagrams are shown in Figure 7.58b and c, respectively.

Example 7.20

Solve the Example 7.19 by treating *A* and *D* as hinged supports instead of fixed supports, as shown in Figure 7.59.

Solution:

The frame is unsymmetrical and subjected to lateral sway. Therefore, the frame must be analysed both with non-sway and sway. Let us, first analyse the frame subjected to sway.

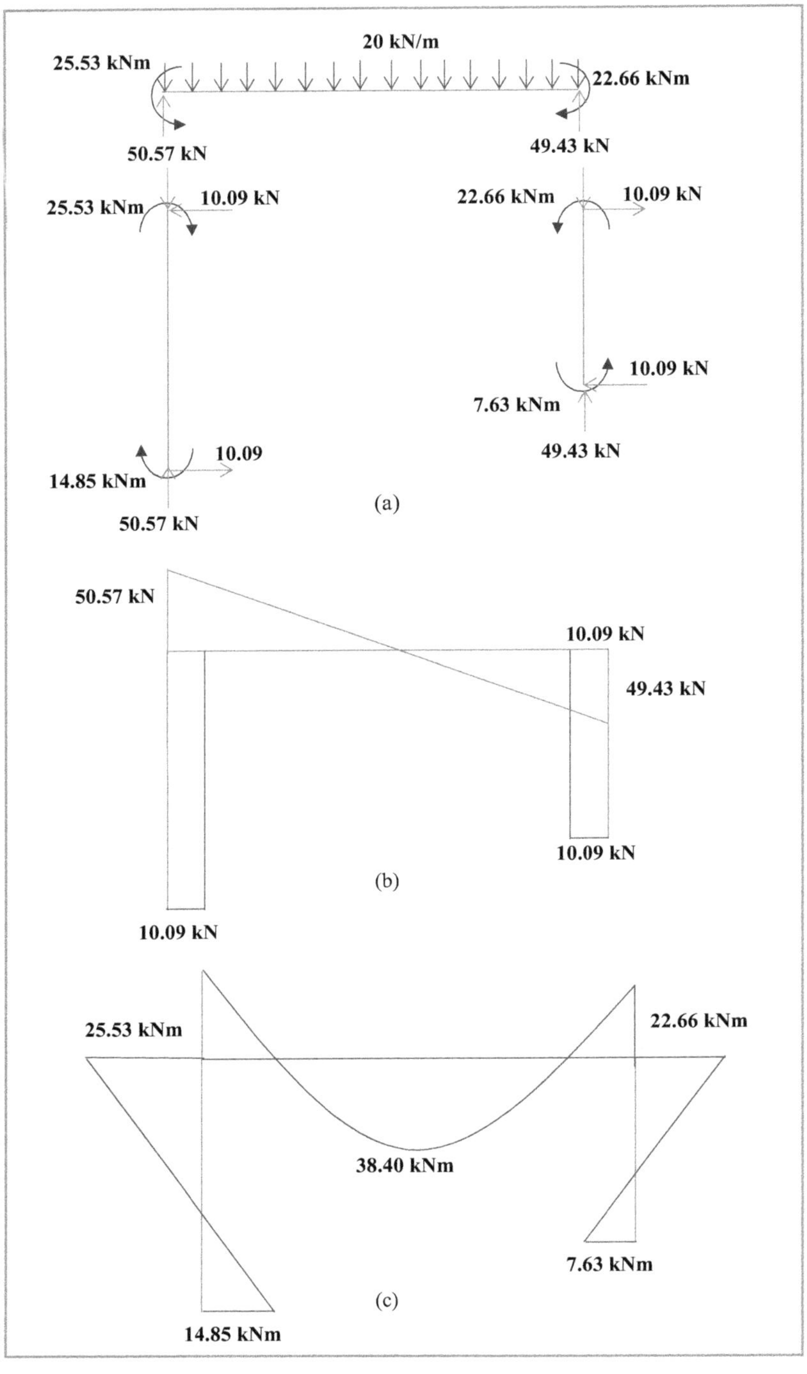

FIGURE 7.58 (a) Free body diagram, (b) shear force diagram, and (c) bending moment diagram.

WITHOUT SWAY ANALYSIS

Neglect the side sway. Also neglect the rotations at *A* and *D* by applying the modified stiffness factors for members *AB* and *CD*, as the moment at these ends are zero due to hinge supports. Therefore, consider the rotations at *B* and *C* as unknown rotations in the frame and calculate the fixed-end moments.

FIXED-END MOMENTS

$$MF_{AB} = MF_{BA} = MF_{CD} = MF_{DC} = 0$$

$$MF_{BC} = -MF_{CB} = -\frac{20 \times 5^2}{12} = -41.667 \text{ kNm}$$

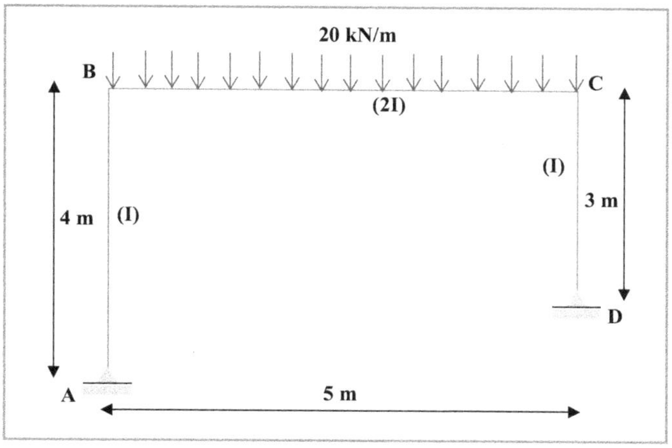

FIGURE 7.59 Rigid frame *ABCD* with unequal column heights.

STIFFNESS FACTORS

Joint *B*:

$$K_{BA} = \frac{3EI}{L} = \frac{3EI}{4} = 0.75EI$$

$$K_{BC} = \frac{4EI}{L} = \frac{4E(2I)}{5} = 1.6EI$$

Joint *C*:

$$K_{CB} = \frac{4EI}{L} = \frac{4E(2I)}{5} = 1.6EI$$

$$K_{CD} = \frac{3EI}{L} = \frac{3EI}{3} = EI$$

DISTRIBUTION FACTORS

Joint *B*:

$$(DF)_{BA} = \frac{0.75EI}{0.75EI + 1.6EI} = 0.319$$

$$(DF)_{BC} = \frac{1.6EI}{0.75EI + 1.6EI} = 0.681$$

Joint *C*:

$$(DF)_{CB} = \frac{1.6EI}{1.6EI + EI} = 0.615$$

$$(DF)_{CD} = \frac{EI}{1.6EI + EI} = 0.385$$

MOMENT DISTRIBUTION TABLE

The iterative process of moment distribution and COMs in each cycle is presented in moment distribution Table 7.23.

FRAME WITH SIDE SWAY ANALYSIS

The free body diagrams of the columns *AB* and *CD* are shown in Figure 7.60. Now apply the horizontal force equilibrium equation.

$$\sum F_X = 0 \Rightarrow H_A + H_D = 0$$

$$\Rightarrow -\left(\frac{M_{BA}}{4}\right) - \left(\frac{M_{CD}}{3}\right) = 0$$

$$\Rightarrow -\left(\frac{19.424}{4}\right) - \left(\frac{-23.994}{3}\right) = 0$$

$$\Rightarrow 3.142 \text{ kN} \neq 0$$

Thus, $\sum F_X = 0$ is not satisfied by the solution obtained previously. To satisfy the equilibrium condition $\sum F_X = 0$, the frame should be analysed separately for the "pure sway" loading case, as shown in Figure 7.60b, wherein the frame is subjected to a lateral load of 3.142 kN (←) acting towards the left.

FIXED-END MOMENTS

Using Figure 7.60c, the fixed-end moments are

$$MF_{BA} = \frac{3EI\delta}{L^2} = \frac{3EI\delta}{16} = 0.1875EI\delta$$

$$MF_{CD} = \frac{3EI\delta}{L^2} = \frac{3EI\delta}{9} = 0.3333EI\delta$$

$$MF_{BC} = MF_{CB} = 0$$

Now we shall carry out the moment distribution with the above fixed-end moments, as carried out earlier. For convenience, let us take $MF_{BA} = 1{,}875$ kNm and $MF_{CD} = 3{,}333$ kNm and then apply a multiplication factor=0.0001$EI\delta$ to the final results.

TABLE 7.23
Moment Distribution Table (Without Sway)

Joint	A	B		C		D
Member	AB	BA	BC	CB	CD	DC
DF	—	0.319	0.681	0.615	0.385	—
COF	0.000	0.000	0.500	0.500	0.000	0.000
FEM	0.000	0.000	−41.667	41.667	0.000	0.000
Balance and distribution		13.298	28.369	−25.641	−16.026	
COM			−12.821	14.185		
Balance and distribution		4.092	8.729	−8.729	−5.456	
COM			−4.364	4.364		
Balance and distribution		1.393	2.972	−2.686	−1.679	
COM			−1.343	1.486		
Balance and distribution		0.429	0.914	−0.914	−0.571	
COM			−0.457	0.457		
Balance and distribution		0.146	0.311	−0.281	−0.176	
COM			−0.141	0.156		
Balance and distribution		0.045	0.096	−0.096	−0.060	
COM			−0.048	0.048		
Balance and distribution		0.015	0.033	−0.029	−0.018	
COM			−0.015	0.016		
Balance and distribution		0.005	0.010	−0.010	−0.006	
COM			−0.005	0.005		
Balance and distribution		0.002	0.003	−0.003	−0.002	
Final end moments (kNm)	0.000	19.424	−19.424	23.994	−23.994	0.000

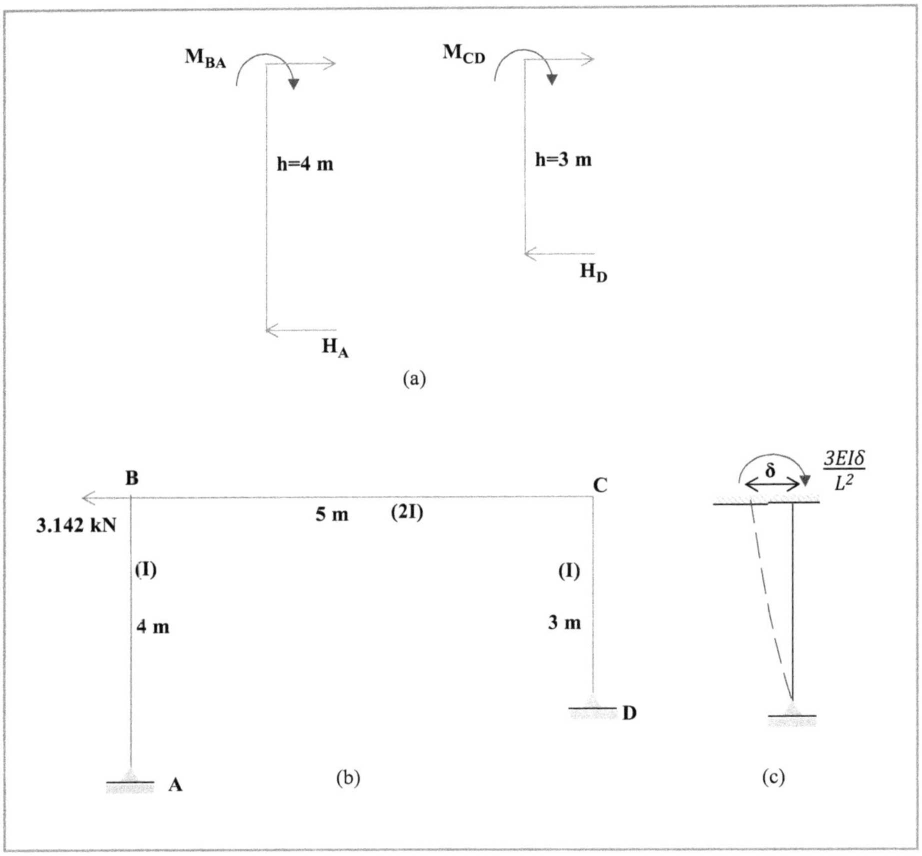

FIGURE 7.60 (a) Free body diagram of columns, (b) rigid frame *ABCD* with horizontal load, and (c) fixed-end moments.

MOMENT DISTRIBUTION

The distribution factors calculated in the non-sway analysis can be used. The iterative process of the moment distribution and COMs in each cycle is conveniently carried out in the moment distribution (Table 7.24).

To get final δ, we use $\sum F_X = 0$.

$$\sum F_X = H_A + H_D + P = 0$$

$$\Rightarrow -\left(\frac{M_{BA}}{h}\right) - \left(\frac{M_{CD}}{h}\right) + P = 0$$

$$\Rightarrow \left(\frac{1{,}572.14}{4} + \frac{2{,}175.298}{3}\right) \times 0.0001EI\delta = P = 3.142$$

$$\therefore EI\delta = 28.1 \text{ kNm}^3$$

FINAL END MOMENTS DUE TO SWAY

The final end moments due to side sway can be calculated by my multiplying the end moments obtained in Table 7.24 with the above known value of $EI\delta$, obtained above. Therefore, the moments due to sway analysis are as follows:

$$M_{BA} = 1{,}572.14 \times (28.1 \times 0.0001) = 4.42 \text{ kNm}$$

$$M_{BC} = -1{,}572.14 \times (28.1 \times 0.0001) = -4.42 \text{ kNm}$$

$$M_{CB} = -2{,}175.298 \times (28.1 \times 0.0001) = -6.11 \text{ kNm}$$

$$M_{CD} = 2{,}175.298 \times (28.1 \times 0.0001) = 6.11 \text{ kNm}$$

NET MEMBER MOMENTS

The net member end moments can be calculated by summing the final end moments arrived at in Table 7.23 under the frame without side sway analysis, and the final end moments obtained in the case of the frame subjected to pure sway analysis, which are as follows:

$$M_{BA} = 19.42 + 4.42 = 23.84 \text{ kNm}$$

$$M_{BC} = -19.42 - 4.42 = -23.84 \text{ kNm}$$

$$M_{CB} = 23.99 - 6.11 = 17.88 \text{ kNm}$$

$$M_{CD} = -23.99 + 6.11 = -17.88 \text{ kNm}$$

After obtaining the net moments, using the free body diagram (Figure 7.61a) and static equilibrium equations, the other support reactions can be calculated. The shear force and bending moment diagrams are shown in Figure 7.61b and c, respectively. The probable elastic curve is shown in Figure 7.61d.

TABLE 7.24
Moment Distribution Table (with Sway)

Joint	*A*	*B*		*C*		*D*
Member	*AB*	*BA*	*BC*	*CB*	*CD*	*DC*
DF	-----	0.319	0.681	0.615	0.385	—
COF	0.000	0.000	0.500	0.500	0.000	0.000
FEM	0.000	1,875.000	0.000	0.000	3,333.000	0.000
Balance and distribution		−598.404	−1,276.596	−2,051.077	−1,281.923	
COM			−1,025.538	−638.298		
Balance and distribution		327.300	698.239	392.799	245.499	
COM			196.399	349.119		
Balance and distribution		−62.681	−133.719	−214.843	−134.277	
COM			−107.421	−66.859		
Balance and distribution		34.283	73.138	41.144	25.715	
COM			20.572	36.569		
Balance and distribution		−6.566	−14.007	−22.504	−14.065	
COM			−11.252	−7.003		
Balance and distribution		3.591	7.661	4.310	2.694	
COM			2.155	3.830		
Balance and distribution		−0.688	−1.467	−2.357	−1.473	
COM			−1.179	−0.734		
Balance and distribution		0.376	0.802	0.451	0.282	
COM			0.226	0.401		
Balance and distribution		−0.072	−0.154	−0.247	−0.154	
End moments (kNm)	0.000	1,572.140	−1,572.140	−2,175.298	2,175.298	0.000

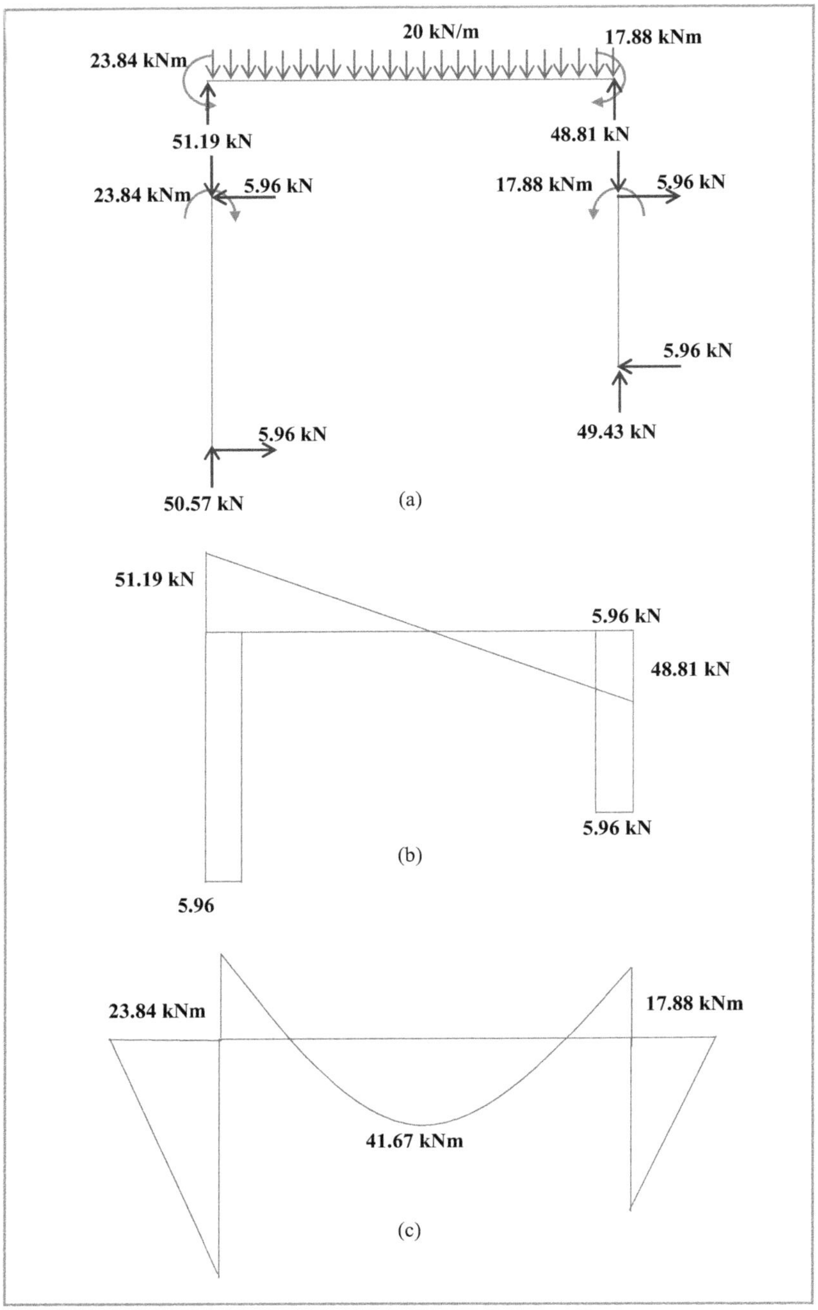

FIGURE 7.61 (a) Free body diagram, (b) shear force diagram, and (c) bending moment diagram.

Example 7.21

Analyse an unsymmetrical rigid joint frame with an inclined column, as shown in Figure 7.62, using the moment distribution method.

Solution:

WITHOUT SIDE SWAY ANALYSIS

Neglect the side sway and consider the rotations at B and C as unknown rotations in the frame and calculate the fixed-end moments.

FIXED-END MOMENTS

$$MF_{AB} = MF_{BA} = MF_{CD} = MF_{DC} = 0$$

$$MF_{BC} = -MF_{CB} = -\frac{20 \times 5^2}{12} = -41.667 \text{ kNm}$$

STIFFNESS FACTORS

Joint B:

$$K_{BA} = \frac{4EI}{L} = \frac{4EI}{4} = EI$$

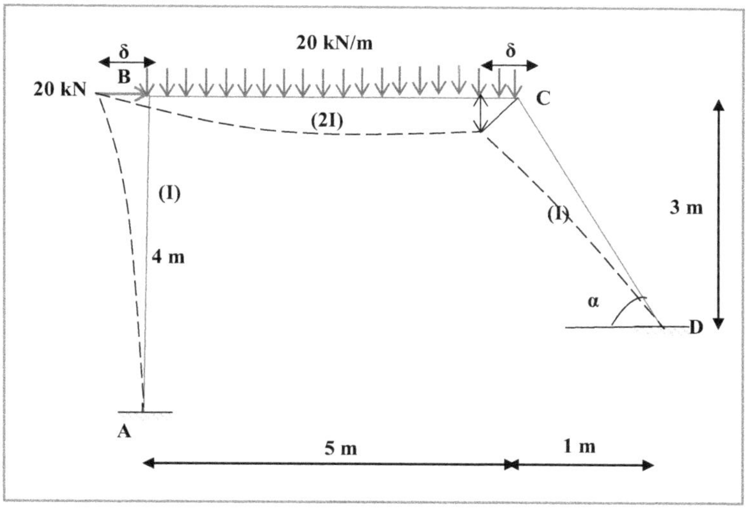

FIGURE 7.62 Rigid frame *ABCD* with an inclined column.

$$K_{BC} = \frac{4EI}{L} = \frac{4E(2I)}{5} = 1.6EI$$

Joint C:

$$K_{CB} = \frac{4EI}{L} = \frac{4E(2I)}{5} = 1.6EI$$

$$K_{CD} = \frac{4EI}{L} = \frac{4EI}{3.16} = 1.265EI$$

DISTRIBUTION FACTORS

Joint B:

$$(DF)_{BA} = \frac{EI}{EI + 1.6EI} = 0.385$$

$$(DF)_{BC} = \frac{1.6EI}{EI + 1.6EI} = 0.615$$

Joint C:

$$(DF)_{CB} = \frac{1.6EI}{1.6EI + 1.265EI} = 0.558$$

$$(DF)_{CD} = \frac{1.265EI}{1.6EI + 1.265EI} = 0.442$$

MOMENT DISTRIBUTION TABLE

The moment distribution and COMs in each cycle are presented in the moment distribution Table 7.25.

FRAME WITH SIDE SWAY ANALYSIS

The free body diagrams of columns *AB* and *CD* are shown in Figure 7.63a. Now apply the horizontal force equilibrium equation.

$$\sum F_X = 0 \Rightarrow -H_A + H_D - P = 0$$

$$\Rightarrow -\left(\frac{M_{AB} + M_{BA}}{4}\right) + \left(\frac{M_{CD} + M_{DC} + 50.78 \times 1}{3}\right)$$

$$-20 = 0$$

$$\Rightarrow -\left(\frac{11.213 + 22.425}{4}\right) + \left(\frac{26.328 + 13.164 + 50.78}{3}\right)$$

$$-20 = 0$$

$$\Rightarrow 21.68 - 20 = 1.68 \text{ kN} \neq 0$$

Thus, $\sum F_X = 0$ is not satisfied by the solution obtained previously. To satisfy the equilibrium condition $\sum F_X = 0$, it is required to analyse the frame separately for the "pure sway" loading case, as shown in Figure 7.63b, where the frame is subjected to a lateral load of 1.68 kN (←) acting towards the left.

FIXED-END MOMENTS

The fixed-end moments are calculated using the free body diagram shown in Figure 7.64.

$$MF_{AB} = MF_{BA} = -\frac{6EI\delta}{4^2} = -0.375EI\delta \text{ kNm}$$

$$MF_{BC} = MF_{CB} = -\frac{6EI\delta\cot(\alpha)}{5^2} = -0.08EI\delta \text{ kNm}$$

$$MF_{CD} = MF_{DC} = -\frac{6EI\delta / \sin\alpha}{3.162^2} = -0.6329EI\delta \text{ kNm}$$

MOMENT DISTRIBUTION TABLE

Now we shall carry out the moment distribution with the above fixed-end moments. For convenience, let us take

TABLE 7.25
Moment Distribution Table (Without Side Sway)

Joint	A	B		C		D
Member	*AB*	*BA*	*BC*	*CB*	*CD*	*DC*
DF	1.000	0.385	0.615	0.558	0.442	
COF	0.500	0.000	0.500	0.500	0.000	0.500
FEM	0.000	0.000	–41.667	41.667	0.000	0.000
Balance and distribution		16.026	25.641	–23.263	–18.404	
COM	8.013		–11.631	12.821		–9.202
Balance and distribution		4.474	7.158	–7.158	–5.663	
COM	2.237		–3.579	3.579		–2.831
Balance and distribution		1.376	2.202	–1.998	–1.581	
COM	0.688		–0.999	1.101		–0.790
Balance and distribution		0.384	0.615	–0.615	–0.486	
COM	0.192		–0.307	0.307		–0.243
Balance and distribution		0.118	0.189	–0.172	–0.136	
COM	0.059		–0.086	0.095		–0.068
Balance and distribution		0.033	0.053	–0.053	–0.042	
COM	0.017		–0.026	0.026		–0.021
Balance and distribution		0.010	0.016	–0.015	–0.012	
COM	0.005		–0.007	0.008		–0.006
Balance and distribution		0.003	0.005	–0.005	–0.004	
COM	0.001		–0.002	0.002		–0.002
Balance and distribution	0.0005	0.001	0.001	–0.001	–0.001	–0.0005
Final end moments (kNm)	11.213	22.425	–22.425	26.328	–26.328	–13.164

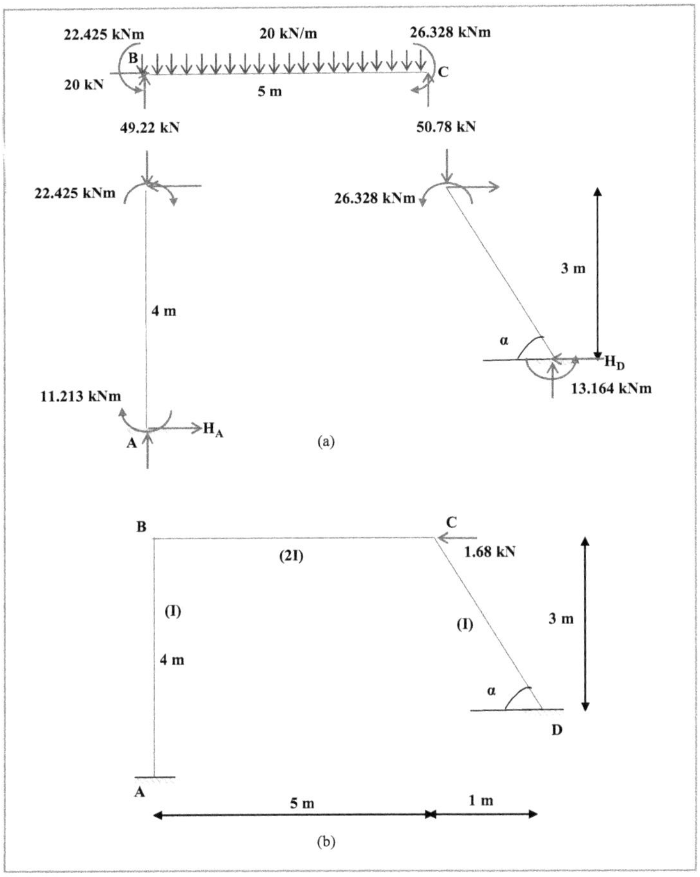

FIGURE 7.63 (a) Free body diagram and (b) rigid frame under pure sway condition.

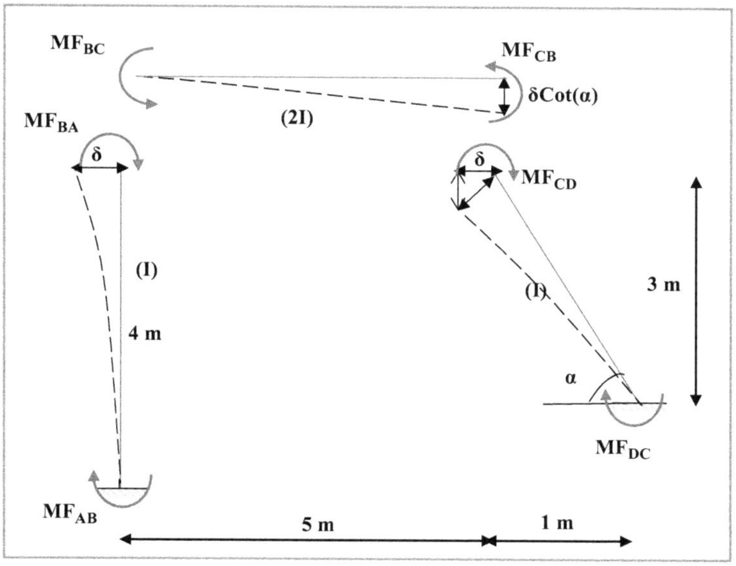

FIGURE 7.64 Fixed-end moments.

TABLE 7.26
Moment Distribution Table (with Side Sway)

Joint	A	B		C		D
Member	AB	BA	BC	CB	CD	DC
DF		0.385	0.615	0.558	0.442	
COF	0.500	0.000	0.500	0.500	0.000	0.500
FEM	3,750.000	3,750.000	−800.000	−800.000	6,329.000	6,329.000
Balance and distribution		−1,134.615	−1,815.385	−3,086.862	−2,442.138	
COM	−567.308		−1,543.431	−907.692		−1,221.069
Balance and distribution		593.627	949.804	506.768	400.924	
COM	296.814		253.384	474.902		200.462
Balance and distribution		−97.455	−155.929	−265.140	−209.762	
COM	−48.728		−132.570	−77.964		−104.881
Balance and distribution		50.988	81.581	43.528	34.437	
COM	25.494		21.764	40.791		17.218
Balance and distribution		−8.371	−13.393	−22.774	−18.017	
COM	−4.185		−11.387	−6.697		−9.009
Balance and distribution		4.380	7.007	3.739	2.958	
COM	2.190		1.869	3.504		1.479
Balance and distribution		−0.719	−1.150	−1.956	−1.548	
COM	−0.359		−0.978	−0.575		−0.774
Balance and distribution		0.376	0.602	0.321	0.254	
COM	0.188		0.161	0.301		0.127
Balance and distribution	−0.031	−0.062	−0.099	−0.168	−0.133	−0.066
Final end moments (kNm)	3,454.075	3,158.149	−3,158.149	−4,095.975	4,095.975	5,212.487
End moments (no sway)	11.213	22.425	−22.425	26.328	−26.328	−13.164
End moments (with sway) (kNm)	1.108	1.013	−1.013	−1.314	1.314	1.672
Final end moments (kNm)	12.321	23.438	−23.438	25.014	−25.014	−11.492

$MF_{AB} = MF_{BA} = -3{,}750$ kNm, $MF_{BC} = MF_{CB} = -80$ kNm and and $MF_{CD} = MF_{DC} = -6{,}329$ kNm and then apply a multiplication factor=$0.0001EI\delta$ to the final results.

The moment distribution and COMs in each cycle are presented in the moment distribution Table 7.26.

To get the final δ, we use $\sum F_X = 0$.

$$\sum F_X = H_A + H_D + P = 0$$

$$\Rightarrow -\left(\frac{M_{BA}+M_{AB}}{4}\right)-\left(\frac{M_{CD}+M_{DC}}{3}\right)-\left(\frac{V_D\times 1}{3}\right)+1.68=0$$

$$\Rightarrow -\left(\begin{array}{l}\frac{3,454.075+3,158.149}{4}+\frac{4,095.975+5,212.487}{3.16}\\+\frac{3,158.149+4,095.975}{5}\times\frac{1}{3}\end{array}\right)$$

$$\times 0.0001EI\delta = P = 1.68$$

$$\therefore EI\delta = 3.206 \text{ kNm}^3$$

FINAL END MOMENTS DUE TO SWAY

Using the above known value of $EI\delta$, the end moments due to side sway can be obtained by multiplying the end moments obtained in Table 7.25, as shown below.

$$M_{AB} = 3,454.075\times(3.206\times 0.0001) = 1.108 \text{ kNm}$$

$$M_{BA} = 3,158.149\times(3.206\times 0.0001) = 1.013 \text{ kNm}$$

$$M_{BC} = -3,158.149\times(3.206\times 0.0001) = -1.013 \text{ kNm}$$

$$M_{CB} = -4,095.975\times(3.206\times 0.0001) = -1.314 \text{ kNm}$$

$$M_{CD} = 4,095.975\times(3.206\times 0.0001) = 1.314 \text{ kNm}$$

$$M_{DC} = 5,212.487\times(3.206\times 0.0001) = 1.672 \text{ kNm}$$

RESULTANT MOMENTS

The net moments are obtained by summing the moments obtained in the case of sway and non-sway analysis, and are shown below, as well as in Table 7.25.

$$M_{AB} = 11.213 + 1.108 = 12.321 \text{ kNm}$$

$$M_{BA} = 22.425 + 1.013 = 23.438 \text{ kNm}$$

$$M_{BC} = -22.425 - 1.013 = -23.438 \text{ kNm}$$

$$M_{CB} = 26.328 - 1.314 = 25.014 \text{ kNm}$$

$$M_{CD} = -26.328 + 1.314 = -25.014 \text{ kNm}$$

$$M_{DC} = -13.164 + 1.672 = -11.492 \text{ kNm}$$

FINAL END SHEARS

The end shears can be calculated using the equilibrium equations as shown in Figure 7.65a. The shear force and bending moment diagrams are shown in Figure 7.65b and c, respectively. The probable elastic curve is shown in Figure 7.65d.

Example 7.22

Analyse the unsymmetrical rigid joint frame *ABCD* with both inclined columns, as shown in Figure 7.66, using the moment distribution method. (The same has been solved in Example 6.16 using the SD method.)

Solution:

The kinematic indeterminacy of the given frame is 3, i.e., θ_B, θ_C and Δ. First, let us carry out the analysis without side sway.

ANALYSIS WITHOUT SIDE SWAY

Neglect the side sway and consider the rotations at *B* and *C* as unknown rotations in the frame and calculate the fixed-end moments.

FIXED-END MOMENTS

Using Table 6.1, the fixed-end moments can be determined as

$$MF_{AB} = MF_{BA} = MF_{CD} = MF_{DC} = 0$$

$$MF_{BC} = \frac{-5wL^2}{96} = \frac{-5\times 25\times 3^2}{96} = -11.719 \text{ kNm} = -MF_{CB}$$

STIFFNESS FACTORS

Joint *B*

$$K_{BA} = \frac{4EI}{L} = \frac{4EI}{3.605} = 1.109EI$$

$$K_{BC} = \frac{4EI}{L} = \frac{4E(3I)}{3} = 4EI$$

Joint *C*

$$K_{CB} = \frac{4EI}{L} = \frac{4E(3I)}{3} = 4EI$$

$$K_{CD} = \frac{4EI}{L} = \frac{4E(2I)}{5} = 1.6EI$$

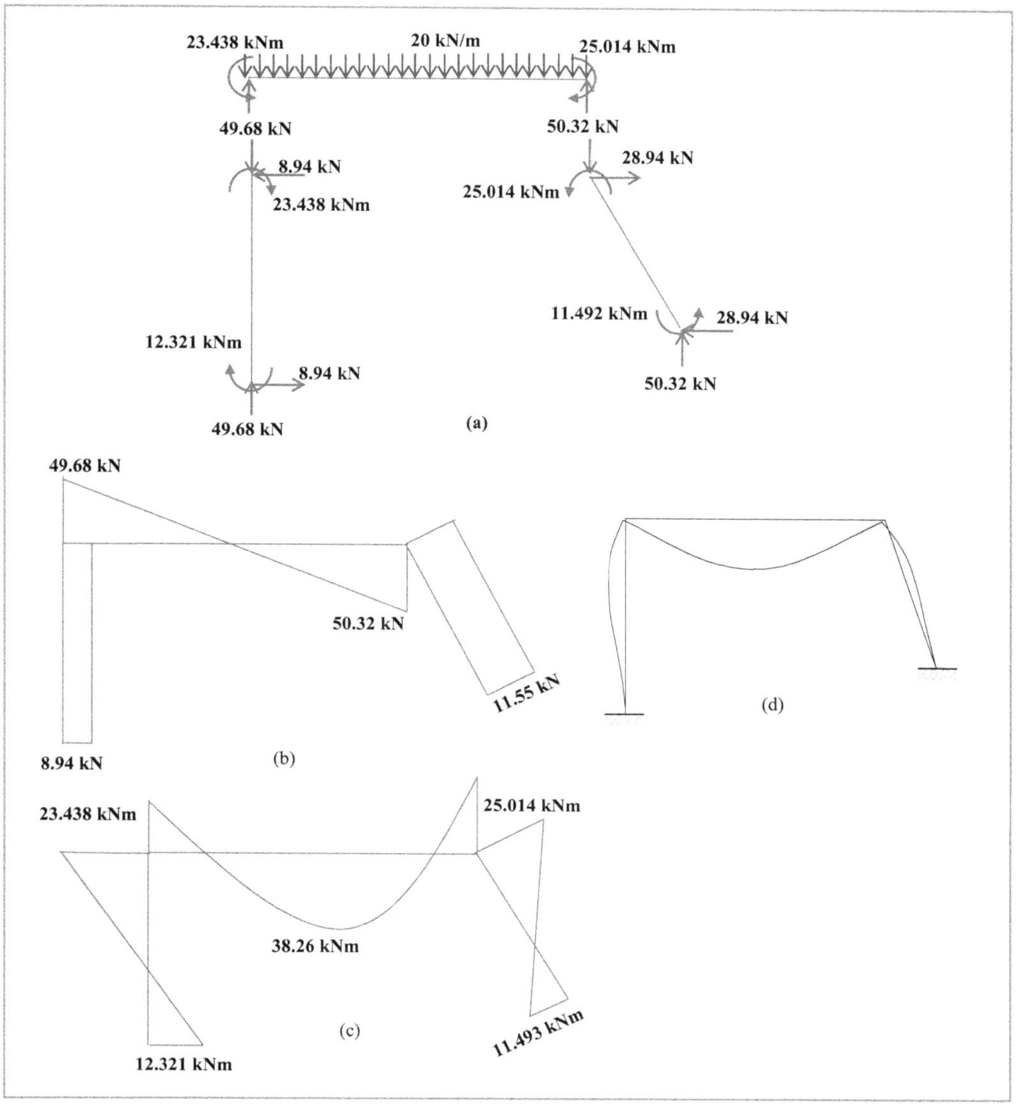

FIGURE 7.65 (a) Free body diagram, (b) shear force diagram, (c) bending moment diagram, and (d) probable elastic curve.

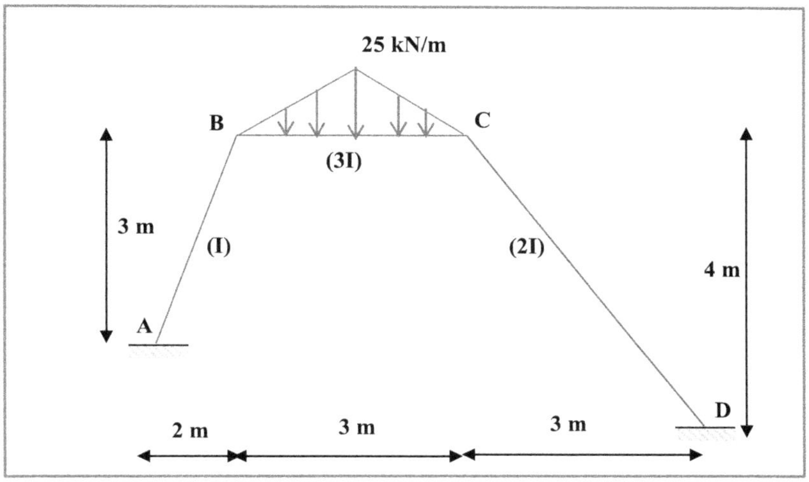

FIGURE 7.66 Unsymmetrical rigid frame *ABCD* with inclined columns.

DISTRIBUTION FACTORS

Joint *B*

$$(DF)_{BA} = \frac{1.109EI}{1.109EI + 4EI} = 0.217$$

$$(DF)_{BC} = \frac{4EI}{1.109EI + 4EI} = 0.783$$

Joint *C*

$$(DF)_{CB} = \frac{4EI}{4EI + 1.6EI} = 0.714$$

$$(DF)_{CD} = \frac{1.6EI}{4EI + 1.6EI} = 0.286$$

MOMENT DISTRIBUTION TABLE

The moment distribution and COMs in each cycle are presented in the moment distribution Table 7.27.

FRAME WITH SIDE SWAY ANALYSIS

The free body diagrams of columns *AB*, *BC*, and *CD* are shown in Figure 7.67. Now apply the horizontal force equilibrium equation.

$$\sum F_X = 0 \Rightarrow H_A + H_D = 0$$

$$V_{BC} = \left(0.5 \times 1.5 \times 25 - (5.415 - 4.014) \times \frac{1}{3}\right) = 18.283 \text{ kN}$$

$$V_{CB} = \left(0.5 \times 1.5 \times 25 + (5.415 - 4.014) \times \frac{1}{3}\right) = 19.217 \text{ kN}$$

$$\Rightarrow -\left(\frac{M_{AB} + M_{BA} + V_{BC} \times 2}{3}\right) + \left(\frac{M_{CD} + M_{DC} + V_{CB} \times 3}{4}\right) = 0$$

$$\Rightarrow -\left(\frac{2.007 + 4.014 + 18.283 \times 2}{3}\right)$$

$$+\left(\frac{5.415 + 2.708 + 19.217 \times 3}{4}\right) = 0$$

$$\Rightarrow +2.247 \text{ kN} \neq 0$$

Thus, $\sum F_X = 0$ is not satisfied by the solution obtained above. To satisfy the equilibrium condition $\sum F_X = 0$, the frame must be analysed separately for the pure sway loading case, as shown in Figure 7.68, wherein the frame is subjected to a horizontal load of 2.247 kN (←) acting towards the left at either *B* or *C*.

FIXED-END MOMENTS

The fixed-end moments can be calculated using Table 6.1 for the free body diagrams shown in Figure 7.69.

TABLE 7.27
Moment Distribution Table (Without Side Sway)

Joint	A	B		C		D
Member	*AB*	*BA*	*BC*	*CB*	*CD*	*DC*
DF	—	0.217	0.783	0.714	0.286	—
COF	0.500	0.000	0.500	0.500	0.000	0.500
FEM	0.000	0.000	−11.719	11.719	0.000	0.000
Balance and distribution		2.545	9.174	−8.371	−3.348	
COM	1.272		−4.185	4.587		−1.674
Balance and distribution		0.909	3.276	−3.276	−1.311	
COM	0.454		−1.638	1.638		−0.655
Balance and distribution		0.356	1.282	−1.170	−0.468	
COM	0.178		−0.585	0.641		−0.234
Balance and distribution		0.127	0.458	−0.458	−0.183	
COM	0.064		−0.229	0.229		−0.092
Balance and distribution		0.050	0.179	−0.164	−0.065	
COM	0.025		−0.082	0.090		−0.033
Balance and distribution		0.018	0.064	−0.064	−0.026	
COM	0.009		−0.032	0.032		−0.013
Balance and distribution		0.007	0.025	−0.023	−0.009	
COM	0.003		−0.011	0.013		−0.005
Balance and distribution		0.002	0.009	−0.009	−0.004	
COM	0.001		−0.004	0.004		−0.002
Balance and distribution	0.0005	0.001	0.004	−0.003	−0.001	−0.0005
Final end moments (kNm)	2.007	4.014	−4.014	5.415	−5.415	−2.708

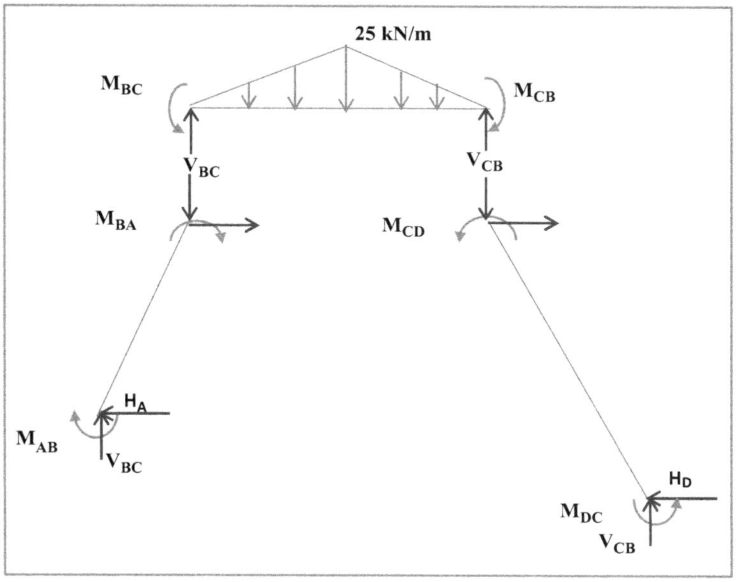

FIGURE 7.67 Free body diagram.

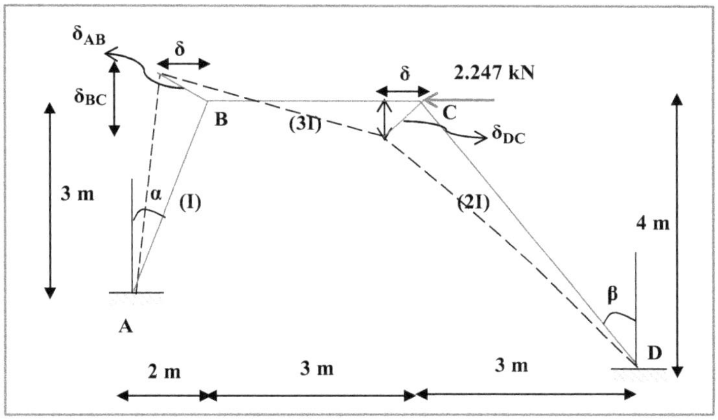

FIGURE 7.68 Frame subjected to pure sway conditions.

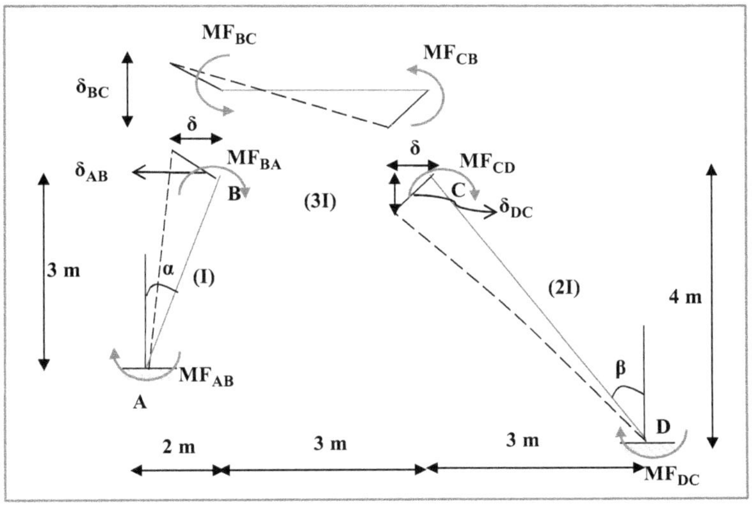

FIGURE 7.69 Fixed-end moments.

$$\delta_{AB} = \frac{\delta}{\cos\alpha} = \frac{\delta}{3/3.605} = 1.201\delta$$

$$\delta_{CD} = \frac{\delta}{\cos\beta} = \frac{\delta}{3/5} = 1.667\delta$$

$$\delta_{BC} = \delta\tan\alpha + \delta\tan\beta = \delta\times\frac{3}{4} + \delta\times\frac{2}{3} = 1.416\delta$$

$$MF_{AB} = MF_{BA} = \frac{6EI\delta_{AB}}{L_{AB}^2} = \frac{6EI(1.201\delta)}{3.605^2} = 0.554EI\delta$$

$$MF_{BC} = MF_{CB} = -\frac{6EI\delta_{BC}}{L_{BC}^2} = \frac{6E(3I)(1.416\delta)}{3^2} = -2.832EI\delta$$

$$MF_{CD} = MF_{DC} = -\frac{6EI\delta_{BC}}{L_{BC}^2} = \frac{6E(2I)(1.667\delta)}{5^2} = 0.8EI\delta$$

MOMENT DISTRIBUTION

For convenience, let us assume arbitrary moment values of $MF_{AB}=MF_{BA}=5{,}540$ kNm, $MF_{BC}=MF_{CB}=-28{,}320$ kNm and $MF_{CD}=MF_{DC}=8{,}000$ kNm. For corrected values, the results are to be multiplied by $0.0001EI\delta$. The iterative procedure of moments distribution in each cycle is shown in Table 7.28. The resultant member end moments are also shown in Table 7.28.

To get the final δ, we can use $\sum F_X = 0$ (Figure 7.70).

$$\sum F_X = H_A + H_D + P = 0$$

$$\Rightarrow -\left(\frac{M_{BA} + M_{AB} + V_{BC}\times 2}{3}\right) - \left(\frac{M_{CD} + M_{DC} + V_{CB}\times 3}{4}\right)$$

$$+2.247 = 0$$

$$V_{BC} = -V_{CB} = \left((9{,}459.069 + 11{,}788.367)\times\frac{1}{3}\right) = 7{,}082.48 \text{ kN}$$

$$\Rightarrow -\left(\begin{array}{c}\dfrac{7{,}499.535 + 9{,}459.069 + 7{,}082.48\times 2}{3} \\ +\dfrac{11{,}788.367 + 9{,}894.183 + 7{,}082.48\times 3}{4}\end{array}\right)\times 0.0001EI\delta$$

$$= -2.247$$

$$\therefore EI\delta = 1.064 \text{ kNm}^3$$

TABLE 7.28
Moment Distribution Table (with Pure Sway)

Joint	*A*	*B*		*C*		*D*
Member	*AB*	*BA*	*BC*	*CB*	*CD*	*DC*
DF	1.000	0.217	0.783	0.714	0.286	
COF	0.500	0.000	0.500	0.500	0.000	0.500
FEM	5,540.000	5,540.000	−28,320.000	−28,320.000	8,000.000	8,000.000
Balance and distribution		4,946.797	17,833.203	14,514.286	5,805.714	
COM	2,473.398		7,257.143	8,916.602	0.000	2,902.857
Balance and distribution		−1,575.927	−5,681.216	−6,369.001	−2,547.600	
COM	−787.963		−3,184.501	−2,840.608	0.000	−1,273.800
Balance and distribution		691.531	2,492.969	2,029.006	811.602	
COM	345.766		1,014.503	1,246.485	0.000	405.801
Balance and distribution		−220.305	−794.198	−890.346	−356.138	
COM	−110.152		−445.173	−397.099	0.000	−178.069
Balance and distribution		96.672	348.501	283.642	113.457	
COM	48.336		141.821	174.251		56.728
Balance and distribution		−30.797	−111.024	−124.465	−49.786	
COM	−15.399		−62.232	−55.512		−24.893
Balance and distribution		13.514	48.718	39.651	15.861	
COM	6.757		19.826	24.359		7.930
Balance and distribution		−4.305	−15.520	−17.399	−6.960	
COM	−2.153		−8.700	−7.760		−3.480
Balance and distribution	0.945	1.889	6.811	5.543	2.217	1.109
End moments (kNm)	7,499.535	9,459.069	−9,459.069	−11,788.367	11,788.367	9,894.183
Final End Moments						
End moments (no sway)	2.007	4.014	−4.014	5.415	−5.415	−2.708
End moments (with sway) (kNm)	0.80	1.0	−1.0	−1.25	1.25	1.05
Final end moments (kNm)	2.807	5.014	−5.014	4.165	−4.165	−1.658

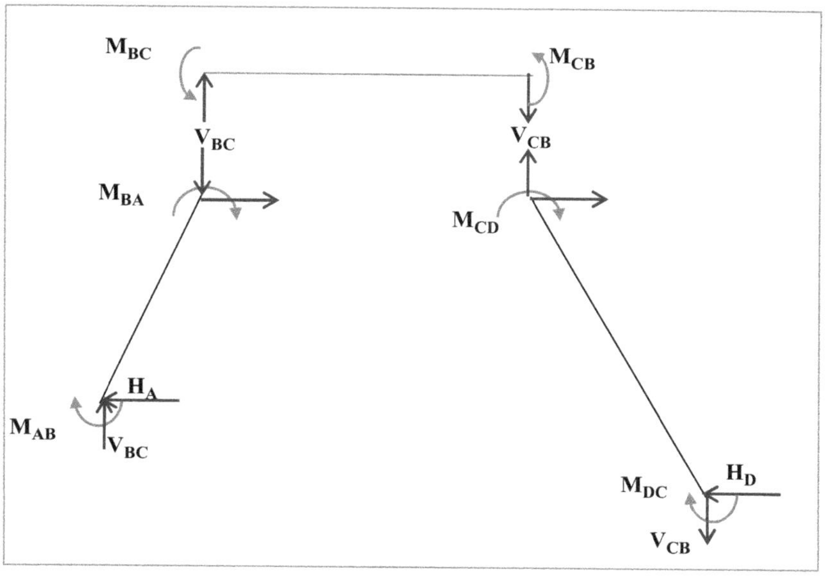

FIGURE 7.70 Free body diagram.

FINAL END MOMENTS DUE TO SWAY

The final sway moments can be obtained by multiplying the end moments from Table 7.28 with the $EI\delta$ calculated as above. Thus, the final end moments are shown below, as well as in Table 7.28.

$$M_{AB} = 7{,}499.535 \times (1.064 \times 0.0001) = 0.80 \text{ kNm}$$

$$M_{BA} = 9{,}459.069 \times (1.064 \times 0.0001) = 1.0 \text{ kNm}$$

$$M_{BC} = -9{,}459.069 \times (1.064 \times 0.0001) = -1.0 \text{ kNm}$$

$$M_{CB} = -11{,}788.367 \times (1.064 \times 0.0001) = -1.25 \text{ kNm}$$

$$M_{CD} = 11{,}788.367 \times (1.064 \times 0.0001) = 1.25 \text{ kNm}$$

$$M_{DC} = 9{,}894.183 \times (1.064 \times 0.0001) = 1.05 \text{ kNm}$$

RESULTANT END MOMENTS

The resultant moments can be calculated by summing the moments obtained in the case of non-sway and sway as follows. Also, the resultant moments are shown in Table 7.28.

$$M_{AB} = 2.007 + 0.80 = 2.807 \text{ kNm}$$

$$M_{BA} = 4.014 + 1.0 = 5.014 \text{ kNm}$$

$$M_{BC} = -4.014 - 1.0 = -5.014 \text{ kNm}$$

$$M_{CB} = 5.415 - 1.25 = 4.165 \text{ kNm}$$

$$M_{CD} = -5.415 + 1.25 = -4.165 \text{ kNm}$$

$$M_{DC} = -2.708 + 1.05 = -1.658 \text{ kNm}$$

FINAL END SHEARS

Once the net member end moments are obtained, the member end shear forces can be calculated using the static equilibrium equations, as shown in the free body diagram in Figure 7.71a. The shear force and bending moment diagrams are shown in Figure 7.71b and c, respectively. The probable elastic curve is shown in Figure 7.71d.

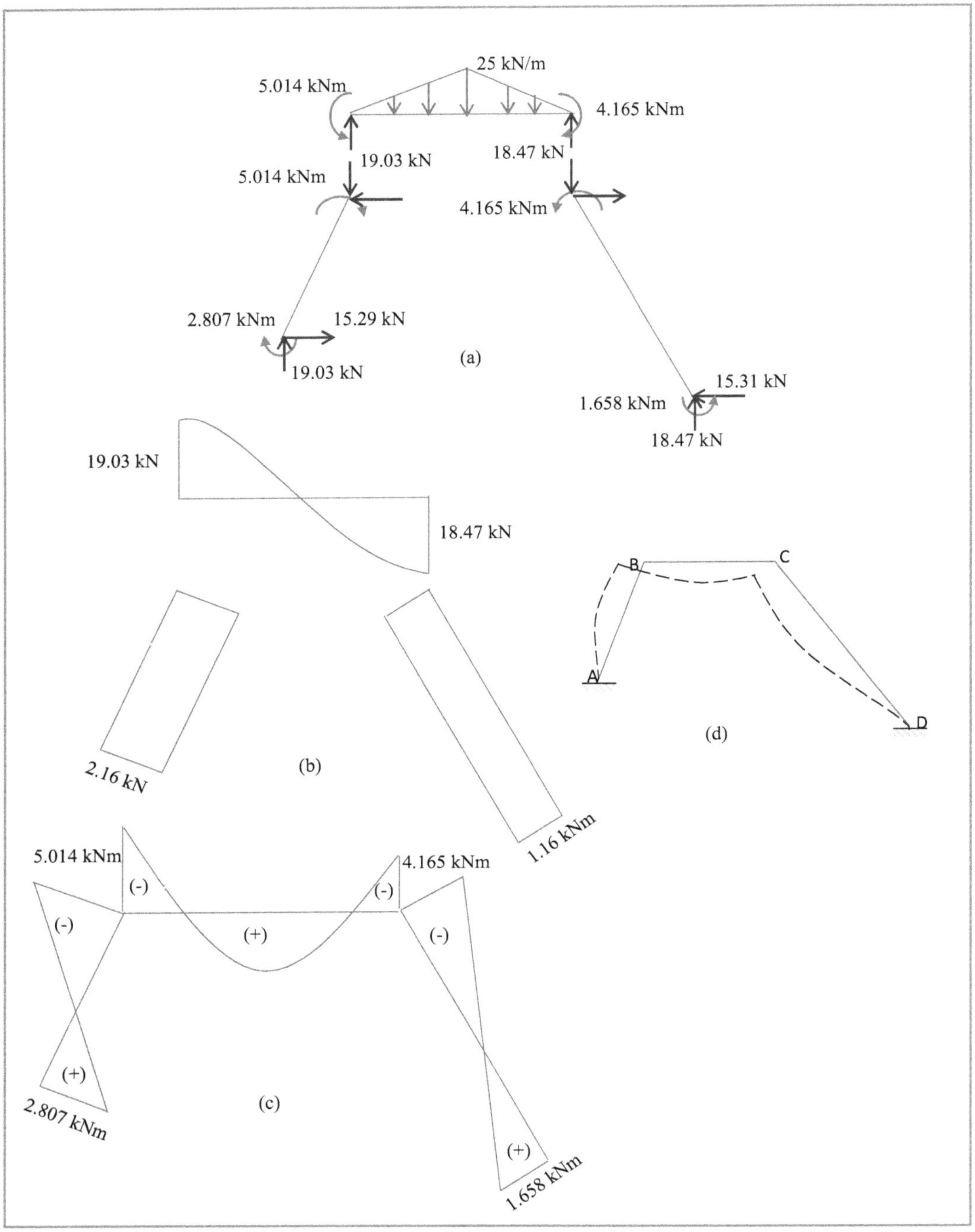

FIGURE 7.71 (a) Free body diagram, (b) shear force diagram, (c) bending moment diagram, and (d) probable deflected shape.

7.7 PROBLEMS FOR PRACTICE

7.7.1 How can the moment distribution method be categorised as a displacement method, even though the unknowns are not the displacements?

7.7.2 Compare the flexural stiffness of the beam elements with different support conditions at the far ends when the near ends are subjected to a couple M in a clockwise direction.

7.7.3 Define the terms (a) stiffness factor, (b) joint stiffness, (c) distribution factor, and (d) COF.

7.7.4 Explain the step-by-step procedure involved in the multi-cycle moment distribution method.

7.7.5 How do you verify whether the final end moments obtained in the moment distribution method are correct?

7.7.6 Explain the advantages and disadvantages of the moment distribution method compared to the slope deflection method.

7.7.7 Analyse the two-span continuous beam *ABC* as shown in Figure 7.72, using the moment distribution method. Draw the shear force diagram, bending moment diagrams, and the deflected shape. Assume *EI* is constant.

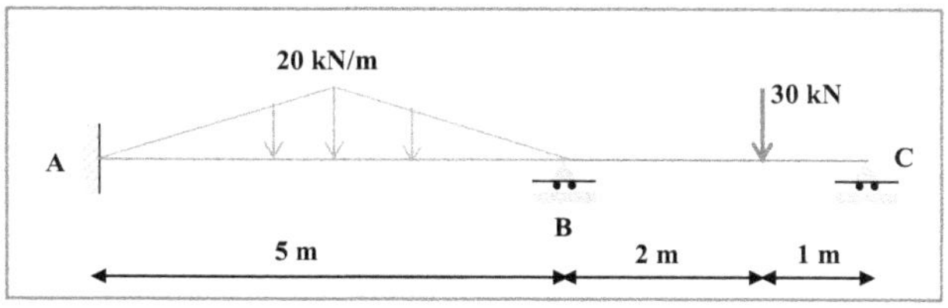

FIGURE 7.72

7.7.8 Re-analyse the two-span continuous beam *ABC*, as shown in Figure 7.72, if support *A* is hinged instead of fixed, using the moment distribution method. Also, draw the shear force diagram, bending moment diagrams, and the deflected shape. Assume constant flexural rigidity.

7.7.9 Re-analyse the two-span continuous beam *ABC*, as shown in Figure 7.72, if support *A* is guided fixed instead of fixed using the moment distribution method. Also, draw the shear force diagram, bending moment diagram, and the deflected shape. Assume constant flexural rigidity.

7.7.10 Evaluate the support reactions of the continuous beam shown in Figure 7.73, using the moment distribution method. Assume flexural rigidity is constant. Draw the shear force diagram, bending moment diagram, and the deflected shape.

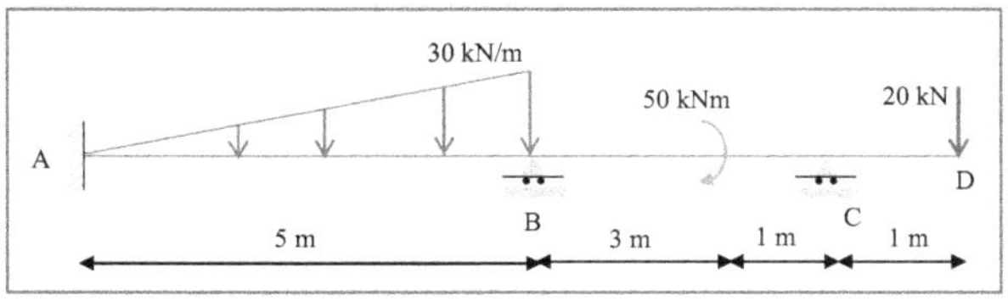

FIGURE 7.73

7.7.11 Analyse the symmetrical three-span continuous beam shown in Figure 7.74, using the moment distribution method. Draw the shear force diagram, bending moment diagram, and the deflected shape.

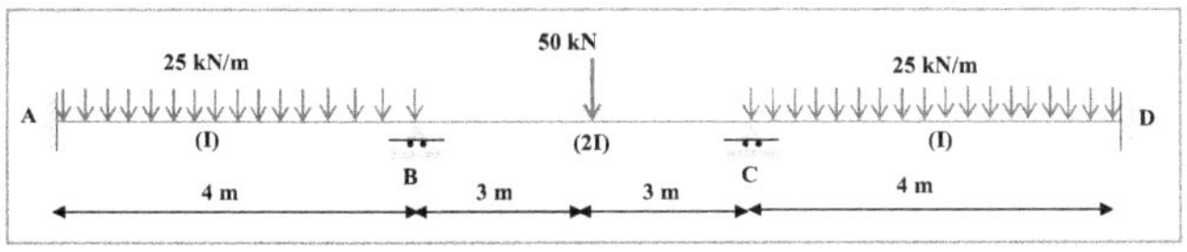

FIGURE 7.74

7.7.12 A three-span continuous beam *ABCDE* with overhang as shown in Figure 7.75. Analyse the beam using the moment distribution method. Construct the shear force diagram, bending moment diagram, and the deflected shape. Assume *EI* is constant.

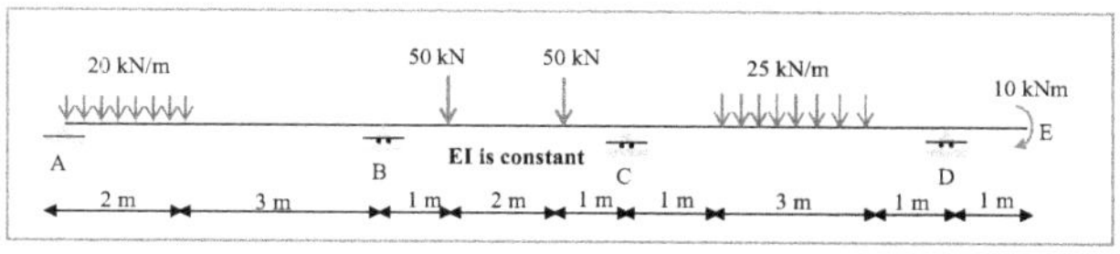

FIGURE 7.75

7.7.13 Analyse the beam shown in Figure 7.73 using the MD method, if support *A* is rotates by 0.002 rad in the clockwise direction. Assume $EI = 138{,}375$ kNm^2. Construct the shear force and bending moment diagrams.

7.7.14 Re-analyse the beam shown in Figure 7.73 using the MD method, if support *B* settles down by 10 mm. Assume $EI = 138{,}375$ kNm^2. Construct the shear force and bending moment diagrams.

7.7.15 Analyse the box frame *ABCD* as shown in Figure 7.76 using the MD method. Take advantage of the symmetry. Draw the shear force and bending moment diagrams.

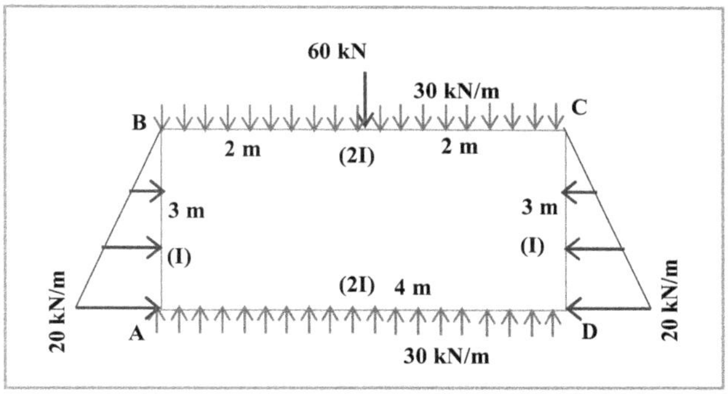

FIGURE 7.76

7.7.16 Analyse the frame shown in Figure 7.77 using the MD method. Construct the shear force, bending moment diagram, and the deflected shape.

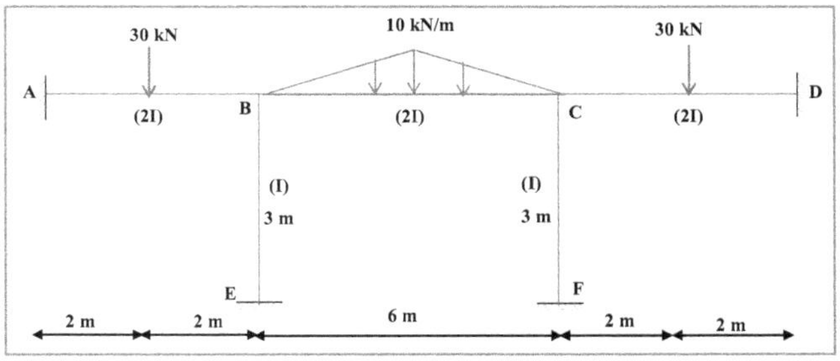

FIGURE 7.77

7.7.17 Re-analyse the frame shown in Figure 7.77 using the MD method, if the supports *A*, *E*, *F*, and *D* are hinged. Construct the shear force and bending moment diagrams and the deflected shape.

7.7.18 Analyse the frame shown in Figure 7.78 using the MD method. Draw the shear force and bending moment diagrams. Also, plot the probable deflected shape.

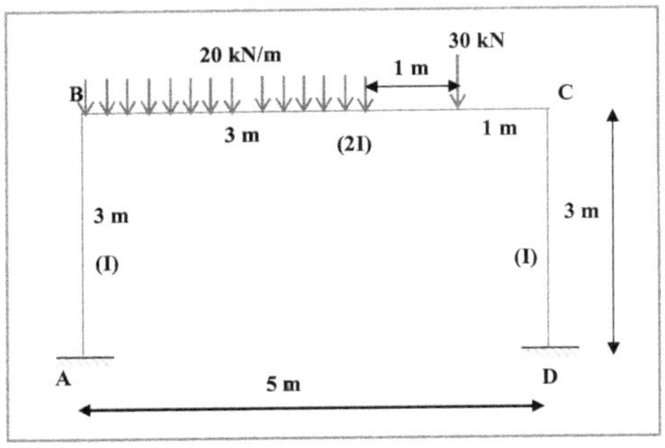

FIGURE 7.78

7.7.19 Re-analyse the frame given in Figure 7.78 using the MD method, in the absence of external loading, if support A settles down by 10 mm. Construct the shear force and bending moment diagrams. Assume $EI = 138{,}375$ kNm^2.

7.7.20 Analyse the unsymmetrical rigid joint frame shown in Figure 7.79 using the MD method. Draw the shear force diagram, bending moment diagram, and the probable deflected shape.

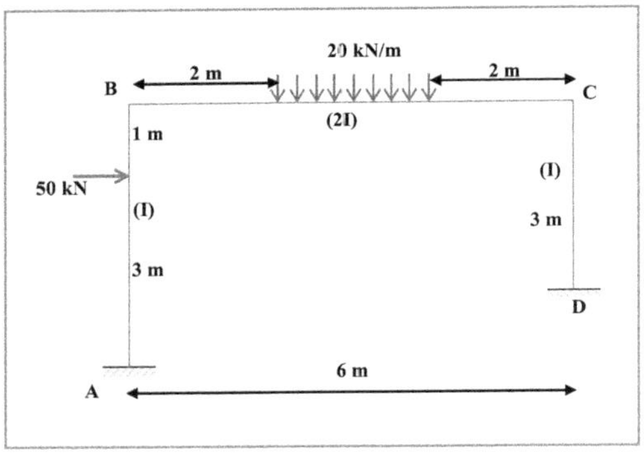

FIGURE 7.79

7.7.21 Analyse the two-bay frame shown in Figure 7.80 by using the MD method. Construct the shear force and bending moment diagrams. Assume EI is constant.

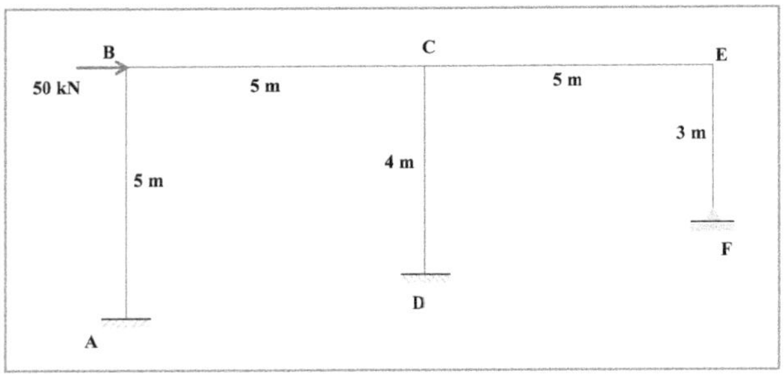

FIGURE 7.80

7.7.22 Analyse the two-storey frame shown in Figure 7.81 using the MD method. Construct the shear force and bending moment diagrams. Also draw the deflected shape.

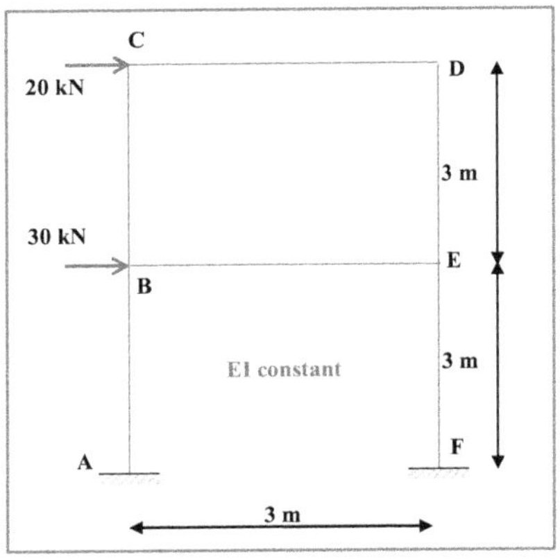

FIGURE 7.81

7.8 ANSWER KEY

Q.No.	Figure No.	Answers at Key Locations
7.7.7	7.72	$M_A = -28.87$ kNm, $M_B = -20.33$ kNm
7.7.8	7.72	$M_B = -29.28$ kNm, $R_A = 19.15$ kN, $R_B = 50.61$ kN, $R_C = 10.24$ kN
7.7.9	7.72	$M_A = -49.09$ kNm, $M_B = -75.91$ kNm, $R_A = 0$ kN, $R_B = 85.3$ kN, $R_C = 5.34$ kN
7.7.10	7.73	$M_A = -42.30$ kNm, $M_B = -2.65$ kNm, $M_C = -20$ kNm $R_A = 32.93$ kN
7.7.11	7.74	$M_A = -32.01$ kNm, $M_B = -35.83$ kNm, $R_A = 49.06$ kN, $R_B = 75.93$ kN
7.7.12	7.75	$M_B = -23.84$ kNm, $M_C = -51.20$ kNm, $R_A = 27.23$ kN, $R_B = 55.93$ kN, $R_C = 102.58$ kN, $R_D = 29.26$ kN
7.7.13	7.73	$M_A = 149.50$ kNm, $M_{BA} = -M_{BC} = 55.46$ kNm, $R_A = -15.99$ kN, $R_B = 87.36$ kN, $R_C = 23.64$ kN
7.7.14	7.73	$M_A = -352.84$ kNm, $M_{BA} = -M_{BC} = -289.68$ kNm, $R_A = 153.51$ kN, $R_B = -168.43$ kN, $R_C = 109.92$ kN
7.7.15	7.76	$M_A = -17.08$ kNm, $M_B = -35.86$ kNm, $M_C = -35.86$ kNm, $M_D = -17.08$ kNm
7.7.16	7.77	$M_A = -14.46$ kNm, $M_E = 0.69$ kNm, $R_A = 14.46$ kN, $R_E = 30.54$ kN
7.7.17	7.77	$R_A = 9.87$ kN, $R_E = 35.12$ kN, $M_{BA} = 20.50$ kNm, $M_{BC} = -19.46$ kNm, $M_{BE} = -1.04$ kNm
7.7.18	7.78	$M_A = 11.77$ kNm, $M_D = -11.96$ kNm, $R_A = 47.99$ kN, $R_D = 42.01$ kN
7.7.19	7.78	$M_A = 103.38$ kNm $= M_D$, $R_A = -51.69$ kN $= -R_D$
7.7.20	7.79	$M_A = -34.14$ kNm, $M_D = -49.98$ kNm, $R_A = 14.06$ kN, $R_D = 25.93$ kN
7.7.21	7.80	$M_A = -35.16$ kNm, $M_D = -59.44$ kNm, $R_A = -9.91$ kN, $R_D = -1.62$ kN, $R_F = 11.53$ kN
7.7.22	7.81	$M_A = -44.88$ kNm, $M_F = -44.7$ kNm, $RA = -40.14$ kN, $R_F = 40.14$ kN

Note: The answers are provided based on the analysis using structural analysis software. The results from the exact solution may differ slightly (decimals), but these differences are generally negligible.

FURTHER READING

Bhavikatti, S. S. (2013). *Structural Analysis-II*, Vikas Publishing House, New Delhi, ISBN 978-9325968806.

Derucher, K., Putcha, C., and Kim, U. (2013). *Indeterminate Structural Analysis*, The Edwin Mellen Press, New York, ISBN 978-0773444706.

Hibbeler, R. C. (2022). *Structural Analysis*, Pearson Education (Singapore) Pte. Ltd, Delhi, ISBN 978-9354497841.

Khalfallah, S. (2018). *Structural Analysis 2: Statically Indeterminate Structures*, John Wiley & Sons, Inc., Hoboken, NJ, ISBN 978-1786303394.

Leet, K. Uang, C.-M. and Lanning, J. (2018). *Fundamentals of Structural Analysis* (6 E), McGraw Hill Education Pvt Ltd, ISBN 978-1260477245.

Menon, D. (2023). *Structural Analysis*, Narosa Publishing House Pvt Ltd, New Delhi, ISBN 978-8184877342.

Reddy, C. S. (2011). *Basic Structural Analysis*, 3 E, Tata McGraw-Hill Education Pvt Ltd, New Delhi, ISBN 978-0070702769.

Sahoo, D. R. (2023). *Indeterminate Structural Analysis*, PHI Learning Pvt Ltd, Delhi, ISBN 9789354432958.

Shah, H. J. and Junarkar, S. B. (2015). *Mechanics of Structures*, Vol. II (24 E), Charotar Publishing House, Anand, ISBN 978-9385039027

Thandavmoorthy, T. S. (2011). *Structural Analysis*, Oxford University Press, ISBN 978-0198069188, 978-0198069189.

Vijayanand, M., Muthu, K. U., Narendra, H., and Janardhana, M. (2019). *Indeterminate Structural Analysis*, Dreamtech Press Publisher, New Delhi, ISBN 978-9389307177.

Wang, C. K. (1953). *Statically Indeterminate Structures,* McGraw Hill Higher Education, New York, ISBN 978-0070681309.

Williams, A. (2009). *Structural Analysis in Theory and Practices*, Elsevier, Burlington, ISBN 978-1-85617-550-0.

8 Kani's Method

8.1 INTRODUCTION

Among the various displacement methods discussed earlier, the slope deflection method is an exact method of analysis. However, its main drawback is that displacements must be explicitly calculated using equilibrium equations before determining the end moments in various beam elements. This makes the calculations cumbersome, especially when the kinematic degree of indeterminacy is high. The moment distribution method addresses these issues by simplifying the calculation procedure, particularly for problems involving only rotational degrees of freedom. The method is straightforward and precise when only one rotational degree of freedom (i.e., moment distribution at a single joint) is required. Even for multi-cycle moment distribution, involving multiple joint rotations, the procedure remains relatively simple, though the solution may be approximate if acceptable convergence is not achieved. However, the process becomes tedious when multiple moment distribution tables and solutions to horizontal force (shear) equilibrium equations are required due to the presence of lateral displacements (sway degrees of freedom).

In 1947, Professor Gasper Kani of Germany proposed an elegant and systematic approach for analysing rigid joint frames. This method addressed the shortcomings of the moment distribution method, particularly for problems involving sway degrees of freedom, by providing a solution using a single iteration table. Initially developed for the analysis of multi-storey frames, Kani's method can be applied to all types of beams and rigid joint plane frames. It integrates the procedures of slope deflection equations (in a different form) and the iterative process of the moment distribution method (also in a different form).

In Kani's method, the primary unknowns are rotational contributions (moments) and sway contributions (moments). These quantities are calculated using appropriate moment equilibrium and force equilibrium equations through a numerical technique resembling the Gauss–Seidel method for solving simultaneous linear equations. Once the rotational and sway moments are determined, the final end moments of various members can be easily obtained by substituting these values into slope deflection equations, as in the slope deflection method. This additional step differentiates Kani's method from the moment distribution method for problems involving sway degrees of freedom. However, compared to the moment distribution method, Kani's method is more efficient for such problems, as the entire solution can be achieved in a single iteration table.

This chapter explores the application of Kani's method, starting with problems involving only rotational degrees of freedom (no sway) and then extending to problems involving both rotational and sway degrees of freedom.

8.2 BEAMS SUBJECTED TO ONLY ROTATIONAL DEGREES OF FREEDOM

As discussed in the Introduction, similar to the moment distribution method, Kani's method can also be easily applied to beams and frames subjected solely to rotational degrees of freedom. In cases where other degrees of freedom are involved, these problems can be addressed by implementing appropriate modifications to the stiffness and fixed-end moments. Let us now formulate Kani's equations for such problems.

8.2.1 Kani's Analysis Procedure Formulation When Far End Fixed

Consider a typical prismatic beam AB whose span is L and flexural rigidity EI. Using the slope deflection equation, the end moment M_{AB} at the end A can be expressed in terms of the fixed end moment MF_{AB}, and the end rotations θ_A and θ_B (assumed to be positive in the clockwise direction),

$$M_{AB} = MF_{AB} + \frac{4EI}{L}\theta_A + \frac{2EI}{L}\theta_B \tag{8.1}$$

In the above equation, let us define the unknown rotation moments at the near end A as $M^1_{AB} = \frac{2EI}{L}\theta_A$ and at the far end B as $M^1_{BA} = \frac{2EI}{L}\theta_B$. Therefore, equation 8.1 can be expressed as

$$M_{AB} = MF_{AB} + 2M^1_{AB} + M^1_{BA} \tag{8.2}$$

Similarly, the end moment at B can be written as

$$M_{BA} = MF_{BA} + 2M^1_{BA} + M^1_{AB} \tag{8.3}$$

So,

The final end moment = Fixed end moment + 2 (rotation moment at near end) + 1 (rotation moment at far end)

8.2.2 Far End is Hinged/Pinned Support

When the far end is a hinged or pinned support, equation 8.1 can be modified as (taking advantage of condensed kinematic indeterminacy)

$$M_{AB} = MF_{AB} + \frac{3EI}{L}\theta_A = MF_{AB} + 2M^1_{AB} \tag{8.4}$$

where the rotation contribution $M^1_{AB} = \frac{1}{2}\times\frac{3EI}{L}\theta_A$ and $M^1_{BA} = 0$

DOI: 10.1201/9781003562955-8

It should be noted that MF_{AB} should be modified by treating the far end as hinged (propped cantilever), with the end moment at the pinned end (far end) being zero, i.e., $M_{BA} = 0$, unless there is some external moment acting at B.

8.2.3 Far End is Guided-Fixed Support

When the far end is a guided-fixed support, we can see that the beam behaves as a cantilever under the end rotation A. Therefore, we can modify equation 8.1 as

$$M_{AB} = MF_{AB} + \frac{EI}{L}\theta_A = MF_{AB} + 2M^1_{AB} \quad (8.5)$$

Where the rotation contribution $M^1_{AB} = \frac{1}{2} \times \frac{EI}{L}\theta_A$ and $M^1_{BA} = 0$.

The far-end moment at B can be expressed as

$$M_{BA} = MF_{BA} - \frac{EI}{L}\theta_A = MF_{BA} - 2M^1_{AB} \quad (8.6)$$

Equations 8.2–8.6 can be expressed more generally using Kani's format. Considering the typical beam element Oi whose far ends are A, B, C, and D (as shown in Figure 8.1), then the end moment at O can be written in a general form as

$$M_{Oi} = MF_{Oi} + 2M^1_{Oi} + M^1_{iO} \quad (8.7)$$

$M^1_{Oi} = \frac{1}{2} \times \frac{4EI}{L}\theta_O$ For far end is continuous

$M^1_{Oi} = \frac{1}{2} \times \frac{3EI}{L}\theta_O$ For far end is hinged

$M^1_{Oi} = \frac{1}{2} \times \frac{EI}{L}\theta_O$ For far end is guided fixed

$M^1_{iO} = \frac{2EI}{L}\theta_i$ For far end is continuous

$M^1_{iO} = 0$ For far end is fixed or guided support

$M^1_{iO} = 0$ For far end is pinned or roller support as the rotation at this end is neglected

If the rotation contributions or moments are generated, the end moments can be determined using equation 8.7 without calculating the end rotations, as Kani suggested a simple process for this purpose.

Considering the joint moment equilibrium at O,

$$\sum_i M_{Oi} = 0 \Rightarrow \sum_i MF_{Oi} + 2\sum_i M^1_{Oi} + \sum_i M^1_{iO} = 0$$

$$\sum_i M^1_{Oi} = \left(-\frac{1}{2}\right)\left[\sum_i MF_{Oi} + \sum_i M^1_{iO}\right] \quad (8.8)$$

For each member meeting at joint O,

$$\sum_i M^1_{Oi} = \frac{1}{2} \times \left(\sum_i k_i\right)\theta_O \quad (8.9)$$

Where $k_i = \frac{4EI}{L}$ for general beam element

$k_i = \frac{3EI}{L}$ for far end i pinned

$k_i = \frac{EI}{L}$ for far end i guided fixed

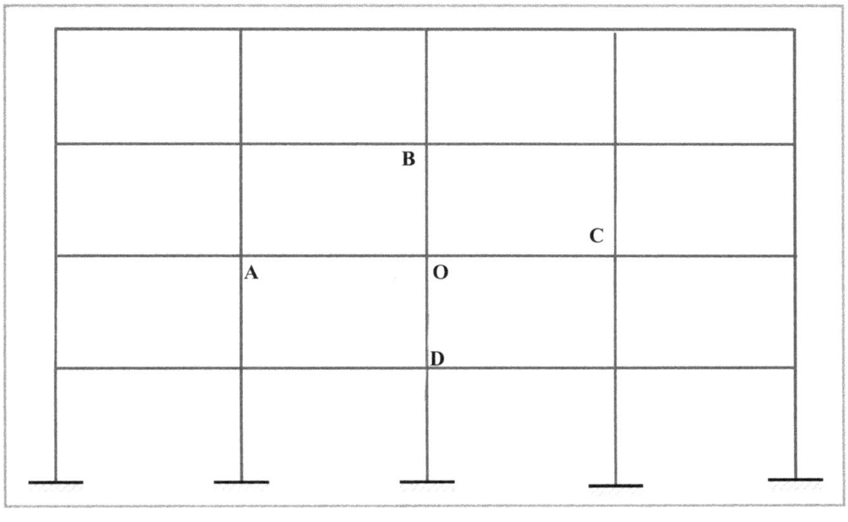

FIGURE 8.1 A typical frame.

$$\theta_O = \frac{2\sum_i M_{Oi}^1}{\sum_i k_i} \quad (8.10)$$

For any individual member,

$$M_{Oi}^1 = \frac{1}{2} k_i \theta_O \quad (8.11)$$

Since θ_O is the same for all the members meeting at joint O, substituting equation 8.10 into equation 8.11, we obtain

$$M_{Oi}^1 = \frac{1}{2} k_i \times \frac{2\sum_i M_{Oi}^1}{\sum_i k_i} \Rightarrow \left[\frac{k_i}{\sum_i k_i}\right] \sum_i M_{Oi}^1 \quad (8.12)$$

The term within the parenthesis is the same as the distribution factor defined in the moment distribution method. Substituting equation 8.8 into equation 8.12, we get

$$M_{Oi}^1 = \left[\frac{k_i}{\sum_i k_i}\right]\left[\left(-\frac{1}{2}\right)\left[\sum_i MF_{Oi} + \sum_i M_{iO}^1\right]\right]$$

$$M_{Oi}^1 = \left(-\frac{1}{2}\right)\left[\frac{k_i}{\sum_i k_i}\right]\left[\left[\sum_i MF_{Oi} + \sum_i M_{iO}^1\right]\right] \quad (8.13)$$

The expression $\left(-\frac{1}{2}\right)\left[\frac{k_i}{\sum_i k_i}\right]$ is called the rotation factor (RF) for any member meeting at joint O.

$$\therefore \text{Rotation factor } (\text{RF})_i = \left(-\frac{1}{2}\right)\left[\frac{k_i}{\sum_i k_i}\right]$$

$$= \frac{-1}{2}(\text{distribution factor}) \quad (8.13a)$$

It should be noted that the sum of all rotation factors at a joint equals −0.5. Equation 8.13 represents the rotation contributions at any joint, obtained from the distributed sum of all fixed end moments of various connecting members at the joint $\left(\sum_i MF_{Oi}\right)$, plus the sum of far-end rotational moments according to their rotation factors. In this equation, the rotation factors $(\text{RF})_i$, and the fixed end moments $\sum_i MF_{Oi}$ are known constants, while the far-end rotation contributions are unknowns that need to be calculated iteratively. To begin the problem, all far-end rotation contributions M_{iO}^1 are initially assumed to be zero, and then calculate the values of M_{Oi}^1 at joint O, using equation 8.13. Subsequently, we proceed to the adjacent joint and calculate improved estimates of the rotation contributions using equation 8.13 based on the rotation contribution values obtained in the previous cycle. We continue this procedure with progressively refined estimates of the rotation contributions in every succeeding cycle until the desired accuracy or convergence is attained. This procedure is similar to the operation carried out in the moment distribution method, i.e., balancing and carryover operation in the moment distribution method. For convenience, this procedure can be performed in a tabular format, as demonstrated in the numerical examples in the subsequent sections.

8.2.3.1 General Procedure for Beams and Frames Subjected to Rotational Degrees of Freedom

The analysis of indeterminate beams and frames subjected to only the rotational degrees of freedom using Kani's method is outlined below.

i. **Calculate the kinematic indeterminacy of the beam.**
   Fix all the joints (arrest all the degrees of freedom) and calculate the fixed end moments due to the external loads for each span using the expressions provided in Table 6.1. If the far end of a span is hinged or supported on a roller, calculate the modified fixed end moment, considering that the moment at the hinge is zero.
ii. **Determine the stiffness factors** for each connecting member using the appropriate equations:
   if the far end is fixed, use $K = 4EI/L$,
   if the far end is hinged, use $K = 3EI/L$,
   if the far end has a guided support, use $K = EI/L$,
   For symmetrical spans, use $K = EI/(L/2) = 2EI/L$
iii. **Calculate the rotation factors** of each joint using equation 8.13a. It should be noted that the rotation factor for hinge or roller support is −1/2, and for fixed support, it is zero.
iv. **To start Kani's iterative cycles at any joint**, initially assume the rotational contributions of the far ends of the connecting members of that joint to be zero and calculate the rotation contributions of the near ends of each connecting member using equation 8.13. This procedure is to be continued for all the joints. While calculating the rotation contributions of the members at the next joint, use the rotation contributions obtained previously at the far ends (if available), if not available, assume them to be zero, and continue this procedure at all the joints.
v. **To calculate the second cycle rotational contributions** of each member, use the rotational contributions of the members obtained in first cycle as initial values, and continue the same procedure for all the joints as explained in (iv).

vi. **Continue the iterative procedure** (Kani's cycles) until the difference between two consecutive rotational contributions is negligible.
vii. **Calculate the final member end moments**, using equation 8.7.
viii. Once the final member end moments are obtained, the **other internal forces and support reactions** can be determined with the help of the free body diagrams and using the static equilibrium equations.
ix. Finally **sketch the shear force and bending moment** diagrams.

The above procedure is demonstrated in the following examples in a table format more conveniently.

8.3 NUMERICAL EXAMPLES ON BEAMS

Example 8.1

Analyse the two-span continuous beam shown in Figure 8.2, using Kani's method. Assume *EI* is constant. The same has been solved in Example 6.1, using SD method.

Solution:

The kinematic indeterminacy of the beam is one, i.e., rotation at *B*. Let us fix all the joints and calculate the fixed end moments using the expressions provided in Table 6.1.

FIXED END MOMENTS

$$MF_{AB} = -\frac{Wab^2}{L^2} = -\frac{30\times 2\times 3^2}{5^2} = -21.66 \text{ kNm}$$

$$MF_{BA} = \frac{Wba^2}{L^2} = \frac{30\times 3\times 2^2}{5^2} = 14.40 \text{ kNm}$$

$$MF_{BC} = -\frac{wL^2}{12} = -\frac{15\times 5^2}{12} = -31.25 \text{ kNm}$$

$$MF_{CB} = \frac{wL^2}{12} = \frac{15\times 5^2}{12} = 31.25 \text{ kNm}$$

STIFFNESS FACTORS

Joint *B*

$$K_{BA} = \frac{4EI}{L} = \frac{4EI}{5} = 0.8EI$$

$$K_{BC} = \frac{4EI}{L} = \frac{4EI}{5} = 0.8EI$$

ROTATION FACTORS

$$(\text{RF})_i = \frac{-1}{2}(\text{DF})_i = \left(\frac{-1}{2}\right)\frac{K_i}{\sum_{i=1}^{n} K_i}$$

Joint	Member	*K*	$\sum K$	DF	RF
B	*BA*	$\frac{4EI}{5} = 0.8EI$		0.5	−0.25
	BC	$\frac{4EI}{5} = 0.8EI$	0.8EI+0.8EI=1.6EI	0.5	−0.25

Note: Sum of the rotation factors at each joint equal to −0.5

RESULTANT RESTRAINT (FIXED END) MOMENTS

$$MF_i = \sum MF_{ij}$$

$$MF_B = MF_{BA} + MF_{BC} = 14.40 - 31.25 = -16.85 \text{ kNm}$$

KANI'S ITERATION CYCLES

Rotation contributions: At joint *B*

Rotation contribution

$$M_{BA}^1 = \text{RF}\left(\begin{array}{l}\sum (\text{Fixed end moments at joint } B) \\ +\sum (\text{Rotation contribution from far end})\end{array}\right)$$

Initially assume all rotational contributions to be zero at all joints which will indirectly mean that θ_i=0

$$M_{AB}^1 = M_{CB}^1 = 0 \,(\text{since } A \text{ and } C \text{ are fixed ends})$$

$$M_{BA}^1 = (\text{RF})_{BA}\left[MF_B + \left(M_{AB}^1 + M_{CB}^1\right)\right]$$
$$= -0.25[-16.85 + (0+0)] = 4.213$$

$$M_{BC}^1 = (\text{RF})_{BC}\left[MF_B + \left(M_{AB}^1 + M_{CB}^1\right)\right]$$
$$= -0.25[-16.85 + (0+0)] = 4.213$$

The ends *A* and *C* are fixed, hence, the rotational contributions are zero. Therefore, no further cycles are required.

FINAL END MOMENTS

Final end moment = Fixed end moment
+ 2(Near end rotational contribution)
+ 1(Far end rotational contribution)

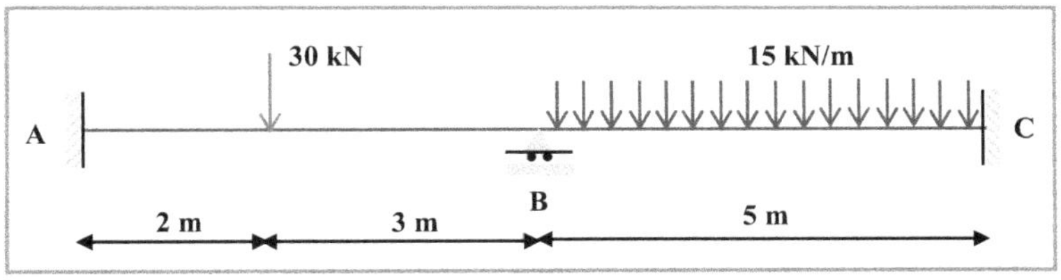

FIGURE 8.2 Two-span continuous beam.

TABLE 8.1
Kani's Iteration Cycles and Final End Moments

Joint	A	B		C
Member	*AB*	*BA*	*BC*	*CB*
Rotation factor	0.000	−0.250	−0.250	0.000
FEM	−21.660	14.400	−31.250	31.250
Sum of FEMs		−16.850		31.250
Rotation contribution		4.213	4.213	
		4.213	4.213	
FEM	−21.660	14.400	−31.250	31.250
2×near end contribution	0.000	8.425	8.425	0.000
1×far end contribution	4.213	0.000	0.000	4.213
Final end moments (kNm)	−17.448	22.825	−22.825	35.463

$$M_{AB} = MF_{AB} + 2\left(M_{AB}^1\right) + 1\left(M_{BA}^1\right) = -21.66 + 2(0) + 1(4.213)$$
$$= -17.448 \text{ kNm}$$
$$M_{BA} = MF_{BA} + 2\left(M_{BA}^1\right) + 1\left(M_{AB}^1\right) = 14.40 + 2(4.213) + 1(0)$$
$$= 22.825 \text{ kNm}$$
$$M_{BC} = MF_{BC} + 2\left(M_{BC}^1\right) + 1\left(M_{CB}^1\right) = -31.25 + 2(4.213) + 1(0)$$
$$= -22.825 \text{ kNm}$$
$$M_{CB} = MF_{CB} + 2\left(M_{CB}^1\right) + 1\left(M_{BC}^1\right) = 31.25 + 2(0) + 1(4.213)$$
$$= 35.463 \text{ kNm}$$

The above calculation is more conveniently shown in Table 8.1.

FINAL END SHEARS

Once the member end moments are known from the table, using the static equilibrium equations the member end shears are calculated, as shown in the free body diagram (Figure 8.3a). The shear force and bending moment diagrams are shown in Figure 8.3b and c, respectively.

Example 8.2

Analyse the three-span continuous beam *ABCD* as shown in Figure 8.4, using Kani's method. Draw the shear force and bending moment diagrams.

Solution:

The kinematic indeterminacy of the beam is 03, i.e., the rotations at *B*, *C*, and *D*. Arrest these DoF by assuming the fixed ends and calculate the fixed end moments.

FIXED END MOMENTS

Using Table 6.1, the fixed end moments are given as follows:

$$MF_{AB} = -\frac{Wab^2}{L^2} = -\frac{50 \times 3 \times 1^2}{4^2} = -9.375 \text{ kNm}$$

$$MF_{BA} = \frac{Wba^2}{L^2} = \frac{50 \times 1 \times 3^2}{4^2} = 28.125 \text{ kNm}$$

$$MF_{BC} = -\frac{wL^2}{12} = -\frac{30 \times 6^2}{12} = -90 \text{ kNm} = -MF_{CB}$$

$$MF_{CD} = MF_{DC} = \frac{M}{4} = \frac{80}{4} = 20 \text{ kNm}$$

ROTATION FACTORS (SHOWN IN TABLE)

$$(RF)_i = \frac{-1}{2}(DF)_i = \left(\frac{-1}{2}\right)\frac{K_i}{\sum_{i=1}^{n} K_i}$$

Joint	Member	K	$\sum K$	DF	RF
B	*BA*	$\frac{4EI}{4} = EI$	$EI + 1.33EI = 2.33EI$	0.429	−0.214
	BC	$\frac{4E(2I)}{6} = 1.33EI$		0.571	−0.286
C	*CB*	$\frac{4E(2I)}{6} = 1.33EI$	$1.33EI + EI = 2.33EI$	0.571	−0.286
	CD	$\frac{4EI}{4} = EI$		0.429	−0.214

RESULTANT RESTRAINT (FIXED END) MOMENTS

$$MF_i = \sum MF_{ij}$$
$$MF_B = MF_{BA} + MF_{BC} = 28.125 - 90 = -61.875 \text{ kNm}$$
$$MF_C = MF_{CB} + MF_{CD} = 90 + 20 = 110 \text{ kNm}$$

KANI'S ITERATION CYCLES: FIRST CYCLE

Rotation contributions: Joint *B*

Rotation contribution

$$M_{BA}^1 = \text{RF}\begin{pmatrix} \sum (\text{Fixed end moments at joint } B) \\ +\sum (\text{Rotation contribution from far end}) \end{pmatrix}$$

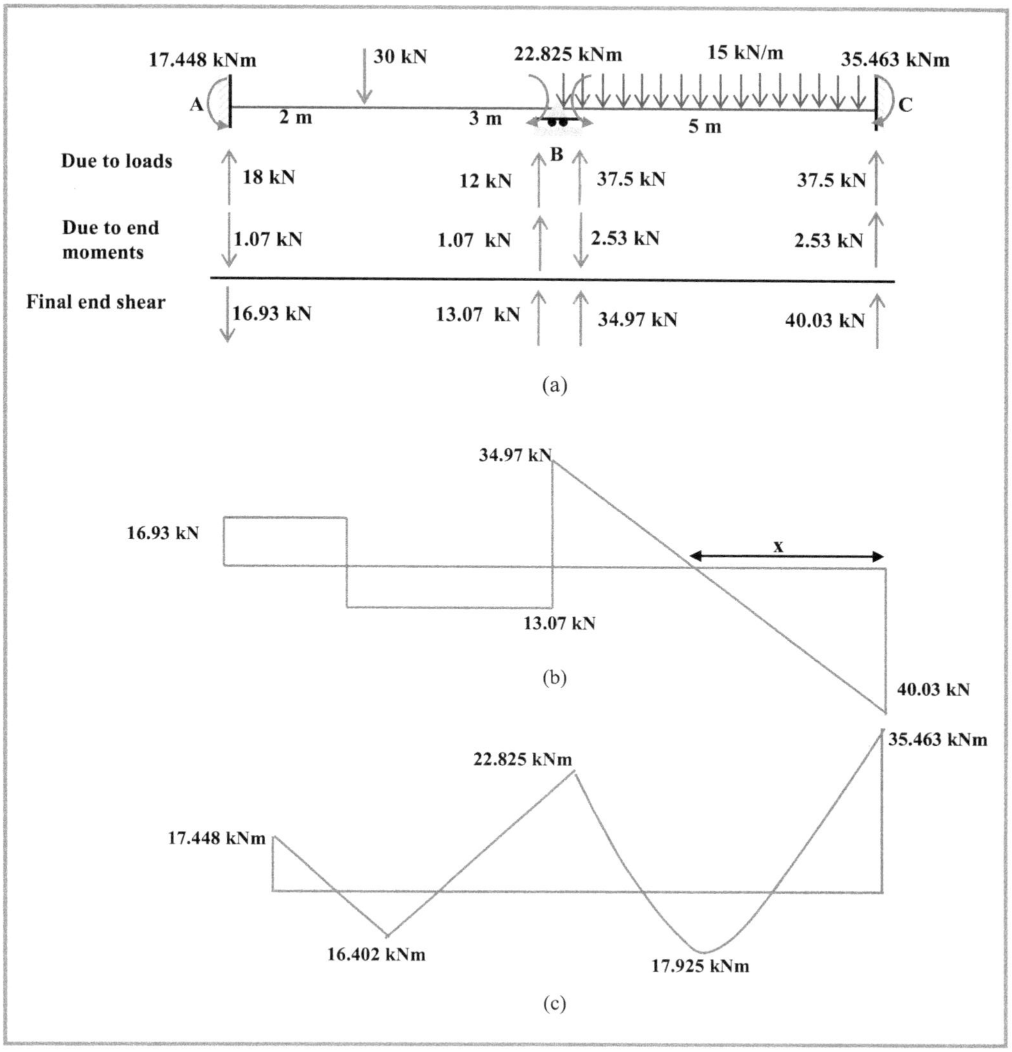

FIGURE 8.3 (a) Free body diagram, (b) shear force diagram, and (c) bending moment diagram.

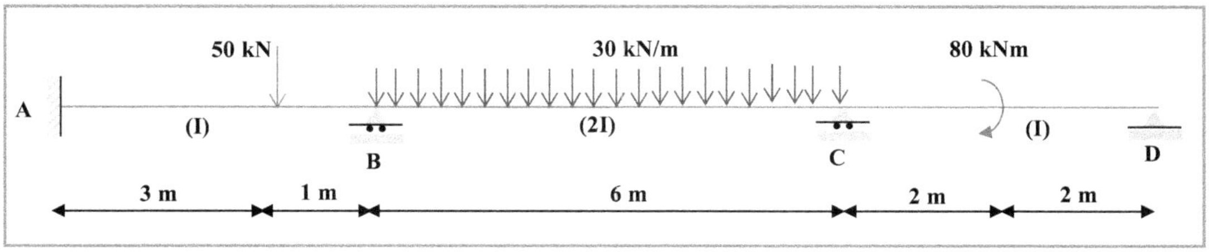

FIGURE 8.4 A three-span continuous beam.

Initially, all far-end rotational contributions of the connecting members are assumed to be zero, which indirectly means that $\theta_i = 0$

$$M^1_{AB} = 0 \text{ (since } A \text{ is fixed end)}$$

$$M^1_{BA} = (\text{RF})_{BA}\left[MF_B + \left(M^1_{AB} + M^1_{CB}\right)\right]$$

$$= -0.214\left[-61.875 + (0+0)\right] = 13.241$$

$$M^1_{BC} = (\text{RF})_{BC}\left[MF_B + \left(M^1_{AB} + M^1_{CB}\right)\right]$$

$$= -0.286\left[-61.875 + (0+0)\right] = 17.696$$

$$M^1_{CB} = (\text{RF})_{CB}\left[MF_C + \left(M^1_{BC} + M^1_{DC}\right)\right]$$

$$= -0.286\left[110 + (17.696+0)\right] = -36.521$$

$$M_{CD}^1 = (\mathrm{RF})_{CD}\left[MF_C + \left(M_{BC}^1 + M_{DC}^1\right)\right]$$

$$= -0.214\left[110 + (17.696 + 0)\right] = -27.327$$

$$M_{DC}^1 = (\mathrm{RF})_{DC}\left[MF_D + \left(M_{CD}^1\right)\right] = -0.5\left[20 + (-27.327)\right] = 3.663$$

SECOND CYCLE

For calculating the second cycle rotational contributions of each member, use the rotational contributions of the members obtained in the first cycle.

$$M_{AB}^1 = 0$$

$$M_{BA}^1 = (\mathrm{RF})_{BA}\left[MF_B + \left(M_{AB}^1 + M_{CB}^1\right)\right]$$

$$= -0.214\left[-61.875 + (0 - 36.521)\right] = 21.057$$

$$M_{BC}^1 = (\mathrm{RF})_{BC}\left[MF_B + \left(M_{AB}^1 + M_{CB}^1\right)\right]$$

$$= -0.286\left[-61.875 + (0 - 36.521)\right] = 28.141$$

$$M_{CB}^1 = (\mathrm{RF})_{CB}\left[MF_C + \left(M_{BC}^1 + M_{DC}^1\right)\right]$$

$$= -0.286\left[110 + (28.141 + 3.663)\right] = -40.556$$

$$M_{CD}^1 = (\mathrm{RF})_{CD}\left[MF_C + \left(M_{BC}^1 + M_{DC}^1\right)\right]$$

$$= -0.214\left[110 + (28.141 + 3.663)\right] = -30.346$$

$$M_{DC}^1 = (\mathrm{RF})_{DC}\left[MF_D + \left(M_{CD}^1\right)\right]$$

$$= -0.5\left[20 + (-30.346)\right] = 5.173$$

The above procedure should continue until the difference between two consecutive rotational contributions is negligible.

FINAL END MOMENTS

Final end moment = Fixed end moment + 2 (Near end rotational contribution) +1(Far end rotational contribution)

$$M_{AB} = MF_{AB} + 2\left(M_{AB}^1\right) + 1\left(M_{BA}^1\right)$$

$$= -9.375 + 2(0) + 1(22.121) = 12.746 \text{ kNm}$$

$$M_{BA} = MF_{BA} + 2\left(M_{BA}^1\right) + 1\left(M_{AB}^1\right)$$

$$= 28.125 + 2(22.121) + 1(0) = 72.367 \text{ kNm}$$

The cyclic operation of rotation contributions in each cycle and the final end moments are conveniently calculated in Table 8.2.

The member end shears, shear force diagram, and bending moment diagrams are shown in Figure 8.5a–c, respectively.

ALTERNATE SOLUTION

Let us take advantage of the known moment, i.e., zero moment at *D* due to the hinge support, and use the modified stiffness and modified fixed end moments for span *CD*. Therefore, the kinematic indeterminacy is limited to 02, i.e., rotations at *B* and *C* only.

FIXED END MOMENTS

$$MF_{AB} = -\frac{Wab^2}{L^2} = -\frac{50 \times 3 \times 1^2}{4^2} = -9.375 \text{ kNm}$$

$$MF_{BA} = \frac{Wba^2}{L^2} = \frac{50 \times 1 \times 3^2}{4^2} = 28.125 \text{ kNm}$$

$$MF_{BC} = -\frac{wL^2}{12} = -\frac{30 \times 6^2}{12} = -90 \text{ kNm} = -MF_{CB}$$

TABLE 8.2
Kani's Iteration Cycles and Final End Moments

Joint	A	B		C		D
Member	AB	BA	BC	CB	CD	DC
RF	0.000	−0.214	−0.286	−0.286	−0.214	−0.500
FEM	−9.375	28.125	−90.000	90.000	20.000	20.000
Sum of FEMs		−61.875		110.000		20.000
Rotation contribution		13.241	17.696	−36.521	−27.327	3.663
		21.057	28.141	−40.556	−30.346	5.173
		21.920	29.295	−41.318	−30.916	5.458
		22.083	29.513	−41.462	−31.024	5.512
		22.114	29.554	−41.489	−31.044	5.522
		22.120	29.562	−41.494	−31.048	5.524
		22.121	29.564	−41.495	−31.049	5.524
		22.121	29.564	−41.495	−31.049	5.524
FEM	−9.375	28.125	−90.000	90.000	20.000	20.000
2×near end contribution	0.000	44.242	59.128	−82.991	−62.098	11.049
1×far end contribution	22.121	0.000	−41.495	29.564	5.524	−31.049
Final end moments (kNm)	12.746	72.367	−72.367	36.573	−36.573	0.000

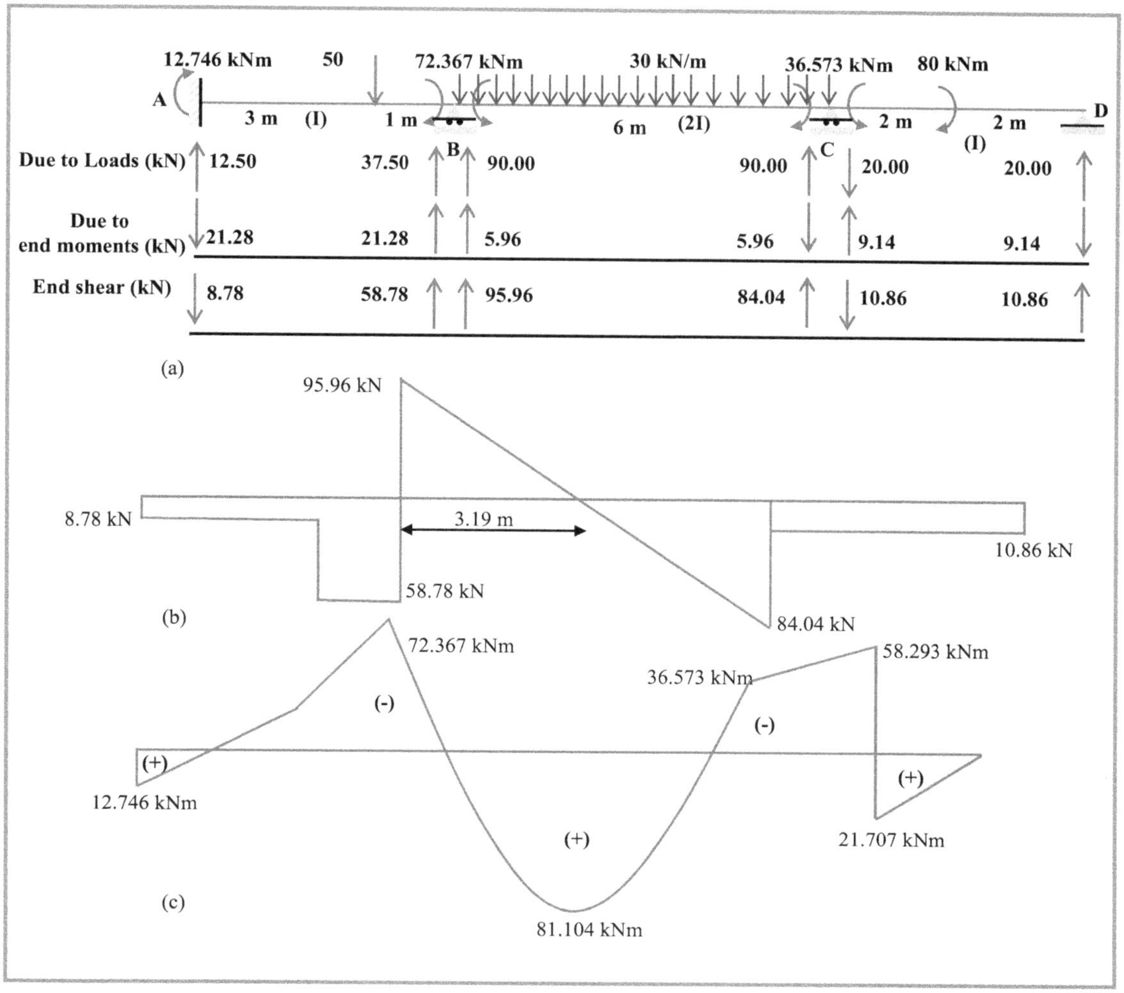

FIGURE 8.5 (a) Free body diagram, (b) shear force diagram, and (c) bending moment diagram.

$$MF_{CD} = \frac{M}{4} - \frac{M}{8} = \frac{M}{8} = \frac{80}{8} = 10 \text{ kNm}$$

$$MF_{DC} = 0$$

ROTATION FACTORS (SHOWN IN TABLE)

$$(\text{RF})_i = \frac{-1}{2}(\text{DF})_i = \left(\frac{-1}{2}\right)\frac{K_i}{\sum_{i=1}^{n} K_i}$$

Joint	Member	K	$\sum K$	DF	RF
B	*BA*	$\frac{4EI}{4} = EI$	$EI + 1.33EI = 2.33EI$	0.429	−0.214
	BC	$\frac{4E(2I)}{6} = 1.33EI$		0.571	−0.286
C	*CB*	$\frac{4E(2I)}{6} = 1.33EI$	$1.33EI + 0.75EI = 2.08EI$	0.639	−0.320
	CD	$\frac{3EI}{4} = 0.75EI$		0.361	−0.180

KANI'S ROTATION CYCLE

Kani's iterations are shown in Table 8.3.

FINAL END MOMENTS

Final end moment = Fixed end moment
+2(Near end rotational contribution)
+1(Far end rotational contribution)

The final end moments are conveniently calculated in Table 8.3.

Example 8.3

Re-analyse Example 8.2 by taking *D* as fixed support instead of hinged support, as shown in Figure 8.6. Draw the shear force and bending moment diagrams.

Solution:

Since there is no change in the geometry and loading conditions, the fixed end moments and rotation factors remain the same as those calculated in Example 8.2. Therefore, they are used in this case. The calculation of rotation contributions in each Kani's cycle, along with the final end moments, is presented in Table 8.4.

TABLE 8.3
Kani's Iteration Cycles and Final End Moments

Joint	A	B		C		D
Member	*AB*	*BA*	*BC*	*CB*	*CD*	*DC*
RF	0.000	−0.214	−0.286	−0.320	−0.180	
FEM	−9.375	28.125	−90.000	90.000	10.000	0.000
Sum of FEMs		−61.875		100.000		
Rotation contribution		13.259	17.679	−37.657	−21.182	
		21.328	28.438	−41.100	−23.119	
		22.066	29.421	−41.415	−23.296	
		22.134	29.511	−41.444	−23.312	
		22.140	29.520	−41.446	−23.314	
		22.140	29.520	−41.446	−23.314	
FEM	−9.375	28.125	−90.000	90.000	10.000	0.000
2 × near end contribution	0.000	44.281	59.041	−82.893	−46.627	0.000
1 × far end contribution	22.140	0.000	−41.447	29.520	0.000	0.000
Final end moments (kNm)	12.765	72.406	−72.406	36.627	−36.627	0.000

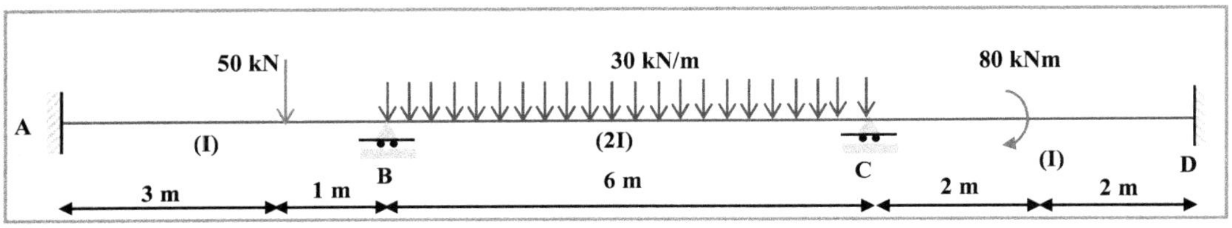

FIGURE 8.6 A three-span continuous beam.

TABLE 8.4
Kani's Iteration Cycles and Final End Moments

Joint	A	B		C		D
Member	*AB*	*BA*	*BC*	*CB*	*CD*	*DC*
RF	0.000	−0.214	−0.286	−0.286	−0.214	
FEM	−9.375	28.125	−90.000	90.000	20.000	20.000
Sum of FEMs		−61.875		110.000		
Rotation contributions		13.259	17.679	−36.480	−27.360	
		21.076	28.101	−39.458	−29.593	
		21.714	28.952	−39.701	−29.775	
		21.766	29.022	−39.720	−29.790	
		21.770	29.027	−39.722	−29.792	
		21.771	29.028	−39.722	−29.792	
FEM	−9.375	28.125	−90.000	90.000	20.000	20.000
2×near end contribution	0.000	43.542	58.056	−79.444	−59.583	0.000
1×far end contribution	21.771	0.000	−39.722	29.028	0.000	−29.792
Final end moments (kNm)	12.396	71.667	−71.667	39.583	−39.583	−9.792

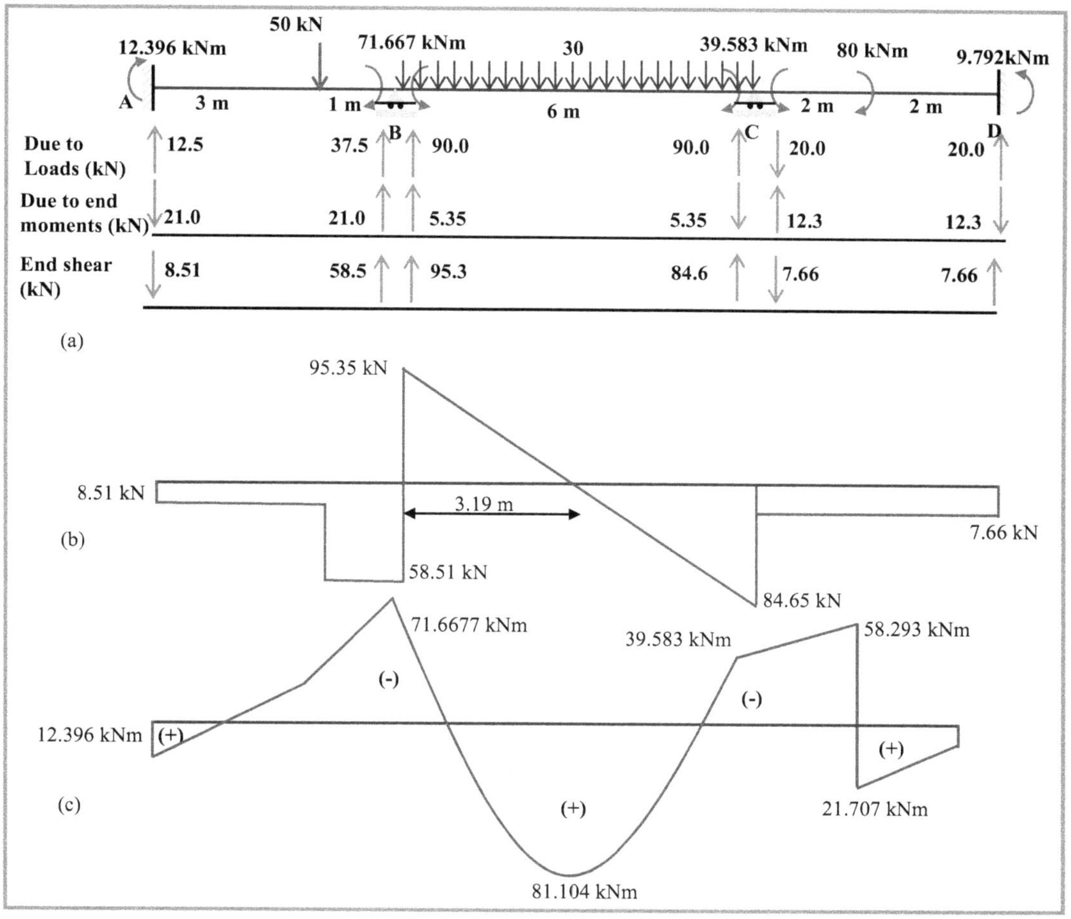

FIGURE 8.7 (a) Free body diagram, (b) shear force diagram, and (c) bending moment diagram.

FINAL END MOMENTS

Final end moment = Fixed end moment
+2(Near end rotational contribution)
+1(Far end rotational contribution)

These moments are calculated conveniently in the above Table 8.4. The member end shears, shear force, and bending moment diagrams are shown in Figure 8.7a–c, respectively.

Example 8.4

Analyse the three-span continuous beam with one side overhang as shown in Figure 8.8, using Kani's method. Draw the shear force and bending moment diagrams.

Solution:

FIXED END MOMENTS

The moment at *A* is zero due to the hinge support. The cantilever portion *DE* is statically determinate, and its effect can be represented as a moment of 20 kNm acting at *D* in the clockwise direction. Therefore, due to the known moments at *A* and *D*, the rotations at these supports can be neglected by appropriately modifying the stiffness and fixed end moments. As a result, the kinematic indeterminacy of the beam is reduced to two, i.e., rotations at *B* and *C* only.

$$MF_{BA} = \frac{3WL}{16} = \frac{3 \times 40 \times 4}{16} = 30 \text{ kNm}$$

$$MF_{BC} = -\frac{WL^2}{12} = -\frac{12 \times 6^2}{12} = -36 \text{ kNm} = -MF_{CB}$$

$$MF_{CD} = \left(\frac{-Wab^2}{L^2} - \frac{Wba^2}{2L^2}\right) + \frac{M_0}{2}$$

$$= \left(\frac{-60 \times 2 \times 4^2}{6^2} - \frac{60 \times 4 \times 2^2}{2 \times 6^2}\right) + \frac{20}{2} = -56.667 \text{ kNm}$$

$$MF_{DC} = M_0 = 20 \text{ kNm}$$

ROTATION FACTORS (SHOWN IN TABLE)

$$(\text{RF})_i = \frac{-1}{2}(\text{DF})_i = \left(\frac{-1}{2}\right)\frac{K_i}{\sum_{i=1}^{n} K_i}$$

Joint	Member	K	$\sum K$	DF	RF
B	BA	$\frac{3EI}{4}=0.75EI$	$0.75EI+2EI=2.75EI$	0.272	−0.136
	BC	$\frac{4E(3I)}{6}=2EI$		0.728	−0.364
C	CB	$\frac{4E(3I)}{6}=2EI$	$2EI+EI=3EI$	0.667	−0.333
	CD	$\frac{3E(2I)}{6}=EI$		0.333	−0.167

Rotational contributions, along with Kani's rotational cycles, and final end moments, are presented in Table 8.5.

FINAL END MOMENTS

$$\text{Final end moment} = \text{Fixed end moment} + 2(\text{Near end rotational contribution}) + 1(\text{Far end rotational contribution})$$

The final end moments are conveniently calculated and shown in the above table.

FINAL MEMBER END SHEARS

With the above known member end moments, the member end shears are calculated using the static equilibrium equations as shown in the free body diagram (Figure 8.9a). The shear force and bending moment diagrams are shown in Figure 8.9b and c, respectively.

Example 8.5

Analyse the continuous beam shown in Figure 8.10 using Kani's method. Construct the shear force and bending moment diagrams.

Solution:

Due to the known moment at A, i.e., zero moment, the rotation at A can be neglected by adopting the modified fixed end moment and stiffness for member BA. Hence, the kinematic indeterminacy of the beam is limited to 02, i.e., rotations at B and C.

FIXED END MOMENTS

$$MF_{BA}=\frac{wL^2}{20}+\frac{1}{2}\frac{wL^2}{30}=\frac{wL^2}{15}=\frac{10\times 6^2}{15}=24\text{ kNm}$$

$$MF_{BC}=\frac{M}{4}=\frac{48}{4}=12\text{ kNm}=MF_{CB}$$

$$MF_{CD}=-\frac{11}{192}wL^2=-\frac{11\times 16\times 6^2}{192}=-33\text{ kNm}$$

$$MF_{DC}=\frac{5}{192}wL^2=\frac{5\times 16\times 6^2}{192}=15\text{ kNm}$$

ROTATION FACTORS (SHOWN IN TABLE)

$$(\text{RF})_i=\frac{-1}{2}(\text{DF})_i=\left(\frac{-1}{2}\right)\frac{K_i}{\sum_{i=1}^{n}K_i}$$

TABLE 8.5
Kani's Iteration Cycles and Final End Moments

Joints	A	B		C		D	
Member	AB	BA	BC	CB	CD	DC	DE
RF	0.000	−0.136	−0.364	−0.333	−0.167	—	—
FEM	0.000	30.000	−36.000	36.000	−56.667	20.000	−20.000
Sum of FEMs		−6.000		−20.667			
Rotation contribution		0.816	2.184	6.155	3.087		
		−0.021	−0.056	6.901	3.461		
		−0.123	−0.328	6.991	3.506		
		−0.135	−0.361	7.002	3.512		
		−0.136	−0.365	7.004	3.512		
		−0.136	−0.365	7.004	3.512		
FEM	0.000	30.000	−36.000	36.000	−56.667	20.00	−20.000
2×near end contribution	—	−0.273	−0.731	14.008	7.025	—	
1×far end contribution	—	0.000	7.004	−0.365	0.000	—	
Final end moments (kNm)	0.000	29.727	−29.727	49.642	−49.642	20.000	−20.000

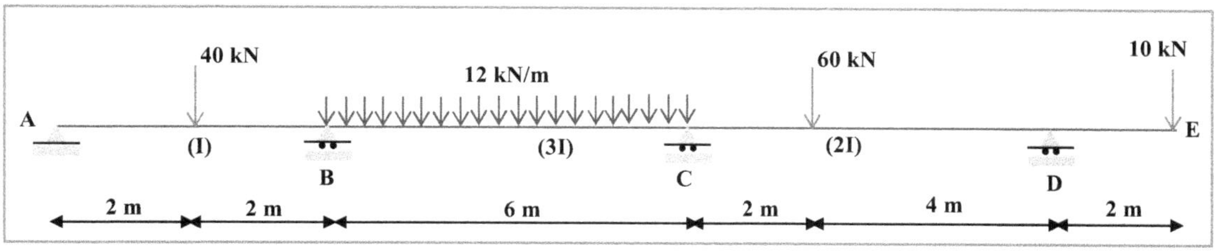

FIGURE 8.8 A three-span continuous beam with overhang.

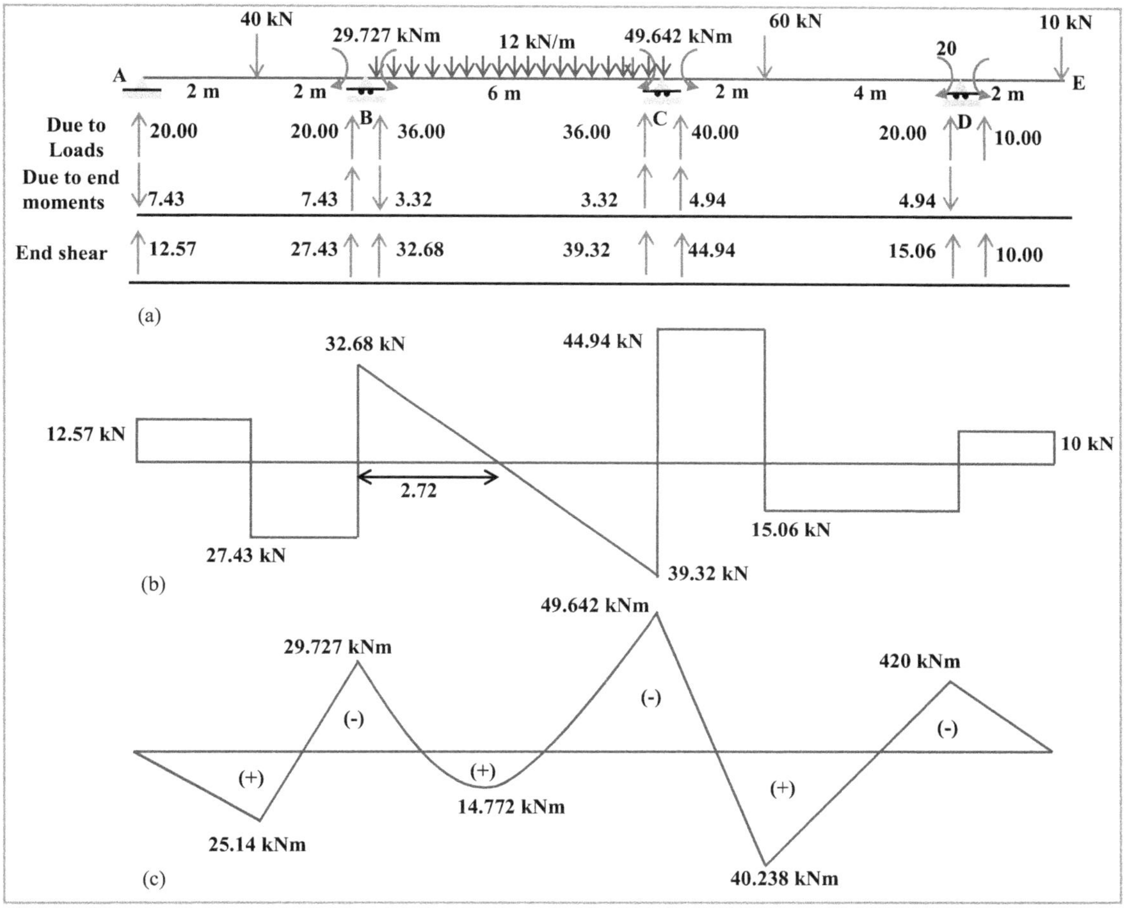

FIGURE 8.9 (a) Free body diagram, (b) shear force diagram, and (c) bending moment diagram.

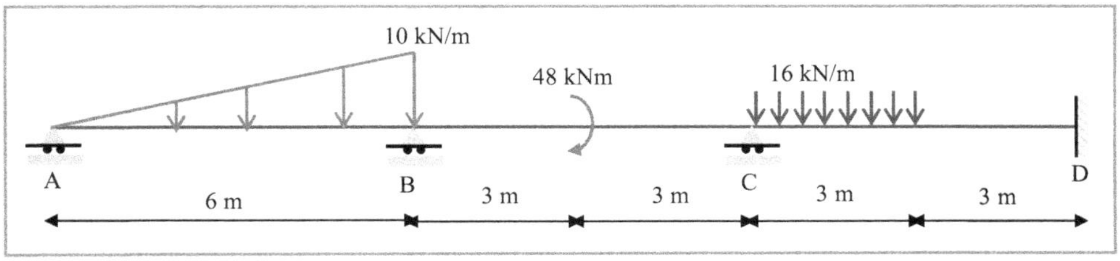

FIGURE 8.10 Three-span continuous beam *ABCD*.

TABLE 8.6
Kani's Iteration Cycles and Final End Moments

Joint	A	B		C		D
Member	*AB*	*BA*	*BC*	*CB*	*CD*	*DC*
RF	0.000	−0.214	−0.286	−0.250	−0.250	
FEMs	0.000	24.000	12.000	12.000	−33.000	15.000
Sum of FEMs		36.000		−21.000		
Rotation contribution		−7.704	−10.296	7.824	7.824	
		−9.378	−12.534	8.383	8.383	
		−9.498	−12.694	8.423	8.423	
		−9.507	−12.705	8.426	8.426	
		−9.507	−12.706	8.426	8.426	
		−9.507	−12.706	8.426	8.426	
FEM		24.000	12.000	12.000	−33.000	15.000
2 × near end contribution		−19.015	−25.412	16.853	16.853	0.000
1 × far end contribution		0.000	8.426	−12.706	0.000	8.426
Final end moments (kNm)	0.000	4.985	−4.985	16.147	−16.147	23.426

Joint	Member	K	$\sum K$	DF	RF
B	*BA*	$\frac{3EI}{6}=0.5EI$	$0.5EI+0.667EI=1.167EI$	0.428	−0.214
	BC	$\frac{4EI}{6}=0.667EI$		0.572	−0.286
C	*CB*	$\frac{4EI}{6}=0.667EI$	$0.667EI+0.667EI=1.334EI$	0.5	−0.25
	CD	$\frac{4EI}{6}=0.667EI$		0.5	−0.25

Rotational contributions, along with Kani's rotational cycles and final end moments, are presented in Table 8.6.

FINAL END MOMENTS

$$\begin{aligned}\text{Final end moment} &= \text{Fixed end moment}\\ &\quad +2(\text{Near end rotational contribution})\\ &\quad +1(\text{Far end rotational contribution})\end{aligned}$$

The final end moments are calculated and presented in the above table using the above equation.

FINAL END SHEARS

After determining the final member end moments (from the table), the member end shears are calculated using the static equilibrium equations and are shown in Figure 8.11a. The shear force and bending moment diagrams are shown in Figure 8.11b and c, respectively.

Example 8.6

Using Kani's method, analyse the two-span continuous beam, one end of which is fixed and the other end has guided support, as shown in Figure 8.12. Assume *EI* is constant. (The same problem has been solved in Example 7.5 using the MD method.)

Solution:

By using modified stiffness and modified fixed end moments for span *BC* (refer to 7.7 in the MD method), the deflection at the guided support can be neglected. Hence, the kinematic indeterminacy can be limited to one, i.e., rotation at *B*.

FIXED END MOMENTS

$$MF_{AB}=-\frac{WL}{8}=-\frac{25\times 4}{8}=-12.5\text{ kNm}=-MF_{BA}$$

$$MF_{BC}=-\frac{WL^2}{12}=-\frac{15\times 4^2}{12}=-20\text{ kNm}$$

$$MF_{CB}=-\left(\frac{WL^2}{8}-\frac{WL^2}{12}\right)=-\left(\frac{WL^2}{24}\right)=-\frac{15\times 4^2}{24}=-10\text{ kNm}$$

ROTATION FACTORS

Joint	Member	K	$\sum K$	DF	RF
	BA	$\frac{4EI}{4}=EI$		0.667	−0.333
B	*BC*	$\frac{EI}{2}=0.5EI$	$EI+0.5EI=1.5EI$	0.333	−0.167

Rotational contributions, along with Kani's rotational cycles, and final end moments are presented in Table 8.7.

FINAL END MOMENTS

$$\begin{aligned}\text{Final end moment} &= \text{Fixed end moment}\\ &\quad +2(\text{Near end rotational contribution})\\ &\quad +1(\text{Far end rotational contribution})\end{aligned}$$

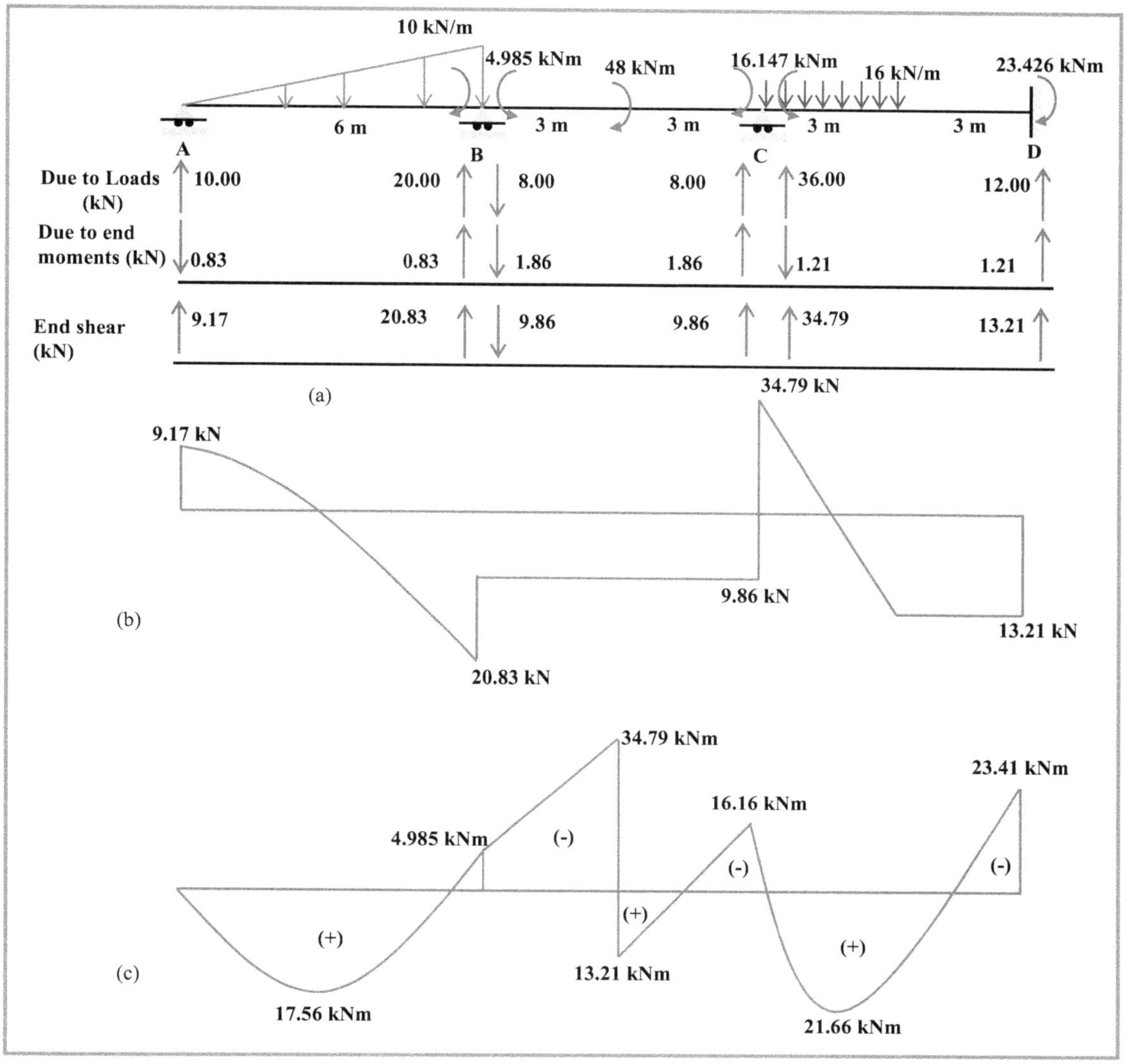

FIGURE 8.11 (a) Free body diagram, (b) shear force diagram, and (c) bending moment diagram.

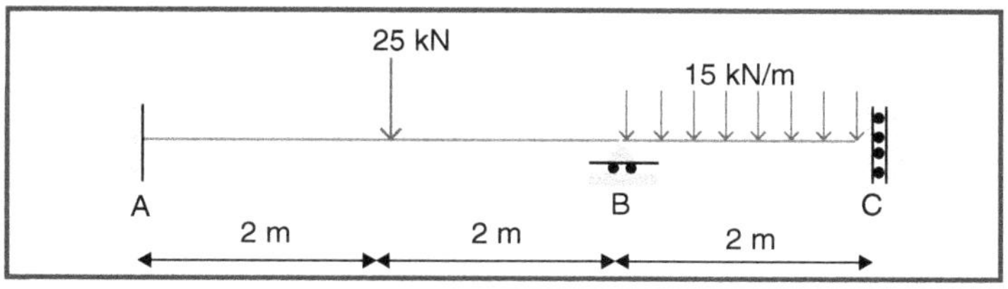

FIGURE 8.12 Two-span continuous beam *ABC* with guided support at *C*.

The final end moments are calculated in the table 8.7 conveniently. The member end shears are shown in Figure 8.13a. The shear force and bending moment diagrams are shown in Figure 8.13b and c, respectively.

Example 8.7

A two-span continuous beam is shown in Figure 8.14, in which support *A* is subjected to a clockwise rotational slip of 0.002 rad. Analyse the beam using Kani's method. Assume $EI = 75{,}000$ kNm². (The same has been solved in Example 7.7 using the MD method.)

Solution:

FIXED END MOMENTS

Span *AB* is subjected to no external load, but the support *A* rotates clockwise by 0.002 rad, which develops the end moments at *A* and *B*, as shown in Figure 8.15.

TABLE 8.7
Kani's Iteration Cycles and Final End Moments

Joint	A	B		C
Member	*AB*	*BA*	*BC*	*CB*
RF	0.000	−0.333	−0.167	0.000
FEMs	−12.500	12.500	−20.000	−10.000
Sum of FEMs		−7.500		−10.000
Rotation contribution		2.500	1.250	0.000
		2.500	1.250	0.000
2×near end contribution	0.000	5.000	2.500	0.000
1×far end contribution	2.500	0.000	0.000	−2.500
Final end moments (kNm)	−10.000	17.500	−17.500	−12.500

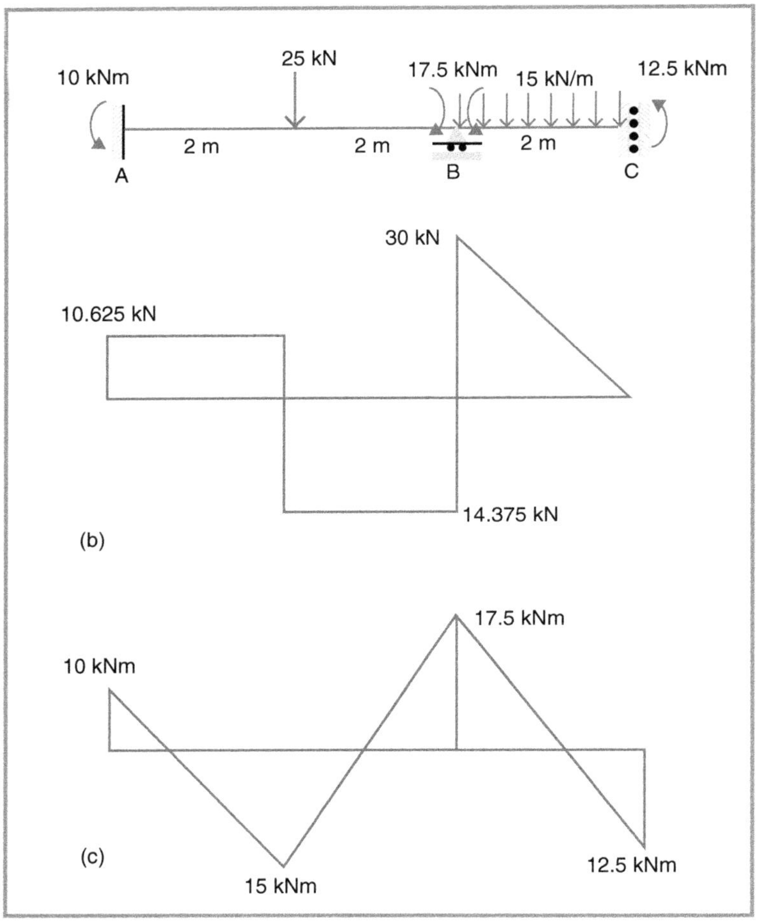

FIGURE 8.13 (a) Free body diagram, (b) shear force diagram, and (c) bending moment diagram.

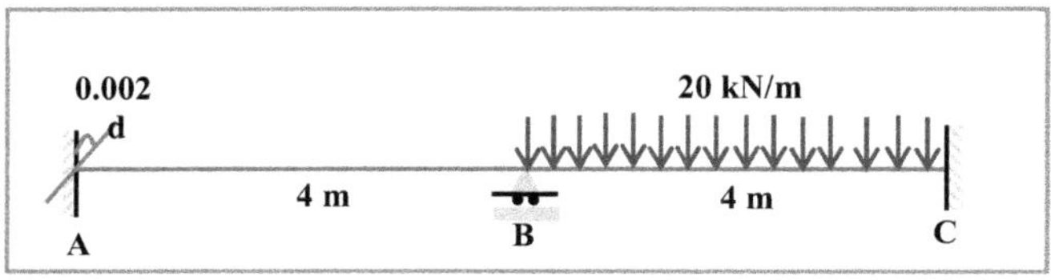

FIGURE 8.14 Two-span continuous beam *ABC*.

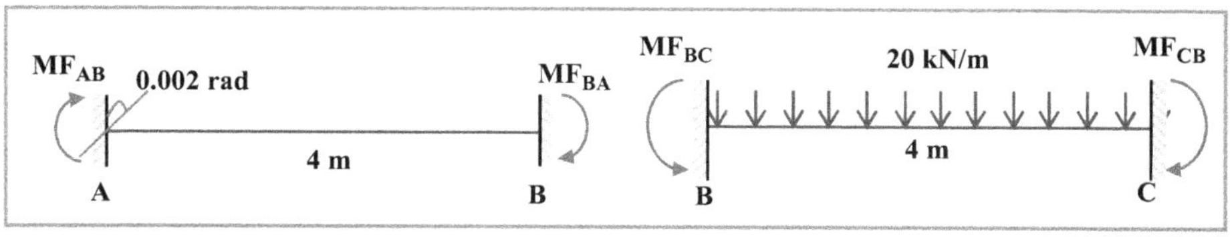

FIGURE 8.15 Fixed end moments.

TABLE 8.8
Kani's Iteration Cycles and Final End Moments

Joint	*A*	*B*		*C*
Member	*AB*	*BA*	*BC*	*CB*
RF	0.000	−0.250	−0.250	0.000
FEMs	150.000	75.000	−26.670	26.670
Sum of FEMs		48.330		26.670
Rotation contribution		−12.083	−12.083	0.000
		−12.083	−12.083	0.000
		−12.083	−12.083	0.000
2×near end contribution	0.000	−24.165	−24.165	0.000
1×far end contribution	−12.083	0.000	0.000	−12.083
Final end moments (kNm)	137.918	50.835	−50.835	14.588

$$MF_{AB} = \frac{4EI\theta}{L} = \frac{4\times 75,000\times 0.002}{4} = 150 \text{ kNm}$$

$$MF_{BA} = \frac{2EI\theta}{L} = \frac{2\times 75,000\times 0.002}{4} = 75 \text{ kNm}$$

$$MF_{BC} = -\frac{wL^2}{12} = -\frac{20\times 4^2}{12} = -26.67 \text{ kNm} = -MF_{CB}$$

ROTATION FACTORS

Joint	**Member**	***K***	$\sum K$	**DF**	**RF**
B	*BA*	$\frac{4EI}{4} = EI$	*EI*+*EI*=2*EI*	0.50	−0.25
	BC	$\frac{4EI}{4} = EI$		0.50	−0.25

Rotational contributions, Kani's rotational cycles, and final end moments are presented in Table 8.8.

FINAL END MOMENTS

Final member end moment=Fixed end moment+2 (Near-end rotational contribution)+1(Far-end rotational contribution)

Using this, the final member end moments are calculated in the above table. The member end shears are shown in the free body diagram in Figure 8.16a. The shear force and bending moment diagrams are shown in Figure 8.16b and c, respectively.

Example 8.8

Analyse the beam shown in Figure 8.17, if support *B* settles down by 30 mm using Kani's method. Assume *EI*=104,335.6 kNm². The same has been solved in Example 7.8 using the MD method.

Solution:

The fixed end moments due to the external loads are calculated using Table 6.1, while the fixed end moments resulting from the yielding of supports are calculated separately. These are then added together to obtain the net fixed end moments (refer to Example 7.10 in the MD method).

Due to the External Loads

$$(MF_{AB})_L = -\frac{22.5\times 7.2^2}{12} = -97.2 \text{ kNm} = (MF_{BA})_L$$

$$(MF_{BC})_L = (MF_{CB})_L = (MF_{CD})_L = (MF_{DC})_L = 0$$

Due to the Support Settlement at *B*

$$(MF_{AB})_s = (MF_{BA})_s = -\frac{6\times 104,335.6\times 0.03}{7.2^2}$$

$$= -362.276 \text{ kNm}$$

$$(MF_{BC})_s = (MF_{CB})_s = \frac{6\times 104,335.6\times 0.03}{6^2} = 521.678 \text{ kNm}$$

$$(MF_{CD})_s = (MF_{DC})_s = 0$$

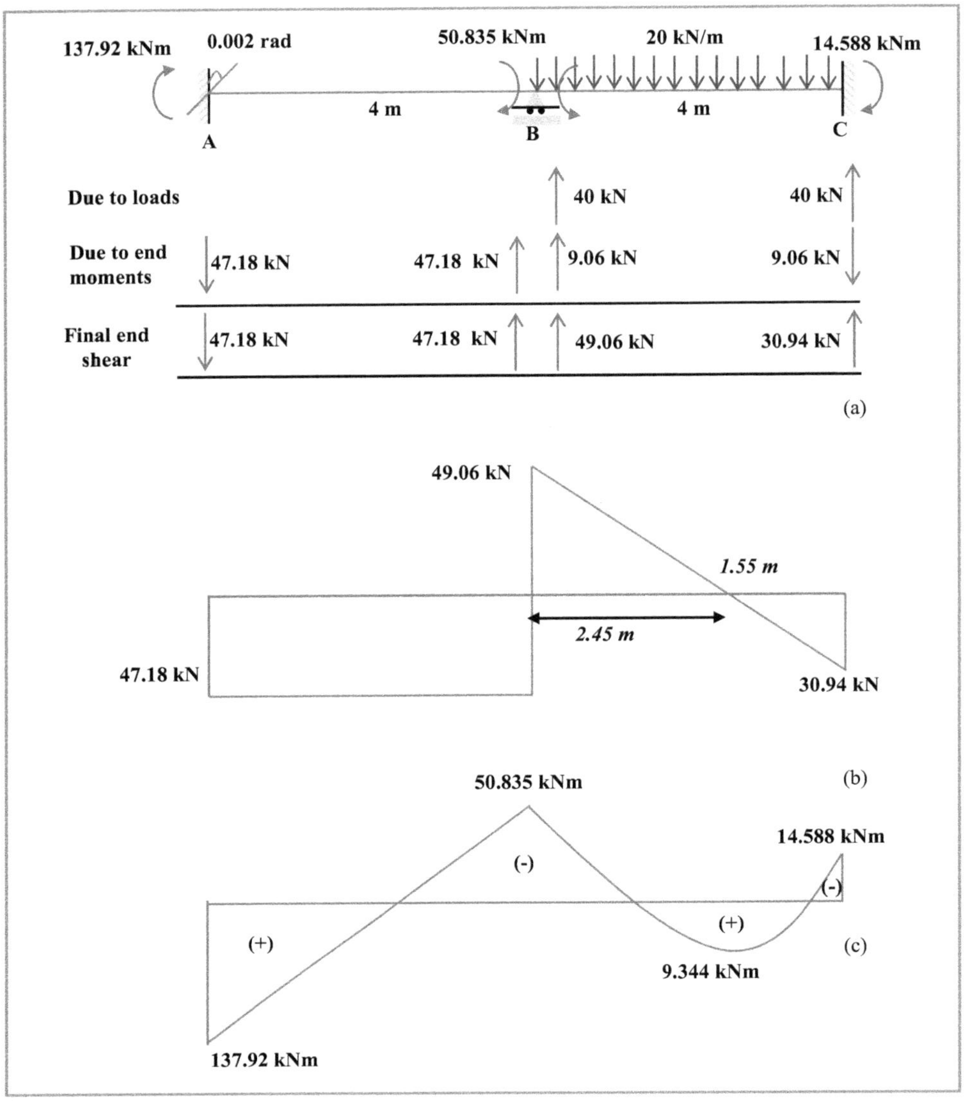

FIGURE 8.16 (a) Free body diagram, (b) shear force diagram, and (c) bending moment diagram.

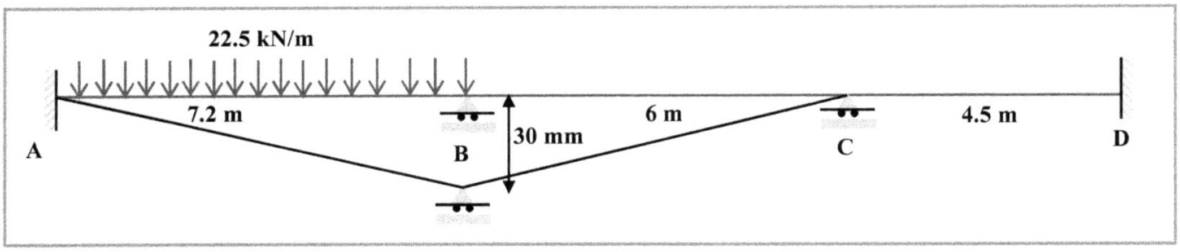

FIGURE 8.17 Three-span continuous beam with support settlement.

NET FIXED END MOMENTS

$$MF_{AB} = -97.2 - 362.276 = -459.479 \text{ kNm}$$

$$MF_{BA} = -97.2 + 362.276 = -265.076 \text{ kNm}$$

$$MF_{BC} = MF_{CB} = 521.678 \text{ kNm}$$

$$MF_{CD} = MF_{DC} = 0$$

ROTATION FACTORS (SHOWN IN TABLE)

$$(\text{RF})_i = \frac{-1}{2}(\text{DF})_i = \left(\frac{-1}{2}\right)\frac{K_i}{\sum_{i=1}^{n} K_i}$$

Joint	Member	K	$\sum K$	DF	RF
B	BA	$\frac{4EI}{7.2}=0.556EI$	$0.556EI+0.667EI=1.223EI$	0.455	−0.227
	BC	$\frac{4EI}{6}=0.667EI$		0.545	−0.273
C	CB	$\frac{4EI}{6}=0.667EI$	$0.667EI+0.889EI=1.556EI$	0.428	−0.214
	CD	$\frac{4EI}{4.5}=0.889EI$		0.572	−0.286

Rotational contributions, Kani's rotational cycles, and the final member end moments are presented in Table 8.9.

FINAL END MOMENTS

Final end moment = Fixed end moment

+2(Near end rotational contribution)

+1(Far end rotational contribution)

Using this expression, the final member end moments are calculated in the table 8.9. The free body diagram in Figure 8.18a shows the member end shears, while Figure 8.18b and c shows the shear force and bending moment diagrams, respectively.

Example 8.9

Analyse the three-span continuous beam shown in Figure 8.19 subjected to the given external loading. Furthermore, supports B and C have yielded down by 5 and 10 mm, respectively. Use Kani's method for the analysis. Assume the flexural rigidity EI=80,000 kNm². Draw the shear force and bending moment diagrams. (The same problem was solved in Example 7.9 using the MD method.)

Solution:

Since DE is statically determinate, the moment at D is known. Using the modified fixed end moments and stiffness for span CD, the rotation at D can be neglected. Hence, the unknown rotations are only at B and C. The fixed end moments due to the external loads and the settlement of supports are calculated separately, and then added together to obtain the net fixed end moments, as shown below.

FIXED END MOMENTS

Due to the External Loads

$$(MF_{AB})_L = -\frac{30\times 8^2}{12} = -160 \text{ kNm} = -(MF_{BA})_L$$

$$(MF_{BC})_L = -\frac{100\times 6}{8} - \frac{5}{192}\times 20\times 6^2 = -93.75 \text{ kNm}$$

$$(MF_{CB})_L = \frac{100\times 6}{8} + \frac{11}{192}\times 20\times 6^2 = 116.25 \text{ kNm}$$

$$(MF_{CD})_L = -\frac{150\times 2\times 4^2}{6^2} - \frac{1}{2}\times\frac{150\times 2^2\times 4}{6^2} + \frac{50}{2}$$

$$= -141.67 \text{ kNm}$$

$$M_{DC} = 50 \text{ kNm}$$

Due to the Settlement of Supports

$$(MF_{AB})_s = (MF_{BA})_s = -\frac{6EI\Delta_B}{L^2} = -\frac{6\times 4\times 80{,}000}{8^2}\times\frac{5}{1{,}000}$$

$$= -150 \text{ kNm}$$

$$(MF_{BC})_s = (MF_{CB})_s = -\frac{6EI(\Delta_C-\Delta_B)}{L^2}$$

$$= \frac{6\times 3\times 80{,}000}{6^2}\times\frac{(10-5)}{1{,}000} = -200 \text{ kNm}$$

TABLE 8.9
Kani's Iteration Cycles and Final End Moments

Joint	A	B		C		D
Member	AB	BA	BC	CB	CD	DC
RF	0.000	−0.227	−0.273	−0.214	−0.286	
FEMs	−459.476	−265.076	521.678	521.678	0.000	0.000
Sum of FEMs	−459.476	256.602		521.678		0.000
Rotation contribution		−58.319	−69.982	−96.792	−129.056	
		−36.320	−43.585	−102.449	−136.598	
		−35.035	−42.042	−102.779	−137.039	
		−34.960	−41.952	−102.798	−137.065	
		−34.955	−41.946	−102.800	−137.066	
		−34.955	−41.946	−102.800	−137.066	
2×near end contribution	0.000	−69.910	−83.892	−205.599	−274.133	0.000
1×far end contribution	−34.955	0.000	−102.800	−41.946	0.000	−137.066
Final end moments (kNm)	−494.431	−334.986	334.986	274.133	−274.133	−137.066

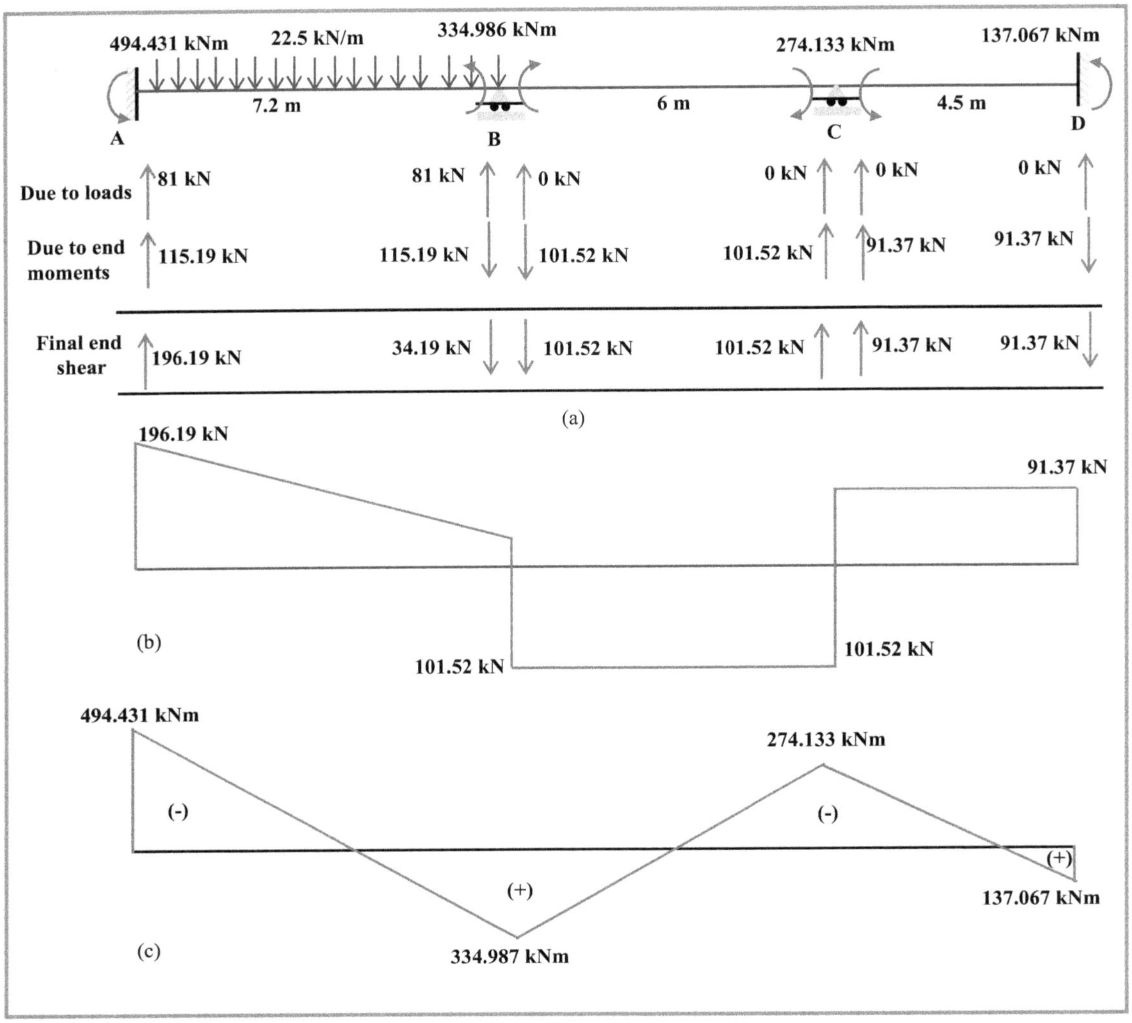

FIGURE 8.18 (a) Free body diagram, (b) shear force diagram, and (c) bending moment diagram.

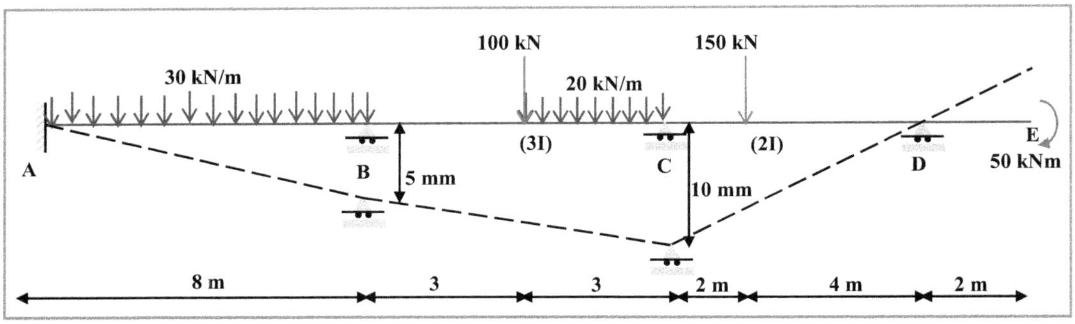

FIGURE 8.19 Three-span continuous beam with supports settlement.

$$(MF_{CD})_s = \frac{3EI\Delta_C}{L^2} = \frac{3 \times 2 \times 80,000}{6^2} \times \frac{10}{1,000} = 133.33 \text{ kNm}$$

NET FIXED END MOMENTS

$$MF_{AB} = -160 - 150 = -310 \text{ kNm}$$

$$MF_{BA} = 160 - 150 = 10 \text{ kNm}$$

$$MF_{BC} = -93.75 - 200 = -293.75 \text{ kNm}$$

$$MF_{CB} = 116.25 - 200 = -83.75 \text{ kNm}$$

$$MF_{CD} = -141.67 + 133.33 = -8.33 \text{ kNm}$$

TABLE 8.10
Kani's Iteration Cycles and Final End Moments

Joint	A	B		C		D	
Member	*AB*	*BA*	*BC*	*CB*	*CD*	*DC*	*DE*
RF	0.000	−0.250	−0.250	−0.333	−0.167		
FEMs	−310.000	10.000	−293.750	−83.750	−8.330	50.000	−50.00
Sum of FEMs	−310.000	−283.750		−92.080		50.000	−50.00
Rotation contribution		70.938	70.938	7.048	3.524		
		69.176	69.176	7.635	3.817		
		69.029	69.029	7.684	3.842		
		69.017	69.017	7.688	3.844		
		69.016	69.016	7.688	3.844		
		69.015	69.015	7.688	3.844		
		69.015	69.015	7.688	3.844		
2×near end contribution	0.000	138.031	138.031	15.376	7.688	0.000	
1×far end contribution	69.015	0.000	7.688	69.015	0.000	0.000	
Final end moments (kNm)	−240.985	148.031	−148.031	0.642	−0.642	50.000	−50.00

ROTATION FACTORS (SHOWN IN TABLE)

$$(\mathrm{RF})_i = \frac{-1}{2}(\mathrm{DF})_i = \left(\frac{-1}{2}\right)\frac{K_i}{\sum_{i=1}^{n} K_i}$$

Joint	Member	K	$\sum K$	DF	RF
B	*BA*	$\frac{4E(4I)}{8} = 2EI$	$2EI + 2EI = 4EI$	0.5	−0.25
	BC	$\frac{4E(3I)}{6} = 2EI$		0.5	−0.25
C	*CB*	$\frac{4E(3I)}{6} = 2EI$	$2EI + EI = 3EI$	0.667	−0.333
	CD	$\frac{3E(2I)}{6} = EI$		0.333	−0.167

Rotational contributions, Kani's rotational cycles, and the final end moments are presented in Table 8.10. The member end shears are shown in the free body diagram (Figure 8.20a). The shear force and bending moment diagrams are shown in Figure 8.20b and c, respectively.

8.4 RIGID JOINT PLANE FRAMES (NON-SWAY)

The analysis of rigid joint plane frames using Kani's method is exactly similar to the analysis of indeterminate beams, as explained in the previous section and examples. However, the key difference is that in the case of beams, only two members meet at a joint, whereas in the case of frames, more than two members may meet at a joint. Therefore, the rotation contributions of all the members connected to a joint must be considered. This is demonstrated in the examples.

Example 8.10

A rigid joint plane frame is shown in Figure 8.21. Compute the support reactions using Kani's method. Construct the shear force and bending moment diagrams. Assume *EI* is constant. (The same was solved in Example 6.10 using the SD method.)

Solution:

FIXED END MOMENTS

The degree of kinematic indeterminacy of the frame, in general, is 3, but due to the zero moment at *E* due to the hinge support, the rotation at *E* can be ignored by using the modified stiffness and fixed end moments for span *CE*. As a result, the kinematic indeterminacy is reduced to 2, i.e., θ_B and θ_C. The fixed end moments are calculated using Figure 8.22 and Table 6.1, as shown below.

$$MF_{AB} = -MF_{BA} = -\frac{WL}{8} = -\frac{10 \times 4}{8} = -5 \text{ kNm}$$

$$MF_{BC} = -MF_{CB} = -\frac{5}{96}wL^2 = -\frac{5}{96} \times 20 \times 4^2 = -16.67 \text{ kNm}$$

$$MF_{CE} = \frac{Mb}{L^2}(2a - b) - \frac{1}{2}\frac{Ma}{L^2}(2b - a)$$

$$= \frac{10 \times 1}{3^2}(2 \times 2 - 1) - \frac{1}{2}\frac{10 \times 2}{3^2}(2 \times 1 - 2)$$

$$= 3.33 \text{ kNm}$$

$$MF_{CD} = MF_{DC} = 0$$

ROTATION FACTORS

$$(\mathrm{RF})_i = \frac{-1}{2}(\mathrm{DF})_i = \left(\frac{-1}{2}\right)\frac{K_i}{\sum_{i=1}^{n} K_i}$$

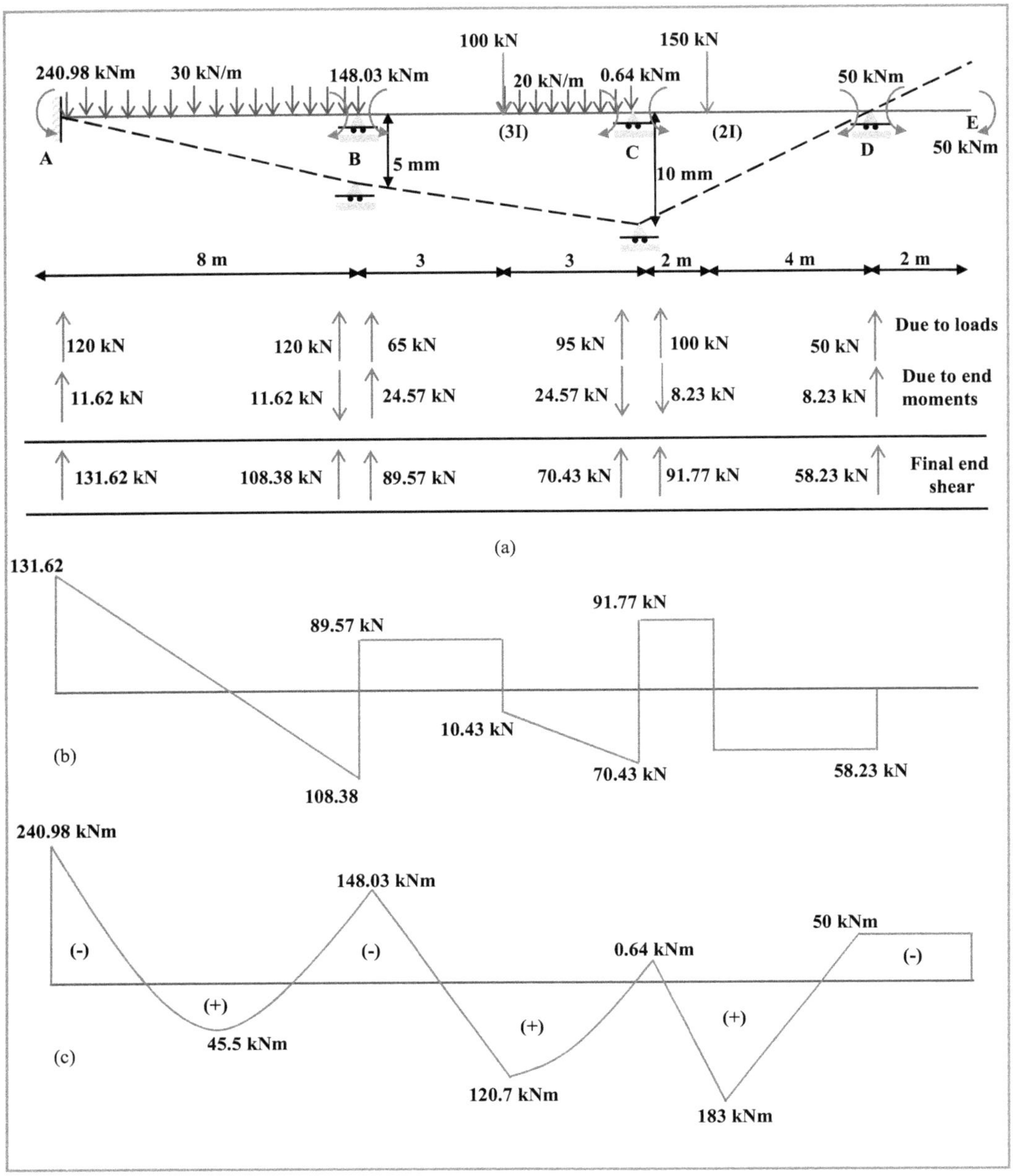

FIGURE 8.20 (a) Free body diagram, (b) shear force diagram, and (c) bending moment diagram.

Rotational contributions, along with Kani's rotational cycles and the final end moments, are presented in Table 8.11.

Joint	Member	K	$\sum K$	DF	RF
B	BA	$\frac{4EI}{4} = EI$	$EI + EI = 2EI$	0.5	−0.25
	BC	$\frac{4EI}{4} = EI$		0.5	−0.25
C	CB	$\frac{4EI}{4} = EI$	$EI + EI + EI = 3EI$	0.333	−0.167
	CD	$\frac{4EI}{4} = EI$		0.333	−0.167
	CE	$\frac{3EI}{3} = EI$		0.333	−0.167

FINAL END SHEARS

With the known final end moments from the table 8.11 and using the static equilibrium equations, the member end shears are calculated and are shown in Figure 8.23a. The shear force and bending moment diagrams are shown in Figure 8.23b and c, respectively.

Example 8.11

Analyse (without taking advantage of symmetry) the symmetrically loaded frame, as shown in Figure 8.24a, using Kani's method. The same has been solved in Example 7.13 using the MD method.

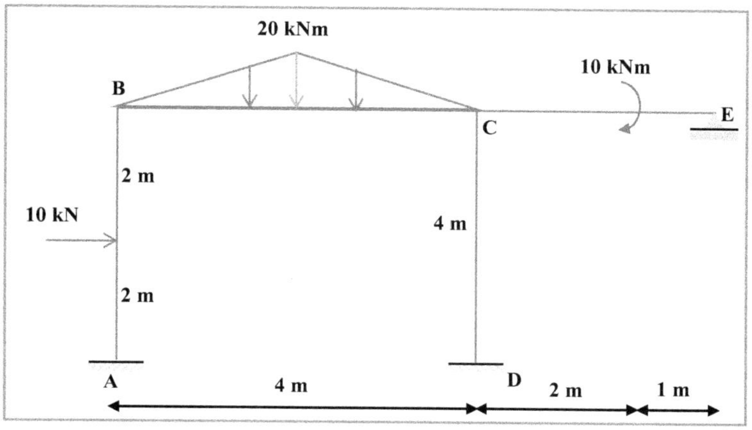

FIGURE 8.21 Rigid joint frame.

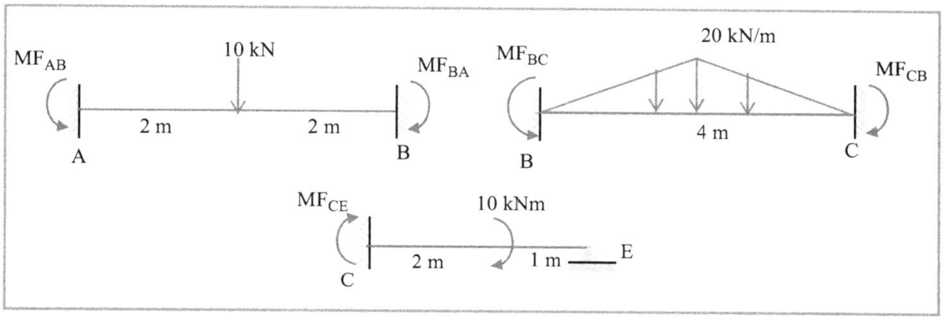

FIGURE 8.22 Free body diagrams.

TABLE 8.11
Kani's Iteration Cycles and Final End Moments

Joint	A	B		C			D	E
Member	*AB*	*BA*	*BC*	*CB*	*CD*	*CE*	*DC*	*EC*
RF	0.000	−0.250	−0.250	−0.167	−0.167	−0.167	0.000	—
FEMs	−5.000	5.000	−16.670	16.670	0.000	3.330	0	0
Sum of FEMs	−5.000	−11.670		20.000			0	0
Rotation contribution		2.918	2.918	−3.820	−3.820	−3.820		
		3.872	3.872	−3.979	−3.979	−3.979		
		3.912	3.912	−3.985	−3.985	−3.985		
		3.914	3.914	−3.986	−3.986	−3.986		
		3.914	3.914	−3.986	−3.986	−3.986		
2×near end contribution	0.000	7.828	7.828	−7.971	−7.971	−7.971	0.000	
1×far end contribution	3.914	0.000	−3.986	3.914	0.000	0.000	−3.986	
Final end moments (kNm)	−1.086	12.828	−12.828	12.613	−7.971	−4.641	−3.986	0.000

Solution:

The frame is symmetrical with respect to the vertical centroid axis. Therefore, only half of the frame is sufficient for consideration in the analysis, as shown in Figure 8.24b, provided the appropriate modifications are made to the stiffness factors and fixed end moments.

FIXED END MOMENTS

$$MF_{AB} = -\frac{wL^2}{30} = -\frac{20 \times 3^2}{30} = -6 \text{ kNm}$$

$$MF_{BA} = \frac{wL^2}{20} = \frac{20 \times 3^2}{20} = 9 \text{ kNm}$$

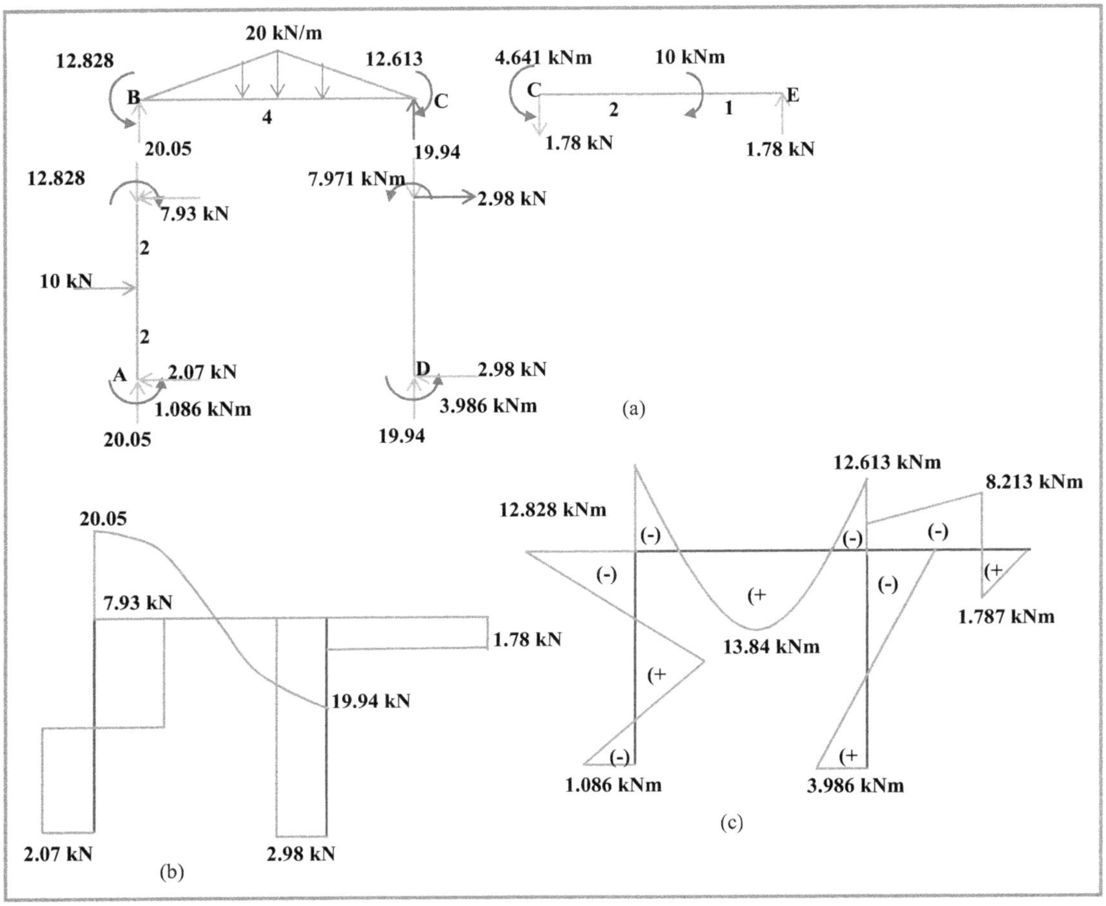

FIGURE 8.23 (a) Free body diagram, (b) shear force diagram, and (c) bending moment diagram.

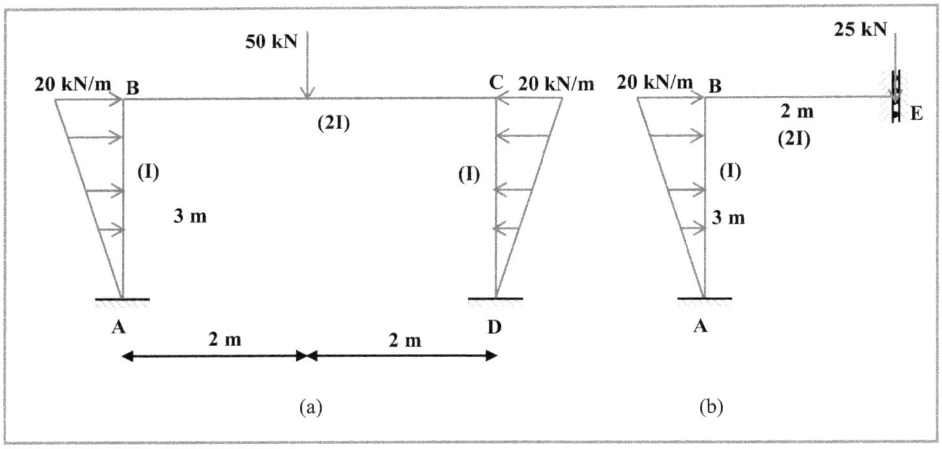

FIGURE 8.24 (a) Symmetrical frame *ABCD* and (b) half portion (*ABE*) of symmetrical frame *ABCD*.

$$MF_{BE} = -\frac{WL}{8} = -\frac{50 \times 4}{8} = -25 \text{ kNm}$$

$$MF_{BE} = -\frac{WL}{8} = -\frac{50 \times 4}{8} = -25 \text{ kNm (Mid span moment)}$$

ROTATION FACTORS

$$(\text{RF})_i = \frac{-1}{2}(\text{DF})_i = \left(\frac{-1}{2}\right)\frac{K_i}{\sum_{i=1}^{n} K_i}$$

Joint	Member	K	$\sum K$	DF	RF
	BA	$\frac{4EI}{3}=1.333EI$	$1.333EI+EI$ $=2.333EI$	0.571	−0.286
B	BE	$\frac{2E(2EI)}{4}=EI$		0.429	−0.214

The rotational contributions, along with Kani's rotational cycles and the final end moments, are presented in Table 8.12.

FINAL END SHEARS

The member end shears, shear force, and bending moment diagrams are shown in Figure 8.25a–c, respectively.

Example 8.12

Analyse the frame shown in Figure 8.26 using Kani's method. Construct the shear fore and bending moment diagrams. This has been solved in Example 7.10 using the moment distribution method.

TABLE 8.12
Kani's Iteration Cycles and Final End Moments

Joint	A	B		E
Member	AB	BA	BE	EB
RF	0.000	−0.286	−0.214	0.000
FEMs	−6.000	9.000	−25.000	−25.000
Sum of FEMs		−16.000		−25.000
Rotation contribution		4.576	3.424	0.000
		4.576	3.424	0.000
2×near end contribution	0.000	9.152	6.848	0.000
1×far end contribution	4.576	0.000	0.000	−6.848
Final end moments (kNm)	−1.424	18.152	−18.152	−31.848

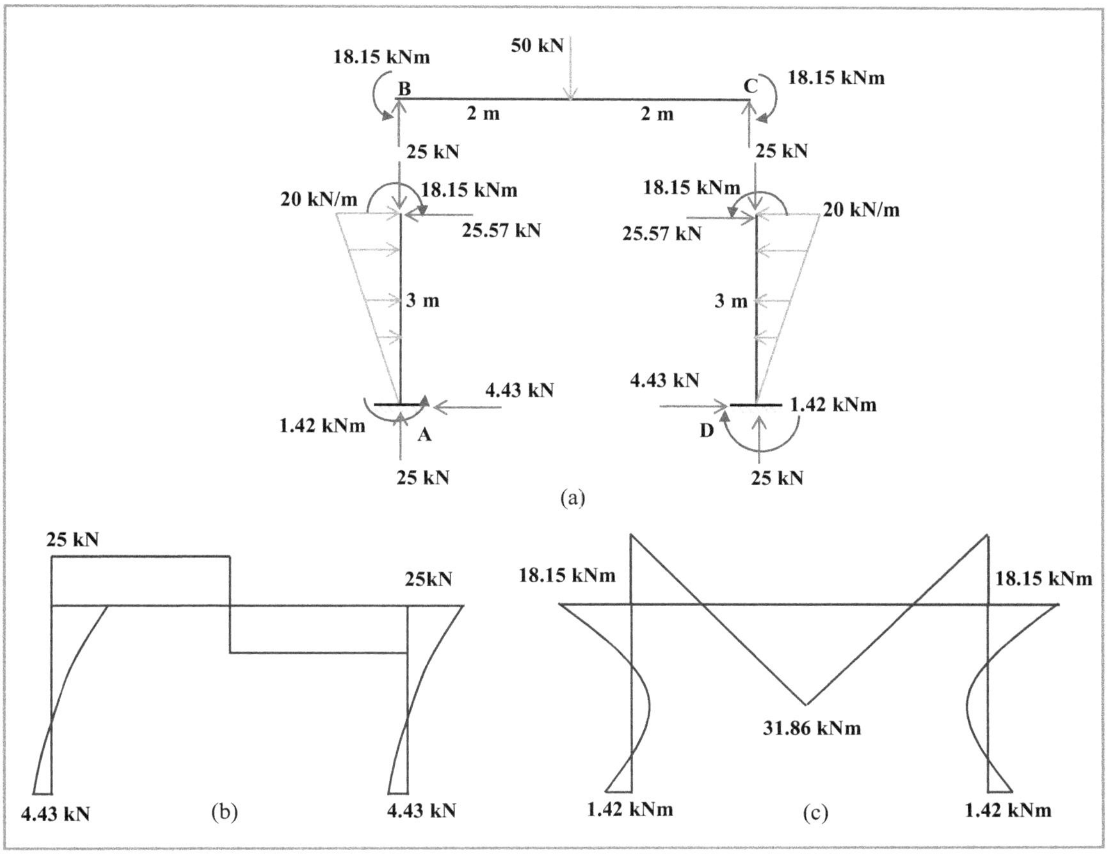

FIGURE 8.25 (a) Free body diagram, (b) shear force diagram, and (c) bending moment diagram.

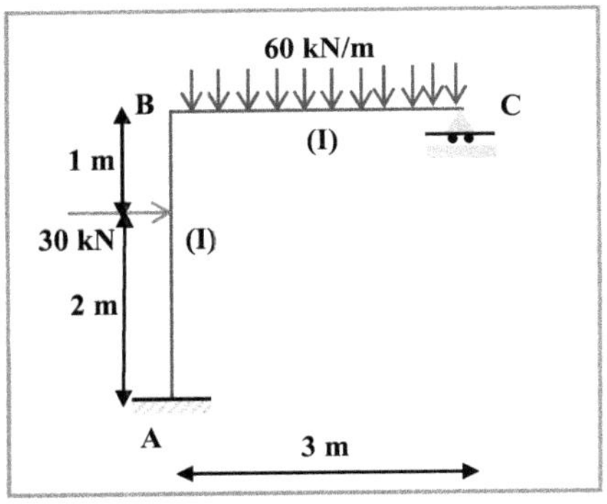

FIGURE 8.26 A simple frame *ABC*.

TABLE 8.13
Kani's Iteration Cycles and Final End Moments

Joint	*A*	*B*		*E*
Member	*AB*	*BA*	*BC*	*EC*
RF	0.000	−0.125	−0.375	–
FEMs	−40.000	−20.000	−67.500	0
Sum of FEMs	−40.000	−87.500		0
Rotation contribution		10.938	32.813	
		10.938	32.813	
2×near end contribution	0.000	21.875	65.625	0
1×far end contribution	−21.875	0.000	0.000	0
Final end moments (kNm)	−61.875	1.875	−1.875	0.000

Solution:

As explained in Example 6.21, the kinematic indeterminacy is limited to 1, i.e., θ_B, as unknown rotation.

FIXED END MOMENTS

The calculation of the fixed end moments is explained in Example 7.12 and these are

$$MF_{AB} = -\frac{30\times 2\times 4(2+4)}{6^2} = -40 \text{ kNm}$$

$$MF_{BA} = -(30\times 3 - 40 - 30\times 1) = -20 \text{ kNm}$$

$$MF_{BC} = -\frac{60\times 3^2}{8} = -67.5 \text{ kNm}$$

ROTATION FACTORS

$$(\text{RF})_i = \frac{-1}{2}(\text{DF})_i = \left(\frac{-1}{2}\right)\frac{K_i}{\sum_{i=1}^{n} K_i}$$

Joint	Member	*K*	$\sum K$	DF	RF
B	*BA*	$\frac{EI}{3} = 0.333EI$	$0.333EI+$ $EI = 1.333EI$	0.249	−0.125
	BE	$\frac{3EI}{3} = EI$		0.751	−0.375

The rotational contributions along with Kani's rotational cycles and final end moments, are presented in Table 8.13. The member end shears are shown in Figure 8.27a. The shear force and the bending moment diagrams are shown in Figure 8.27b and c, respectively.

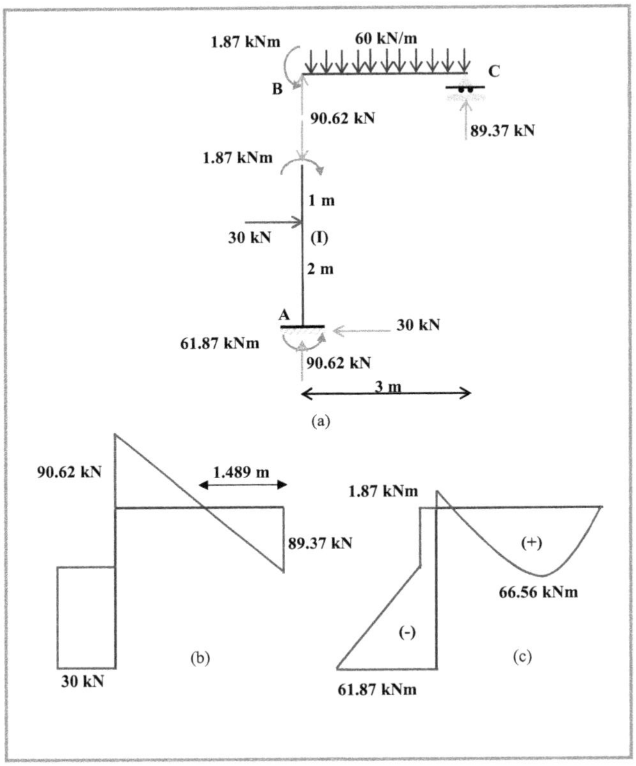

FIGURE 8.27 (a) Free body diagram, (b) shear force diagram, and (c) bending moment diagram.

Example 8.13

Analyse the symmetrical frame shown in Figure 8.28 using Kani's method. Draw the shear force and bending moment diagrams.

Solution:

The given frame is symmetrical with respect to the centre of *BC*. Therefore, only half of the frame needs (Figure 8.29) to be considered for the analysis, making appropriate modifications to the stiffness factors and fixed end moments. Furthermore, the moment at *A* is zero due to simple support. Hence, the problem can be further simplified by incorporating the modifications in the stiffness factors and fixed end moments.

FIXED END MOMENTS

Using Table 6.1, the fixed end moments are calculated as shown below:

$$MF_{BA} = \frac{wL^2}{8} = \frac{30\times 4^2}{8} = 60 \text{ kNm}$$

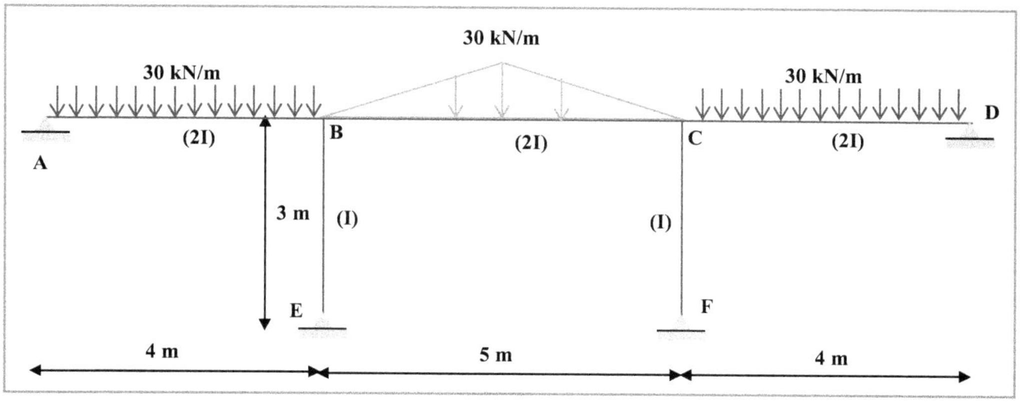

FIGURE 8.28 Symmetrical rigid joint frame.

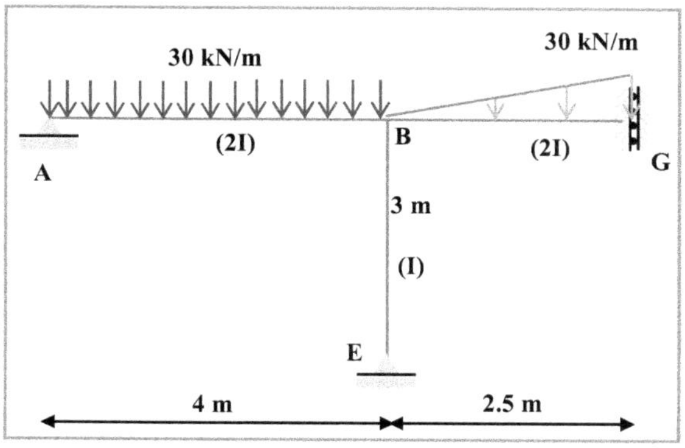

FIGURE 8.29 Half portion of rigid joint frame

$$MF_{BG} = -\frac{5wL^2}{96} = -\frac{5\times 30\times 5^2}{96} = -39.06 \text{ kNm}$$

$$MF_{GB} = -\left(\frac{wL^2}{8} - \frac{5wL^2}{96}\right) = -\frac{wL^2}{32} = -23.43 \text{ kNm}$$

ROTATION FACTORS

$$(\text{RF})_i = \frac{-1}{2}(\text{DF})_i = \left(\frac{-1}{2}\right)\frac{K_i}{\sum_{i=1}^{n} K_i}$$

Joint	Member	K	$\sum K$	DF	RF
B	BA	$\frac{3E(2I)}{4} = 1.5EI$	$1.5EI + 0.8EI + EI = 3.3EI$	0.454	−0.227
	BG	$\frac{2E(2I)}{5} = 0.8EI$		0.243	−0.121
	BE	$\frac{3EI}{3} = EI$		0.303	−0.152

The rotational contributions along with Kani's rotational cycles and the final end moments are presented in Table 8.14.

TABLE 8.14
Kani's Iteration Cycles and Final End Moments

Joint	*A*		*B*		*G*
Member	*AB*	*BA*	*BE*	*BG*	*GB*
RF	0.000	−0.227	−0.152	−0.121	0.000
FEMs	0.000	60.000	0.000	−39.063	−23.430
Sum of FEMs		20.937			−23.430
Rotation contribution		−4.758	−3.172	−2.538	0.000
		−4.758	−3.172	−2.538	0.000
2×near end contribution	0.000	−9.517	−6.345	−5.076	0.000
1×far end contribution	0.000	0.000	0.000	0.000	−5.076
Final end moments (kNm)	0.000	50.483	−6.345	−44.139	−18.354

MEMBER END SHEARS

With the known final member end moments and using the static equilibrium equations, the member end shears are shown in the free body diagram in Figure 8.30a. The shear force and bending moment diagrams are shown in Figure 8.30b and c, respectively.

Example 8.14

Analyse the symmetrical two-storey frame shown in Figure 8.31 using Kani's method. Draw the shear force and bending moment diagrams.

Solution:

The frame is symmetrical with respect to the vertical centre line passing through the beams. Therefore, only half of the frame (Figure 8.31a) needs to be considered for the analysis by using appropriate modifications in the stiffness and fixed end moments.

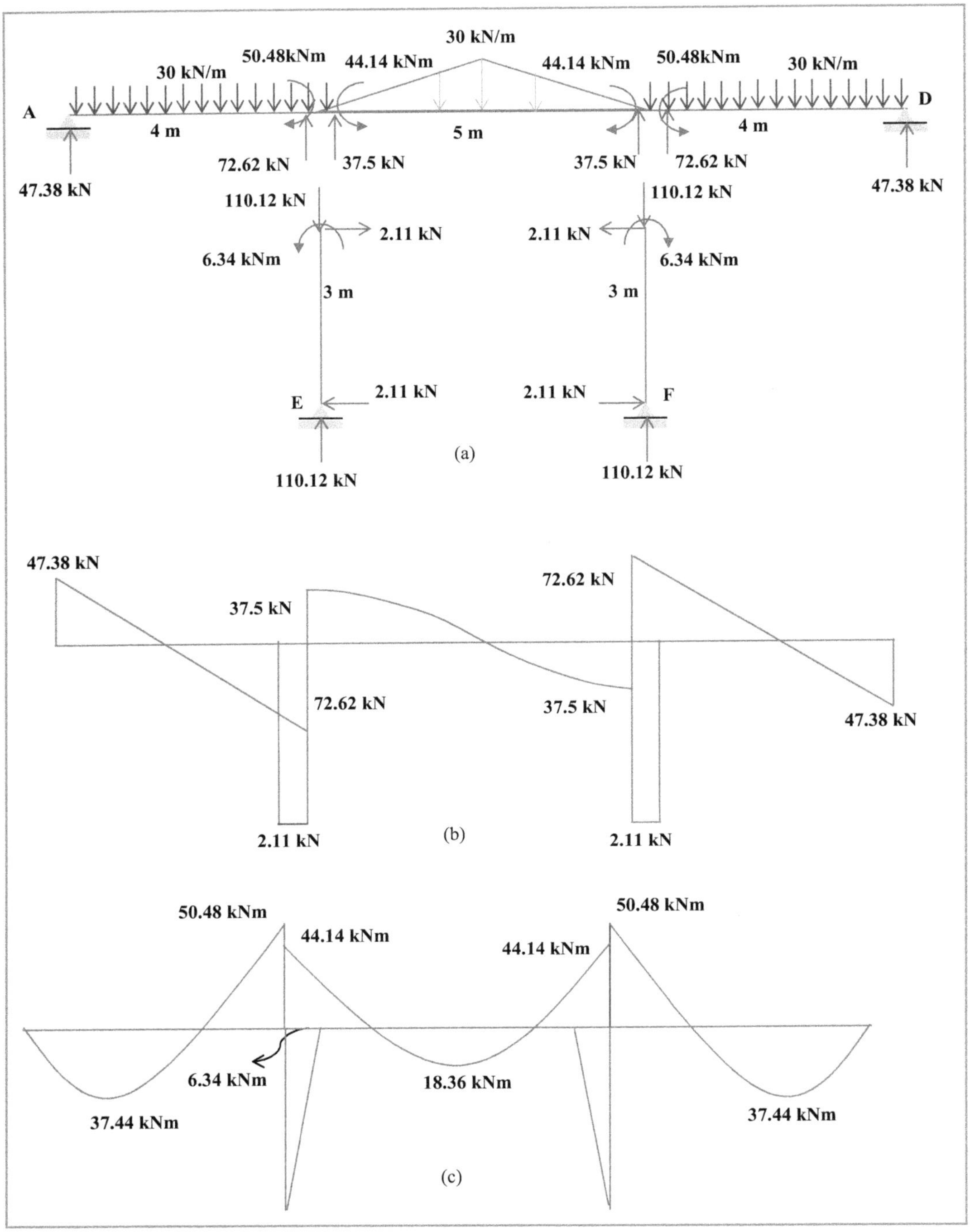

FIGURE 8.30 (a) Free body diagram, (b) shear force diagram, and (c) bending moment diagram.

FIXED END MOMENTS

$$MF_{BE^1} = -\frac{wL^2}{12} = -\frac{50 \times 5^2}{12} = -104.16 \text{ kNm}$$

$$MF_{CD^1} = -\frac{wL^2}{12} = -\frac{30 \times 5^2}{12} = -62.50 \text{ kNm}$$

$$MF_{E^1B} = -\left(\frac{wL^2}{8} - \frac{wL^2}{12}\right) = -\frac{wL^2}{24} = -\frac{50 \times 5^2}{24} = -52.08 \text{ kNm}$$

$$MF_{D^1C} = -\left(\frac{wL^2}{8} - \frac{wL^2}{12}\right) = -\frac{wL^2}{24} = -\frac{30 \times 5^2}{24} = -31.25 \text{ kNm}$$

$$MF_{AB} = MF_{BA} = MF_{BC} = MF_{CB} = 0$$

ROTATION FACTORS

$$(\text{RF})_i = \frac{-1}{2}(\text{DF})_i = \left(\frac{-1}{2}\right)\frac{K_i}{\sum_{i=1}^{n} K_i}$$

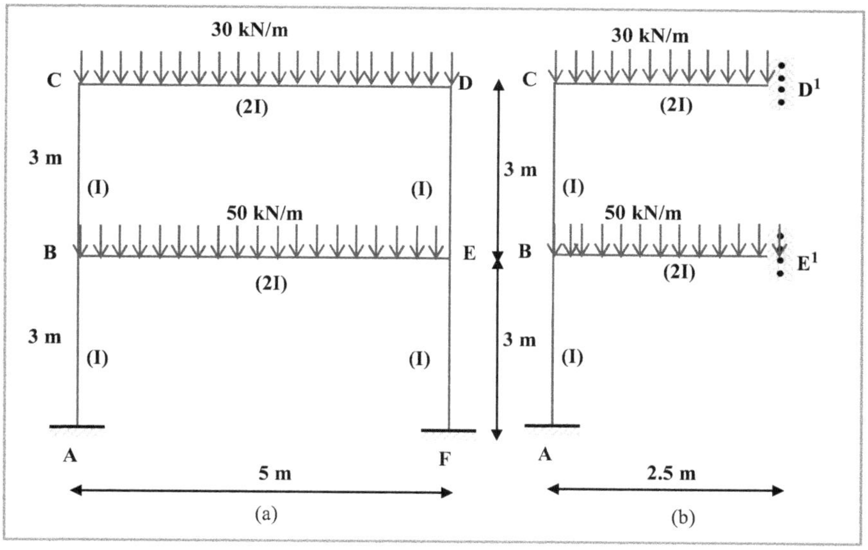

FIGURE 8.31 (a) Symmetrical rigid joint frame and (b) half rigid joint frame.

TABLE 8.15
Kani's Iteration Cycles and Final End Moments

Joint	A	B			C		D	E
Member	*AB*	*BA*	*BC*	*BE*¹	*CB*	*CD*¹	*D*¹*C*	*E*¹*B*
RF	—	−0.192	−0.192	−0.116	−0.312	−0.188	—	—
FEMs	0.000	0.000	0.000	−104.16	0.000	−62.500	−52.08	−31.25
Sum of FEMs		−104.160			−62.500			
Rotation contribution		19.999	19.999	12.083	13.260	11.750		
		17.453	17.453	10.544	14.055	8.469		
		17.300	17.300	10.452	14.102	8.498		
		17.291	17.291	10.447	14.105	8.499		
		17.291	17.291	10.446	14.105	8.499		
2×near end contribution	0.000	34.581	34.581	20.893	28.211	16.999	0	0
1×far end contribution	17.290	0.000	14.105	0.000	17.290	0.000	8.499	10.446
Final end moments (kNm)	17.290	34.581	48.686	−83.267	45.501	−45.501	−43.581	−20.804

Joint	Member	K	$\sum K$	DF	RF
B	*BA*	$\frac{4E(I)}{3}=1.33EI$	$1.33EI+0.8$ $EI+1.33$ $EI=3.46EI$	0.384	−0.192
	BC	$\frac{4E(I)}{3}=1.33EI$		0.384	−0.192
	*BE*¹	$\frac{2E(2I)}{5}=0.8EI$		0.232	−0.116
C	*CB*	$\frac{4E(I)}{3}=1.33EI$	$1.33EI+0.8$ $EI=2.13EI$	0.624	0.312
	*CD*¹	$\frac{2E(2I)}{5}=0.8EI$		0.376	0.188

The rotational contributions along with Kani's rotational cycles, and the final end moments are presented in Table 8.15.

The member end shears, shear force, and bending moment diagrams are shown, respectively, in Figure 8.32a–c.

8.5 RIGID JOINT PLANE FRAMES (SIDE SWAY)

In the previous section, we studied the problems involving frames without side sway. In this section frames subjected to lateral sway are discussed. If a frame is unsymmetrical in terms of loading, supporting conditions or geometry, it will be subjected to lateral displacements (sway). This lateral sway induces additional moments and shears in the columns. These additional moments are referred to as sway or displacement contributions. Kani provided formulations for such frames, classified into two categories:

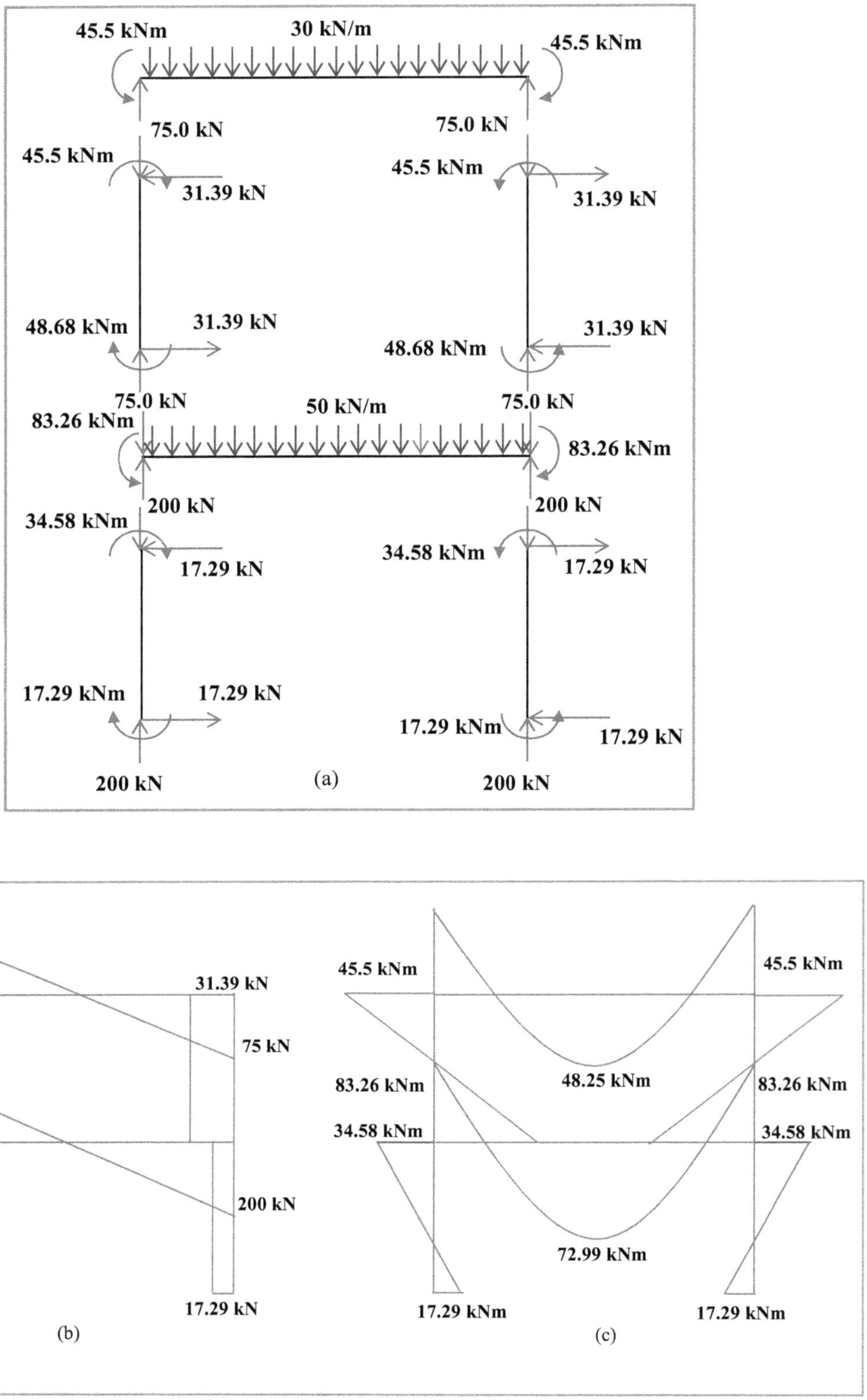

FIGURE 8.32 (a) Free body diagram, (b) shear force diagram, and (c) bending moment diagram.

1. All columns in a storey are of equal height
2. All columns in a storey are of different height

8.5.1 All Columns Are of Equal Height in a Storey

Consider a multi-storey frame as shown in Figure 8.33, where the columns in each storey are of equal height and are fixed at the base, as shown in the figure. The total kinematic indeterminacy is equal to one rotational degree of freedom at each joint and one lateral displacement at each floor level (assuming the axial deformations are neglected). Therefore, there are 15 unknown joint rotations and 3 lateral displacements. The lateral sway induces chord rotations only in the columns, not in the beams.

$$M_{OA} = MF_{OA} + \frac{4EI}{L}\theta_O + \frac{2EI}{L}\theta_A - \frac{6EI}{L}\varphi_{OA} \quad (8.14)$$

where θ_O and θ_A are the end rotations and $\varphi_{OA} = \frac{\Delta_{OA}}{L}$ is the chord rotation, which is assumed to act in a clockwise direction as positive.

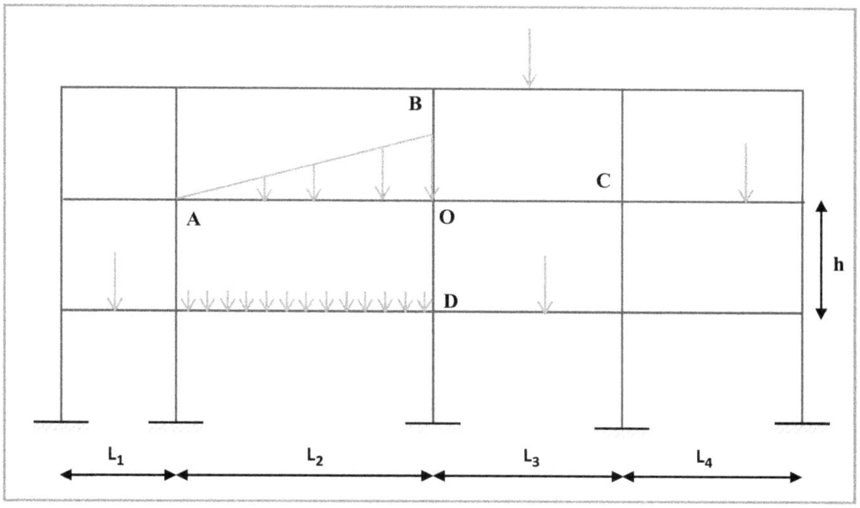

FIGURE 8.33 Multi-storey frame subjected to lateral sway.

In the above figure for any column element, Oi, we can write the above equation in Kani's format as

$$M_{Oi} = MF_{Oi} + 2M'_{Oi} + M'_{iO} + M''_{Oi} \quad (8.15)$$

where the additional term M''_{Oi} due to the sway effect is referred to as the sway moment or sway contribution in the column Oi, which is given by

$$M'_{Oi} = -\frac{6EI_c}{h^2{}_i}\Delta = -\frac{6EI_c}{h_i}\varphi_{Oi} \quad \text{where } \varphi_{Oi} = \frac{\Delta}{h_i} \quad (8.16)$$

where EI_c is the flexural rigidity of column I, h_i is the height of column i, and Δ is the lateral sway.

Normally, the fixed end moments in columns are zero, as the wind load is assumed to be acting at the joints.

Therefore, the moment at the near end is

$$M_{Oi} = 2M'_{Oi} + M'_{iO} + M''_{Oi} \quad (8.17)$$

Similarly, for the far end, we can write

$$M_{iO} = 2M'_{iO} + M'_{Oi} + M''_{iO} \quad (8.18)$$

By adding equations 8.17 and 8.18, we obtain

$$M_{Oi} + M_{iO} = 3M'_{Oi} + M'_{iO} + M''_{Oi} + M''_{iO}$$

However, due to lateral sway, the end moments in a column are equal i.e. $M''_{Oi} = M''_{iO}$

Therefore,

$$M_{Oi} + M_{iO} = 3M'_{Oi} + M'_{iO} + 2M''_{Oi} \quad (8.19)$$

Applying the summation for all columns in a storey, then

$$\sum M_{Oi} + \sum M_{iO} = 3\sum M'_{Oi} + 3\sum M'_{iO} + 2\sum M''_{Oi}$$

$$\sum M''_{Oi} = \frac{1}{2}\left[\sum M_{Oi} + \sum M_{iO} - 3\sum M'_{Oi} - 3\sum M'_{iO}\right] \tag{8.20}$$

Consider the moment equilibrium at joint O and i, the far end of the beam or column connected at joint O, as shown in the figure below.

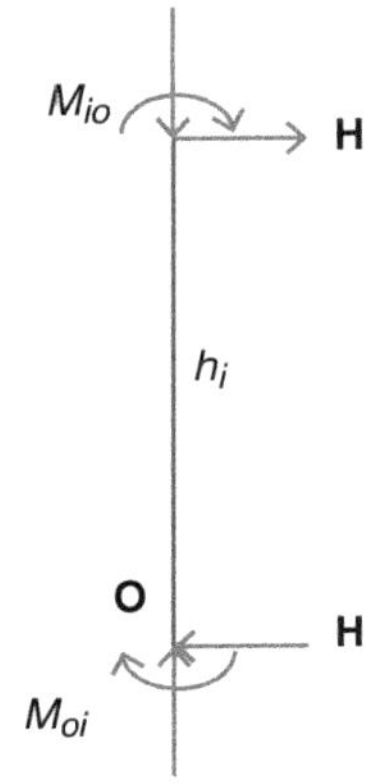

$$\sum_i M_{Oi} = 0 \Rightarrow H \times h_i = -\left(M_{Oi} + M_{iO}\right)$$

Let us apply this for all the columns in a storey, where S_i is the storey shear $= \sum H$

Then,

$$S_i h_i = -\sum \left(M_{Oi} + M_{iO}\right) \tag{8.21}$$

From equations 8.20 and 8.21, we get

$$\sum M''_{Oi} = -\frac{S_i h_i}{2} - \frac{3}{2}\left[\sum M'_{Oi} + \sum M'_{iO}\right]$$

$$\sum M''_{Oi} = -\frac{3}{2}\left[\frac{S_i h_i}{3} + \left[\sum M'_{Oi} + \sum M'_{iO}\right]\right] \quad (8.22$$

From equation 8.16,

$$M''_{Oi} = -\frac{6EI_c}{h^2_i}\Delta = -1.5\frac{k}{h_i}\Delta \quad \text{where } k = \frac{4EI}{h_i}$$

$$\Rightarrow \sum M''_{Oi} = -\frac{1.5\Delta}{h_i}\sum k$$

$$\frac{M''_{Oi}}{\sum M''_{Oi}} = \frac{k}{\sum k} \quad \text{(or)} \quad M''_{Oi} = \frac{k}{\sum k}\sum M''_{Oi} \tag{8.23}$$

Now substituting equation 8.22 into equation 8.23, we get

$$M''_{Oi} = \frac{k}{\sum k}\left(-\frac{3}{2}\left[\frac{S_i h_i}{3} + \left[\sum M'_{Oi} + \sum M'_{iO}\right]\right]\right)$$

$$M''_{Oi} = -\frac{3}{2}\frac{k}{\sum k}\left(\left[\frac{S_i h_i}{3} + \left[\sum M'_{Oi} + \sum M'_{iO}\right]\right]\right)$$

$$M''_{Oi} = (\mathrm{SF})_i\left(\left[\frac{S_i h_i}{3} + \left[\sum M'_{Oi} + \sum M'_{iO}\right]\right]\right)$$

Where $(\mathrm{SF})_i = -\frac{3}{2}\frac{k}{\sum k}$ is called the displacement factor or sway factor

(8.24a)

It should be noted that the sum of all the sway factors at a joint is equal to $-\frac{3}{2}$ and the term $\frac{S_i h_i}{3}$ is called the storey moment.

Therefore, the displacement contribution to a column

$M''_{Oi} = (\mathrm{SF})_i$ [Storey moment + Rotation contribution of columns from top and bottom]

When displacement contributions are to be taken into account, the rotation contributions calculated in equation 8.13 can take the form as

$$M^1_{Oi} = \left(-\frac{1}{2}\right)\left[\frac{k_i}{\sum_i k_i}\right]\left[\sum_i MF_{Oi} + \sum_i\left(M^1_{iO} + M''_{Oi}\right)\right] \tag{8.25}$$

Therefore, equations 8.24 and 8.25, will assist us in evolving the distribution procedure that will account for the sway effect.

Step-by-step procedure to solve the problems is as follows:

i. **Calculate the fixed end moment** for all the members
ii. **Find the rotation factors** for all the members meeting at each joint using equation 8.13a
iii. **Calculate the sway/displacement factors** for all the columns in each storey using equation 8.24a
iv. **Start Kani's iterations**: Assume the unknown rotation and sway contributions are initially zero in Kani's table.
v. **Apply Kani's procedure**: Use equation 8.25 to find the rotation contributions joint by joint until all joints are completed. While applying equation 8.25, as and when the rotation contributions are available, those values should be used in the subsequent cycles.
vi. **Calculate displacement contributions**: Use equation 8.24 to compute the displacement contributions of columns in each storey.
vii. **Update contributions**: After completing steps v and vi, use the calculated rotation and displacement

contributions at all joints to update these values in the next cycle.

viii. **Iterate until convergence**: Repeat the above procedure until the rotation contributions and displacement contributions remain the same in two consecutive cycles, indicating convergence.

ix. Calculate the final end moment using equation 8.14:

Final end moment = [Fixed end moment + 2 × Near end contribution
+ 1 × Far end rotation contribution
+ Displacement contribution]

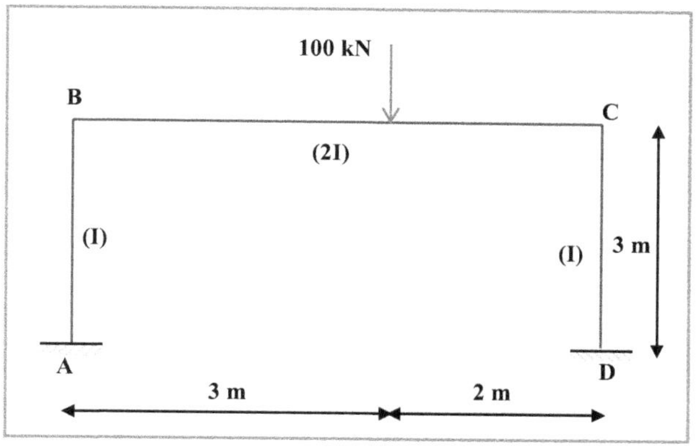

FIGURE 8.34 Rigid frame *ABCD* with unsymmetrical loading.

Example 8.15

Analyse the rigid joint frame ABCD shown in Figure 8.34 using Kani's method. Draw the shear force and bending moment diagrams.

Solution:

The kinematic indeterminacy of the given frame is 03, i.e., rotation at *B* and *C* (assumed positive in the clockwise direction) and the translational displacement of *BC* from left to right.

FIXED END MOMENTS

$$MF_{BC} = -\frac{Wab^2}{L^2} = -\frac{100 \times 3 \times 2^2}{5^2} = -48 \text{ kNm}$$

$$MF_{CB} = \frac{Wba^2}{L^2} = \frac{100 \times 2 \times 3^2}{5^2} = 72 \text{ kNm}$$

$$MF_{AB} = MF_{BA} = MF_{CD} = MF_{DC} = 0$$

ROTATION FACTORS (RF)

$$(\text{RF})_i = \frac{-1}{2}(\text{DF})_i = \left(\frac{-1}{2}\right)\frac{K_i}{\sum_{i=1}^{n} K_i}$$

Joint	Member	K	$\sum K$	DF	RF
B	BA	$\frac{4E(I)}{3} = 1.33EI$	1.33EI+1.6 EI=2.93EI	0.453	−0.227
	BC	$\frac{4E(2I)}{5} = 1.6EI$		0.547	−0.273
C	CB	$\frac{4E(2I)}{5} = 1.6EI$	1.6EI+1.33 EI=2.93EI	0.547	−0.273
	CD	$\frac{4E(I)}{3} = 1.33EI$		0.453	−0.227

Displacement/sway factors (SF): $(\text{SF})_i = -\frac{3}{2}\frac{k}{\sum k}$

Storey	Member	K	$\sum K$	DF	SF
A	AB	$\frac{4E(I)}{3} = 1.33EI$	1.33EI+1.33 EI=2.66EI	0.5	−0.75
	DC	$\frac{4E(I)}{3} = 1.33EI$		0.5	−0.75

KANI'S DISTRIBUTION

FIRST CYCLE

To begin, assume all rotational contributions are zero.

Sum of the fixed end moments at B, $\sum MF_B = MF_{BA} + MF_{BC} = 0 + (-48) = -48$ kNm

Sum of the fixed end moments at C, $\sum MF_C = MF_{CB} + MF_{CD} = 72 + (0) = 72$ kNm

Storey moment $\frac{S_1 h_1}{3} = 0$ (since $\sum H = 0$)

Rotation contributions/moments

$$M^1_{Oi} = \left(-\frac{1}{2}\right)\left[\frac{k_i}{\sum_i k_i}\right]\left[\sum_i MF_{Oi} + \sum_i M^1_{iO} + M''_{Oi}\right]$$

$$M^1_{AB} = M^1_{DC} = 0 \text{ (since } A \text{ and } D \text{ are fixed ends)}$$

$$M^1_{BA} = (\text{RF})_{BA}\left[MF_B + \left(M^1_{AB} + M^1_{CB}\right) + M''_{AB}\right]$$

$$= -0.227 \times [-48 + 0 + 0 + 0] = 10.896$$

$$M^1_{BC} = (\text{RF})_{BC}\left[MF_B + \left(M^1_{AB} + M^1_{CB}\right) + M''_{AB}\right]$$

$$= -0.273 \times [-48 + 0 + 0 + 0] = 13.104$$

$$M_{CB}^{1} = (\text{RF})_{CB}\left[MF_C + \left(M_{BC}^{1} + M_{DC}^{1}\right) + M_{DC}''\right]$$

$$= -0.273 \times [72 + 13.104 + 0 + 0] = -23.233$$

$$M_{CD}^{1} = (\text{RF})_{CD}\left[MF_C + \left(M_{BC}^{1} + M_{DC}^{1}\right) + M_{DC}''\right]$$

$$= -0.227 \times [72 + 13.104 + 0 + 0] = -19.319$$

DISPLACEMENT CONTRIBUTIONS

$$M_{Oi}'' = (\text{SF})_i\left(\left[\frac{S_i h_i}{3} + \left[\sum M_{Oi}' + \sum M_{iO}''\right]\right]\right)$$

$$M_{AB}'' = (\text{SF})_{AB}\left(\left[\frac{S_i h_i}{3} + \left[\sum M_{BA}' + \sum M_{CD}'\right]\right]\right)$$

$$M_{AB}'' = M_{CD}'' = -0.75 \times (0 + 10.896 - 19.319) = 6.317$$

SECOND CYCLE

Now use the above values of rotation contributions and sway contributions in the second cycle of Kani's distribution.

$$M_{BA}^{1} = (\text{RF})_{BA}\left[MF_B + (M_{AB}^{1} + M_{CB}^{1}) + M_{AB}''\right]$$

$$= -0.227 \times [-48 + 0 - 23.233 + 6.317] = 14.736$$

$$M_{BC}^{1} = (\text{RF})_{BC}\left[MF_B + (M_{AB}^{1} + M_{CB}^{1}) + M_{AB}''\right]$$

$$= -0.273 \times [-48 + 0 - 23.233 + 6.317] = 17.722$$

$$M_{CB}^{1} = (\text{RF})_{CB}\left[MF_C + (M_{BC}^{1} + M_{DC}^{1}) + M_{DC}''\right]$$

$$= -0.273 \times [72 + 17.722 + 0 + 6.317] = -26.219$$

$$M_{CD}^{1} = (\text{RF})_{CD}\left[MF_C + (M_{BC}^{1} + M_{DC}^{1}) + M_{DC}''\right]$$

$$= -0.227 \times [72 + 17.722 + 0 + 6.317] = -21.801$$

DISPLACEMENT CONTRIBUTIONS

$$M_{AB}'' = M_{CD}'' = (\text{SF})_{AB}\left(\left[\frac{S_i h_i}{3} + \left[\sum M_{BA}' + \sum M_{CD}'\right]\right]\right)$$

$$M_{AB}'' = M_{CD}'' = -0.75 \times (0 + 14.736 - 21.801) = 5.299$$

The above procedure is continued until the rotation contributions and displacement contributions remain the same in two consecutive cycles, i.e., convergence is attained. The whole Kani's distribution cycles are presented in Table 8.16.

FINAL END MOMENTS

Final end moments = Fixed end moment + 2 × Near end rotation contribution
+Far end rotation contribution + Sway contribution

The calculation of final end moments is presented in Table 8.16.

The free body diagram (Figure 8.35a) shows the stress resultants, and the shear force and bending moment diagrams are shown in Figure 8.35b and c, respectively.

Example 8.16

Analyse the portal frame *ABCD*, which is subjected to a horizontal load of 50 kN at *B*, and the beam *BC* is subjected to a triangular load, as shown in Figure 8.36, using Kani's method. Also, draw the shear force and bending moment diagrams.

Solution:

The total kinematic indeterminacy of the given frame is 03, i.e., rotational degrees of freedom at the joints *B* and *C*, and the translational displacement of *BC*.

FIXED END MOMENTS

$$MF_{BC} = -MF_{CB} = -\frac{5wL^2}{96} = -\frac{5 \times 50 \times 6^2}{96} = -93.75 \text{ kNm}$$

$$MF_{AB} = MF_{BA} = MF_{CD} = MF_{DC} = 0$$

ROTATION FACTORS (RF)

$$(\text{RF})_i = \frac{-1}{2}(\text{DF})_i = \left(\frac{-1}{2}\right)\frac{K_i}{\sum_{i=1}^{n} K_i}$$

Joint	Member	K	$\sum K$	DF	RF
B	BA	$\frac{4E(I)}{3} = 1.33EI$	$1.33EI + 1.33EI = 2.93EI$	0.5	−0.25
	BC	$\frac{4E(2I)}{6} = 1.33EI$		0.5	−0.25
C	CB	$\frac{4E(2I)}{6} = 1.33EI$	$1.336EI + 1.33EI = 2.93EI$	0.5	−0.25
	CD	$\frac{4E(I)}{3} = 1.33EI$		0.5	−0.25

Displacement/sway factors (SF): $(\text{SF})_i = -\frac{3}{2}\frac{k}{\sum k}$

Storey	Member	K	$\sum K$	DF	SF
A	AB	$\frac{4E(I)}{3} = 1.33EI$	$1.33EI + 1.33EI = 2.66EI$	0.5	−0.75
	DC	$\frac{4E(I)}{3} = 1.33EI$		0.5	−0.75

$$\text{Storey moment} = \frac{S_i h_i}{3} = \frac{h_i}{3}\sum H = \frac{3 \times 50}{3} = 50 \text{ kNm}$$

TABLE 8.16
Kani's Iteration Cycles Table

Joint	A	B		C		D
Member	AB	BA	BC	CB	CD	DC
RF	0.000	−0.227	−0.273	−0.273	−0.227	
FEM	0.000	0.000	−48.000	72.000	0.000	0.000
Sum of FEMs	0.000	−48.000		72.000		0.000
Rotation contribution		10.896	13.104	−23.233	−19.319	
		14.736	17.722	−26.219	−21.801	
		15.645	18.815	−26.239	−21.818	
		15.801	19.003	−26.108	−21.709	
		15.817	19.022	−26.059	−21.668	
		15.815	19.020	−26.046	−21.658	
		15.814	19.018	−26.044	−21.656	
		15.813	19.018	−26.044	−21.656	
Displacement contribution		−0.750			−0.750	
		6.317			6.317	
		5.299			5.299	
		4.630			4.630	
		4.431			4.431	
		4.388			4.388	
		4.382			4.382	
		4.382			4.382	
		4.382			4.382	
2×near end contribution	0.000	31.627	38.036	−52.088	−43.311	0.000
1×far end contribution	15.813	0.000	−26.044	19.018	0.000	−21.656
Displacement contribution		4.382			4.382	
Final end moments (kNm)	20.195	36.009	−36.008	38.930	−38.930	−17.274

KANI'S DISTRIBUTION CYCLES

ROTATION CONTRIBUTIONS

Sum of the fixed end moments at B, $\sum MF_B = MF_{BA} + MF_{BC} = 0 + (-93.75) = -93.75\ \text{kNm}$

Sum of the fixed end moments at C, $\sum MF_C = MF_{CB} + MF_{CD} = 93.75 + (0) = 93.75\ \text{kNm}$

Storey moment $\dfrac{S_1 h_1}{3} = 50\ \text{kNm}$

ROTATION CONTRIBUTIONS/MOMENTS

$$M_{Oi}^{1} = \left(-\frac{1}{2}\right)\left[\frac{k_i}{\sum_i k_i}\right]\left[\sum_i MF_{Oi} + \sum_i M_{iO}^{1} + M_{Oi}''\right]$$

DISPLACEMENT CONTRIBUTIONS

$$M_{Oi}'' = (\text{SF})_i\left(\left[\frac{S_i h_i}{3}\right] + \left[\sum M_{Oi}' + \sum M_{iO}'\right]\right)$$

As explained in the previous problem, the rotation contributions of all the members are initially assumed to be zero to find the rotations in the first cycle using the above equations. The complete Kani's distribution is presented in Table 8.17.

FINAL END MOMENTS

Final end moments = Fixed end moments
+ 2 × near end rotation contribution
+ Far end rotation contribution
+ Sway contribution

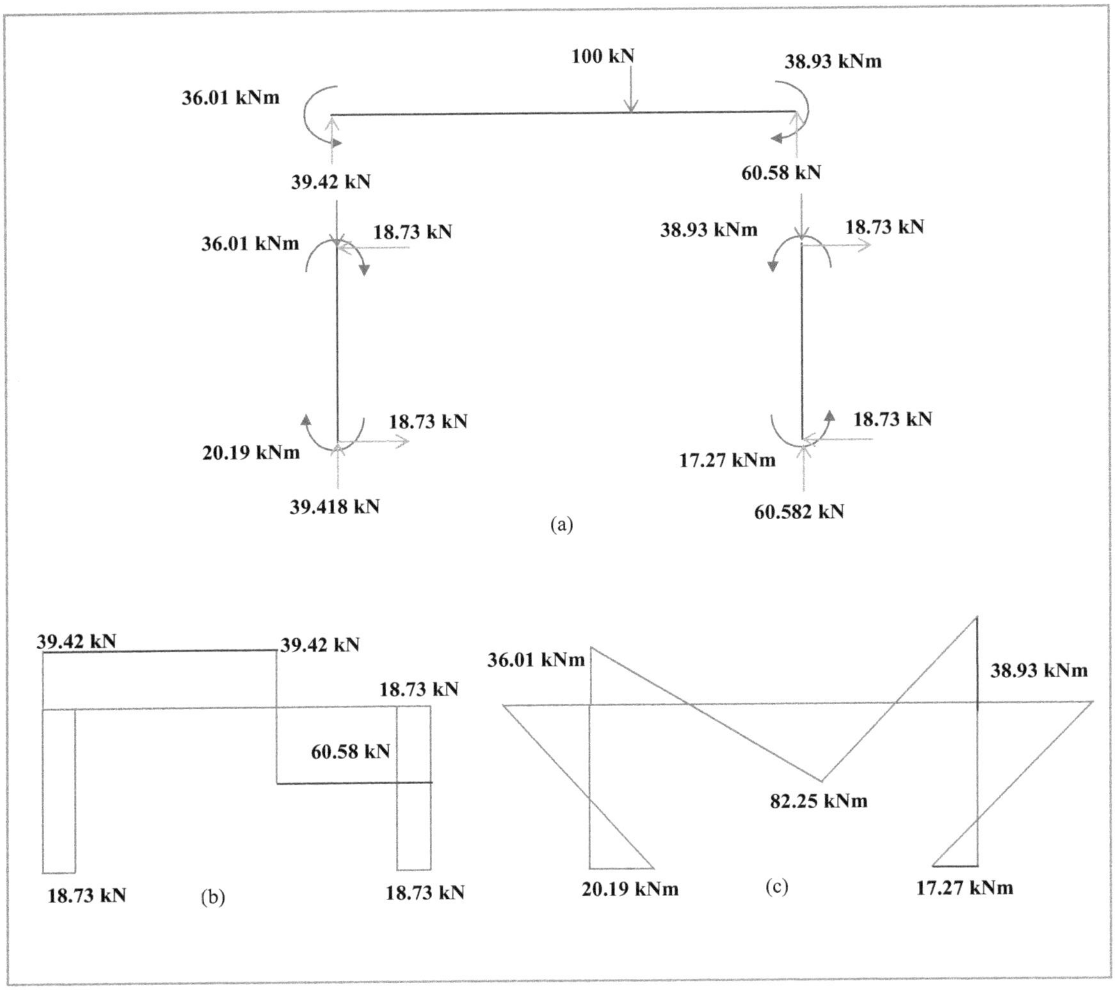

FIGURE 8.35 (a) Free body diagram, (b) shear force diagram, and (c) bending moment diagram.

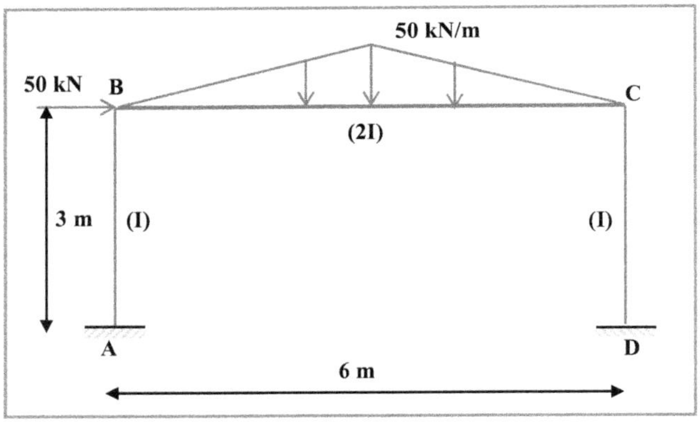

FIGURE 8.36 Rigid frame *ABCD* with horizontal load at *B*.

All final end moments are calculated using the above expression, and these are shown in Table 8.17.

The stress resultants are shown in the free body diagram (Figure 8.37a), and the shear force and bending moment diagrams are shown in Figure 8.37b and c, respectively.

Example 8.17

Analyse the two-storey frame subjected to an unsymmetrical load as shown in Figure 8.38 using Kani's method. Also, draw the shear force and bending moment diagrams. Assume the flexural rigidity of the beams is twice that of the columns.

Solution:

The given multi-storey frame is subjected to four rotational degrees of freedom, one at each joint *B*, *C*, *D*, and *E*, and two translational displacements, one at each floor, i.e., *BC* and *CD*, respectively.

TABLE 8.17
Kani's Iteration Cycles and Final End Moments

Joint	A	B		C		D
Member	*AB*	*BA*	*BC*	*CB*	*CD*	*DC*
RF	0.000	–0.250	–0.250	–0.250	–0.250	
FEM	0.000	0.000	–93.750	93.750	0.000	0.000
Sum of FEMs	0.000	–93.750		93.750		0.000
Rotation contribution		23.438	23.438	–29.297	–29.297	
		39.038	39.038	–24.921	–24.921	
		41.690	41.690	–21.838	–21.838	
		41.994	41.994	–20.839	–20.839	
		41.989	41.989	–20.593	–20.593	
		41.972	41.972	–20.544	–20.544	
		41.966	41.966	–20.536	–20.536	
		41.965	41.965	–20.536	–20.536	
Storey moment (kNm)		**50.000**				
Displacement contribution		–0.750			–0.750	
		–33.105			–33.105	
		–48.088			–48.088	
		–52.389			–52.389	
		–53.367			–53.367	
		–53.547			–53.547	
		–53.571			–53.571	
		–53.573			–53.573	
		–53.572			–53.572	
2×near end contribution	0.000	83.929	83.929	–41.071	–41.071	0.000
1×far end contribution	41.965	0.000	–20.536	41.965	0.000	–20.536
Displacement contribution		–53.572			–53.572	
Final end moments (kNm)	–11.607	30.358	–30.356	94.644	–94.643	–74.107

FIXED END MOMENTS

$$MF_{BE} = -MF_{EB} = -\frac{wL^2}{12} = -\frac{20\times 6^2}{12} = -60 \text{ kNm}$$

$$MF_{CD} = -MF_{DC} = -\frac{wL^2}{12} = -\frac{10\times 6^2}{12} = -30 \text{ kNm}$$

$$MF_{AB} = MF_{BA} = MF_{BC} = MF_{CB} = MF_{DE}$$

$$= MF_{ED} = MF_{EF} = MF_{FE} = 0$$

STOREY MOMENTS

$$\text{Top storey moment} = \frac{S_2 h_2}{3} = \frac{20\times 3}{3} = 20 \text{ kNm}$$

$$\text{Bottom storey moment} = \frac{S_1 h_1}{3} = \frac{50\times 3}{3} = 50 \text{ kNm}$$

ROTATION FACTORS (RF)

$$(\text{RF})_i = \frac{-1}{2}(\text{DF})_i = \left(\frac{-1}{2}\right)\frac{K_i}{\sum_{i=1}^{n} K_i}$$

Joint	Member	K	∑K	DF	RF
B	*BA*	$\frac{4E(I)}{3} = 1.33EI$	1.33*EI*+1.33 *EI*+1.33 *EI*=3.99*EI*	0.333	–0.167
	BC	$\frac{4E(I)}{3} = 1.33EI$		0.333	–0.167
	BE	$\frac{4E(2I)}{6} = 1.33EI$		0.333	–0.167
C	*CB*	$\frac{4E(I)}{3} = 1.33EI$	1.33*EI*+1.33 *EI*=2.66*EI*	0.5	–0.25
	CD	$\frac{4E(2I)}{6} = 1.33EI$		0.5	–0.25
D	*DC*	$\frac{4E(2I)}{6} = 1.33EI$	1.33*EI*+1.33 *EI*=2.66*EI*	0.5	–0.25
	DE	$\frac{4E(I)}{3} = 1.33EI$		0.5	–0.25
E	*ED*	$\frac{4E(I)}{3} = 1.33EI$	1.33*EI*+1.33 *EI*+1.33 *EI*=3.99*EI*	0.333	–0.167
	EB	$\frac{4E(2I)}{6} = 1.33EI$		0.333	–0.167
	EF	$\frac{4E(I)}{3} = 1.33EI$		0.333	–0.167

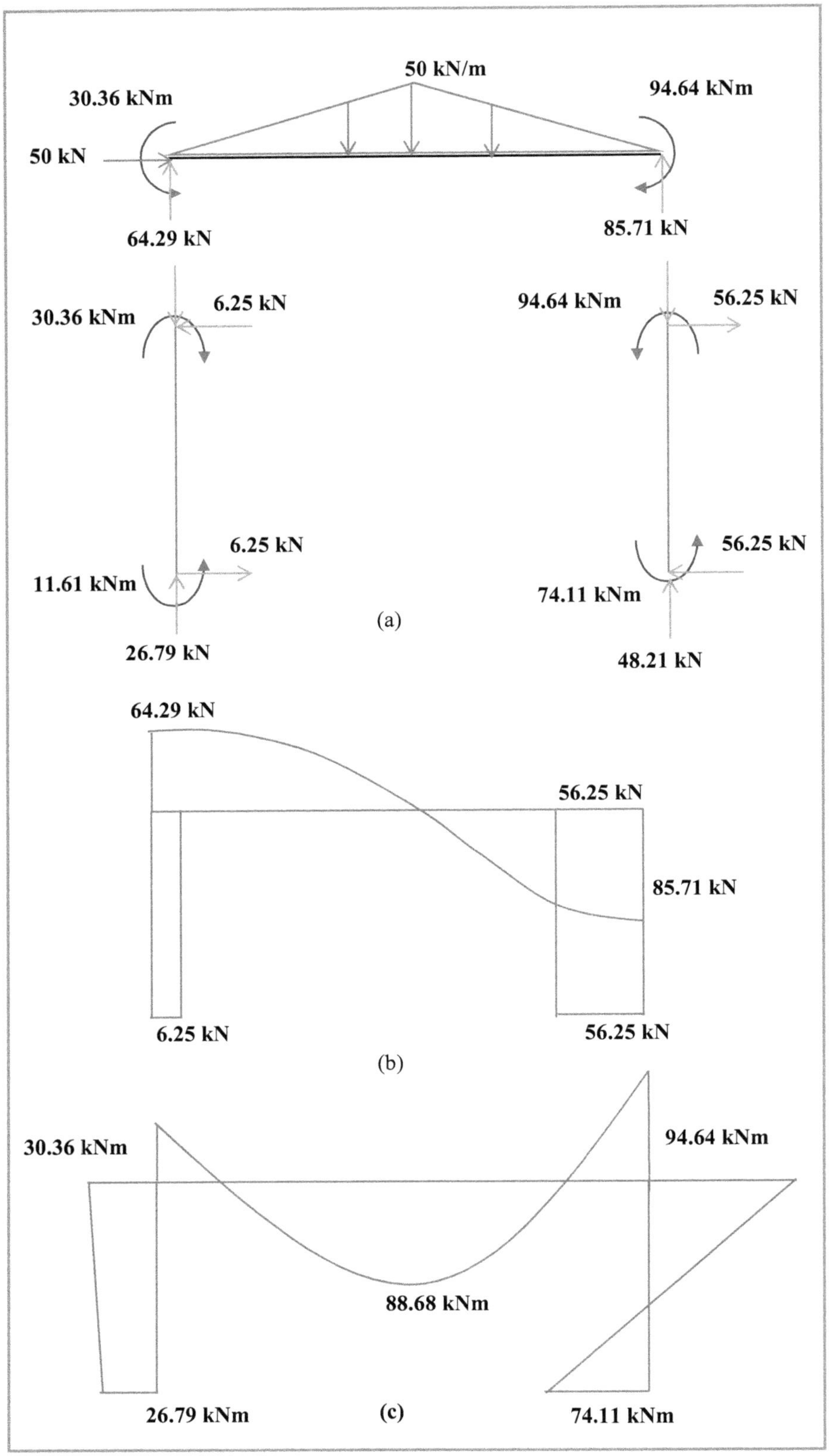

FIGURE 8.37 (a) Free body diagram (b) shear force diagram and (c) bending moment diagram.

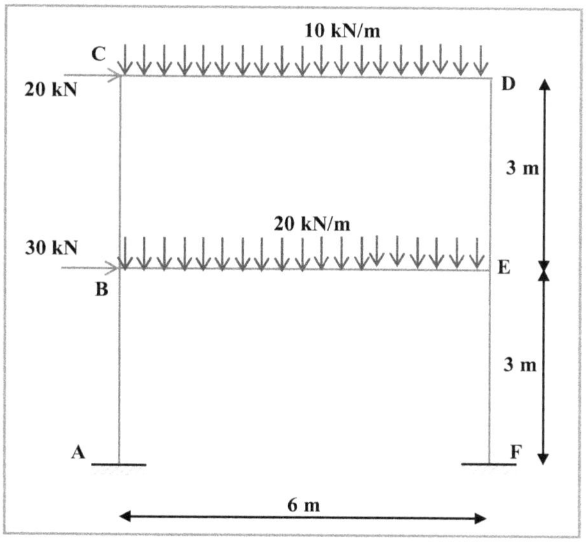

FIGURE 8.38 Two-storey frame.

Displacement/sway factors (SF): $(\mathrm{SF})_i = -\dfrac{3}{2}\dfrac{k}{\sum k}$

Storey	Member	K	$\sum K$	DF	SF
Top	AB	$\dfrac{4E(I)}{3}=1.33EI$	$1.33EI+1.33EI=2.66EI$	0.5	–0.75
	EF	$\dfrac{4E(I)}{3}=1.33EI$		0.5	–0.75
Bottom	BC	$\dfrac{4E(I)}{3}=1.33EI$	$1.33EI+1.33EI=2.66EI$	0.5	–0.75
	DE	$\dfrac{4E(I)}{3}=1.33EI$		0.5	–0.75

KANI'S DISTRIBUTION

First Cycle

To begin initially assume all rotation contributions are zero.

Sum of the fixed end moments at B, $\sum MF_B = MF_{BA} + MF_{BC} + MF_{BE} = -60$ kNm

Sum of the fixed end moments at C, $\sum MF_B = MF_{CB} + MF_{CD} = -30$ kNm

Sum of the fixed end moments at E, $\sum MF_E = MF_{EB} + MF_{ED} + MF_{EF} = 60$ kNm

Sum of the fixed end moments at D, $\sum MF_D = MF_{DC} + MF_{DE} = 30$ kNm

Rotation contributions/moments

$$M_{Oi}^{1} = (\mathrm{RF})_i\left[\left[\sum_i MF_{Oi} + \sum_i (M_{iO}^{1} + M_{Oi}'')\right]\right]$$

$$M_{AB}^{1} = M_{FE}^{1} = 0\,(\text{since } A \text{ and } F \text{ are fixed ends})$$

Joint C

$$M_{CB}^{1} = (\mathrm{RF})_{CB}\left[MF_C + (M_{BC}^{1} + M_{DC}^{1}) + (M_{BC}'')\right]$$
$$= -0.25\times[-30+0+0+0] = 7.5$$
$$M_{CD}^{1} = (\mathrm{RF})_{CD}\left[MF_C + (M_{BC}^{1} + M_{DC}^{1}) + (M_{BC}'')\right]$$
$$= -0.25\times[-30+0+0+0] = 7.5$$

Joint D

$$M_{DC}^{1} = (\mathrm{RF})_{DC}\left[MF_D + (M_{CD}^{1} + M_{ED}^{1}) + M_{DE}''\right]$$
$$= -0.25\times[30+7.5+0+0] = -9.375$$
$$M_{DE}^{1} = (\mathrm{RF})_{DE}\left[MF_D + (M_{CD}^{1} + M_{ED}^{1}) + M_{DE}''\right]$$
$$= -0.25\times[30+7.5+0+0] = -9.375$$

Joint E

$$M_{ED}^{1} = (\mathrm{RF})_{ED}\left[MF_E + (M_{DE}^{1} + M_{BE}^{1} + M_{FE}^{1}) + M_{DE}'' + M_{FE}''\right]$$
$$= -0.167\times[60-9.375+0+0+0+0] = -8.454$$
$$M_{EB}^{1} = (\mathrm{RF})_{EB}\left[MF_E + (M_{DE}^{1} + M_{BE}^{1} + M_{FE}^{1}) + M_{DE}'' + M_{FE}''\right]$$
$$= -0.167\times[60-9.375+0+0+0+0] = -8.454$$
$$M_{EF}^{1} = (\mathrm{RF})_{EF}\left[MF_E + (M_{DE}^{1} + M_{BE}^{1} + M_{FE}^{1}) + M_{DE}'' + M_{FE}''\right]$$
$$= -0.167\times[60-9.375+0+0+0+0] = -8.454$$

Joint *B*

$$M_{BC}^{1} = (\mathrm{RF})_{BC}\left[MF_B + (M_{EB}^{1} + M_{CB}^{1} + M_{AB}^{1}) + M_{BC}'' + M_{AB}''\right]$$
$$= -0.167\times[-60-8.454+7.5+0+0+0] = 10.179$$
$$M_{BE}^{1} = (\mathrm{RF})_{BE}\left[MF_B + (M_{EB}^{1} + M_{CB}^{1} + M_{AB}^{1}) + M_{BC}'' + M_{AB}''\right]$$
$$= -0.167\times[-60-8.454+7.5+0+0+0] = 10.179$$
$$M_{BA}^{1} = (\mathrm{RF})_{BA}\left[MF_B + (M_{EB}^{1} + M_{CB}^{1} + M_{AB}^{1}) + M_{BC}'' + M_{AB}''\right]$$
$$= -0.167\times[-60-8.454+7.5+0+0+0] = 10.179$$

DISPLACEMENT CONTRIBUTIONS

$$M_{Oi}'' = (\mathrm{SF})_i\left(\left[\frac{S_i h_i}{3}\right] + \left[\sum M_{Oi}' + \sum M_{iO}'\right]\right)$$

Bottom Storey

$$M_{AB}'' = (\mathrm{SF})_{AB}\left(\left[\frac{S_b h_b}{3}\right] + \left[\sum M_{AB}' + \sum M_{FE}'\right]\right)$$

$$M_{AB}'' = M_{CD}'' = -0.75\times(50+10.179+0-8.454+0) = -38.794$$

Top Storey

$$M''_{BC} = (\text{SF})_{BC}\left(\left[\frac{S_t h_t}{3} + \left[\sum M'_{BC} + \sum M'_{ED}\right]\right]\right)$$

$$M''_{BC} = M''_{DE} = -0.75 \times (20 + 10.179 + 7.5 - 8.454 - 9.375)$$

$$= -14.888$$

Now, use the above values of rotation contributions and sway contributions in Kani's second cycle. The procedure is repeated until the rotation contributions and displacement contributions remain the same in two consecutive cycles, i.e., until convergence is attained. The complete Kani distribution cycles are presented in Table 8.18.

FINAL END MOMENTS

The free body diagram (Figure 8.39a) shows the internal stress resultants, and Figure 8.39b and c, respectively, shows the shear force and bending moment diagrams.

8.5.2 Columns with Unequal Heights

In certain situations, the foundations for column bases are provided at different levels, resulting in unequal column heights on the ground floor. Therefore, suitable modifications must be applied to the sway moment contribution, i.e., equation 8.24, to account for these varying heights on the ground floor. For convenience, let us consider a reference storey height, h_{ref} (which can be chosen as the largest column height on the ground floor), and define a height factor C_i for the ith column in the ground storey. Then,

$$C_i = \frac{h_{\text{ref}}}{h_i} \tag{8.26}$$

Substituting equation 8.26 into the derivation of the sway moment equation 8.22, we obtain the following modifications in the sway moment equations:

$$\sum M''_{Oi} = -\frac{S_i h_{\text{ref}}}{2 \times C_i} - \frac{3}{2}\left[\sum M'_{Oi} + \sum M'_{iO}\right]$$

$$\Rightarrow \sum C_i M''_{Oi} = -\frac{S_i h_{\text{ref}}}{2} - \frac{3}{2}\left[\sum C_i (M'_{Oi} + M'_{iO})\right] \tag{8.27}$$

But

$$M''_{Oi} = -\frac{6EI_c}{h_{\text{ref}}^2}\Delta = -1.5\frac{k}{h_{\text{ref}}} C_i \Delta \quad \text{where } k = \frac{4EI}{h_{\text{ref}}}$$

$$\Rightarrow \sum M''_{Oi} = -\frac{1.5\Delta}{h_{\text{ref}}} \sum kC_i^2$$

$$\frac{M''_{Oi}}{\sum M''_{Oi}} = \frac{kC_i}{\sum kC_i^2} \quad \text{(or)} \quad M''_{Oi} = \frac{kC_i}{\sum kC_i^2} \sum M''_{Oi} \tag{8.28}$$

Now substituting equation 8.27 into equation 8.28, we obtain

$$M''_{Oi} = -\frac{3}{2}\frac{kC_i}{\sum kC_i^2}\left(\left[\frac{S_i h_i}{3} + \left[\sum C_{Oi}(M'_{Oi}) + C_{iO}(M'_{iO})\right]\right]\right)$$

$$M''_{Oi} = (\text{SF})_i\left(\left[\frac{S_i h_i}{3} + \left[\sum C_{Oi}(M'_{Oi}) + C_{iO}(M'_{iO})\right]\right]\right) \tag{8.29}$$

where the sway factor or displacement factor takes the form as

$$(\text{SF})_i = -\frac{3}{2}\frac{kC_i}{\sum_i kC_i^2} \tag{6.30}$$

Using the above modifications in the displacement factors and displacement contributions, we can analyse the frames of unequal heights. This is illustrated in the following Example 8.18.

Example 8.18

Analyse the unsymmetrical frame *ABCD* as shown in Figure 8.40 using Kani's method. Draw the SFD and BMD. This has been solved in Example 7.19 using the moment distribution method.

Solution:

The kinematic indeterminacy of the frame is 03, i.e., the rotational degrees of freedom at *B* and *C*, and the translational displacement of *BC*.

FIXED END MOMENTS

$$MF_{AB} = MF_{BA} = MF_{CD} = MF_{DC} = 0$$

$$MF_{BC} = -MF_{CB} = -\frac{20 \times 5^2}{12} = -41.667 \text{ kNm}$$

STOREY MOMENTS

$$\text{Storey moment} = \frac{S_1 h_1}{4} = 0 \text{ kNm}$$

ROTATION FACTORS (RF)

$$(\text{RF})_i = \frac{-1}{2}(\text{DF})_i = \left(\frac{-1}{2}\right)\frac{K_i}{\sum_{i=1}^{n} K_i}$$

Joint	Member	*K*	$\sum K$	DF	RF
B	*BA*	$\frac{4E(I)}{4} = 1.0EI$	$1.0EI + 1.6EI = 2.6EI$	0.384	–0.192
	BC	$\frac{4E(2I)}{5} = 1.6EI$		0.616	–0.308
C	*CB*	$\frac{4E(2I)}{5} = 1.6EI$	$1.6EI + 1.33EI = 2.93EI$	0.546	–0.273
	CD	$\frac{4E(I)}{3} = 1.33EI$		0.454	–0.227

DISPLACEMENT/SWAY FACTORS (SF)

The columns are of different heights; hence, choose the height of column AB as the reference height, i.e., $h_{ref} = h_{AB} = 3$ m

$$C_{CD} = 1; \quad C_{AB} = \frac{h_{ref}}{h_{AB}} = \frac{3}{4} = 0.75$$

$$(\text{SF})_i = -\frac{3}{2}\frac{kC_i}{\sum C_i^2 k}$$

TABLE 8.18
Kani's Iteration Cycles and Final End Moments

Joint	A	B			C		D		E			F
Member	*AB*	*BA*	*BC*	*BE*	*CB*	*CD*	*DC*	*DE*	*ED*	*EB*	*EF*	*FE*
RF	0.000	−0.167	−0.167	−0.167	−0.250	−0.250	−0.250	−0.250	−0.167	−0.167	−0.167	0.000
FEM	0.000	0.000	0.000	−60.000	0.000	−30.000	30.000	0.000	0.000	60.000	0.000	0.000
Sum of FEMs	0.000	−60.000				−30.000		30.000		60.000		0.000
Rotation contribution		10.179	10.179	10.179	7.500	7.500	−9.375	−9.375	−8.454	−8.454	−8.454	
		17.481	17.481	17.481	11.021	11.021	−4.420	−4.420	−2.017	−2.017	−2.017	
		21.316	21.316	21.316	12.122	12.122	−2.139	−2.139	0.886	0.886	0.886	
		23.150	23.150	23.150	12.490	12.490	−1.059	−1.059	2.177	2.177	2.177	
		23.997	23.997	23.997	12.619	12.619	−0.557	−0.557	2.750	2.750	2.750	
		24.381	24.381	24.381	12.667	12.667	−0.328	−0.328	3.006	3.006	3.006	
		24.554	24.554	24.554	12.685	12.685	−0.224	−0.224	3.119	3.119	3.119	
		24.631	24.631	24.631	12.693	12.693	−0.178	−0.178	3.170	3.170	3.170	
Top storey moment (kNm)			20.000									
Displacement contribution			−0.750					−0.750				
			−14.888					−14.888				
			−31.549					−31.549				
			−39.138					−39.138				
			−42.568					−42.568				
			−44.107					−44.107				
			−44.795					−44.795				
			−45.101					−45.101				
			−45.237					−45.237				
Bottom storey moments (kNm)		50.000										
Displacement contribution		**−0.750**									**−0.750**	
		−38.794									−38.794	
		−49.098									−49.098	
		−54.151									−54.151	
		−56.495									−56.495	
		−57.560									−57.560	
		−58.040									−58.040	
		−58.255									−58.255	
		−58.351									−58.351	
FEM	0.000	0.000	0.000	−60.000	0.000	−30.000	30.000	0.000	0.000	60.000	0.000	0.000
2×near end contribution	0.000	49.263	49.263	49.263	25.385	25.385	−0.356	−0.356	6.339	6.339	6.339	0.000
1×far end contribution	24.631	0.000	12.693	3.170	24.631	−0.178	12.693	3.170	−0.178	24.631	0.000	3.170
Displacement contribution	−58.351	−58.351	−45.237		−45.237			−45.237	−45.237		−58.351	−58.351
Final end moments (kNm)	−33.719	−9.148	16.719	−7.568	4.790	−4.792	42.417	−42.423	−39.076	90.091	−52.012	−55.181

Storey	Member	k	C_i	$\sum C_i^2 k$	SF
1	AB	$\frac{4EI}{4}=1.00$	0.75	1.8955	−0.594
	CD	$\frac{4EI}{3}=1.333$	1.0		−1.055

KANI'S DISTRIBUTION CYCLES

Rotation Contributions

Using the following equation the rotation contributions can be determined. Initially assume all the rotation contributions are zeros.

$$M_{Oi}^{1}=(\text{RF})_i\left[\sum_i MF_{Oi}+\sum_i (M_{iO}^{1}+M_{Oi}'')\right]$$

Joint B:

$$M_{BA}^{1}=(\text{RF})_{BA}\left[\sum_i MF_B+\sum_i (M_{AB}^{1}+M_{CB}^{1}+M_{AB}'')\right]$$

$$M_{BA}^{1}=-0.192\left[-41.667+(0+0+0)\right]=8.000$$

$$M_{BC}^{1}=(\text{RF})_{BC}\left[\sum_i MF_B+\sum_i (M_{AB}^{1}+M_{CB}^{1}+M_{AB}'')\right]$$

$$C=-0.308\left[-41.667+(0+0+0)\right]=12.833$$

Joint C:

$$M_{CB}^{1}=(\text{RF})_{CB}\left[\sum_i MF_C+\sum_i (M_{BC}^{1}+M_{DC}^{1}+M_{CD}'')\right]$$

$$M_{CB}^{1}=-0.273\left[41.667+(12.833+0+0)\right]=14.879$$

$$M_{CD}^{1}=(\text{RF})_{CD}\left[\sum_i MF_C+\sum_i (M_{BC}^{1}+M_{DC}^{1}+M_{CD}'')\right]$$

$$M_{CD}^{1}=-0.227\left[41.667+(12.833+0+0)\right]=12.372$$

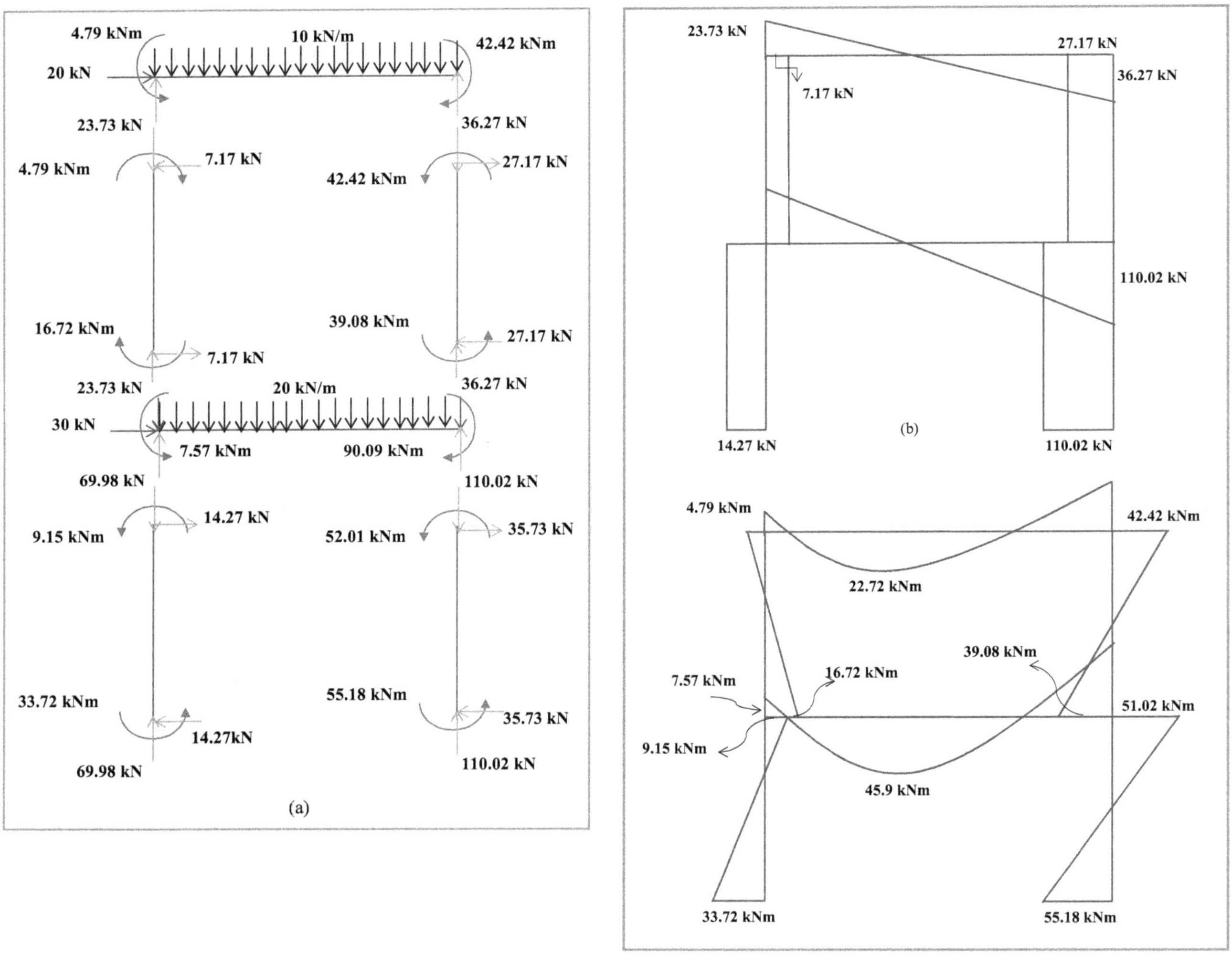

FIGURE 8.39 (a) Free body diagram, (b) shear force diagram, and (c) bending moment diagram.

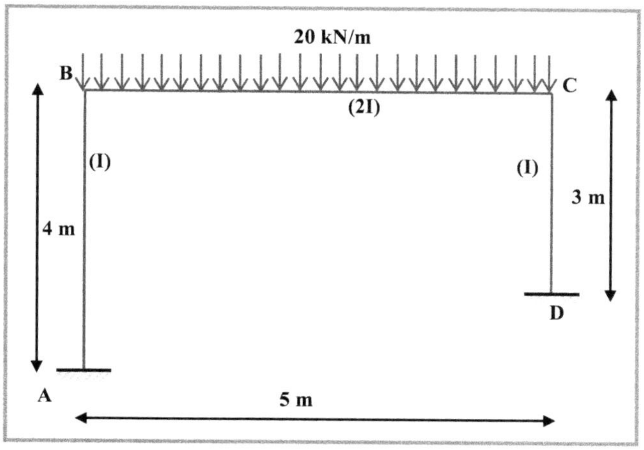

FIGURE 8.40 Rigid frame *ABCD* with unequal height of columns.

Displacement Contributions

$$M''_{Oi} = (\text{SF})_i \left(\left[\frac{S_i h_i}{3} + \left[\sum C_{Oi} (M'_{Oi}) + C_{iO} (M'_{iO}) \right] \right] \right)$$

For AB, $M''_{AB} = (\text{SF})_{AB} \left(\left[\frac{S_i h_i}{3} + \left[\sum C_{AB} (M'_{BA}) + C_{DC} (M'_{CD}) \right] \right] \right)$

$$M''_{AB} = -0.594\left[0 + \left(0.75 \times 8.0 + 1.0 \times (-12.372)\right)\right] = 3.785$$

For DC, $M''_{DC} = (SF)_{DC} \left(\left[\frac{S_i h_i}{3} + \left[\sum C_{AB} (M'_{BA}) + C_{DC} (M'_{CD}) \right] \right] \right)$

$$M''_{DC} = -1.055\left[0 + \left(0.75 \times 8.0 + 1.0 \times (-12.372)\right)\right] = 6.722$$

The above process will be continued until the convergence is achieved. The iteration process is presented in Table 8.19.

The member end forces, shear force and bending moment diagrams are shown in Figure 8.41a–c, respectively.

TABLE 8.19
Kani's iteration cycles and final end moments

Joint	*A*	*B*		*C*		*D*
Member	*AB*	*BA*	*BC*	*CB*	*CD*	*DC*
RF	0.000	−0.192	−0.308	−0.273	−0.227	
FEM	0.000	0.000	−41.667	41.667	0.000	0.000
Sum of FEMs	0.000	−41.667		41.667		0.000
Rotation contributions		8.000	12.833	−14.879	−12.372	
		10.130	16.250	−17.647	−14.673	
		10.581	16.974	−18.047	−15.006	
		10.659	17.098	−18.079	−15.033	
		10.669	17.114	−18.075	−15.029	
		10.669	17.115	−18.071	−15.026	
		10.669	17.114	−18.071	−15.026	
		10.669	17.114	−18.070	−15.025	
Ci		0.750			1.000	
Displacement contribution		−0.594			−1.055	
		3.785			6.722	
		4.203			7.465	
		4.200			7.459	
		4.181			7.426	
		4.174			7.414	
		4.173			7.411	
		4.172			7.411	
		4.172			7.410	
FEM	0.000	0.000	−41.667	41.667	0.000	0.000
2×near end contribution	0.000	21.337	34.228	−36.141	−30.051	0.000
1×far end contribution	10.669	0.000	−18.070	17.114	0.000	−15.025
Displacement contribution	4.172	4.172			7.410	7.410
Final end moments (kNm)	14.841	25.509	−25.509	22.640	−22.641	−7.615

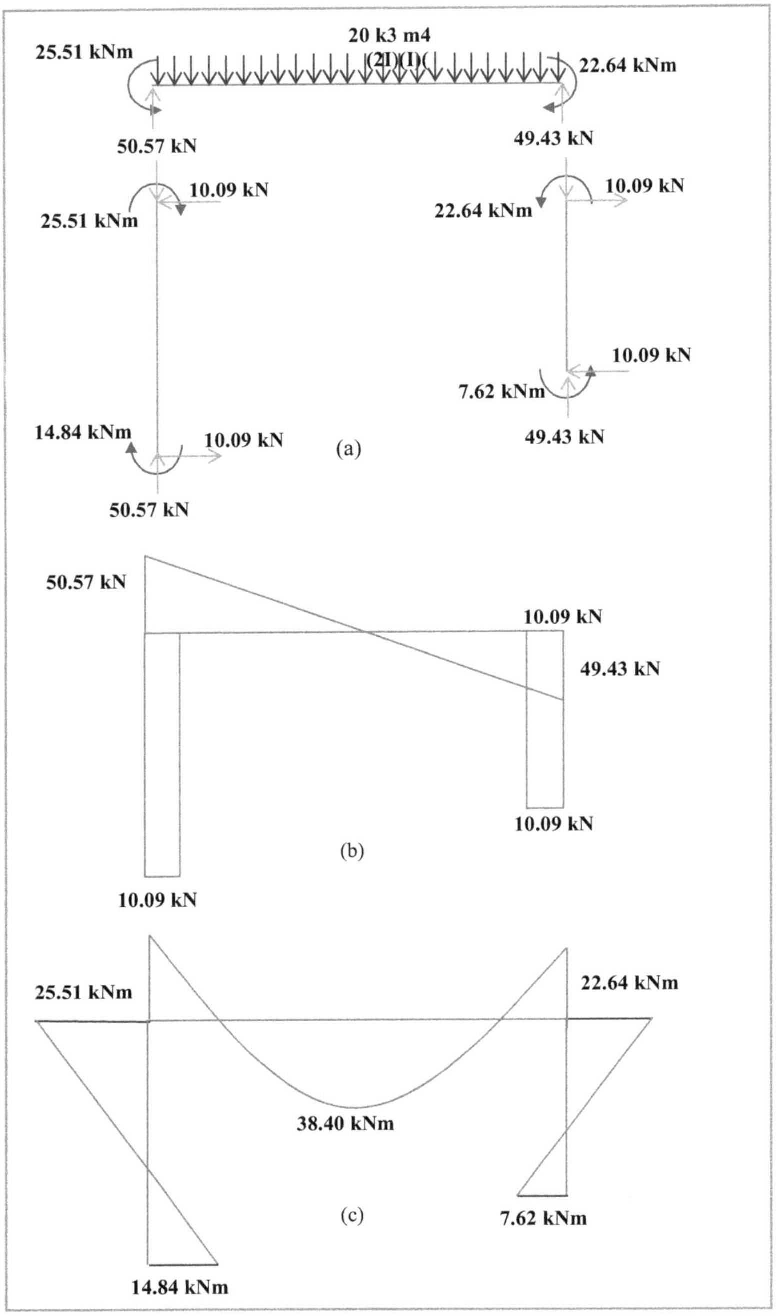

FIGURE 8.41 (a) Free body diagram (b) shear force diagram and (c) bending moment diagram.

8.6 PROBLEMS FOR PRACTICE

8.6.1 The primary unknowns in Kani's method are member end moments. Explain how this method considered to be a displacement method.

8.6.2 Explain the difference between Kani's method and the moment distribution method?

8.6.3 Is Kani's method an exact method or an approximate method? Discuss.

8.6.4 Explain the terms (a) rotation factor, (b) displacement factor, (c) storey shear, and (d) storey moment.

8.6.5 Explain the step-by-step procedure for the analysis of problems subjected to non-sway and sway analysis.

8.6.6 Analyse the three-span continuous beam *ABCD* shown in Figure 8.42 using Kani's method. Draw the shear force and bending moment diagrams.

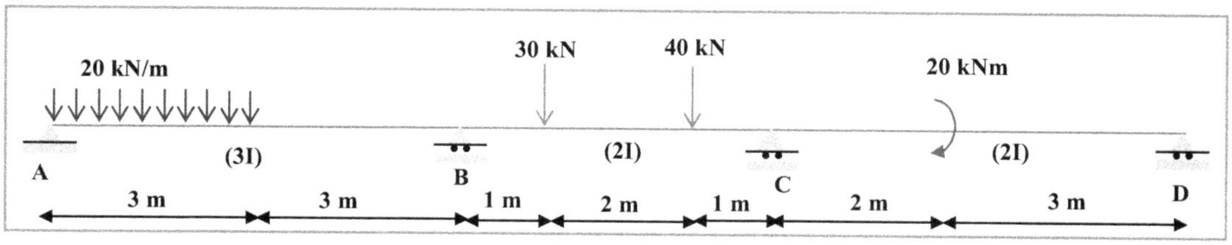

FIGURE 8.42

8.6.7 Re-analyse the beam shown in Figure 8.42, with end *A* as a fixed support, instead of a hinged support, using Kani's method. Construct the shear force and bending moment diagrams.

8.6.8 Re-analyse the beam shown in Figure 8.42, with end *A* as a guided support instead of a hinged support using Kani's method. Construct the shear force and bending moment diagrams.

8.6.9 Analyse the three-span continuous beam *ABCDE* with overhang, as shown in Figure 8.43, using Kani's method. Draw the shear force and bending moment diagrams. Assume *EI* is constant.

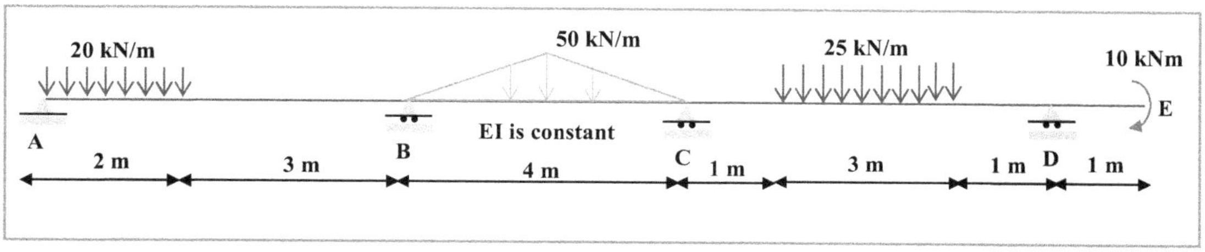

FIGURE 8.43

8.6.10 Using Kani's method, analyse the three-span continuous beam *ABCDE* with overhang, as shown in Figure 8.44, considering 5 and 10 mm settlements at supports *B* and *C* respectively. Draw the shear force and bending moment diagrams. Assume $EI = 138{,}375$ kNm^2.

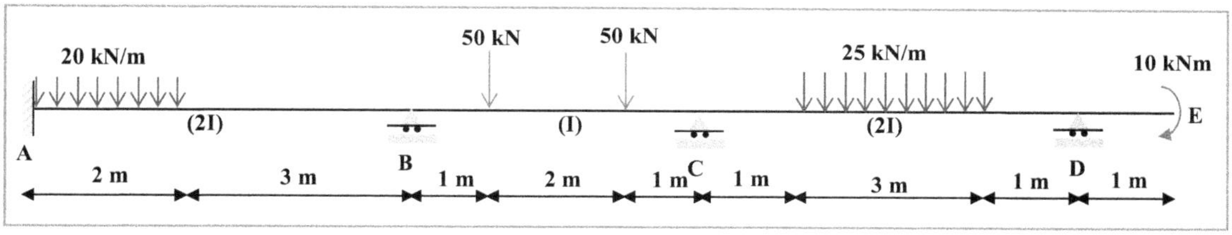

FIGURE 8.44

8.6.11 Re-analyse the beam shown in Figure 8.44 by neglecting the external loads and considering that support *A* rotates by 0.002 rad clockwise and support *C* settles by 10 mm. Assume $EI = 138{,}375$ kNm^2. Construct the shear force and bending moment diagrams.

8.6.12 Analyse the beam shown in Figure 8.45 using Kani's method, taking advantage of symmetry. Draw the shear force and bending moment diagrams.

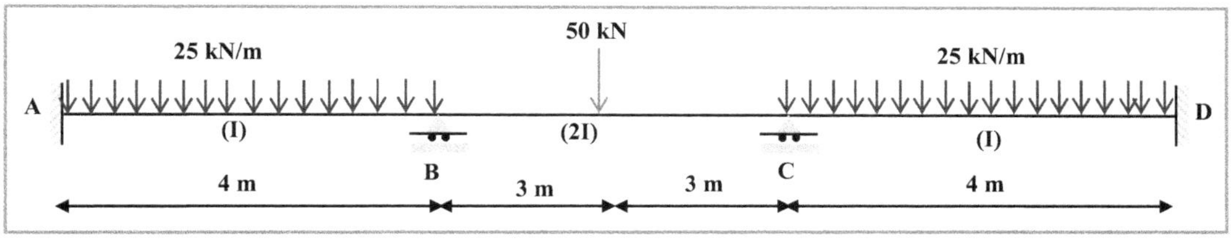

FIGURE 8.45

8.6.13 Analyse the rigid joint frame shown in Figure 8.46 using Kani's method, taking advantage of symmetry. Sketch the shear force and bending moment diagrams.

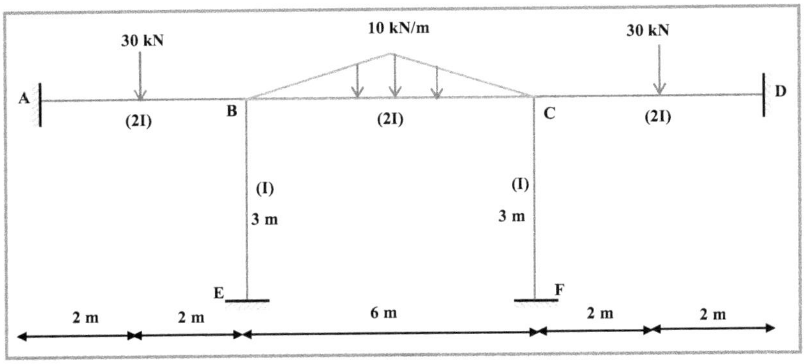

FIGURE 8.46

8.6.14 Analyse the rigid joint frame shown in Figure 8.47 using Kani's method. Sketch the shear force and bending moment diagrams.

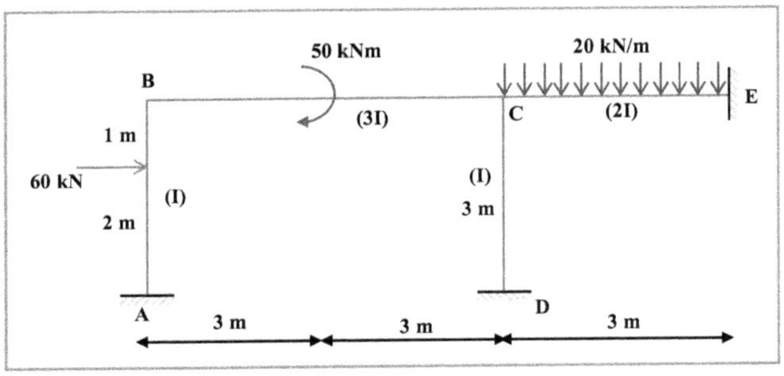

FIGURE 8.47

8.6.15 Analyse the multi-storey rigid joint plane frame shown in Figure 8.48 using Kani's method, taking advantage of symmetry. Draw the shear force and bending moment diagrams. Assume EI is constant.

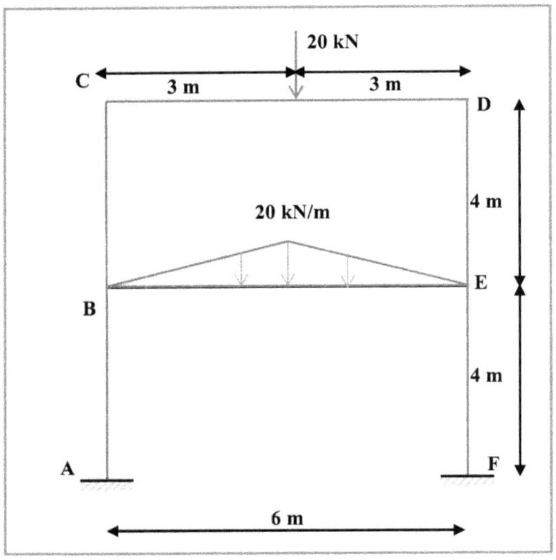

FIGURE 8.48

8.6.16 Analyse the rigid joint plane frame shown in Figure 8.49 subjected to horizontal loading using Kani's method. Sketch the shear force and bending moment diagrams.

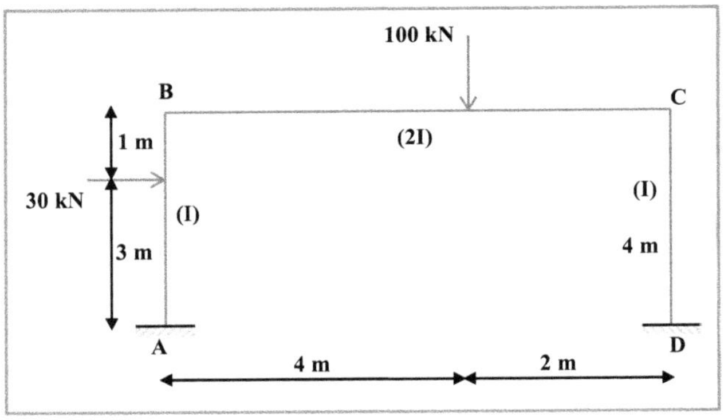

FIGURE 8.49

8.6.17 Analyse the multi-storey frame subjected to horizontal loading, as shown in Figure 8.50, using Kani's method. Sketch the shear force and bending moment diagrams.

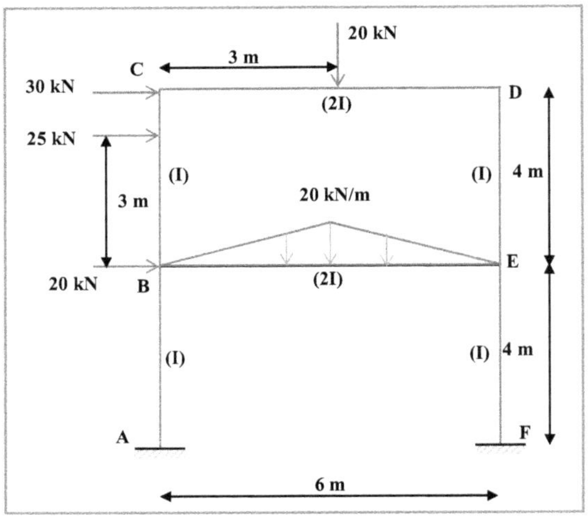

FIGURE 8.50

8.6.18 Analyse the multi-storey frame subjected to horizontal loading, as shown in Figure 8.51, using Kani's method. Draw the shear force and bending moment diagrams.

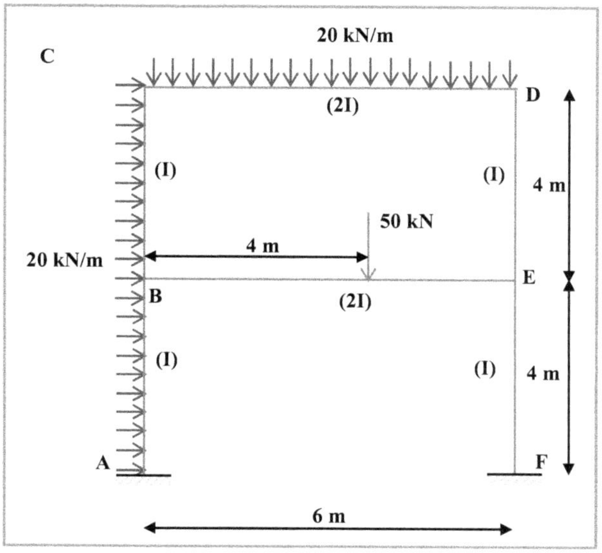

FIGURE 8.51

8.6.19 A rigid joint frame *ABCD* with unequal column heights, as shown in Figure 8.52. Analyse the frame using Kani's method. Sketch the shear force and bending moment diagrams.

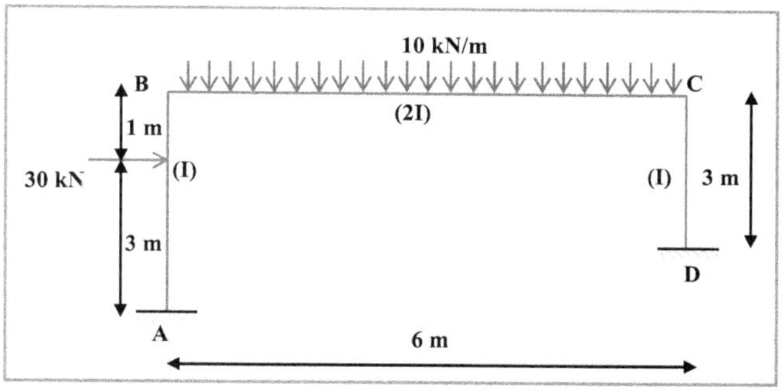

FIGURE 8.52

8.6.20 Analyse the rigid joint frame *ABCD* with hinged bases as shown in Figure 8.53 using Kani's method. Draw the shear force and bending moment diagrams.

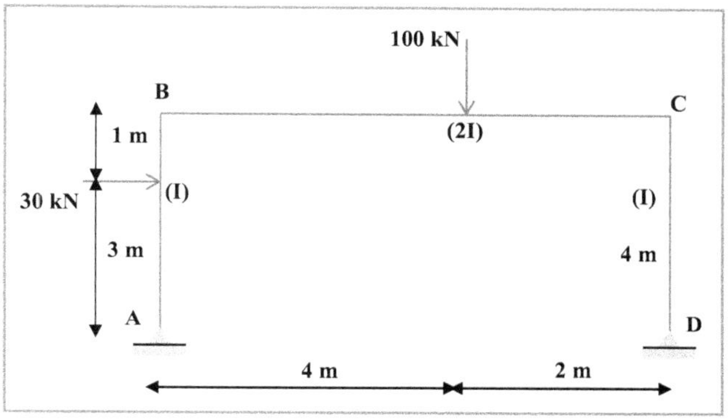

FIGURE 8.53

8.7 ANSWER KEY

Problem No.	Answer Key at Salient Locations
8.6.6	$M_{BA}=-M_{BC}=36.15$ kNm, $R_A=38.97$ kN, $R_B=59.97$ kN, $R_C=29.12$ kN, $R_D=1.93$ kN
8.6.7	$M_A=-37.51$ kNm, $M_{BA}=-M_{BC}=134.81$ kNm, $R_B=129.01$ kN, $R_C=-5.25$ kN, $R_D=6.25$ kN
8.6.8	$M_A=-135.19$ kNm, $M_{BA}=-M_{BC}=26.30$ kNm, $R_A=46.86$ kN, $R_B=49.08$ kN, $R_C=32.56$ kN, $R_D=1.49$ kN
8.6.9	$M_{BA}=-M_{BC}=26.12$ kNm, $R_A=26.78$ kN, $R_B=56.40$ kN, $R_C=103.01$ kN, $R_D=28.82$ kN
8.6.10	$M_A=-150.75$ kNm, $M_{BA}=-M_{BC}=82.25$ kNm, $R_A=45.69$ kN, $R_B=108.78$ kN, $R_C=-14.11$ kN, $R_D=74.63$ kN
8.6.11	$M_A=369.61$ kNm, $M_{BA}=-M_{BC}=77.26$ kNm, $R_A=-89.37$ kN, $R_B=114.56$ kN, $R_C=-29.89$ kN, $R_D=4.69$
8.6.12	$M_A=-32.01$ kNm, $M_B=-35.83$ kNm, $R_A=49.06$ kN, $R_B=75.93$ kN
8.6.13	$M_A=-14.46$ kNm, $M_E=0.69$ kNm, $R_A=14.46$ kN, $R_E=30.54$ kN
8.6.14	$M_A=-22.11$ kNm, $M_{BA}=9.76$ kNm$=M_{BC}$, $M_E=19.05$ kNm, $R_A=-7.52$ kN, $R_D=33.73$ kN
8.6.15	$M_A=7.36$ kNm$=M_D$, $R_A=40$ kN$=R_F$
8.6.16	$M_A=-13.33$ kNm, $M_D=-41.83$ kNm, $R_A=27.53$ kN, $R_D=72.47$ kN
8.6.17	$M_A=-81.40$ kNm, $M_F=-95.97$ kNm, $M_{CD}=36.43$ kNm$=-M_{CB}$, $R_A=-12.93$ kN, $R_F=92.93$ kN
8.6.18	$M_A=-158.58$ kNm, $M_F=-139.27$ kNm, $M_{CD}=5.36$ kNm$=-M_{CB}$, $R_A=19.64$ kN, $R_F=150.36$ kN
8.6.19	$M_A=-15.73$ kNm, $M_D=-31.77$ kNm, $R_A=26.59$ kN, $R_D=33.41$ kN
8.6.20	$M_{BA}=-5.20$ kNm$=-M_{BC}$, $R_A=18.33$ kN, $R_D=81.67$ kN

Note: The answers are provided based on the analysis using structural analysis software. The results from the exact solution may differ slightly (decimals), but these differences are generally negligible.

FURTHER READING

Bhavikatti, S. S. (2013). *Structural Analysis-II*, Vikas Publishing House, New Delhi, ISBN 978-9325968806.

Menon, D. (2023). *Structural Analysis*, Narosa Publishing House Pvt Ltd, New Delhi, ISBN: 978-8184877342.

Reddy, C. S. (2011). *Basic Structural Analysis* (3 E), Tata McGraw-Hill Education Pvt Ltd, New Delhi, ISBN 978-0070702769.

Thandavmoorthy, T. S. (2011). *Structural Analysis*, Oxford University Press, New York, ISBN 978-0198069188, 987-0198069189.

Vijayanand, M., Muthu, K. U., Narendra, H., and Janardhana, M (2019). *Indeterminate Structural Analysis*, Dreamtech Press Publisher, Delhi, ISBN 978-9389307177.

9 Matrix Methods
Flexibility Matrix Method

9.1 INTRODUCTION

The basic concepts of the force and displacement methods of analysis are discussed in Chapter 1. The force method, also known as the flexibility matrix method or compatibility method, is introduced there. The consistent deformation method, covered in detail in Chapter 3, is also a force method because the compatibility conditions (equations) are derived by superimposing external loads and redundant forces. Another reason it is called the force method is that the primary unknowns in this approach are forces. This method is sometimes referred to as the flexibility method because flexibility coefficients are developed to write the compatibility conditions. The flexibility and stiffness matrix methods provide systematic procedures for solving structures with a high degree of static and kinematic indeterminacy, respectively, making them well-suited for computer program development. The analysis of indeterminate beams, as well as rigid and pin-jointed plane frames, was discussed in Chapter 3, with various problems solved using the compatibility equations in the consistent deformation method. In this chapter, additional problems involving indeterminate beams, rigid-jointed, and pin-jointed plane frames are analysed using the matrix approach.

9.2 FLEXIBILITY MATRIX EQUATION

Consider a fixed beam AB subjected to some random loading, as shown in Figure 9.1a. The static indeterminacy of the beam is 02. Let us select the reactions at support B as redundants and assign co-ordinates to each redundant. Thus, R_1 and R_2 are the redundants at co-ordinates 1 and 2, respectively, as shown in Figure 9.1a. Remove the redundants from the original beam to get the basic determinate structure. The displacements at the co-ordinates, due to external loads and redundants in the basic determinate beam, are shown in Figure 9.1b and c. According to the principle of superposition, the net displacement at any point in the determinate structure equal to the sum of the displacements in the basic determinate structure due to external loads and redundants. Based on the compatibility condition, we can write

$$\left.\begin{aligned}\Delta_1 &= \Delta_{1L} + \Delta_{1R}\\ \Delta_2 &= \Delta_{2L} + \Delta_{2R}\end{aligned}\right\} \tag{9.1}$$

where

Δ_1, and Δ_2 are the displacements at the co-ordinates in the indeterminate or original structure or beam

Δ_{1L}, and Δ_{2L} are the displacements at the co-ordinates in the basic determinate structure due to external loads

Δ_{1R}, and Δ_{2R} are the displacements at the co-ordinates in the basic determinate structure due to redundants.

Equation 9.1 can be expressed in a matrix form as

$$[\Delta]=[\Delta_L]+[\Delta_R] \tag{9.2}$$

Apply unit forces corresponding to each redundant in the direction of the redundant in the basic determinate structure, and find the flexibility coefficients at the co-ordinates, as shown in Figure 9.1d and e.

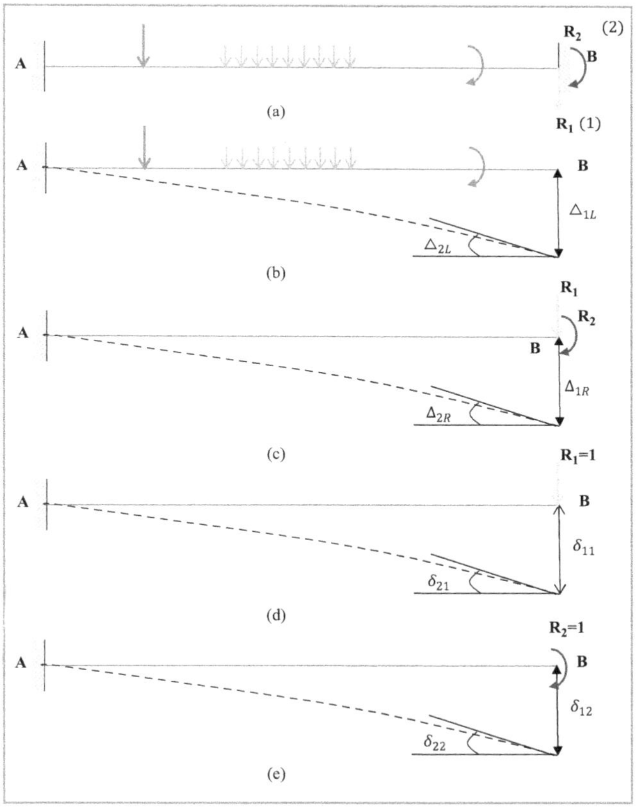

FIGURE 9.1 Fixed beam ($I_S=2$) subjected to external loads and redundants, and a primary beam subjected to loading and unit load at R_1 and R_2. (a) Indeterminate fixed beam with selected redundants. (b) Basic determinate beam with external loads. (c) Primary beam with redundant R_1 and R_2. (d) Primary beam with unit load at R_1 i.e. $R_1=1$. (e) Primary beam with unit load at R_2 i.e. $R_2=1$.

DOI: 10.1201/9781003562955-9

Using the principle of superposition, we can write

$$\left.\begin{aligned}\Delta_{1R} &= R_1\delta_{11} + R_2\delta_{12} \\ \Delta_{2R} &= R_1\delta_{21} + R_2\delta_{22} = 0\end{aligned}\right\} \quad (9.3)$$

where

$\delta_{11}, \delta_{12}, \delta_{21}$ and δ_{22}... are the flexibility coefficients due to unit loads at co-ordinates1 and 2

R_1 and R_2 are the redundants

Substituting equation 9.3 into equation 9.1,

$$\left.\begin{aligned}\Delta_1 &= \Delta_{1L} + R_1\delta_{11} + R_2\delta_{12} = 0 \\ \Delta_2 &= \Delta_{2L} + R_1\delta_{21} + R_2\delta_{22} = 0\end{aligned}\right\} \quad (9.4)$$

Equation 9.4 may be written in matrix form as

$$\begin{bmatrix} \Delta_1 \\ \Delta_2 \end{bmatrix} = \begin{bmatrix} \Delta_{1L} \\ \Delta_{2L} \end{bmatrix} + \begin{bmatrix} \delta_{11} & \delta_{12} \\ \delta_{21} & \delta_{22} \end{bmatrix} \begin{bmatrix} R_1 \\ R_2 \end{bmatrix} \quad (9.5)$$

or simply we can write

$$[\Delta] = [\Delta_L] + [\delta][R] \quad (9.6)$$

or

$$[R] = [\delta]^{-1}[[\Delta] - [\Delta_L]] \quad (9.7)$$

If the net displacements at the co-ordinates in the indeterminate or original structure is zero, i.e., $\Delta_1 = \Delta_2 = 0$, then equation 9.7 is reduced to

$$[R] = [\delta]^{-1}[-\Delta_L] \quad (9.8)$$

Equations 9.6–9.8 are called the flexibility matrix equations. These equations can be used for any degree of static indeterminacy. The applications of these equations to statically indeterminate beams, frames, and trusses are discussed in the subsequent sections.

9.2.1 Procedure to Solve the Problems

i. Determine the degree of static indeterminacy of the given structure and identify the redundant (unknown) reactions as R_1 R_2, R_3, ... etc., Assign the co-ordinates (1), (2), (3), ... etc., to each selected unknown reaction (redundant) in any chosen direction. While selecting the redundants, ensure that the structure remains stable.
ii. Release the selected redundants from the original structure by modifying the appropriate supporting conditions to make the structure statically determinate and stable. This modified structure is referred to as the **basic structure or released structure.**
iii. Next, develop the **elements of the flexibility matrix** $[\delta]$. The flexibility matrix is a square matrix of size equal to the degree of static indeterminacy. For instance, to develop the first column elements of the flexibility matrix, apply a unit force at co-ordinate (1) in its specified direction (without any force at other co-ordinates) on the released structure. Then, calculate the resulting displacements at all co-ordinates in the released structure. Repeat the process to develop the other columns elements of the flexibility matrix.
iv. Identify any displacements arising from the yielding of supports at the co-ordinates in the original structure and represent these as the **external displacements matrix** $[\Delta]$.
v. Formulate the **redundant force matrix** $[\boldsymbol{R}]$ for the unknown reactions.
vi. Substitute the matrices $[\delta]$, $[\Delta]$, and $[R]$ into the flexibility matrix equation (9.6) and solve for the unknown forces or redundants.
vii. Finally, after determining the redundants, calculate the remaining support reactions and member end forces using the static equilibrium equations, aided by the free body diagram.

9.3 APPLICATION TO BEAMS

9.3.1 Problems without Yielding of Supports

Example 9.1

Analyse the two-span continuous beam shown in Figure 9.2 using the flexibility matrix method. Assume EI is constant. Also, draw the SFD and BMD.

Solution:

The static indeterminacy, $I_S = 5 - 3 = 2$, indicates that the beam is statically indeterminate to the second degree. Select the moments at A and B as the redundants. The basic determinate beam is obtained by releasing these moments. This can be achieved by inserting hinges at supports A and B, allowing the spans AB and BC to behave as simply supported beams. The determinate structure, with co-ordinates corresponding to the redundants, is shown in Figure 9.3.

From equation 9.6 of the flexibility matrix,

$$[\Delta] = [\Delta_L] + [\delta][R]$$

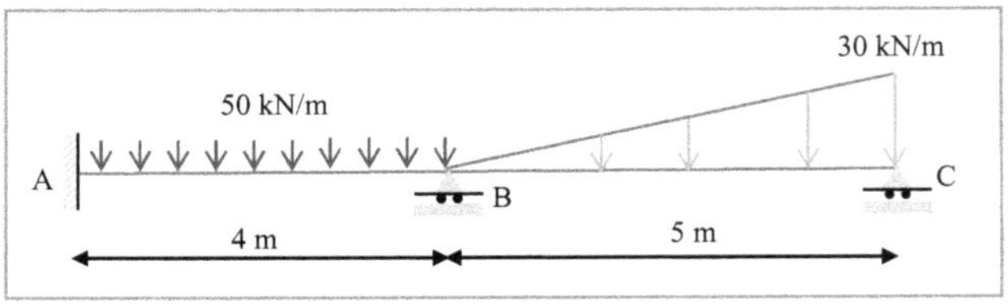

FIGURE 9.2 Two-span continuous beam.

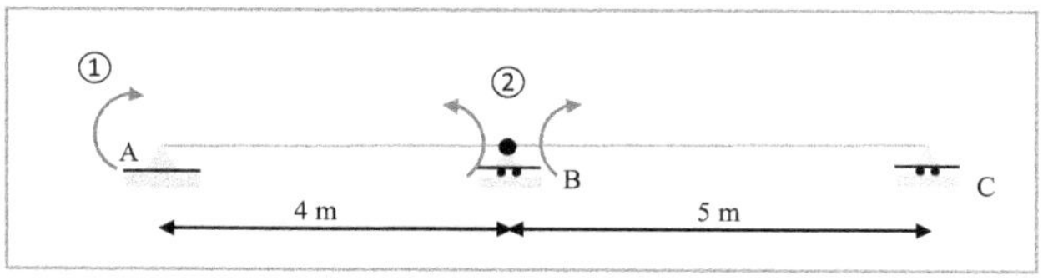

FIGURE 9.3 Basic or released structure with co-ordinates corresponding to the redundants.

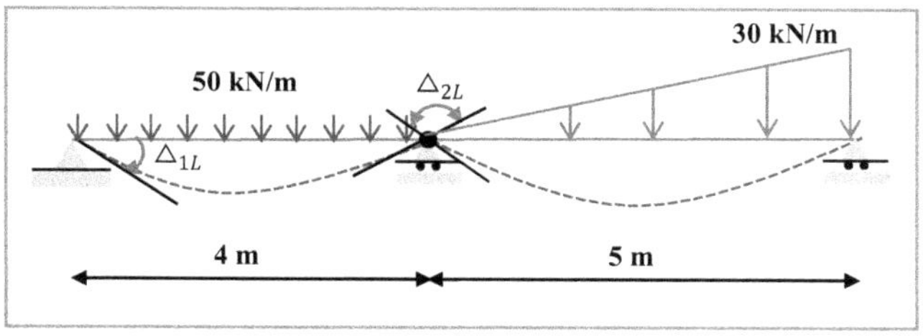

FIGURE 9.4 Displacements due to external loads at co-ordinates.

Let Us Calculate the Elements of Each Matrix

$[\Delta]=\begin{bmatrix}\Delta_1\\ \Delta_2\end{bmatrix}=\begin{bmatrix}0\\ 0\end{bmatrix}$ (∵for the continuity of the beam, the net flexural external displacements at co-ordinates 1 and 2 in the original beam must be zero)

$[\Delta_L]=\begin{bmatrix}\Delta_{1L}\\ \Delta_{2L}\end{bmatrix}$ are displacements due to loads corresponding to the co-ordinates ① and ② in the released or basic determinate beam. The deflection curve due to the external loads is shown in Figure 9.4. These displacements can be computed using the standard expressions provided in Table 3.2.

$$\Delta_{1L}=\frac{WL^3}{24EI}=\frac{50\times4^3}{24EI}=\frac{133.33}{EI}$$

$$\Delta_{2L}=\frac{wL^3}{24EI}+\frac{7wL^3}{360EI}=\frac{50\times4^3}{24EI}+\frac{70\times30\times5^3}{360EI}=\frac{206.25}{EI}$$

Flexibility Matrix $[\delta]=\begin{bmatrix}\delta_{11} & \delta_{12}\\ \delta_{21} & \delta_{22}\end{bmatrix}$: To develop the flexibility matrix, apply a unit force successively at co-ordinates 1 and 2 in the basic structure, and compute the resulting displacements (i.e., flexibility coefficients) at all co-ordinates.

First column elements of the flexibility matrix: To generate the first column elements of the flexibility matrix, apply a unit force (moment) at co-ordinate ① in its direction on the released beam. The resulting elastic curve due to this unit force at co-ordinate 1 is shown in Figure 9.5a. Calculate the displacements at all co-ordinates in the released structure using Table 3.2.

$$\delta_{11}=\frac{L}{3EI}=\frac{4}{3EI}$$

$$\delta_{21}=\frac{L}{6EI}=\frac{4}{6EI}$$

Second column elements of the flexibility of matrix: Similarly, to obtain the second column elements of the flexibility matrix, apply a unit force (moment) at co-ordinate ② in its specified direction on the released beam. The resulting deflected shape is shown in Figure 9.5b. Compute the displacements at all co-ordinates using Table 3.2.

$$\delta_{12}=\frac{L}{6EI}=\frac{4}{6EI}$$

$$\delta_{22}=\left(\frac{L}{3EI}\right)_{AB}+\left(\frac{L}{3EI}\right)_{BC}=\frac{4}{3EI}+\frac{5}{3EI}=\frac{3}{EI}$$

So the flexibility matrix $[\delta]=\frac{1}{EI}\begin{bmatrix}4/3 & 4/6\\ 4/6 & 3\end{bmatrix}$

Redundant force matrix $[R]=\begin{bmatrix}R_1\\ R_2\end{bmatrix}=\begin{bmatrix}M_A\\ M_B\end{bmatrix}$

Substituting the above into equation 9.6,

$$\begin{bmatrix}0\\ 0\end{bmatrix}=\begin{bmatrix}133.33/EI\\ 206.25/EI\end{bmatrix}+\frac{1}{EI}\begin{bmatrix}4/3 & 4/6\\ 4/6 & 3\end{bmatrix}\begin{bmatrix}M_A\\ M_B\end{bmatrix}$$

By solving the above, we obtain

$$\begin{bmatrix}M_A\\ M_B\end{bmatrix}=\begin{bmatrix}4/3 & 4/6\\ 4/6 & 3\end{bmatrix}^{-1}\begin{bmatrix}-133.33\\ -206.25\end{bmatrix}$$

$$=\frac{1}{3.56}\begin{bmatrix}3 & -4/6\\ -4/6 & 4/3\end{bmatrix}\begin{bmatrix}-133.33\\ -206.25\end{bmatrix}=\begin{bmatrix}-73.73\\ -52.27\end{bmatrix}\text{kNm}$$

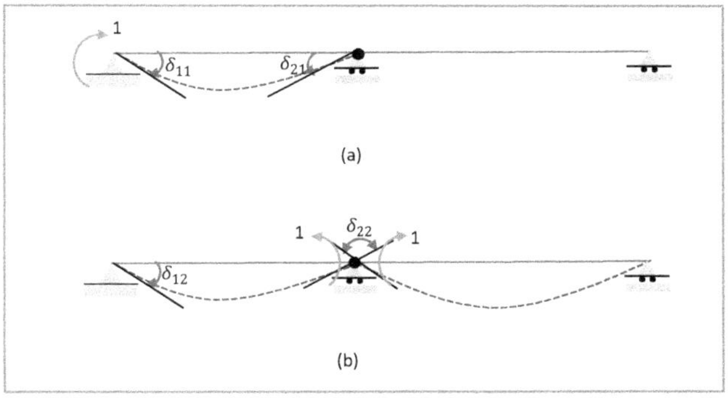

FIGURE 9.5 (a) Unit force (couple) applied at co-ordinate 1 and (b) unit couple at co-ordinate 2.

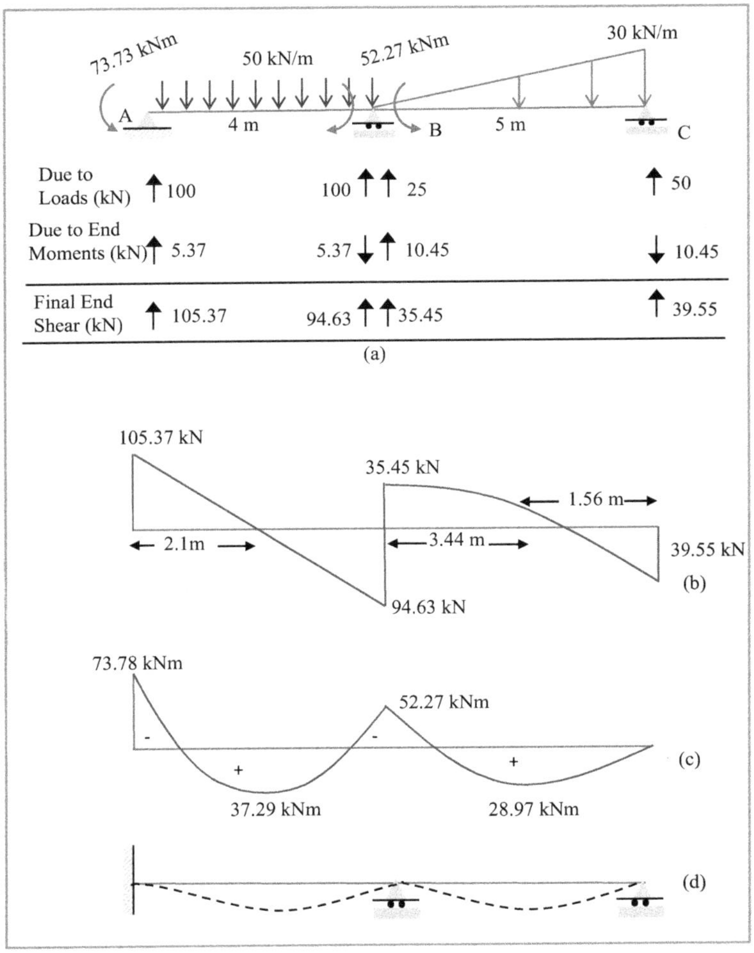

FIGURE 9.6 (a) Free body diagram, (b) shear force diagram, (c) bending moment diagram, and (d) probable deflected shape.

Other support reactions: With the known values of the moments at *A* and *B*, the remaining end reactions can be determined using the free body diagram (Figure 9.6a) and the static equilibrium equations. The shear force diagram, bending moment diagram and probable deflected shape are shown in Figure 9.6b–d, respectively.

Example 9.2

Analyse the continuous beam shown in Figure 9.7 using the Flexibility matrix method. Assume *EI* is constant.

Solution:

The static indeterminacy of the beam, $I_S = 5 - 3 = 2$, indicates that the beam is indeterminate to the second degree. Select the moments at *A* and *B* as the redundants, and assign co-ordinates 1 and 2, respectively, to them. The released beam can be obtained by inserting the hinges at *A* and *B*, allowing the spans *AB* and *BC* to behaves as simply supported beams. The released beam, with the co-ordinates corresponding to the redundants, is shown in Figure 9.8.

The equation of the flexibility matrix in equation 9.6 is

$$[\Delta] = [\Delta_L] + [\delta][R]$$

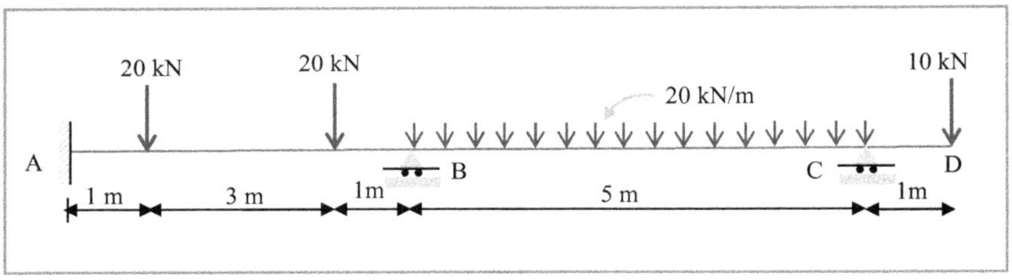

FIGURE 9.7 Continuous beam ABCD.

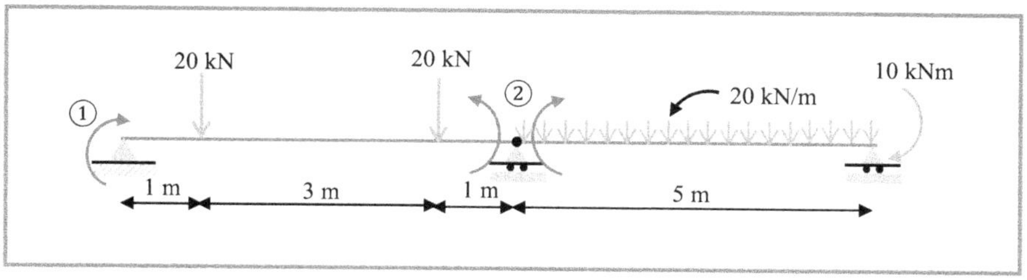

FIGURE 9.8 Released beam with co-ordinates and external loads.

where

$$[\Delta]=\begin{bmatrix}0\\0\end{bmatrix}$$ (since due to the continuity, the net displacements at A and B are zero)

$$[R]=\begin{bmatrix}M_A\\M_B\end{bmatrix}$$

$$[\delta]=\begin{bmatrix}\delta_{11} & \delta_{12}\\ \delta_{21} & \delta_{22}\end{bmatrix}$$ is the flexibility matrix. To develop this, a unit load is to be applied successively at the co-ordinates 1 and 2 on the released beam and compute the resulting displacements at all co-ordinates, as shown below.

First column elements of the flexibility matrix: To develop the first column elements of the flexibility matrix, apply a unit force (moment) at co-ordinate ① in its specified direction on the released beam (Figure 9.9a), and determine the resulting displacements at all co-ordinates on the released beam, using Table 3.2.

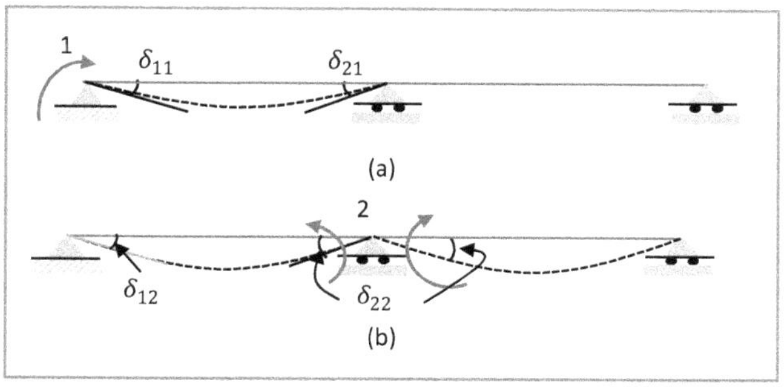

FIGURE 9.9 (a) Unit force (couple) applied at co-ordinate 1 and (b) unit couple at co-ordinate 2.

$$\delta_{11}=\frac{L}{3EI}=\frac{5}{3EI}$$

$$\delta_{21}=\frac{L}{6EI}=\frac{5}{6EI}$$

Second column elements of the flexibility of matrix: To generate the second column elements of the flexibility matrix, apply a unit force (moment) at co-ordinates ② in its specified direction on the basic determinate beam (Figure 9.9b), and find the displacements at all co-ordinates on the released beam using Table 3.2.

$$\delta_{12}=\frac{L}{6EI}=\frac{5}{6EI}=\delta_{21}$$

$$\delta_{22}=\left(\frac{L}{3EI}\right)_{AB}+\left(\frac{L}{3EI}\right)_{BC}$$

$$=\frac{5}{3EI}+\frac{5}{3EI}=\frac{10}{3EI}$$

$$[\delta]=\frac{1}{EI}\begin{bmatrix}5/3 & 5/6\\ 5/6 & 10/3\end{bmatrix}$$

$$[\Delta_L]=\begin{bmatrix}\Delta_{1L}\\ \Delta_{2L}\end{bmatrix}$$ are the displacements due to the external loads at co-ordinates on the released beam. The resulting deflected shape due to external loads is shown in Figure 9.10. The displacements due to the loads are calculated using the expressions provided in Table 3.2, as shown below.

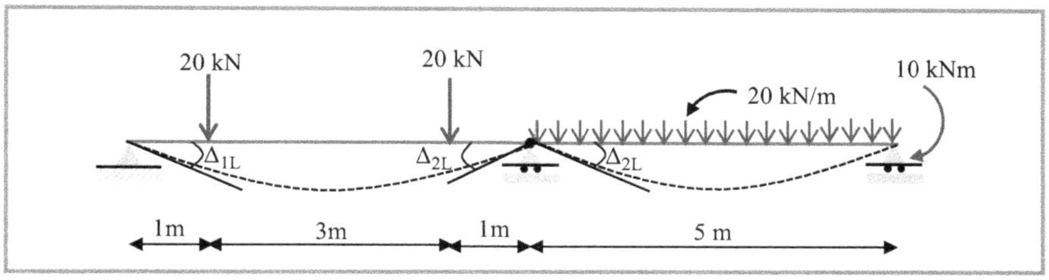

FIGURE 9.10 Displacements due to external loads at co-ordinates.

$$\Delta_{1L} = \frac{W_1 ab}{6EIL}(L+b) + \frac{W_2 ab}{6EIL}(L+b)$$

$$= \frac{20\times1\times4}{6EI\times5}(5+4) + \frac{20\times4\times1}{6EI\times5}(5+1)$$

$$= \frac{40}{EI}$$

$$\Delta_{2L} = \frac{40}{EI} + \frac{20\times5^3}{24EI} - \frac{10\times5}{6EI} = \frac{135.85}{EI}$$

$$[\Delta_L] = \begin{bmatrix} 40/EI \\ 135.83/EI \end{bmatrix}$$

Substituting the above values into the flexibility matrix equation, we obtain

$$\begin{bmatrix} 0 \\ 0 \end{bmatrix} = \begin{bmatrix} 40/EI \\ 135.83/EI \end{bmatrix} + \frac{1}{EI}\begin{bmatrix} 5/3 & 5/6 \\ 5/6 & 10/3 \end{bmatrix}\begin{bmatrix} M_A \\ M_B \end{bmatrix}$$

By solving the above, we obtain

$$\begin{bmatrix} M_A \\ M_B \end{bmatrix} = \begin{bmatrix} 5/3 & 5/6 \\ 5/6 & 10/3 \end{bmatrix}^{-1}\begin{bmatrix} -40 \\ -135.83 \end{bmatrix}$$

$$= \frac{1}{4.86}\begin{bmatrix} 10/3 & -5/6 \\ -5/6 & 5/3 \end{bmatrix}\begin{bmatrix} -40 \\ -135.83 \end{bmatrix}$$

$$= \begin{bmatrix} -4.14 \\ -39.72 \end{bmatrix}\text{kNm}$$

Final end shears: With the known values of M_A and M_B, the shear force at the supports is calculated using the static equilibrium equations with the help of the free body diagram, as shown in Figure 9.11a. The shear force, bending moment diagrams and probable deflected shape are shown in Figure 9.11b–d, respectively.

Example 9.3

Analyse the continuous beam *ABC* shown in Figure 9.12 using the flexibility matrix method. Assume *EI* is constant (the same has been solved using the strain energy method in Example 5.2).

Solution:

The beam is statically indeterminate to the second degree. Select the moments at *A* and *B* as redundants. The released beam is obtained by releasing these moments from the original beam. This can be achieved by inserting the hinges at the supports *A* and *B*, which allows the spans *AB* and *BC* to behave as simply supported beams. The released beam, with co-ordinates corresponding to the redundants, is shown in Figure 9.13.

The flexibility matrix equation 9.6 is

$$[\Delta] = [\Delta_L] + [\delta][R]$$

Since there are no yielding of supports and due to the continuity of the beam, the displacements in the indeterminate beam at co-ordinates are zero. That is

$$[\Delta] = \begin{bmatrix} 0 \\ 0 \end{bmatrix}$$

$[\Delta_L] = \begin{bmatrix} \Delta_{1L} \\ \Delta_{2L} \end{bmatrix}$ are the displacements due to the loads at co-ordinates on the released beam. The deflected shape of the released beam under the loads is shown in Figure 9.14. These displacements are calculated using Table 3.2, as shown below.

$$\Delta_{1L} = \frac{ML}{24EI} = \frac{10\times4}{24EI} = \frac{1.67}{EI}$$

$$\Delta_{2L} = \frac{ML}{24EI} + \frac{WL^3}{24EI} = \frac{10\times4}{24EI} + \frac{20\times4^3}{24EI} = \frac{55}{EI}$$

$$\begin{bmatrix} \Delta_{1L} \\ \Delta_{2L} \end{bmatrix} = \begin{bmatrix} \dfrac{1.67}{EI} \\ \dfrac{55}{EI} \end{bmatrix}$$

Flexibility Matrix $[\delta] = \begin{bmatrix} \delta_{11} & \delta_{12} \\ \delta_{21} & \delta_{22} \end{bmatrix}$: The elements of the flexibility matrix can be obtained by successively applying the unit forces at the co-ordinates on the released beam, as presented below.

First column elements of the flexibility matrix: To obtain the elements of the first column of the flexibility matrix, apply a unit force (couple) at co-ordinate ① in the direction of it on the released structure, and compute the resulting displacements at all co-ordinates, as shown in Figure 9.15a.

$$\delta_{12} = \frac{L}{3EI} = \frac{4}{3EI}$$

$$\delta_{22} = \frac{-L}{6EI} = \frac{-4}{6EI}$$

Second column elements: Apply a unit force at ② in its specified direction on the released beam, as shown in Figure 9.15b, and calculate the resulting displacements at all co-ordinates.

$$\delta_{21} = \frac{-L}{6EI} = \frac{-4}{6EI} = \delta_{12}$$

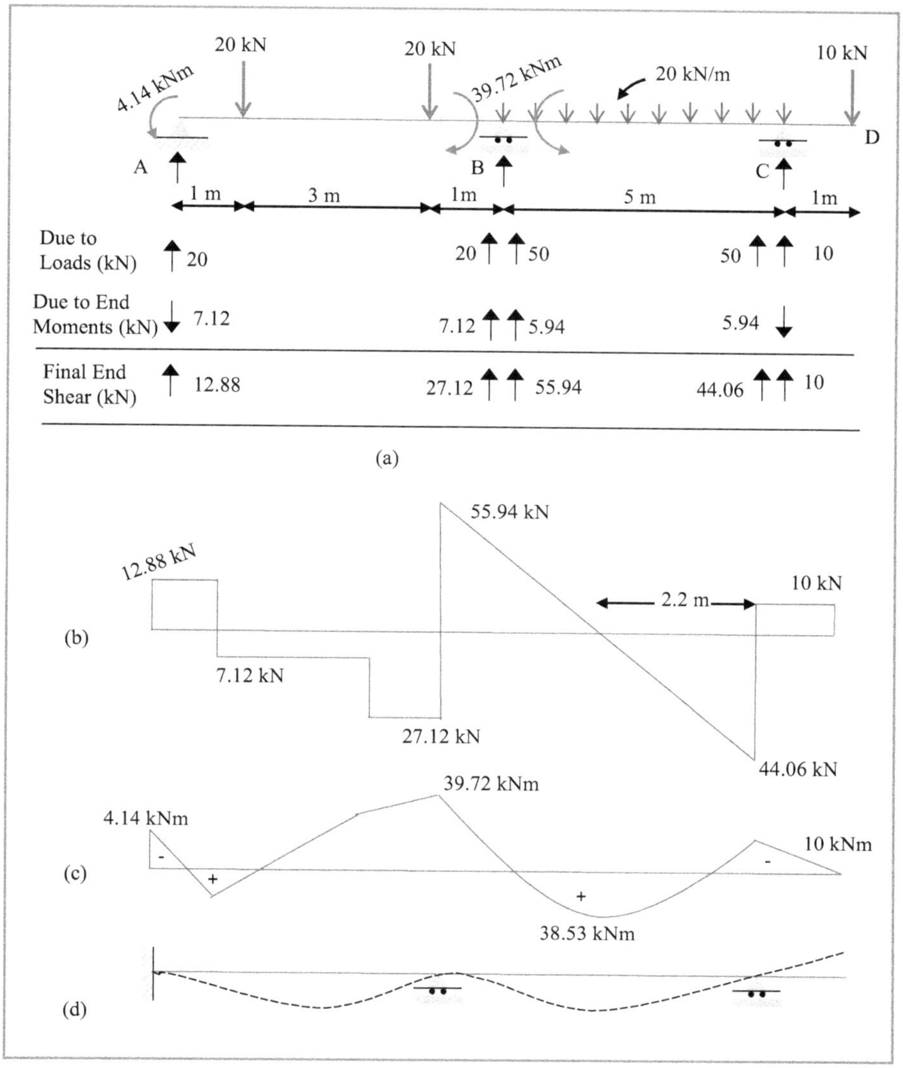

FIGURE 9.11 (a) Free body diagram, (b) shear force diagram (c) bending moment diagram, and (d) probable deflected shape.

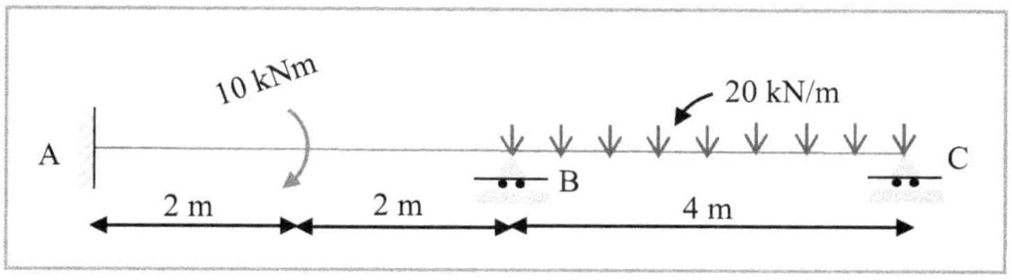

FIGURE 9.12 Two-span continuous beam.

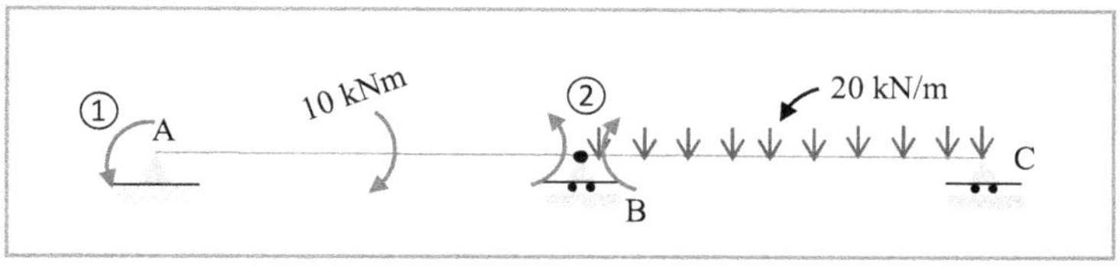

FIGURE 9.13 Released beam with co-ordinates and loads.

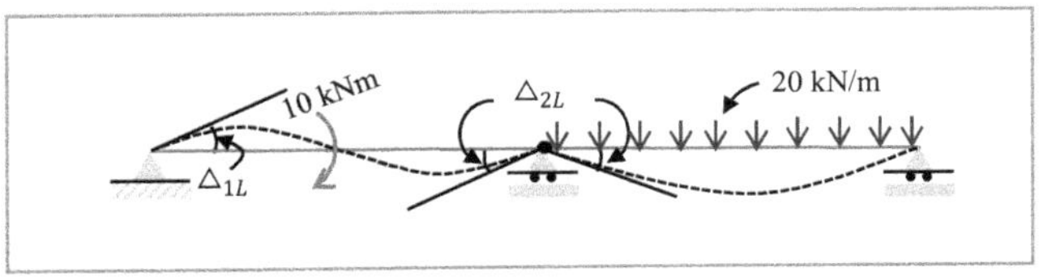

FIGURE 9.14 Displacement due to loads.

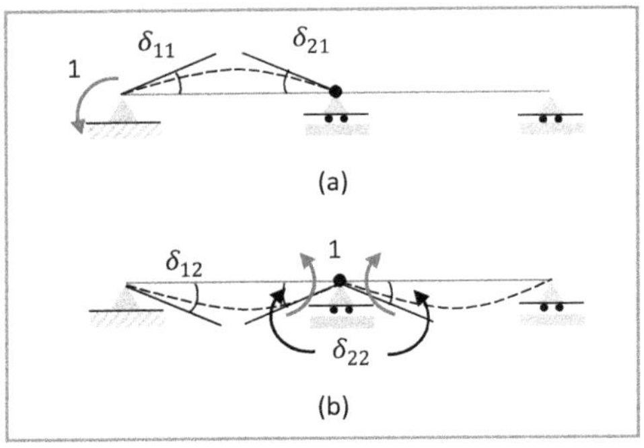

FIGURE 9.15 (a) Unit force at co-ordinate 1 and (b) unit force at co-ordinate 2.

$$\delta_{22} = \left(\frac{L}{3EI}\right)_{AB} + \left(\frac{L}{3EI}\right)_{BC}$$

$$= \frac{4}{3EI} + \frac{4}{3EI} = \frac{8}{3EI}$$

$$[\delta] = \frac{1}{EI}\begin{bmatrix} 4/3 & -4/6 \\ -4/6 & 8/3 \end{bmatrix}$$

Substituting all the above values into equation 9.6,

$$\begin{bmatrix} 0 \\ 0 \end{bmatrix} = \begin{bmatrix} 1.67/EI \\ 55/EI \end{bmatrix} + \frac{1}{EI}\begin{bmatrix} 4/3 & -4/6 \\ -4/6 & 8/3 \end{bmatrix}\begin{bmatrix} M_A \\ M_B \end{bmatrix}$$

By solving the above equation, we obtain

$$\begin{bmatrix} M_A \\ M_B \end{bmatrix} = \begin{bmatrix} -13.22 \\ -23.93 \end{bmatrix} \text{kNm}$$

Final end shears: The free body diagram is shown in Figure 9.16a. With the known values of the moments at *A* and *B*, using the static equilibrium equations the other support reactions can be determined. The shear force and bending moment diagrams and probable deflected shapes are shown in Figure 9.16b–d, respectively.

Example 9.4

A symmetrical three-span continuous beam is shown in Figure 9.17. Find the support reactions, and hence draw the shear force and bending moment diagrams.

Solution:

The beam is symmetric with respect to the centre of span *BC*. Therefore, the slope at the centre of span *BC* is zero, resulting in zero shear force at that point. To simulate this condition, a sliding guided support can be introduced at the centre of *BC*, reducing the static indeterminacy to one, as shown in Figure 9.18a. Let us select the moment at B^I as redundant, and assign it co-ordinate 1. The released beam, with the loads and co-ordinate corresponding to the redundant, is shown in Figure 9.18b.

Equation 9.6 of the flexibility matrix is

$$[\Delta] = [\Delta_L] + [\delta][R]$$

$$[\Delta] = [0] \;(\text{Due to symmetry the slope at } B^I \text{ is zero.})$$

Δ_{1L} = slope at B^I due to loads on the released beam. Let us calculate this using the unit load method and by applying the area moment multiplication technique. The M_L and m_1 diagrams are shown in Figure 9.19.

$$\Delta_{1L} = \int M_L m_1 \frac{dx}{EI}$$

$$= \frac{1}{EI}\left(\frac{2}{3}\times 100\times 4\times\frac{1}{2}\right)$$

$$+\frac{1}{EI}\left[\frac{4}{6}(-150\times 2\times 1) + \frac{1}{2EI}\left(\frac{1}{2}\times(-150)\times 3\times 1\right)\right]$$

$$= \frac{-179.16}{EI}$$

$$\delta_{11} = \int m_1 m_1 \frac{dx}{EI} = \frac{1}{EI}\left[\frac{4}{6}\times(1\times 2\times 1)\right] + \frac{1}{2EI}(1\times 3\times 1) = \frac{2.83}{EI}$$

By substituting the above into equation 9.6

$$0 = \frac{-179.16}{EI} + \frac{2.83}{EI}\times M_{B^I} \Rightarrow M_{B^I} = \frac{179.16}{2.83} = 63.3\,kNm$$

Support reactions/end shears: After knowing the moment at B^1, using the static equilibrium equations and with the help of the free body diagram, the other support reactions/end shears are calculated, as shown in Figure 9.20a. The shear force and bending moment diagrams are shown in Figure 9.20b and c, respectively.

Example 9.5

Analyse the continuous beam ***ABCD*** with an internal hinge at ***C***, as shown in Figure 9.21. Also, draw the shear force and bending moment diagrams. Assume *EI* is constant.

Solution:

The static indeterminacy, $I_S = 5-(2+1) = 2$. Select the moments at *A* and *D* as redundants. The released beam can be obtained by introducing the hinge supports at *A* and *B*, as shown in Figure 9.22.

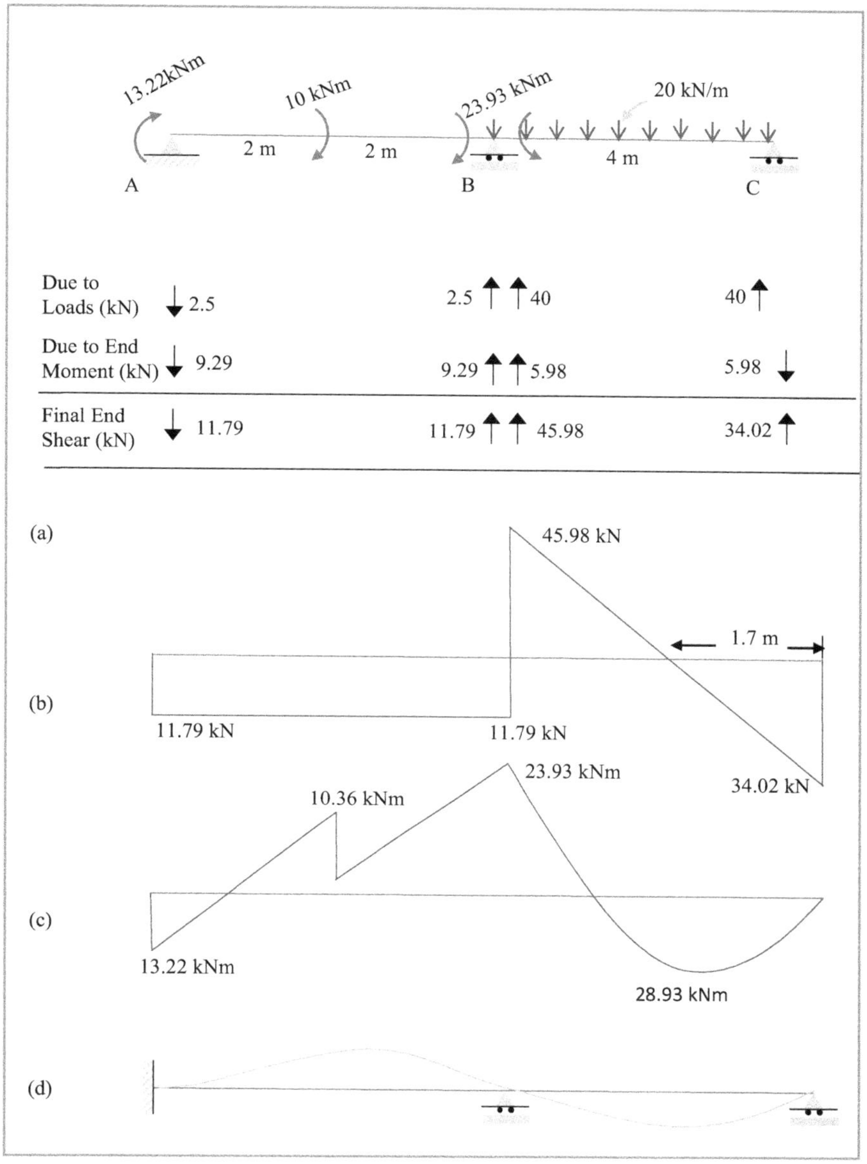

FIGURE 9.16 (a) Free body diagram, (b) shear force diagram, (c) bending moment diagram, and (d) probable deflected shape.

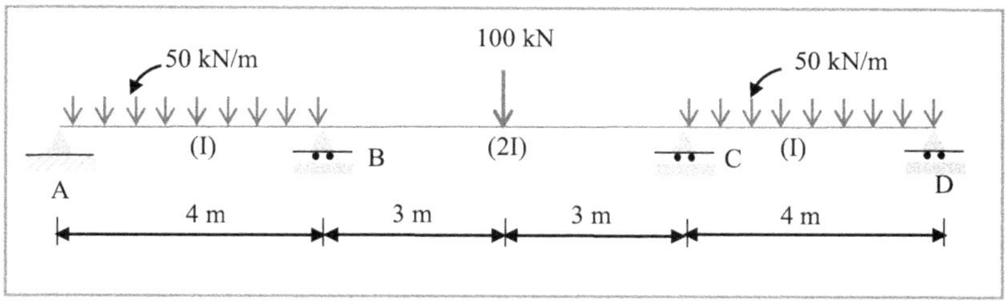

FIGURE 9.17 Symmetrical three-span continuous beam.

The co-ordinates assigned to the redundants are also shown in the figure.

Equation 9.6 of the flexibility matrix is

$$[\Delta]=[\Delta_L]+[\delta][R]$$

where $[\Delta]=\begin{bmatrix}\Delta_1\\\Delta_2\end{bmatrix}=\begin{bmatrix}0\\0\end{bmatrix}$, as there are no rotations at *A* and *D* in the continuous beam.

$[\Delta_L]$ and $[\delta]$ are can be calculated using the unit load method.

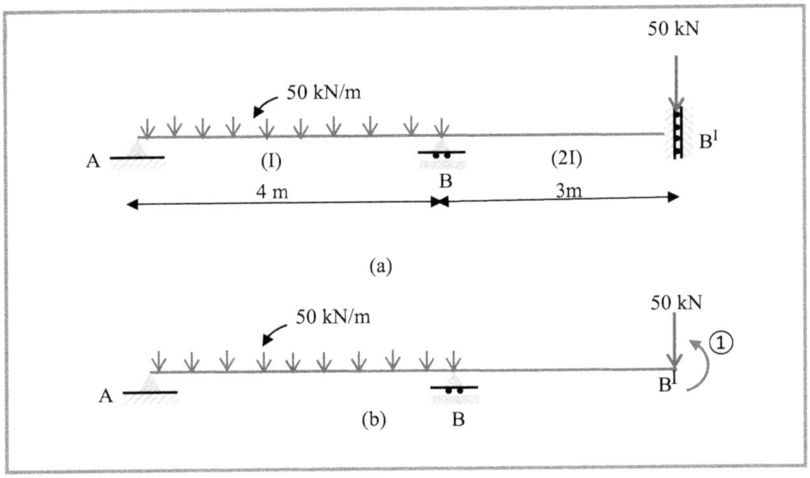

FIGURE 9.18 (a) Half portion of the beam ABB^I due to symmetry and (b) released beam ABB^1 with co-ordinates and loads.

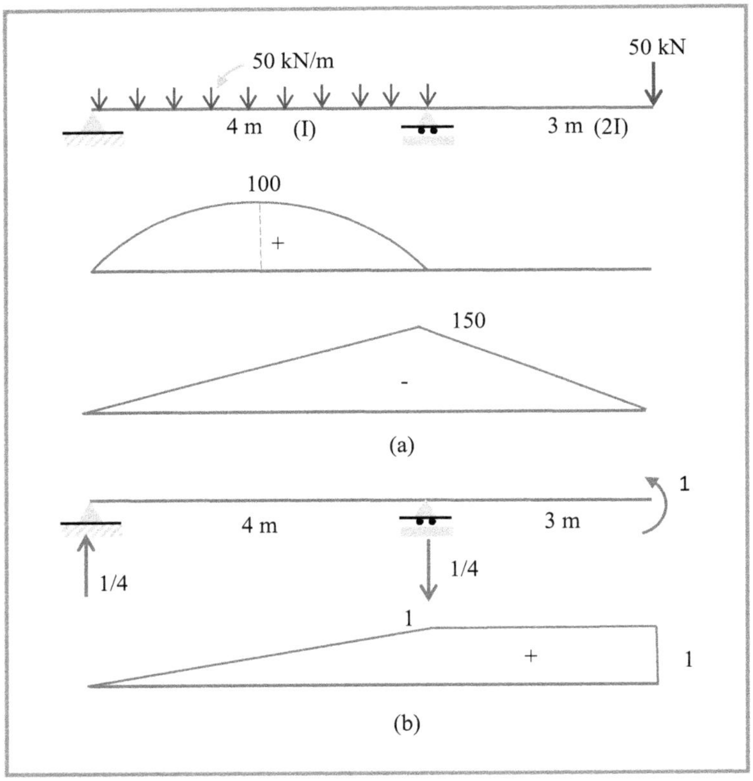

FIGURE 9.19 (a) M_L diagram due to loads and (b) m_1 diagram due to unit force at ①.

$$[\Delta_L] = \begin{bmatrix} \Delta_{1L} \\ \Delta_{2L} \end{bmatrix}$$

$$\Delta_{1L} = \int M_L m_1 \frac{dx}{EI}$$

$$\Delta_{2L} = \int M_L m_2 \frac{dx}{EI}$$

where M_L is the bending moment due to loads, and m_1 and m_2 are the bending moments due to unit forces (moments) applied at the co-ordinates ① and ②, respectively, as shown in Figure 9.23.

Similarly, the displacements (flexibility coefficient) due to unit forces are also calculated using the unit load method and are presented in Table 9.1.

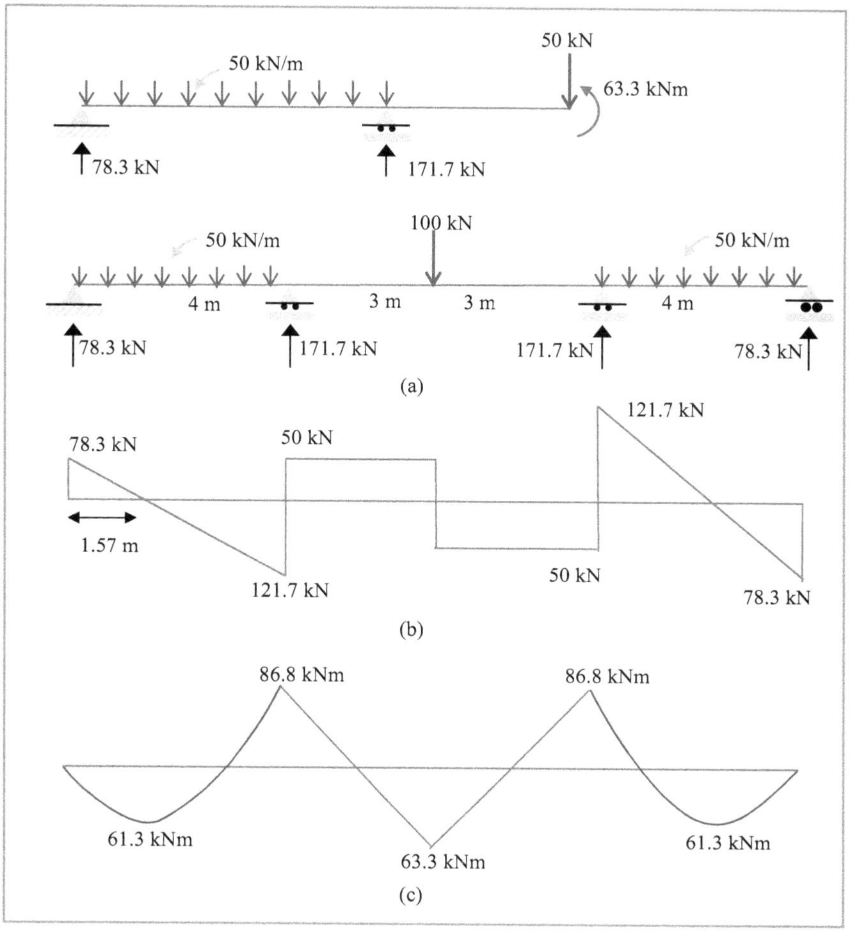

FIGURE 9.20 (a) Free body diagram, (b) shear force, and (c) bending moment diagram.

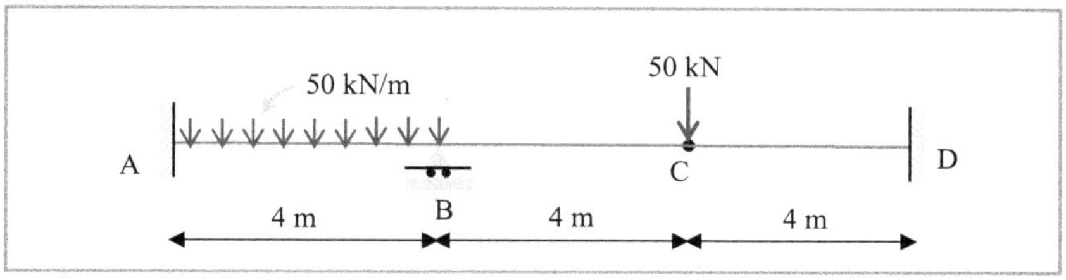

FIGURE 9.21 Continuous beam *ABCD* with internal hinge at *C*.

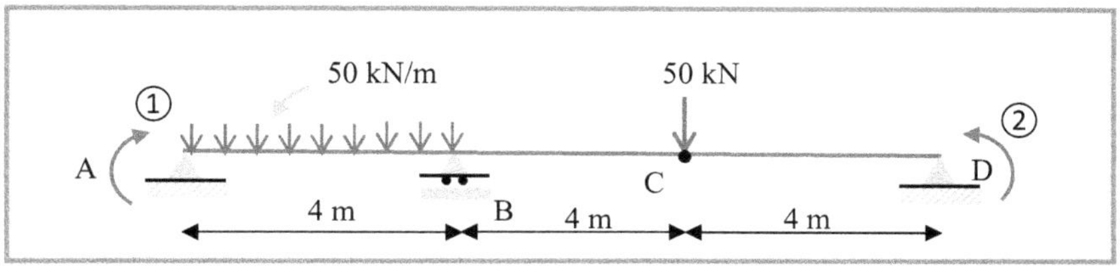

FIGURE 9.22 Released beam with loads and co-ordinates.

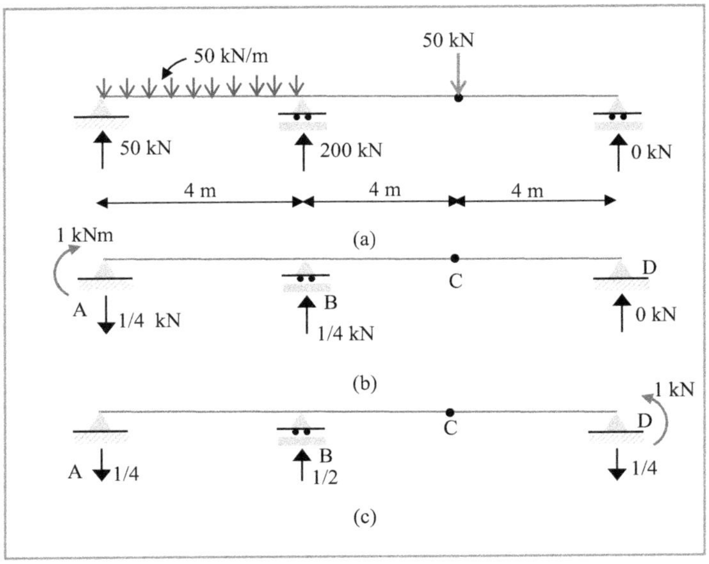

FIGURE 9.23 Free body diagrams due to (a) external loads, (b) unit force (couple) applied at co-ordinate ①, and (c) unit force (couple) applied at co-ordinate ② on the released beam.

TABLE 9.1
Moment Expressions Due to External Loads and Unit Forces

Segment/FBD	Limits (m)	Moment (kNm) Due to Loads	Unit Force at ①	Unit Force at ②
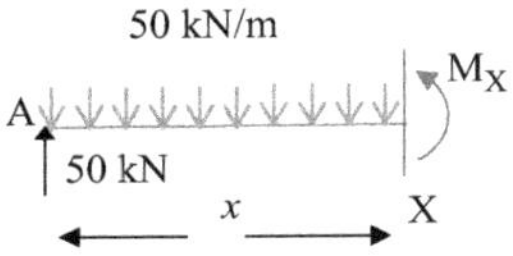	0–4	$50x^2 - 25x^2$		
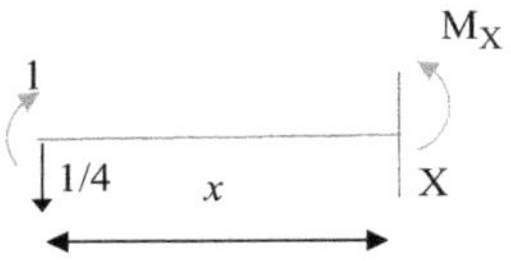	0–4		$1-\frac{x}{4}$	
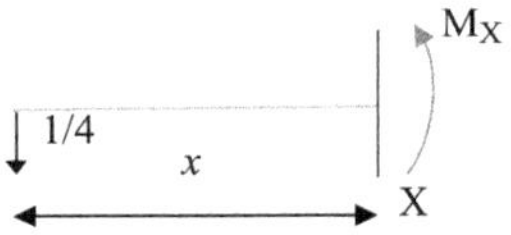	0–4			$-\frac{x}{4}$
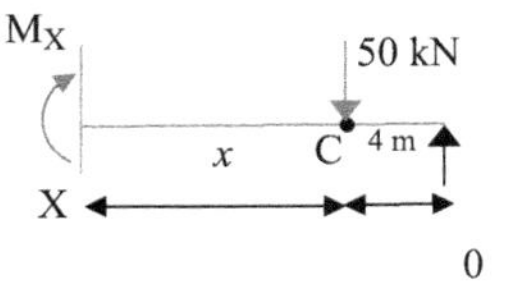	0–4	$-50x$		
BC 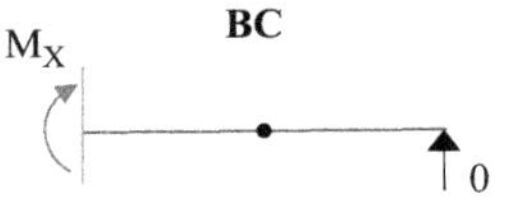	0–4		0	
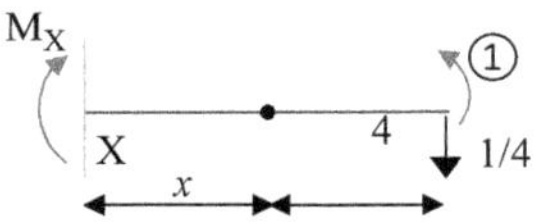	0–4			$1-\frac{1}{4}(x+4)=\frac{-x}{4}$
CD 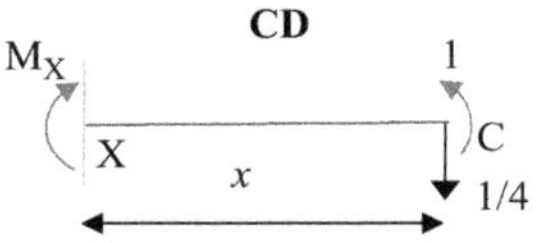	0–4	0	0	$1-\frac{x}{4}$

$$\Delta_{1L}=\int_0^4 (50x-25x^2)\left(1-\frac{x}{4}\right)\frac{dx}{EI}$$

$$=\frac{1}{EI}\left[50\frac{x^2}{2}-25\frac{x^3}{3}-50\frac{x^3}{12}+25\frac{x^4}{16}\right]_0^4=0$$

$$\Delta_{2L}=\int_0^4 (50x-25x^2)\left(-\frac{x}{4}\right)\frac{dx}{EI}+\int_0^4 (-50x)\left(-\frac{x}{4}\right)\frac{dx}{EI}$$

$$=\frac{1}{EI}\left[(-50)\frac{x^3}{12}+25\frac{x^4}{16}\right]_0^4+\frac{1}{EI}\left[50\frac{x^3}{12}\right]_0^4$$

$$=\frac{400}{EI}$$

$$\delta_{11}=\int_0^4 \left(1-\frac{x}{4}\right)\left(1-\frac{x}{4}\right)\frac{dx}{EI}=\frac{1}{EI}\left[x-\frac{x^2}{4}+\frac{x^3}{48}\right]_0^4=\frac{1.33}{EI}$$

$$\delta_{12}=\delta_{21}=\int_0^4 \left(1-\frac{x}{4}\right)\left(-\frac{x}{4}\right)\frac{dx}{EI}=\frac{1}{EI}\left[-\frac{x^2}{8}+\frac{x^3}{48}\right]_0^4=\frac{-0.67}{EI}$$

$$\delta_{22}=2\int_0^4 \left(-\frac{x}{4}\right)\left(-\frac{x}{4}\right)\frac{dx}{EI}+\int_0^4 \left(1-\frac{x}{4}\right)\left(1-\frac{x}{4}\right)\frac{dx}{EI}$$

$$=\frac{2}{EI}\left[\frac{x^3}{48}\right]_0^4+\frac{1.33}{EI}=\frac{4}{EI}$$

Substituting into equation 9.6,

$$\begin{bmatrix}0\\0\end{bmatrix}=\begin{bmatrix}-266.67/EI\\400/EI\end{bmatrix}+\frac{1}{EI}\begin{bmatrix}1.33 & -0.67\\-0.67 & 4\end{bmatrix}\begin{bmatrix}M_A\\M_B\end{bmatrix}$$

By solving the above equation, we obtain

$$\begin{bmatrix}M_A\\M_B\end{bmatrix}=\begin{bmatrix}-55\\-109.22\end{bmatrix}\text{kNm}$$

Other support reactions: With the known values of the moments at A and D, using the static equilibrium equations, the remaining support reactions can be determined, as shown in Figure 9.24a. The shear force and bending moment diagrams and probable deflected shape are shown in Figure 9.24b–d, respectively.

Example 9.6

Analyse the non-prismatic beam shown in Figure 9.25 using the flexibility matrix method. Draw the shear force and bending moment diagrams.

Solution:

The beam is statically indeterminate to the second degree. Let us select R_B and M_B as redundants, and assign the co-ordinates 1 and 2, respectively, to them. The released beam is obtained by releasing these redundants, i.e., by removing the fixed support at D.

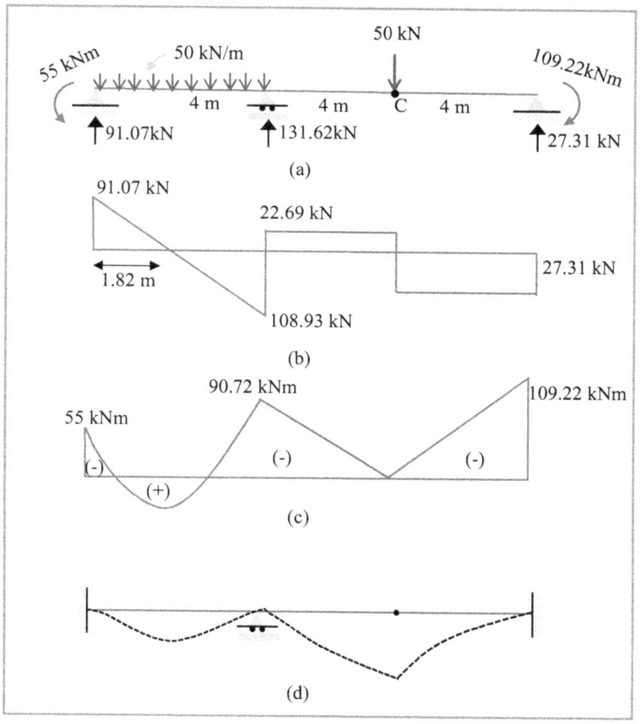

FIGURE 9.24 (a) Free body diagram, (b) shear force diagram, (c) bending moment diagram, and (d) probable deflected shape.

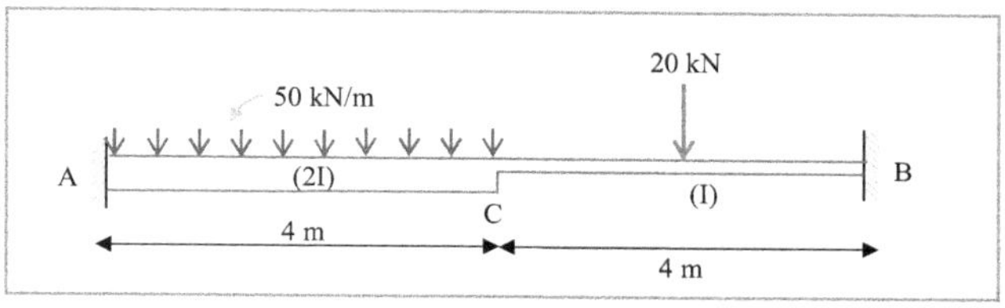

FIGURE 9.25 Indeterminate non-prismatic beam *ABC*.

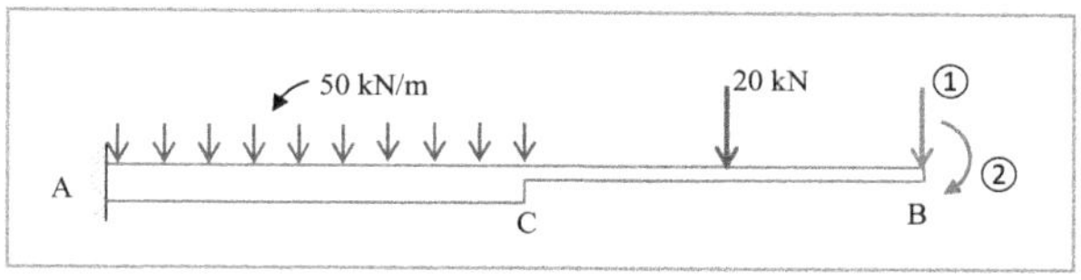

FIGURE 9.26 Released beam with loads and co-ordinates.

The basic determinate beam with the applied loads and co-ordinates corresponding to the redundants is shown in Figure 9.26.

The flexibility matrix equation 9.6 is

$$[\Delta]=[\Delta_L]+[\delta][R]$$

where $[\Delta]=\begin{bmatrix}\Delta_1\\\Delta_2\end{bmatrix}=\begin{bmatrix}0\\0\end{bmatrix}$, as there is no yielding of supports.

$$[R]=\begin{bmatrix}R_B\\M_B\end{bmatrix}$$

$[\Delta_L]$ and $[\delta]$ can be calculated using the unit load method and the area moment multiplication technique. The M_L diagram due to the applied loads, as well as the m_1 and m_2 diagrams resulting from the unit forces applied at co-ordinates 1 and 2, respectively, are shown in Figure 9.27.

$$\Delta_{1L} = \int M_L m_1 \frac{dx}{EI}$$

$$\Delta_{2L} = \int M_L m_2 \frac{dx}{EI}$$

$$\Delta_{1L} = \int M_L m_1 \frac{dx}{EI}$$

$$= \frac{-1}{2EI}\begin{bmatrix} \frac{1}{3}\times 4\times 400\times 7+\frac{4}{6} \\ \times\left[120(2\times 8+4)+40(2\times 4+8)\right] \end{bmatrix}$$

$$+\frac{2}{6EI}\left[40(2\times 4+2)\right]$$

$$= \frac{3{,}013.33}{EI}$$

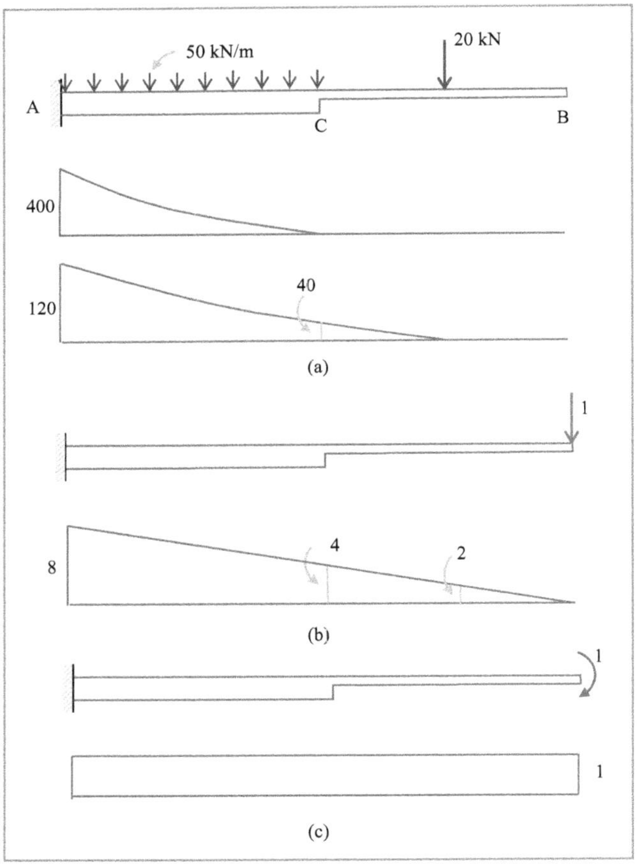

FIGURE 9.27 M_L, m_1, and m_2 diagrams. (a) M_L diagrams due to loads. (b) m_1 diagram due to unit force at coordinate 1. (c) m_2 diagram due to unit force at coordinate 2.

$$\Delta_{2L} = \frac{1}{2EI}\left[\frac{1}{3}\times 4\times 400\times 1+\frac{4}{2}(120+40)\times 1\right]$$

$$+\frac{1}{EI}\left(\frac{1}{2}\times 2\times 40\times 1\right)$$

$$= \frac{466.67}{EI}$$

$$\delta_{11} = \int m_1 m_1 \frac{dx}{EI} = \frac{1}{2EI}\left[\frac{4}{6}\{8(2\times 8+4)+4(2\times 4+8)\}\right]$$

$$+\frac{1}{EI}\left(\frac{1}{2}\times 4\times 4\times 2.67\right)$$

$$= \frac{96.02}{EI}$$

$$\delta_{12} = \delta_{21} = \int m_1 m_2 \frac{dx}{EI} = \frac{1}{2EI}\left[\frac{4}{2}(8+4)\times 1\right]$$

$$+\frac{1}{EI}\left[\frac{1}{2}\times 4\times 4\times 1\right]$$

$$= \frac{20}{EI}$$

$$\delta_{22} = \frac{1}{2EI}[1\times 4\times 1]+\frac{1}{EI}[1\times 4\times 1] = \frac{6}{EI}$$

On substituting the above values into equation 9.6,

$$\begin{bmatrix} 0 \\ 0 \end{bmatrix} = \begin{bmatrix} 3{,}013.33/EI \\ 466.67/EI \end{bmatrix} + \frac{1}{EI}\begin{bmatrix} 96.02 & 20 \\ 20 & 6 \end{bmatrix}\begin{bmatrix} R_B \\ M_B \end{bmatrix}$$

$$\Rightarrow \begin{bmatrix} R_B \\ M_B \end{bmatrix} = \begin{bmatrix} -49.67\text{ kN} \\ 87.76\text{ kNm} \end{bmatrix}$$

Other support reactions: The free body diagram shows the other support reactions in Figure 9.28a. The shear force and bending moment diagrams and the probable deflected shape are shown in Figure 9.28b–d, respectively.

9.3.2 Problems with Yielding of Supports

In the previous problems, to ensure the continuity of the structure, the displacements $[\triangle]$ at co-ordinates 1, 2, …, i assigned to the redundants were considered zero. However, if the supports yield, the displacements becomes known quantities at co-ordinates $i+1$, $i+2$, … assigned to the redundants. Therefore, in the case of yielding of supports, the $[\triangle]$ matrix can be written as

$$[\Delta] = \begin{bmatrix} \Delta_1 = 0 \\ \Delta_2 = 0 \\ \vdots \\ \Delta_i = 0 \\ \Delta_{i+1} \\ \Delta_{i+2} \\ \vdots \\ \Delta_n \end{bmatrix}$$

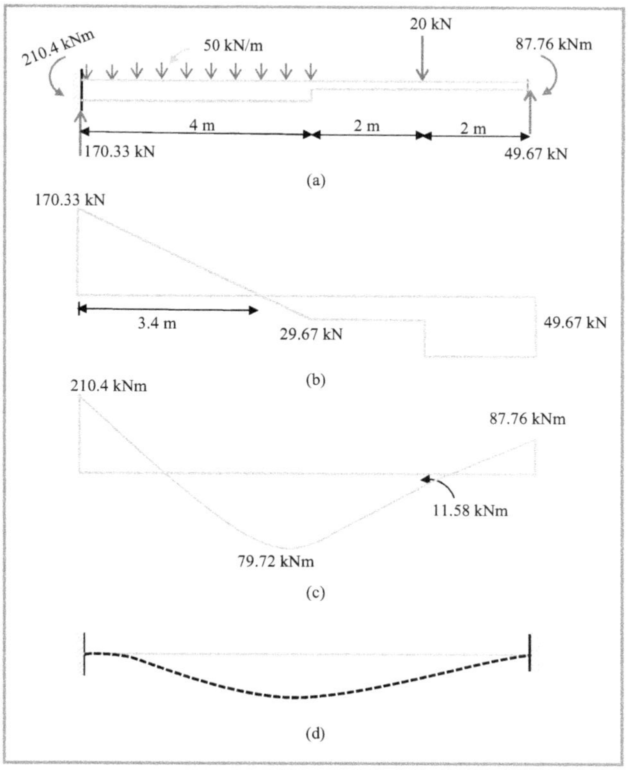

FIGURE 9.28 (a) Free body diagram, (b) shear force diagram, (c) bending moment diagram, and (d) probable deflected shape.

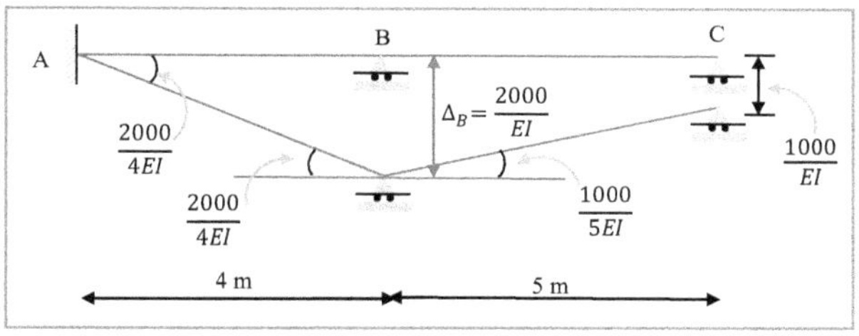

FIGURE 9.29 Continuous beam with yielding supports.

Let us now solve some numerical problems involving yielding of supports in the following examples.

Example 9.7

Re-analyse the continuous beam solved in Example (9.1), if the supports B and C settle down by $\frac{2{,}000}{EI}$ and $\frac{1{,}000}{EI}$, respectively, instead of the external load (Figure 9.29).

Solution:

The static indeterminacy, $I_S = 4 - 2 = 2$. Select M_A and M_B as redundants. The released beam with co-ordinates, as shown in Figure 9.30.

Equation 9.6 of flexibility matrix is

$$[\Delta] = [\Delta_L] + [\delta][R]$$

where $[\Delta]$ = *displacements at co-ordinates in the original beam.*

$$= \begin{bmatrix} \Delta_1 \\ \Delta_2 \end{bmatrix} = \begin{bmatrix} 2{,}000/4EI \\ -\frac{2{,}000}{4EI} - \frac{1{,}000}{5EI} \end{bmatrix} = \begin{bmatrix} 500/EI \\ -700/EI \end{bmatrix}$$

$$[\Delta_L] = \begin{bmatrix} \Delta_{1L} \\ \Delta_{2L} \end{bmatrix} = \begin{bmatrix} 0 \\ 0 \end{bmatrix}$$ (No external loads on the beam)

$[\delta]$ = *same as* Example 9.1 *as the selected redundants remain same.*

$$= \frac{1}{EI}\begin{bmatrix} 4/3 & 4/6 \\ 4/6 & 3 \end{bmatrix}$$

$$[R] = \begin{bmatrix} M_A \\ M_B \end{bmatrix}$$

On substituting into equation 9.9,

$$\begin{bmatrix} \frac{500}{EI} \\ \frac{-700}{EI} \end{bmatrix} = \begin{bmatrix} 0 \\ 0 \end{bmatrix} + \frac{1}{EI}\begin{bmatrix} 4/3 & 4/6 \\ 4/6 & 3 \end{bmatrix}\begin{bmatrix} M_A \\ M_B \end{bmatrix}$$

By solving the above equation, we obtain

$$\begin{bmatrix} M_A \\ M_B \end{bmatrix} = \begin{bmatrix} -553.13 \\ 356.25 \end{bmatrix} \text{kNm}$$

Note: Alternatively, the reactions R_B and R_C can be chosen as redundants, so that the $[\triangle]$ Matrix would be

$$[\Delta] = \begin{bmatrix} \Delta_1 \\ \Delta_2 \end{bmatrix} = \begin{bmatrix} 2{,}000/EI \\ 1{,}000/EI \end{bmatrix}$$

$[\delta] = \begin{bmatrix} \delta_{11} & \delta_{12} \\ \delta_{21} & \delta_{22} \end{bmatrix}$ can be obtained by using the unit load method and the area moment multiplication technique. The released beam with co-ordinates is shown in Figure 9.31.

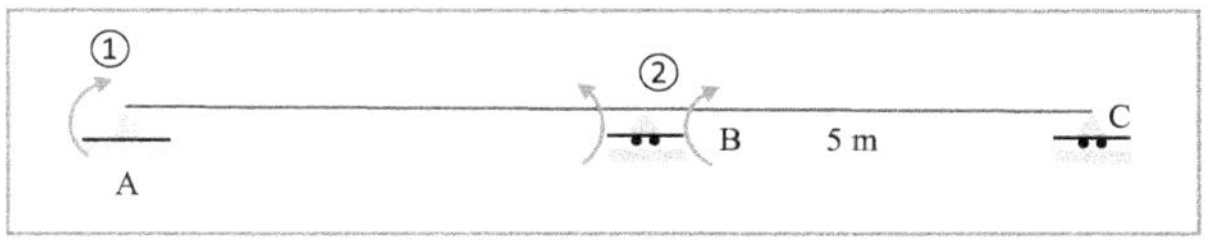

FIGURE 9.30 Released beam with co-ordinates corresponding to redundants M_A and M_B.

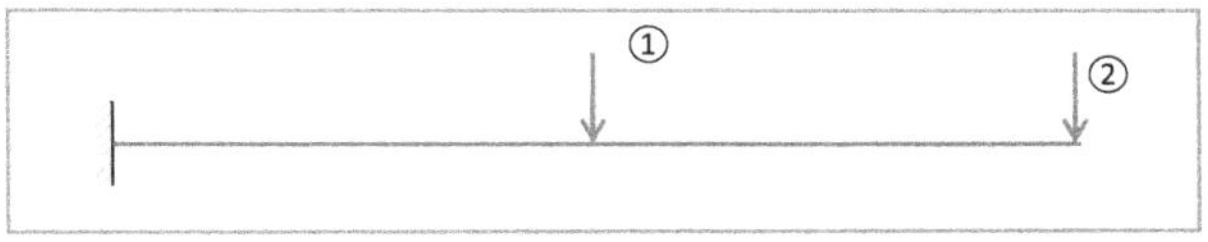

FIGURE 9.31 Released beam with co-ordinates corresponding to redundants R_B and R_C.

Elements of [δ]: To obtain the first column elements of the flexibility matrix, apply a unit load at ① in its specified direction, and calculate the displacement at all the co-ordinates in the released beam. Similarly, to get the second column elements, apply a unit force at ② and compute the displacements at ① and ②. The resulting m_1 and m_2 diagrams due to the unit forces at the co-ordinates are shown in Figure 9.32.

$$\delta_{11}=\int m_1 m_1\frac{dx}{EI}=\frac{1}{EI}\left(\frac{1}{2}\times4\times4\times2.67\right)$$

$$=\frac{21.33}{EI}$$

$$\delta_{12}=\delta_{21}=\int m_1 m_2\frac{dx}{EI}=\frac{1}{EI}\left[\frac{4}{6}(4(2\times9+5)+0)\right]$$

$$=\frac{61.33}{EI}$$

$$\delta_{22}=\int m_2 m_2\frac{dx}{EI}=\frac{1}{EI}\left[\frac{1}{2}\times9\times9\times6\right]=\frac{243}{EI}$$

Substituting the above values into equation 9.6

$$\begin{bmatrix}\dfrac{2{,}000}{EI}\\ \dfrac{1{,}000}{EI}\end{bmatrix}=\begin{bmatrix}0\\0\end{bmatrix}+\frac{1}{EI}\begin{bmatrix}21.33 & 61.33\\ 61.33 & 243\end{bmatrix}\begin{bmatrix}R_B\\R_C\end{bmatrix}$$

By solving, we obtain

$$\begin{bmatrix}R_B\\R_C\end{bmatrix}=\begin{bmatrix}298.68\\-71.27\end{bmatrix}\text{kN}$$

The shear force and bending moment diagrams and the probable deflected shape are shown in Figure 9.33.

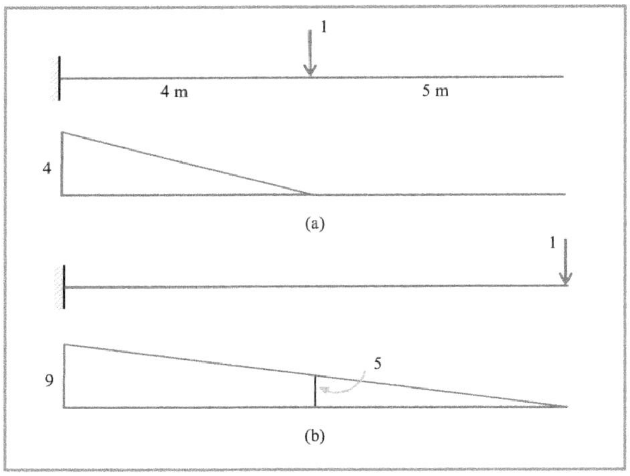

FIGURE 9.32 (a) m_1 diagram and (b) m_2 diagram due to unit forces at coordinates 1 and 2 respectively.

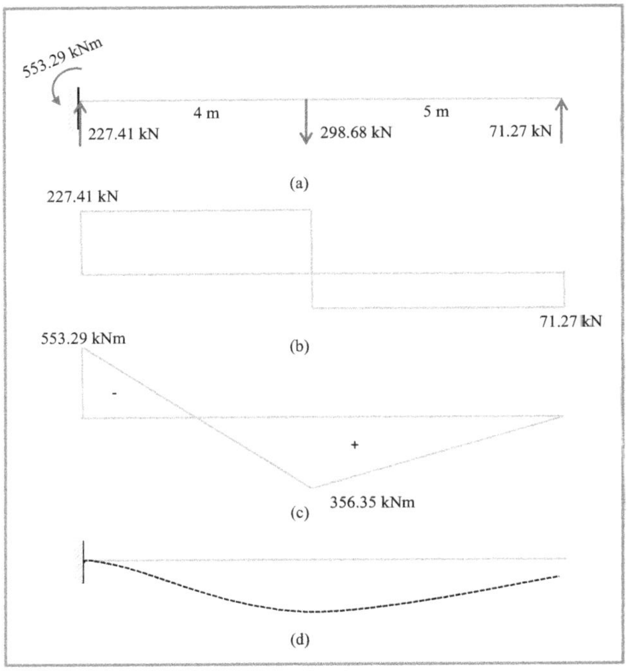

FIGURE 9.33 (a) Free body diagram, (b) shear force diagram, (c) bending moment diagram, and (d) probable deflected shape.

Example 9.8

Analyse the beam shown in Figure 9.34, if the support *A* rotates by 0.001 rad in a clockwise direction and supports *B* settle down by 10 mm. Assume $EI = 135\times10^3$ kNm².

Solution:

The static indeterminacy, $I_S = 2$. Select the moment at *A* and the vertical reaction at *B* as the redundants, and assign co-ordinates to these redundants, as shown in Figure 9.35. The released beam is obtained by replacing the fixed support at *A* with a hinged support and removing the support at *B*. The released beam, along with the co-ordinates and applied loads, is shown in Figure 9.35.

Equation 9.6 of the flexibility matrix is

$$[\Delta]=[\Delta_L]+[\delta][R]$$

where

$$[\Delta]=\begin{bmatrix}\Delta_1\\ \Delta_2\end{bmatrix}\text{ displacements corresponding}$$

to the coordinates in original beam.

$$=\begin{bmatrix}0.001\\0.01\end{bmatrix}$$

$[\Delta_L]$ and [δ] matrix can be calculated using the unit load method and the area moment multiplication technique, as presented below. The M_L, m_1, and m_2 diagrams due to the applied loads and unit forces at co-ordinates 1 and 2, respectively, are shown in Figure 9.36a–c.

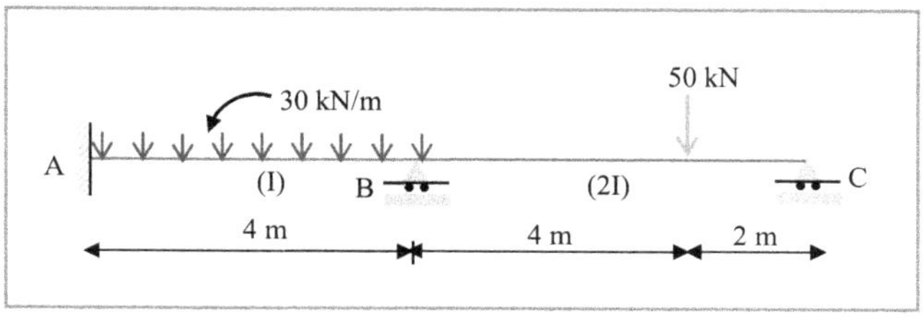

FIGURE 9.34 Two-span continuous beam.

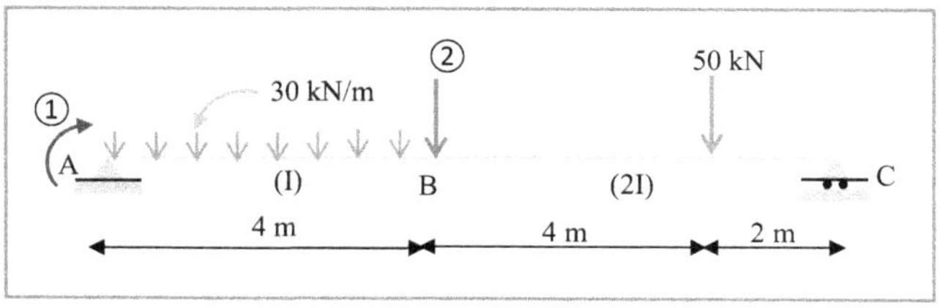

FIGURE 9.35 Released beam with co-ordinates and loads.

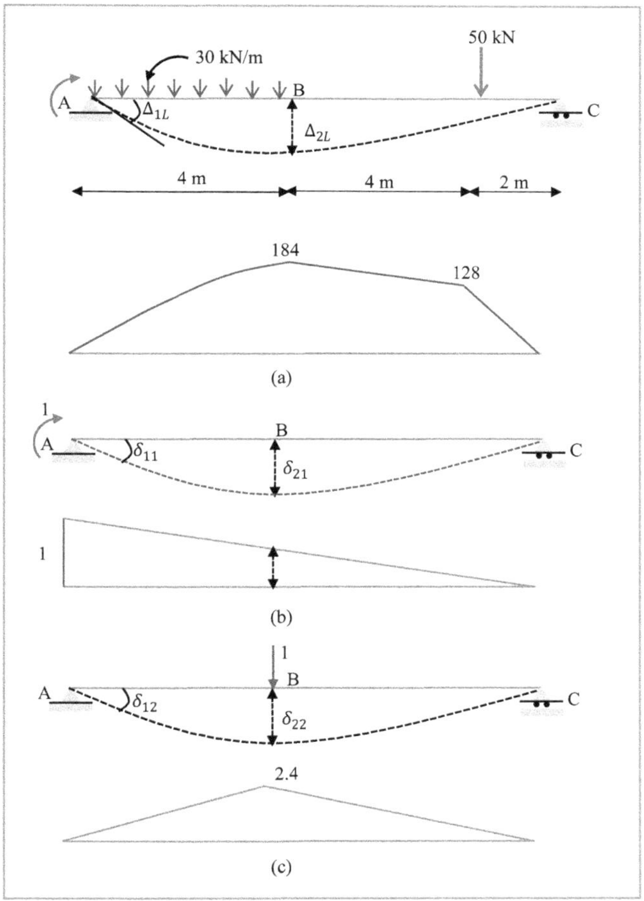

FIGURE 9.36 (a) M_L diagram due to loads, (b) m_1 diagram due to unit force at co-ordinate 1, and (c) m_2 diagram due to unit force at co-ordinate at 2.

$$\Delta_{1L} = \int M_L m_1 \frac{dx}{EI}$$

$$= \frac{1}{EI}\left[\left(\frac{2}{3}\times 4\times 184\times\frac{7.5}{10}\right)\right]$$

$$+\frac{1}{2EI}\left[\begin{array}{l}\frac{4}{6}\left[184\left(2\times\frac{6}{10}+\frac{2}{10}\right)+128\left(2\times\frac{2}{10}+\frac{6}{10}\right)\right]\\ +\frac{2}{6}\left[128\times\left(2\times\frac{2}{10}+0\right)\right]\end{array}\right]$$

$$= \frac{368}{EI}+\frac{128.53}{EI}+\frac{8.53}{EI}=\frac{505.06}{EI}$$

$$\Delta_{2L} = \int M_L m_2 \frac{dx}{EI}$$

$$= \frac{1}{EI}\left[\left(\frac{2}{3}\times 4\times 184\times 1.5\right)\right]$$

$$+\frac{1}{2EI}\left[\begin{array}{l}\frac{4}{6}[184(2\times 2.4+0.8)+128(2\times 0.8+2.4)]\\ +\frac{2}{6}[128\times(2\times 0.8+0)]\end{array}\right]$$

$$= \frac{736}{EI}+\frac{514.13}{EI}+\frac{34.13}{EI}=\frac{1{,}284.26}{EI}$$

$$\delta_{11} = \int m_1 m_1 \frac{dx}{EI} = \frac{1}{EI}\left[\frac{4}{6}\left\{1\times\left(2\times 1+\frac{6}{10}\right)+\frac{6}{10}\left(2\times\frac{6}{10}+1\right)\right\}\right]$$

$$+\frac{1}{2EI}\left[\frac{6}{6}\left(\frac{6}{10}\times 2\times\frac{6}{10}+0\right)\right]$$

$$= \frac{2.61}{EI}+\frac{0.36}{EI}=\frac{2.97}{EI}$$

$$\delta_{12} = \delta_{21} = \int m_1 m_2 \frac{dx}{EI} = \frac{1}{EI}\left[0+\frac{4}{6}\left(0+2.4\left(2\times\frac{6}{10}+1\right)\right)\right]$$

$$+\frac{1}{2EI}\left[\frac{6}{6}\left[2.4\left(2\times\frac{6}{10}+0\right)+0\right]\right]$$

$$= \frac{3.51}{EI}+\frac{1.44}{EI}=\frac{4.96}{EI}$$

$$\delta_{22} = \int m_2 m_2 \frac{dx}{EI} = \frac{1}{EI}\left[\frac{4}{6}\{0+2.4\times(2\times 2.4+0)\}\right]$$

$$+\frac{1}{2EI}\left[\frac{6}{6}\{2.4(2\times 2.4+0)\}\right]$$

$$= \frac{7.68}{EI}+\frac{5.76}{EI}=\frac{13.44}{EI}$$

Substituting the above values into equation 9.6, we have

$$\begin{bmatrix} 0.001 \\ 0.01 \end{bmatrix} = \begin{bmatrix} \dfrac{505.06}{EI} \\ \dfrac{1{,}284.26}{EI} \end{bmatrix} + \frac{1}{EI}\begin{bmatrix} 2.97 & 4.96 \\ 4.96 & 13.44 \end{bmatrix}\begin{bmatrix} M_A \\ R_B \end{bmatrix}$$

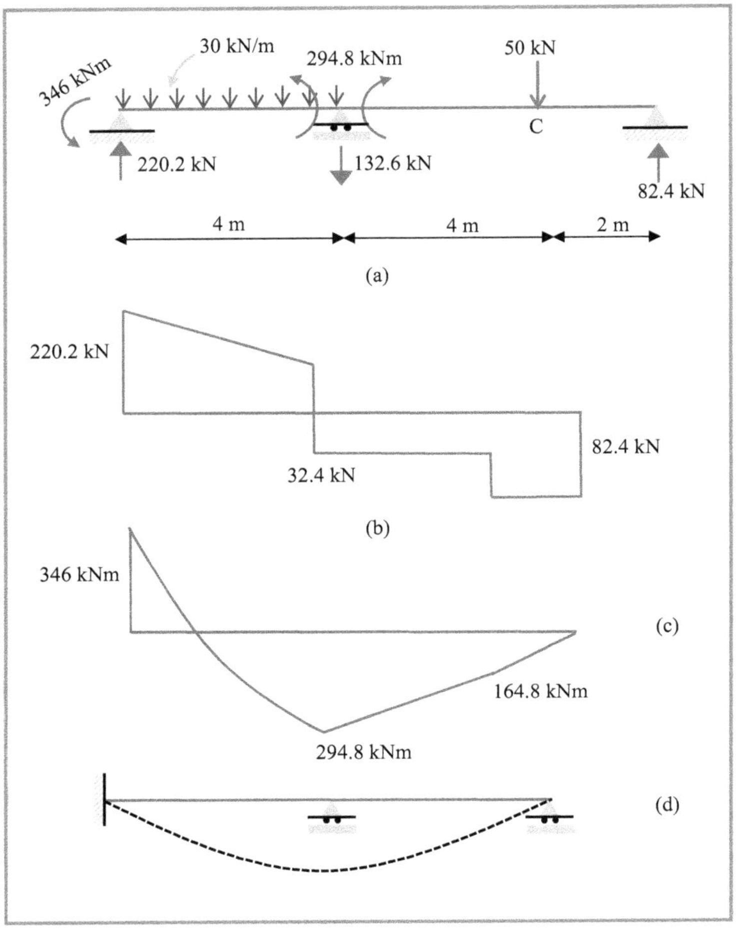

FIGURE 9.37 (a) Free body diagram, (b) shear force diagram, (c) bending moment diagram, and (d) probable deflected shape.

By solving the above equation, we obtain

$$\begin{bmatrix} M_A \\ R_B \end{bmatrix} = \begin{bmatrix} -346.04 \\ 132.59 \end{bmatrix} = \begin{bmatrix} -346 \text{ kNm} \\ 132.6 \text{ kN} \end{bmatrix}$$

Other support reactions: With the obtained values of M_A and R_B, using the static equilibrium equations, the remaining unknown reactions can be determined and are shown in Figure 9.37a. The shear force and bending moment diagrams are shown in Figure 9.37b and c, respectively. The probable deflected shape is shown in Figure 9.37d.

9.4 APPLICATIONS TO RIGID-JOINTED PLANE FRAMES

Example 9.9

Analyse the rigid-jointed plane frame ABCD shown in Figure 9.38 using the flexibility matrix method. Assume *EI* is constant. Draw the shear force and bending moment diagrams. Also, draw the probable deflected shape. (The same has been solved in the Energy Method Example 4.2.)

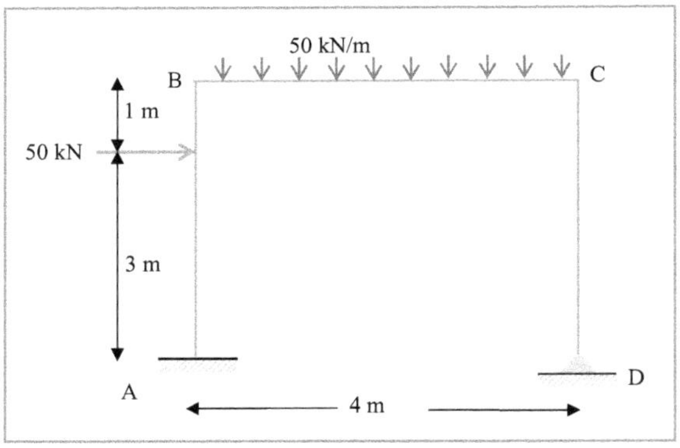

FIGURE 9.38 Rigid-jointed plane frame *ABCD*.

Solution:

The static indeterminacy of the frame is $I_s = 5 - 3 = 2$. Select the reactions at *D* (H_D and V_D) as the redundants. The basic determinate frame is obtained by removing the support *D*, thereby releasing the two reaction components. The basic determinate frame, with co-ordinates corresponding to the chosen redundants and the external loads, is shown in Figure 9.39.

Equation 9.6 of the flexibility Matrix is

$$[\Delta] = [\Delta_L] + [\delta][R]$$

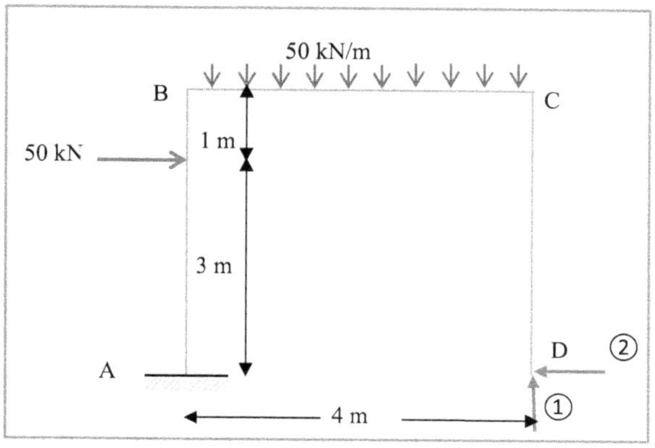

FIGURE 9.39 Basic determinate frame with redundants and external loads.

$$\text{where}\,[\Delta]=\begin{bmatrix}\Delta_1\\ \Delta_2\end{bmatrix}=\begin{bmatrix}0\\ 0\end{bmatrix}\ (\because \text{No yielding of supports})$$

$[\Delta_L]=\begin{bmatrix}\Delta_{1L}\\ \Delta_{2L}\end{bmatrix}$ are the displacements at co-ordinates ① & ② due to the external loads on the basic determinate structure.

$[\delta]=\begin{bmatrix}\delta_{11} & \delta_{12}\\ \delta_{21} & \delta_{22}\end{bmatrix}$ are the elements of the flexibility matrix due to the unit loads successively applied at co-ordinates 1 and 2 on the determinate frame.

Both $[\Delta_L]$ and $[\delta]$ can be calculated using the unit load method and the area moment multiplication technique. For this purpose, the loading diagrams, along with M_L, m_1, and m_2 diagrams, are shown in Figure 9.40a–d, respectively.

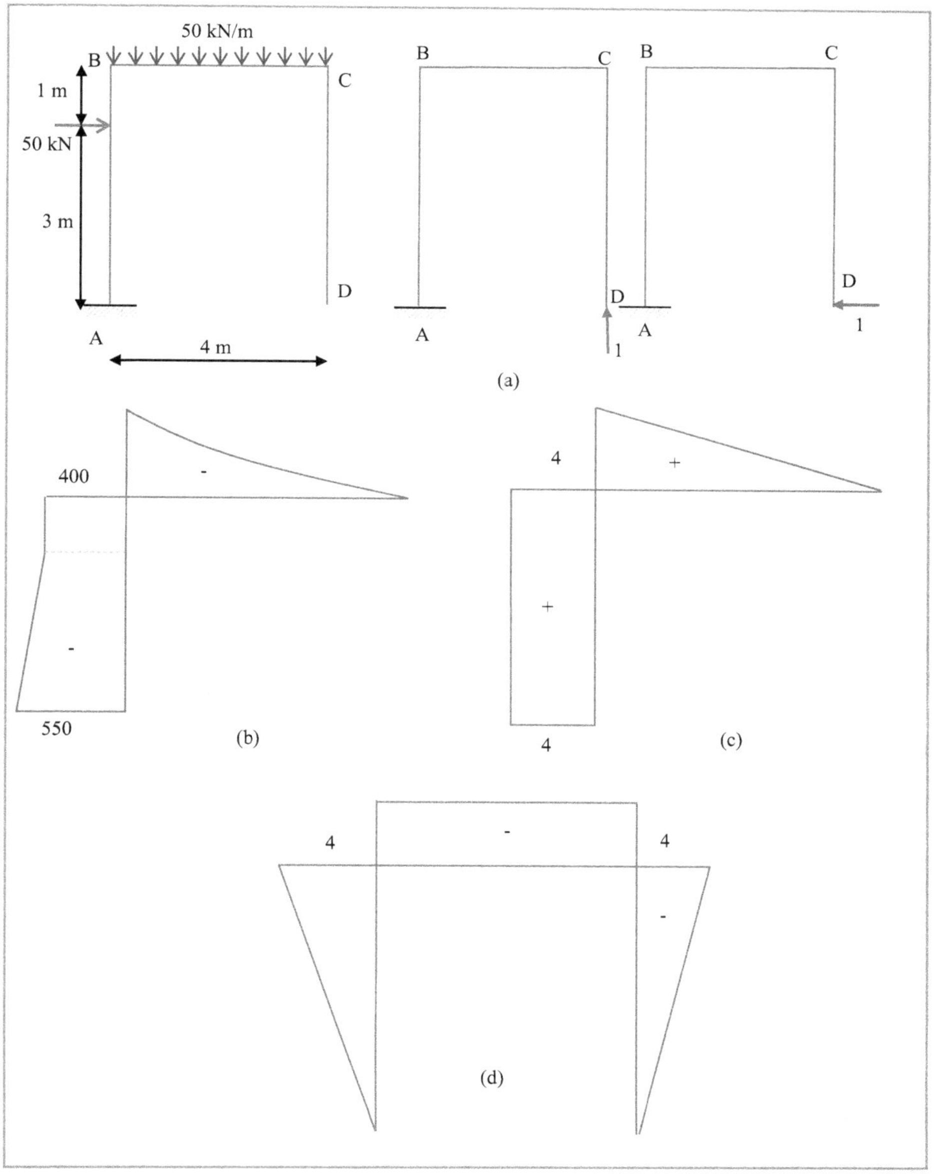

FIGURE 9.40 (a) Basic frame with external loads and unit forces at co-ordinates 1 and 2, (b) M_L diagram, (c) m_1 diagram, and (d) m_2 diagram.

$$\Delta_{1L} = \int M_L m_1 \frac{dx}{EI} = \sum \frac{A_L \overline{m_1}}{EI}$$ where $\overline{m_1}$ is the ordinate of the m_1 diagram corresponding to the c.g. of the M_L diagram.

$$= \frac{1}{EI}\left[\left(\frac{1}{3}\times 4\times(-400)\times 3\right)+\left(-400\times 1+\frac{3}{2}(-400-550)\right)\times 4\right]$$

$$= \frac{-8{,}900}{EI}$$

$$\Delta_{2L} = \int M_L m_2 \frac{dx}{EI} = \sum \frac{A_L \overline{m_2}}{EI}$$

$$= \frac{1}{EI}\begin{bmatrix}\left(\frac{1}{3}\times 4\times(-400)\times(-4)\right)+ \\ \left(-400\times 1+\frac{3}{2}(-400-550)\right)\times(-1.87)\end{bmatrix}$$

$$= \frac{5{,}546.1}{EI}$$

$$\left[\text{c.g. } \bar{x} = \frac{400\times 3.5+1425\times 1.42}{400+1{,}425} = 1.87 \text{ m from bottom}\right]$$

$$\delta_{11} = \int m_1 m_1 \frac{dx}{EI} = \sum \frac{A_{m1}\overline{m_1}}{EI}$$

$$= \frac{1}{EI}\left[\left(\frac{1}{2}\times 4\times 4\times 2.67\right)+(4\times 4\times 4)\right] = \frac{85.36}{EI}$$

$$\delta_{12} = \delta_{21} = \frac{1}{EI}\left[\left(\frac{1}{2}\times 4\times(-4)\times 4\right)+(4\times(-4)\times 2)\right] = \frac{-64}{EI}$$

$$\delta_{22} = \frac{1}{EI}\left[\left(\frac{1}{2}\times 4\times 4\times 2.67\right)+(4\times 4\times 4)+\left(\frac{1}{2}\times 4\times 4\times 2.67\right)\right]$$

$$= \frac{106.72}{EI}$$

Equation 9.6 of the Flexibility Matrix is

$$[\Delta] = [\Delta_L] + [\delta][R]$$

Substituting the above values into equation 9.6, we have

$$\begin{bmatrix} 0 \\ 0 \end{bmatrix} = \begin{bmatrix} \frac{-8{,}900}{EI} \\ \frac{5{,}546.1}{EI} \end{bmatrix} + \frac{1}{EI}\begin{bmatrix} 85.36 & -64 \\ -64 & 106.72 \end{bmatrix}\begin{bmatrix} V_D \\ H_D \end{bmatrix}$$

By solving the above, we obtain

$$\begin{bmatrix} V_D \\ H_D \end{bmatrix} = \begin{bmatrix} 118.65 \\ 19.18 \end{bmatrix} \text{kN}$$

Final end reactions: The final end reactions can be calculated using the equilibrium equations, with the help of free body diagrams, as shown in Figure 9.41a. The axial force, shear force, and bending moment diagrams along with the probable deflected shape are shown in Figure 9.41b–e, respectively.

Example 9.10

Analyse the rigid-jointed plane frame shown in Figure 9.42 using the flexibility matrix method. Draw the shear force and bending moment diagrams.

Solution:

The static indeterminacy of the rigid-jointed frame is $I_s = 5-3 = 2$. Select V_A and H_A as the redundants. The released structure can be obtained by removing the selected redundants from the original beam and assigning the co-ordinates to the selected redundants, as shown in Figure 9.43.

Equation 9.6 of the flexibility matrix is

$$[\Delta] = [\Delta_L] + [\delta][R]$$

where

$$[\Delta] = \begin{bmatrix} \Delta_1 \\ \Delta_2 \end{bmatrix} = \begin{bmatrix} 0 \\ 0 \end{bmatrix} \text{ (No yielding of supports.)}$$

$$[R] = \begin{bmatrix} R_1 \\ R_2 \end{bmatrix} = \begin{bmatrix} V_A \\ H_A \end{bmatrix} \text{ (Redundant forces)}$$

The $[\Delta_L]$ and $[\delta]$ represents the displacements due to the external loads and unit forces applied at the co-ordinates on the basic determinate structure. These can be calculated using the unit load method and the area moment multiplication technique with the help of M_L, m_1, and m_2 diagrams corresponding to external loads and unit forces, respectively, as shown in Figure 9.44.

$$\Delta_{1L} = \int M_L m_1 \frac{dx}{EI} = \sum \frac{A_L \overline{m_1}}{EI}$$

$$= \left[\frac{1}{2EI}\left(\frac{-6\times 300}{4}\right)\times 4.8\right] + \frac{1}{EI}[-300\times 4\times 6]$$

$$= \frac{-1{,}080}{EI} - \frac{7{,}200}{EI} = \frac{-8{,}280}{EI}$$

$$\Delta_{2L} = \frac{1}{2EI}\left[\frac{300\times 6}{4}\times 3\right] + \frac{1}{EI}(300\times 4\times 1)$$

$$= \frac{675}{EI} + \frac{1{,}200}{EI} = \frac{1{,}875}{EI}$$

$$\delta_{11} = \int m_1 m_1 \frac{dx}{EI} = \frac{1}{2EI}\left[\frac{1}{2}\times 6\times 6\times 4\right] + \frac{1}{EI}\left(\frac{1}{2}\times 6\times 4\times 6\right)$$

$$= \frac{180}{EI}$$

$$\delta_{12} = \delta_{21} = \int m_1 m_2 \frac{dx}{EI} = \frac{1}{2EI}[(18)\times(-3)] + \frac{1}{EI}[6\times 4\times(-1)]$$

$$= \frac{-51}{EI}$$

$$\delta_{22} = \int m_2 m_2 \frac{dx}{EI} = \frac{1}{EI}\left[\left(\frac{1}{2}\times 3\times 3\times 2\times 2\right)+\left(\frac{1}{2}\times 1\times 1\times 0.67\right)\right]$$

$$+ \frac{1}{2EI}[3\times 6\times 3] = \frac{45.34}{EI}$$

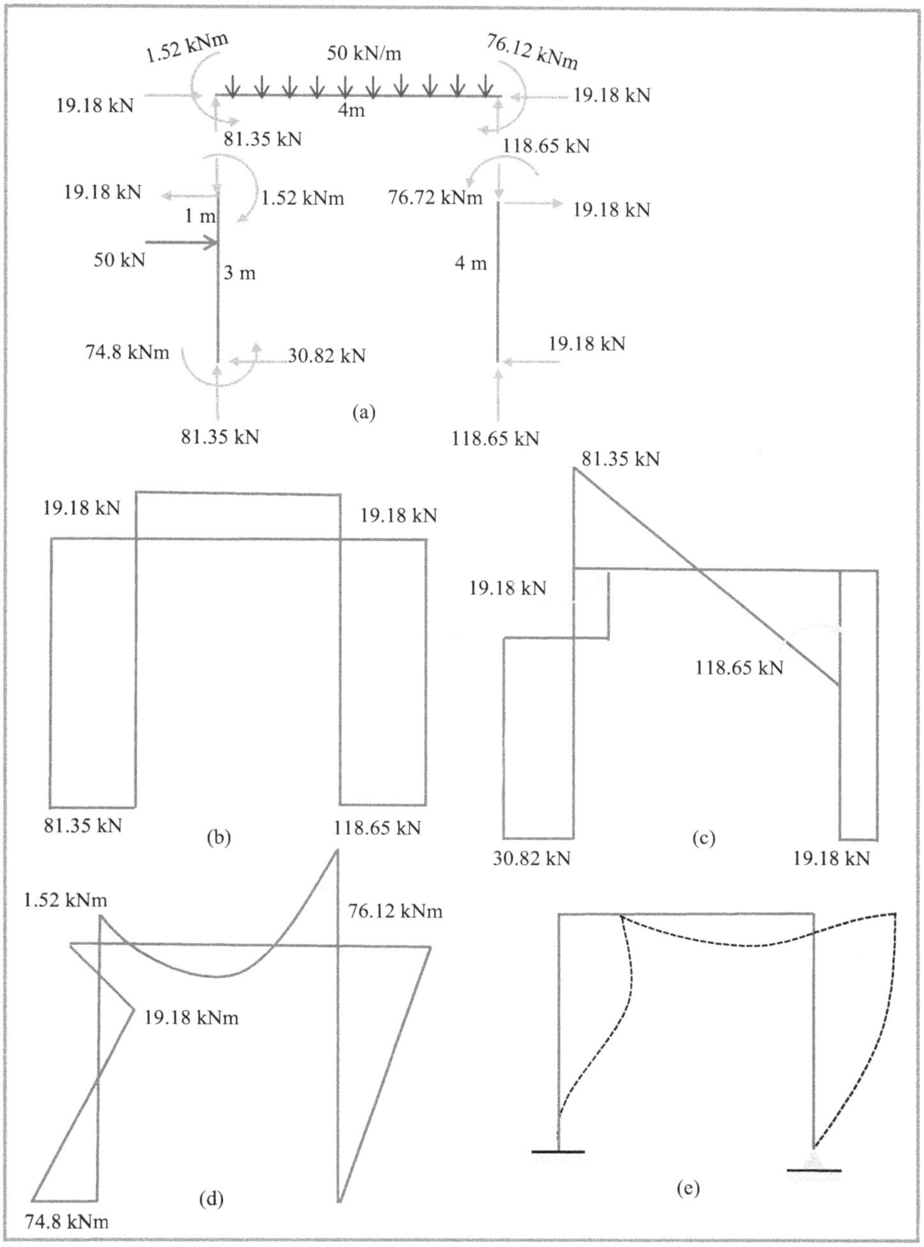

FIGURE 9.41 (a) Free body diagram, (b) axial force diagram, (c) shear force diagram, (d) bending moment diagram, and (e) probable deflected shape.

Substituting the above values into equation 9.6, we have

$$\begin{bmatrix} 0 \\ 0 \end{bmatrix} = \begin{bmatrix} \dfrac{-8{,}280}{EI} \\ \dfrac{1{,}875}{EI} \end{bmatrix} + \frac{1}{EI}\begin{bmatrix} 180 & -51 \\ -51 & 45.34 \end{bmatrix}\begin{bmatrix} V_A \\ H_A \end{bmatrix}$$

By solving the above equation, we obtain,

$$\begin{bmatrix} V_A \\ H_A \end{bmatrix} = \begin{bmatrix} 50.32 \\ 15.25 \end{bmatrix} \text{kN}$$

Final end reactions: After obtaining the reactions V_A and H_A, the remaining reactions can be calculated using the static equilibrium equations with the help of the free body diagram, as shown in Figure 9.45a. The axial force, shear force, and bending moment diagrams are shown in Figure 9.45b–d, respectively. The probable deflected shape is shown in Figure 9.45e.

Example 9.11

Analyse the symmetrical frame shown in Figure 9.46 using the flexibility matrix method. Draw the probable deflected shape, shear force, and bending moment diagrams.

Solution:

The static indeterminacy of the frame is $I_s = 6 - 3 = 3$. By taking advantage of symmetry, the static indeterminacy can be reduced to 2. As shown in Figure 9.47a, the

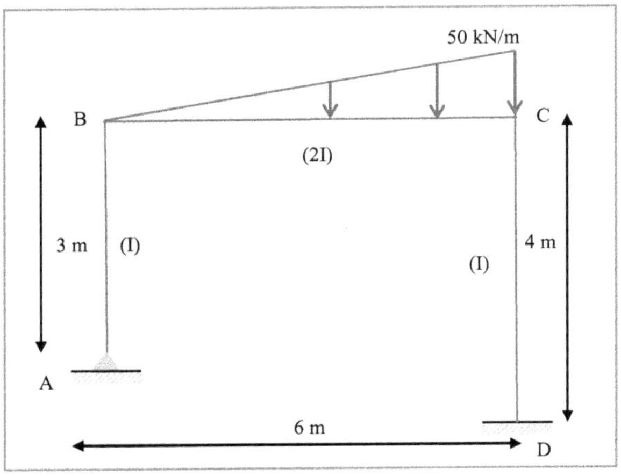

FIGURE 9.42 ABCD rigid-jointed frame.

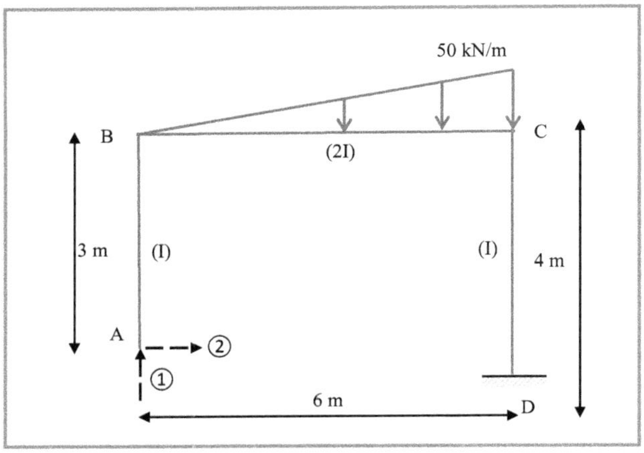

FIGURE 9.43 Released frame with loads and co-ordinates corresponding to the redundants.

slope at the centre of *BC* is zero, implying that the *SF* at the centre is also zero. Therefore, select M_E and H_E as the redundants. The determinate structure is obtained by releasing M_E and H_E and assigning co-ordinates to them, as shown in Figure 9.47b.

Equation 9.6 of the flexibility Matrix is

$$[\Delta] = [\Delta_L] + [\delta][R]$$

where

$$[\Delta] = \begin{bmatrix} \Delta_1 \\ \Delta_2 \end{bmatrix} = \begin{bmatrix} 0 \\ 0 \end{bmatrix}$$

$$[R] = \begin{bmatrix} M_E \\ H_E \end{bmatrix}$$

$[\Delta_L] = \begin{bmatrix} \Delta_{1L} \\ \Delta_{2L} \end{bmatrix}$ represents the displacements at co-ordinates due to the applied loads on the basic or determinate structure.

$[\delta] = \begin{bmatrix} \delta_{11} & \delta_{12} \\ \delta_{21} & \delta_{22} \end{bmatrix}$ represents the flexibility coefficients due to unit forces at co-ordinates.

Both $[\Delta_L]$ and $[\delta]$ can be calculated using the unit load method and the area moment multiplication technique. The basic frame, subjected to external loads and unit forces at co-ordinates 1 and 2, is shown in Figure 9.48a. The BM diagram due to external loads (M_L), and the BM diagrams due to unit loads (m_1 and m_2) are shown in Figure 9.48b–e, and can be used to calculate the displacements.

$$\Delta_{1L} = \int M_L m_1 \frac{dx}{EI}$$

$$= \frac{1}{2EI}\left[\left(\frac{1}{3} \times 3 \times 225 \times 1\right) + \left(\frac{1}{2} \times 3 \times 150 \times 1\right)\right]$$

$$+ \frac{1}{EI}\left[(3 \times 225 \times 1) + (3 \times 150 \times 1)\right]$$

$$= \frac{1{,}350}{EI}$$

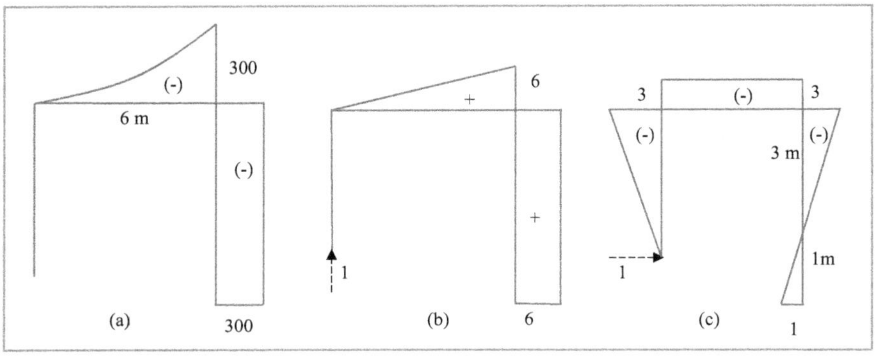

FIGURE 9.44 (a) M_L diagram, (b) m_1 diagram, and (c) m_2 diagram.

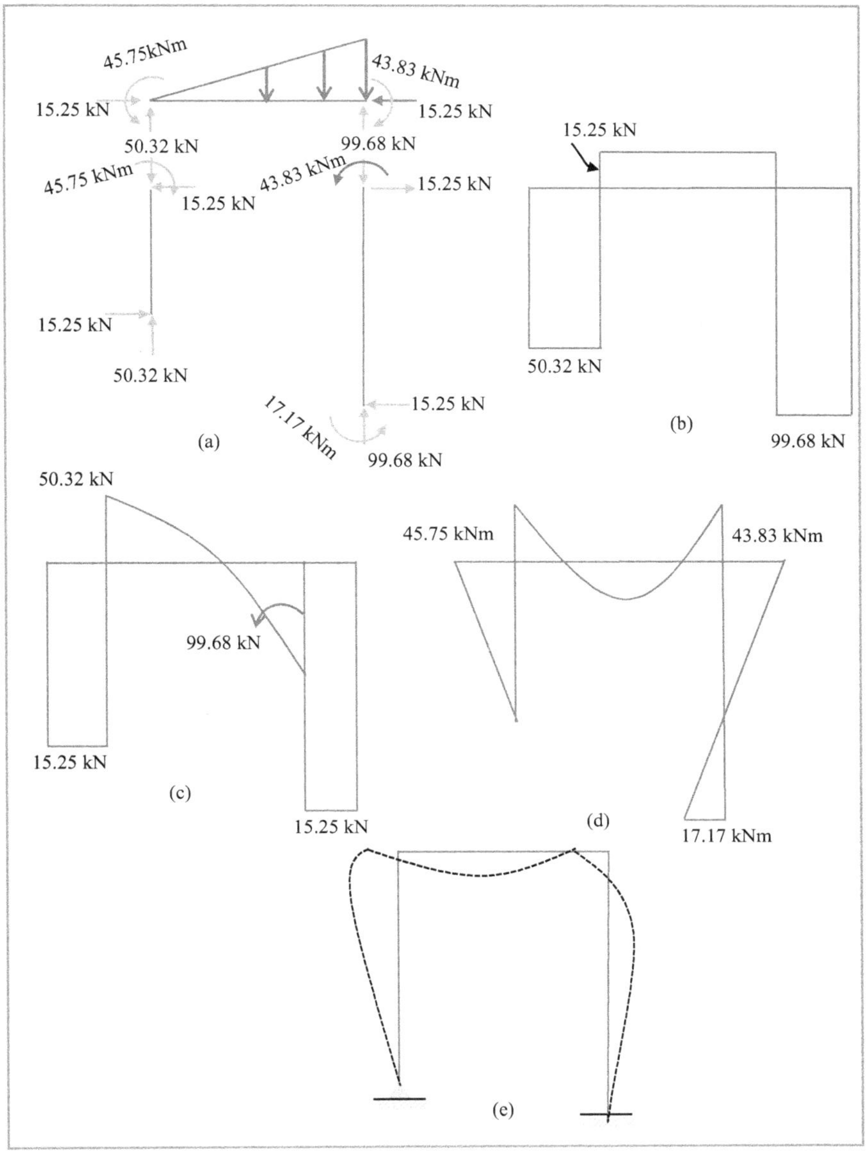

FIGURE 9.45 (a) Free body diagram, (b) axial force diagram, (c) shear force diagram, (d) bending moment diagram, and (e) probable deflected shape.

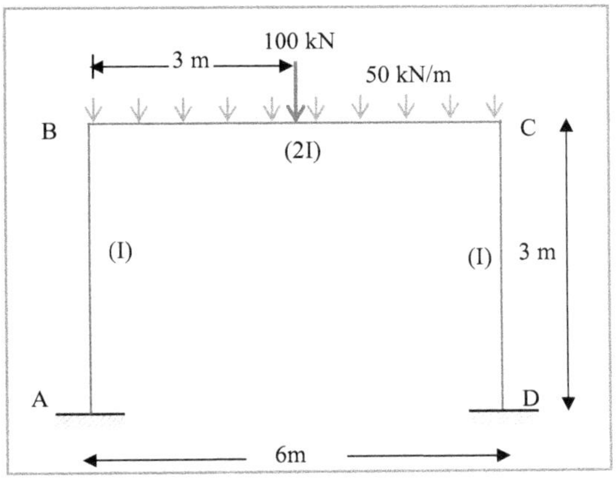

FIGURE 9.46 Symmetrical rigid-jointed frame.

$$\Delta_{2L} = \int M_L m_2 \frac{dx}{EI} = \frac{1}{2EI}\left[\frac{1}{3} \times 3 \times 225 \times 0\right]$$

$$+\frac{1}{EI}\left[(225 \times 3 + 150 \times 3)(1.5)\right] = \frac{1{,}687.5}{EI}$$

$$\delta_{11} = \int m_1 m_1 \frac{dx}{EI} = \frac{1}{2EI}[1 \times 3 \times 1] + \frac{1}{EI}[1 \times 3 \times 1] = \frac{4.5}{EI}$$

$$\delta_{12} = \delta_{21} = \int m_1 m_2 \frac{dx}{EI} = \frac{1}{EI}\left[\frac{1}{2} \times 3 \times 3 \times 1\right] = \frac{4.5}{EI}$$

$$\delta_{22} = \int m_2 m_2 \frac{dx}{EI} = \frac{1}{EI}\left[\frac{1}{2} \times 3 \times 3 \times 2\right] = \frac{9}{EI}$$

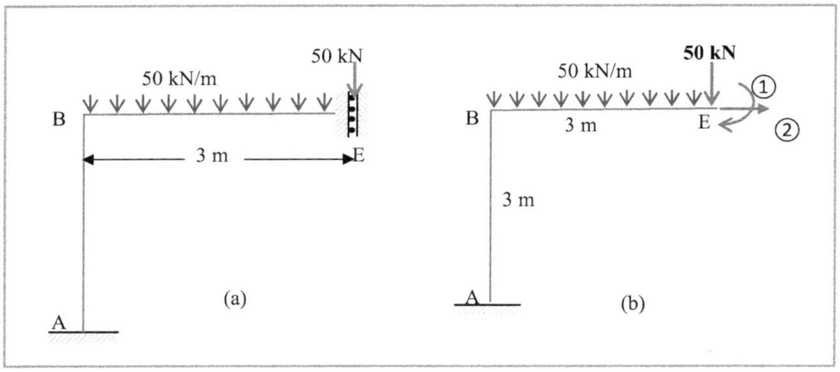

FIGURE 9.47 (a) Half portion of the frame due to symmetry and (b) basic determinate frame with co-ordinates and loads.

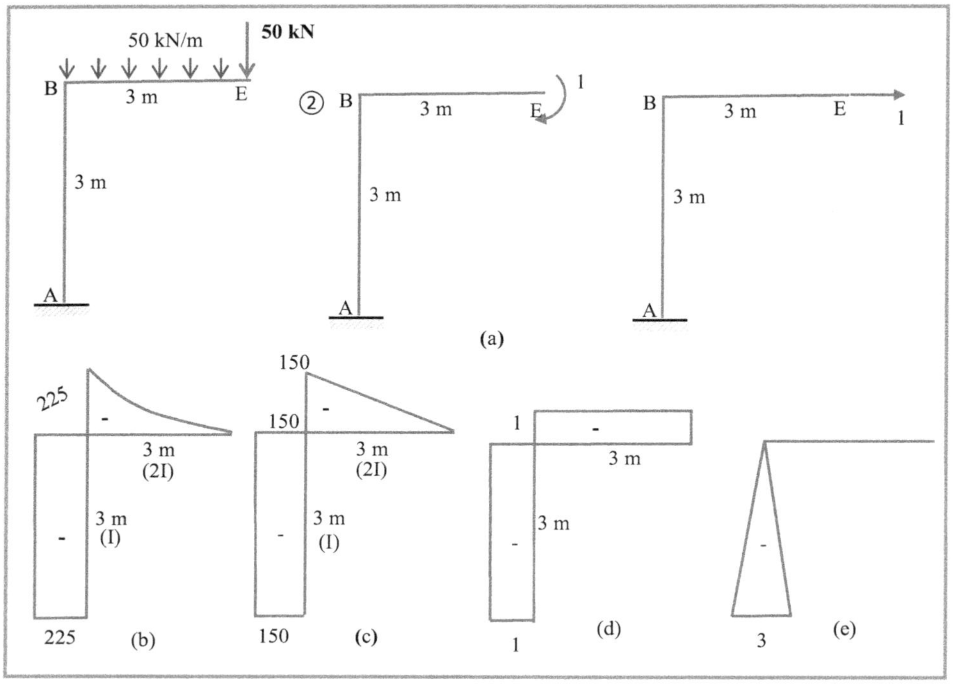

FIGURE 9.48 (a) Basic frame with loads and unit forces at co-ordinates, (b and c) M_L diagram, (d) m_1 diagram, and (e) m_2 diagram.

Substituting the above values into equation 9.6,

$$\begin{bmatrix} 0 \\ 0 \end{bmatrix} = \begin{bmatrix} \dfrac{1{,}350}{EI} \\ \dfrac{1{,}687.5}{EI} \end{bmatrix} + \frac{1}{EI}\begin{bmatrix} 4.5 & 4.5 \\ 4.5 & 9 \end{bmatrix}\begin{bmatrix} M_E \\ H_E \end{bmatrix}$$

By solving the above equation, we obtain

$$\begin{bmatrix} M_E \\ H_E \end{bmatrix} = \begin{bmatrix} -225 \text{ kNm} \\ -75 \text{ kN} \end{bmatrix}$$

Final end reactions: After determining M_E and H_E and using the free body diagrams (Figure 9.49a), along with the static equilibrium equations, the other support reactions can be calculated. The shear force, bending moment diagrams, and probable deflected shapes are shown in Figure 9.49b–e, respectively.

Example 9.12

Analyse the rigid-jointed plane frame shown in Figure 9.50, using the flexibility matrix method. Draw the *AFD*, *SFD*, and *BMD*. Also, draw the probable deflected shape. Assume *EI* is constant.

Solution:

The static indeterminacy is $I_s = 5 - 3 = 2$. Select V_D and H_D as the redundants. The basic or determinate structure can be obtained by releasing the selected redundants V_D and H_D by removing the hinged support at *D*. Assign co-ordinates to these redundants, as shown in Figure 9.51.

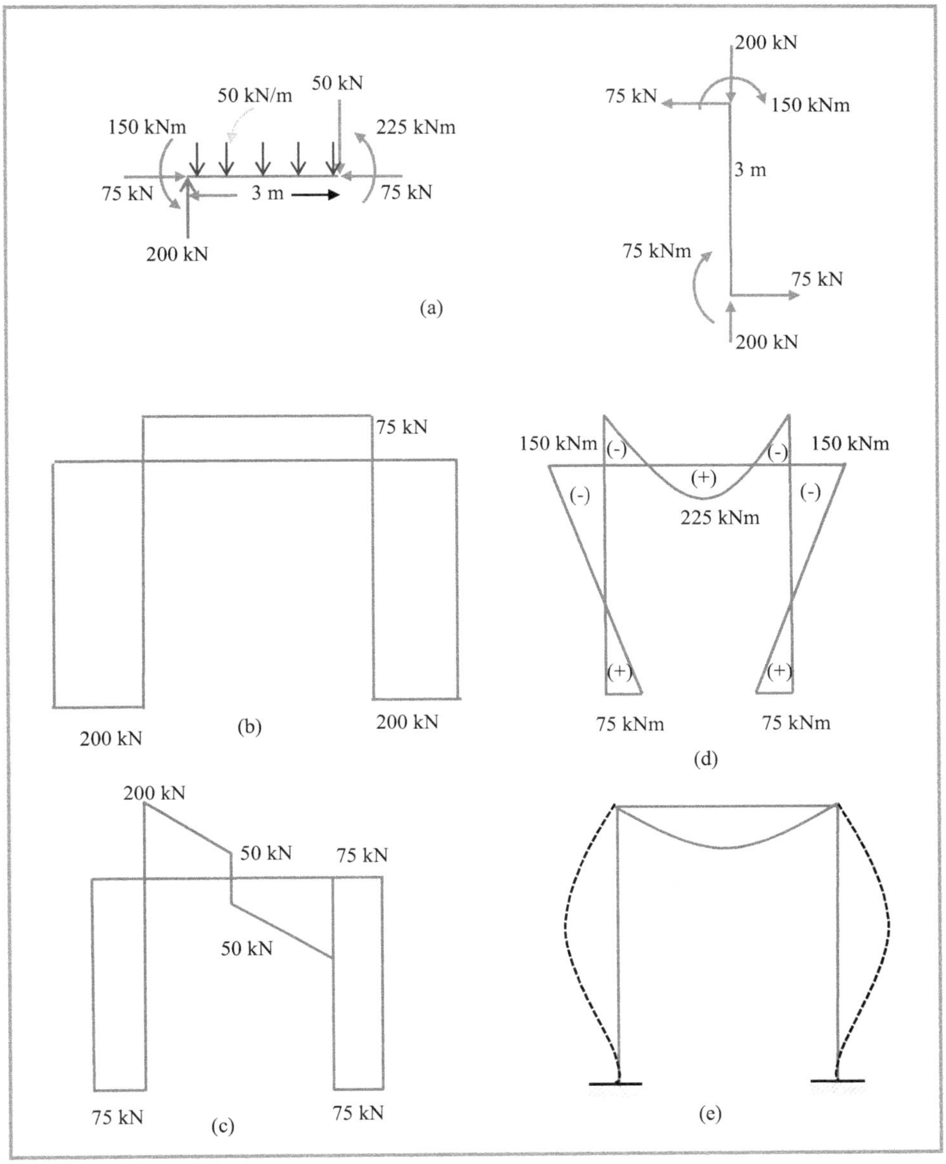

FIGURE 9.49 (a) Free body diagram, (b) axial force diagram, (c) shear force diagram, (d) bending moment diagram, and (e) probable deflected shape.

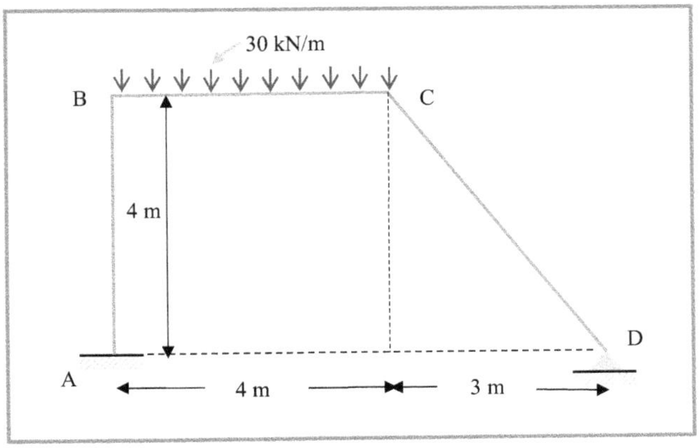

FIGURE 9.50 Indeterminate rigid-jointed frame ABCD.

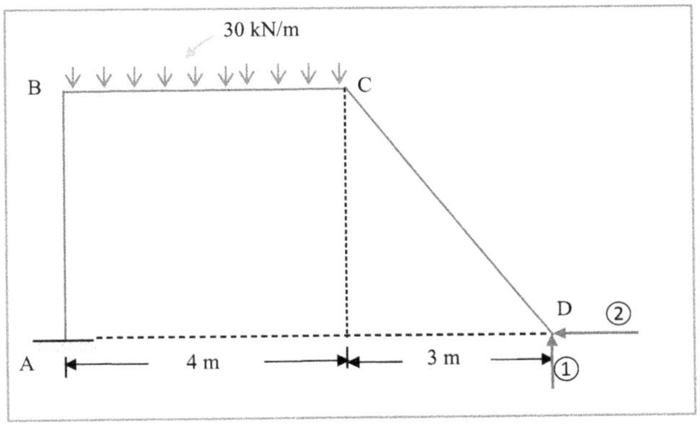

FIGURE 9.51 Basic determinate frame with co-ordinates assigned to redundants.

Equation 9.6 of the flexibility matrix is

$$[\Delta]=[\Delta_L]+[\delta][R]$$

where

$$[\Delta]=\begin{bmatrix}0\\0\end{bmatrix}$$ (Since no yielding of supports.)

$$[R]=\begin{bmatrix}V_D\\H_D\end{bmatrix}$$ (Chosen redundants)

$$[\Delta_L]=\begin{bmatrix}\Delta_{1L}\\\Delta_{2L}\end{bmatrix}$$ represents the displacements at co-ordinates due to the external loads on the basic structure.

$$[\delta]=\begin{bmatrix}\delta_{11} & \delta_{12}\\\delta_{21} & \delta_{22}\end{bmatrix}$$ represents the flexibility coefficients.

The $[\Delta_L]$ and $[\delta]$ can be computed using the unit load method and the area moment multiplication technique. The basic determinate frame, subjected to external loads and unit forces at the co-ordinates, is shown in Figure 9.52a for the construction of M_L, m_1, and m_2 diagrams. The BM diagrams due to the external loads and unit forces at the co-ordinates are shown in Figure 9.52b–d, respectively.

$$\Delta_{1L}=\int M_L m_1\frac{dx}{EI}=\frac{1}{EI}\begin{bmatrix}\left(\frac{1}{3}\times 4\times(-240)\times 6\right)\\+((-240)\times 4\times 7)\end{bmatrix}=\frac{-8{,}640}{EI}$$

$$\Delta_{2L}=\int M_L m_2\frac{dx}{EI}=\frac{1}{EI}\begin{bmatrix}\left(\frac{1}{3}\times 4\times(-240)\times(-4)\right)\\+(4\times(-240)(-2))\end{bmatrix}=\frac{3{,}200}{EI}$$

$$\delta_{11}=\int m_1 m_1\frac{dx}{EI}=\frac{1}{EI}\left[\left(\frac{1}{2}\times 3\times 5\times 2\right)+\frac{4}{6}[7(2\times 7+3)\right.$$
$$\left.+3(2\times 3+7)](7\times 4\times 7)\right]=\frac{316.33}{EI}$$

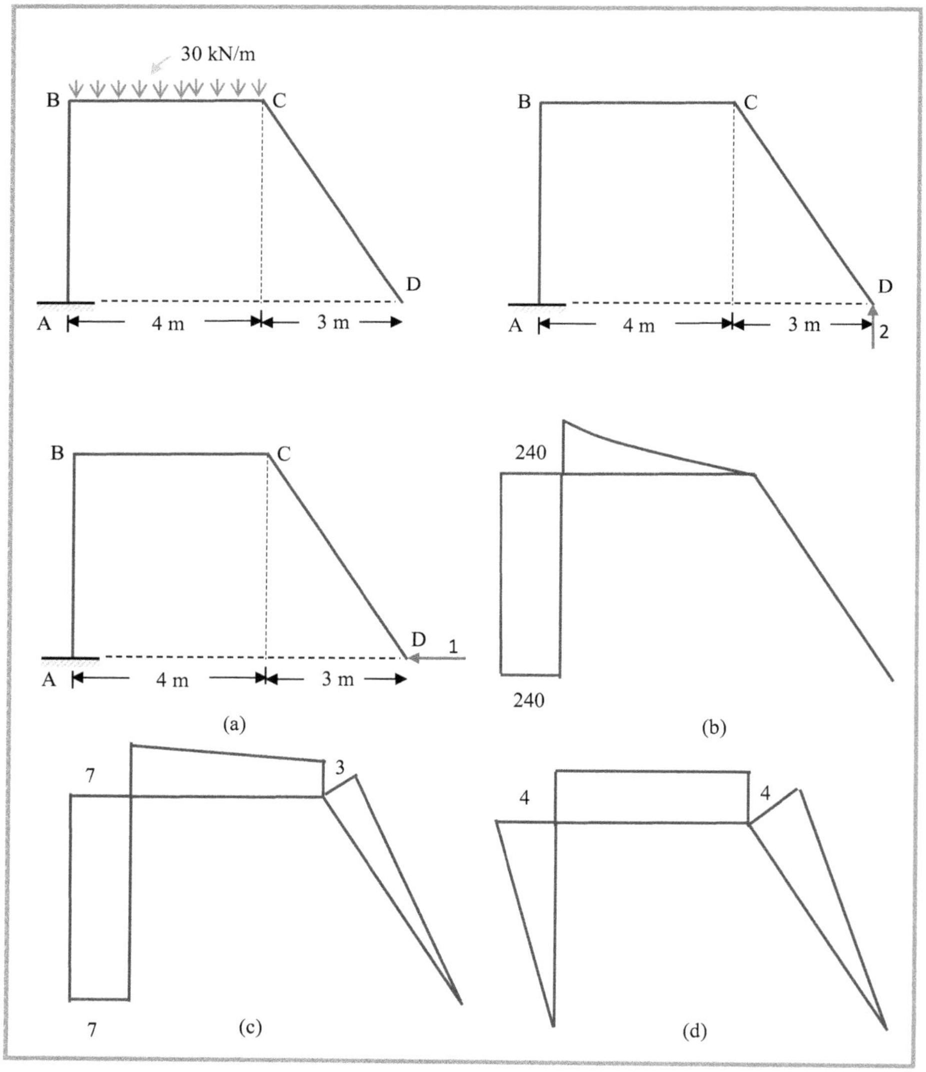

FIGURE 9.52 (a) Basic frame with loads and unit forces at co-ordinates, (b) M_L diagram, (c) m_1 diagram, and (d) m_2 diagram.

$$\delta_{12} = \delta_{21} = \int m_1 m_2 \frac{dx}{EI} = \frac{-1}{EI}\begin{bmatrix} \frac{5}{6}[3(2\times4+0)]+ \\ \left[\frac{4}{2}(3+7)\times4\right]+\left(\frac{1}{2}\times4\times4\times7\right) \end{bmatrix}$$

$$= \frac{-156}{EI}$$

$$\delta_{22} = \int m_2 m_2 \frac{dx}{EI} = \frac{1}{EI}\begin{bmatrix} \frac{5}{6}\{4(2\times4+0)\} \\ +(4\times4\times4)+\left(\frac{1}{2}\times4\times4\times2.67\right) \end{bmatrix}$$

$$= \frac{112}{EI}$$

Substituting the above values into equation 9.6, we have

$$\begin{bmatrix} 0 \\ 0 \end{bmatrix} = \frac{1}{EI}\begin{bmatrix} -8640 \\ 3200 \end{bmatrix} + \frac{1}{EI}\begin{bmatrix} 316.33 & -156 \\ -156 & 112 \end{bmatrix}\begin{bmatrix} V_D \\ H_D \end{bmatrix}$$

By solving the above, we get

$$\begin{bmatrix} V_D \\ H_D \end{bmatrix} = \begin{bmatrix} 42.23 \\ 30.25 \end{bmatrix} \text{kN}$$

Final end reactions: Using the free body diagram shown in Figure 9.53a and applying the static equilibrium equations, we can compute the other support reactions. The shear force and bending moment diagrams are shown in Figure 9.53b–d, respectively. The probable elastic curve is also shown in Figure 9.53e.

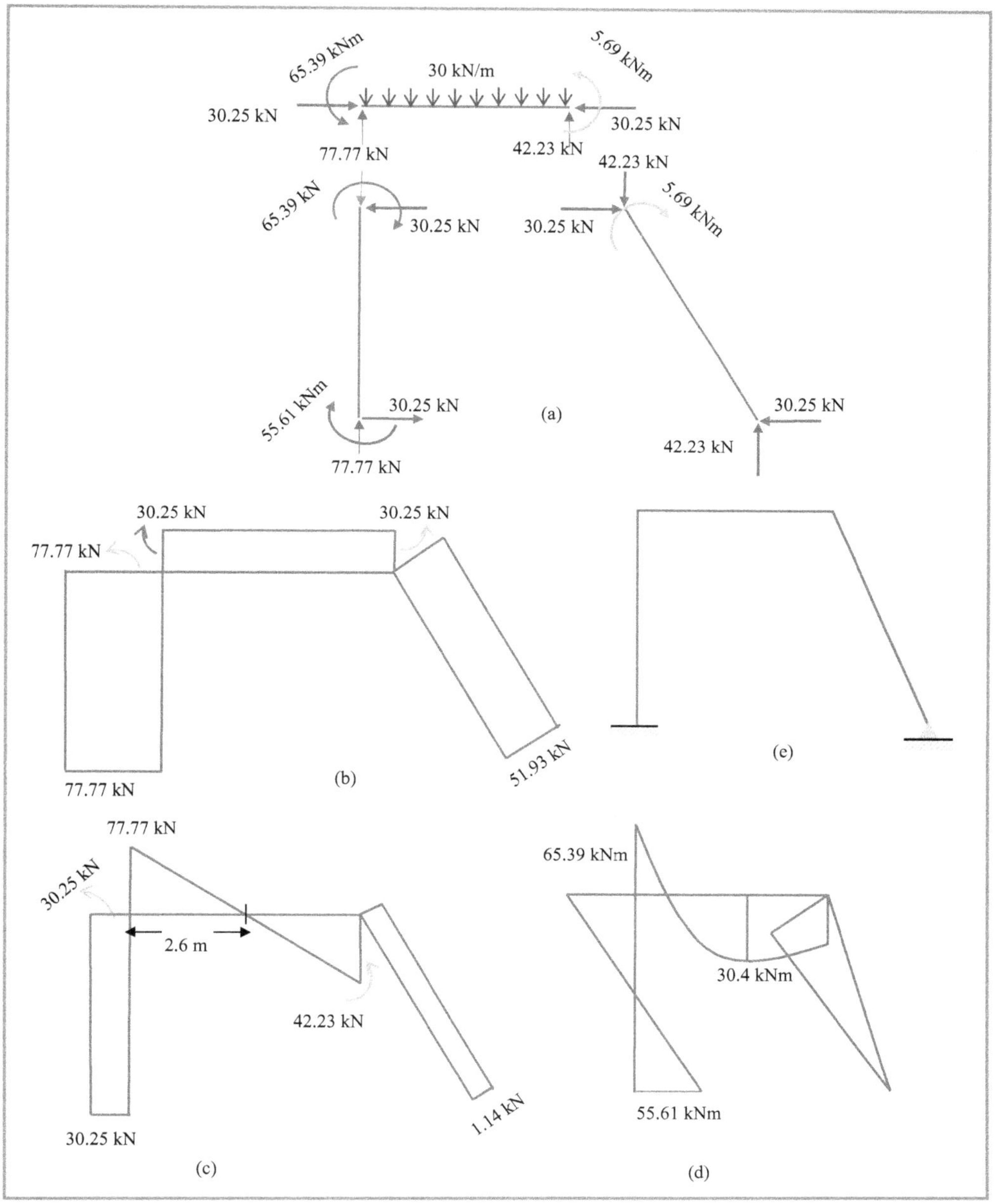

FIGURE 9.53 (a) Free body diagram, (b) axial force diagram, (c) shear force diagram, (d) bending moment diagram, and (e) probable deflected shape.

Example 9.13

Analyse Q. No. 9.11, if support D settles down by 10 mm and rotates 0.002 rad in a clockwise direction instead of external loads, as shown in Figure 9.54. Assume $EI = 135 \times 10^3$ kNm^2.

Solution:

The static indeterminacy is $I_s = 5 - 3 = 2$. Let us select the reactions at D, V_D, and H_D as the redundants, so that the yielding conditions can be directly incorporated into the $[\Delta]$ matrix. The released structure can be obtained by removing the chosen redundants. Assign the co-ordinates to the redundants, as shown in Figure 9.55.

Equation 9.6 of the flexibility Matrix is

$$[\Delta] = [\Delta_L] + [\delta][R]$$

where

$$[\Delta] = \begin{bmatrix} \Delta_1 \\ \Delta_2 \\ \Delta_3 \end{bmatrix} = \begin{bmatrix} 0.01 \\ 0.00 \\ 0.002 \end{bmatrix}$$

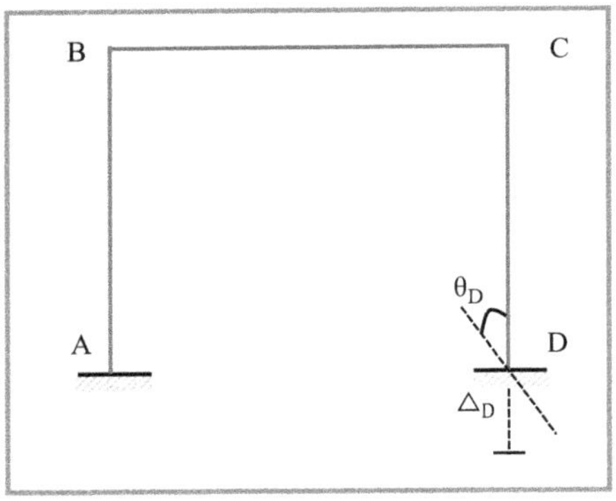

FIGURE 9.54 Rigid frame subjected to yielding of supports.

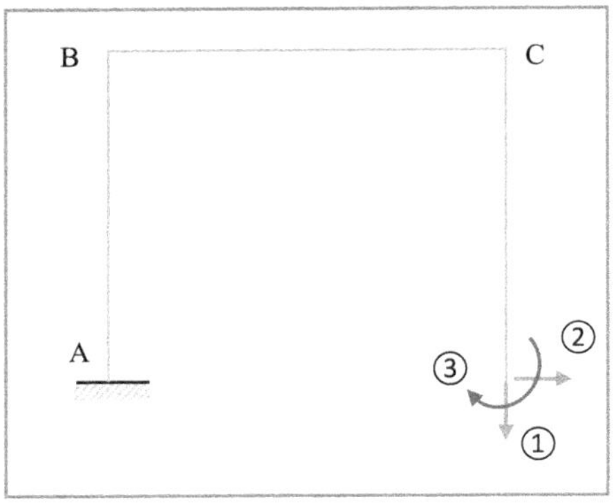

FIGURE 9.55 Released frame.

$$[R] = \begin{bmatrix} V_D \\ H_D \\ M_D \end{bmatrix}; \ [\Delta_L] = \begin{bmatrix} \Delta_{1L} \\ \Delta_{2L} \\ \Delta_{3L} \end{bmatrix}$$

$$= \begin{bmatrix} 0 \\ 0 \\ 0 \end{bmatrix} \text{(since, no external loads acting on the basic structure)}$$

$[\delta] = \begin{bmatrix} \delta_{11} & \delta_{12} & \delta_{13} \\ \delta_{21} & \delta_{22} & \delta_{23} \\ \delta_{31} & \delta_{32} & \delta_{33} \end{bmatrix}$ represents the flexibility coefficients due to unit loads. These can be calculated using the unit load method and the area moment multiplication technique, utilising the m_1, m_2, and m_3 diagrams. The m_1, m_2, and m_3 diagrams are constructed in Figure 9.56b–d for the unit loads applied separately at each co-ordinate, as shown in Figure 9.56a.

$$\delta_{11} = \int m_1 m_1 \frac{dx}{EI} = \frac{1}{2EI}\left[\frac{6}{6}(6(2\times 6+0))\right] + \frac{1}{EI}[6\times 3\times 6] = \frac{144}{EI}$$

$$\delta_{12} = \delta_{21} = \int m_1 m_2 \frac{dx}{EI}$$

$$= \frac{1}{2EI}\left[\frac{-1}{2}\times 6\times 6\times 3\right] + \frac{1}{EI}\left[\frac{-1}{2}\times 3\times 3\times 6\right] = \frac{-54}{EI}$$

$$\delta_{13} = \delta_{31} = \int m_1 m_3 \frac{dx}{EI}$$

$$= \frac{1}{2EI}\left[\frac{1}{2}\times(-6)\times 6\times(-1)\right] + \frac{1}{EI}[3\times(-6)\times(-1)] = \frac{27}{EI}$$

$$\delta_{22} = \int m_2 m_2 \frac{dx}{EI} = \frac{1}{2EI}[3\times 6\times 3] + \frac{1}{EI}\left[2\times\frac{1}{2}\times 3\times 3\times 2\right] = \frac{45}{EI}$$

$$\delta_{23} = \delta_{32} = \int m_2 m_3 \frac{dx}{EI}$$

$$= \frac{1}{2EI}[3\times 6\times(-1)] + \frac{1}{EI}\left[2\times\frac{1}{2}\times 3\times 3\times(-1)\right] = \frac{-18}{EI}$$

$$\delta_{33} = \int m_3 m_3 \frac{dx}{EI} = \frac{1}{2EI}[1\times 6\times 1] + \frac{1}{EI}[2\times 1\times 3\times 1] = \frac{9}{EI}$$

$$\begin{bmatrix} 0.01 \\ 0.0 \\ 0.002 \end{bmatrix} = \begin{bmatrix} 0 \\ 0 \\ 0 \end{bmatrix} + \frac{1}{EI}\begin{bmatrix} 252 & -54 & 45 \\ -54 & 45 & -18 \\ 45 & -18 & 9 \end{bmatrix}\begin{bmatrix} V_D \\ H_D \\ M_D \end{bmatrix}$$

$$\begin{bmatrix} V_D \\ H_D \\ M_D \end{bmatrix} = \begin{bmatrix} 8.57 \text{ kN} \\ 60 \text{ kN} \\ 124.29 \text{ kNm} \end{bmatrix}$$

Final reactions: With the known values of V_D, H_D, and M_D, with the help of the free body diagram and using the

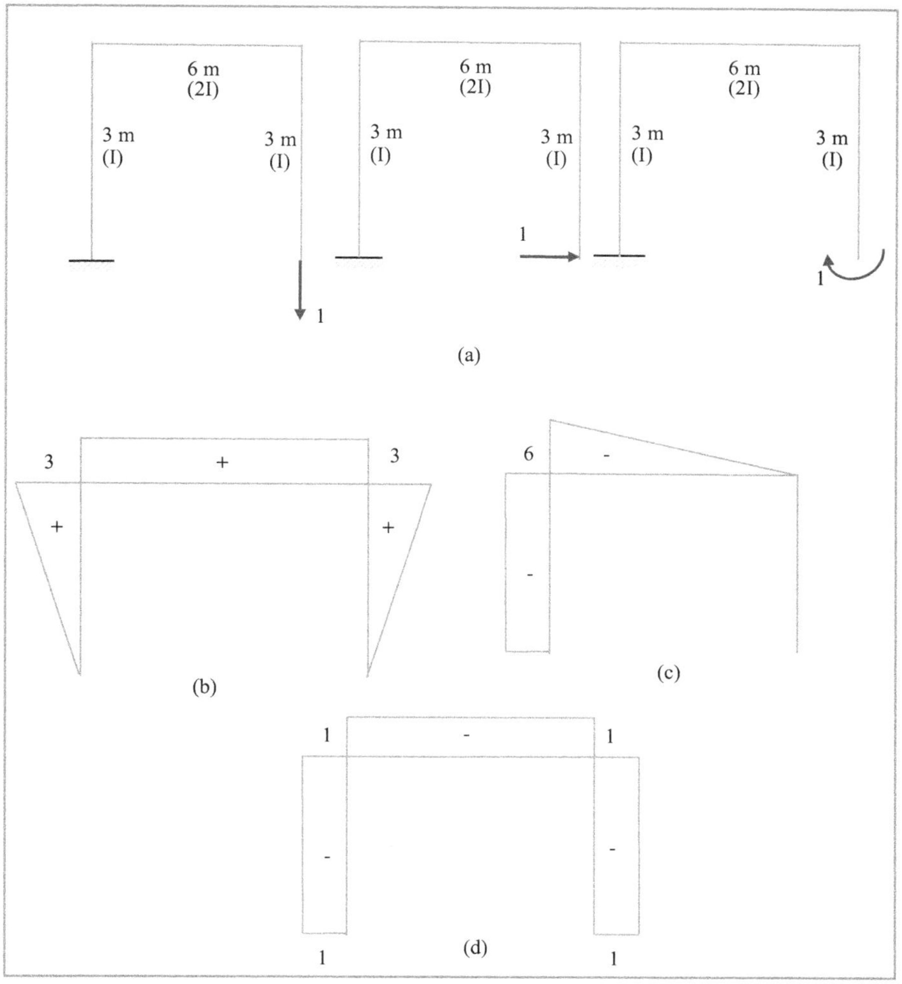

FIGURE 9.56 (a) Unit loads at co-ordinates 1, 2, and 3, (b) m_1 diagram, (c) m_2 diagram, and (d) m_3 diagram.

static equilibrium equations, the other support reactions can be determined, as shown in the in Figure 9.57a. The shear force and bending moment diagrams are shown in Figure 9.57b–c, respectively. The probable deflected shape is also shown in Figure 9.57d.

Example 9.14

Analyse the rigid-jointed bent shown in Figure 9.58 using the flexibility matrix method. Draw the shear force and bending moment diagrams. Assume EI is constant.

Solution:

The static indeterminacy is $I_S = 5 - 3 = 2$, indicating that the frame is statically indeterminate to the second degree. Let us select the moments at B and C as the redundants. The basic determinate structure can be obtained by introducing hinges at B and C, thereby releasing the moment at these points. Assign co-ordinates 1 and 2 to the redundants, as shown in Figure 9.59.

Equation 9.6 of the flexibility matrix is

$$[\Delta] = [\Delta_L] + [\delta][R]$$

where

$$[\Delta] = \begin{bmatrix} \Delta_1 \\ \Delta_2 \end{bmatrix} = \begin{bmatrix} 0 \\ 0 \end{bmatrix} \text{ (Since no yielding of supports)}$$

$$[R] = \begin{bmatrix} M_B \\ M_C \end{bmatrix} \text{(Redundants moments.)}$$

The $[\Delta_L]$ and $[\delta]$ represents the displacements due to external loads and the flexibility coefficients at co-ordinates in the basic frame, respectively. The basic or determinate frame, subjected to external loads and unit forces at the co-ordinates, is shown in Figure 9.60a, for the construction of M_L, m_1, and m_2 diagrams. $[\Delta_L]$ and $[\delta]$ can be calculated using the unit load method and the area moment multiplication technique with the help of M_L, m_1, and m_2 diagrams, as shown in Figure 9.60b–d, respectively.

$$\Delta_{1L} = \int M_L m_1 \frac{dx}{EI} = \frac{1}{EI}\left[\frac{1}{2} \times 5 \times 60 \times 0.47\right] = \frac{70.5}{EI}$$

$$\Delta_{2L} = \int M_L m_2 \frac{dx}{EI} = \frac{1}{EI}\left[\frac{1}{2} \times 5 \times 60 \times 0.53\right] = \frac{79.5}{EI}$$

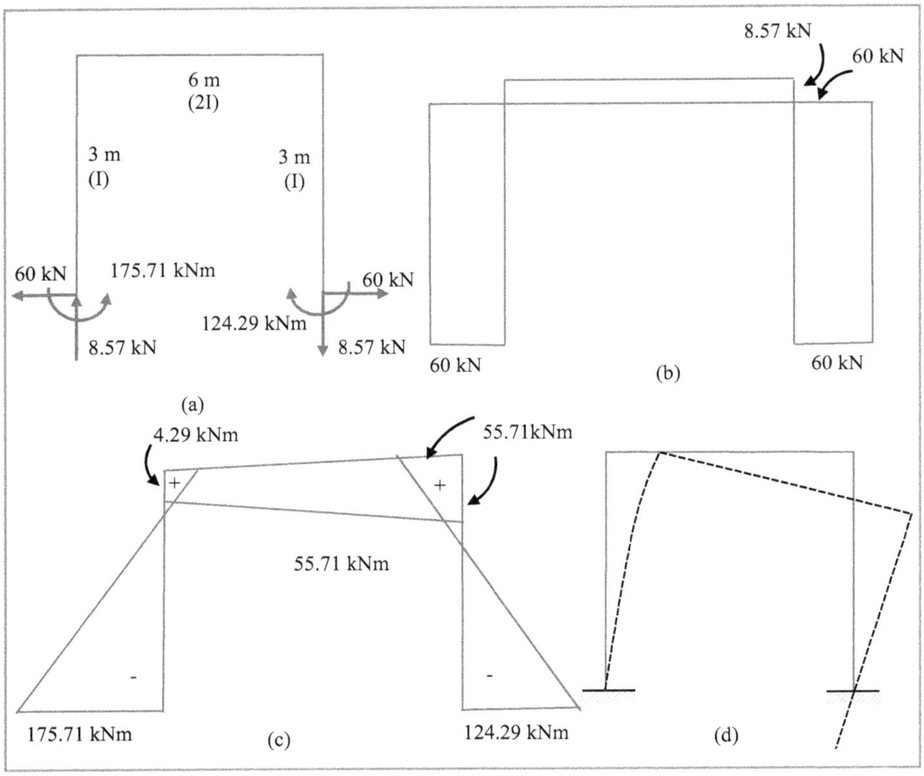

FIGURE 9.57 (a) Free body diagram, (b) shear force diagram, (c) bending moment diagram, and (d) probable deflected shape.

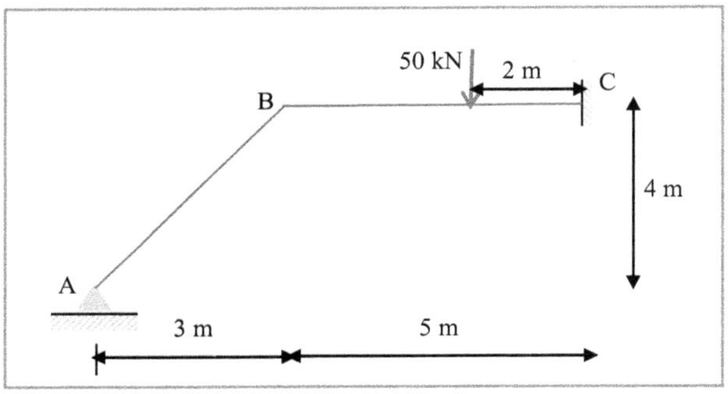

FIGURE 9.58 A rigid bent ABC.

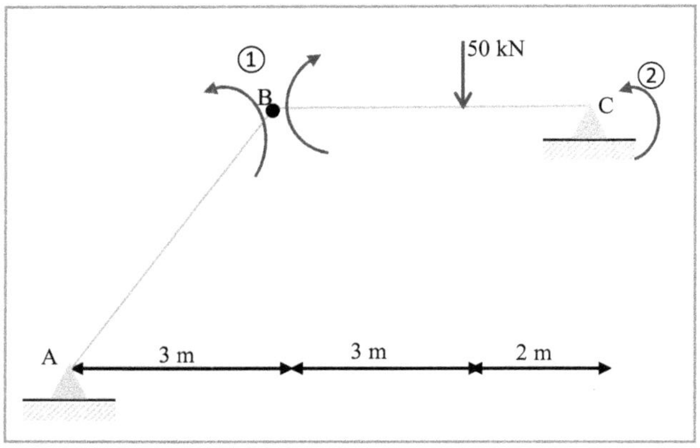

FIGURE 9.59 Determinate frame ABC with co-ordinates and loads.

$$\delta_{11} = \int m_1 m_1 \frac{dx}{EI} = \frac{1}{EI}\left[\frac{5}{6}\times 1\times 2\times 1\times 2\right] = \frac{3.33}{EI}$$

$$\delta_{12} = \delta_{21} = \int m_1 m_2 \frac{dx}{EI} = \frac{1}{EI}\left[\frac{5}{6}[1(2\times 0+1)]+0\right] = \frac{0.83}{EI}$$

$$\delta_{22} = \int m_2 m_2 \frac{dx}{EI} = \frac{1}{EI}\left[\frac{1}{2}\times 5\times 1\times 0.67\right] = \frac{1.67}{EI}$$

By substituting the above into equation 9.6,

$$\begin{bmatrix} 0 \\ 0 \end{bmatrix} = \begin{bmatrix} 70.5/EI \\ 79.5/EI \end{bmatrix} + \frac{1}{EI}\begin{bmatrix} 3.11 & 0.83 \\ 0.83 & 1.67 \end{bmatrix}\begin{bmatrix} M_B \\ M_C \end{bmatrix}$$

By solving the above equation, we get

$$\begin{bmatrix} M_B \\ M_C \end{bmatrix} = \begin{bmatrix} -10.62 \\ -42.32 \end{bmatrix} \text{kNm}$$

Final end reactions: After determining the moments at *B* and *C*, and using the static equilibrium equations, the other support reactions can be calculated, as shown in the free body diagram in Figure 9.61a. The axial force, shear force, and bending moment diagrams are shown in Figure 9.61b–d, respectively. The probable deflected shape is shown in Figure 9.61e.

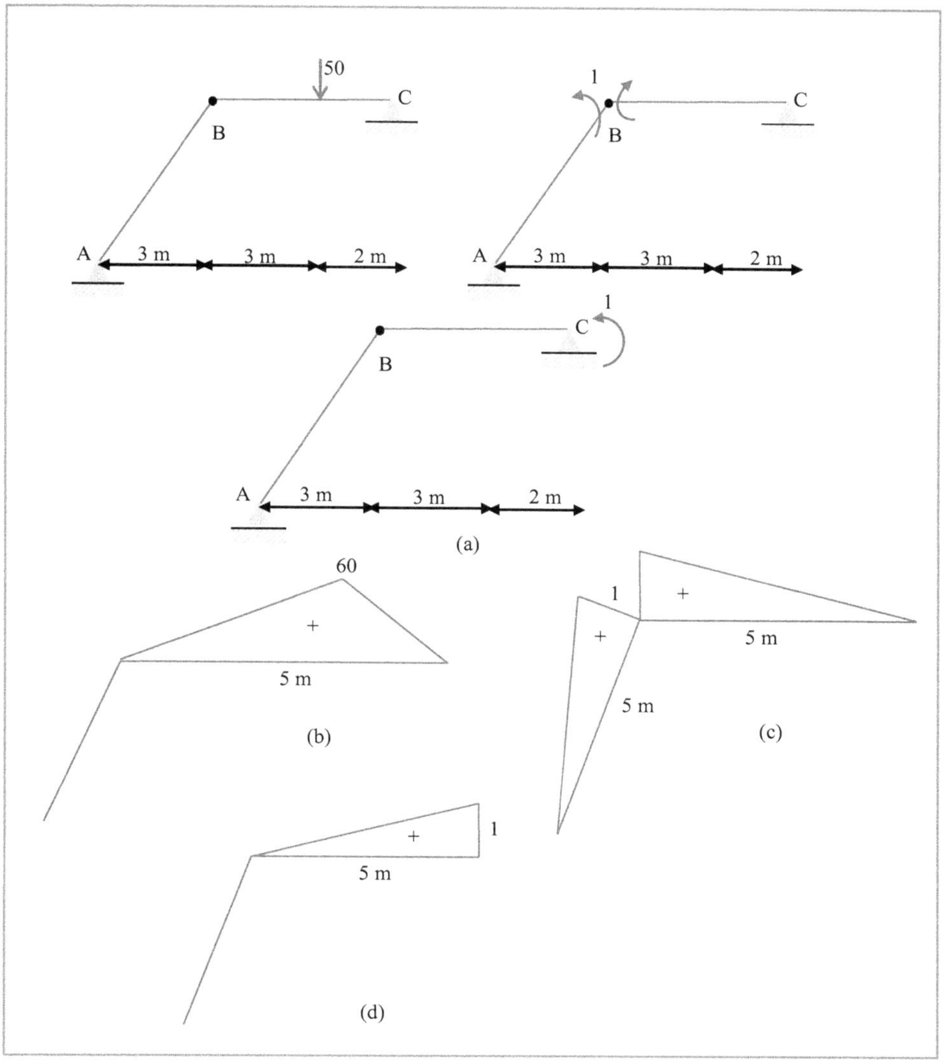

FIGURE 9.60 (a) Unit loads at co-ordinates 1, 2, 3, (b) M_L diagram, (c) m_1 diagram, and (d) m_2 diagram.

Example 9.15

Analyse the rigid-jointed frame shown in Figure 9.62. Draw the shear force and bending moment diagrams. Assume $EI = 135 \times 10^3$ kNm².

Solution:

The static indeterminacy of the rigid frame is $I_S = 6 - 3 = 3$. Select the reactions at *B* and *C* as the redundants. The basic or determinate structure is obtained by releasing the redundants, i.e., by removing the supports at *B* and *C*, and assign co-ordinates to them, as shown in Figure 9.63.

Equation 9.6 of the flexibility matrix is

$$[\Delta] = [\Delta_L] + [\delta][R]$$

where

$$[\Delta] = \begin{bmatrix} 0 \\ 0 \\ 0 \end{bmatrix} \text{(Due to no yielding of supports)}$$

$$[R] = \begin{bmatrix} V_B \\ V_C \\ H_C \end{bmatrix} \text{(Redundants forces)}$$

The $[\Delta_L]$ and $[\delta]$ represents the displacements due to the external loads and the flexibility coefficients resulting from unit forces at the co-ordinates in the basic determinate structure. These can be calculated by using the unit load method and area moment multiplication technique with the help of M_L, m_1, m_2, and m_3 diagrams, constructed for the external loads and unit forces at co-ordinates 1, 2, and 3, respectively, as shown in Figure 9.64.

$$\Delta_{1L} = \int M_L m_1 \frac{dx}{EI}$$

$$= \frac{1}{2EI}\left[\left(\frac{1}{3} \times 10 \times (-2{,}500) - \frac{1}{3} \times 5 \times (-625)\right) \times 3.03\right]$$

$$+ \frac{1}{EI}\left[\frac{-5}{2}(2{,}500 + 2{,}750) \times 5\right]$$

$$= \frac{-11{,}046.88}{EI} + \left(\frac{-65{,}525}{EI}\right) = \frac{-76{,}671.88}{EI}$$

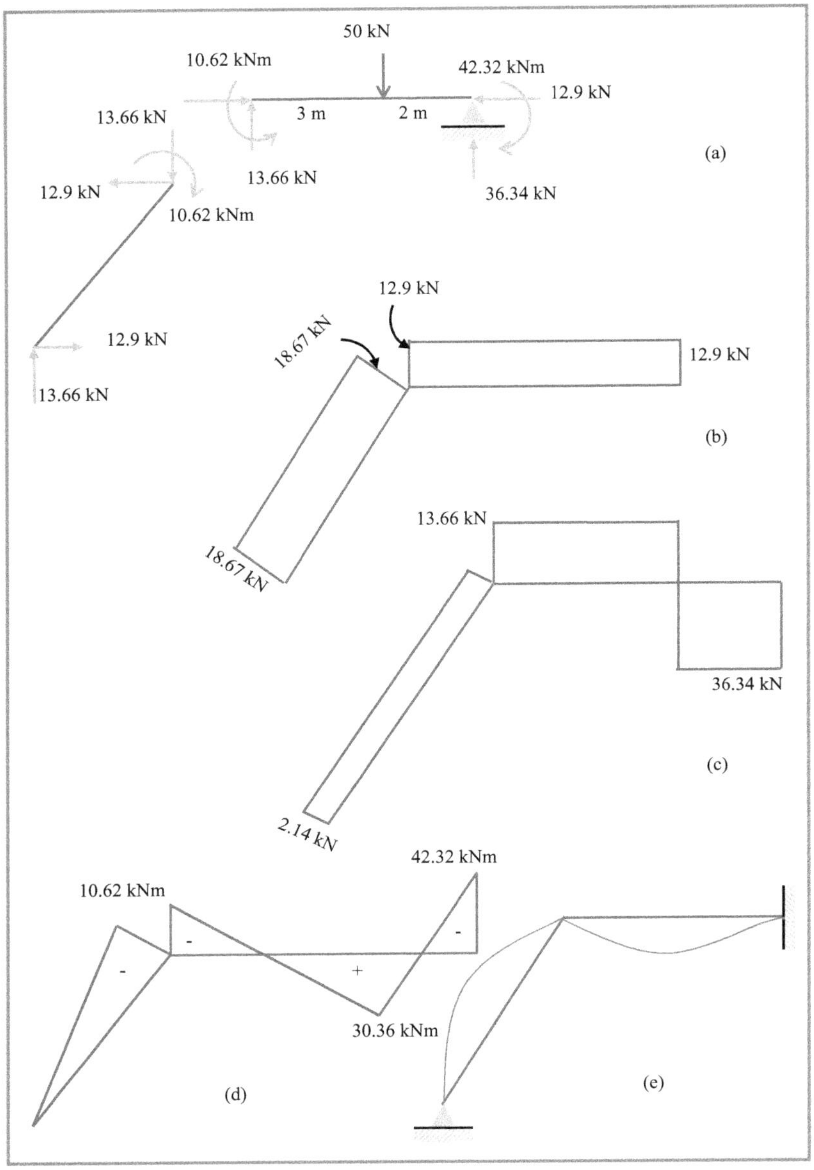

FIGURE 9.61 (a) Free body diagram, (b) axial force diagram, (c) shear force diagram, (d) bending moment diagram, and (e) probable deflected shape.

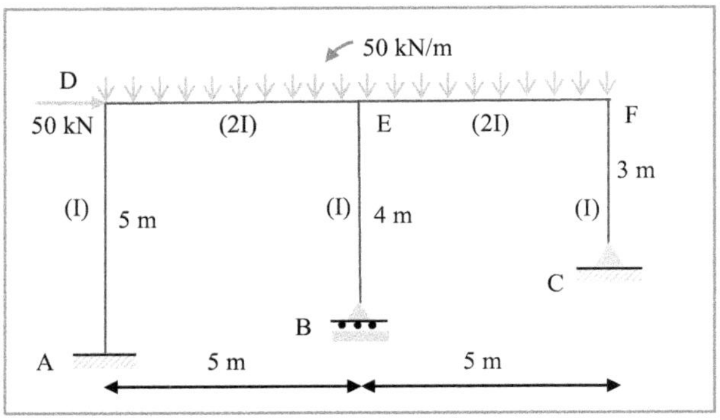

FIGURE 9.62 Indeterminate two-bay frame.

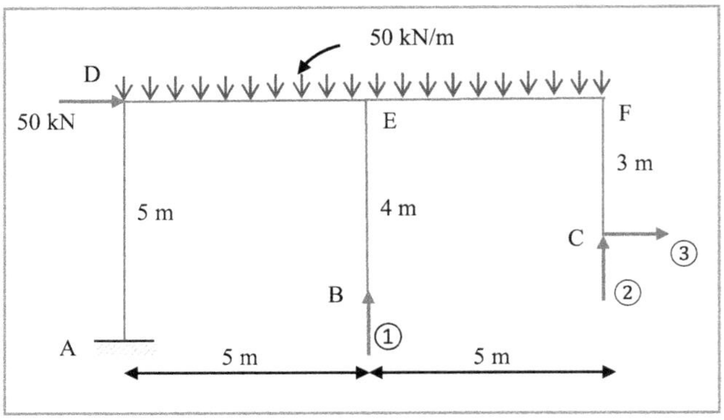

FIGURE 9.63 Basic frame with loads and co-ordinates.

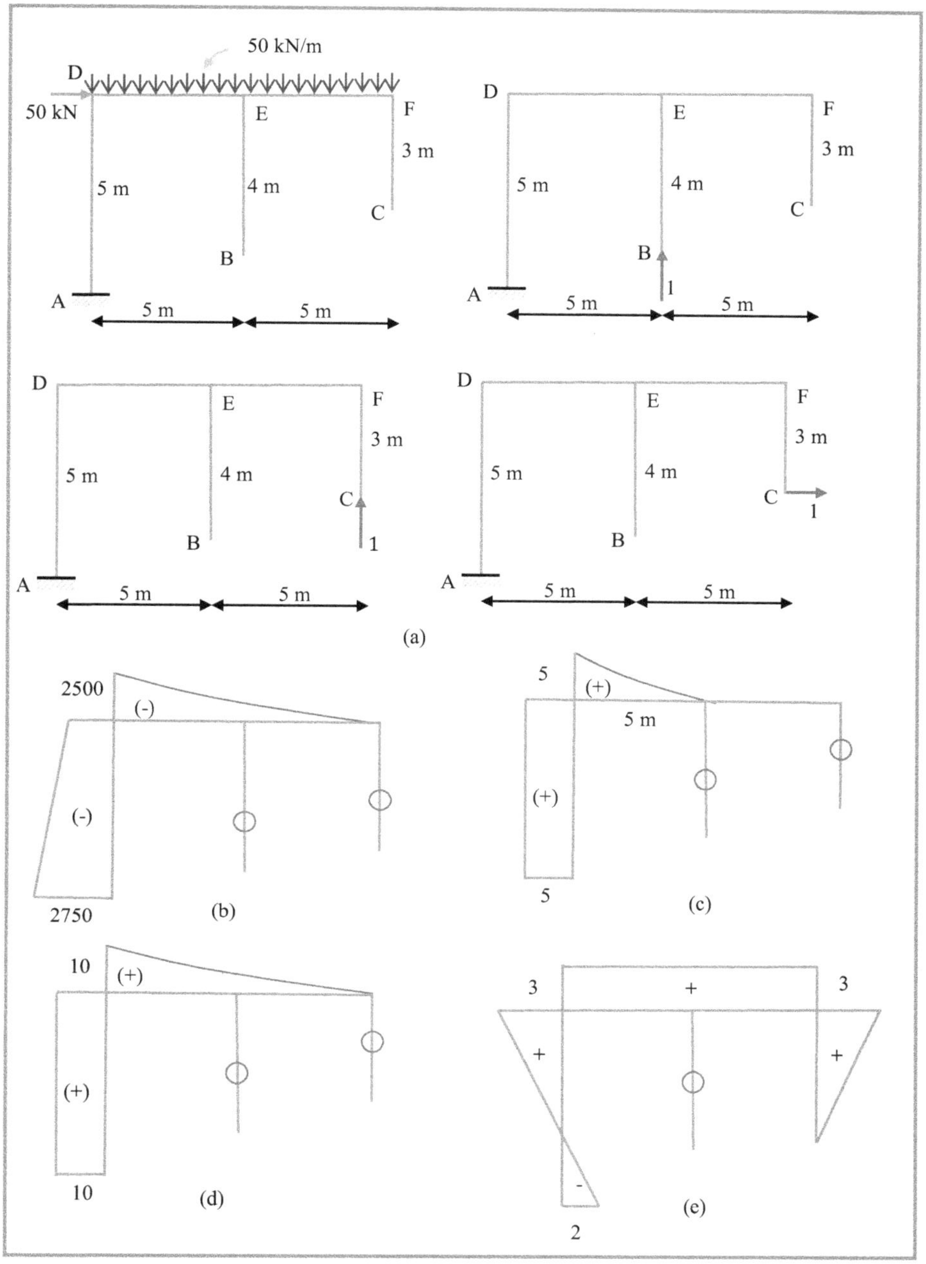

FIGURE 9.64 (a) Determinate frame with external loads and unit loads at co-ordinates 1, 2, and 3, (b) M_L diagram, (c) m_1 diagram, (d) m_2 diagram, and (e) m_3 diagram.

$$\Delta_{2L} = \frac{1}{2EI}\left[\frac{1}{3}\times 10\times(-2{,}500)\times 7.5\right]$$

$$+\frac{1}{EI}\left[\frac{5}{2}(-2{,}500-2{,}750)\times(10)\right] = \frac{-162{,}500}{EI}$$

$$\Delta_{3L} = \int M_L m_3 \frac{dx}{EI} = \frac{1}{2EI}\left[\frac{1}{3}\times 10\times(-2{,}500)\times 3\right]$$

$$+\frac{1}{EI}\left[\frac{5}{2}(-2{,}500-2{,}750)\times 0.46\right]$$

$$= \frac{-18{,}537.5}{EI}$$

$$\delta_{11} = \int m_1 m_1 \frac{dx}{EI} = \frac{1}{2EI}\left[\frac{5}{6}(5\times 2\times 5)\right] + \frac{1}{EI}[5\times 5\times 5] = \frac{145.83}{EI}$$

$$\delta_{12} = \delta_{21} = \int m_1 m_2 \frac{dx}{EI} = \frac{1}{2EI}\left[\frac{5}{6}(5(2\times 10+5))\right]$$

$$+\frac{1}{EI}[5\times 5\times 10] = \frac{302.08}{EI}$$

$$\delta_{13} = \delta_{31} = \int m_1 m_3 \frac{dx}{EI} = \frac{1}{2EI}\left[\frac{1}{2}\times 5\times 5\times 3\right]$$

$$+\frac{1}{EI}[5\times 5\times 0.5] = \frac{31.25}{EI}$$

$$\delta_{22} = \int m_2 m_2 \frac{dx}{EI} = \frac{1}{2EI}\left[\frac{10}{6}\times 10\times 2\times 10\right]$$

$$+\frac{1}{EI}[10\times 5\times 10] = \frac{666.67}{EI}$$

$$\delta_{23} = \delta_{32} = \int m_2 m_3 \frac{dx}{EI} = \frac{1}{2EI}\left[\frac{1}{2}\times 10\times 10\times 3)\right]$$

$$+\frac{1}{EI}[10\times 5\times 0.5] = \frac{100}{EI}$$

$$\delta_{33} = \int m_3 m_3 \frac{dx}{EI} = \frac{1}{2EI}[10\times 3\times 3]$$

$$+\frac{1}{EI}\left[\left(\frac{1}{2}\times 3\times 3\times 2\times 7\right)+\left(\frac{1}{2}\times 2\times 2\times 1.33\right)\right] = \frac{65.66}{EI}$$

On substitution of the above values into equation 9.6, we have

$$\begin{bmatrix} 0 \\ 0 \\ 0 \end{bmatrix} = \frac{1}{EI}\begin{bmatrix} -76{,}671.88 \\ -162{,}500 \\ -18{,}537.5 \end{bmatrix}$$

$$+\frac{1}{EI}\begin{bmatrix} 145.83 & 302.08 & 31.25 \\ 302.08 & 666.67 & 100 \\ 31.25 & 100 & 65.66 \end{bmatrix}\begin{bmatrix} V_B \\ V_C \\ H_C \end{bmatrix}$$

By solving the above equation, we obtain

$$\begin{bmatrix} V_B \\ V_C \\ H_C \end{bmatrix} = \begin{bmatrix} 281.23 \\ 121.89 \\ -37.16 \end{bmatrix} \text{kN}$$

Remaining reactions: After knowing the values of reactions at B and C, and using the static equilibrium equations, the remaining support reactions can be calculated, as shown in Figure 9.65a. The shear force and bending moment diagrams are shown in Figure 9.65b–c, respectively. The probable deflected shape is shown in Figure 9.65d.

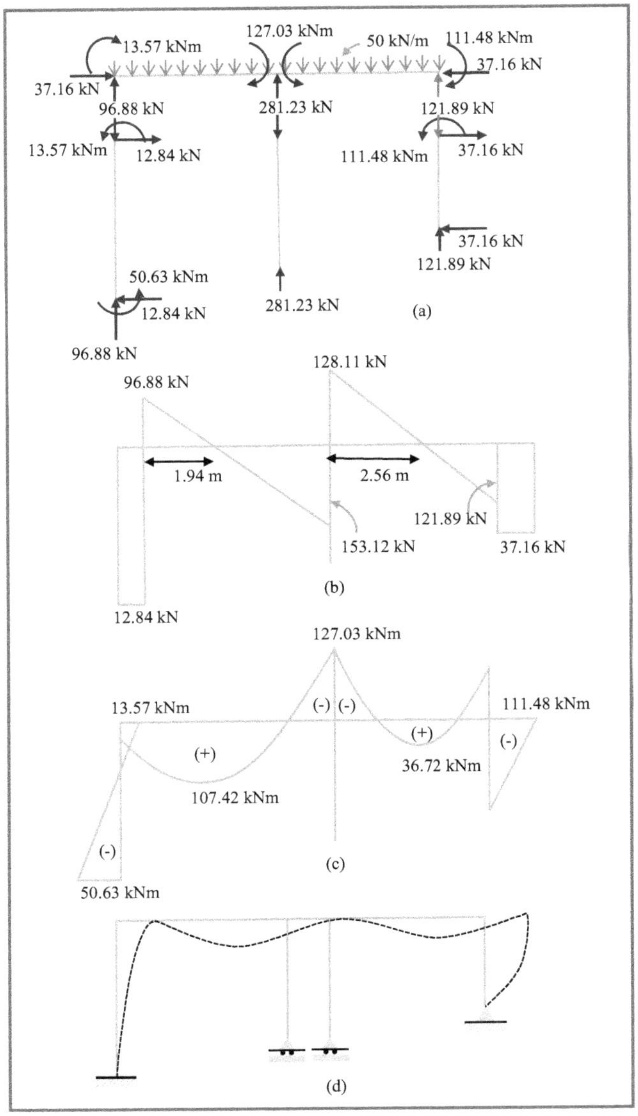

FIGURE 9.65 (a) Free body diagram, (b) shear force diagram, (c) bending moment diagram, and (d) probable deflected shape.

Example 9.16

Analyse the two-storey single-bay frame shown in Figure 9.66a. Draw the shear force and bending moment diagrams.

Solution:

The frame is symmetric with respect to the centre line of CD & BE. Therefore, half of the frame needs to be considered for the analysis with appropriate support conditions at the centre of CD and BE. Since, the frame is symmetric, the slope at the centre of CD and BE is zero, making the shear force at these points also zero. As a result, guided fixed supports are placed at these locations, as shown in Figure 9.66b, reducing the static indeterminacy to $I_S = 6-3 = 3$. Select the moments at C', B' and the horizontal reaction at A as the redundants. The basic determinate structure is obtained by releasing the chosen redundants, i.e., by replacing the guided and hinged supports with roller supports at A, B^1, and C^1 and assigning co-ordinates to them as shown in Figure 9.67.

Equation 9.6 of the flexibility matrix is

$$[\Delta] = [\Delta_L] + [\delta][R]$$

where

$$[\Delta] = \begin{bmatrix} 0 \\ 0 \\ 0 \end{bmatrix}$$

$$[R] = \begin{bmatrix} M_{B'} \\ M_{C'} \\ H_A \end{bmatrix}$$

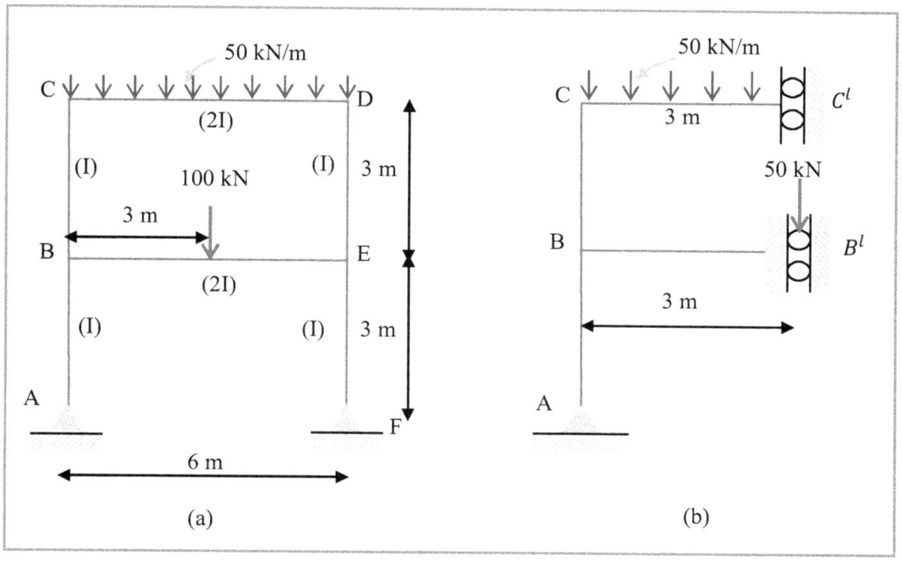

FIGURE 9.66 (a) Symmetrical two-storey frame and (b) half portion of frame.

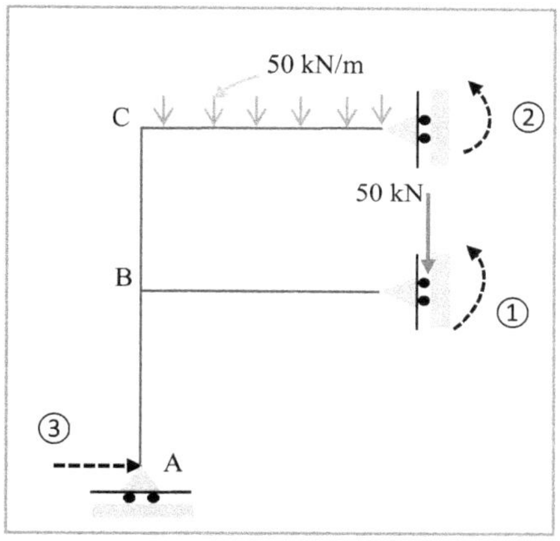

FIGURE 9.67 Basic determinate frame with co-ordinates and loads.

The $[\Delta_L]$ and $[\delta]$ can be calculated using the unit load method and the area moment multiplication technique with the help of M_L, m_1, m_2, and m_3 diagrams, as shown in Figure 9.68. The basic frame subjected to external loads and unit loads at all the co-ordinates are is shown in Figure 9.68 for the construction of M_L, m_1, m_2, and m_3 diagrams.

$$\Delta_{1L} = \int M_L \times m_1 \frac{dx}{EI}$$

$$= \frac{1}{2EI}\left[\left(\frac{1}{3}\times 3\times(-225)(0)+\frac{1}{2}\times 3\times(-150)\right)\times 1\right]$$

$$+\frac{1}{EI}\left[\frac{1}{2}\times 3\times(-1)\times(25)\right]$$

$$= \frac{-150}{EI}$$

$$\Delta_{2L} = \int M_L m_2 \frac{dx}{EI} = \frac{1}{2EI}\left[\frac{1}{3}\times 3\times(-225)\times(+1)\right]$$

$$+\frac{1}{EI}\left[\frac{1}{2}\times 3\times 1\times(-100)\right]$$

$$= \frac{-112.5}{EI}+\frac{-150}{EI} = \frac{-262.5}{EI}$$

$$\Delta_{3L} = \int M_L m_3 \frac{dx}{EI} = \frac{1}{EI}\left[\frac{1}{2}\times 3\times(-3)\times(+25)\right] = \frac{-112.5}{EI}$$

$$\delta_{11} = \int m_1 m_1 \frac{dx}{EI} = \frac{1}{2EI}[(1\times 3\times 1)]+\frac{1}{EI}\left[\frac{1}{2}\times(-1)\times 3\times\left(\frac{-2}{3}\right)\right] = \frac{2.5}{EI}$$

$$\delta_{12} = \delta_{21} = \int m_1 m_2 \frac{dx}{EI} = \frac{1}{EI}\left[\frac{3}{6}(-1(2\times 0+1))\right] = \frac{-0.5}{EI}$$

$$\delta_{13} = \delta_{31} = \int m_1 m_3 \frac{dx}{EI} = \frac{1}{EI}\left[-1\times\frac{1}{2}\times 3\times(-2)\right] = \frac{3}{EI}$$

$$\delta_{22} = \int m_2 m_2 \frac{dx}{EI} = \frac{1}{2EI}[1\times 3\times 1]+\frac{1}{EI}\left[\frac{3}{6}\times 1\times 2\times 1\right] = \frac{2.5}{EI}$$

$$\delta_{23} = \delta_{32} = \int m_2 m_3 \frac{dx}{EI} = \frac{1}{EI}\left[\frac{3}{6}(1\times(2\times 0+(-3)))\right] = \frac{-1.5}{EI}$$

$$\delta_{33} = \int m_3 m_3 \frac{dx}{EI} = \frac{1}{EI}\left[\frac{1}{2}\times 3\times 3\times 2\times 2\right] = \frac{18}{EI}$$

By substituting the above values into equation 9.6, we have

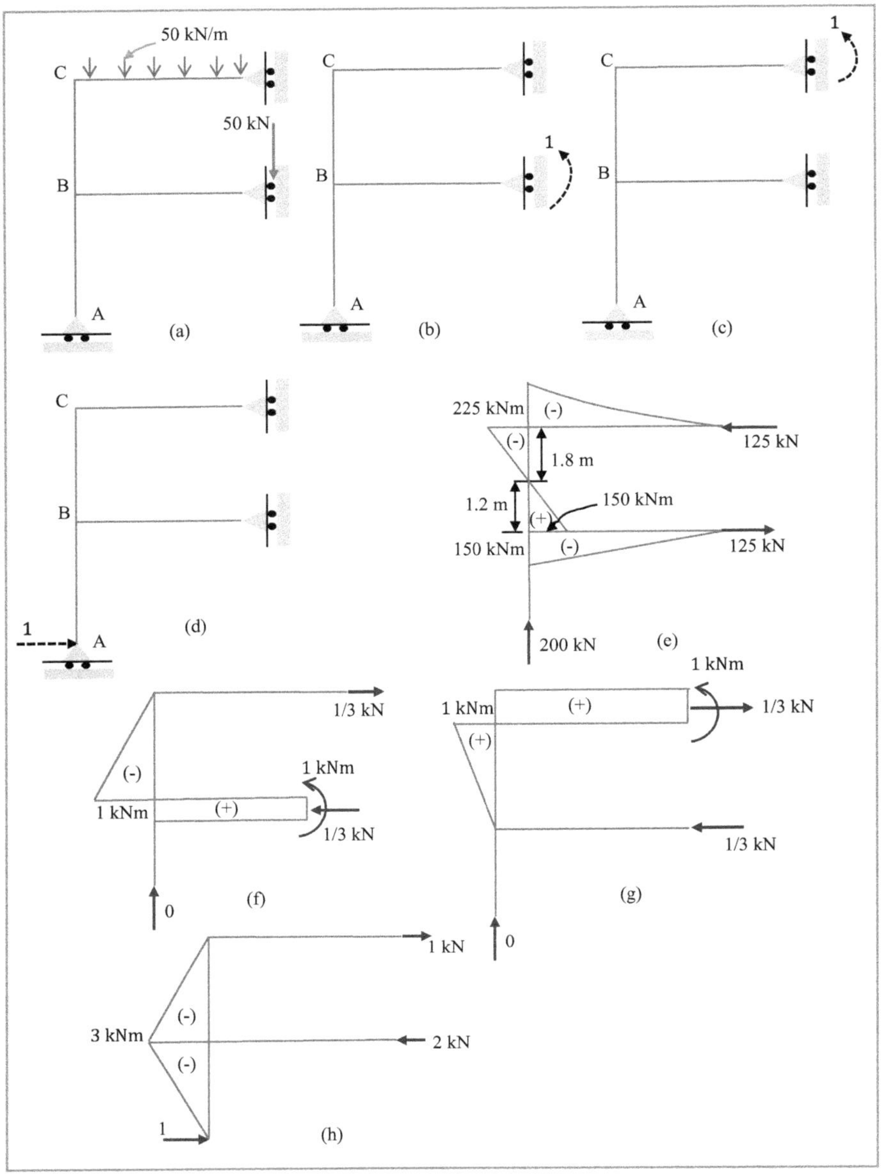

FIGURE 9.68 (a–d) Determinate frame with external loads and unit loads at co-ordinates 1, 2, and 3, respectively, (e) M_L diagram, (f) m_1 diagram, (g) m_2 diagram, and (h) m_3 diagram.

$$\begin{bmatrix} 0 \\ 0 \\ 0 \end{bmatrix} = \frac{1}{EI}\begin{bmatrix} -150 \\ -262.5 \\ -112.5 \end{bmatrix} + \frac{1}{EI}\begin{bmatrix} 2.5 & -1 & 3 \\ -1 & 2.5 & -1.5 \\ 3 & -1.5 & 18 \end{bmatrix}\begin{bmatrix} M_{B'} \\ M_{C'} \\ H_A \end{bmatrix}$$

After solving the above equation, we get

$$\begin{bmatrix} M_{B'} \\ M_{C'} \\ H_A \end{bmatrix} = \begin{bmatrix} 81\text{ kNm} \\ 123\text{ kNm} \\ 3\text{ kN} \end{bmatrix}$$

The shear force and bending moment diagrams are shown in Figure 9.69.

9.5 APPLICATIONS TO PIN-JOINTED PLANE FRAMES (TRUSSES)

Example 9.17

Analyse the pin-jointed plane frame shown in Figure 9.70 using the flexibility matrix method. Assume the axial rigidity of all the members is the same. The same has been solved using the energy method in Example 4.16. Assume tension is +ve and compression is −ve.

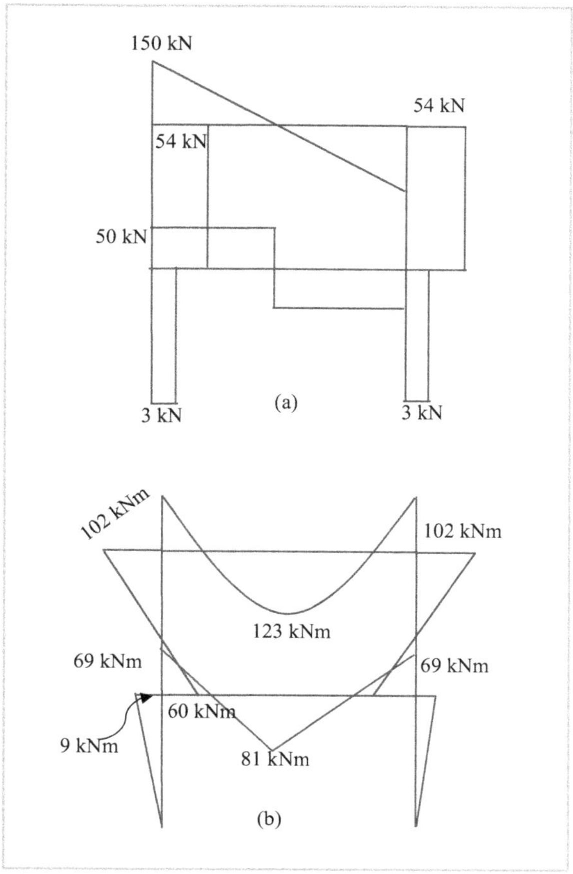

FIGURE 9.69 (a) Shear force diagram and (b) bending moment diagram.

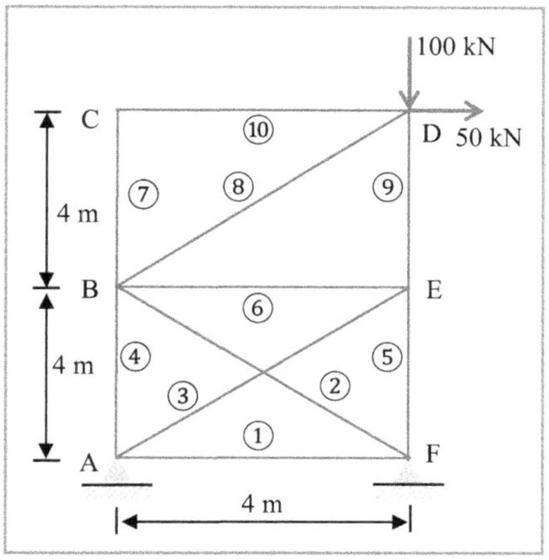

FIGURE 9.70 Statically indeterminate truss.

Solution:

The static indeterminacy $I_s = m - 2j + R$

$$= 10 - 2 \times 6 + 4 = 2$$

That is, the truss is statically indeterminate to the second degree (externally and internally indeterminate to the first degree). Let us select the horizontal reaction

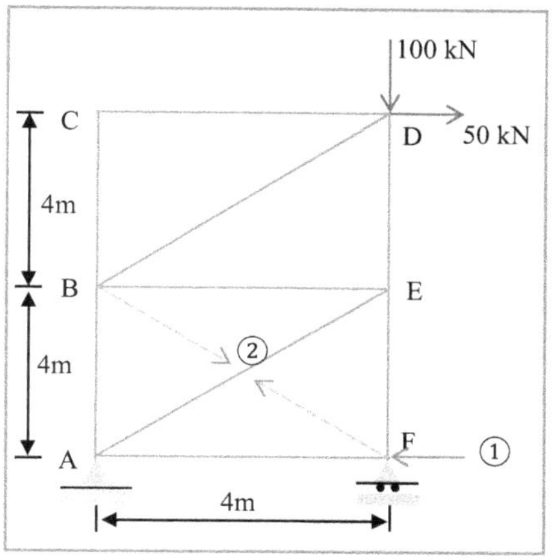

FIGURE 9.71 Basic truss with co-ordinates assigned to the selected redundants.

at support *F* and the force in member *BF* as the redundants. The truss can be made externally determinate by replacing the hinged support at *F* with a roller support, thereby releasing the horizontal reaction at *F*. Similarly, the truss can be made internally determinate by providing the cut in the member *FB*. The basic frame can be obtained by releasing the chosen redundants and assigning the co-ordinates ① and ②, respectively, to the horizontal reaction at *F* and the force in member *BF*, as shown in Figure 9.71.

$[\Delta] = \begin{bmatrix} \Delta_1 \\ \Delta_2 \end{bmatrix} = \begin{bmatrix} 0 \\ 0 \end{bmatrix}$ (∵ No initial displacements at ① and ② in the original frame)

$[R] = \begin{bmatrix} H_F \\ F_2 \end{bmatrix}$ are redundant forces.

$[\Delta_L] = \begin{bmatrix} \Delta_{1L} \\ \Delta_{2L} \end{bmatrix}$ represents the displacements at co-ordinates ① and ② due to the external loads on the released or basic or determinate frame.

$[\delta] = \begin{bmatrix} \delta_{11} & \delta_{12} \\ \delta_{21} & \delta_{22} \end{bmatrix}$ indicates the displacements at co-ordinates ① and ② due to the unit forces applied at ① and ② simultaneously on the released frame (Figure 9.72).

These displacements can be calculated using the unit load method, as shown below.

$$\Delta_{1L} = \sum F f_1 \frac{L}{AE}; \quad \Delta_{2L} = \sum F f_2 \frac{L}{AE};$$
$$\delta_{11} = \sum f_1 f_1 \frac{L}{AE}; \quad \delta_{12} = \delta_{21} = \sum f_1 f_2 \frac{L}{AE}; \quad (9.9)$$
$$\delta_{22} = \sum f_2 f_2 \frac{L}{AE};$$

where *F*, f_1, and f_2 are the forces in various members due to the applied loads and unit forces applied at ① and ②, respectively, and are presented in Table 9.2.

Substituting the values from Table 9.2 into equation 9.9, we have

TABLE 9.2
Forces in Members Due to External Loads and Unit Forces

Member	L (m)	F	f_1	f_2	Ff_1L	Ff_2L	f_1^2L	f_1f_2L	f_2f_2L
1	4	0	−1	−0.707	0	0	4	2.828	2.00
2	5.66	0	0	1	0	0	0	0	5.66
3	5.66	70.75	0	1	0	400	0	0	5.66
4	4	50	0	−0.707	0	−141.4	0	0	2.00
5	4	−200	0	−0.707	0	565.6	0	0	2.00
6	4	−50	0	−0.707	0	141.4	0	0	2.00
7	4	0	0	0	0	0	0	0	0
8	5.66	70.75	0	0	0	0	0	0	0
9	4	−150	0	0	0	0	0	0	0
10	4	0	0	0	0	0	0	0	0
Σ					0	965.6	4	2.828	19.32

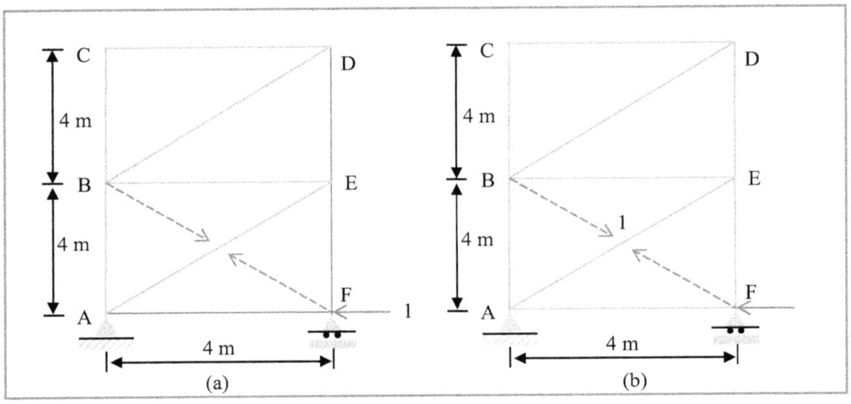

FIGURE 9.72 (a) unit load applied at co-ordinate 1 and (b) unit load applied at co-ordinate 2.

$$\Delta_{1L} = \sum Ff_1 \frac{L}{AE} = 0;$$

$$\Delta_{2L} = \sum Ff_2 \frac{L}{AE} = \frac{965.6}{AE};$$

$$\delta_{11} = \sum f_1 f_1 \frac{L}{AE} = \frac{4}{AE};$$

$$\delta_{12} = \delta_{21} = \sum f_1 f_2 \frac{L}{AE} = \frac{2.828}{AE};$$

$$\delta_{22} = \sum f_2 f_2 \frac{L}{AE} = \frac{19.32}{AE};$$

On substituting the above values into equation 9.6, we have

$$\therefore \begin{bmatrix} 0 \\ 0 \end{bmatrix} = \begin{bmatrix} 0 \\ \dfrac{965.6}{AE} \end{bmatrix} + \frac{1}{AE}\begin{bmatrix} 4 & 2.828 \\ 2.828 & 19.32 \end{bmatrix}\begin{bmatrix} H_F \\ F_2 \end{bmatrix}$$

By solving the above equation, we obtain

$$\begin{bmatrix} H_F \\ F_2 \end{bmatrix} = \begin{bmatrix} 39.41 \\ -55.75 \end{bmatrix} \text{kN}$$

Final member end forces: With the known values of H_F and F_2, the forces in other members can be calculated simply by adding the force due to the external loads and the redundants as given by the following equation, and the forces are shown below.

$$\text{Net force} = F + H_F \times f_1 + F_2 \times f_2$$

$F_1 = 0$; $F_2 = -55.75$ kN; $F_3 = 15$ kN;

$F_4 = 89.41$ kN; $F_5 = -160.58$ kN

$F_6 = -10.58$ kN; $F_7 = 0$ kN; $F_8 = 70.75$ kN;

$F_9 = -150$ kN; $F_{10} = 0$ kN

Example 9.18

A pin-jointed plane frame shown in Figure 9.73 using the flexibility matrix method. The cross-sectional area of each top and bottom chord member is 2,000 mm²,

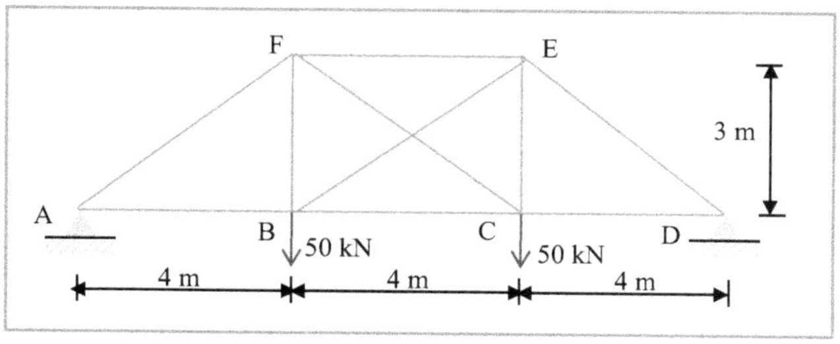

FIGURE 9.73 Indeterminate pin-jointed plane frame.

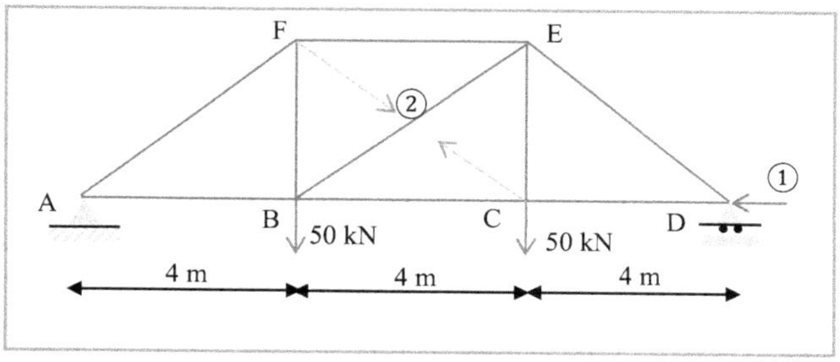

FIGURE 9.74 Basic determinate truss with co-ordinates assigned to redundants.

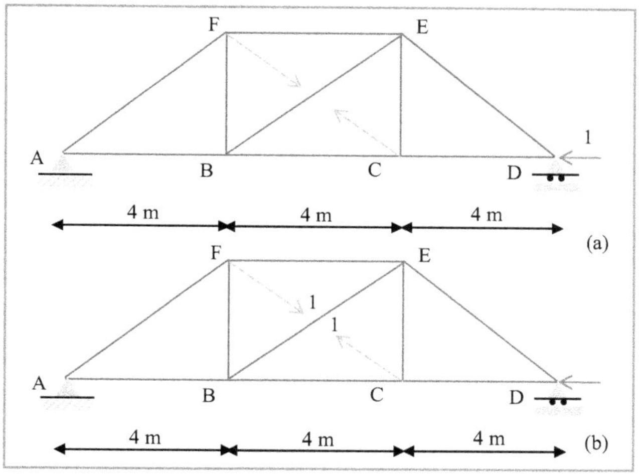

FIGURE 9.75 (a) Unit load applied at co-ordinate 1 and (b) unit load applied at co-ordinate 2.

and the web member is 1,000 mm². Analyse the frame in each of the following cases, respectively.

a. Due to external loads only.
b. If the member *CF* is too long by 4 mm, and assume $E = 2.05\times10^8$ kN/ m^2.
c. If the member *CF* undergoes a rise in temperature of 72.73°C, and assume the coefficient of thermal expansion $\alpha = 11\times10^{-6}$/°C.
d. If the pin-jointed plane frame shown in Figure 9.73 undergoes a uniform rise in temperature of 50°C.

Solution:

The frame is internally and externally indeterminate to the first degree. Therefore, select the horizontal reaction at *D* (H_D) and the force in member *CF* (F_{CF}) (member subjected to fabrication error and rise in temperature) as the redundants. The released truss can be obtained by releasing the redundant forces (i.e., by replacing the hinged support with a roller support at *D*, and by making the cut in the member *CF*) and assigning co-ordinates to the redundants, as shown in Figure 9.74. The unit loads applied at the co-ordinates are shown in Figure 9.75.

Case-(a): Only the external loads are acting on the truss

∴ Equation 9.6 of the flexibility matrix is

$$[\Delta] = [\Delta_L] + [\delta][R]$$

where

$$[\Delta] = \begin{bmatrix} \Delta_1 \\ \Delta_2 \end{bmatrix} = \begin{bmatrix} 0 \\ 0 \end{bmatrix}$$

$$[R] = \begin{bmatrix} H_D \\ F_{CF} \end{bmatrix}$$

$[\Delta_L]$ and $[\delta]$ are the displacements at co-ordinates due to the external loads and unit forces applied at co-ordinates ① and ② on the released structure. These can be calculated using the unit load method. The forces in various members due to the external loads and unit forces applied at ① and ② are calculated and are presented in Table 9.3.

Substituting the values from the above table into the following equations:

$$\Delta_{1L} = \sum F_i f_1 \frac{L}{AE} = \frac{-400}{2.05\times10^2} = -1.95 \text{ mm};$$

$$\Delta_{2L} = \sum F_i f_2 \frac{L}{AE} = \frac{-180}{2.05\times10^2} = -0.878 \text{ mm};$$

$$\delta_{11} = \sum f_1 f_1 \frac{L}{AE} = \frac{6}{2.05\times10^2} = 0.0292 \text{ mm};$$

$$\delta_{12} = \delta_{21} = \sum f_1 f_2 \frac{L}{AE} = \frac{1.6}{2.05\times10^2} = 0.0078 \text{ mm};$$

$$\delta_{22} = \sum f_2 f_2 \frac{L}{AE} = \frac{14.72}{2.05\times10^2} = 0.0718 \text{ mm};$$

On substitution of the above values into equation 9.6, we have,

$$\therefore \begin{bmatrix} 0 \\ 0 \end{bmatrix} = \begin{bmatrix} -1.95 \\ -0.878 \end{bmatrix} + \begin{bmatrix} 0.0292 & 0.0078 \\ 0.0078 & 0.0718 \end{bmatrix} \begin{bmatrix} H_D \\ F_{CF} \end{bmatrix}$$

By solving the above equation, we get

$$\therefore \begin{bmatrix} H_D \\ F_{CF} \end{bmatrix} = \begin{bmatrix} 65.41 \\ 5.12 \end{bmatrix} \text{kN}$$

TABLE 9.3
Forces in Members Due to External Loads and Unit Forces

Member	$\left(\frac{L}{A}\right)$ (mm)	F (kN)	f_1	f_2	$Ff_1\frac{L}{A}$ kN/mm	$Ff_2\frac{L}{A}$ kN/mm	$f_1^2\frac{L}{A}$	$f_2^2\frac{L}{A}$	$f_1 f_2\frac{L}{A}$
AB	2	66.67	−1	0	−133.34	0	2.0	0	0
BC	2	66.67	−1	−0.8	−133.34	−106.67	2.0	1.28	1.6
CD	2	66.67	−1	0	−133.34	0	2.0	0	0
DE	5	−83.33	0	0	0	0	0	0	0
BE	5	0	0	1.0	0	0	0	5	0
EF	2	−66.67	0	−0.8	0	106.67	0	1.28	0
BF	3	50	0	−0.6	0	−90	0	1.08	0
AF	5	−83.33	0	0	0	0	0	0	0
CF	5	0	0	1.0	0	0	0	5	0
CE	3	50	0	−0.6	0	−90	0	1.08	0
				Σ	−400	−180	6	14.72	1.6

Final member end forces: Knowing the values of H_D and F_{CF}, the other member forces can be calculated by adding the forces caused by the applied loads and the redundants, as shown in column 2 of Table 9.4 using the following equation.

$$\text{Net force} = F + H_D f_1 + F_{CF} f_2$$

Case-(b): If the member *CF* is too long by 4 mm in the absence of external loads.

$$\therefore [\Delta_L] = \begin{bmatrix} \Delta_{1L} \\ \Delta_{2L} \end{bmatrix} = \begin{bmatrix} 0 \\ 0 \end{bmatrix}$$ since no external loads on the truss

The member (redundant) *CF* is too long by 4 mm, so it would be under compression. Hence, $\Delta_2 = -4$ mm and $\Delta_1 = 0$.

Substituting the above into equation 9.6, we have

$$\therefore \begin{bmatrix} 0 \\ -4 \end{bmatrix} = \begin{bmatrix} 0 \\ 0 \end{bmatrix} + \begin{bmatrix} 0.0292 & 0.0078 \\ 0.0078 & 0.0718 \end{bmatrix} \begin{bmatrix} H_D \\ F_{CF} \end{bmatrix}$$

By solving the above equation, we obtain

$$\therefore \begin{bmatrix} H_D \\ F_{CF} \end{bmatrix} = \begin{bmatrix} 15.33 \\ -57.38 \end{bmatrix} \text{kN}$$

Net member forces: Knowing the values of H_D and F_{CF}, the forces in the other members can be calculated using the following equation, and the calculated member forces are presented in column 3 of Table 9.4.

$$\text{Net force} = H_D f_1 + F_{CF} f_2$$

Case-(c): If the member *CF* undergoes a rise in temperature of 72.73°C and $\alpha = 11 \times 10^{-6}$/°C.

A free thermal expansion of the redundant member *CF* is

$$\Delta_{CF} = \Delta_2 = L\alpha\Delta T = 5{,}000 \times 11 \times 10^{-6} \times 72.73 = 4 \text{ mr}$$

TABLE 9.4
Final Member End Forces

	Force (kN)		
Member	**Case (a)**	**Case (b)**	**Case (d)**
AB	1.26	−15.33	−232.78
BC	−2.84	30.57	−212.56
CD	1.26	−15.33	−232.78
DE	−83.33	0	0
EF	−70.77	45.9	20.22
BE	5.12	−57.38	−25.28
CE	46.93	34.43	15.17
FC	5.12	−57.38	−25.28
BF	46.93	34.43	15.17
AF	−83.33	0	0

Consequently, the member *CF* is too long by 4 mm, causing the same initial lack of fit as in case (b). Hence, the forces due to the rise in temperature of the member *CF* are the same as obtained in case (b), and are presented in column (3) of Table 9.4.

Case-(d): If the pin-jointed plane frame undergoes a uniform rise in temperature of 50°C.

As the truss is statically indeterminate externally, even a uniform change in temperature will cause internal forces in the members.

∴ The free thermal expansion of the bottom chord *AD*

$$\Delta_{AD} = L\alpha\Delta T = 12 \times 11 \times 10^{-6} \times 50 = 0.0066 \text{ m} = 6.6 \text{ mm}$$

$$\therefore \Delta_1 = 6.6 \text{ mm and } \Delta_2 = 0$$

$$[\Delta_L] = \begin{bmatrix} \Delta_{1L} \\ \Delta_{2L} \end{bmatrix} = \begin{bmatrix} 0 \\ 0 \end{bmatrix}$$ as no external loads on the truss.

Substituting the above values into equation 9.6, we have

$$\therefore \begin{bmatrix} 6.6 \\ 0 \end{bmatrix} = \begin{bmatrix} 0 \\ 0 \end{bmatrix} + \begin{bmatrix} 0.0292 & 0.0078 \\ 0.0078 & 0.0718 \end{bmatrix} \begin{bmatrix} H_D \\ F_{CF} \end{bmatrix}$$

By solving the above equation, we obtain

$$\therefore \begin{bmatrix} H_D \\ F_{CF} \end{bmatrix} = \begin{bmatrix} 232.78 \\ -25.28 \end{bmatrix} \text{kN}$$

Net forces in the members: By knowing the above H_D and F_{CF}, the other member forces can be calculated using the following equation, and are presented in column (4) of Table 9.4.

$$\text{Net force} = H_D f_1 + F_{CF} f_2$$

Example 9.19

Analyse the pin-jointed plane frame shown in Figure 9.76. Assume the flexibility of the members $= BC = DE = EF = 20$ mm/MN, members $CD = BE = 15$ mm/MN and members $AE = BF = BD = CE = 10$ mm/MN.

Solution:

The static Indeterminacy, $I_S = m - 2j + R$

$$= 10 - 2 \times 6 + 4 = 2$$

∴ Select the forces in the members *CE* (F_{CE}) and *BF* (F_{BF}) as the redundants. Release the forces in members *CE* and *BF* by introducing cuts to obtain the basic or determinate truss. Assign co-ordinates 1 and 2 to the chosen redundants, respectively, as shown in Figure 9.77.

The displacements due to the applied loads and unit forces at co-ordinates can be calculated using the unit load method. The member forces due to the applied loads and unit forces applied successively at co-ordinates ① and ② (Figure 9.78) can be determined, and are presented in Table 9.5.

Equation 9.9 of the flexibility matrix is

$$[\Delta] = [\Delta_L] + [\delta][R]$$

where

$$[\Delta] = \begin{bmatrix} \Delta_1 \\ \Delta_2 \end{bmatrix} = \begin{bmatrix} 0 \\ 0 \end{bmatrix}$$

$$[R] = \begin{bmatrix} F_{BF} \\ F_{CE} \end{bmatrix}$$

$$\Delta_{1L} = \sum F_i f_1 \frac{L}{AE} = 5.75 \text{ mm;}$$

$$\Delta_{2L} = \sum F_i f_2 \frac{L}{AE} = 5.455 \text{ mm;}$$

$$\delta_{11} = \sum f_1 f_1 \frac{L}{AE} = 0.044 \text{ mm;}$$

$$\delta_{12} = \delta_{21} = \sum f_1 f_2 \frac{L}{AE} = 0.0096 \text{ mm;}$$

$$\delta_{22} = \sum f_2 f_2 \frac{L}{AE} = 0.05468 \text{ mm;}$$

Substituting the above values into equation 9.8, we have

$$\therefore \begin{bmatrix} F_{BF} \\ F_{CE} \end{bmatrix} = \begin{bmatrix} 0.044 & 0.0096 \\ 0.0096 & 0.05468 \end{bmatrix}^{-1} \begin{bmatrix} -5.75 \\ -5.455 \end{bmatrix}$$

By solving the above equation, we obtain

$$\therefore \begin{bmatrix} F_{BF} \\ F_{CE} \end{bmatrix} = \begin{bmatrix} -113.25 \\ -79.88 \end{bmatrix} \text{kN}$$

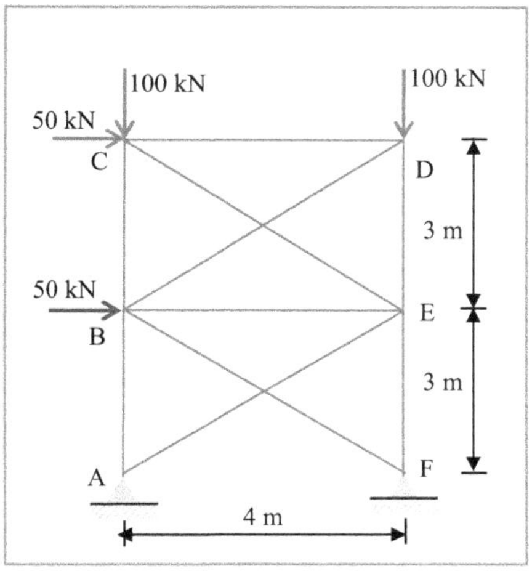

FIGURE 9.76 Indeterminate truss.

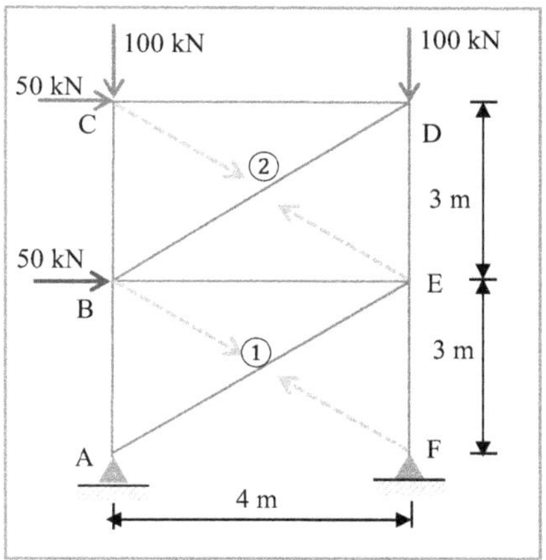

FIGURE 9.77 Basic determinate truss with loads and co-ordinates assigned to the redundants.

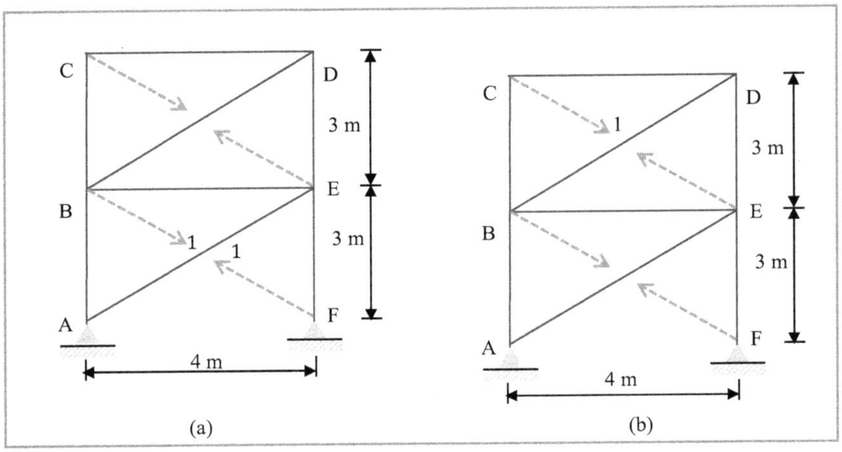

FIGURE 9.78 (a) unit load applied at co-ordinate 1 and (b) unit load applied at co-ordinate 2.

TABLE 9.5
Forces in Members Due to External Loads and Unit Forces

Member	$\left(\frac{L}{A}\right)$ (mm/kN)	F (kN)	f_1	f_2	$Ff_1\frac{L}{AE}$ kN/mm	$Ff_2\frac{L}{AE}$ kN/mm	$f_1^2\frac{L}{AE}$	$f_2^2\frac{L}{AE}$	$f_1f_2\frac{L}{AE}$
AB	0.02	−62.5	−0.6	0	0.75	0	0.0072	0.0	0
BC	0.02	−100	0	−0.6	0	1.38	0	0.00828	0
CD	0.015	−50	0	−0.8	0	0.6	0	0.0096	0
DE	0.02	−137.5	0	−0.6	0	1.65	0	0.0072	0
EF	0.02	−212.5	−0.6	0	2.55	0	0.0072	0	0
AE	0.01	125	1.0	0	1.25	0	0.01	0	0
BD	0.01	62.5	0	1.0	0	0.625	0	0.01	0
CE	0.01	0	0	1.0	0	0	0	0.01	0
BF	0.01	0	1.0	0	0	0	0.01	0	0
BE	0.015	−100	−0.8	−0.8	1.2	1.2	0.0096	0.0096	0.0096
				Σ	5.75	5.455	0.044	0.05468	0.0096

Net forces in the members:

$$\text{Net force} = F_i + f_1(F_{BF}) + f_2(F_{CE})$$

$F_{AB} = 5.45$ kN; $F_{BC} = -52.07$ kN; $F_{CD} = 13.90$ kN;

$F_{DE} = -89.57$ kN;

$F_{EF} = -144.55$ kN; $F_{AE} = 11.75$ kN; $F_{BD} = -17.38$ kN;

$F_{BE} = 54.50$ kN;

$$F_{CE} = 79.88 \text{ kN}; \quad F_{BF} = -113.25 \text{ kN};$$

Example 9.20

Analyse the pin-jointed plane frame shown in Figure 9.79. Assume the axial rigidity of each member is 4.1×10^5 kN.

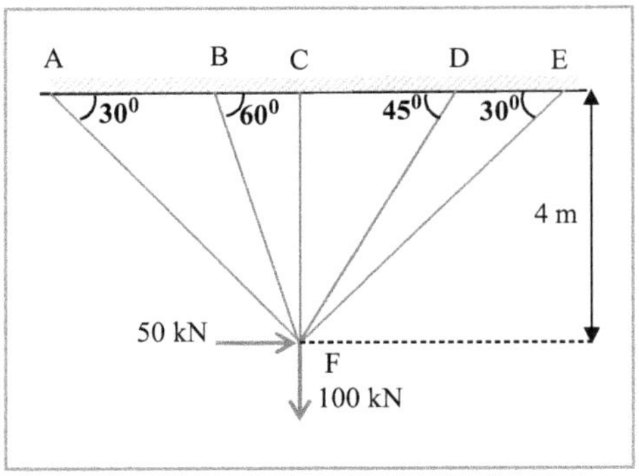

FIGURE 9.79 Indeterminate truss.

Solution:

At joint *F*, there are five unknown member forces, but only two equilibrium equations are available. Hence, the degree of static indeterminacy is $I_S = 5 - 2 = 3$. Therefore, select the forces in members *AF*, *EF*, and *BF*

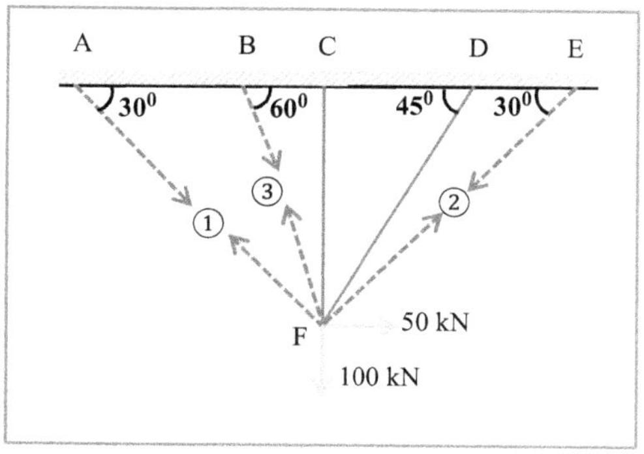

FIGURE 9.80 Released truss with co-ordinates and loads.

as the redundants. The released frame can be obtained by introducing the cut sections to release these member forces and assigning co-ordinates ①, ②, and ③ to them, respectively, as shown in Figure 9.80.

The displacements due to the external loads, and the unit forces applied successively at co-ordinates can be calculated using the unit load method. The member forces (F) due to the external loads and the unit forces (f_1, f_2, f_3) can be calculated using Figure 9.81, and the calculated values are presented in Table 9.6.

Equation 9.6 of the flexibility matrix is

$$[\Delta]=[\Delta_L]+[\delta][R]$$

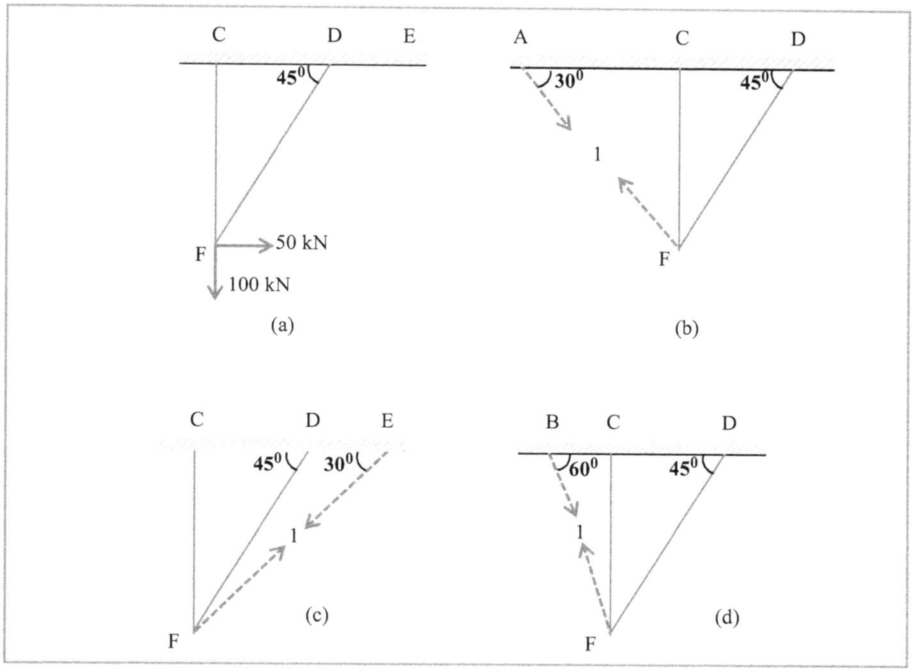

FIGURE 9.81 (a) Basic determinate truss, (b) unit load applied at co-ordinate 1, (c) unit load applied at co-ordinate 2, and (d) unit load applied at co-ordinate 3.

TABLE 9.6
Forces in Members Due to External Loads and Unit Forces

Member	(L) (m)	F_i (kN)	f_1	f_2	f_3	$F_i f_1 L$	$F_i f_2 L$	$F_i f_3 L$	$f_1^2 L$	$f_2^2 L$	$f_3^2 L$	$f_1 f_2 L$	$f_1 f_3 L$	$f_3 f_2 L$	Net Forces (kN)
AF	5.66	0	1.0	0	0	0	0	0	5.66	0	0	0	0	0	−35.98
BF	4.62	0	0	0	1.0	0	0	0	0	0	4.62	0	0	0	−43.29
CF	4.0	150	−1.414	0.363	−1.366	−848.4	217.8	−819.6	7.99	0.53	7.46	−2.05	7.73	−1.98	−38.14
DF	5.66	−70.71	1.0	−1.22	0.707	−400.22	488.27	−282.96	5.66	8.42	2.83	−6.91	4.0	−4.89	−2.09
EF	8.0	0	0	1.0	0	0	0	0	0	8	0	0	0	0	−5.09
Σ						−1,248.62	706.07	−1,102.55	19.32	16.95	14.91	−8.96	11.73	−6.87	

where

$$[\Delta]=\begin{bmatrix}\Delta_1\\ \Delta_2\end{bmatrix}=\begin{bmatrix}0\\ 0\end{bmatrix}$$

$$[R]=\begin{bmatrix}F_{AF}\\ F_{EF}\\ F_{BF}\end{bmatrix}$$

Substituting the values from the table in the following equations

$$\Delta_{1L}=\sum F_i f_1\frac{L}{AE}=\frac{-1,248.62}{AE}$$

$$\Delta_{2L}=\sum F_i f_2\frac{L}{AE}=\frac{706.07}{AE}$$

$$\Delta_{3L}=\sum F_i f_3\frac{L}{AE}=\frac{-1,102.55}{AE}$$

$$\delta_{11}=\sum f_1 f_1\frac{L}{AE}=\frac{19.32}{AE}$$

$$\delta_{12}=\delta_{21}=\sum f_1 f_2\frac{L}{AE}=\frac{-8.96}{AE}$$

$$\delta_{22}=\sum f_2 f_2\frac{L}{AE}=\frac{16.95}{AE}$$

$$\delta_{33}=\sum f_3 f_3\frac{L}{AE}=\frac{14.91}{AE}$$

$$\delta_{13}=\delta_{31}=\sum f_1 f_3\frac{L}{AE}=\frac{11.73}{AE}$$

$$\delta_{23}=\delta_{32}=\sum f_2 f_3\frac{L}{AE}=\frac{-6.87}{AE}$$

By substituting the above values into equation 9.8, we have

$$\therefore\begin{bmatrix}F_{AF}\\ F_{EF}\\ F_{BF}\end{bmatrix}=\begin{bmatrix}19.32 & -8.96 & 11.73\\ -8.96 & 16.95 & -6.87\\ 11.73 & -6.87 & 14.91\end{bmatrix}^{-1}\begin{bmatrix}1,248.62\\ -706.07\\ 1,102.55\end{bmatrix}$$

By solving the above equation, we obtain

$$\therefore\begin{bmatrix}F_{BF}\\ F_{CE}\end{bmatrix}=\begin{bmatrix}35.98\\ -5.09\\ 43.29\end{bmatrix}\text{kN}$$

Net forces in the members: After determining the forces in members *AF, EF,* and *BF*, we can calculate the other member forces using the following equation, and these are presented in column 16 of Table 9.6.

$$\text{Net force}=F_i+f_1(F_{AF})+f_2(F_{EF})+f_3(F_{BF})$$

9.6 PROBLEMS FOR PRACTICE

9.6.1 Develop the flexibility matrix for a general beam element.

9.6.2 Write the properties of the flexibility matrix.

9.6.3 Analyse the continuous beam shown in Figure 9.82 using the flexibility matrix method. Draw the shear force and bending moment diagrams.

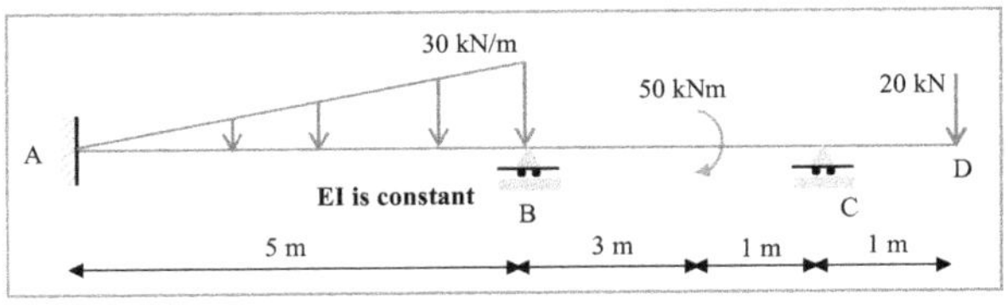

FIGURE 9.82

9.6.4 Analyse the continuous beam shown in Figure 9.83. Construct the shear force and bending moment diagrams. Assume EI is constant.

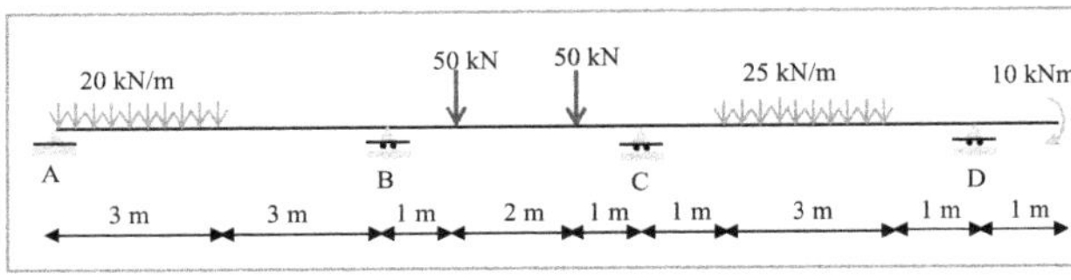

FIGURE 9.83

9.6.5 Analyse the three-span continuous beam shown in Figure 9.84 by taking advantage of symmetry. Sketch the shear force, and bending moment diagrams.

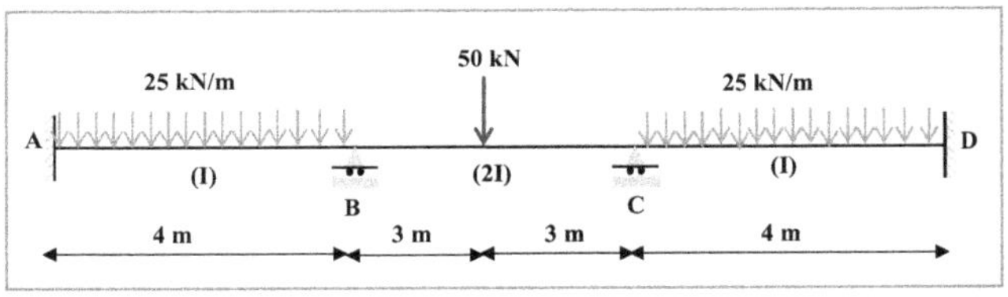

FIGURE 9.84

9.6.6 Compute the support reactions of the beam shown in Figure 9.84, if the supports *B* and *C* settle down by 10 mm each in the absence of external loads. Assume $EI = 135\times10^3$ kNm².

9.6.7 Analyse the beam shown in Figure 9.85 if support A rotates by 0.002 radians in anti-clockwise direction and support *B* settles down by 10 mm. Assume $EI = 138,375$ kNm².

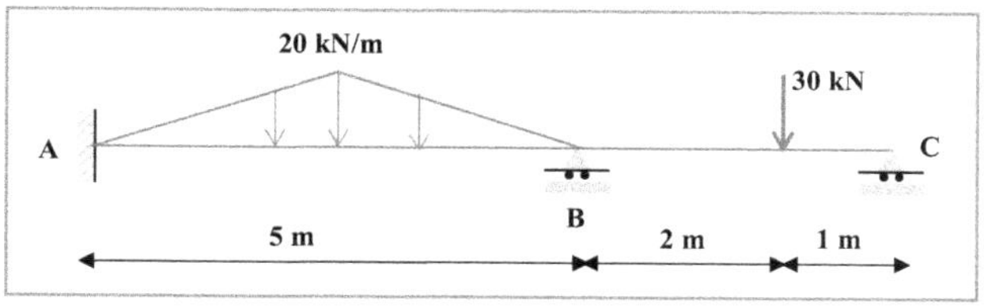

FIGURE 9.85

9.6.8 Find the support reactions of a non-prismatic beam shown in Figure 9.86. Also, draw the shear force and bending moment diagrams.

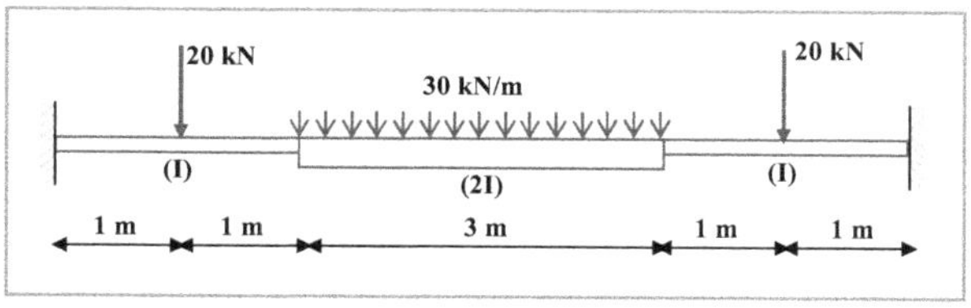

FIGURE 9.86

9.6.9 A rigid-jointed plane frame with inclined legs is shown in Figure 9.87. Analyse the frame using the flexibility matrix method. Construct the axial force, shear force, and bending moment diagrams. Also, draw the elastic curve.

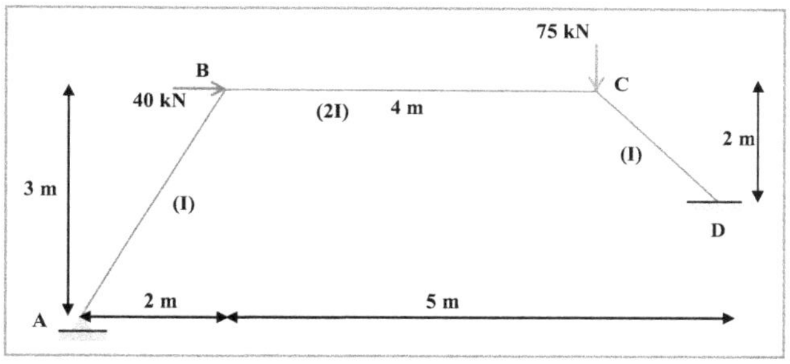

FIGURE 9.87

9.6.10 Re-analyse the rigid-jointed plane frame shown in Figure 9.87 using the flexibility matrix method, if the support D settle down by 10 mm and rotates in clockwise direction by 0.002 radians, in the absence of the external loads. Assume $EI=85{,}000$ kNm^2.

9.6.11 A pitched portal frame ABCD is shown in Figure 9.88. Compute the support reactions using the flexibility matrix method. Draw the axial force, shear force and bending moment diagrams. Also draw the deflected shape.

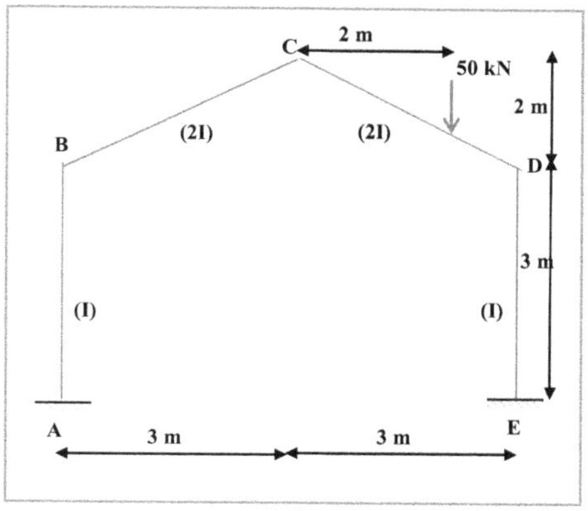

FIGURE 9.88

9.6.12 Analyse the pin-jointed plane frame shown in Figure 9.89 using the flexibility matrix method. Assume all members have the same axial rigidity.

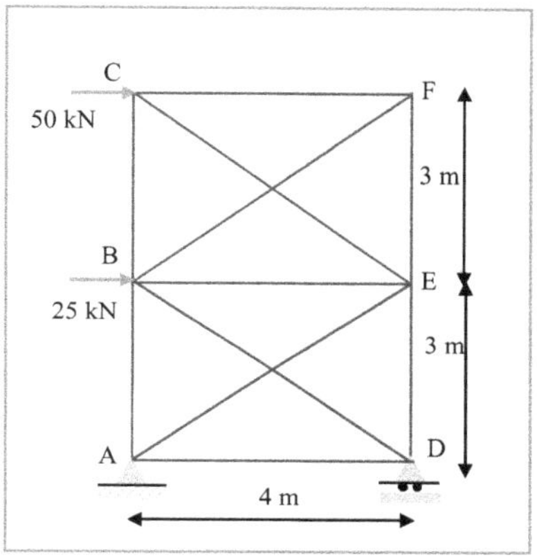

FIGURE 9.89

9.6.13 A plane braced frame is shown in Figure 9.90. The member BF in the frame is initially short in length by 2.5 mm. Analyse the truss for all member end forces. Assume the cross-sectional area of all members is constant and is equal to 800 mm^2, and the modulus of elasticity $E=200$ kN/mm^2. Also, analyse the truss if member BF is subjected to a temperature rise of 20°C. $\alpha = 11 \times 10^{-6}$/°C.

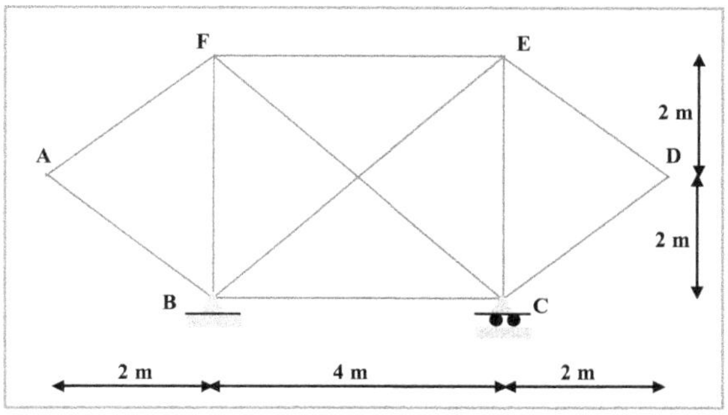

FIGURE 9.90

9.7 ANSWER KEY

Q. No.	Fig. No.	Answers at Key Points
9.6.3	9.8.2	$M_{A=}-42.27$ kNm; $R_A=32.92$ kN; $R_B=25.24$ kN
9.6.4	9.8.3	$M_B=-36.39$ kNm; $M_C=-48.42$ kNm; $R_A=38.94$ kN; $R_B=68.05$ kN; $R_C=98.19$ kN
9.6.5	9.8.4	$M_A=-32.11$ kNm; $M_B=-35.82$ kNm; $R_A=49.07$ kN; $R_B=75.93$ kN
9.6.6	9.8.4	$M_A=-391.95$ kNm; $M_B=169.82$ kNm; $R_A=190.44$ kN; $R_B=-65.44$ kN
9.6.7	9.8.5	$M_A=-581.43$ kNm; $M_B=426.09$ kNm; $R_A=226.51$ kN; $R_C=162.03$ kN
9.6.8	9.8.6	$M_A=-581.43$ kNm$=-M_B$; $R_A=65$ kN$=R_B$
9.6.9	9.87	$R_A=-0.58$ kN, $R_D=75.58$ kN, $M_D=-1.63$ kNm, $M_{BA}=-M_{BC}=0.94$ kNm
9.6.10	9.87	$M_D=-23.55$ kNm; $V_A=3.00$ kN; $H_A=2.5$ kN
9.6.11	9.88	$M_A=10.20$ kNm; $M_E=5.64$ kNm; $V_A=7.54$ kN; $V_E=42.45$ kN; $H_A=6.45$ kN; $H_E=-6.45$ kN
9.6.12	9.89	$V_A=93.75$ kN; $H_A=-75$ kN; $V_D=-93.75$ kN; $F_{AB}=62.15$ kN; $F_{AE}=52.66$ kN; $F_{AF}=32.87$ kN; $F_{BC}=23.43$ kN; $F_{CD}=18.75$ kN
9.6.13	9.90	(a) $F_{AB}=F_{AF}=F_{DE}=F_{DC}=0$; $F_{BF}=F_{BC}=F_{EF}=F_{EC}=10.36$ kN; $F_{CF}=F_{BE}=-14.65$ kN (b) $F_{AB}=F_{AF}=F_{DE}=F_{DC}=0$; $F_{BF}=F_{BC}=F_{EF}=F_{EC}=-3.65$ kN; $F_{CF}=F_{BE}=5.16$ kN

Note: The answers are provided based on the analysis using structural analysis software. The results from the exact solution may differ slightly (decimals), but these differences are generally negligible.

FURTHER READING

Bhavikatti, S. S. (2011). *Matrix Methods of Structural Analysis*, I.K. International Publishing House Pvt. Limited. ISBN 9789381141359, 9381141355.

Leet, K. (2020). Uang, C.-M. and Lanning, J. *Fundamentals of Structural Analysis* (6th E), McGraw Hill Education Pvt Ltd, ISBN (13) 9781260477245.

Pandit, G. S. and Gupta, S. P. (2008). *Structural Analysis: A Matrix Approach*, McGraw Hill Education, New Delhi, ISBN 978-0070667358.

Reddy, C. S. (2011). *Basic Structural Analysis* (3rd E), Tata McGraw-Hill Education Pvt Ltd, New Delhi, ISBN (13) 978-0-07–070276-9.

Sahoo, D. R. (2021). *Indeterminate Structural Analysis*, PHI Learning Pvt Ltd, New Delhi, ISBN 9789354432958.

Shah, H. J and Junarkar, S. B. (2015). *Mechanics of Structures – Vol. II*, (24th E) Charotar Publishing House, Anand. ISBN 9789385039027.

Vijayanand, M., Muthu, K. U., Narendra, H. and Janardhana, M. (2019). *Indeterminate Structural Analysis*, Dreamtech Press Publisher, India, ISBN 978-9389307177.

10 Matrix Methods
Stiffness Matrix Method

10.1 INTRODUCTION

The flexibility matrix method was discussed in detail in Chapter 9. The flexibility of a structure is defined as the displacement produced by a unit force. Stiffness is the converse of flexibility and can be defined as the force developed due to a unit displacement. Let us consider a structural element of length L, cross-sectional area A, and modulus of elasticity E, with one end fixed and the other end free, having axial, transverse, and rotational degrees of freedom (DoFs) as shown in Figure 10.1a. The stiffness of the structural member with respect to each DoF can be defined as follows.

Axial displacement: Let P is the axial force applied at co-ordinate 1, as shown in Figure 10.1b, then the axial displacement at co-ordinate 1 is given by Equation 10.1, as

$$\Delta_1 = \frac{PL}{AE} \tag{10.1}$$

By definition, the axial stiffness of the member is the force developed due to a unit displacement at co-ordinate 1. Hence, substituting $\Delta_1 = 1$, into equation 10.1, then

$$\text{Axial stiffness } P = k_{11} = \frac{AE}{L} \tag{10.2}$$

Equation 10.2 provides the axial stiffness, which is relevant to pin-jointed plane frames (i.e., trusses that carry only axial forces). In general, axial displacements are relatively small compared to flexural displacements and are therefore often neglected in the analysis of rigid-jointed structures.

Transverse displacement: Let us give the displacement Δ_2 at co-ordinate 2 without any displacement at other co-ordinates, as shown in Figure 10.1c. The force developed at co-ordinate 2, when the far end is fixed, is given by equation 10.3, as

$$W = \frac{12EI\Delta_2}{L^3} \tag{10.3}$$

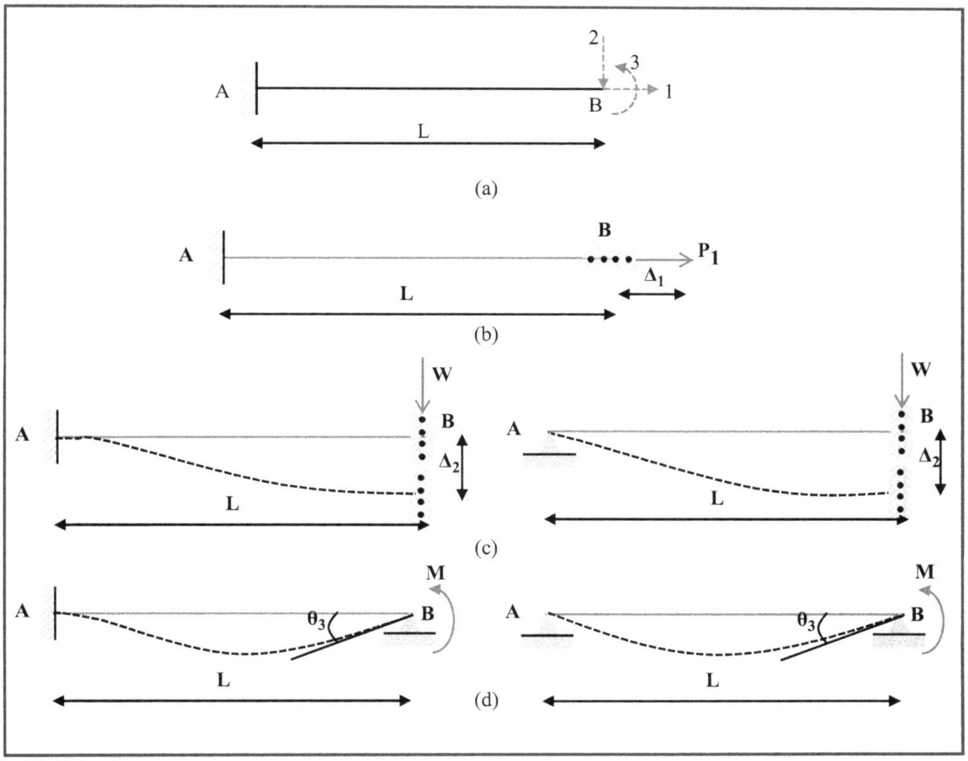

FIGURE 10.1 (a) Member AB with one end fixed and the other end free, (b) axial displacement at 1, (c) transverse displacement at 2, and (d) flexural displacement at 3 in a beam element.

DOI: 10.1201/9781003562955-10

By definition, the transverse stiffness of a member is the force developed due to a unit displacement at co-ordinate 2. Substituting $\Delta_2 = 1$ into equation 10.3,

$$\text{The transverse stiffness } W = k_{22} = \frac{12EI}{L^3} \quad (10.4)$$

When the far end is hinged at A, to produce the displacement Δ_2 at co-ordinate 2 without displacements at the other co-ordinates, the force required at co-ordinate 2 is given by equation 10.4.

$$W = \frac{3EI\Delta_2}{L^3} \quad (10.5)$$

Hence, by definition, to determine the transverse stiffness, substitute $\Delta_2 = 1$ into equation 10.5,

$$\text{Transverse stiffness } W = k_{22} = \frac{3EI}{L^3} \quad (10.6)$$

Rotational or flexural displacement: Let the force M (moment) required at co-ordinate 3 to produce rotational displacement at co-ordinate 3, without displacements at the other co-ordinates (Figure 10.1d), be given by equation 10.7

$$M = \frac{4EI\theta_3}{L} \quad (10.7)$$

By definition, to get the rotational or flexural stiffness, substitute $\theta_3 = 1$ into equation 10.7, yielding

$$\text{Rotational stiffness } M = k_{33} = \frac{4EI}{L} \quad (10.8)$$

Equation 10.8 is applicable when the far end A is fixed. If the far end is hinged at A, then the force required at co-ordinate 3 to produce the displacement θ_3 at co-ordinate 3, without any displacement at the other co-ordinates, is given by equation 10.9:

$$M = \frac{3EI\theta_3}{L} \quad (10.9)$$

By definition, substituting $\theta_3 = 1$ into equation 10.9 gives the flexural stiffness as

$$k_{33} = \frac{3EI}{L} \quad (10.10)$$

Equation 10.10 gives the rotational or flexural stiffness when the far end is hinged.

10.2 FORMULATION OF STIFFNESS MATRIX EQUATION OR FORCE DISPLACEMENT RELATIONSHIP

In the stiffness method (displacement method or equilibrium method), the degree of kinematic indeterminacy (DoFs) of the structure is determined. In general the unknown joint or support displacements are considered as the independent DoFs. To formulate the stiffness equation, let us consider a beam with a finite degree of kinematic indeterminacy (I_k=4), as shown in Figure 10.2a.

In this case, the unknown displacements are the rotations at each support (Δ_1, Δ_2, Δ_3, Δ_4), and co-ordinates 1, 2, 3, and 4 are assigned to them, respectively, as shown in Figure 10.2a. The structure is then made completely restrained by locking (fixing) all the supports and joints so that no displacements are possible at the co-ordinates in the structure. Let the forces required at these co-ordinates to make the displacements Δ_1, Δ_2, Δ_3, and Δ_4 zero be P_{1L}, P_{2L}, P_{3L}, and P_{4L}, respectively. Now unlock (release) all the supports and joints successively, allowing the displacements Δ_1, Δ_2, Δ_3, and Δ_4 at co-ordinates 1, 2, 3, and 4, respectively. Let the forces required to attain these displacements (Δ_1, Δ_2, Δ_3, Δ_4) at co-ordinates 1, 2, 3, and 4 be $P_{1\Delta}$, $P_{2\Delta}$, $P_{3\Delta}$, and $P_{4\Delta}$, respectively. Finally, let the external forces acting at co-ordinates 1, 2, 3, and 4 in the kinematically indeterminate structure be P_1, P_2, P_3, and P_4, respectively. For the equilibrium of the structure, we can write

$$\left.\begin{aligned} P_1 &= P_{1L} + P_{1\Delta} \\ P_2 &= P_{2L} + P_{2\Delta} \\ P_3 &= P_{3L} + P_{3\Delta} \\ P_4 &= P_{4L} + P_{4\Delta} \end{aligned}\right\} \quad (10.11)$$

Using the principle of superposition (Figure 10.2c–f), we can write

$$\left.\begin{aligned} P_{1\Delta} &= k_{11}\Delta_1 + k_{12}\Delta_2 + k_{13}\Delta_3 + k_{14}\Delta_4 \\ P_{2\Delta} &= k_{21}\Delta_1 + k_{22}\Delta_2 + k_{23}\Delta_3 + k_{24}\Delta_4 \\ P_{3\Delta} &= k_{31}\Delta_1 + k_{32}\Delta_2 + k_{33}\Delta_3 + k_{34}\Delta_4 \\ P_{4\Delta} &= k_{41}\Delta_1 + k_{42}\Delta_2 + k_{43}\Delta_3 + k_{44}\Delta_4 \end{aligned}\right\} \quad (10.12)$$

Substituting equation 10.12 into equation 10.11, we can write

$$\left.\begin{aligned} P_1 &= P_{1L} + k_{11}\Delta_1 + k_{12}\Delta_2 + k_{13}\Delta_3 + k_{14}\Delta_4 \\ P_2 &= P_{2L} + k_{21}\Delta_1 + k_{22}\Delta_2 + k_{23}\Delta_3 + k_{24}\Delta_4 \\ P_3 &= P_{3L} + k_{31}\Delta_1 + k_{32}\Delta_2 + k_{33}\Delta_3 + k_{34}\Delta_4 \\ P_4 &= P_{4L} + k_{41}\Delta_1 + k_{42}\Delta_2 + k_{43}\Delta_3 + k_{44}\Delta_4 \end{aligned}\right\} \quad (10.13)$$

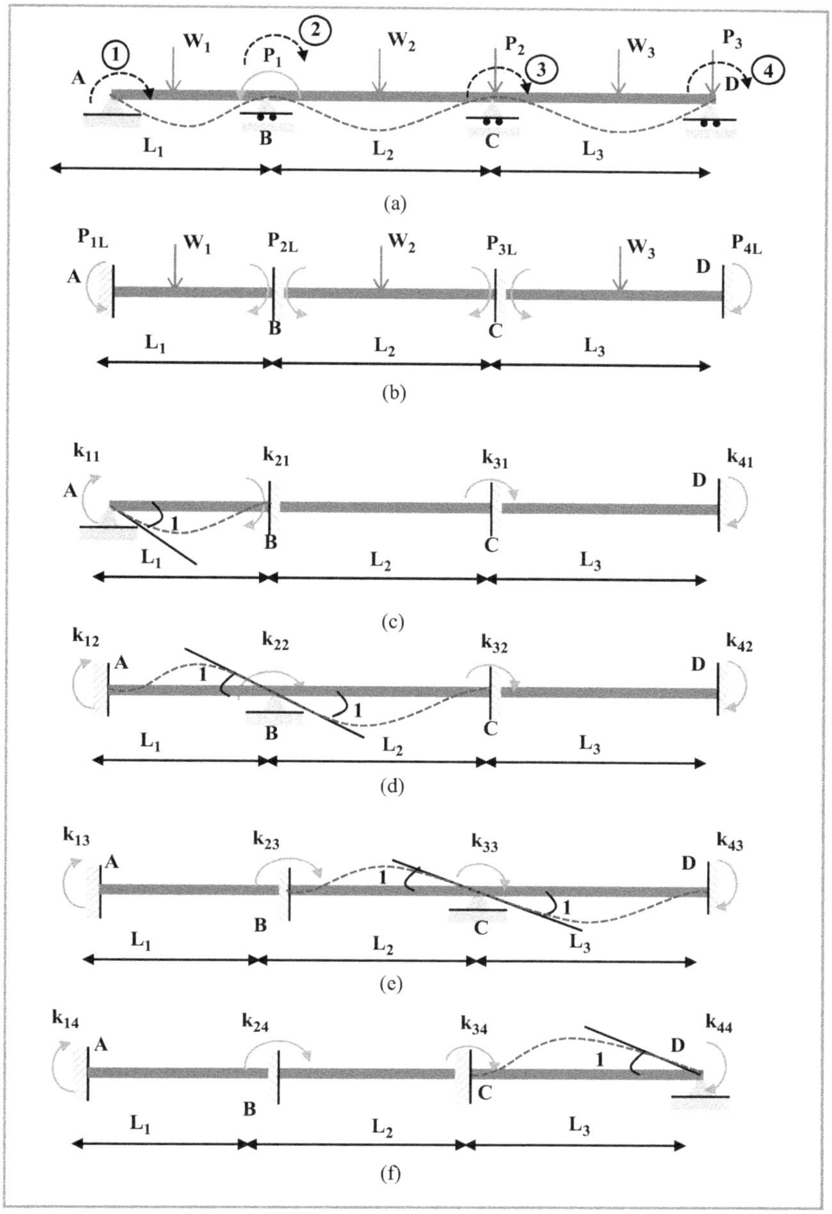

FIGURE 10.2 (a) Kinematically indeterminate beam, (b) restrained beam, (c) unit displacement at co-ordinate 1, (d) unit displacement at co-ordinate 2, (e) unit displacement at co-ordinate 3, and (f) unit displacement at co-ordinate 3.

Equation 10.13 can be written in matrix form as

$$\begin{bmatrix} P_1 \\ P_2 \\ P_3 \\ P_4 \end{bmatrix} = \begin{bmatrix} P_{1L} \\ P_{2L} \\ P_{3L} \\ P_{4L} \end{bmatrix} + \begin{bmatrix} k_{11} & k_{12} & k_{13} & k_{14} \\ k_{21} & k_{22} & k_{23} & k_{24} \\ k_{31} & k_{32} & k_{33} & k_{34} \\ k_{41} & k_{42} & k_{43} & k_{44} \end{bmatrix} \begin{bmatrix} \Delta_1 \\ \Delta_2 \\ \Delta_3 \\ \Delta_4 \end{bmatrix} \tag{10.14}$$

Simply we can write equation 10.14 as

$$[P] = [P_L] + [K][\Delta] \tag{10.15}$$

Equation 10.15 is called the force–displacement relationship or the stiffness matrix equation. By solving this equation, the unknown displacements at the joints and supports can be obtained.

$$[\Delta] = [K]^{-1}\big[[P] - [P_L]\big] \tag{10.16}$$

After knowing the displacements at the joints, the bending moments in the member ends can be calculated using the slope–deflection equation (equation 6.4),

If there are no external forces acting at the co-ordinates, i.e., $P_1 = P_2 = P_3 = P_4 = 0$, then equation 10.16 takes the form as

$$[\Delta] = -[K]^{-1}\big[[+P_L]\big] \tag{10.17}$$

If the external forces act only at co-ordinates, i.e., $P_{1L}=P_{2L}=P_{3L}=P_{4L}=0$, then equation 10.16 takes the form as

$$[\Delta]=[K]^{-1}[[P]] \tag{10.18}$$

If there is yielding of supports present in the structure, calculate the additional fixed end forces (moments) due to these effects, and designate them as P_{1Y}, P_{2Y}, P_{3Y}, P_{4Y} etc. or simply as P_Y, then equation 10.15 takes the form

$$[P]=\{[P_L]+[P_Y]\}+[K][\Delta] \tag{10.19}$$

or equation 10.16 takes the form

$$[\Delta]=[K]^{-1}\left[[P]-\{[P_L]+[P_Y]\}\right] \tag{10.20}$$

The displacement method is applicable to both statically determinate and indeterminate structures. Since the displacements are the primary unknowns in this method, it is referred to as the displacement method. The method is also referred to as the stiffness matrix method because the stiffness matrix of the structure is developed to calculate the unknown displacements and write the equilibrium equations. Additionally, the method is known as the equilibrium method, as equilibrium equations are used to determine the unknown displacements in the structure. The application of the stiffness matrix method to beams, rigid-jointed plane frames, and pin-jointed plane frames is demonstrated with examples in the following sections.

10.2.1 Procedure to Solve the Problems

i. **Determine the degree of kinematic indeterminacy**: Find the kinematic indeterminacy (unknown joint displacements) of the given structure and identify them as Δ_1, Δ_2, Δ_3, … etc. Assign the co-ordinates (1), (2), (3), … etc., to each unknown DoF in the chosen direction.
ii. **Restrained Structure:** Make the structure fully restrained by locking all the joints (preventing all DoFs by fixing all the joints). This structure is called the restrained structure. If the far end supports are hinged or roller, we may neglect the corresponding DoFs and modify the stiffness and fixed end moments accordingly. Furthermore, use modified slope–deflection equations for the calculation of the member end moments.
iii. **Fixed end moments:** Calculate the fixed end moments due to external loads $[P_L]$ using the expressions provided in Table 6.1. If there is any yielding of supports, determine the fixed end moments caused by the yielding of supports $[P_Y]$. Then calculate the resultant fixed end moments due to loads and yielding of supports by adding them algebraically $[[P_L]+[P_Y]]$. Assume clockwise moments are positive.
iv. **Stiffness matrix development**: Develop the elements of the stiffness matrix $[K]$, which is a square matrix of size equal to the degree of kinematic indeterminacy. To compute the first column elements of the stiffness matrix, apply a unit displacement at co-ordinate (1) in its specified direction, keeping all other co-ordinates fixed (no displacements). Calculate the resulting forces at all co-ordinates in the restrained structure. Repeat this procedure for each co-ordinate to develop all columns of the stiffness matrix.
v. **External force matrix:** Identify any forces acting at the co-ordinates in the original structure and write them as the external force matrix $[P]$
vi. **Unknown displacement matrix**: Write the unknown displacement matrix $[\Delta]$
vii. **Solve the stiffness matrix equation**: Substitute all the above matrices into the stiffness matrix equation (equation 10.15) and solve for the unknown displacements.
viii. **Member end moments**: Once the displacements are determined, calculate the member end moments using the appropriate slope–deflection equations.
ix. **Support reactions**: Finally, calculate the other support reactions (member end shears) using the static equilibrium equations.

10.2.2 Development of Stiffness Matrix for a General Beam Element

Consider a general beam element of length L with 6 DoFs, as shown in Figure 10.3(i)a. Assume the flexural rigidity (EI) and cross-sectional area (A) are constant. Let us develop the elements of the stiffness matrix. To do so, consider the restrained structure by fixing all ends.

First column elements: To generate the first column elements, apply a unit axial displacement at co-ordinate (1) without any displacement at other co-ordinates (Figure 10.3(i)b) and find the forces developed at all the co-ordinates.

$$K_{11}=\frac{AE}{L};\quad K_{21}=K_{31}=K_{51}=K_{61}=0;\quad K_{41}=-\frac{AE}{L}$$

Second column elements: To develop the second column elements of the stiffness matrix, apply a unit translational displacement at co-ordinate (2) while keeping all other

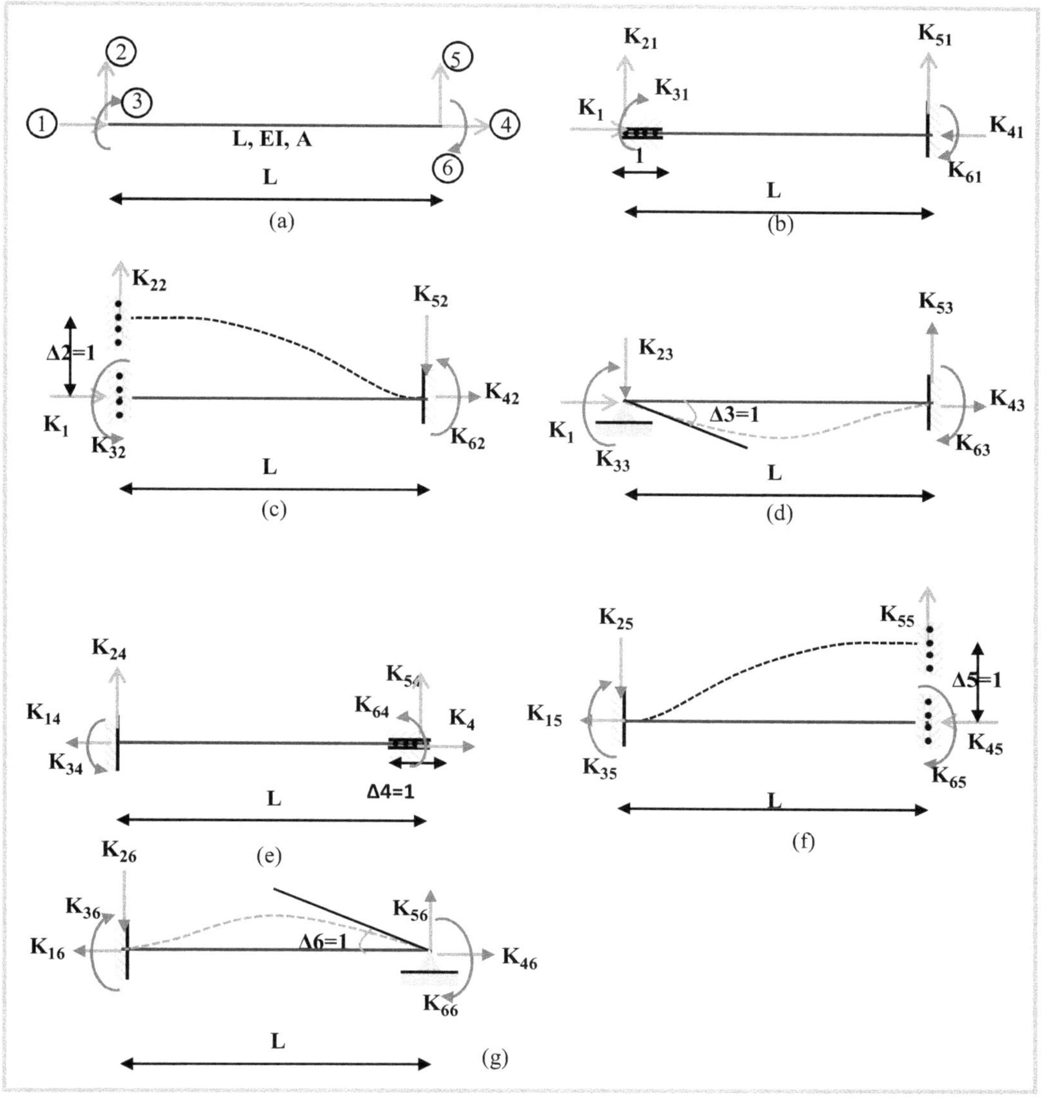

FIGURE 10.3 (a) Beam element with co-ordinates, (b) first column elements, (c) second column elements, (d) third column elements, (e) fourth column elements, (f) fifth column elements, and (g) sixth column elements.

co-ordinates restrained, as shown in Figure 10.3(i)c, and determine the forces generated at all the co-ordinates.

$$K_{12} = 0 = K_{42}$$

$$K_{22} = \frac{12EI}{L^3};\ K_{32} = -\frac{6EI}{L^2};\ K_{52} = -\frac{12EI}{L^3};\ K_{62} = -\frac{6EI}{L^2};$$

Third column elements: The third column elements of the stiffness matrix can be obtained by applying a unit rotational displacement at co-ordinate (3) without any displacement at other co-ordinates (Figure 10.3(i)d) and finding the forces produced at all the co-ordinates.

$$K_{13} = 0 = K_{43}$$

$$K_{23} = -\frac{6EI}{L^2};\quad K_{33} = \frac{4EI}{L};\quad K_{53} = \frac{6EI}{L^2};\quad K_{63} = \frac{2EI}{L};$$

Fourth column elements: To obtain the fourth column elements of the stiffness matrix, apply a unit axial displacement at co-ordinate (4) while keeping all other co-ordinates restrained (Figure 10.3(i)e), and find the forces developed at all the co-ordinates.

$$K_{14} = -\frac{AE}{L};\quad K_{24} = K_{34} = K_{54} = K_{64} = 0;\quad K_{44} = \frac{AE}{L}$$

Fifth column elements: The fifth column elements of the stiffness matrix can be generated by applying a unit translational displacement at co-ordinate (5) without any displacement at other co-ordinates (Figure 10.3(i)f), and determining the forces generated at all the co-ordinates.

$$K_{12} = 0 = K_{42}$$

$$K_{25} = -\frac{12EI}{L^3};\quad K_{35} = \frac{6EI}{L^2};\quad K_{55} = \frac{12EI}{L^3};\quad K_{65} = \frac{6EI}{L^2};$$

Sixth column elements: The sixth column elements of the stiffness matrix can be obtained by applying a unit rotational displacement at co-ordinate (6) without any displacement at other co-ordinates (Figure 10.3(i)g), and finding the forces produced at all the co-ordinates.

$$K_{16} = 0 = K_{46}$$

$$K_{26} = -\frac{6EI}{L^2};\quad K_{36} = \frac{2EI}{L};\quad K_{56} = \frac{6EI}{L^2};\quad K_{66} = \frac{4EI}{L};$$

Let us arrange these elements in matrix form.

$$[K] = \begin{bmatrix} AE/L & 0 & 0 & -AE/L & 0 & 0 \\ 0 & 12EI/L^3 & -6EI/L^2 & 0 & -12EI/L^3 & -6EI/L^2 \\ 0 & -6EI/L^2 & 4EI/L & 0 & 6EI/L^2 & 2EI/L \\ -AE/L & 0 & 0 & AE/L & 0 & 0 \\ 0 & -12EI/L^3 & 6EI/L^2 & 0 & 12EI/L^3 & 6EI/L^2 \\ -0 & -6EI/L^2 & 2EI/L & 0 & 6EI/L^2 & 4EI/L \end{bmatrix}$$

If the beam is assumed to be inextensible, i.e., axial deformations are neglected, the stiffness matrix can be obtained by simply eliminating the elements corresponding to the co-ordinates 1 and 4.

$$[K] = \begin{bmatrix} 12EI/L^3 & -6EI/L^2 & -12EI/L^3 & -6EI/L^2 \\ -6EI/L^2 & 4EI/L & 6EI/L^2 & 2EI/L \\ -12EI/L^3 & 6EI/L^2 & 12EI/L^3 & 6EI/L^2 \\ -6EI/L^2 & 2EI/L & 6EI/L^2 & 4EI/L \end{bmatrix}$$

Stiffness matrix for a truss element: To get the elements of the stiffness matrix for the truss element, consider only the elements corresponding to the co-ordinates 1 and 4. That is

$$[K] = \begin{bmatrix} AE/L & -AE/L \\ -AE/L & AE/L \end{bmatrix}$$

The application of the stiffness matrix method to beams, rigid-jointed, and pin-jointed plane frames is demonstrated in the following sections.

10.3 APPLICATION TO BEAMS

Example 10.1

Analyse the two-span continuous beam, as shown in Figure 10.4 using the stiffness matrix method. Assume EI is constant (this was solved in Example 6.1 using the slope–deflection method.)

Solution:

The unknown joint displacement is only the rotation at support *B* ($\theta_B = \Delta_1$), i.e., the kinematic indeterminacy, I_K=1. Co-ordinate 1 can be assigned to the unknown DoF, θ_B, as shown in Figure 10.5.

The restrained beam can be obtained by locking joint *B*, as shown in Figure 10.6. Hence, the fixed end moments due to the external loads are

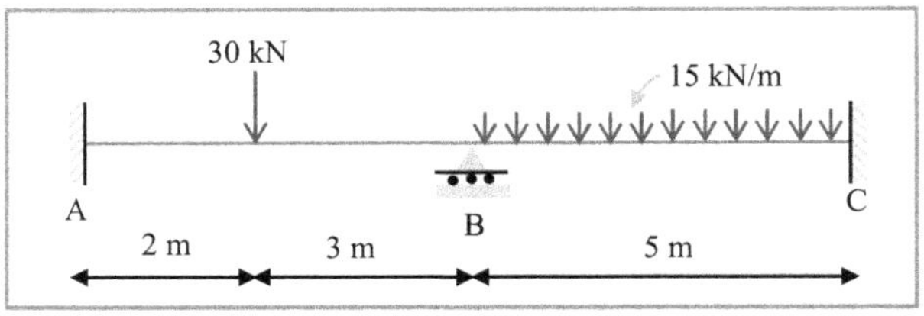

FIGURE 10.4 Two span continuous beam.

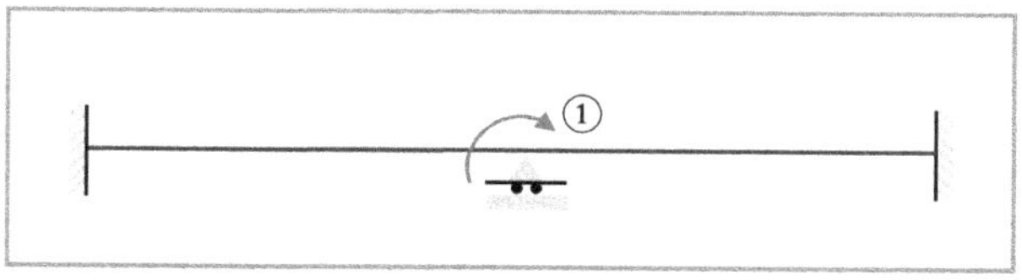

FIGURE 10.5 Co-ordinate 1 assigned to θ_B.

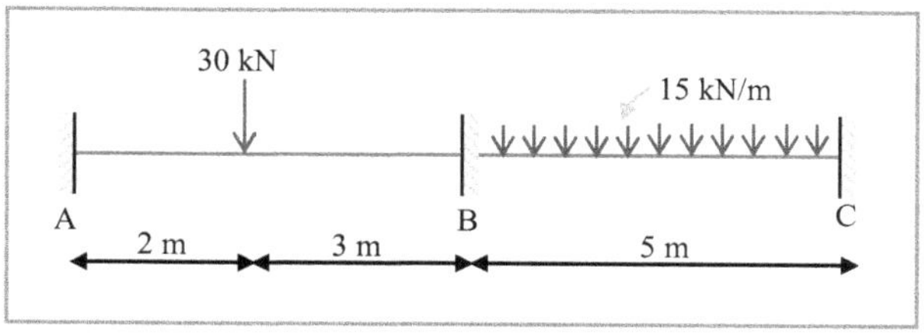

FIGURE 10.6 Restrained beam.

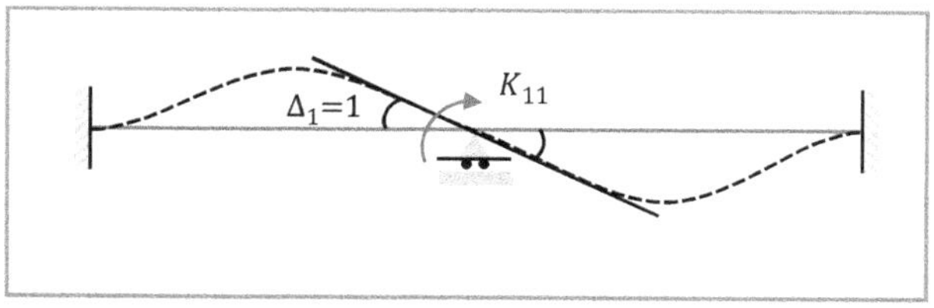

FIGURE 10.7 Unit displacement at ①.

$$MF_{AB} = \frac{-30 \times 3^2 \times 2}{5^2} = -21.66 \text{ kNm}$$

$$MF_{BA} = \frac{30 \times 2^2 \times 3}{5^2} = 14.40 \text{ kNm}$$

$$MF_{BC} = -MF_{CB} = \frac{-15 \times 5^2}{12} = -31.25 \text{ kNm}$$

∴ Forces due to the applied loads at co-ordinate ① are

$$P_L = 14.40 - 31.25 = -16.85 \text{ kNm}$$

Stiffness matrix with respect to the co-ordinates: To obtain an element of the stiffness matrix, apply a unit displacement at co-ordinate ① in its chosen direction, and determine the force developed at co-ordinate ①, as shown in Figure 10.7.

$$K_{11} = \left(\frac{4EI}{L}\right)_{AB} + \left(\frac{4EI}{L}\right)_{BC}$$

$$= \left(\frac{4}{5} + \frac{4}{5}\right)EI = \frac{8EI}{5}$$

Since, there is no external load at co-ordinate ① in the original beam, $P_1 = 0$

∴ Equation 10.15 of the stiffness matrix method is

$$[P] = [P_L] + [K][\Delta]$$

Substituting the values of $P_L, K_{11},$ and P_1, into equation 10.15,

$$0 = -16.85 + \frac{8EI}{5}(\Delta_1)$$

$$\therefore \Delta_1 = \frac{10.53}{EI}(\curvearrowright) = \theta_B$$

Knowing the rotational displacement at B, the final member end moments can be calculated using the slope–deflection equation (equation 6.4), as shown below.

$$M_{AB} = MF_{AB} + \frac{2EI}{L}(2\theta_A + \theta_B - 3\varphi)$$

$$= -21.66 + \frac{2EI}{5}\left(\frac{10.53}{EI}\right) = -17.448 \text{ kNm}$$

$$M_{BA} = MF_{BA} + \frac{2EI}{L}(\theta_A + 2\theta_B - 3\varphi)$$

$$= 14.40 + \frac{2EI}{5}\left(2 \times \frac{10.53}{EI}\right) = 22.825 \text{ kNm}$$

$$M_{BC} = MF_{BC} + \frac{2EI}{L}(2\theta_B + \theta_C - 3\varphi)$$

$$= -31.25 + \frac{2EI}{5}\left(2 \times \frac{10.53}{EI}\right) = -22.825 \text{ kNm}$$

$$M_{CB} = MF_{CB} + \frac{2EI}{L}(\theta_B + 2\theta_C - 3\varphi)$$

$$= 31.25 + \frac{2EI}{5}\left(\frac{10.53}{EI}\right) = 35.462 \text{ kNm}$$

Final end shears: Using the above member end moments, with the help of the free body diagram shown in Figure 10.8a, and using the static equilibrium equations, the final member end shears can be calculated. The shear force, bending moment diagrams, and the probable deflected shape are shown in Figure 10.8b–d, respectively.

Example 10.2

Analyse the continuous beam ABCD as shown in Figure 10.9 using the stiffness matrix method. Assume EI is constant.

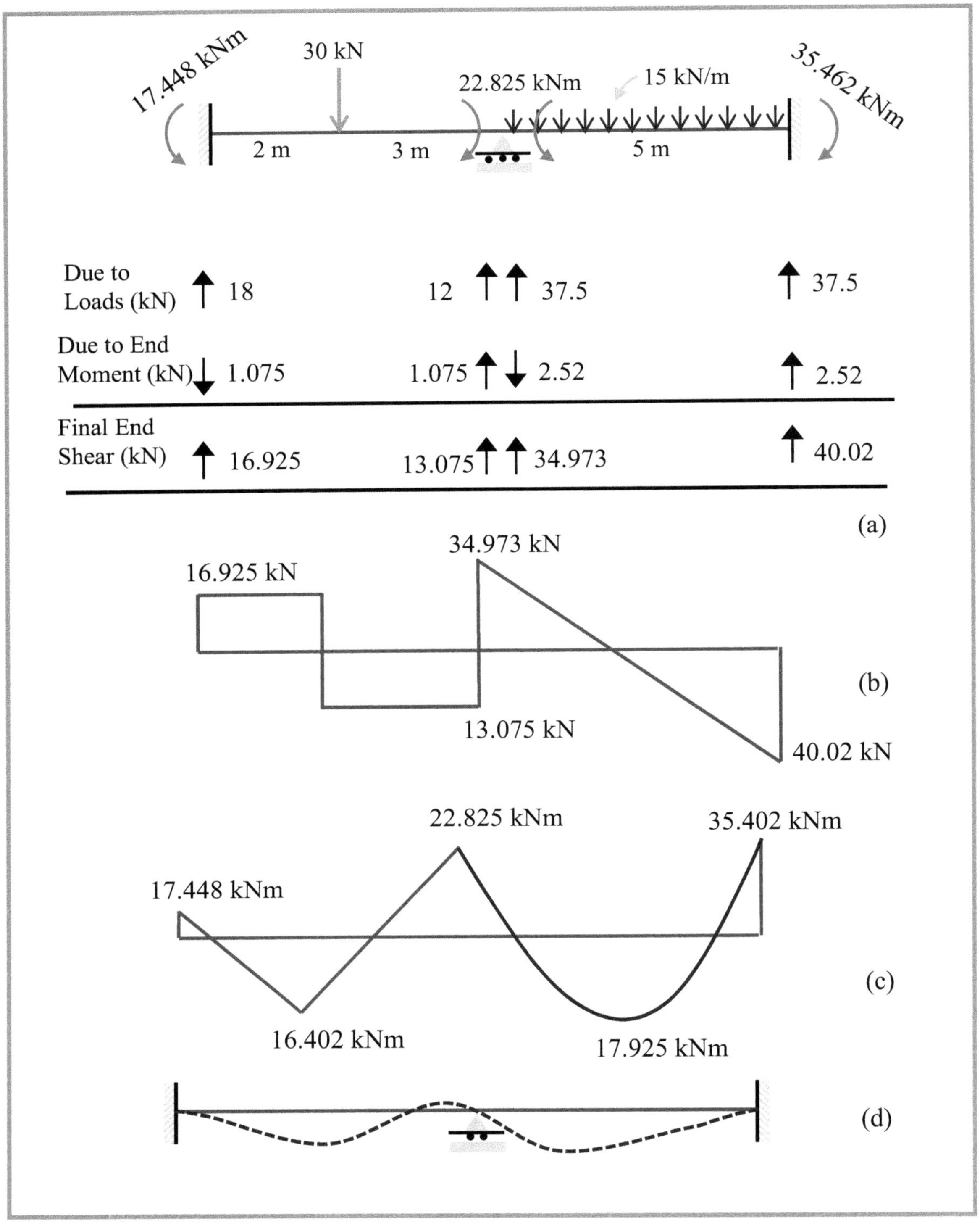

FIGURE 10.8 (a) Free body diagram, (b) shear force diagram, (c) bending moment diagram, and (d) probable deflected shape.

Solution

The overhang portion AB is determinate, and its effect can be represented as a concentrated moment at B, as shown in Figure 10.10a. Therefore, the number of unknown DoF, i.e., kinematic indeterminacy (I_k), is 2: θ_B and θ_C. The co-ordinates can be assigned to the rotations at B and C as ① and ②, respectively, as shown in Figure 10.10b. The restrained structure can be obtained by locking joints B and C, as shown in Figure 10.10c.

∴ Equation 10.15 of the stiffness matrix method is

$$[P]=[P_L]+[P_Y]+[K][\Delta]$$

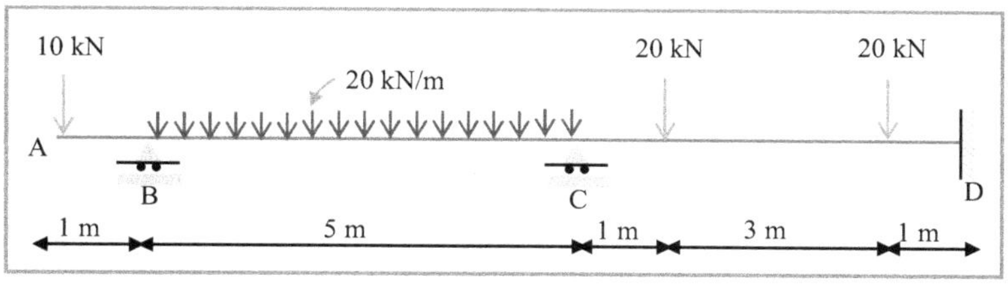

FIGURE 10.9 Two-span continuous beam with one side overhang.

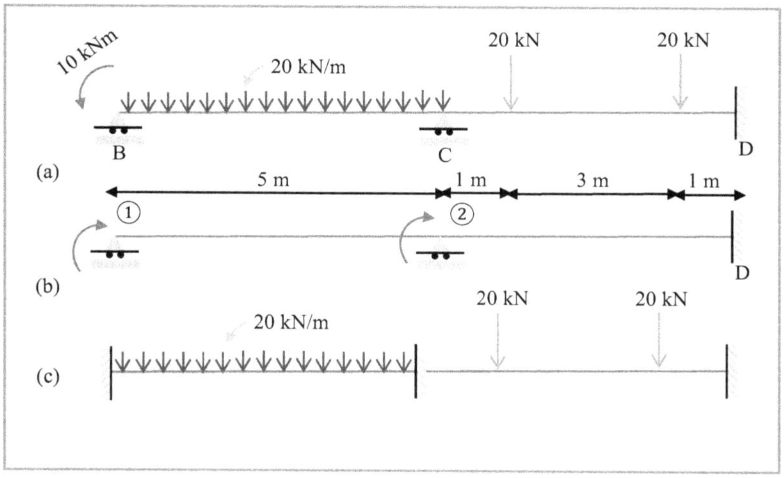

FIGURE 10.10 (a) Equivalent continuous beam, (b) co-ordinates assigned to θ_B and θ_C, and (c) restrained structure.

where $[P]=\begin{bmatrix} -10 \\ 0 \end{bmatrix}$ $[\because 10 \text{ kNm is acting at } B]$

$[P_L]=\begin{bmatrix} P_{1L} \\ P_{2L} \end{bmatrix}$ Fixed end moments due to the applied loads.

$[P_Y]=\begin{bmatrix} 0 \\ 0 \end{bmatrix}$ $(\because$ No yielding of supports$)$

$[\Delta]=\begin{bmatrix} \Delta_1=\theta_B \\ \Delta_2=\theta_C \end{bmatrix}$ $(\because$ Unknown DoF$)$

$[K]=\begin{bmatrix} K_{11} & K_{12} \\ K_{21} & K_{22} \end{bmatrix}$ is the stiffness matrix.

Fixed end moments: The fixed end moments can be calculated using the standard expressions provided in Table 6.1.

$$\text{MF}_{BC} = -\text{MF}_{CB} = \frac{-20\times 5^2}{12} = -41.67 \text{ kNm}$$

$$\text{MF}_{CD} = -\text{MF}_{DC} = \frac{-20\times 1\times 4^2}{5^2} - \frac{-20\times 4\times 1^2}{5^2} = -16 \text{ kNm}$$

Therefore, $[P_L]=\begin{bmatrix} -41.67 \\ 25.67 \end{bmatrix}$

ELEMENT OF THE STIFFNESS MATRIX

First column elements: The elements of the first column of the stiffness matrix can be obtained by applying a unit displacement at co-ordinate 1 in its chosen direction on the restrained beam and calculating the forces developed at all the co-ordinates, as shown in Figure 10.11a.

$$K_{11} = \frac{4EI}{L} = \frac{4}{5}EI$$

$$K_{21} = \frac{2EI}{L} = \frac{2}{5}EI$$

Second column elements of the stiffness matrix: Apply a unit displacement at co-ordinate 2 in its chosen direction without any displacement at co-ordinate 1 on the restrained beam and obtain the forces developed at both co-ordinates 1 and 2, as shown in Figure 10.11b.

$$K_{12} = \frac{2EI}{L} = \frac{2}{5}EI$$

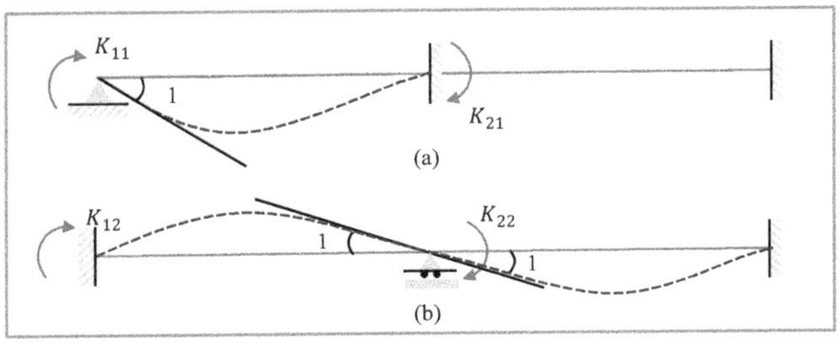

FIGURE 10.11 (a) Unit displacement at co-ordinate 1 and (b) unit displacement at co-ordinate 2.

$$K_{22}=\left(\frac{4EI}{L}\right)_{BC}+\left(\frac{4EI}{L}\right)_{CD}=\frac{4}{5}EI+\frac{4}{5}EI=\frac{8EI}{5}$$

$$\therefore [K]=EI\begin{bmatrix} 4/5 & 2/5 \\ 2/5 & 8/5 \end{bmatrix}$$

Substituting the above into equation 10.15, we have

$$\begin{bmatrix} -10 \\ 0 \end{bmatrix}=\left[\begin{bmatrix} -41.67 \\ 25.67 \end{bmatrix}+\begin{bmatrix} 0 \\ 0 \end{bmatrix}\right]\begin{bmatrix} 4/5 & 2/5 \\ 2/5 & 8/5 \end{bmatrix}EI\begin{bmatrix} \Delta_1 \\ \Delta_2 \end{bmatrix}$$

By solving the above equation, we obtain

$$\therefore \begin{bmatrix} \Delta_1 \\ \Delta_2 \end{bmatrix}=\begin{bmatrix} 54.41/EI \\ -29.64/EI \end{bmatrix}=\begin{bmatrix} \theta_B \\ \theta_C \end{bmatrix}$$

Final member end moments: Knowing the displacements θ_B and θ_C, the final member end moments can be calculated by using slope–deflection equations, as shown below.

$$M_{BC}=\text{MF}_{BC}+\frac{2EI}{L}(2\theta_B+\theta_C-3\varphi)$$

$$=-41.67+\frac{2EI}{5}\left(2\times\frac{54.41}{EI}-\frac{29.64}{EI}\right)=-10\text{ kNm}$$

$$M_{CB}=41.67+\frac{2EI}{5}\left(\frac{54.41}{EI}-\frac{2\times 29.64}{EI}\right)=39.72\text{ kNm}$$

$$M_{CD}=-16+\frac{2EI}{5}\left(\frac{2\times(-29.64)}{EI}\right)=-39.72\text{ kNm}$$

$$M_{DC}=16+\frac{2EI}{5}\left(\frac{(-29.64)}{EI}\right)=4.14\text{ kNm}$$

Final end reactions: Using the known end moments and static equilibrium equations with the help of a free body diagram, the member end shears can be calculated, as shown in Figure 10.12a. The shear force and bending moment diagrams are shown in Figure 10.12b and c, respectively. The probable deflected shape is shown in Figure 10.12d.

Example 10.3

Analyse the three-span continuous beam *ABCD* as shown in Figure 10.13 using the stiffness matrix method. Assume *EI* is constant.

Solution

The kinematic indeterminacy (I_k) is 3, corresponding to the rotations at supports *A*, *B*, and *C* (θ_A, θ_B, and θ_C). However, since the moment at end *A* is zero, θ_A can be neglected. Hence, the kinematic indeterminacy is reduced to two, i.e., θ_B and θ_C, provided appropriate modifications are made to account for span *AB* in the fixed end moments, stiffness, and slope–deflection equations. The co-ordinates can be assigned to θ_B and θ_C, as shown in Figure 10.14a. The restrained structure is shown in Figure 10.14b.

Equation 10.15 of the stiffness matrix is

$$[P]=[[P_L]+[P_Y]]+[K][\Delta]$$

where $[P_L]$ are the fixed moments due to the applied loads in the restrained structure.

$$\text{MF}_{BA}=\frac{wL^2}{20}+\frac{1}{2}\frac{wL^2}{30}=\frac{wL^2}{15}=\frac{10\times 6^2}{15}$$

$$=24\text{ kNm (modified fixed end moment)}$$

$$\text{MF}_{BC}=\frac{M}{4}=\frac{48}{4}=12\text{ kNm}=\text{MF}_{CB}$$

$$\text{MF}_{CD}=-\frac{11}{192}wL^2=-\frac{11\times 16\times 6^2}{192}=-33\text{ kNm}$$

$$\text{MF}_{DC}=\frac{5}{192}wL^2=\frac{5\times 16\times 6^2}{192}=15\text{ kNm}$$

$$[P_L]=\begin{bmatrix} P_{1L} \\ P_{2L} \end{bmatrix}=\begin{bmatrix} 24+12=36 \\ -33+12=-21 \end{bmatrix}\text{kNm}$$

$$[P_Y]=\begin{bmatrix} P_{1Y} \\ P_{2Y} \end{bmatrix}=\begin{bmatrix} 0 \\ 0 \end{bmatrix}\quad(\because \text{No yielding of supports})$$

$$[K]=\text{stiffness matrix}=\begin{bmatrix} K_{11} & K_{12} \\ K_{21} & K_{22} \end{bmatrix}$$

First column elements of the stiffness matrix: To obtain the first column elements of the stiffness matrix, apply a unit displacement at co-ordinate ① in its chosen direction and without any displacement at other co-ordinates on the restrained structure. Determine the forces developed at all the co-ordinates, as shown in Figure 10.15a.

$$K_{11}=\left(\frac{3EI}{L}\right)_{AB}+\left(\frac{4EI}{L}\right)_{BC}=\frac{3}{6}EI+\frac{4}{6}EI=\frac{7EI}{6}$$

$$K_{21}=\frac{2EI}{L}=\frac{2}{6}EI$$

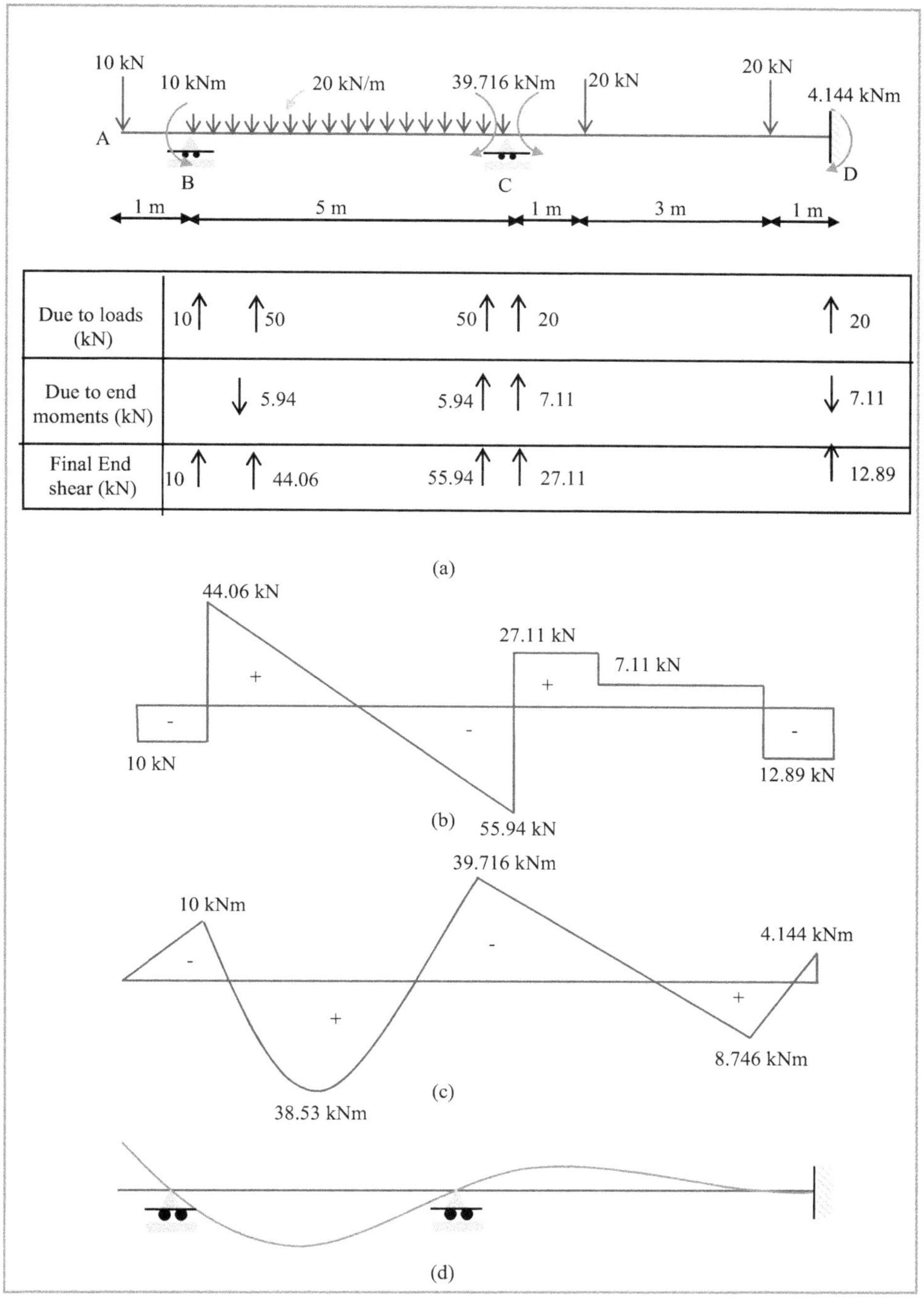

FIGURE 10.12 (a) Free body diagram, (b) shear force diagram, (c) bending moment diagram, and (d) probable deflected shape.

Second column elements of the stiffness matrix: To generate the second column elements of the stiffness matrix, apply a unit displacement at ② in its chosen direction, without displacement at ① on the restrained beam, and obtain the forces developed at co-ordinates ① and ②, as shown in Figure 10.15b.

$$K_{12} = \frac{2EI}{L} = \frac{2}{6}EI$$

$$K_{22} = \left(\frac{4EI}{L}\right)_{BC} + \left(\frac{4EI}{L}\right)_{CD} = \frac{4}{6}EI + \frac{4}{6}EI = \frac{8EI}{6}$$

Substituting the above values into equation 10.15,

$$\begin{bmatrix} 0 \\ 0 \end{bmatrix} = \left[\begin{bmatrix} 36 \\ -21 \end{bmatrix} + \begin{bmatrix} 0 \\ 0 \end{bmatrix}\right] + \begin{bmatrix} 7 & 2 \\ 2 & 8 \end{bmatrix} \frac{EI}{6} \begin{bmatrix} \Delta_1 = \theta_B \\ \Delta_2 = \theta_C \end{bmatrix}$$

By solving the above equation, we obtain

$$\therefore \begin{bmatrix} \theta_B \\ \theta_C \end{bmatrix} = \begin{bmatrix} -38.07/EI \\ 25.26/EI \end{bmatrix}$$

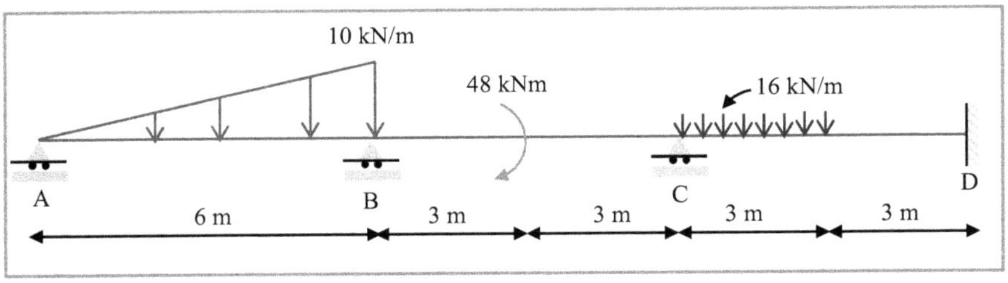

FIGURE 10.13 A three-span continuous beam.

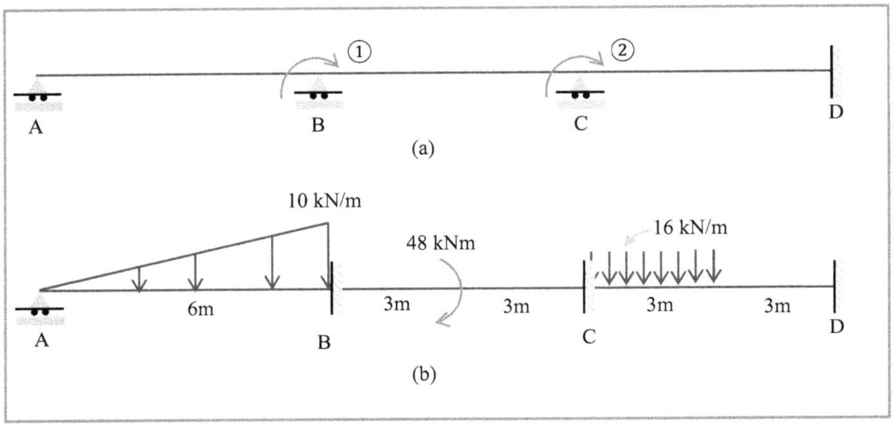

FIGURE 10.14 (a) Co-ordinates assigned to θ_B and θ_C and (b) restrained beam.

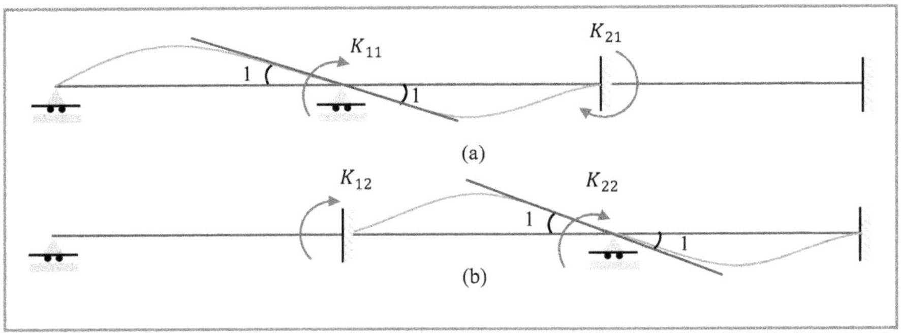

FIGURE 10.15 (a) Unit displace at co-ordinate 1 and (b) unit displacement at co-ordinate 2.

Final member end moments: Knowing the displacements θ_B and θ_C, the final member end moments can be calculated using the slope–deflection equations as shown below.

$$M_{BA} = 24 + \frac{3EI}{6}\left(\frac{(-38.07)}{EI}\right) = 4.96 \text{ kNm}$$

$$M_{BC} = 12 + \frac{4EI}{6}\left(2 \times \frac{(-38.07)}{EI} + \frac{25.26}{EI}\right) = -4.96 \text{ kNm}$$

$$M_{CB} = 12 + \frac{2EI}{6}\left(\frac{-38.07}{EI} + 2 \times \frac{25.26}{EI}\right) = 16.16 \text{ kNm}$$

$$M_{CD} = -33 + \frac{2EI}{6}\left(2 \times \frac{25.26}{EI}\right) = -16.16 \text{ kNm}$$

$$M_{DC} = 15 + \frac{2EI}{6}\left(\frac{25.26}{EI}\right) = 23.42 \text{ kNm}$$

Final member end shears: With the known member end moments, and using the equilibrium equations, the final member end shears can be calculated as shown in the free body diagram in Figure 10.16a.

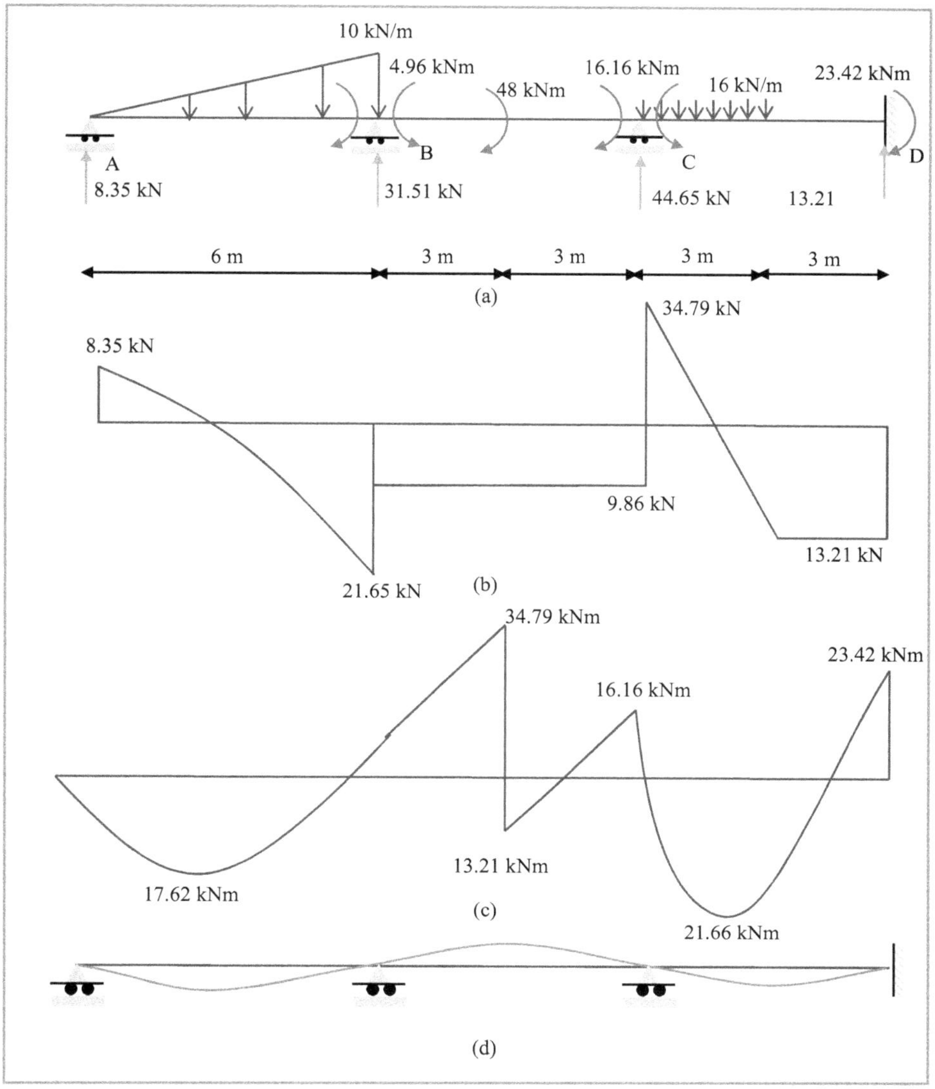

FIGURE 10.16 (a) Free body diagram, (b) shear force diagram, (c) bending moment diagram, and (d) probable deflected shape.

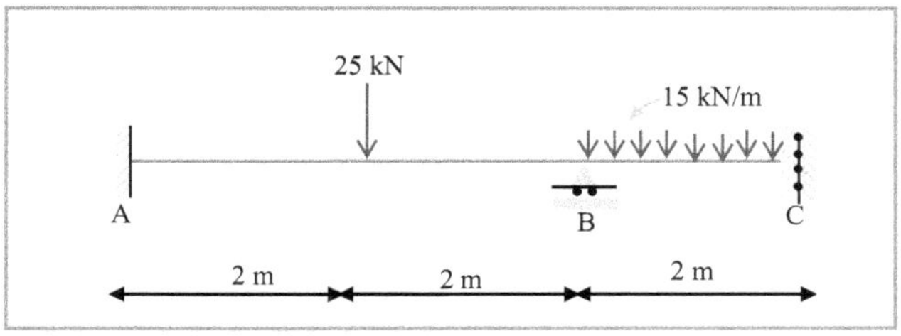

FIGURE 10.17 Two-span continuous beam with guided support.

Example 10.4

Analyse the two-span continuous beam whose one end is fixed and the other end is a guided support, as shown in Figure 10.17 using the stiffness matrix method. Assume *EI* is constant.

Solution

The kinematic indeterminacy of the beam, $I_K = 2$ i.e. θ_B and Δ_C. The co-ordinates can be assigned to these unknown DoF as ① and ②, respectively, as shown in Figure 10.18a. The restrained beam can be obtained by preventing the θ_B and Δ_C that is by introducing fixity at *B* and *C*, as shown in Figure 10.18b.

Equation 8.15 of the stiffness matrix method is

$$[P]=\left[[P_L]+[P_Y]\right]+[K][\Delta]$$

where

$$[P]=\begin{bmatrix} 0 \\ 0 \end{bmatrix} \quad [\because \text{No external forces at coordinates in the original beam}]$$

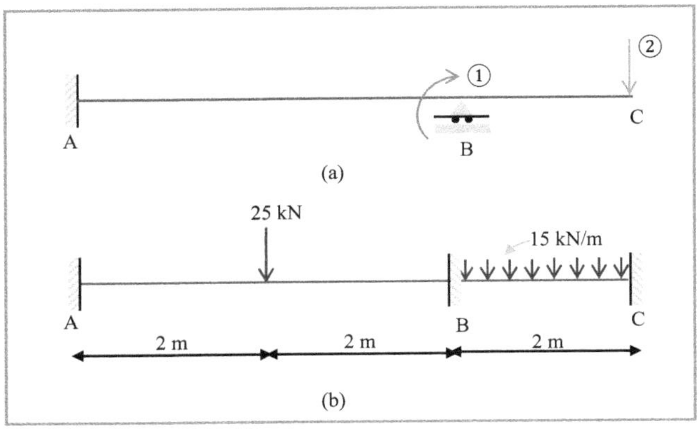

FIGURE 10.18 (a) Co-ordinates assigned to θ_B and Δ_C and (b) restrained beam.

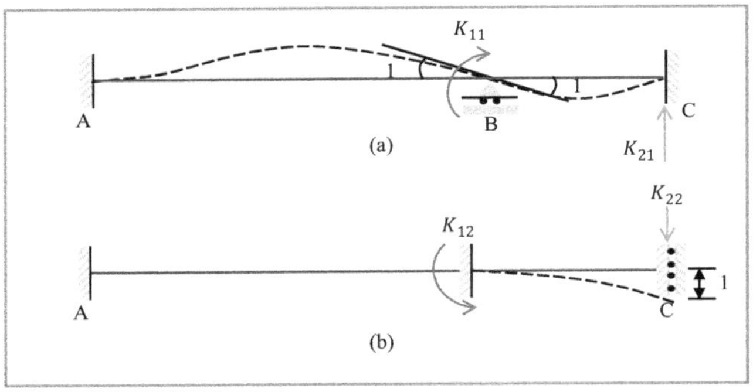

FIGURE 10.19 (a) First column elements and (b) second column elements.

$$[P_Y]=\begin{bmatrix}0\\0\end{bmatrix}\quad (\because \text{No yielding of supports})$$

$$[P_L]=\begin{bmatrix}P_{1L}\\P_{2L}\end{bmatrix}$$ are the fixed end moments due to the external loads

$$[K]=\begin{bmatrix}K_{11} & K_{12}\\K_{21} & K_{22}\end{bmatrix}$$ is the stiffness matrix

Fixed end moments: The fixed end moments can be calculated using the expressions given in Table 6.1 as below

$$MF_{AB}=-MF_{BA}=\frac{-25\times 4}{8}=-12.5\text{ kNm}$$

$$MF_{BC}=-MF_{CB}=\frac{-15\times 2^2}{12}=-5\text{ kNm}$$

$$V_{CB}=-\frac{WL}{2}=\frac{-15\times 2}{2}=-15\text{ kN}$$

$$\therefore\begin{bmatrix}P_{1L}\\P_{2L}\end{bmatrix}=\begin{bmatrix}12.5-5\\-15\end{bmatrix}=\begin{bmatrix}7.5\\-15\end{bmatrix}$$

STIFFNESS MATRIX

First column elements of the stiffness matrix: To obtain the first column elements of the stiffness matrix, apply a unit displacement at co-ordinate ① in its chosen direction, without displacement at other co-ordinates on the restrained beam, and determine the forces developed at all co-ordinates, as shown in Figure 10.19a.

$$K_{11}=\left(\frac{4EI}{L}\right)_{AB}+\left(\frac{4EI}{L}\right)_{BC}=\frac{4}{4}EI+\frac{4}{2}EI=3EI$$

$$K_{21}=\frac{-6EI}{L^2}=\frac{-6}{2^2}EI=-1.5EI$$

Second column elements of the stiffness matrix: The second column elements can be obtained by applying a unit displacement at ②, without any displacement at ①, and finding the forces at all co-ordinates in the restrained beam, as shown in Figure 10.19b.

$$K_{12}=\frac{-6EI}{L^2}=\frac{-6}{2^2}EI=-1.5EI$$

$$K_{22}=\frac{12EI}{L^3}=\frac{12EI}{2^3}=1.5EI$$

$$[K]=EI\begin{bmatrix}3 & -1.5\\-1.5 & 1.5\end{bmatrix}$$

Substituting all the values into equation 10.15,

$$\begin{bmatrix}0\\0\end{bmatrix}=\left[\begin{bmatrix}7.5\\-15\end{bmatrix}+\begin{bmatrix}0\\0\end{bmatrix}\right]+EI\begin{bmatrix}3 & -1.5\\-1.5 & 1.5\end{bmatrix}\begin{bmatrix}\theta_B\\\Delta_C\end{bmatrix}$$

By solving the above equation, we obtain

$$\therefore\begin{bmatrix}\theta_B\\\Delta_C\end{bmatrix}=\begin{bmatrix}5/EI\\15/EI\end{bmatrix}$$

Final member end moments: After knowing the values of θ_B and Δ_C, using the slope–deflection equations, the final member end moments can be calculated as below.

$$M_{AB}=-12.5+\frac{2EI}{4}\left(\frac{5}{EI}\right)=-10\text{ kNm}$$

$$M_{BA}=12.5+\frac{2EI}{4}\left(2\times\frac{5}{EI}\right)=17.5\text{ kNm}$$

$$M_{BC}=-5+\frac{2EI}{4}\left(2\times\frac{5}{EI}-\frac{3}{2}\times\frac{15}{EI}\right)=17.5\text{ kNm}$$

$$M_{CB}=5+\frac{2EI}{2}\left(\frac{5}{EI}-\frac{3\times 15}{2EI}\right)=-12.5\text{ kNm}$$

Final member end shears: The final member end shears can be calculated using the known end moments and equilibrium equations with the help of the free body diagram,

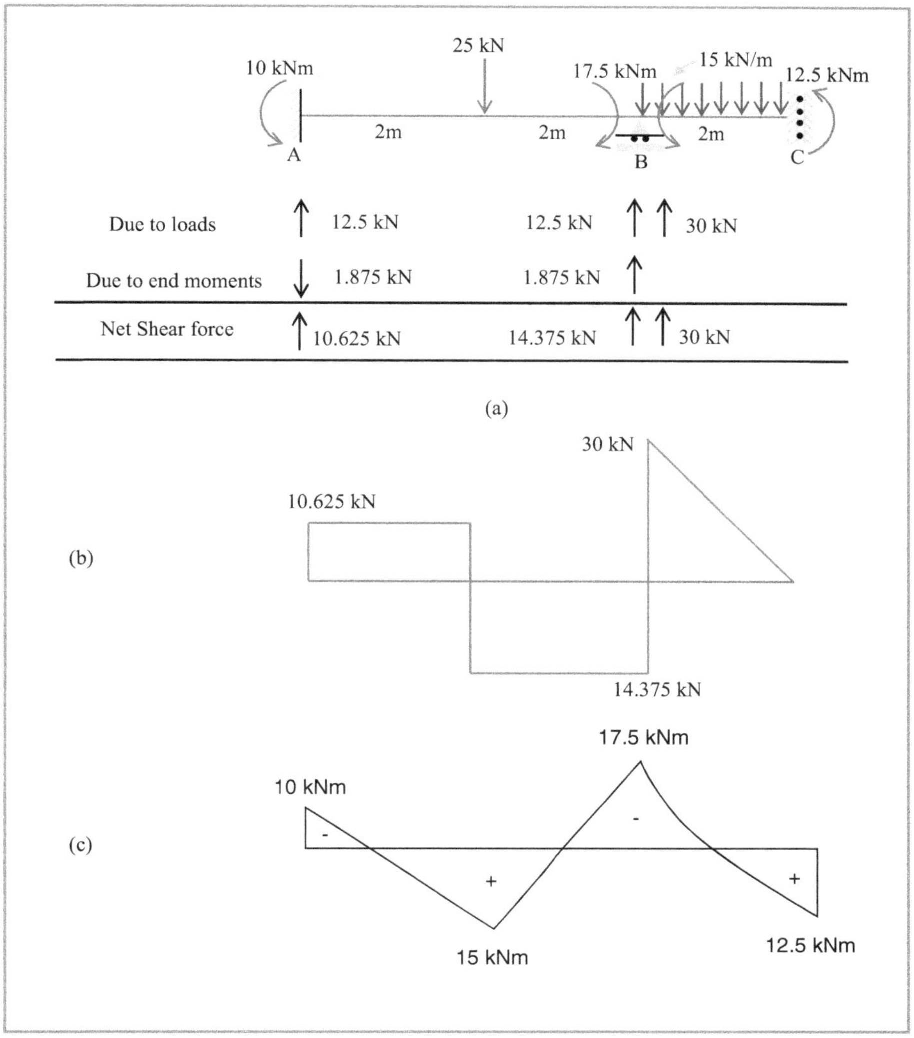

FIGURE 10.20 (a) Free body diagram, (b) shear force diagram, and (c) bending moment diagram.

as shown in Figure 10.20a. The shear force and bending moment diagrams are shown in Figure 10.20b and c, respectively.

Alternate Solution: By using the modified fixed end moments and modified slope–deflection equations for span BC, Δ_C may be ignored. Hence, the kinematic indeterminacy is reduced to θ_B only. To apply the modified fixed end moments and slope–deflection equations for BC, the beam $ABCB^1A^{1'}$ should be visualised as symmetrical with respect to C, where ABC is the left half of $ABCB'A'$, as shown in Figure 10.21a. The moment at C is equivalent to the mid-span moment in the span BCB^1.

Fixed end moments: The fixed end moments (Figure 10.21b) can be calculated using the expressions provided in Table 6.1.

$$\mathrm{MF}_{AB} = -\frac{WL}{8} = -\frac{25\times 4}{8} = -12.5 \text{ kNm} = -\mathrm{MF}_{BA}$$

$$\mathrm{MF}_{BC} = -\frac{wL^2}{12} = -\frac{15\times 4^2}{12} = -20 \text{ kNm}$$

$$\mathrm{MF}_{CB} = -\left(\frac{wL^2}{8} - \frac{wL^2}{12}\right) = -\left(\frac{wL^2}{24}\right) = -\frac{15\times 4^2}{24} = -10 \text{ kNm}$$

Stiffness coefficient:

$$K_{11} = \left(\frac{4EI}{L}\right)_{BA} + \left(\frac{EI}{L}\right)_{BC} = \frac{4}{4}EI + \frac{1}{2}EI = 1.5EI$$

Substituting the above values into equation 10.15,

$$0 = [-20 + 12.5] + (1.5EI)(\theta_B)$$

By solving, we get

$$\theta_B = \frac{5}{EI}$$

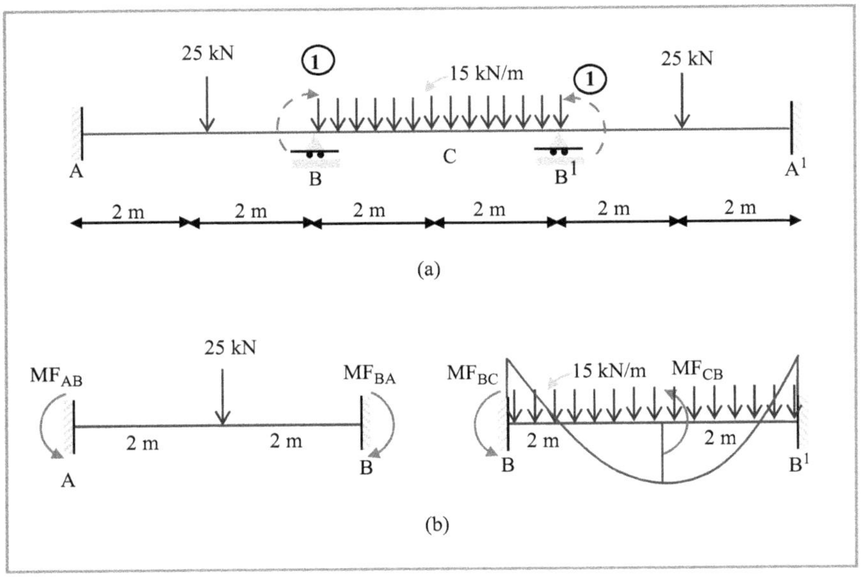

FIGURE 10.21 (a) Visualised as symmetrical beam $ABCB^1A^1$ and (b) fixed end moments.

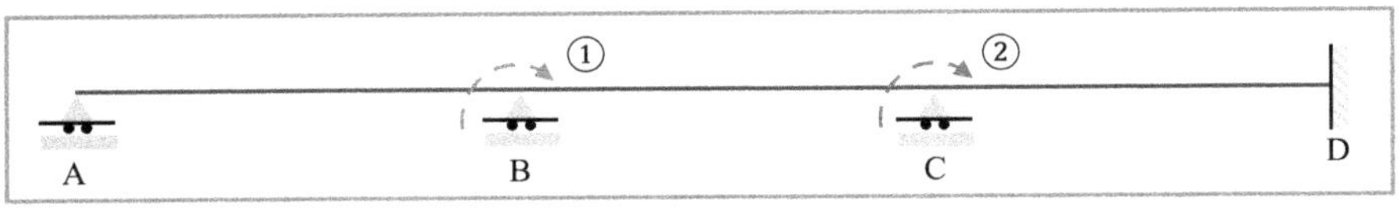

FIGURE 10.22 Co-ordinates assigned to θ_B and θ_C.

Final end moments: After knowing the rotation at *B*, using the slope–deflection equations, the member end moments can be calculated, as shown below.

$$M_{AB} = -12.5 + \frac{2EI}{4}\left(\frac{5}{EI}\right) = -10 \text{ kNm}$$

$$M_{BA} = 12.5 + \frac{2EI}{4}\left(2\times\frac{5}{EI}\right) = 17.5 \text{ kNm}$$

$$M_{BC} = \text{MF}_{BC} + \frac{EI}{L}(\theta_B) = -20 + \frac{EI}{2}\left[\frac{5}{EI}\right] = -17.5 \text{ kNm}$$

$$M_{CB} = \text{MF}_{CB} - \frac{EI}{L}(\theta_B) = -10 - \frac{EI}{2}\left(\frac{5}{EI}\right) = -12.5 \text{ kNm}$$

Example 10.5

Re-analyse Example 10.3, if the supports *B* and *C* are settling down by 2,000/*EI* kNm and 1,000/*EI* kNm, respectively, in addition to the external loads.

Solution

The beam in this example is the same as that in example 10.3, except that supports *B* and *C* undergoes some pre-specified settlements. Hence, the same co-ordinates ① and ② may be chosen, as shown in Figure 10.22.

The fixed end moments due to the applied loads, as calculated in example 10. 3, are

$$\text{MF}_{BA} = 24 \text{ kNm}$$

$$\text{MF}_{BC} = \text{MF}_{CB} = 12 \text{ kNm}$$

$$\text{MF}_{CD} = -33 \text{ kNm}$$

$$\text{MF}_{DC} = 15 \text{ kNm}$$

Fixed end moments due to the settlement of supports:

$$\text{MF}^Y_{BA} = \frac{-6EI\Delta_B}{L^2} + \frac{3EI\Delta_C}{L^2} = \frac{-3EI}{6^2}\times\left(\frac{2{,}000}{EI}\right) = -166.67 \text{ kNm}$$

$$\text{MF}^Y_{BC} = \frac{6EI}{L^2}(\Delta_B - \Delta_C) + \frac{3EI\Delta}{L^2} = \frac{6EI}{6^2}\times\left(\frac{1{,}000}{EI}\right) = 166.67 \text{ kNm}$$

$$\text{MF}^Y_{CB} = \frac{6EI}{L^2}(\Delta_B - \Delta_C) = \frac{6EI}{6^2}\times\left(\frac{1{,}000}{EI}\right) = 166.67 \text{ kNm}$$

$$\text{MF}^Y_{CD} = \text{MF}^Y_{DC} = \frac{6EI}{L^2}(\Delta_C) = \frac{6EI}{6^2}\times\left(\frac{1000}{EI}\right) = 166.67 \text{ kNm}$$

The force matrix corresponding to the co-ordinates is

$$\begin{bmatrix} P_{1L} + P_{1Y} \\ P_{2L} + P_{2Y} \end{bmatrix} = \begin{bmatrix} 24 + 12 - 166.67 + 166.67 \\ 12 - 33 + 166.67 + 166.67 \end{bmatrix}$$

$$= \begin{bmatrix} 36 \\ 312.34 \end{bmatrix} \text{kNm}$$

STIFFNESS MATRIX

Since the co-ordinates are the same as those selected in example 10.3, the stiffness matrix remains unchanged and is given by

$$[K] = EI\begin{bmatrix} 7 & 2 \\ 2 & 8 \end{bmatrix}$$

$$\begin{bmatrix} P_1 \\ P_2 \end{bmatrix} = \begin{bmatrix} 0 \\ 0 \end{bmatrix}$$ (∵ No external forces at coordinates in the original beam)

On substituting the above values into equation 10.19, i.e.,

$$[P] = [[P_L] + [P_Y]] + [K][\Delta] \quad (10.19)$$

$$\begin{bmatrix} 0 \\ 0 \end{bmatrix} = \begin{bmatrix} 36 \\ 312.34 \end{bmatrix} + \frac{EI}{6}\begin{bmatrix} 7 & 2 \\ 2 & 8 \end{bmatrix}\begin{bmatrix} \theta_B \\ \theta_C \end{bmatrix}$$

By solving the above, we obtain

$$\therefore \begin{bmatrix} \theta_B \\ \theta_C \end{bmatrix} = \begin{bmatrix} 38.847/EI \\ -243.967/EI \end{bmatrix}$$

Final end moments: With the known values of θ_B and θ_C and using the slope–deflection equations, the final member end moments can be calculated, as shown below.

$$M_{BA} = \text{MF}_{AB} + \frac{3EI}{L}(\theta_B - \varphi)$$

OR

$$= (\text{MF}_{AB})_L + \text{MF}^Y_{AB} + \frac{3EI}{L}(\theta_B)$$

$$= 24 - 166.67 + \frac{3EI}{6}(38.847) = -123.24 \text{ kNm}$$

$$M_{BC} = (\text{MF}_{BC})_L + (\text{MF}^Y_{BC}) + \frac{2EI}{L}(2\theta_B + \theta_C)$$

$$= 12 + 166.67 + \frac{2EI}{6}\left[2\left(\frac{38.847}{EI}\right) - \frac{243.967}{EI}\right] = 123.24 \text{ kNm}$$

$$M_{CB} = 12 + 166.67 + \frac{2EI}{6}\left[\frac{38.847}{EI} - \frac{2 \times 243.967}{EI}\right] = 28.97 \text{ kNm}$$

$$M_{CD} = -33 + 166.67 + \frac{2EI}{6}\left[-\frac{2 \times 243.967}{EI}\right] = -28.97 \text{ kNm}$$

$$M_{DC} = 15 + 166.67 + \frac{2EI}{6}\left[-\frac{243.967}{EI}\right] = 100.34 \text{ kNm}$$

Final end shears: With the known member end moments, using the free body diagram and the static equilibrium equations, the member end shear can be calculated, as shown in Figure 10.23a. The shear force and bending moment diagrams are shown in Figure 10.23b and c, respectively. The probable deflected shape is shown in Figure 10.23d.

Example 10.6

A three-span continuous beam with an internal hinge at *B*, as shown in Figure 10.24. Analyse the beam and draw the shear force and bending moment diagrams. The same has been solved in Example 6.7 using the slope–deflection method.

Solution

The kinematic indeterminacy of the beam is 4 i.e. $\theta_{BA}, \theta_{BC}, \Delta_B$ and θ_C. However, by applying appropriate modifications to the fixed end moments and the slope–deflection equation for spans *BA* and *BC*, θ_{BA} and θ_{BC} may be ignored; as a result, the kinematic indeterminacy is reduced to 2, i.e., θ_C and Δ_B. Let us assign the co-ordinates to Δ_B and θ_C as ① and ②, respectively, as shown in Figure 10.25.

Fixed end moments due to loads: The fixed end moments can be calculated using the expressions provided in Table 6.1, as shown below.

$$\text{MF}_{AB} = \text{MF}_{CB} = 0 \quad (\because \text{No loads on spans } AB \text{ and } BC)$$

$$\text{MF}_{CD} = \text{MF}_{DC} = \frac{10}{4} = 2.5 \text{ kNm}$$

Since there is no yielding of supports in the beam, the additional fixed end moments are zero, i.e.,

$$P_{1Y} = P_{2Y} = 0$$

$$\therefore P_{1L} = 0$$

$$P_{2L} = 2.5 \text{ kNm}$$

STIFFNESS MATRIX

First column elements of the stiffness matrix: To obtain the first column elements of the stiffness matrix, apply a unit displacement at co-ordinate ① in its chosen direction, without any displacement at ②, and find the forces at all co-ordinates, as shown in Figure 10.26a.

$$K_{11} = \left(\frac{3EI}{L^3}\right)_{AB} + \left(\frac{3EI}{L^3}\right)_{BC} = \frac{3EI}{3^3} + \frac{3EI}{3^3} = \frac{6EI}{27} = \frac{2EI}{9}$$

$$K_{21} = \frac{3EI}{L^2} = \frac{3EI}{9} = \frac{EI}{3}$$

Second column elements of the stiffness matrix: The second column elements of the stiffness matrix can be obtained by applying a unit displacement at ② without any displacement at ①, and finding the forces developed at all co-ordinates, as shown in Figure 10.26b.

$$K_{12} = \frac{3EI}{L^2} = \frac{3EI}{9} = \frac{EI}{3}$$

$$K_{22} = \left(\frac{3EI}{L}\right)_{BC} + \left(\frac{4EI}{L}\right)_{CD} = \frac{3EI}{3} + \frac{4EI}{6} = \frac{5EI}{3}$$

Equation 10.19 of the stiffness matrix is

$$[P] = [[P_L] + [P_Y]] + [K][\Delta]$$

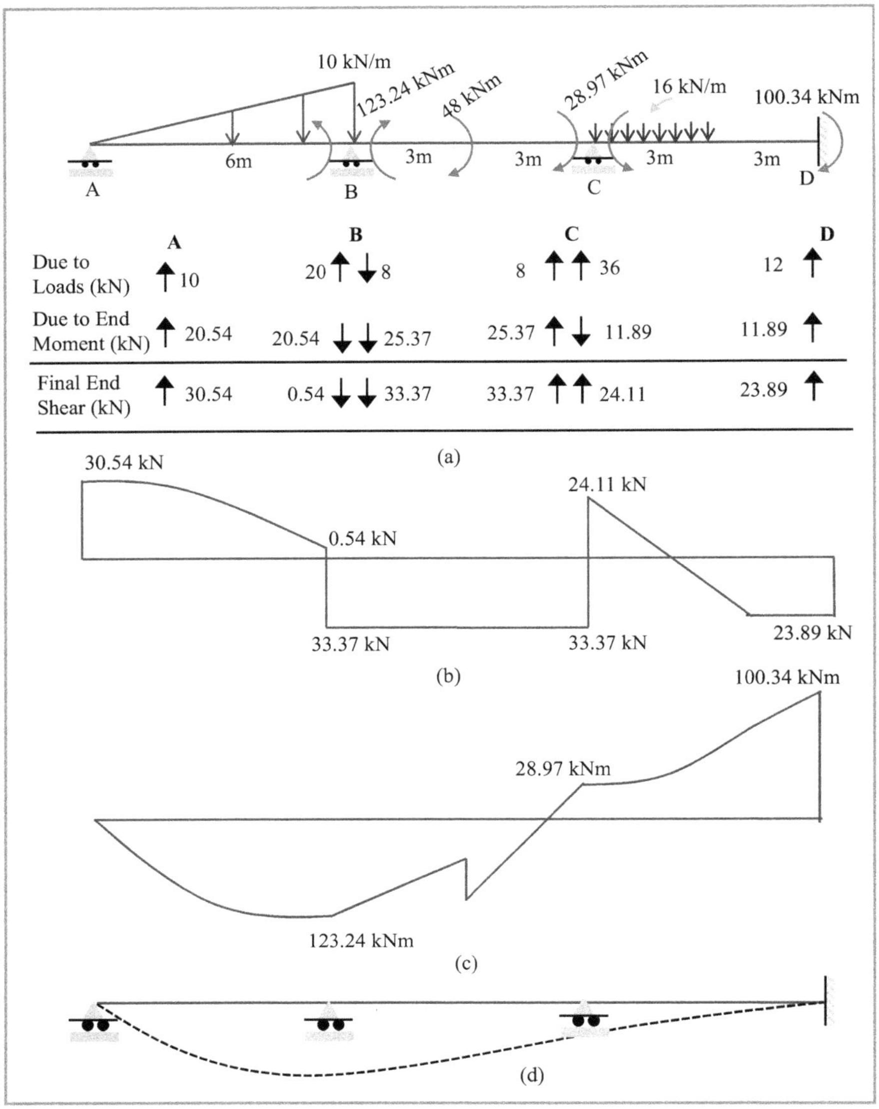

FIGURE 10.23 (a) Free body diagram, (b) shear force diagram, and (c) bending moment diagram and (d) probable deflected shape.

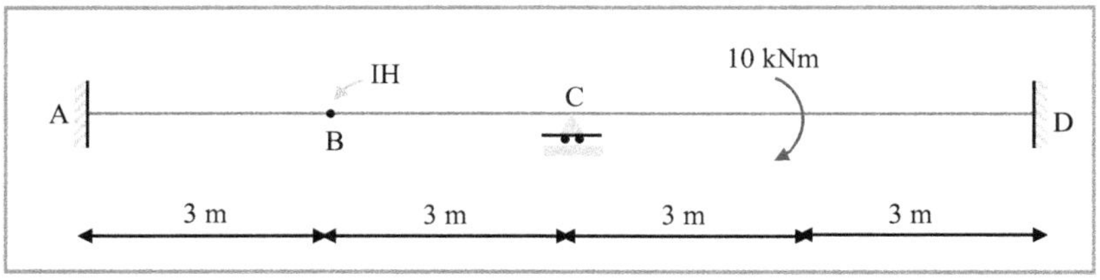

FIGURE 10.24 Continuous beam with internal hinge.

By substituting the above values into equation 10.19,

$$\begin{bmatrix} 0 \\ 0 \end{bmatrix} = \begin{bmatrix} 0 \\ 2.5 \end{bmatrix} + EI \begin{bmatrix} 1/3 & 1/3 \\ 1/3 & 5/3 \end{bmatrix} \begin{bmatrix} \Delta_B \\ \theta_C \end{bmatrix}$$

By solving the above equation, we obtain

$$\therefore \begin{bmatrix} \Delta_B \\ \theta_C \end{bmatrix} = \begin{bmatrix} 3.21/EI \\ -2.14/EI \end{bmatrix}$$

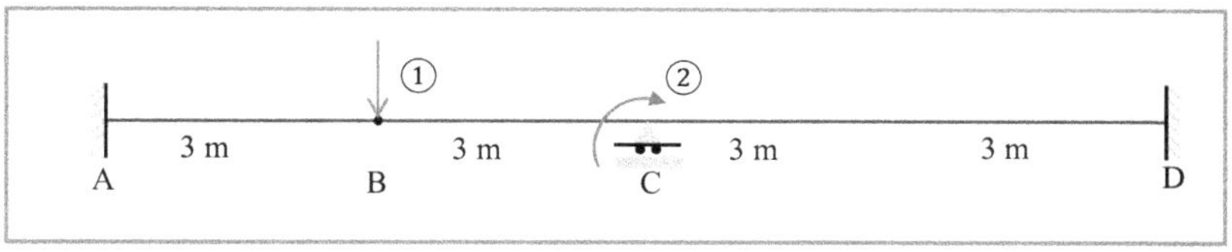

FIGURE 10.25 Co-ordinates assigned to Δ_B *and* θ_C.

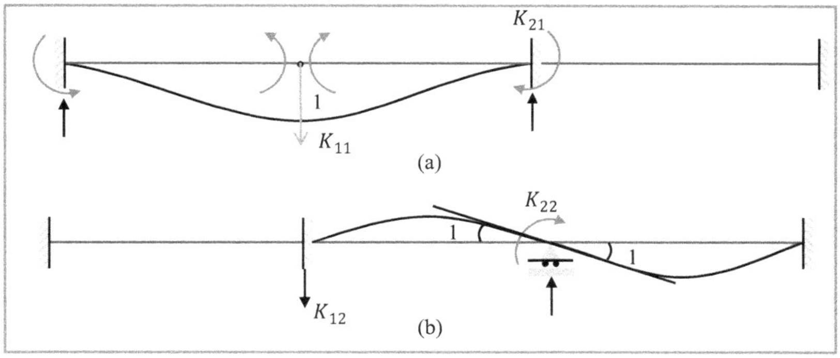

FIGURE 10.26 (a) Unit displacement at ① and (b) unit displacement at co-ordinate 2.

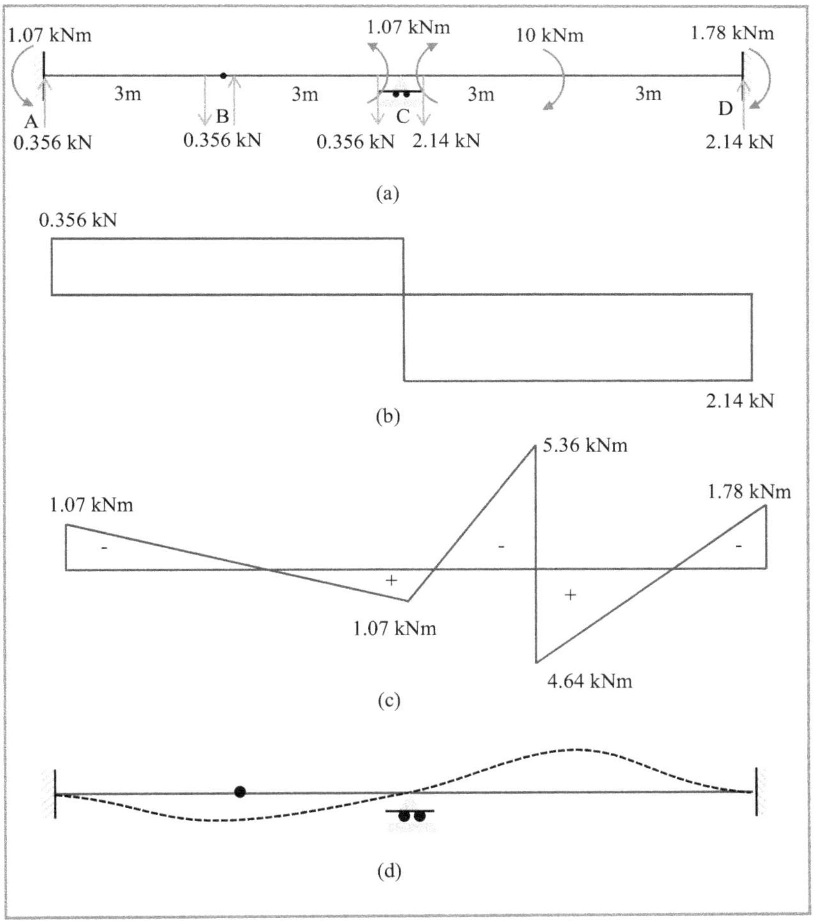

FIGURE 10.27 (a) Free body diagram, (b) shear force diagram, (c) bending moment diagram, and (d) probable deflected shape.

Final end moments: With the known values of Δ_B and θ_C, and the slope–deflection equations, the final member end moments can be calculated, as shown below.

$$M_{AB} = MF_{AB} + \frac{3EI}{L}(\theta_A - \varphi) = 0 + \frac{3EI}{3}\left(0 - \frac{3.21}{3EI}\right) = -1.07 \text{ kNm}$$

$$M_{CB} = MF_{CB} + \frac{3EI}{L}(\theta_C - \varphi) = MF_{CB} + \frac{3EI}{L}\left(\theta_C - \frac{\Delta_B}{L}\right)$$

$$= 0 + \frac{3EI}{3}\left(\frac{-2.14}{EI} + \frac{3.21}{3EI}\right) = -1.07 \text{ kNm}$$

$$M_{CD} = MF_{CD} + \frac{2EI}{L}(2\theta_C + \theta_D - 3\varphi)$$

$$= 2.5 + \frac{2EI}{6}[-2.14 \times 2] = 1.07 \text{ kNm}$$

$$M_{DC} = MF_{DC} + \frac{2EI}{L}(\theta_C + 2\theta_D - 3\varphi)$$

$$= 2.5 + \frac{2EI}{6}[-2.14] = 1.78 \text{ kNm}$$

Final member end shears: With the known values of the member end moments, the member end shears can be calculated using the equilibrium equations and with the help of the free body diagram, is shown in Figure 10.27a. The shear force and bending moment diagrams are shown in Figure 10.27b and c, respectively. The probable elastic curve is shown in Figure 10.27d.

Example 10.7

Analyse the non-prismatic fixed beam *ABC* is shown in Figure 10.28. Draw the shear force and bending moment diagrams. The same has been solved in Example 6.9 using the slope–deflection method.

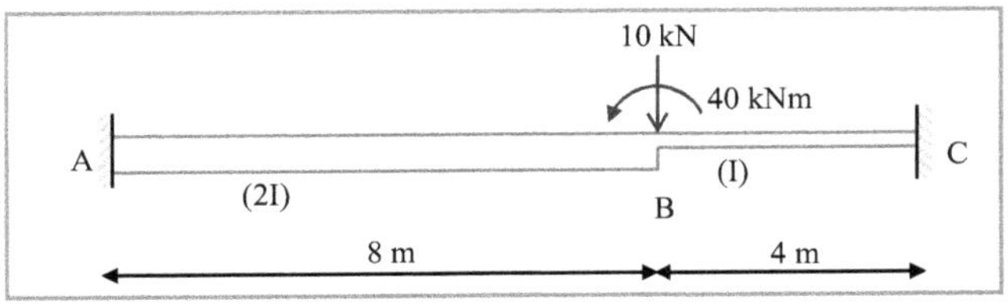

FIGURE 10.28 Non-prismatic fixed beam.

Solution

Since the ends are fixed, the beam is kinematically indeterminate. Therefore, to solve this problem, a node can be considered at *B*, so that the kinematic indeterminacy can be considered as $I_K = 2$ *i.e.* θ_B and Δ_B. The co-ordinates ① and ② can be assigned to θ_B and Δ_B, respectively, as shown in Figure 10.29a. The restrained beam can be obtained by preventing the DoF, i.e., θ_B and Δ_B as shown in Figure 10.29b.

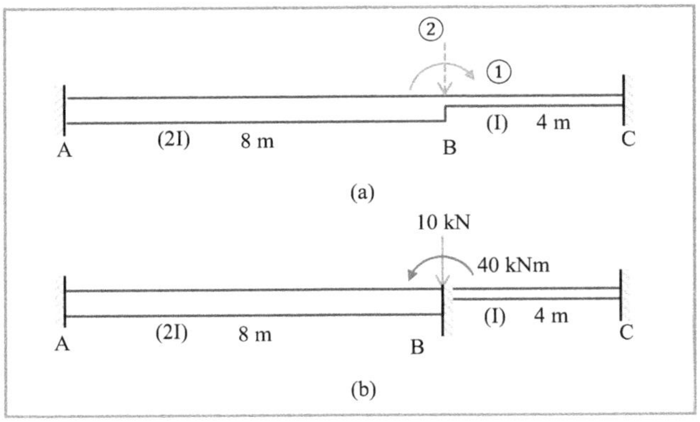

FIGURE 10.29 (a) Co-ordinates assigned to θ_B and Δ_C and (b) restrained beam.

Fixed end moments: Since there are no external loads on *AB* and *BC* in the restrained beam, the fixed end moments are zero.

$$\therefore MF_{AB} = MF_{BA} = MF_{BC} = MF_{CB} = 0$$

Similarly, as no supports are yielding, the additional fixed end moments due to the yielding of supports are also zero.

$$\therefore \begin{bmatrix} P_{1L} + P_{1Y} \\ P_{2L} + P_{2Y} \end{bmatrix} = \begin{bmatrix} 0 \\ 0 \end{bmatrix}$$

$$\begin{bmatrix} P_1 \\ P_2 \end{bmatrix} = \begin{bmatrix} -40 \\ 10 \end{bmatrix} \quad (\because \text{loads are acting at the coordinates})$$

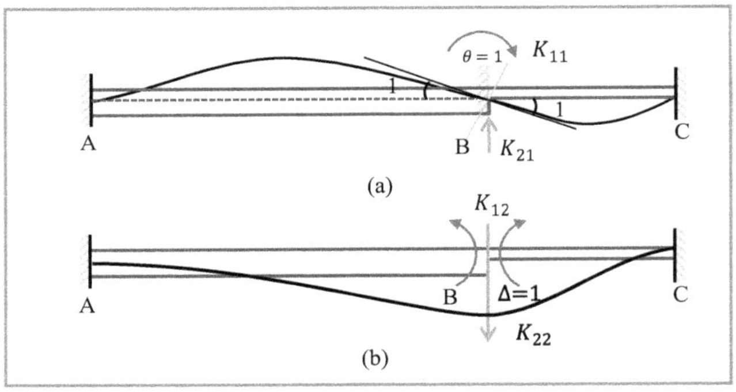

FIGURE 10.30 (a) Unit displacement at ① and (b) unit displacement at 2.

STIFFNESS MATRIX

First column elements: The first column elements of the stiffness matrix can be obtained by applying a unit displacement at co-ordinate ① in its chosen direction and finding the forces developed at all the co-ordinates in the restrained beam, as shown in Figure 10.30a.

$$K_{11} = \left(\frac{4EI}{L}\right)_{AB} + \left(\frac{4EI}{L}\right)_{BC} = \frac{4E(2I)}{8} + \frac{4EI}{4} = 2EI$$

$$K_{21} = -\left(\frac{6EI}{L^2}\right)_{AB} + \left(\frac{6EI}{L^2}\right)_{BC} = \frac{-6E(2I)}{8^2} + \frac{6EI}{4^2} = \frac{3EI}{16}$$

Second column elements of stiffness matrix: To generate the second column elements of the stiffness matrix, apply a unit displacement at ② in its chosen direction and find the forces at all co-ordinates in the restrained beam, as shown in Figure 10.30b.

$$K_{12} = -\left(\frac{6EI}{L^2}\right)_{AB} + \left(\frac{6EI}{L^2}\right)_{BC} = \frac{-12EI}{8^2} + \frac{6EI}{4^2} = \frac{3EI}{16}$$

$$K_{22} = \left(\frac{12EI}{L^3}\right)_{AB} + \left(\frac{12EI}{L^3}\right)_{BC} = \frac{24EI}{8^3} + \frac{12EI}{4^3} = \frac{15EI}{64}$$

Equation 10.19 of the stiffness matrix method is

$$[P] = [[P_L] + [P_Y]] + [K][\Delta]$$

On substituting the above values into equation 10.19,

$$\begin{bmatrix} -40 \\ 10 \end{bmatrix} = \begin{bmatrix} 0 \\ 0 \end{bmatrix} + EI \begin{bmatrix} 2 & 3/16 \\ 3/16 & 15/64 \end{bmatrix} \begin{bmatrix} \theta_B \\ \Delta_B \end{bmatrix}$$

By solving the above equation, we obtain,

$$\therefore \begin{bmatrix} \theta_B \\ \Delta_B \end{bmatrix} = \begin{bmatrix} -25.94/EI \\ 63.42/EI \end{bmatrix}$$

Final End Moments

With the known values of Δ_B and θ_C, and the slope–deflection equations, the final end moments can be calculated, as shown below.

$$M_{AB} = 0 + \frac{2E(2I)}{8}\left[\frac{-25.94}{EI} - \frac{3\times 63.42}{8EI}\right] = -24.86 \text{ kNm}$$

$$M_{BA} = 0 + \frac{2E(2I)}{8}\left[2\times\left(\frac{-25.94}{EI}\right) - \frac{3\times 63.42}{8EI}\right] = -37.84 \text{ kNm}$$

$$M_{BC} = 0 + \frac{2EI}{4}\left[2\times\left(\frac{-25.94}{EI}\right) + \frac{3\times 63.42}{4EI}\right] = -2.16 \text{ kNm}$$

$$M_{CB} = 0 + \frac{2EI}{4}\left[\frac{-25.94}{EI} + \frac{3\times 63.42}{4EI}\right] = 10.81 \text{ kNm}$$

The member end shear forces are shown in Figure 10.31a. The shear force and bending moment diagrams are shown in Figure 10.31b and c, respectively.

10.4 APPLICATIONS TO RIGID-JOINTED PLANE FRAMES

The stiffness of various structural members due to different DoFs is discussed in Section 10.1. In the case of rigid-jointed frames, we generally need to consider both the rotational stiffness and translational stiffness of the rigid joints.

Rotational rigid-jointed stiffness: It depends on the stiffness of the members meeting at the joint. It is defined as the moment required to produce a unit rotation of a joint without any translation. For example, consider a rigid joint O at which four members OP, OQ, OR, and OS meet, as shown in Figure 10.32a.

The ends *P*, *R*, and *S* are fixed, while *Q* is a hinged support. Let us assign co-ordinates 1 and 2 to joints *O* and *Q*, respectively. Now, let us apply a unit rotation at co-ordinate 1 while keeping co-ordinate 2 fixed (Figure 10.32b). In this case, the joint rotational stiffness is the sum of the rotational stiffnesses of the members *OP*, *OQ*, *OR*, and *OS*. That is,

$$k_{11} = \frac{4EI_{OP}}{L_{OP}} + \frac{4EI_{OQ}}{L_{OQ}} + \frac{4EI_{OR}}{L_{OR}} + \frac{4EI_{OS}}{L_{OS}}$$

If modified stiffness is considered for member *OQ* (where rotation at *Q* is neglected as shown in Figure 10.32c), then the rotational stiffness of joint O is

$$k_{11} = \frac{4EI_{OP}}{L_{OP}} + \frac{3EI_{OQ}}{L_{OQ}} + \frac{4EI_{OR}}{L_{OR}} + \frac{4EI_{OS}}{L_{OS}}$$

Translational stiffness: It depends on the extensibility of the members meeting at the joint. It is defined as the force required to produce a unit translation (sway) of a rigid joint in the assumed direction without any rotation of the joint. In general, if the members are assumed to be inextensible, the translational stiffness of the joint is finite. However, if the members meeting at the joint are extensible (i.e., able to undergo extension or contraction), the translational stiffness of the joint becomes infinite. This means that sway is not possible in the assumed direction.

10.4.1 Frames Subjected to No Lateral Sway

Example 10.8

A rigid-jointed plane frame is shown in Figure 10.33. Analyse it by using the stiffness matrix method. Draw the shear force and bending moment diagrams. Also, draw the probable deflected shape. Assume *EI* is constant. The same was solved in Example 6.10 using the slope–deflection method.

Solution

As explained in example 6.10, the kinematic indeterminacy is limited to 2, i.e., θ_B and θ_C. Assign co-ordinates ① and ② to the DoFs θ_B and θ_C, respectively, as shown in Figure 10.34a. The restrained frame can be obtained by locking all the joints, as shown in Figure 10.34b. The end *E* is not restrained because the kinematic indeterminacy is limited to two only (modified fixed end moments, stiffness, and SD equations will be used).

Fixed end moments due to loads: Using the expressions given in Table 6.1,

$$MF_{AB} = -MF_{BA} = \frac{-10\times 4}{8} = -5 \text{ kNm}$$

$$MF_{BC} = -MF_{CB} = \frac{-5}{96}\times 20\times 4^2 = -16.67 \text{ kNm}$$

$$MF_{CE} = \frac{10\times 1}{3^2}(2\times 2-1) - \frac{1}{2}\cdot\frac{10\times 2}{3^2}(2\times 1-2) = 3.33 \text{ kNm}$$

$$MF_{CD} = MF_{DC} = 0$$

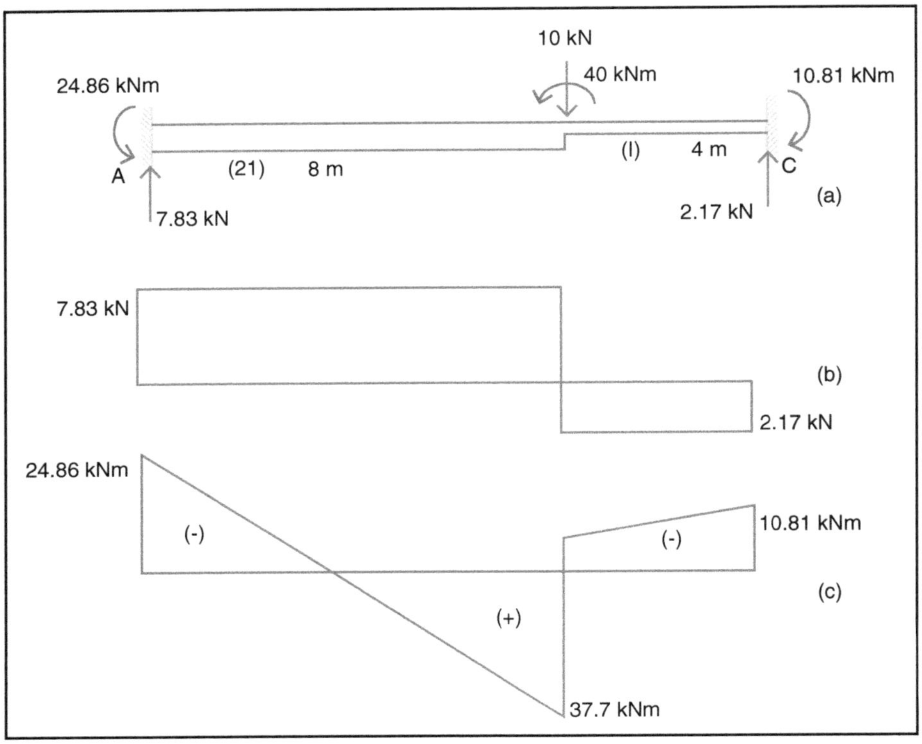

FIGURE 10.31 (a) Free body diagram, (b) shear force diagram, and (c) bending moment diagram.

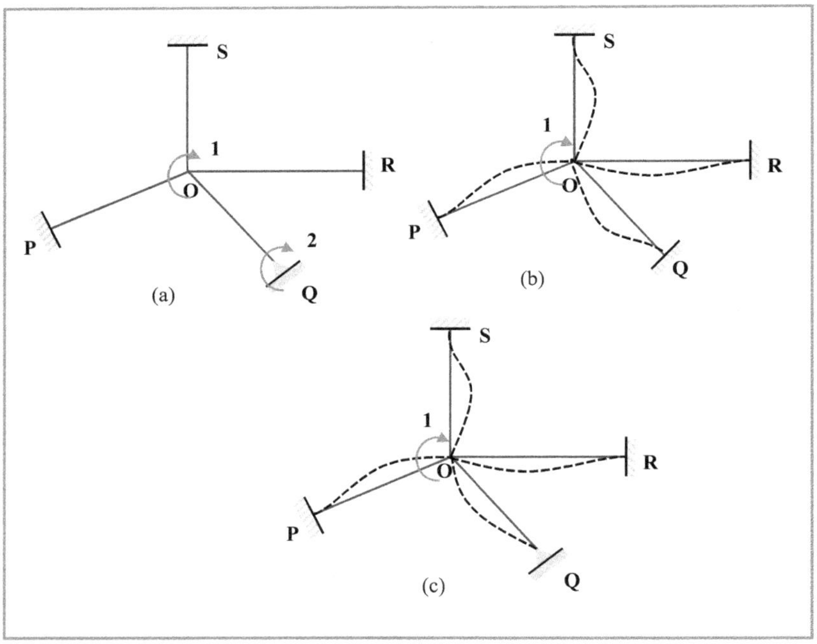

FIGURE 10.32 Rotational stiff of a joint.

Fixed end moments due to the yielding of supports are zero because there is no yielding of supports.

$$\therefore [P_L + P_Y] = \begin{bmatrix} P_{1L} + P_{1Y} \\ P_{2L} + P_{2Y} \end{bmatrix} = \begin{bmatrix} +5 - 16.67 \\ 16.67 + 3.33 \end{bmatrix} = \begin{bmatrix} -11.67 \\ 20 \end{bmatrix}$$

STIFFNESS MATRIX

First column elements: To generate the first column elements, apply a unit displacement (rotational) at co-ordinate

① without any displacement at ② and obtain the forces developed at all co-ordinates, as shown in Figure 10.35a.

$$K_{11} = \left(\frac{4EI}{L}\right)_{AB} + \left(\frac{4EI}{L}\right)_{BC}$$

$$= \frac{4EI}{4} + \frac{4EI}{4} = 2EI$$

$$K_{21} = \frac{2EI}{L} = \frac{2EI}{4} = \frac{EI}{2}$$

Second column elements: The second column elements can be obtained by applying a unit displacement (rotational) at ② without any displacement at ① and finding the forces at all co-ordinates, as shown in Figure 10.35b.

$$K_{12} = \frac{2EI}{L} = \frac{2EI}{4} = \frac{EI}{2}$$

$$K_{22} = \left(\frac{4EI}{L}\right)_{BC} + \left(\frac{3EI}{L}\right)_{CE} + \left(\frac{4EI}{L}\right)_{CD}$$

$$= \frac{4EI}{4} + \frac{3EI}{3} + \frac{4EI}{4} = 3EI$$

$$\therefore \begin{bmatrix} P_1 \\ P_2 \end{bmatrix} = \begin{bmatrix} 0 \\ 0 \end{bmatrix} \quad (\because \text{No external loads at coordinates})$$

$$[\Delta] = \begin{bmatrix} \Delta_1 \\ \Delta_2 \end{bmatrix} = \begin{bmatrix} \theta_B \\ \theta_C \end{bmatrix}$$

On substituting the above values into equation 10.15,

$$\therefore \begin{bmatrix} 0 \\ 0 \end{bmatrix} = \begin{bmatrix} -11.67 \\ 20 \end{bmatrix} + EI \begin{bmatrix} 2 & \frac{1}{2} \\ \frac{1}{2} & 3 \end{bmatrix} \begin{bmatrix} \theta_B \\ \theta_C \end{bmatrix}$$

By solving the above equation, we obtain

$$\therefore \begin{bmatrix} \theta_B \\ \theta_C \end{bmatrix} = \begin{bmatrix} 7.827/EI \\ -7.97/EI \end{bmatrix}$$

Final member end moments: With the known values of θ_B and θ_C, and using the slope–deflection equations, the final member end moments can be calculated, as shown below.

$$M_{AB} = -5 + \frac{2EI}{4}\left[2(0) + \frac{7.827}{EI} - 3(0)\right] = -1.086 \text{ kNm}$$

$$M_{BA} = 5 + \frac{2EI}{4}\left[2 \times \frac{7.827}{EI}\right] = 12.82 \text{ kNm}$$

$$M_{BC} = -16.67 + \frac{2EI}{4}\left[2 \times \left(\frac{7.827}{EI}\right) - \frac{7.97}{EI}\right] = -12.82 \textit{ kNm}$$

$$M_{CB} = 16.67 + \frac{2EI}{4}\left[\frac{7.827}{EI} - 2 \times \frac{7.97}{EI}\right] = 12.61 \textit{ kNm}$$

$$M_{CD} = 0 + \frac{2EI}{4}\left[-2 \times \frac{7.97}{EI}\right] = -7.97 \text{ kNm}$$

$$M_{CE} = 3.33 + \frac{3EI}{3}\left[-\frac{7.97}{EI}\right] = -4.64 \text{ kNm}$$

$$M_{DC} = 0 + \frac{2EI}{4}\left[-\frac{7.97}{EI}\right] = -3.98 \text{ kNm}$$

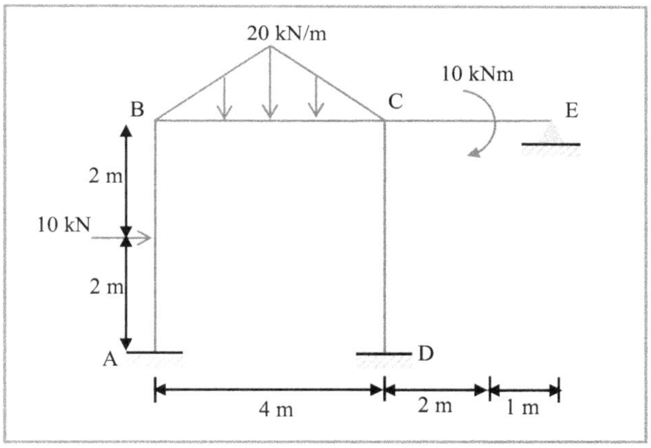

FIGURE 10.33 Rigid-jointed plane frame.

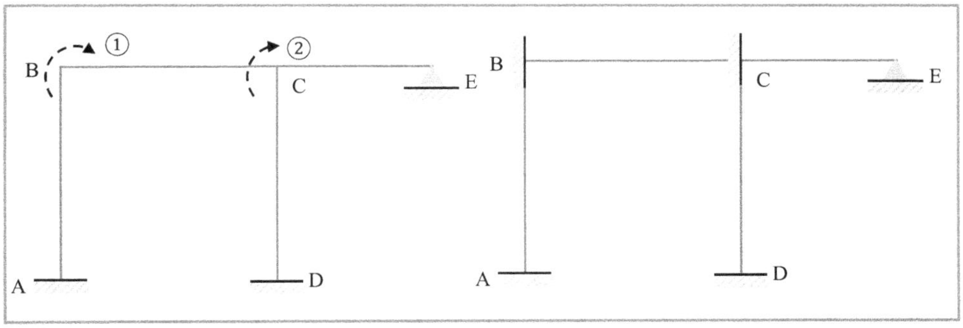

FIGURE 10.34 (a) Co-ordinates assigned to θ_B and θ_C and (b) restrained frame.

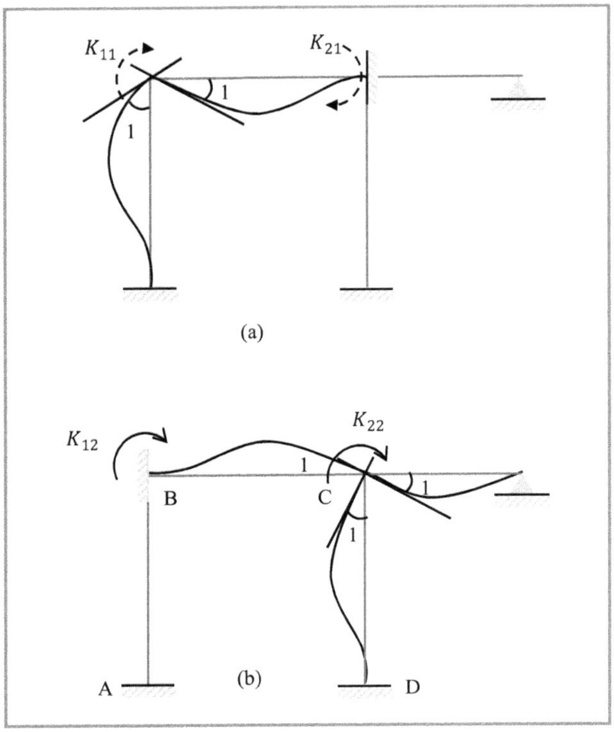

FIGURE 10.35 (a) development of 1st column elements and (b) development of 2nd column elements of stiffness matrix.

Shear force and bending moment diagram: After determining the member end moments, the other support reactions and member end shears can be calculated using the static equilibrium equations, as shown in Figure 10.36a. The shear force and bending moment diagrams are presented in Figure 10.36b and c. The probable elastic curve is shown in Figure 10.36d.

Example 10.9

Analyse the symmetric frame shown in Figure 10.37. Take advantage of symmetry.

Solution

The kinematic indeterminacy of the frame is 4. However, as explained in Example 6.11, the frame is symmetric with respect to the centre of *BC*, so only half of the frame needs to be considered for the analysis, as shown in Figure 10.38a. As a result, the kinematic indeterminacy is reduced to 3, i.e., θ_A, θ_B, and Δ_E. Furthermore, the moment at *A* is zero, so θ_A may be neglected by applying modifications to the stiffness, fixed end moments, and the slope–deflection equation for span *AB*. Similarly, Δ_E can be neglected by applying appropriate modifications to the stiffness and the slope–deflection equation for span *BE*. Therefore, the kinematic indeterminacy is reduced to one, i.e., θ_B. A co-ordinate can be assigned to θ_B as ①, as shown in Figure 10.38b.

Fixed end moments due to loads: Using Table 6.1, the fixed end moments can be calculated as

$$\text{MF}_{BA} = 0$$

$$\text{MF}_{BE} = \frac{-WL}{8} = \frac{-100\times 4}{8} = -50 \text{ kNm}$$

$$\text{MF}_{EB} = -\left(\frac{WL}{4} - \frac{WL}{8}\right) = -\frac{WL}{8} = -50 \text{ kNm}$$

Additional fixed end moments due to the yielding of supports are zero.

$$\therefore P_{1L} + P_{1Y} = -50 + 0 = -50 \text{ kNm}$$

Stiffness Matrix: *To get the stiffness,* give a unit displacement (rotational displacement) at co-ordinate ①, as shown in Figure 10.39, and find the force developed at co-ordinate 1.

$$K_{11} = \left(\frac{3EI}{L}\right)_{AB} + \left(\frac{EI}{L}\right)_{BE}$$

$$= \frac{3EI}{3} + \frac{E(2I)}{2} = 2EI$$

Substituting the above into equation 10.15

$$0 = -50 + 2EI \times \theta_B$$

$$\Rightarrow \theta_B = \frac{25}{EI}$$

Final member end moments: After *knowing* the value of θ_B, using the slope–deflection equations the member end moments can be calculated, as presented below.

$$M_{BA} = \text{MF}_{BA} + \frac{3EI}{L}(\theta_B - \varphi) = 0 + \frac{3EI}{3}\left(\frac{25}{EI} - 0\right) = 25 \text{ kNm}$$

$$M_{BE} = \text{MF}_{BE} + \frac{EI}{L}(\theta_B) = -50 + \frac{2EI}{2}\left(\frac{25}{EI}\right) = -25 \text{ kNm}$$

$$M_{EB} = \text{MF}_{EB} - \frac{EI}{L}(\theta_B) = -50 - \frac{2EI}{2}\left(\frac{25}{EI}\right) = -75 \text{ kNm}$$

Shear force and bending moment diagrams: Using the above results and the free body diagram shown in Figure 10.40a, the other reactions are calculated using the static equilibrium equations, as shown in Figure 10.40a. The shear force and bending moment diagrams are presented in Figure 10.40b and c, respectively. The probable deflected shape is shown in Figure 10.40d.

Example 10.10

Analyse the two-storey symmetrical frame shown in Figure 10.41. Draw the shear force and bending moment diagrams.

Solution

The kinematic indeterminacy of the frame is 4, i.e., $\theta_B, \theta_E, \theta_D$ and θ_C. However, due to symmetry with

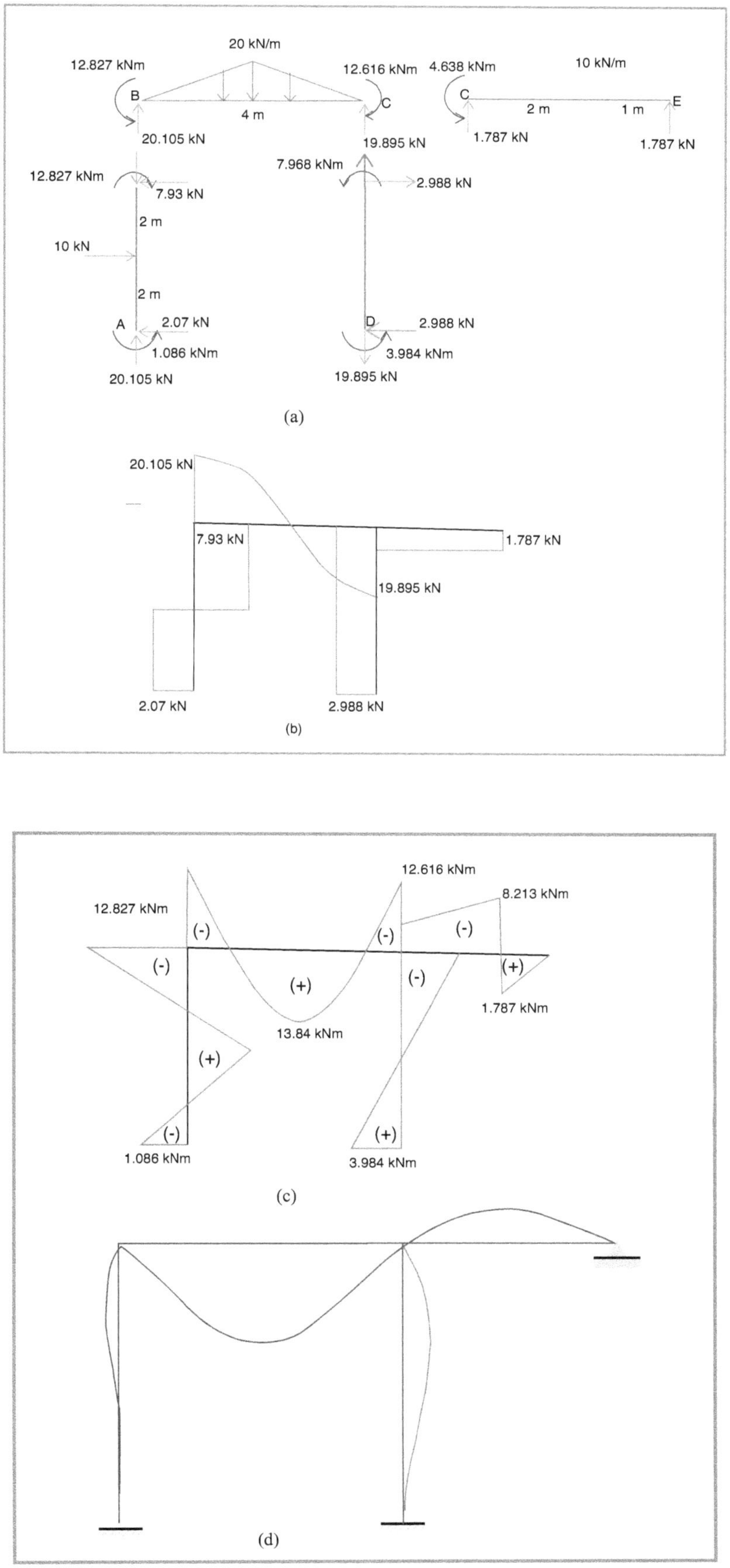

FIGURE 10.36 (a) Free body diagram, (b) shear force diagram, (c) bending moment diagram, and (d) probable elastic curve.

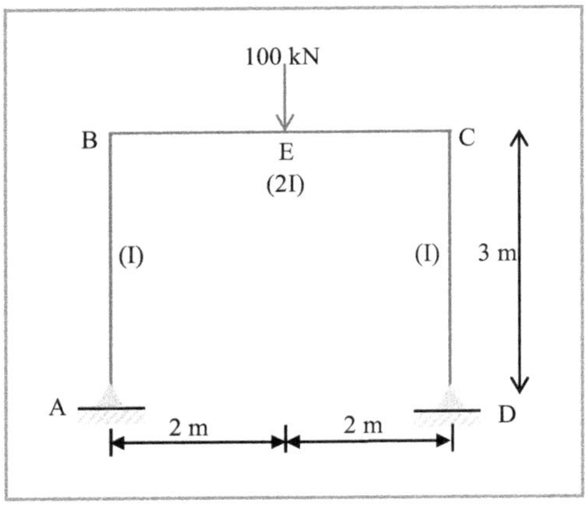

FIGURE 10.37 Symmetrical rectangular frame.

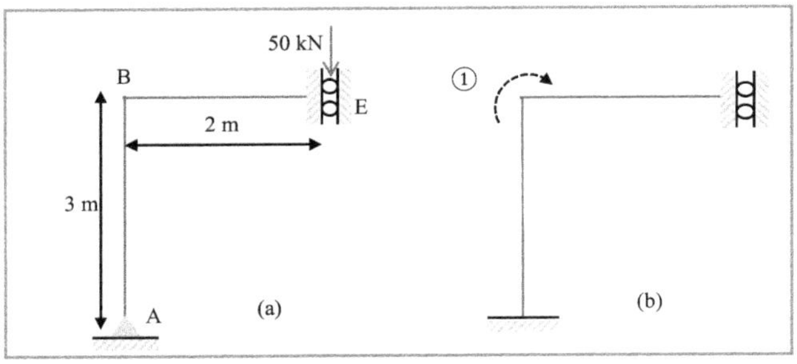

FIGURE 10.38 (a) Half portion of the frame due to symmetry and (b) co-ordinates assigned to θ_B.

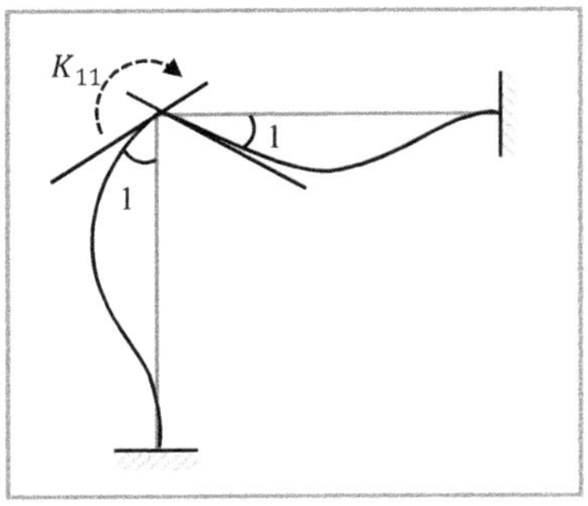

FIGURE 10.39 Elements of stiffness.

respect to *BE* and *CD*, $\theta_C = \theta_D$ and $\theta_B = \theta_E$, only half of the frame needs to be considered for the analysis, as shown in Figure 10.42a. Also, $\Delta_{B^1} = \Delta_{C^1}$ can be neglected by using appropriate modifications to the stiffness, fixed end moments, and the slope–deflection equations for spans BB^1 and CC^1. As a result, the kinematic indeterminacy can be limited to 2, (θ_B and θ_C). Co-ordinates are assigned to the θ_B and θ_C, as shown in Figure 10.42b.

Fixed end moments due to the loads: Using the expressions provided in Table 6.1, the fixed end moments can be calculated as

$$MF_{AB} = MF_{BA} = MF_{BC} = MF_{CB} = 0$$

$$MF_{BC^1} = \left(\frac{50 \times 6^2}{12} + \frac{50 \times 6}{8}\right) = -187.5\ \text{kNm}$$

$$MF_{B'B} = -\left(\frac{wL^2}{24} + \frac{WL}{8}\right) = -\left(\frac{50 \times 6^2}{24} + \frac{50 \times 6}{8}\right) = -112.5\ \text{kNm}$$

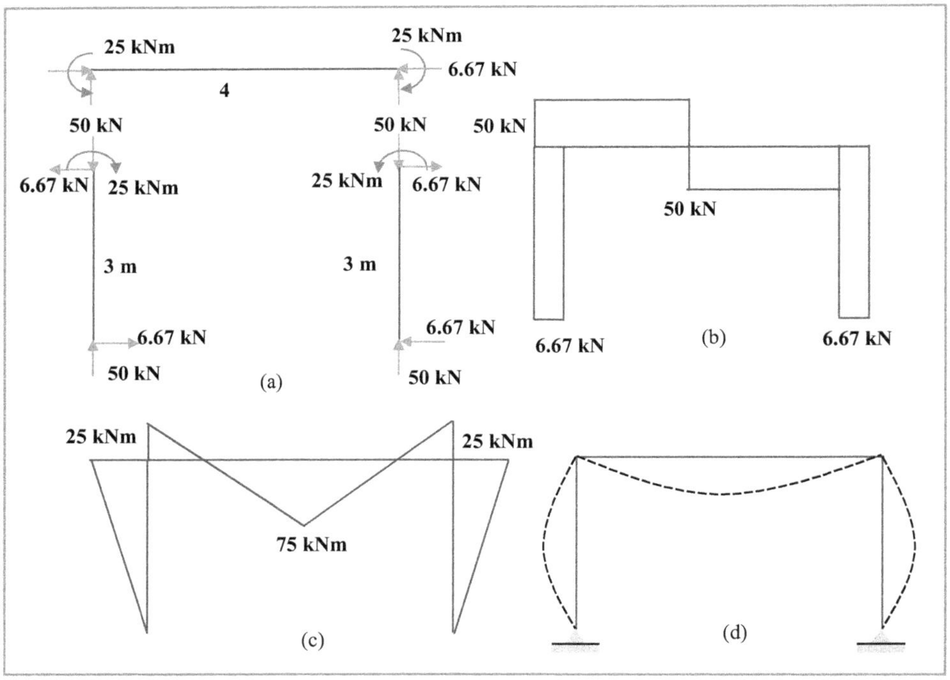

FIGURE 10.40 (a) Free body diagram, (b) shear force diagram, (c) bending moment diagram, and (d) probable deflected shape.

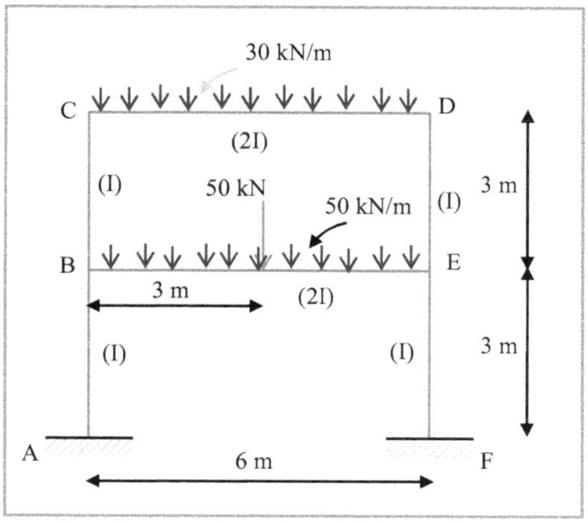

FIGURE 10.41 Two-storey frame.

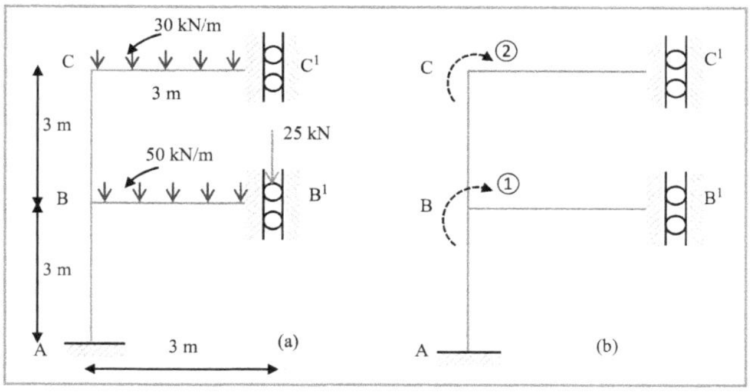

FIGURE 10.42 Half portion of frame due to symmetry and (b) co-ordinates assigned to θ_B and θ_C.

$$\mathrm{MF}_{CD} = \frac{-30 \times 6^2}{12} = -90 \text{ kNm}$$

$$\mathrm{MF}_{C'C} = \frac{-wL^2}{24} = -45 \text{ kNm}$$

As there are no yielding of supports, $P_{1Y} = P_{2Y} = 0$

∴ *The forces corresponding to the co-ordinates* 1 *and* 2 *are*

$$P_{1L} + P_{1Y} = -187.5 + 0 = -187.5 \text{ kNm}$$

$$P_{2L} + P_{2Y} = -90 + 0 = -90 \text{ kNm}$$

Since there are no external forces acting at the co-ordinates,

$$P_1 = P_2 = 0$$

10.4.1.3 Stiffness Matrix

First column elements: To generate the first column elements of the stiffness matrix, apply a unit displacement (rotational) at ① without any displacement at ② in the restrained frame and find the forces developed at all co-ordinates, as shown in Figure 10.43a.

$$K_{11} = \frac{4EI}{3} + \frac{E(2I)}{3} + \frac{4EI}{3} = 3.33EI$$

$$K_{21} = \frac{2EI}{3} = 0.667EI$$

Second column elements: The second column elements of the stiffness matrix can be obtained by applying a unit displacement (rotational) at co-ordinate ② without displacement at co-ordinate ① in the restrained frame and determining the forces developed at all co-ordinates, as shown in Figure 10.43b.

$$K_{12} = \frac{2EI}{3} = 0.667EI$$

$$K_{22} = \frac{4EI}{3} + \frac{E(2I)}{3} = 2EI$$

Substituting the above values into stiffness equation 10.15, we have

$$\therefore \begin{bmatrix} 0 \\ 0 \end{bmatrix} = \begin{bmatrix} -187.5 \\ -90 \end{bmatrix} + EI \begin{bmatrix} 3.33 & 0.667 \\ 0.667 & 2 \end{bmatrix} \begin{bmatrix} \theta_B \\ \theta_C \end{bmatrix}$$

$$\Rightarrow \begin{bmatrix} \theta_B \\ \theta_C \end{bmatrix} = \begin{bmatrix} 50.67/EI \\ 28.09/EI \end{bmatrix}$$

Final member end moments: With the known values of θ_B and θ_B using the slope–deflection equations, the member end moments can be calculated, as shown below.

$$M_{AB} = 0 + \frac{2EI}{3}\left(\frac{50.67}{EI}\right) = 33.78 \text{ kNm}$$

$$M_{BA} = 0 + \frac{2EI}{3}\left(2 \times \frac{50.67}{EI}\right) = 67.56 \text{ kNm}$$

$$M_{BE} = -187.5 + \frac{2EI}{3}\left(\frac{50.67}{EI}\right) = -153.82 \text{ kNm}$$

$$M_{B^1B} = -112.5 - \frac{2EI}{3}\left(\frac{50.67}{EI}\right) = -146.28 \text{ kNm}$$

$$M_{BC} = 0 + \frac{2EI}{3}\left(2 \times \frac{50.67}{EI} + \frac{28.09}{EI}\right) = 86.26 \text{ kNm}$$

$$M_{CB} = 0 + \frac{2EI}{3}\left(\frac{50.67}{EI} + 2 \times \frac{28.09}{EI}\right) = 71.23 \text{ kNm}$$

$$M_{CD} = -90 + \frac{2EI}{3}\left(\frac{28.09}{EI}\right) = -71.23 \text{ kNm}$$

$$M_{C^1C} = -45 - \frac{2EI}{3}\left(\frac{28.09}{EI}\right) = -63.72 \text{ kNm}$$

Shear force and bending moment diagram: The shear force and bending moment diagrams are shown in Figure 10.44a and b, respectively. The probable deflected shape is shown in Figure 10.44c.

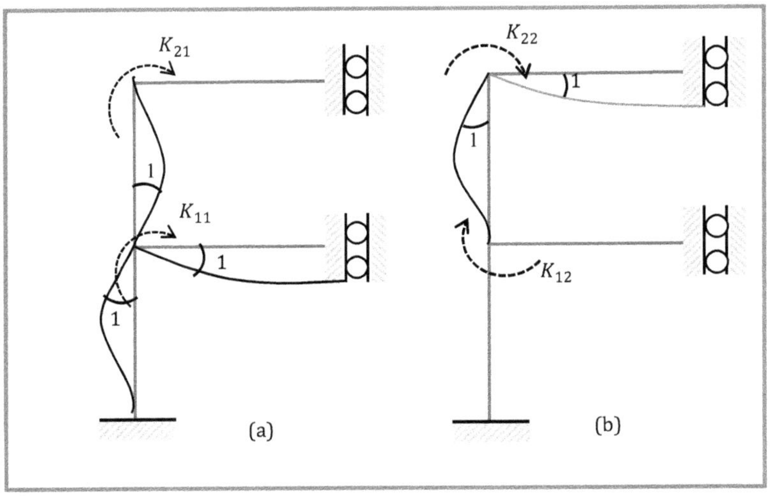

FIGURE 10.43 (a) Stiffness matrix elements of first column and (b) second column.

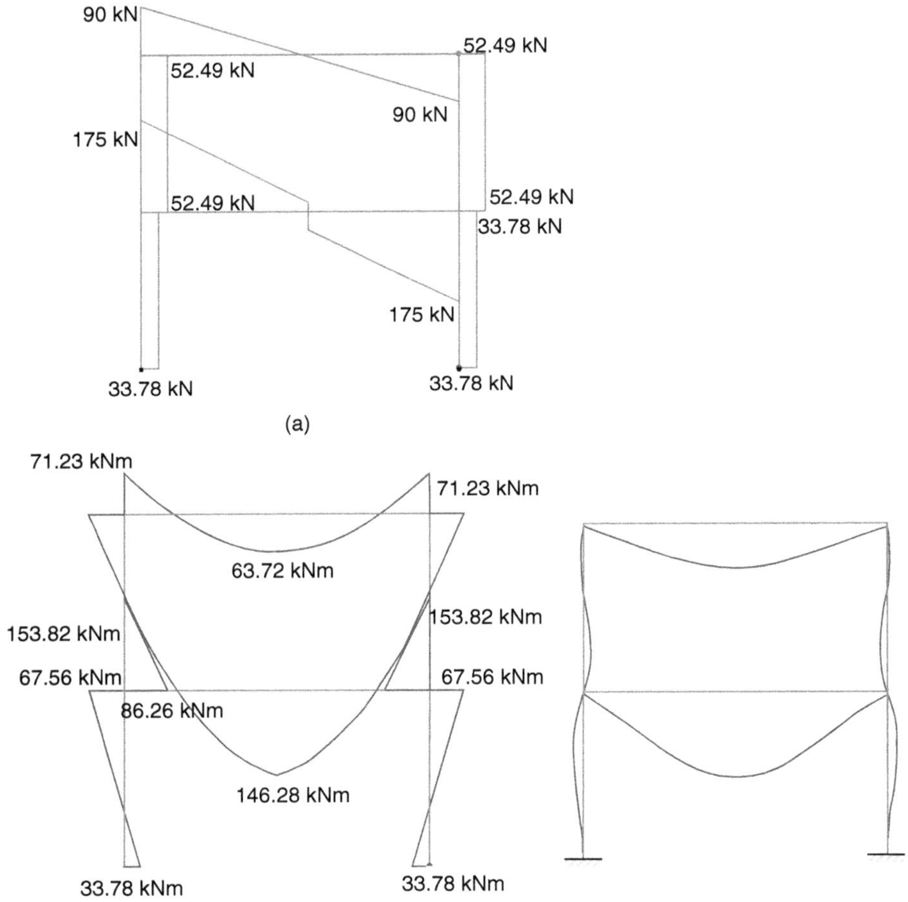

FIGURE 10.44 (a) Shear force diagram, (b) bending moment diagram, and (c) probable deflected shape.

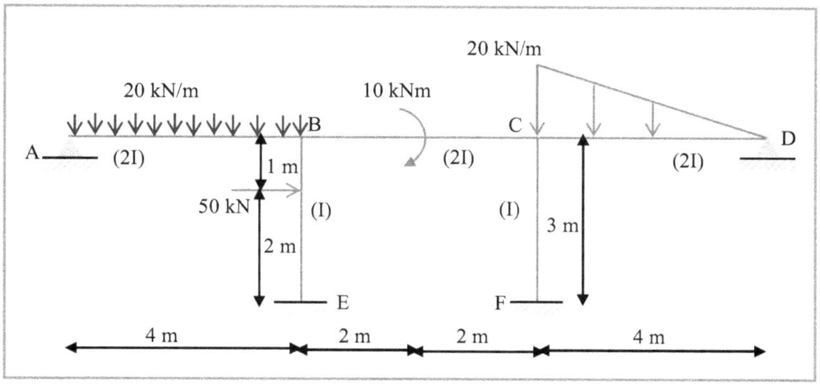

FIGURE 10.45 Rigid-jointed frame.

Example 10.11

Analyse the frame shown in Figure 10.45 using the stiffness matrix method. Draw the shear force and bending moment diagrams. This was solved in Example 6.12 using the SD method.

Solution

In the frame shown in Figure 10.45, *A* and *D* are hinged supports; therefore, the moments at *A* and *D* are zero. We can use the modified stiffness, fixed end moments, and the slope–deflection equations for the spans *AB* and *CD*, so that the rotations at *A* and *D* can be ignored. As a result, the kinematic indeterminacy is limited to 2, i.e., θ_B and θ_C.

Fixed end moments:

$$MF_{BA} = \frac{wL^2}{12} + \frac{wL^2}{24} = \frac{wL^2}{8} = \frac{20\times 4^2}{8} = 40 \text{ kNm}$$

$$MF_{BC} = MF_{CB} = \frac{M}{4} = \frac{10}{4} = 2.5 \text{ kNm}$$

$$MF_{CD} = -\frac{wL^2}{20} - \frac{1}{2}\frac{wL^2}{30} = -\frac{wL^2}{15} = -\frac{20\times 4^2}{15} = -21.33 \text{ kNm}$$

$$MF_{EB} = -\frac{wab^2}{L^2} = -\frac{50\times 2\times 1^2}{3^2} = -11.11 \text{ kNm}$$

$$MF_{BE} = \frac{wba^2}{L^2} = -\frac{50\times 1\times 2^2}{3^2} = 22.22 \text{ kNm}$$

$$MF_{FC} = MF_{CF} = 0$$

As there is no yielding of supports, the additional fixed end moments are zero.

$$\therefore P_{1L} = MF_{BA} + MF_{BE} + MF_{BC} = 40 + 22.22 + 2.5 = 64.72 \text{ kNm}$$

$$P_{2L} = MF_{CB} - MF_{CD} + MF_{CF} = 2.5 - 21.33 = -18.83 \text{ kNm}$$

$$P_{1Y} = 0$$
$$P_{2Y} = 0$$

$$\therefore \begin{bmatrix} P_1 \\ P_2 \end{bmatrix} = \begin{bmatrix} 0 \\ 0 \end{bmatrix} \quad (\because \text{No external loads at coordinates})$$

$$[\Delta] = \begin{bmatrix} \Delta_1 \\ \Delta_2 \end{bmatrix} = \begin{bmatrix} \theta_B \\ \theta_C \end{bmatrix} \quad (\because \text{Unknown DOF})$$

STIFFNESS MATRIX

First column elements: The first column elements of the stiffness matrix can be obtained by applying a unit rotation at co-ordinate 1 in its chosen direction on the restrained structure and finding the forces developed at all co-ordinates, as shown in Figure 10.46a.

Similarly, to get the second column elements of the stiffness matrix, apply a unit rotation at co-ordinate 2, without any displacement at co-ordinate 1, and find the forces developed at all co-ordinates, as shown in Figure 10.46b.

$$K_{11} = \left(\frac{3EI}{L}\right)_{AB} + \left(\frac{4EI}{L}\right)_{BC} + \left(\frac{4EI}{L}\right)_{BE}$$

$$= \frac{6EI}{4} + \frac{8EI}{4} + \frac{4EI}{3} = 4.83\ EI$$

$$K_{21} = K_{12} = \left(\frac{2EI}{L}\right)_{BC} = \frac{2E(2I)}{4} = EI$$

$$K_{22} = \left(\frac{4EI}{L}\right)_{BC} + \left(\frac{3EI}{L}\right)_{CD} + \left(\frac{4EI}{L}\right)_{CF}$$

$$= \frac{8EI}{4} + \frac{6EI}{4} + \frac{4EI}{3} = 4.83\ EI$$

Using the stiffness equation (equation 10.15), we can write

$$\therefore \begin{bmatrix} 0 \\ 0 \end{bmatrix} = \begin{bmatrix} 64.72 \\ -18.83 \end{bmatrix} + EI \begin{bmatrix} 4.83 & 1 \\ 1 & 4.83 \end{bmatrix} \begin{bmatrix} \theta_B \\ \theta_C \end{bmatrix}$$

By solving the above equation, we obtain

$$\begin{bmatrix} \theta_B \\ \theta_C \end{bmatrix} = \begin{bmatrix} -14.83/EI \\ 6.96/EI \end{bmatrix}$$

Final end moments: After knowing the DoF, i.e., θ_B and θ_D, using the slope–deflection equations, the member end moments can be calculated, as shown below.

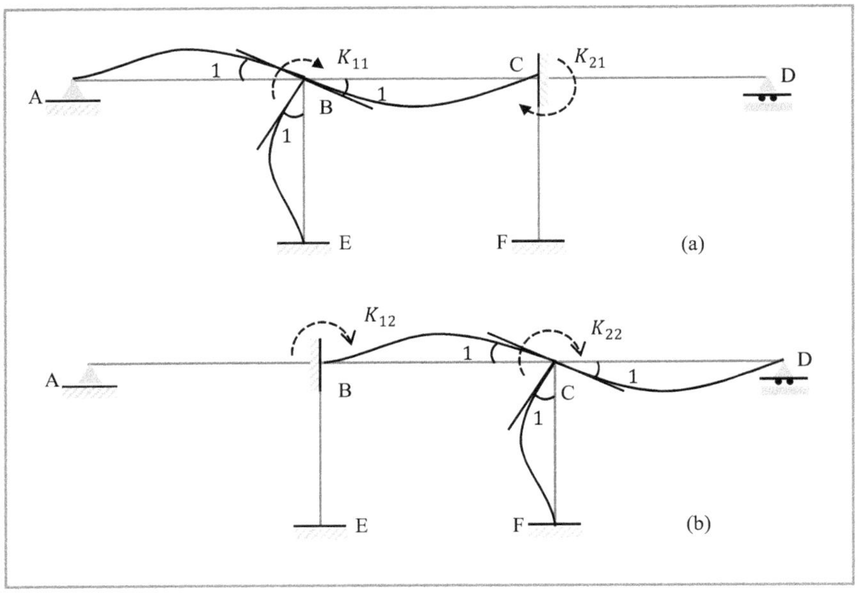

FIGURE 10.46 Stiffness matrix of (a) first column elements and (b) second column elements.

$$M_{BA} = MF_{BA} + \frac{3EI}{L}(\theta_B - \varphi_{AB}) = 40 + \frac{6EI}{4}\left(\frac{-14.83}{EI}\right) = 17.75 \text{ kNm}$$

$$M_{BC} = MF_{BC} + \frac{2EI}{L}(2\theta_B + \theta_C - 3\varphi_{BC})$$

$$= 2.5 + \frac{4EI}{4}\left[2\left(\frac{-14.83}{EI}\right) + \frac{6.96}{EI}\right] = -20.2 \text{ kNm}$$

$$M_{BE} = MF_{BE} + \frac{2EI}{L}(2\theta_B + \theta_E - 3\varphi_{BE})$$

$$= 22.22 + \frac{2EI}{3}\left[2\left(\frac{-14.83}{EI}\right)\right] = 2.45 \text{ kNm}$$

$$M_{CB} = 2.5 + \frac{4EI}{4}\left[2\left(\frac{6.96}{EI}\right) - \frac{14.83}{EI}\right] = 1.6 \text{ kNm}$$

$$M_{CD} = -21.33 + \frac{3E(2I)}{4}\left[\left(\frac{6.96}{EI}\right)\right] = -10.88 \text{ kNm}$$

$$M_{CF} = 0 + \frac{2EI}{3}\left[2\times\left(\frac{6.96}{EI}\right)\right] = 9.28 \text{ kNm}$$

$$M_{FC} = 0 + \frac{2EI}{3}\left[\left(\frac{6.96}{EI}\right)\right] = 4.64 \text{ kNm}$$

$$M_{EB} = -11.11 + \frac{2EI}{3}\left(\frac{-14.83}{EI}\right) = -21 \text{ kNm}$$

Shear force and bending moment diagram: The free body diagram shows the member end forces in Figure 10.47a. The shear force and bending moment diagrams are shown in Figure 10.47b and c, respectively. The probable deflected shape is shown in Figure 10.47d.

10.4.2 Frame Subjected to Lateral Sway

Example 10.12

Analyse the simple frame with unsymmetrical loading as shown in Figure 10.48. Draw the shear force and bending moment diagrams.

Solution

Even though the frame is symmetrical, due to the unsymmetrical loading, the frame is subjected to lateral sway at *BC*. Therefore, the kinematic indeterminacy is three, i.e., the rotations at *B* and *C*, and the lateral sway at *BC*. Let these DoF (θ_B, θ_C and Δ) as unknowns. Let us assign co-ordinates 1, 2, and 3 to these unknown DoFs, respectively, as shown in Figure 10.49a. The restrained structure is obtained by locking all the joints is shown in Figure 10.49b.

Fixed end moments and reactions due to the applied loads are calculated from the free body diagrams, as shown in Figure 10.49c, using the expressions provided in Table 6.1.

$$MF_{BC} = \frac{-wL^2}{20} = \frac{-20\times 6^2}{20} = -36 \text{ kNm}$$

$$MF_{CB} = \frac{WL^2}{30} = \frac{20\times 6^2}{30} = 24 \text{ kNm}$$

$$MF_{AB} = MF_{BA} = MF_{CD} = MF_{DC} = 0$$

As there is no yielding of supports, moments due to these are zero.

$$\therefore [P_L + P_Y] = \begin{bmatrix} P_{1L} + P_{1Y} \\ P_{2L} + P_{2Y} \\ P_{3L} + P_{3Y} \end{bmatrix} = \begin{bmatrix} -36 \\ 24 \\ 0 \end{bmatrix}$$

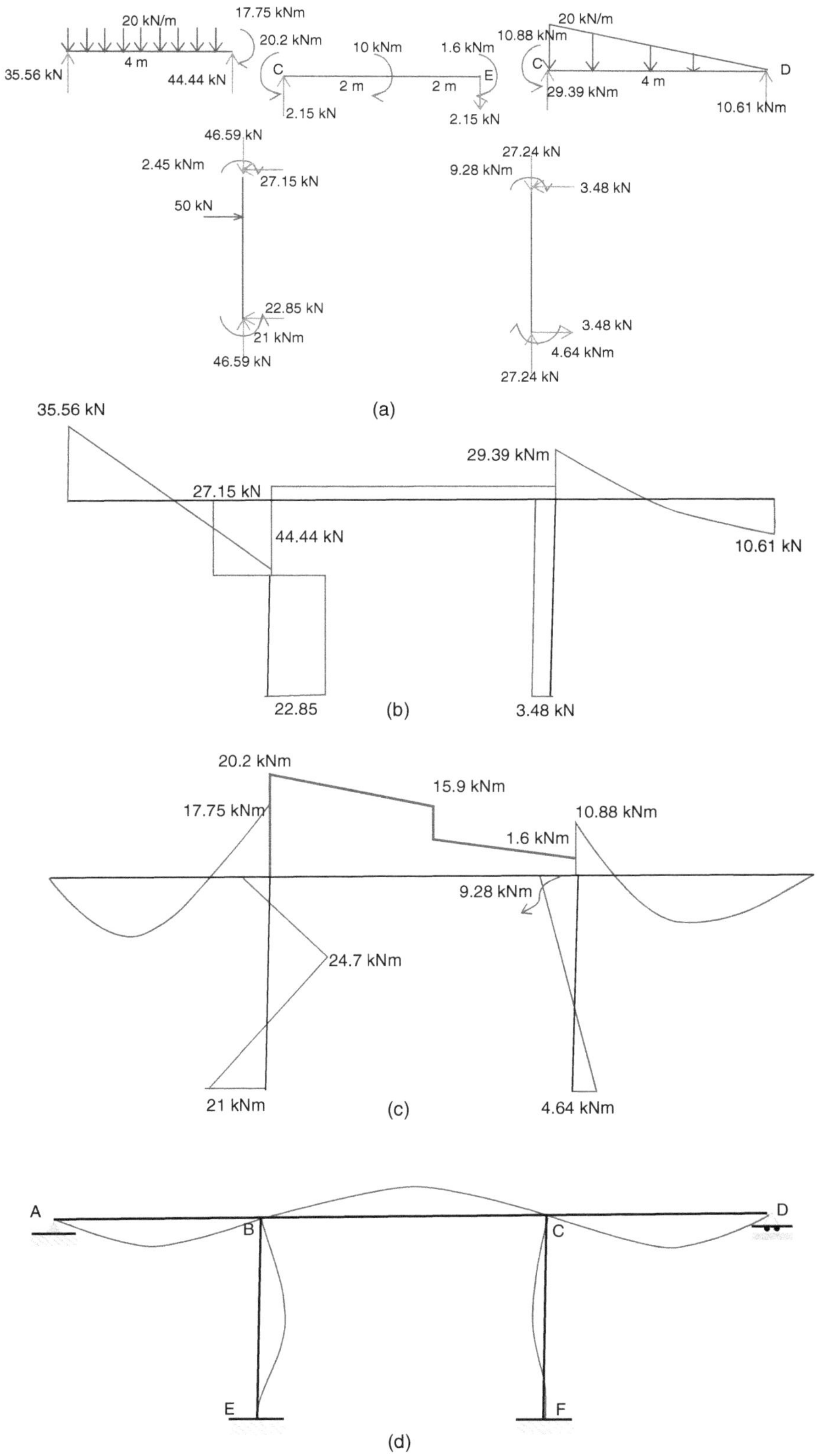

FIGURE 10.47 (a) Free body diagram, (b) shear force diagram, (c) bending moment diagram, and (d) probable elastic curve.

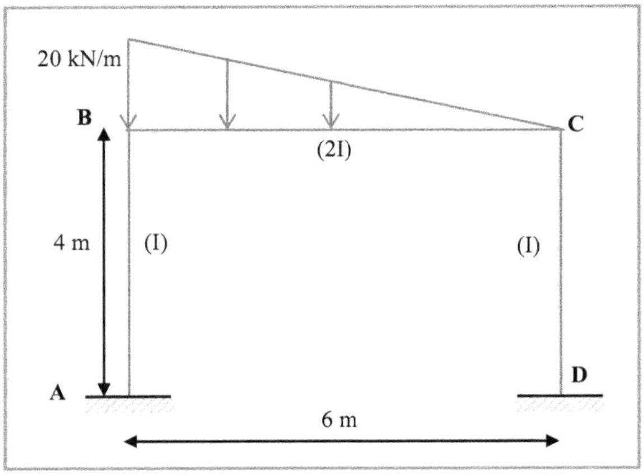

FIGURE 10.48 Rigid-jointed frame subjected to unsymmetrical load.

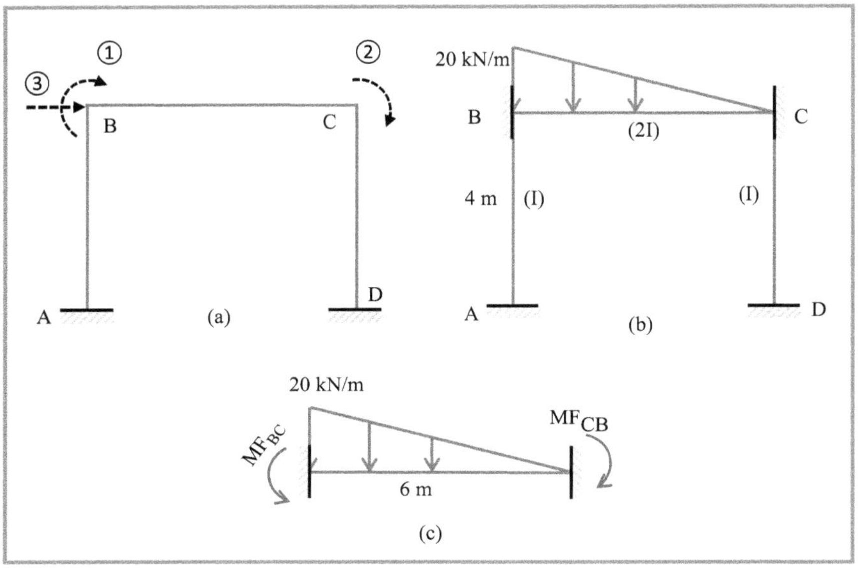

FIGURE 10.49 (a) Restrained frame, (b) co-ordinates assigned to θ_B, θ_C and Δ, and (c) fixed end moments.

Similarly, since there are no external forces acting at the co-ordinates,

$$P_1 = P_2 = P_3 = 0$$

$$[\Delta] = \begin{bmatrix} \Delta_1 \\ \Delta_2 \\ \Delta_3 \end{bmatrix} = \begin{bmatrix} \theta_B \\ \theta_C \\ \Delta \end{bmatrix}$$

STIFFNESS MATRIX

First column elements: The first column elements of the stiffness matrix can be obtained by applying a unit displacement (rotational) at co-ordinate ①, without any displacement at the other co-ordinates on the restrained frame and obtaining the forces at all co-ordinates, as shown in Figure 10.50a. Similarly, the second and third column elements can be obtained by applying a unit displacement at co-ordinates ② (rotational) and ③ (translational) successively and obtaining the forces developed at all co-ordinates in each case separately, as shown in Figure 10.50b and c, respectively.

$$K_{11} = \left(\frac{4EI}{L}\right)_{AB} + \left(\frac{4EI}{L}\right)_{BC} = \frac{4EI}{4} + \frac{8EI}{6} = 2.333\ EI$$

$$K_{21} = K_{12} = \left(\frac{2EI}{L}\right)_{BC} = \frac{4EI}{6} = \frac{2EI}{3}$$

$$K_{13} = K_{31} = -\left(\frac{6EI}{L^2}\right)_{AB} = \frac{-6EI}{4^2} = \frac{-6EI}{16}$$

$$K_{23} = K_{32} = -\left(\frac{6EI}{L^2}\right)_{CD} = \frac{-6EI}{4^2} = \frac{-6EI}{16}$$

$$K_{22} = \left(\frac{4EI}{L}\right)_{BC} + \left(\frac{4EI}{L}\right)_{CD} = \frac{8EI}{6} + \frac{4EI}{4} = 2.333\ EI$$

$$K_{33} = \left(\frac{12EI}{L^3}\right)_{AB} + \left(\frac{12EI}{L^3}\right)_{CD} = \frac{12EI}{4^3} \times 2 = \frac{24EI}{64}$$

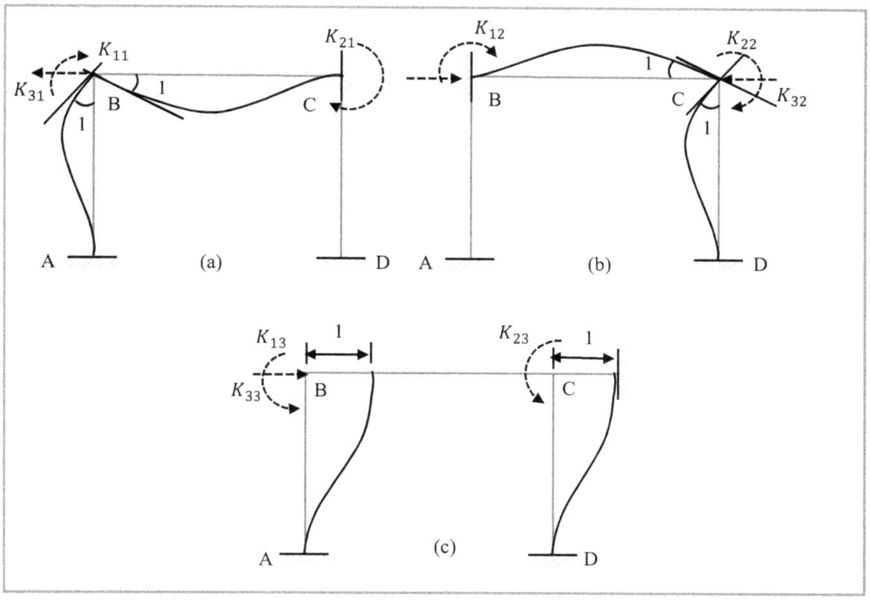

FIGURE 10.50 Elements of stiffness matrix.

Substituting all the values into the stiffness matrix equation 10.15,

$$\begin{bmatrix} 0 \\ 0 \\ 0 \end{bmatrix} = \begin{bmatrix} -36 \\ 24 \\ 0 \end{bmatrix} + EI\begin{bmatrix} 2.33 & 2/3 & -6/16 \\ 2/3 & 2.333 & -6/16 \\ -6/16 & -6/16 & 24/64 \end{bmatrix}\begin{bmatrix} \theta_B \\ \theta_C \\ \Delta \end{bmatrix}$$

By solving the above equation, we obtain

$$\begin{bmatrix} \theta_B \\ \theta_C \\ \Delta \end{bmatrix} = \begin{bmatrix} 20.67/EI \\ -15.33/EI \\ 5.33/EI \end{bmatrix}$$

Final end moments: With the known values of the θ_B, θ_C and Δ, and using the slope–deflection equations, the member end moments can be determined as shown below.

$$M_{AB} = 0 + \frac{2EI}{4}\left[2(0) + \frac{20.67}{EI} - 3\left(\frac{5.33}{EI \times 4}\right)\right] = 8.33 \text{ kNm}$$

$$M_{BA} = 0 + \frac{2EI}{4}\left[0 + 2 \times \frac{20.67}{EI} - 3\left(\frac{5.33}{EI \times 4}\right)\right] = 18.67 \text{ kNm}$$

$$M_{BC} = -36 + \frac{2E(2I)}{6}\left[2 \times \frac{20.67}{EI} - \frac{15.33}{EI} - 0\right] = -18.67 \text{ kNm}$$

$$M_{CB} = 24 + \frac{2E(2I)}{6}\left[\frac{20.67}{EI} - 2 \times \frac{15.33}{EI} - 0\right] = 17.33 \text{ kNm}$$

$$M_{CD} = 0 + \frac{2EI}{4}\left[-2 \times \frac{15.33}{EI} - 3 \times \frac{5.33}{EI \times 4}\right] = -17.33 \text{ kNm}$$

$$M_{DC} = 0 + \frac{2EI}{4}\left[\frac{-15.33}{EI} - \frac{3 \times 5.33}{4EI}\right] = -9.67 \text{ kNm}$$

Shear force and bending moment diagrams: Using the free body diagram and the static equilibrium equations, the other support reactions are determined and are shown in Figure 10.51a. The shear force and bending moment diagrams are shown in Figure 10.51b and c. The probable deflected shape is shown in Figure 10.51d.

Example 10.13

Analyse the frame shown in Figure 10.52 using the stiffness matrix method. Draw the shear force and bending moment diagrams. (This was also solved in Example 6.14 by slope–deflection method.)

Solution

The kinematic indeterminacy of the frame is 3, i.e., θ_B, θ_C at *B* and *C*, and Δ at *BC*. The co-ordinates are assigned to θ_B, θ_C and Δ as 1, 2, and 3, respectively, as shown in Figure 10.53.

Fixed end moments due to loads are calculated using the free body diagrams as shown in Figure 10.54.

$$\text{MF}_{AB} = -\frac{30 \times 3 \times 1^2}{4^2} = -5.625 \text{ kNm}$$

$$\text{MF}_{BA} = \frac{30 \times 3^2 \times 1}{4^2} = 16.875 \text{ kNm}$$

$$\text{MF}_{BC} = -\text{MF}_{CB} = -\frac{50 \times 4^2}{12} = -66.67 \text{ kNm}$$

Net forces at the co-ordinates are

$$P_{1L} = -66.67 + 16.875 = -49.795 \text{ kNm}$$

$$P_{2L} = 66.67 + 0 = 66.67 \text{ kNm}$$

$$P_{3L} = -25.32 + 0 = -25.32 \text{ kNm}$$

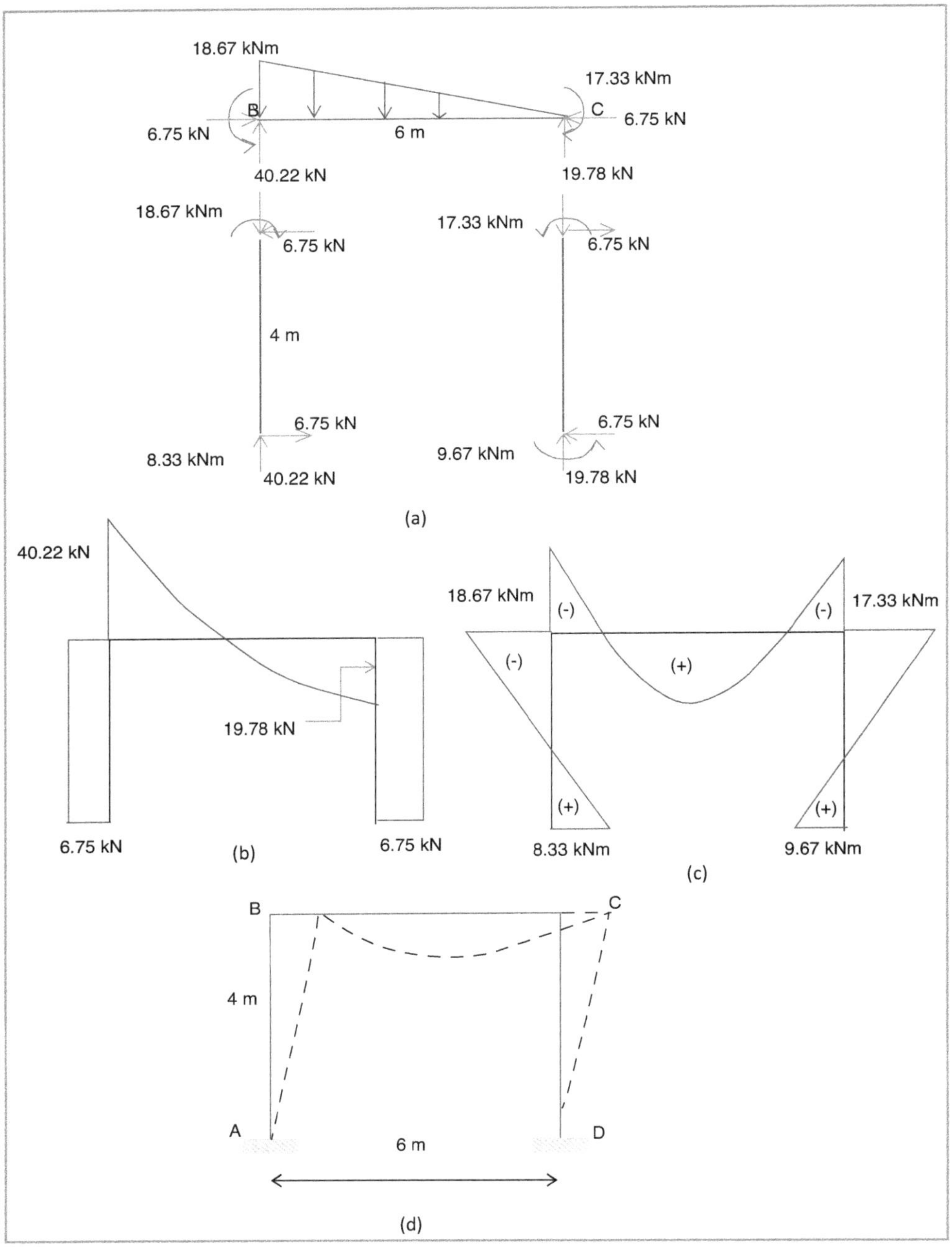

FIGURE 10.51 (a) Free body diagram, (b) shear force diagram, (c) bending moment diagram, and (d) probable elastic curve.

Additional forces due to the yielding of supports are zero, i.e.,

$$P_{1Y} = P_{2Y} = P_{3Y} = 0$$

External factors acting at the co-ordinates are

$$P_1 = P_2 = P_3 = 0$$

Stiffness matrix: To develop the stiffness matrix, apply a unit displacement corresponding to each co-ordinate successively on the restrained structure and find the forces developed at all the co-ordinates each time separately, as shown in Figure 10.55.

$$K_{11} = \left(\frac{4EI}{L}\right)_{AB} + \left(\frac{4EI}{L}\right)_{BC} = \frac{4EI}{4} + \frac{8EI}{4} = 3\ EI$$

$$K_{21} = K_{12} = \left(\frac{2EI}{L}\right)_{BC} = \frac{4EI}{4} = EI$$

$$K_{13} = K_{31} = -\left(\frac{6EI}{L^2}\right)_{AB} = \frac{-6EI}{4^2} = \frac{-6EI}{16}$$

$$K_{23} = K_{32} = -\left(\frac{6EI}{L^2}\right)_{CD} = \frac{-6EI}{25}$$

$$K_{22} = \left(\frac{4EI}{L}\right)_{BC} + \left(\frac{4EI}{L}\right)_{CD} = \frac{8EI}{4} + \frac{4EI}{5} = 2.8\ EI$$

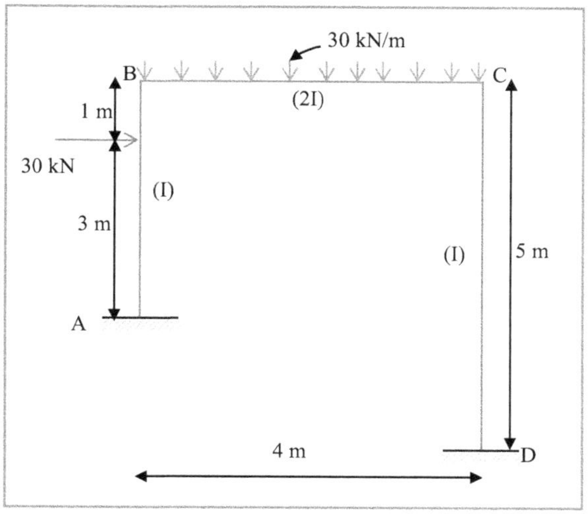

FIGURE 10.52 Unsymmetrical rigid-jointed frame.

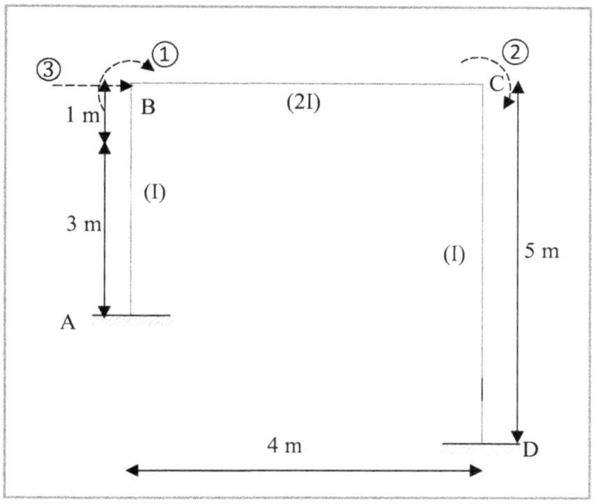

FIGURE 10.53 Co-ordinates assigned to θ_B, θ_C and Δ.

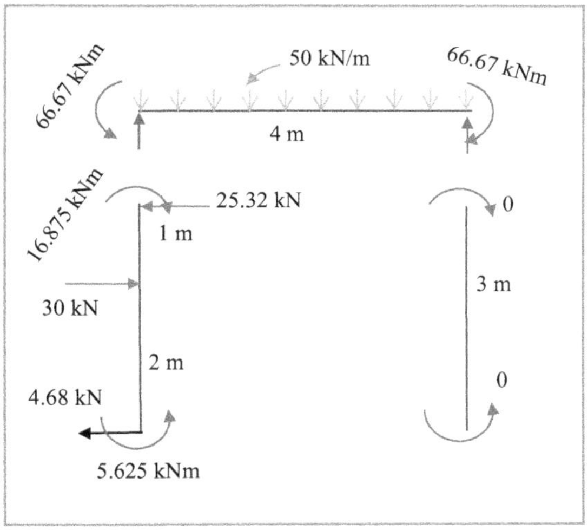

FIGURE 10.54 Force body diagram for fixed end moments.

$$K_{33} = \left(\frac{12EI}{L^3}\right)_{AB} + \left(\frac{12EI}{L^3}\right)_{CD} = \frac{12EI}{64} + \frac{12EI}{125} = 0.2835\ EI$$

$$[K] = EI\begin{bmatrix} 3 & 1 & -6/16 \\ 1 & 2.8 & -6/25 \\ -6/16 & -6/25 & 0.2835 \end{bmatrix}$$

∴ On substituting into stiffness equation 10.15,

$$\begin{bmatrix} 0 \\ 0 \\ 0 \end{bmatrix} = \begin{bmatrix} -49.795 \\ 66.67 \\ -25.32 \end{bmatrix} + EI\begin{bmatrix} 3 & 1 & -6/16 \\ 1 & 2.8 & -6/25 \\ -6/16 & -6/25 & 0.2835 \end{bmatrix}\begin{bmatrix} \theta_B \\ \theta_C \\ \Delta \end{bmatrix}$$

By solving, we obtain

$$\begin{bmatrix} \theta_B \\ \theta_C \\ \Delta \end{bmatrix} = \begin{bmatrix} 40.94/EI \\ -28.18/EI \\ 119.58/EI \end{bmatrix}$$

Final member end moments: By knowing the DoF and with the help of the SD equations, the final member end moments are calculated, as shown below.

$$M_{AB} = -5.625 + \frac{2EI}{4}\left(\frac{40.94}{EI} - \frac{3\times 119.58}{4\times EI}\right) = -30\text{ kNm}$$

$$M_{BA} = 16.875 + \frac{2EI}{4}\left(2\times\frac{40.94}{EI} - \frac{3\times 119.58}{4\times EI}\right) = 12.97\text{ kNm}$$

$$M_{BC} = -66.67 + \frac{2E(2I)}{4}\left(2\times\frac{40.94}{EI} - \frac{28.18}{EI}\right) = -12.97\text{ kNm}$$

$$M_{CB} = 66.67 + \frac{4EI}{4}\left(\frac{40.94}{EI} - \frac{2\times 28.18}{EI}\right) = 51.25\text{ kNm}$$

$$M_{CD} = 0 + \frac{2EI}{5}\left[2\left(\frac{-28.18}{EI}\right) - \frac{3\times 119.58}{5\times EI}\right] = -51.25\text{ kNm}$$

$$M_{DC} = 0 + \frac{2EI}{5}\left[\left(\frac{-28.18}{EI}\right) - \frac{3\times 119.58}{5\times EI}\right] = -39.97\text{ kNm}$$

Shear force and bending moment diagram: The free body diagram and the shear force and bending moment diagrams are shown in Figure 10.56.

Example 10.14

Analyse the frame shown in Figure 10.57. Draw the shear force and bending moment diagrams.

Solution

The kinematic indeterminacy of the frame is 4, i.e., θ_B, θ_C, θ_E and Δ at *BCE*. Assign the co-ordinates to these DoF as shown in Figure 10.58.

The fixed end moments are calculated by assuming all ends are fixed (restrained frame) as shown in the free body diagrams in Figure 10.59.

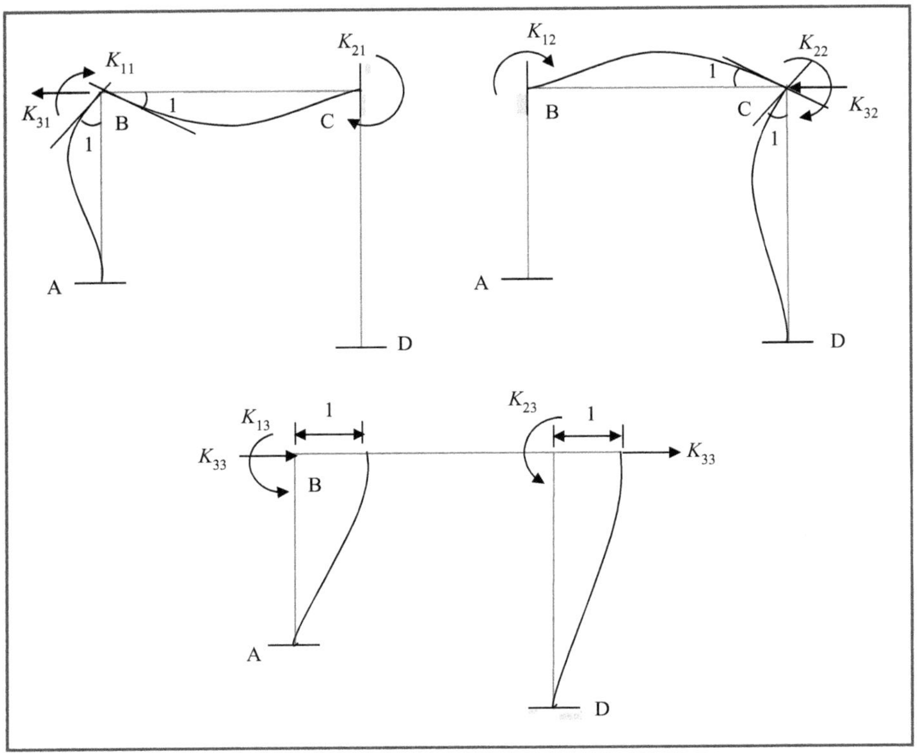

FIGURE 10.55 Development of Elements of stiffness matrix.

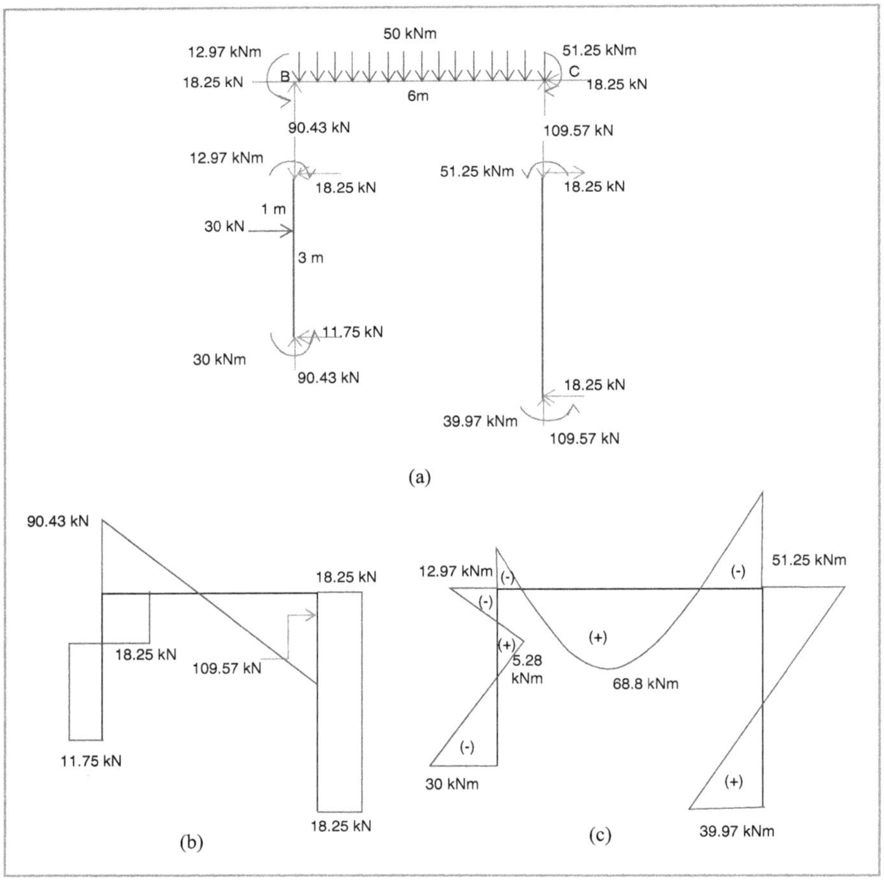

FIGURE 10.56 (a) Free body diagram, (b) shear force diagram, and (c) bending moment diagram.

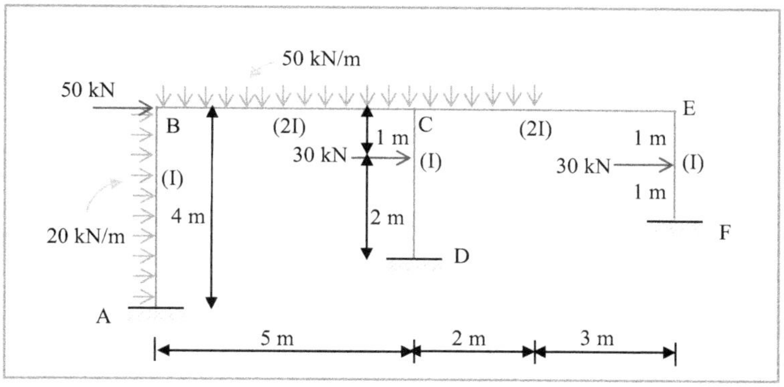

FIGURE 10.57 Two-bay rigid-jointed frame.

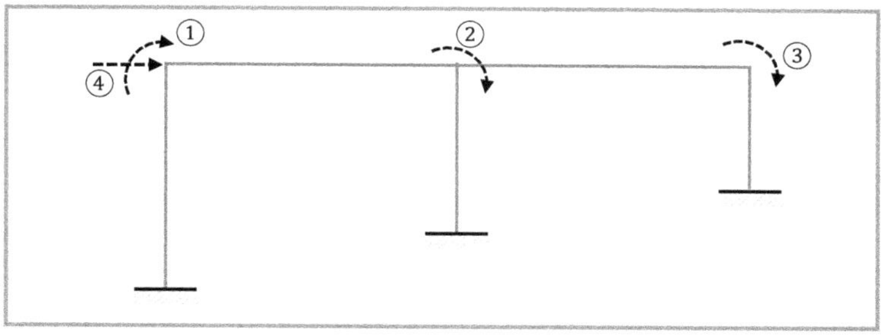

FIGURE 10.58 Co-ordinates assign to DoF.

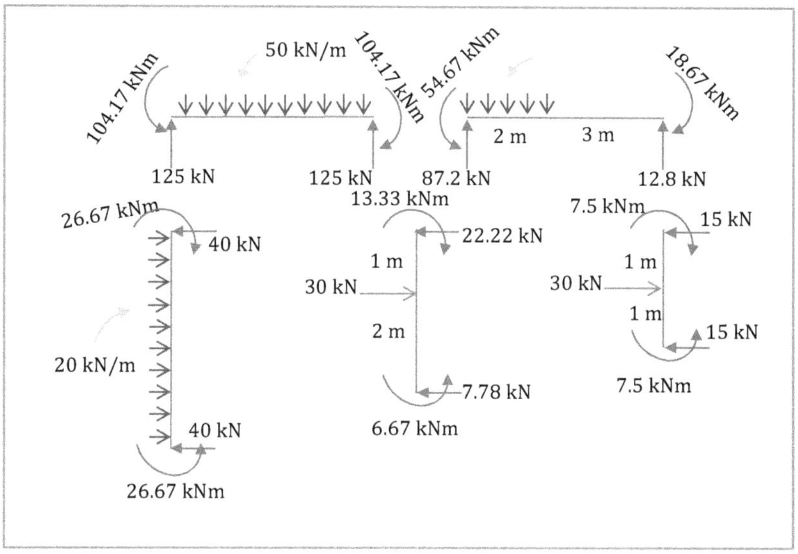

FIGURE 10.59 Free body diagram for fixed end moments and reactions.

As there is no yielding of supports, the additional moments due to the yielding of supports are zero. Therefore, the forces corresponding to the co-ordinates are

$$P_{1L} + P_{1Y} = 26.67 - 104.17 + 0 = -77.5 \text{ kNm}$$

$$P_{2L} + P_{2Y} = 13.33 + 104.17 - 54.67 + 0 = 62.83 \text{ kNm}$$

$$P_{3L} + P_{3Y} = 7.5 + 18.67 + 0 = 26.17 \text{ kNm}$$

$$P_{4L} + P_{4Y} = -40 - 22.22 - 15 + 0 = -77.22 \text{ kNm}$$

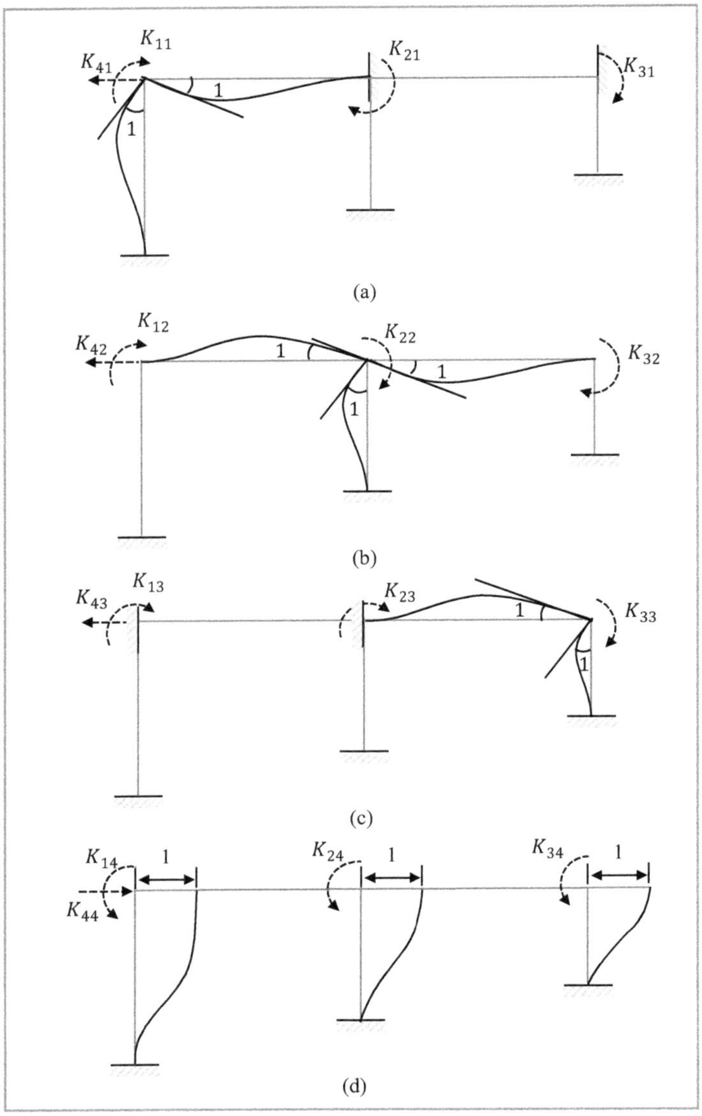

FIGURE 10.60 Elements of stiffness matrix of (a) first column elements, (b) second column elements, (c) third column elements, and (d) fourth column.

External loads acting at the co-ordinates are

$$\begin{bmatrix} P_1 \\ P_2 \\ P_3 \\ P_4 \end{bmatrix} = \begin{bmatrix} 0 \\ 0 \\ 0 \\ 50 \end{bmatrix}$$

STIFFNESS MATRIX

First column elements: To generate the first column elements of the stiffness matrix, apply a unit displacement (rotational) at co-ordinate ① without any displacement at other co-ordinates and find the forces developed at all co-ordinates, as shown in Figure 10.60a.

$$K_{11} = \left(\frac{4EI}{L}\right)_{AB} + \left(\frac{4EI}{L}\right)_{BC} = \frac{4EI}{4} + \frac{4E(2I)}{5} = 2.6\ EI$$

$$K_{21} = K_{12} = \left(\frac{2EI}{L}\right)_{BC} = \frac{2E(2I)}{5} = 0.8\ EI$$

$$K_{31} = 0$$

$$K_{41} = -\left(\frac{6EI}{L^2}\right)_{AB} = \frac{-6EI}{4^2} = -0.375\ EI$$

Second column elements: To develop the second column elements of the stiffness matrix, apply a unit displacement at co-ordinate ② in its chosen direction, without any displacement at other co-ordinates, and calculate the forces developed at all co-ordinates, as shown in Figure 10.60b.

$$K_{12} = K_{21} = 0.8\ EI$$

$$K_{22} = \frac{4E(2I)}{5} \times 2 + \frac{4EI}{3} = 4.533\ EI$$

$$K_{32} = \frac{2E(2I)}{5} + \frac{4EI}{3} = 0.8\ EI$$

$$K_{42} = -\left(\frac{6EI}{L^2}\right)_{CD} = \frac{-6EI}{3^2} = -0.667\ EI$$

Third column elements of the stiffness matrix: To generate the third column elements of the stiffness matrix, give a unit displacement (rotational) at co-ordinate ③, without any displacement at all other co-ordinates, and compute the forces at all co-ordinates, as shown in Figure 10.60c.

$$K_{13} = 0$$

$$K_{23} = \frac{2E(2I)}{5} = 0.8\ EI$$

$$K_{33} = \frac{4E(2I)}{5} + \frac{4EI}{2} = 3.6\ EI$$

$$K_{43} = \frac{-6EI}{2^2} = -1.5\ EI$$

Fourth column elements: To obtain the fourth column elements of the stiffness matrix, apply a unit displacement (translational) at co-ordinate ④, without any displacement at all other co-ordinates, and find the forces at all co-ordinates, as shown in Figure 10.60d.

$$K_{14} = \frac{-6EI}{4^2} = -0.375\ EI$$

$$K_{24} = \frac{-6EI}{3^2} = -0.667EI$$

$$K_{34} = \frac{-6EI}{2^2} = -1.5\ EI$$

$$K_{44} = \frac{12EI}{4^3} + \frac{12EI}{3^3} + \frac{12EI}{2^3} = 2.132EI$$

On substituting the above into equation 10.15,

$$\begin{bmatrix} 0 \\ 0 \\ 0 \\ 50 \end{bmatrix} = \begin{bmatrix} -77.5 \\ 62.83 \\ 26.17 \\ -77.22 \end{bmatrix} + EI \begin{bmatrix} 2.6 & 0.8 & 0 & -0.375 \\ 0.8 & 4.533 & 0.8 & -0.667 \\ 0 & 0.8 & 3.6 & -1.5 \\ -0.375 & -0.667 & -1.5 & 2.132 \end{bmatrix} \begin{bmatrix} \theta_B \\ \theta_C \\ \theta_D \\ \Delta \end{bmatrix}$$

By solving, we obtain

$$\begin{bmatrix} \theta_B \\ \theta_C \\ \theta_E \\ \Delta \end{bmatrix} = \begin{bmatrix} 46.793 \\ -15.137 \\ 31.709 \\ 85.476 \end{bmatrix} \times \frac{1}{EI}$$

Final member end moments: With the known values of DoF, and using the SD equations, the final member end moments can be calculated, as shown below.

$$M_{AB} = -26.67 + \frac{2EI}{4}\left(0 + \frac{46.793}{EI} - \frac{3}{4} \times \frac{85.476}{EI}\right) = -35.33\ \text{kNm}$$

$$M_{BA} = 26.67 + \frac{2EI}{4}\left(2 \times \frac{46.793}{EI} + 0 - \frac{3}{4} \times \frac{85.476}{EI}\right) = 41.41\ \text{kNm}$$

$$M_{BC} = -104.17 + \frac{2E(2I)}{5}\left(2 \times \frac{46.793}{EI} - \frac{15.137}{EI}\right) = -41.41\ \text{kNm}$$

$$M_{CB} = 104.17 + \frac{2E(2I)}{5}\left(\frac{46.793}{EI} - 2 \times \frac{15.137}{EI}\right) = 117.36\ \text{kNm}$$

$$M_{CE} = -54.67 + \frac{2E(2I)}{5}\left(-2 \times \frac{15.137}{EI} + \frac{31.709}{EI}\right) = -53.52\ \text{kNm}$$

$$M_{CD} = 13.33 + \frac{2EI}{3}\left(-2 \times \frac{15.137}{EI} - \frac{3}{3} \times \frac{85.476}{EI}\right) = -63.84\ \text{kNm}$$

$$M_{DC} = -6.67 + \frac{2EI}{3}\left(-\frac{15.137}{EI} - \frac{3}{3} \times \frac{85.476}{EI}\right) = -73.75\ \text{kNm}$$

$$M_{EC} = 18.67 + \frac{2E(2I)}{5}\left(2 \times \frac{31.709}{EI} - \frac{15.137}{EI}\right) = 57.29\ \text{kNm}$$

$$M_{EF} = 7.5 + \frac{2EI}{2}\left(2 \times \frac{31.709}{EI} - \frac{3}{2} \times \frac{85.476}{EI}\right) = -57.29\ \text{kNm}$$

$$M_{FE} = -7.5 + \frac{2EI}{2}\left(\frac{31.709}{EI} - \frac{3}{2} \times \frac{85.476}{EI}\right) = -104\ \text{kNm}$$

Shear force and bending moment diagrams: Using the above results and the static equilibrium equations, the other support reactions (shear forces) are calculated and are shown in free body diagram in Figure 10.61a. The shear force and bending moment diagrams are shown in Figure 10.61b and c, respectively. The probable elastic curve is shown in Figure 10.61d.

Example 10.15

Analyse the two-storey non-symmetrical frame shown in Figure 10.62. Draw the shear force and bending moment diagrams.

Solution

The kinematic indeterminacy of the frame is 6, i.e., θ_B, θ_C, θ_D, θ_E, Δ_{BE} and Δ_{CD}. The co-ordinates can be assigned to these DoF, as shown in Figure 10.63.

Fixed end moments: Since there are no external loads acting on the columns other than at joints, the fixed end moments are

$$MF_{AB} = MF_{BA} = MF_{BC} = MF_{CB} = 0$$

$$MF_{FE} = MF_{EF} = MF_{ED} = MF_{DE} = 0$$

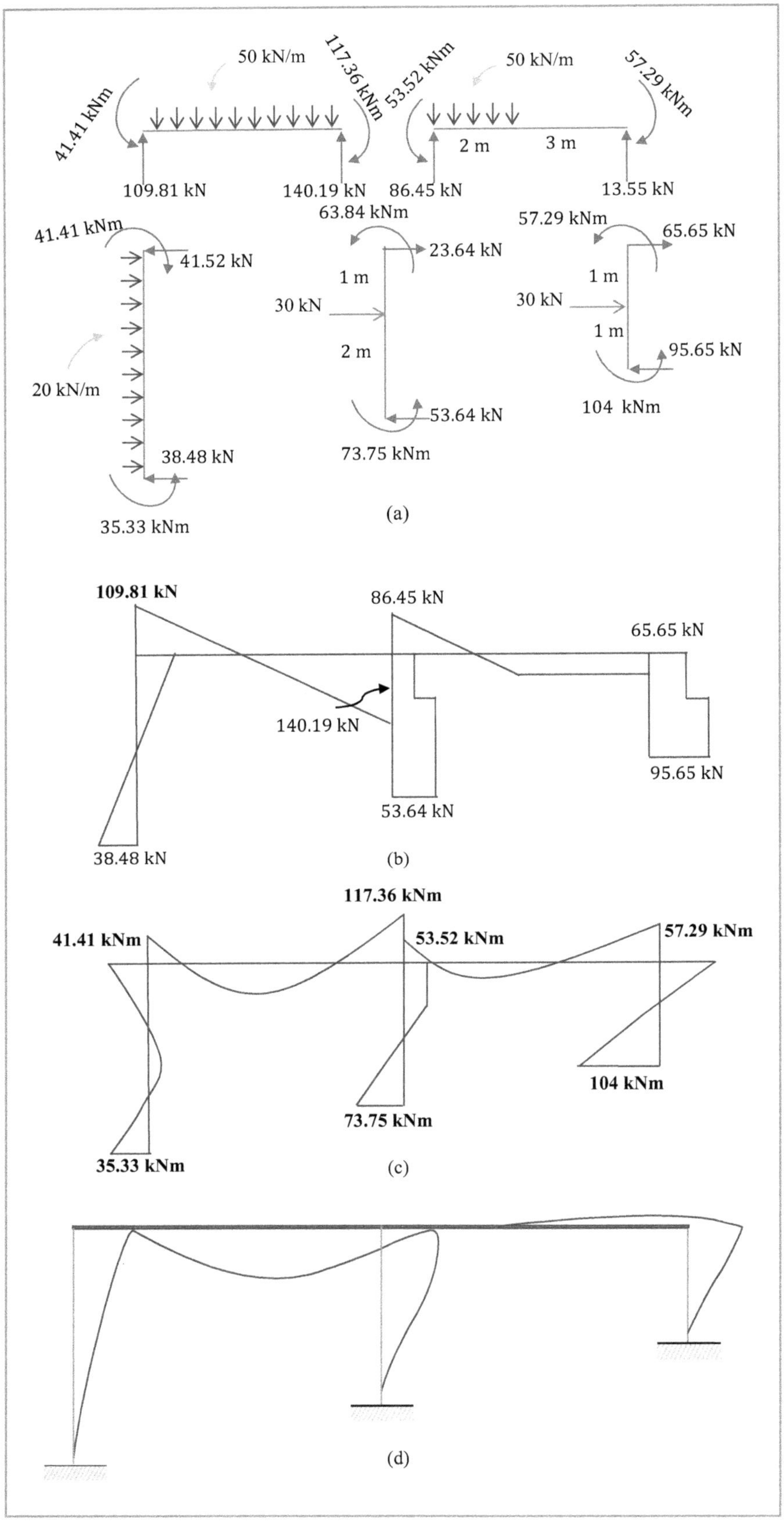

FIGURE 10.61 (a) Free body diagram, (b) shear force diagram, (c) bending moment diagram, and (d) probable deflected shape.

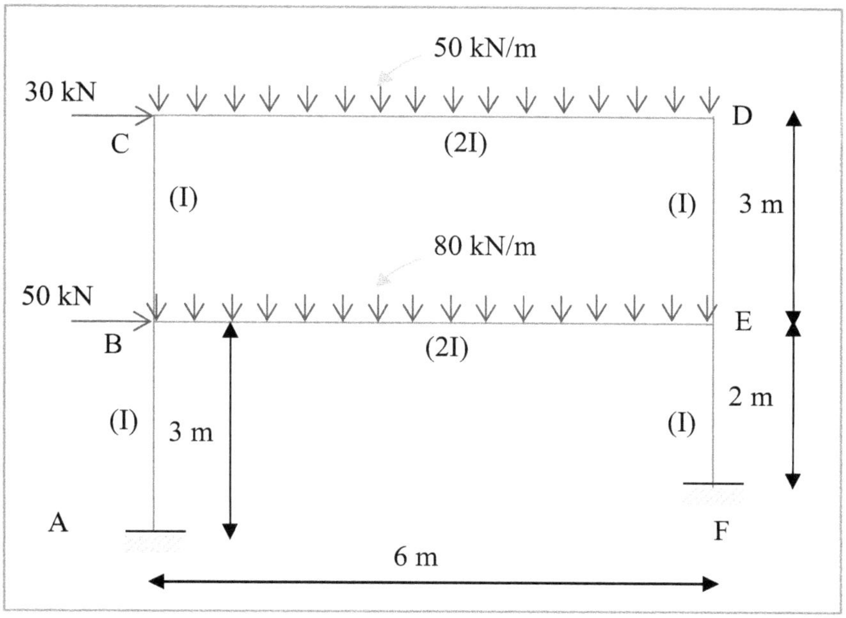

FIGURE 10.62 Two-storey unsymmetrical rigid-jointed frame.

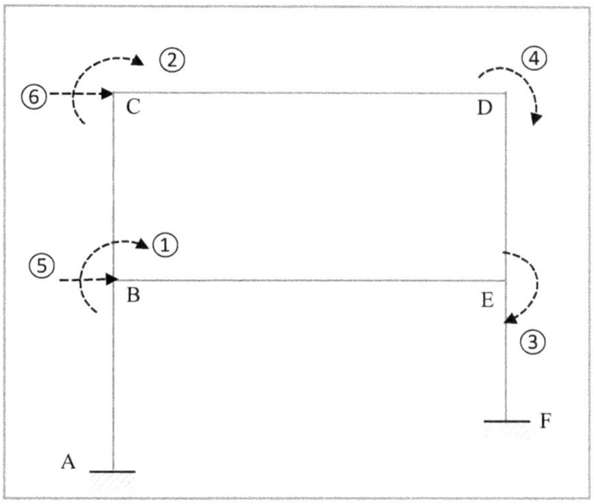

FIGURE 10.63 Co-ordinates assigned to each DoF.

$$MF_{BE} = -MF_{EB} = -\frac{80\times 6^2}{12} = -240 \text{ kNm}$$

$$MF_{CD} = -MF_{DC} = -\frac{50\times 6^2}{12} = -150 \text{ kNm}$$

$$P_{1L} = -240 \text{ kNm}; \quad P_{2L} = -150 \text{ kNm}$$

$$P_{3L} = 240 \text{ kNm}; \quad P_{4L} = 150 \text{ kNm}$$

$$P_{5L} = P_{6L} = 0$$

$$P_{1Y} = P_{2Y} = P_{3Y} = P_{4Y} = P_{5Y} = P_{6Y} = 0 \quad (\because \text{No yielding of supports})$$

$$P_1 = P_2 = P_3 = P_4 = 0 \quad (\because \text{No external loads acting at coordinates})$$

$$P_5 = 50 \text{ kN}, \quad P_6 = 30 \text{ kN}$$

STIFFNESS MATRIX

First column elements: To develop the first column elements of the stiffness matrix, apply a unit displacement at co-ordinate ① in its chosen direction, with no displacement at all other co-ordinates and compute the forces at all co-ordinates, as shown in Figure 10.64a.

$$K_{11} = \frac{4E(2I)}{6} + \frac{4EI}{3} + \frac{4EI}{3} = 4EI$$

$$K_{21} = \frac{2EI}{3} = 0.667EI$$

$$K_{31} = \frac{2E(2I)}{6} = 0.667EI$$

$$K_{41} = 0$$

$$K_{51} = 0$$

$$K_{61} = -\frac{6EI}{3^2} = -0.667EI$$

Similarly, the other column elements of the stiffness matrix can be developed, as shown in Figure 10.66b–f, as presented below.

Second column elements:

$$K_{12} = \frac{2EI}{3} = 0.667EI$$

$$K_{22} = \frac{4E(2I)}{6} + \frac{4EI}{3} = 2.667EI$$

$$K_{32} = 0$$

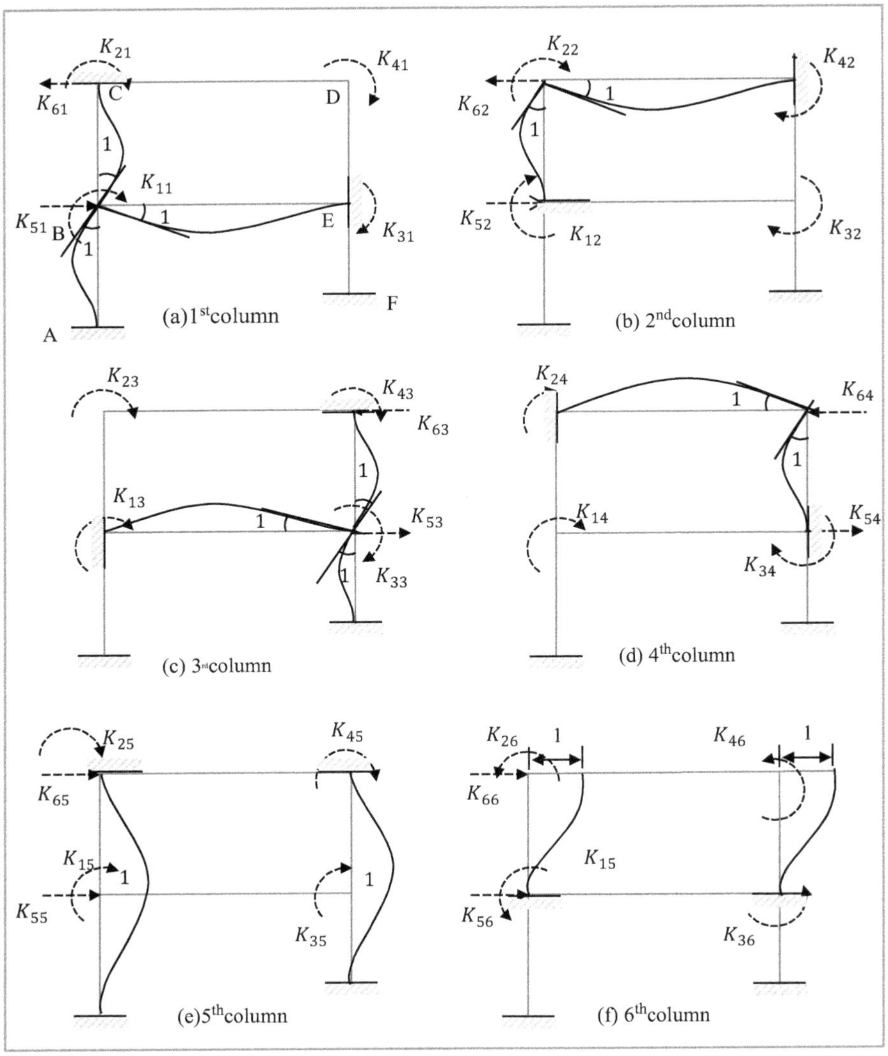

FIGURE 10.64 Development of elements of stiffness matrix.

$$K_{42} = \frac{2E(2I)}{6} = 0.667EI$$

$$K_{52} = \frac{6EI}{3^2} = 0.667EI$$

$$K_{62} = -\frac{6EI}{3^2} = -0.667EI$$

Third column elements:

$$K_{13} = \frac{2E(2I)}{6} = 0.667EI$$

$$K_{23} = 0$$

$$K_{33} = \frac{4E(2I)}{6} + \frac{4EI}{3} + \frac{4EI}{2} = 4.667EI$$

$$K_{43} = \frac{2EI}{3} = 0.667EI$$

$$K_{53} = \frac{6EI}{3^2} - \frac{6EI}{2^2} = -0.833EI$$

$$K_{63} = -\frac{6EI}{3^2} = -0.667EI$$

Fourth column elements:

$$K_{14} = 0$$

$$K_{24} = \frac{2E(2I)}{6} = 0.667EI$$

$$K_{34} = \frac{2EI}{3} = 0.667EI$$

$$K_{44} = \frac{4E(2I)}{6} + \frac{4EI}{3} = 2.667EI$$

$$K_{54} = \frac{6EI}{3^2} = 0.667EI$$

$$K_{64} = -\frac{6EI}{3^2} = -0.667EI$$

Fifth column elements:

$$K_{15} = 0$$

$$K_{25} = \frac{6EI}{3^2} = 0.667EI$$

$$K_{35} = \frac{6EI}{3^2} - \frac{6EI}{2^2} = -0.833EI$$

$$K_{45} = \frac{6EI}{3^2} = 0.667EI$$

$$K_{55} = \frac{12EI}{3^3} \times 3 + \frac{12EI}{2^3} = 2.833EI$$

$$K_{65} = -\frac{12EI}{3^3} \times 2 = -0.888EI$$

Sixth column elements:

$$K_{16} = -\frac{6EI}{3^2} = -0.667EI$$

$$K_{26} = -\frac{6EI}{3^2} = -0.667EI$$

$$K_{36} = -\frac{6EI}{3^2} = -0.667EI$$

$$K_{46} = -\frac{6EI}{3^2} = -0.667EI$$

$$K_{56} = \frac{-12EI}{3^3} \times 2 = -0.888EI$$

$$K_{66} = \frac{12EI}{3^3} \times 2 = 0.888EI$$

By substituting the above values into the stiffness equation 10.15,

$$\begin{bmatrix} 0 \\ 0 \\ 0 \\ 0 \\ 50 \\ 30 \end{bmatrix} = \begin{bmatrix} -240+0 \\ -150+0 \\ 240+0 \\ 150+0 \\ 0 \\ 0 \end{bmatrix} + EI \begin{bmatrix} 4 & 0.667 & 0.667 & 0 & 0 & -0.667 \\ 0.667 & 2.667 & 0 & 0.667 & 0.667 & -0.667 \\ 0.667 & 0 & 4.667 & 0.667 & -0.833 & -0.667 \\ 0 & 0.667 & 0.667 & 2.667 & 0.667 & -0.667 \\ 0 & 0.667 & -0.833 & 0.667 & 2.833 & -0.888 \\ -0.667 & -0.667 & -0.667 & -0.667 & -0.888 & 0.888 \end{bmatrix} \begin{bmatrix} \theta_B \\ \theta_C \\ \theta_E \\ \theta_D \\ \Delta_{BE} \\ \Delta_{CD} \end{bmatrix}$$

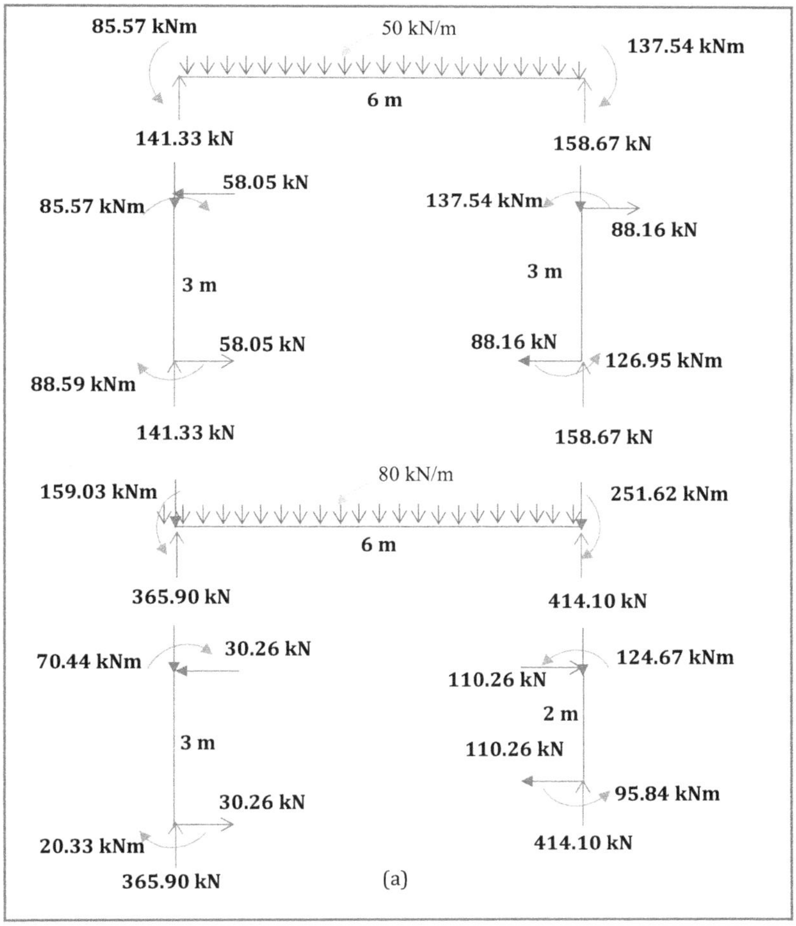

FIGURE 10.65 (a) Free body diagram, (b) shear force diagram, and (c) bending moment diagram and (d) probable deflected shape.

(Continued)

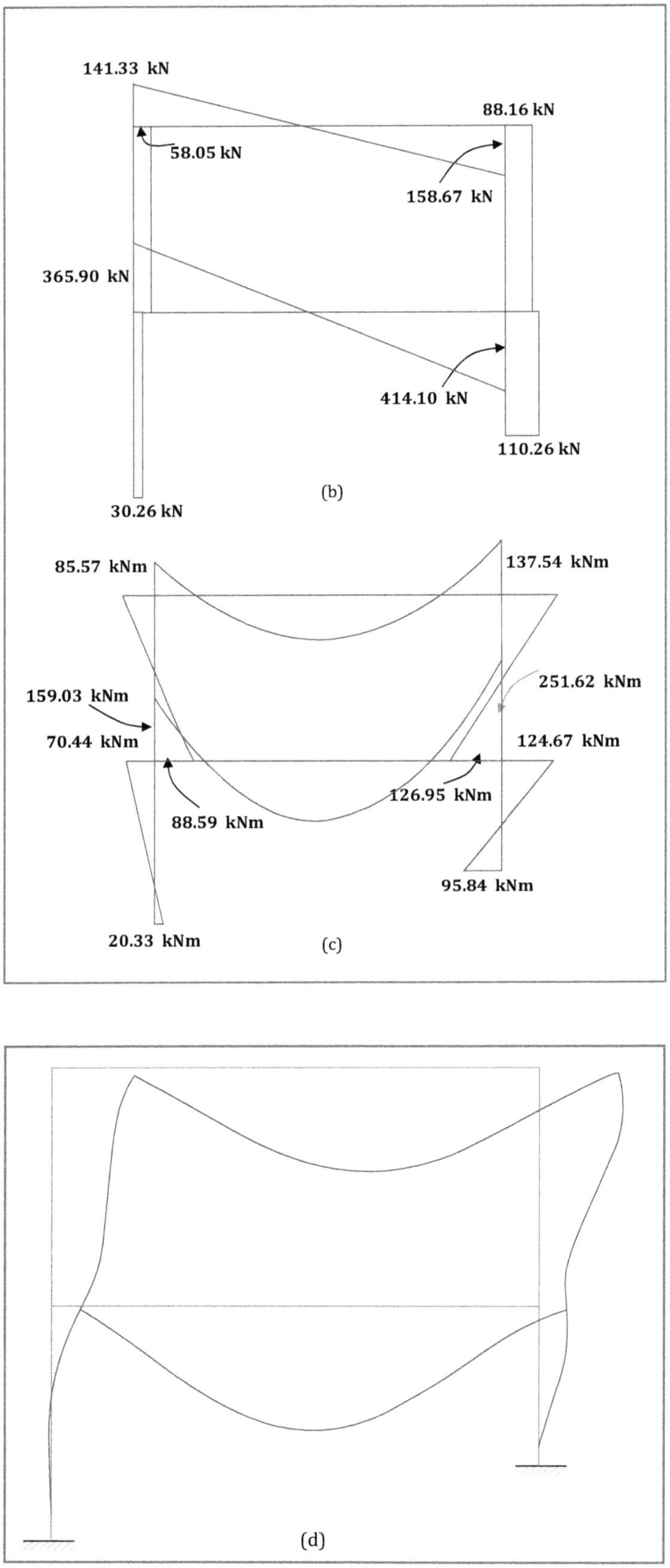

FIGURE 10.65 (*Continued*) (a) Free body diagram, (b) shear force diagram, and (c) bending moment diagram and (d) probable deflected shape.

By solving, we obtain

$$\begin{bmatrix} \theta_B \\ \theta_C \\ \theta_E \\ \theta_D \\ \Delta_{BE} \\ \Delta_{CD} \end{bmatrix} = \frac{1}{EI}\begin{bmatrix} 75.167 \\ 70.648 \\ -28.836 \\ -44.667 \\ 44.669 \\ 132.769 \end{bmatrix}$$

Member end moments: With the known values of the DoF, and using the SD Equations, the member end moments can be calculated, as shown below.

$$M_{AB} = 0 + \frac{2EI}{3}\left(\frac{75.167}{EI} - \frac{3}{3}\times\frac{44.669}{EI}\right) = 20.33 \text{ kNm}$$

$$M_{BA} = 0 + \frac{2EI}{3}\left(2\times\frac{75.167}{EI} - \frac{3}{3}\times\frac{44.669}{EI}\right) = 70.44 \text{ kNm}$$

$$M_{BC} = 0 + \frac{2EI}{3}\left(2\times\frac{75.167}{EI} + \frac{70.648}{EI} - \frac{3}{3}\times\frac{(132.769-44.669)}{EI}\right)$$

$$= 88.59 \text{ kNm}$$

$$M_{CB} = 0 + \frac{2EI}{3}\left(\frac{75.167}{EI} + 2\times\frac{70.648}{EI} - \frac{3}{3}\times\frac{((132.769-44.669))}{EI}\right)$$

$$= 85.57 \text{ kNm}$$

$$M_{BE} = -240 + \frac{2E(2I)}{6}\left(2\times\frac{75.167}{EI} - \frac{28.836}{EI}\right) = -159.03 \text{ kNm}$$

$$M_{EB} = 240 + \frac{2E(2I)}{6}\left(\frac{75.167}{EI} - 2\times\frac{28.836}{EI}\right) = 251.62 \text{ kNm}$$

$$M_{FE} = 0 + \frac{2EI}{2}\left[\frac{-28.836}{EI} - \frac{3}{2}\left(\frac{44.669}{EI}\right)\right] = -95.84 \text{ kNm}$$

$$M_{EF} = 0 + \frac{2EI}{2}\left[2\left(\frac{-28.836}{EI}\right) - \frac{3}{2}\left(\frac{44.669}{EI}\right)\right] = -124.67 \text{ kNm}$$

$$M_{ED} = 0 + \frac{2EI}{3}\left[2\left(\frac{-28.836}{EI}\right) - \frac{44.667}{EI} - \frac{3}{3}\left(\frac{132.769-44.669}{EI}\right)\right]$$

$$= -126.95 \text{ kNm}$$

$$M_{DE} = 0 + \frac{2EI}{3}\left[\frac{-28.836}{EI} - \frac{2\times 44.667}{EI} - \frac{3}{3}\left(\left(\frac{132.769-44.669}{EI}\right)\right)\right]$$

$$= -137.54 \text{ kNm}$$

$$M_{CD} = -150 + \frac{2E(2I)}{6}\left[2\times\frac{70.648}{EI} - \frac{44.667}{EI}\right] = -85.58 \text{ kNm}$$

$$M_{DC} = 150 + \frac{2E(2I)}{6}\left[\frac{70.648}{EI} - \frac{2\times 44.667}{EI}\right] = 137.54 \text{ kNm}$$

Shear force and bending moment diagrams: The free body diagram shows the member end forces in Figure 10.65a. The shear force and bending moment diagrams are shown in Figure 10.65b and c, respectively. The probable deflected shape is shown in Figure 10.65d.

10.5 APPLICATIONS TO PIN-JOINTED PLANE FRAMES

A typical pin-joint '*O*' of a pin-jointed plane frame (truss) is shown in Figure 10.66. Under the action of external loads, the pin-joint offers resistance to translation as the members meeting at joint '*O*' undergo expansion or contraction. Therefore, the translational stiffness of a pin-joint in any chosen direction is defined as the force required to produce a unit translational displacement in that direction. Furthermore, this stiffness depends on the axial stiffness of the members meeting at the joint.

Consider a typical member '*OA*' of a pin-joint '*O*,' as shown in Figure 10.66b. Let the co-ordinates *i* and *j* be selected as positive in the *X* and *Y* directions, respectively. Similarly, let *m* and *n* at point *A* represent the *X* and *Y* directions, respectively, where *X* and *Y* are global co-ordinates.

Development of the stiffness matrix:

a. **Unit displacement along co-ordinate 'i':** Let us apply a unit displacement along '*i*' at co-ordinate '*O*' without any displacement at all other co-ordinates. Now the joint '*O*' moves to '*O*!' as shown in Figure 10.66b. As a result, the members meeting at joint '*O*' are subjected to a change in lengths.

   Let the contraction of member *OA* be *OO*!!=*OO*!. cos *θ*=cos *θ*.

   where '*θ*' is the inclination of the member measured in a counter-clockwise direction from the positive *X*-axis.

   The force required to produce this contraction $= \frac{AE}{L}\cos\theta$.

   ∴ The component of this force along the direction '*i*' $= \frac{AE}{L}\cos^2\theta$.

   Similarly, consider all other members meeting at joint '*O*'.

   It is clear that the net force required at joint '*O*' to displace by a unit magnitude along co-ordinate '*i*' is K_{ii}.

$$\therefore K_{ii} = \sum_{i}^{n}\frac{AE}{L}\cos^2\theta \qquad (10.21)$$

   where $\frac{AE}{L}$ is the axial rigidity (stiffness) of a member meeting at joint '*O*'.

   The component of the force in the direction of '*j*' is $\frac{AE}{L}\cos\theta\cdot\sin\theta$

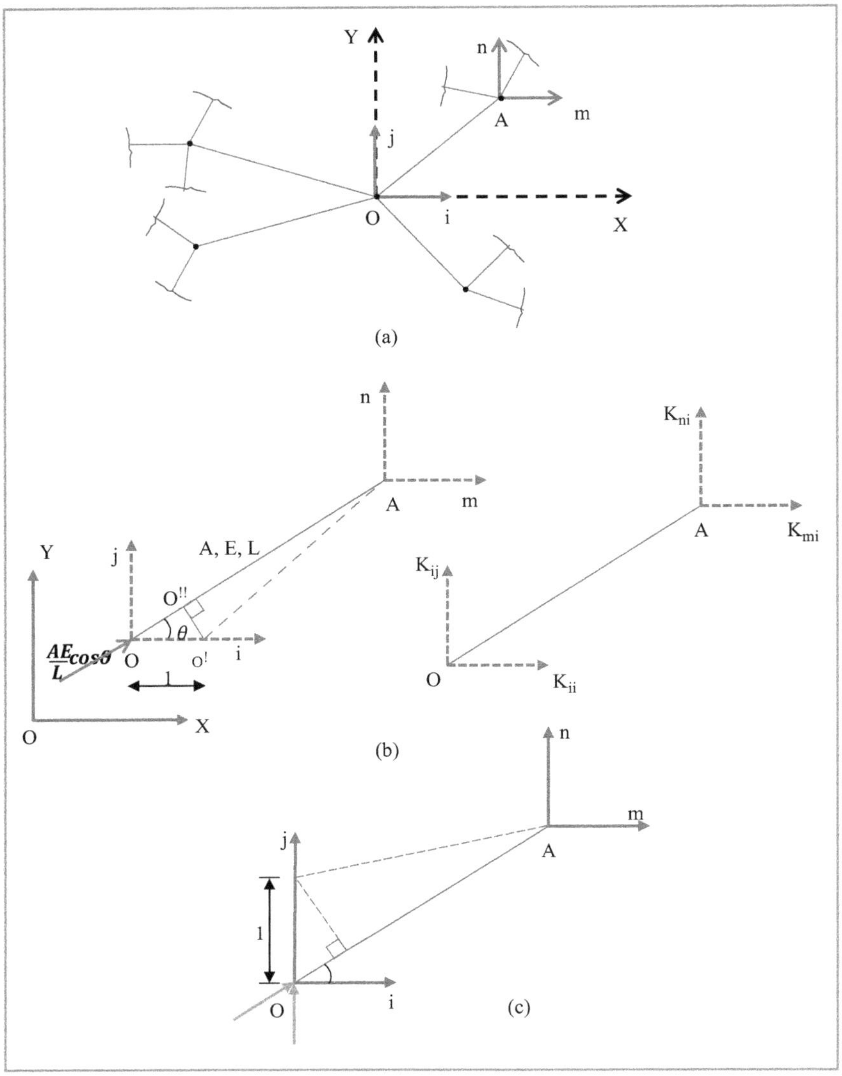

FIGURE 10.66 (a) Typical pin-joint of a plane frames, (b) unit displacement at 'O' along 'i', and (c) unit displacement along 'j' at O.

By considering all the members meeting at joint 'O', the net force developed at co-ordinate 'j' due to a unit displacement at co-ordinate 'i' is K_{ji}

$$K_{ji} = \sum \frac{AE}{L} \sin\theta \cos\theta \tag{10.22}$$

At the other end, the forces at co-ordinates 'm' and 'n' due to a unit displacement at 'i' are opposite to these forces.

$$\therefore K_{mi} = -\frac{AE}{L} \cos^2\theta \tag{10.23}$$

$$K_{ni} = -\frac{AE}{L} \sin\theta \cos\theta \tag{10.24}$$

b. **Unit displacement at co-ordinates 'j':** Let us apply a unit displacement at 'O' in the direction of 'j' without any displacement at the other co-ordinates, as shown in Figure 10.66c. The member OA contracts by $\sin\theta$

Therefore, the compressive force required to produce this contraction $= \frac{AE}{L} \sin\theta$.

The component of this force along the co-ordinate $j = \frac{AE}{L} \sin^2\theta$.

Considering all the members meeting at joint 'O', the net force required at joint 'O' to displace a unit magnitude in the direction of co-ordinate 'j' is given by

$$K_{jj} = \sum \frac{AE}{L} \sin^2\theta \tag{10.25}$$

Similarly, the component of the force along the co-ordinate i by considering all the members meeting at joint 'O' is

$$K_{ij} = K_{ji} = \sum \frac{AE}{L} \sin\theta \cos\theta$$

Also, at the other end, the force developed along the co-ordinates '*m*' and '*n*' due to a unit displacement along '*j*' is opposite to these forces.

$$\therefore K_{mj} = -\frac{AE}{L}\sin\theta\cos\theta \qquad (10.26)$$

$$K_{nj} = -\frac{AE}{L}\sin\theta \qquad (10.27)$$

It should be noted that in the above equations, θ is measured in a counter-clockwise direction with respect to positive direction of the *X*-axis and the co-ordinates *i*, *j*, *m* and *n* are in the positive directions of *X* and *Y* axes at *O* and *A*, respectively.

Sometimes the co-ordinates at a joint may be required to define at a certain inclination with the horizontal or vertical directions, as shown in Figure 10.67.

Let the co-ordinates '*m*' and '*n*' at joint *A* be rotated in an anti-clockwise direction by α with the positive *X*-axis, as shown in Figure 10.67, then the stiffness can be expressed as in equation 10.28:

$$K_{ii} = \sum \frac{AE}{L}\cos^2\theta$$

$$K_{jj} = \sum \frac{AE}{L}\sin^2\theta$$

$$K_{ij} = K_{ji} = \sum \frac{AE}{L}\sin\theta\cos\theta$$

$$K_{mi} = K_{im} = -\frac{AE}{L}\cos\theta\cos(\theta-\alpha)$$

$$K_{ni} = K_{in} = -\frac{AE}{L}\cos\theta\sin(\theta-\alpha)$$

$$K_{mj} = K_{jm} = -\frac{AE}{L}\sin\theta\cos(\theta-\alpha) \qquad (10.28)$$

$$K_{nj} = K_{jn} = -\frac{AE}{L}\sin\theta\sin(\theta-\alpha)$$

$$K_{mm} = \sum \frac{AE}{L}\cos^2(\theta-\alpha)$$

$$K_{nn} = \sum \frac{AE}{L}\sin^2(\theta-\alpha)$$

$$K_{mn} = K_{nm} = \sum \frac{AE}{L}\cos(\theta-\alpha)\sin(\theta-\alpha)$$

Member forces: Consider a typical member '*PQ*' connecting joints *P* and *Q* of a pin-jointed plane frame as shown in Figure 10.68.

Let $P^!$ and $Q^!$ be the displaced positions of *P* and *Q*. Let Δ_{px} and Δ_{py} be the displacement components along *X* and *Y* directions, respectively, at *P*. Similarly, at *Q*, the displacement components along *X* and *Y* directions are Δ_{qx} and Δ_{qy}, respectively.

From Figure 10.68,

The shortening of the member *PQ* at joint $P = (\Delta_{PX}\cos\theta_{PQ} + \Delta_{PY}\sin\theta_{PQ})$

Also, the elongation of the member *PQ* at joint $Q = (\Delta_{QX}\cos\theta_{PQ} + \Delta_{QY}\sin\theta_{PQ})$

$\therefore$ Net elongation of the member *PQ* is $= (\Delta_{QX} - \Delta_{PX})\cos\theta_{PQ} + (\Delta_{QY} - \Delta_{PY})\sin\theta_{PQ}$

Hence, the force in the member *PQ*

$$F_{PQ} = \frac{AE}{L}\left[(\Delta_{QX} - \Delta_{PX})\cos\theta_{PQ} + (\Delta_{QY} - \Delta_{PY})\sin\theta_{PY}\right] \qquad (10.29)$$

A positive sign indicates that the force in member *PQ* is a tensile force. It should be noted that the inclination of the member is always measured in an anti-clockwise direction with respect to the +ve *X*-axis. Using equation 10.29, the forces in the members of a pin-jointed plane frame can be determined.

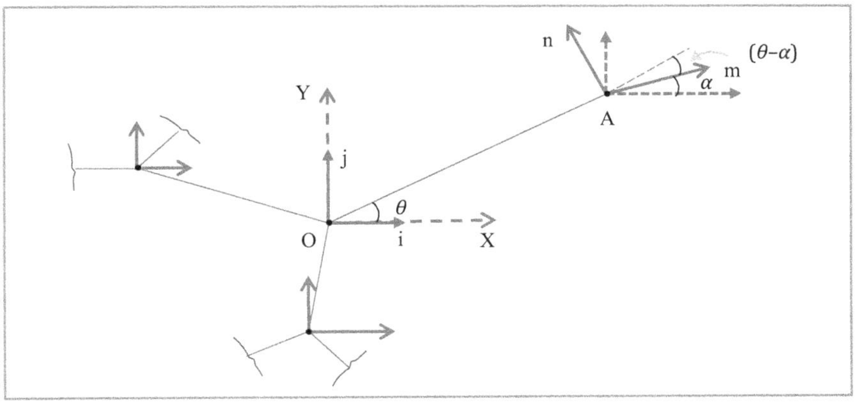

FIGURE 10.67 Co-ordinates at an inclination with horizontal.

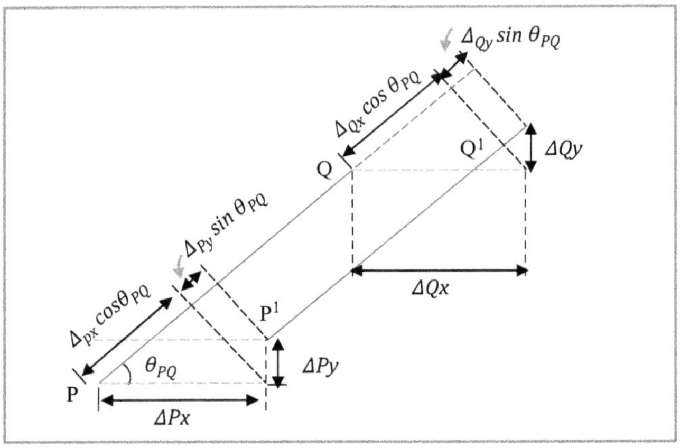

FIGURE 10.68 Member forces in *PQ*.

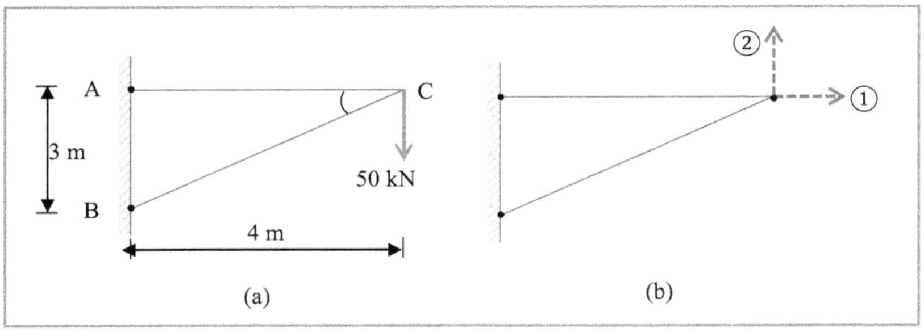

FIGURE 10.69 (a) Truss *ABC* and (b) co-ordinates assigned to DoF.

Example 10.16

A jib crane carrying a vertical load of 50 kN is shown in Figure 10.69a. Compute the displacement of joint 'C'. Hence, calculate the forces in members *AC* and *BC*. The cross-sectional areas of members *AC* and *BC* are 5,000 mm², respectively. Assume $E = 2\times10^5$ N/ mm^2.

Solution

In the case of pin-jointed plane frames, each joint has two DoFs, i.e., the displacement along *X* and *Y* directions. Therefore, the kinematic indeterminacy is 2, i.e., the horizontal and vertical displacements of joint 'C' in this case. Let us assign the co-ordinates ① and ② to these DOF, as shown in Figure 10.69b.

Let Δ_1 and Δ_2 be the displacements at co-ordinates ① and ②, respectively, due to the external loads.

$$\begin{bmatrix} P_1 \\ P_2 \end{bmatrix} = \begin{bmatrix} 0 \\ -50 \end{bmatrix}$$

STIFFNESS MATRIX

The elements of the stiffness matrix with reference to the co-ordinates can be developed by applying a unit displacement successively at co-ordinates ① and ② in their chosen directions, and finding the forces developed at all co-ordinates in each case separately. The computation of these elements of the stiffness matrix is shown in Table 10.1.

$$\therefore K_{11} = \sum \frac{AE}{L}\cos^2\theta = 506$$

$$K_{21} = K_{12} = \sum \frac{AE}{L}\cos\theta\sin\theta = 192$$

$$K_{22} = \sum \frac{AE}{L}\sin^2\theta = 144$$

∴ By substituting all the values into equation 10.15,

TABLE 10.1
Computation of Elements of Stiffness Matrix

Member	$\frac{AE}{L}$ kN/mm	θ	$\cos\theta$	$\sin\theta$	$\frac{AE}{L}\cos^2\theta$ kN/mm	$\frac{AE}{L}\sin^2\theta$ kN/mm	$\frac{AE}{L}\sin\theta\cos\theta$ kN/mm
CA	250	180°	−1	0	250	0	0
CB	400	216.86°	−0.8	−0.60	256	144	192
Σ =					506	144	192

$$\begin{bmatrix} 0 \\ -50 \end{bmatrix} = \begin{bmatrix} 506 & 192 \\ 192 & 144 \end{bmatrix} \begin{bmatrix} \Delta_1 \\ \Delta_2 \end{bmatrix}$$

By solving the above equation, we get

$$\begin{bmatrix} \Delta_1 \\ \Delta_2 \end{bmatrix} = \begin{bmatrix} +0.266 \text{ mm} \\ -0.703 \text{ mm} \end{bmatrix}$$

Final member forces: With the known values of Δ_1 and Δ_2 and using Equation 10.29, the member forces can be calculated as

$$F_{CA} = \frac{AE}{L}\left[(\Delta_{AX} - \Delta_{CX})\cos\theta_{CA} + (\Delta_{AY} - \Delta_{CY})\sin\theta_{CA}\right]$$

$$= 250\left[(0 - 0.266)\cos 180 + (0 + 0.703)\sin 180\right]$$

$$= 66.5 \text{ kN (Tensile)}$$

$$F_{CB} = \frac{AE}{L}\left[(\Delta_{BX} - \Delta_{CX})\cos\theta_{CB} + (\Delta_{BY} - \Delta_{CY})\sin\theta_{CB}\right]$$

$$= 400\left[(0 - 0.266)\cos 216.86 + (0 + 0.703)\sin 216.86\right]$$

$$= -83.55 \text{ kN}(\text{Compression})$$

Example 10.17

Calculate the displacements of joint '*O*' for the pin-jointed frame shown in Figure 10.70a. Also, calculate the forces in the members. The cross-sectional areas of each member are 1,000 mm². Assume $E = 2.05 \times 10^2$ kN/mm². The same was solved in Example No. 9.20 using the flexibility matrix method.

The kinematic indeterminacy, i.e., DoF equal to 2, i.e. Δ_H and Δ_V at '*O*'. The other joints are hinged. The co-ordinates ① and ② can be assigned to these DoF, as shown in Figure 10.70b. The elements of the stiffness matrix with respect to the co-ordinates are calculated symmetrically in a tabular form, as shown in Table 10.2.

$$\therefore K_{11} = \sum \frac{AE}{L}\cos^2\theta = 66.55$$

$$K_{21} = K_{12} = \sum \frac{AE}{L}\cos\theta\sin\theta = -8.12$$

$$K_{22} = \sum \frac{AE}{L}\sin^2\theta = 127.17$$

∴ Substituting all the values into equation 10.15,

$$\begin{bmatrix} 50 \\ -100 \end{bmatrix} = \begin{bmatrix} 66.55 & -8.12 \\ -8.12 & 127.17 \end{bmatrix} \begin{bmatrix} \Delta_1 \\ \Delta_2 \end{bmatrix}$$

By solving the above, we obtain

$$\begin{bmatrix} \Delta_1 \\ \Delta_2 \end{bmatrix} = \begin{bmatrix} +0.661 \text{ mm} \\ -0.744 \text{ mm} \end{bmatrix}$$

Final member forces: With the known values of Δ_1 and Δ_2 and using equation 10.29, the member forces can be calculated as

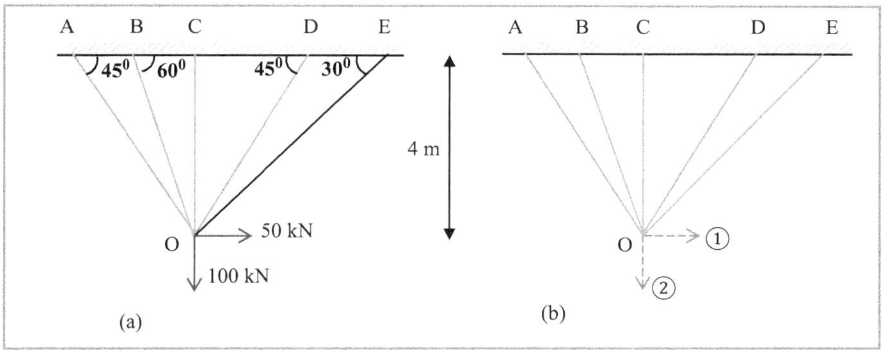

FIGURE 10.70 (a) Pin-jointed frame and (b) co-ordinates assigned to the DoF.

TABLE 10.2
Estimation of the Elements of the Stiffness Matrix

Member	$\frac{AE}{L}$ kN/mm	θ (Degrees)	$\cos\theta$	$\sin\theta$	$\frac{AE}{L}\cos^2\theta$ kN/mm	$\frac{AE}{L}\sin^2\theta$ kN/mm	$\frac{AE}{L}\sin\theta\cos\theta$ kN/mm
OA	36.23	135	−0.707	0.707	18.12	18.12	−18.12
OB	44.38	120	−0.5	0.866	11.10	33.29	−19.22
OC	51.25	90	0	1	0	51.25	0.0
OD	36.23	45	0.707	0.707	18.12	18.12	18.12
OE	25.63	30	0.866	0.5	19.22	6.41	11.10
Σ =					66.55	127.17	−8.12

$$F_{OA} = \frac{AE}{L}\left[(\Delta_{AX} - \Delta_{OX})\cos\theta_{OA} + (\Delta_{AY} - \Delta_{OY})\sin\theta_{OA}\right]$$
$$= 36.23\left[(0 - 0.661)\cos 135 + (0 + 0.744)\sin 135\right]$$
$$= 36 \text{ kN (Tensile)}$$
$$F_{OB} = \frac{AE}{L}\left[(\Delta_{BX} - \Delta_{OX})\cos\theta_{OB} + (\Delta_{BY} - \Delta_{OY})\sin\theta_{OB}\right]$$
$$= 44.38\left[(0 - 0.661)\cos 120 + (0 + 0.744)\sin 120\right]$$
$$= 43.26 \text{ kN (Tensile)}$$
$$F_{OC} = 51.25\left[(0 - 0.661)\cos 90 + (0 + 0.744)\sin 90\right]$$
$$= 38.13 \text{ kN (Tensile)}$$
$$F_{OD} = 36.23\left[(0 - 0.661)\cos 45 + (0 + 0.744)\sin 45\right]$$
$$= 2.12 \text{ kN (Tensile)}$$
$$F_{OE} = 25.63\left[(0 - 0.661)\cos 30 + (0 + 0.744)\sin 30\right]$$
$$= -5.13 \text{ kN (Compression)}$$

Example 10.18

Analyse the pin-jointed plane frame shown in Figure 10.71a. The axial stiffness of each member is 100 kN/mm.

Solution

Joint 'C' moves along the horizontal and vertical directions, whereas joint 'B' moves along the inclined direction, i.e., in the inclined plane *DE*. Therefore, the kinematic indeterminacy=3, i.e., Δ_{HC}, Δ_{VC} and Δ_B. Let them designate as Δ_1, Δ_2 and Δ_3, respectively. The co-ordinates ①, ②, and ③ can be assigned to these DoF, as shown in Figure 10.71b.

$$P_1 = 50 \text{ kN}; \quad P_2 = -100 \text{ kN}; \quad P_3 = 0$$

The stiffness matrix can be calculated using Equation 10.28 as

$$K_{11} = \sum \frac{AE}{L}\cos^2\theta = 100\left[\cos^2\theta_{CA} + \cos^2\theta_{CB}\right]$$
$$= 100\left[\cos^2 225° + \cos^2 300°\right] = 75$$

$$K_{22} = \sum \frac{AE}{L}\sin^2\theta = 100\left[\sin^2\theta_{CA} + \sin^2\theta_{CB}\right]$$
$$= 100\left[\sin^2 225° + \sin^2 300°\right] = 125$$

$$K_{33} = \sum \frac{AE}{L}\cos^2(\theta - \alpha) = 100\begin{bmatrix}\cos^2(\theta_{BC} - \alpha) \\ +\cos^2(\theta_{BA} - \alpha)\end{bmatrix}$$
$$= 100\left[\cos^2(120 - 45)° + \cos^2(180 - 45)°\right] = 56.69$$

$$K_{21} = K_{12} = \sum \frac{AE}{L}\cos\theta\sin\theta = \frac{AE}{L}[\sin\theta_{CA}\cos\theta_{CA}$$
$$+ \sin\theta_{CB}\cos\theta_{CB}]$$
$$= 100[\sin 225\cos 225 + \cos 300\sin 300] = 6.69$$

$$K_{13} = K_{31} = -\frac{AE}{L}\cos\theta\cdot\cos(\theta - \alpha) \qquad \text{for member } CB$$
$$= -100\cdot\cos 300\cdot\cos(300 - 45) = +12.94$$

$$K_{23} = K_{32} = -\frac{AE}{L}\sin\theta\cdot\cos(\theta - \alpha) \qquad \text{for member } CB$$
$$= -100\cdot\sin 300\cdot\cos(300 - 45) = -22.414$$

∴ The stiffness matrix is

$$[K] = \begin{bmatrix} 75 & 6.69 & 12.94 \\ 6.69 & 125 & -22.414 \\ 12.94 & -22.414 & 56.69 \end{bmatrix}$$

On substituting these values into equation 10.15,

$$\begin{bmatrix} 50 \\ -100 \\ 0 \end{bmatrix} = \begin{bmatrix} 75 & 6.69 & 12.94 \\ 6.69 & 125 & -22.414 \\ 12.94 & -22.414 & 56.69 \end{bmatrix}\begin{bmatrix} \Delta_1 \\ \Delta_2 \\ \Delta_3 \end{bmatrix}$$

By solving the above equation, we obtain

$$\begin{bmatrix} \Delta_1 \\ \Delta_2 \\ \Delta_3 \end{bmatrix} = \begin{bmatrix} +0.849 \\ -0.947 \\ -0.568 \end{bmatrix} \text{mm}$$

Now the components of displacements along *X* and *Y* directions are

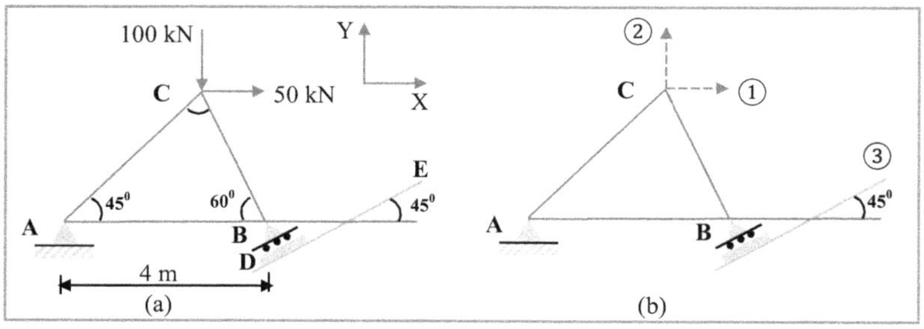

FIGURE 10.71 (a) Pin-jointed plane frame and (b) co-ordinates assigned to the DoF.

$$\Delta_{CX} = 0.849 \text{ mm}; \quad \Delta_{BX} = -0.568 \cdot \cos 45^\circ = -0.402 \text{ mm}; \quad \Delta_{CY} = -0.947 \text{ mm};$$

$$\Delta_{BY} = -0.568 \cdot \sin 45^\circ = -0.402 \text{ mm}; \quad \Delta_{AX} = \Delta_{AY} = 0$$

Final member forces: With the results of the displacement components, and using equation 10.29, the member forces can be calculated as presented below.

$$F_{AC} = \frac{AE}{L}\left[(\Delta_{CX} - \Delta_{AX})\cos\theta_{AC} + (\Delta_{CY} - \Delta_{AY})\sin\theta_{AC}\right]$$

$$= 100\left[(0.849 - 0)\cos 45^\circ + (-0.947 - 0)\sin 45^\circ\right]$$

$$= -6.92 \text{ kN (Compressive)}$$

$$F_{AB} = \frac{AE}{L}\left[(\Delta_{BX} - \Delta_{AX})\cos\theta_{AB} + (\Delta_{BY} - \Delta_{AY})\sin\theta_{AB}\right]$$

$$= 100\left[(-0.402 - 0)\cos 0 + (-0.402 - 0)\sin 0\right]$$

$$= -\ 40.2 \text{ kN (Compressive)}$$

$$F_{BC} = \frac{AE}{L}\left[(\Delta_{CX} - \Delta_{BX})\cos\theta_{BC} + (\Delta_{CY} - \Delta_{BY})\sin\theta_{BC}\right]$$

$$= 100\begin{bmatrix}(0.849 - (-0.402))\cos 120^\circ + (-0.947 - (-0.402)) \\ \sin 120^\circ\end{bmatrix}$$

$$= -109.69 \text{ kN (Compressive)}$$

Example 10.19

Analyse the pin-jointed plane frame shown in Figure 10.72a. The axial stiffness of each member is 70 kN/mm.

Solution

The joints *A* and *D* are hinged supports. The joints *B* and *C* can move horizontally as well as vertically. Therefore, the kinematic indeterminacy, i.e., DoF=4 and we designate them as Δ_1, Δ_2, Δ_3 and Δ_4, respectively. The co-ordinates ①, ②, ③, and ④ can be assigned to these DoF, as shown in Figure 10.72b.

The forces acting at these co-ordinates are

$$P_1 = 50 \text{ kN}; \quad P_2 = 0 \text{ kN}; \quad P_3 = 0; \quad P_4 = -100 \text{ kN}$$

The elements of the stiffness matrix can be calculated using Equation 10.28, as shown in Table 10.3.

$$K_{11} = \sum \frac{AE}{L}\cos^2\theta \qquad \text{(Consider only } AB, BC \text{ and } BD\text{)}$$

$$= 70[0 + 1 + 0.5] = 105$$

$$K_{22} = \sum \frac{AE}{L}\sin^2\theta \qquad \text{(Consider only } AB, BC \text{ and } BD\text{)}$$

$$= 70[0 + 1 + 0.5] = 105$$

$$K_{33} = \sum \frac{AE}{L}\cos^2\theta \qquad \text{(Consider only } BC, AC \text{ and } CD\text{)}$$

$$= 70[0 + 1 + 0.5] = 105$$

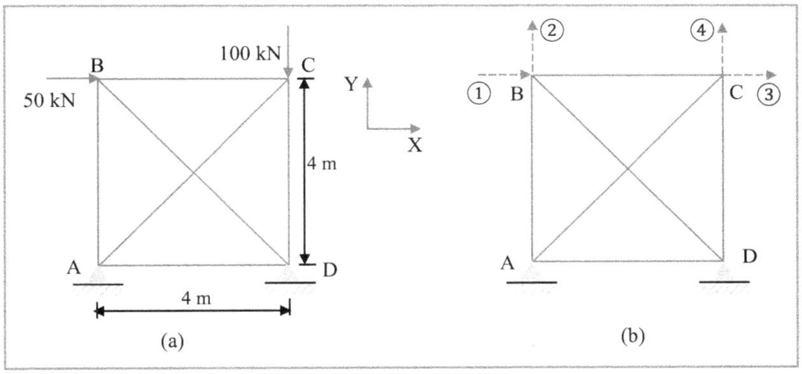

FIGURE 10.72 (a) Pin-jointed plane frame and (b) co-ordinates assigned to the DoF.

TABLE 10.3
Calculation of Stiffness Matrix Elements

Member	θ (Degrees)	$\cos\theta$	$\sin\theta$	$\cos^2\theta$	$\sin^2\theta$	$\sin\theta\cos\theta$
AB	90	0	1.0	0	1.0	0
BC	0	1	0.0	1	0	0
CD	270	0	−1.0	0	1.0	0
AC	45	0.71	0.71	0.5	0.5	0.5
BD	315	0.71	−0.71	0.5	0.5	-0.5
AD	0	1	0	1.0	0.0	0

$$K_{44} = \sum \frac{AE}{L}\sin^2\theta \qquad \text{(Consider only } BC, AC \text{ and } CD\text{)}$$

$$= 70[0+1+0.5] = 105$$

$$K_{21} = K_{12} = \sum \frac{AE}{L}\cos\theta\sin\theta \quad \text{(For members } AB, BC \text{ and } BD\text{)}$$

$$= 70[0+0-0.504] = -35.29$$

$$K_{13} = K_{31} = -\frac{AE}{L}\cos^2\theta \qquad \text{(For member } BC\text{)}$$

$$= -70\times 1 = -70$$

$$K_{23} = K_{32} = -\frac{AE}{L}\sin\theta\cdot\cos\theta \qquad \text{(For member } BC\text{)}$$

$$= 0$$

$$K_{14} = K_{41} = -\frac{AE}{L}\sin\theta\cdot\cos\theta \qquad \text{(For member } BC\text{)}$$

$$= 0$$

$$K_{34} = \sum \frac{AE}{L}\cos\theta\sin\theta \qquad \text{(For members } BC, AC \text{ and } CD\text{)}$$

$$= 70[0+0.504+0] = 35.29$$

$$K_{24} = K_{42} = -\frac{AE}{L}\sin^2\theta \qquad \text{(For member } BC\text{)}$$

$$= 0$$

Substituting these values into equation 10.15

$$\begin{bmatrix} 50 \\ 0 \\ 0 \\ -100 \end{bmatrix} = \begin{bmatrix} 105 & -35.29 & -70 & 0 \\ -35.29 & 105 & 0 & 0 \\ -70 & 0 & 105 & 35.29 \\ 0 & 0 & 35.29 & 105 \end{bmatrix} \begin{bmatrix} \Delta_1 \\ \Delta_2 \\ \Delta_3 \\ \Delta_4 \end{bmatrix}$$

By solving the above, we get

$$\begin{bmatrix} \Delta_1 \\ \Delta_2 \\ \Delta_3 \\ \Delta_4 \end{bmatrix} = \begin{bmatrix} 1.856 \text{ mm} \\ 0.624 \text{ mm} \\ 1.756 \text{ mm} \\ -1.542 \text{ mm} \end{bmatrix} = \begin{bmatrix} \Delta_{BX} \\ \Delta_{BY} \\ \Delta_{CX} \\ \Delta_{CY} \end{bmatrix}$$

Member forces: With the obtained results of Δ_1, Δ_2, Δ_3 and Δ_4, and using Equation 10.29, the member forces can be calculated as

$$F_{AB} = \frac{AE}{L}\left[(\Delta_{BX}-\Delta_{AX})\cos\theta_{AB} + (\Delta_{BY}-\Delta_{AY})\sin\theta_{AB}\right]$$

$$= 70\left[(1.856-0)\cos 90^\circ + (0.624-0)\sin 90^\circ\right]$$

$$= 43.68 \text{ kN (Tensile)}$$

$$F_{BC} = 70\left[(1.756-1.856)\cos 0^\circ + (-1.542-0.624)\sin 0^\circ\right]$$

$$= -7 \text{ kN (Tensile)}$$

$$F_{CD} = 70\left[(0-1.756)\cos 270^\circ + (0+1.542)\sin 270^\circ\right]$$

$$= -107.94 \text{ kN (Compressive)}$$

$$F_{AC} = 70\left[(1.756-0)\cos 45^\circ + (-1.542-0)\sin 45^\circ\right]$$

$$= 10.59 \text{ kN}$$

$$F_{BD} = 70\left[(0-1.856)\cos 315^\circ + (0-0.624)\sin 315^\circ\right]$$

$$= -60.98 \text{ kN}$$

$$F_{AD} = 70\left[(0-0)\cos 0^\circ + (0-0)\sin 0^\circ\right]$$

$$= 0$$

10.6 PROBLEMS FOR PRACTICE

10.6.1 Write the properties of the flexibility matrix and stiffness matrix.

10.6.2 Prove that the flexibility matrix and stiffness matrix are reciprocal to each other.

10.6.3 Explain the difference between the flexibility matrix method and displacement matrix method.

10.6.4 Generate the stiffness matrix for a general beam element

10.6.5 Analyse the continuous beam shown in Figure 10.73 using the stiffness matrix method. Draw the shear force and bending moment diagrams. Also draw the probable elastic curve.

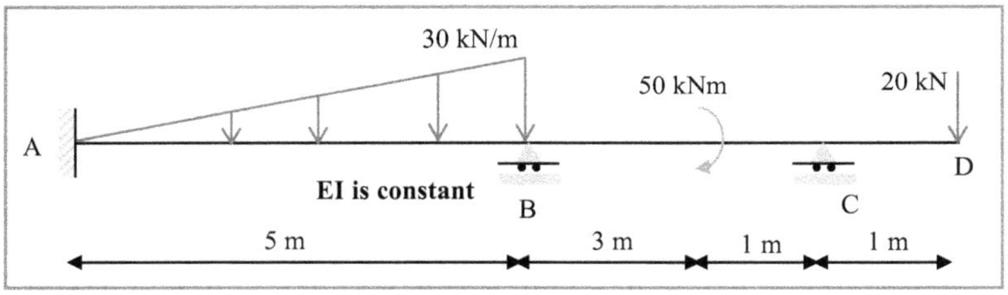

FIGURE 10.73

10.6.6 Analyse the three-span continuous beam shown in Figure 10.74 by neglecting the rotations at *A* and *D*. Construct the shear force and bending moment diagrams. Also, construct the probable deflected shape. Assume *EI* is constant.

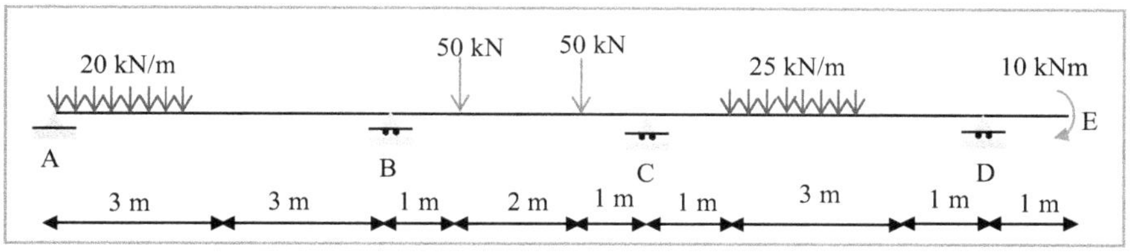

FIGURE 10.74

10.6.7 Analyse the three-span continuous beam shown in Figure 10.75 by taking the advantage of symmetry. Sketch the shear force, bending moment diagrams, and probable deflected shape.

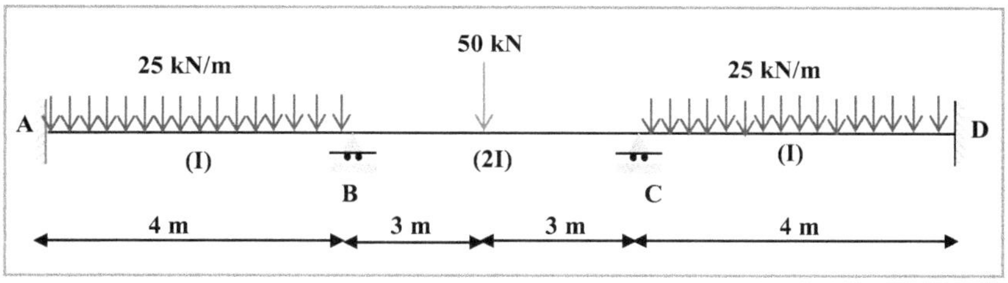

FIGURE 10.75

10.6.8 Calculate the support reactions for the non-prismatic beam shown in Figure 10.76. Also, construct the shear force and bending moment diagrams.

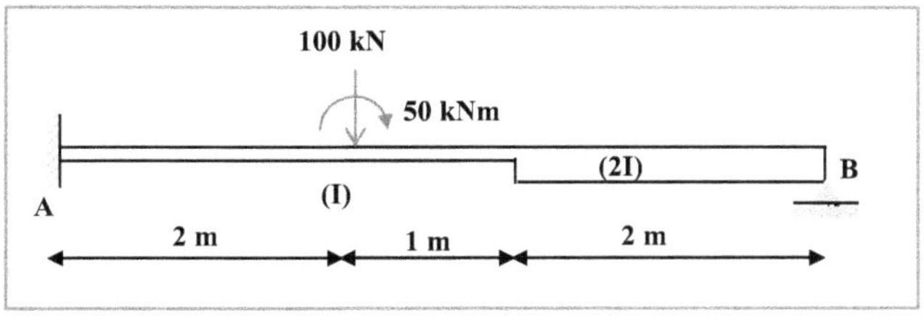

FIGURE 10.76

10.6.9 Analyse the beam shown in Figure 10.77 if support *A* rotates by 0.002 radians in the anti-clockwise direction and support *B* settles down by 10 mm. Assume $EI = 138,375$ kNm^2.

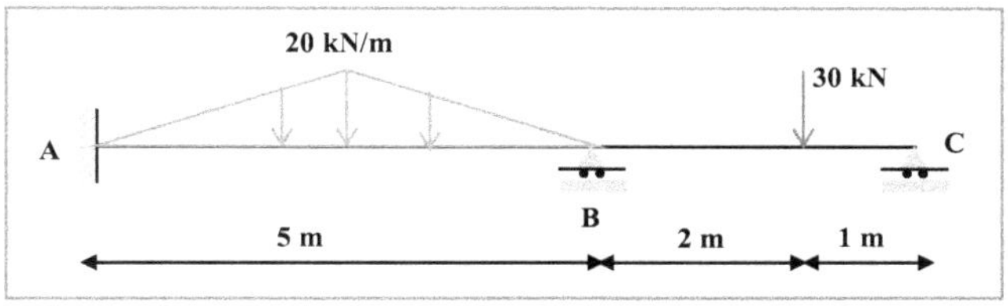

FIGURE 10.77

10.6.10 Determine the support reactions for the beam shown in Figure 10.78 if supports *B* and *C* settle down by 10 and 5 mm, respectively. Assume $EI = 14,660$ kNm^2.

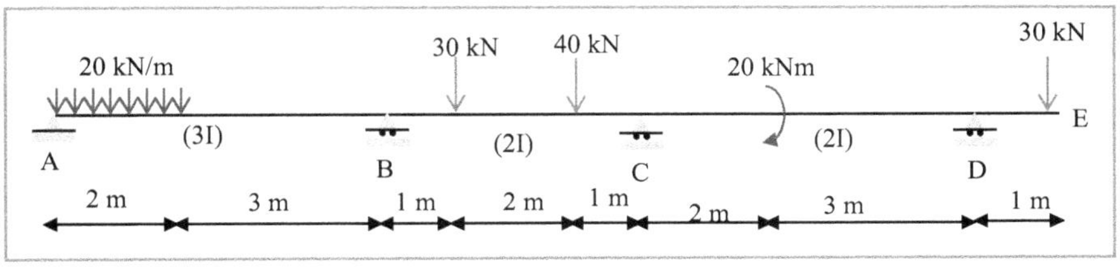

FIGURE 10.78

10.6.11 Analyse the symmetrical rigid-jointed plane frame shown in Figure 10.79 using the stiffness matrix method. Draw the axial force, shear force

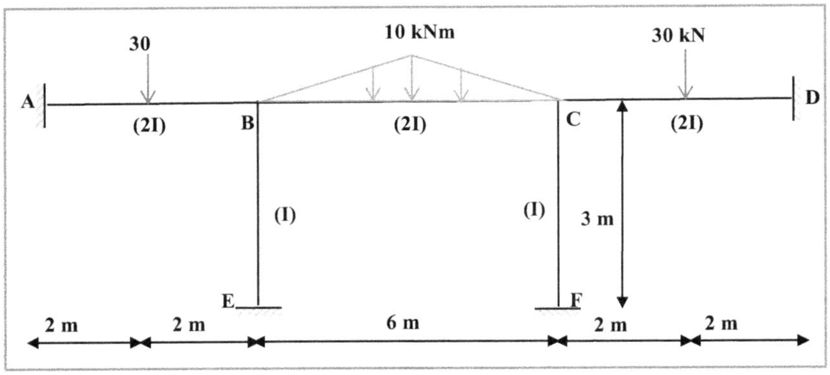

FIGURE 10.79

and bending moment diagrams. Also draw the probable deflected shape.

10.6.12 Determine the support reactions for the frame shown in Figure 10.80. Sketch the axial force,

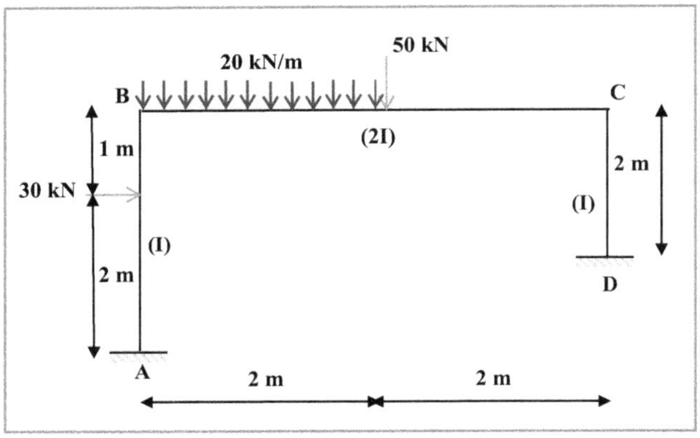

FIGURE 10.80

shear force, and bending moment diagrams. Also, plot the deflected shape.

10.6.13 An unsymmetrical frame with inclined legs is shown in Figure 10.81. Analyse the frame using the stiffness matrix method. Construct the shear

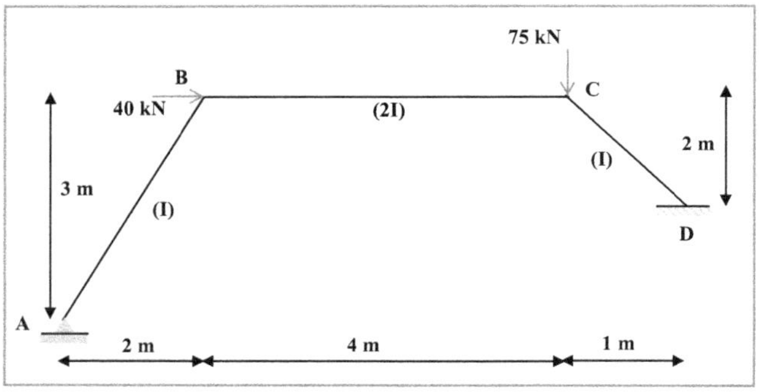

FIGURE 10.81

force and bending moment diagrams. Also, sketch the probable elastic curve.

10.6.14 Analyse the multi-storey frame shown in Figure 10.82. Sketch the shear force and bend-

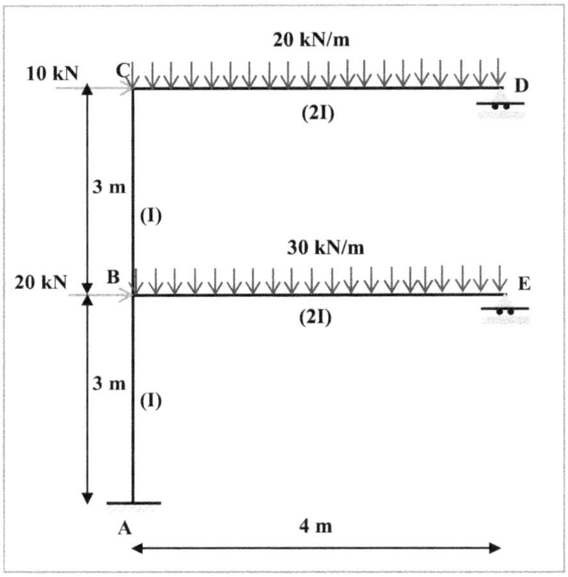

FIGURE 10.82

ing moment diagrams. Also, plot the probable deflected shape. Assume *EI* is constant.

10.6.15 Analyse the gable frame shown in Figure 10.83. Draw the axial force, shear force, and bending moment diagrams. Also, draw the probable deflected shape.

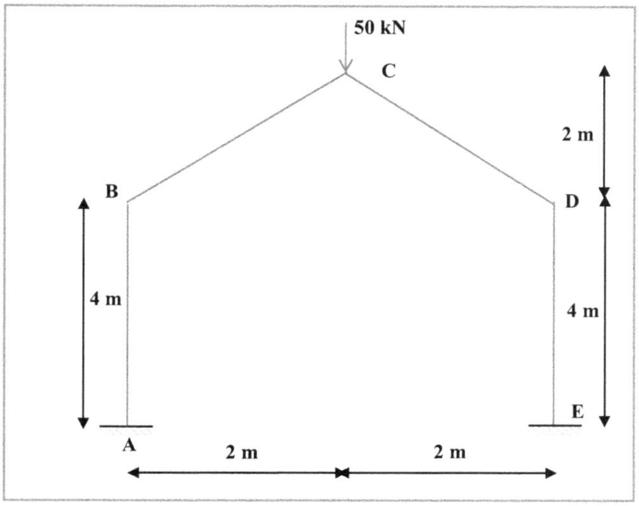

FIGURE 10.83

10.6.16 Figure 10.84 shows a triangular truss subjected to a load of 200 kN at joint *C*. Analyse the truss. The axial stiffness of each member is 100 kN/mm.

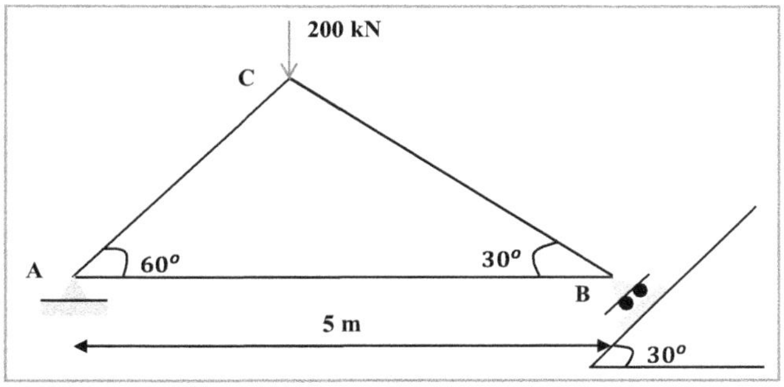

FIGURE 10.84

10.6.17 Analyse the pin-jointed plane frame shown in Figure 10.85, if the member *BD* was made too short by 0.01 m during fabrication. Assume axial rigidity *AE* equal to 8,000 kN.

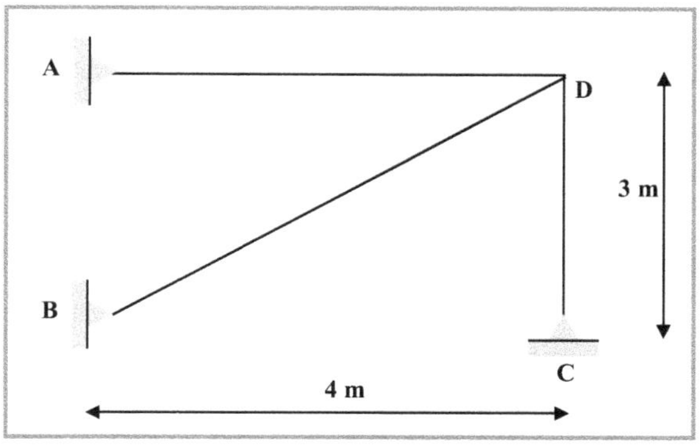

FIGURE 10.85

10.6.18 Determine the displacements at the joints *A* and *C* for the truss shown in Figure 10.86 and hence find the support reactions and member forces. Assume *AE* is constant.

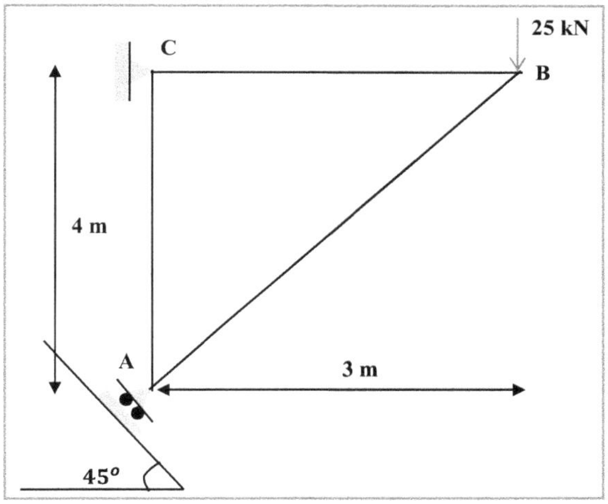

FIGURE 10.86

10.7 ANSWER KEY

Problem No.	Answer Key at Salient Locations
10.6.5	M_A=–42.30 kNm, M_B=–2.65 kNm, M_C=–20 kNm, R_A=32.93 kN
10.6.6	M_B=–23.84 kNm, M_C=–51.20 kNm, R_A=27.23 kN, R_B=55.93 kN, R_C=102.58 kN, R_D=29.26 kN
10.6.7	M_A=–32.01 kNm, M_B=–35.83 kNm, R_A=49.06 kN, R_B=75.93 kN
10.6.8	M_A=–92.58 kNm, R_B=31.48 kN, R_A=68.52 kN
10.6.9	M_A=166.72 kNm, M_{BA}=–M_{BA}=80.84 kNm, R_A=–24.51 kN, R_B=111.46 kN, R_C=–6.95 kN

(Continued)

(Continued)

Problem No.	Answer Key at Salient Locations
10.6.10	M_A=–97.95 kNm, M_{BA} = –M_{BC}=–21.79 kNm, R_A=64.96 kN, R_B=17.78 kN, R_D=36.55 kN
10.6.11	M_A=–14.46 kNm, M_E=0.69 kNm, R_A=14.46 kN, R_E=30.54 kN
10.6.12	R_A=53.38 kN, R_D=36.61 kN, M_A=–5.0 kNm, M_D=–22.18 kNm
10.6.13	R_A=–0.58 kN, R_D=75.58 kN, M_D=–1.63 kNm, M_{BA}=–M_{BC}=0.94 kNm
10.6.14	R_A=86.35 kN, R_E=68.17 kN, R_D=45.47 kN, M_A=–65.42 kNm, M_{CD}=–M_{CB}=37.22 kNm
10.6.15	M_A=12.21 kNm=M_E; R_A=R_E=25 kN; H_A=–H_E=6.96 kN
10.6.16	F_{AC}=–173 kN (C), F_{BC}=–100 kN (C), F_{AB}=58 kN (T)
10.6.17	F_{AD}=–7.41 kN (T), F_{BD}=–9.26 kN (C), F_{CD}=–5.56 kN (C)
10.6.18	Δ_{BV}=243.74/AE; Δ_{BH}=56.25/AE; F_{AB}=–31.25 kN (C); F_{BC}=18.75 kN (T); F_{AC}=6.25 kN (T);

Note: The answers are provided based on the analysis using structural analysis software. The results from the exact solution may differ slightly (decimals), but these differences are generally negligible.

FURTHER READING

Bhavikatti, S. S. (2011). *Matrix Methods of Structural Analysis*, I.K. International Publishing House Pvt. Limited. ISBN 9789381141359, 9381141355.

Derucher, K., Putcha, C. and Kim, U. (2013). *Indeterminate Structural Analysis*, The Edwin Mellen Press, Newyark, ISBN 978-0-7734-4470-6.

Hibbeler, R. C. (2022). *Structural Analysis*, Pearson Education (Singapore) Pte. Ltd., Delhi, ISBN 978-9354497841.

Leet, K., Uang, C.-M. and Lanning, J. (2020). *Fundamentals of Structural Analysis* (6th E), McGraw Hill Education Pvt Ltd, ISBN (13) 9781260477245.

Pandit, G. S. and Gupta, S. P. (2008). *Structural Analysis: A Matrix Approach*, McGraw Hill Education, New Delhi, ISBN 978-0070667358.

Reddy, C. S. (2011). *Basic Structural Analysis* (3rd E), Tata McGraw-Hill Education Pvt Ltd, New Delhi, ISBN (13) 978-0-07-070276-9.

Sahoo, D. R. (2021). *Indeterminate Structural Analysis*, PHI Learning Pvt Ltd, New Delhi, ISBN 9789354432958.

Shah, H. J and Junarkar, S. B. (2015). *Mechanics of Structures – Vol. II*, (24th E), Charotar Publishing House, Anand. ISBN 9789385039027.

Vijayanand, M., Muthu, K. U., Narendra, H. and Janardhana, M. (2019). *Indeterminate Structural Analysis*, Dreamtech Press Publisher, India, ISBN 978-9389307177.

Index

A

analogous column, 119–143
anti-symmetric, 73, 116, 145, 186, 188, 191, 193, 195, 236, 242, 244
applied loads, 2, 18, 20, 24, 76, 119, 148, 325, 326, 328, 334, 349, 353, 365, 367, 368, 374, 388
approximate method, 308
arches, 118
area moment multiplication, 43, 46, 50, 54, 58–60, 64, 320, 326–328, 331, 332, 334, 338, 340, 341, 343, 347
axial, 11, 50, 65, 72, 73, 75, 89–91, 93–98, 100, 101, 106–108, 116, 117, 120, 130, 151, 168, 175, 178, 191, 195, 239, 295, 332, 333, 335, 337, 339, 342, 344, 348, 354, 357, 359, 362–364, 403, 408, 409, 412, 413
axial displacement, 359, 362, 363
axial force, 11, 50, 72, 75, 89, 90, 93, 95–98, 100, 101, 106–107, 116, 117, 130, 175, 191, 332, 333, 335, 337, 339, 342, 344, 357, 359, 412
axial rigidity, 73, 108, 348, 354, 357, 403, 413
axial stiffness, 359, 403, 406, 409, 413
axial, transverse and rotational degrees of freedom, 359

B

basic, 2–4, 31–34, 36, 38, 39, 43, 46, 47, 49–52, 62, 68, 75 –77, 89, 96, 103, 106, 107, 109, 111, 113, 117–129, 131, 133, 134 –140, 145, 146, 201, 265, 312 –315, 317, 325, 330 –332, 334, 336 –338, 340, 341, 343, 344, 346, 347, 349, 351, 353, 355, 358, 414
basic beam, 3, 32, 33, 36, 39, 43, 46, 47, 49, 76, 119, 120
basic determinate, 32, 33, 77, 96, 118–129, 131, 133, 134, 136, 139, 140, 313–315, 317, 325, 330–332, 336–338, 341, 343, 346, 347, 351, 353, 355.
basic determinate beam, 32, 119, 122, 123, 125, 126, 313–315, 317, 325
basic determinate structure, 32, 33, 77, 96, 118, 120, 121, 124, 134, 313, 331, 332, 341, 343, 346
basic structure, 2, 32, 33, 50, 51, 136–138, 314, 315, 338, 340
bays, 11
beam, 1–3, 11, 12, 18–24, 26–29, 31–34, 36–49, 72, 75–78, 80–89, 103–106, 115, 116, 118–126, 136, 138, 142–144, 146–167, 175, 191, 193, 197–199, 202–204, 208–210, 213–216, 218, 219, 221, 223, 225–227, 229, 236, 239, 241, 245, 262, 263, 266–271, 274–284, 296, 298, 308, 309, 313–329, 332, 356, 357, 359–365, 367, 369–376, 378–379, 410, 411
beam element, 149, 191, 197, 202, 262, 266, 267, 356, 359, 362, 363, 410
bending moment, 1, 18, 20–31, 33, 34, 36, 38–48, 50, 52, 54, 55, 57, 59–61, 63, 65, 71, –73, 75, 77, 78, 80–82, 84 –104, 106, 107, 116 –119, 121, 123, 124, 126, 128, 130, 132, 134, 135, 137, 139, 144, 145, 151–156, 158 –162, 164, 166–169, 171–173, 175, 176, 180, 181, 183, 186, 187, 189, 193, 196, 199, 207–209, 212–214, 216, 218, 231, 233, 235–239, 242, 243, 247, 248, 251, 252, 256, 257, 261–264, 269–273, 275–286, 288–294, 297, 298, 300, 302, 304, 306–311, 316, 318–323, 325–328, 330, 332–335, 337, 339, 341–344, 346, 348, 349, 356, 357, 365, 366, 368, 369, 371, 373, 375–380, 382–389, 391–394, 397, 398, 401–403, 410–412
bending moment diagram, 18, 20–31, 33, 34, 36, 38, 48, 50, 52, 54, 55, 57, 59, 61, 63, 65, 71–73, 77, 78, 80, 82, 84–86, 88–91, 93–96, 98, 100–104, 106–107, 116–119, 121, 123, 124, 126, 128, 130, 132, 134, 135, 137, 139, 144, 145, 151–156, 158–162, 164, 166–169, 171–173, 175–176, 180, 181, 183, 186, 187, 189–193, 196–199, 207, 209, 212–214, 216, 218–231, 233, 235–239, 242–243, 247, 248, 251, 252, 256, 257, 261–264, 269–273, 275–286, 288–294, 297–298, 300, 302, 304, 306–311, 316, 318–321, 323, 325–328, 330, 332–335, 337, 339, 341–344, 346, 348–349, 356–357, 365–366, 368–369, 371, 373, 375–380, 382–389, 391–394, 397–398, 401–403, 410–412
bottom chord, 111, 115, 350, 352
boundary conditions, 20, 85
box culvert, 73, 134
box girder bridges, 134

C

cantilever, 3, 33, 35–37, 57, 63, 72, 77, 96, 120 –122, 125, 128, 134, 139, 144, 149, 156, 191, 204, 213, 216, 267, 275
cantilever beam, 3, 33, 35–37, 77, 120–122, 125, 144, 204
carryover factor, 118, 142–144, 203, 210
carryover moment, 203
Castigliano's, 75–77, 89, 101, 106, 114
cells, 04, 11
centroid, 118, 119, 124, 126, 129–132, 134, 135, 139, 140, 231, 287
centroidal axis, 126, 233
centroidal distances, 18, 20, 126
chord rotation, 19, 26, 28, 45, 146, 148, 149, 168, 175, 177, 178, 184, 236, 238, 295.
Clapeyron's, 18–23
clockwise, 45–46, 59, 63, 72, 82, 101, 116, 127, 129, 135, 144, 146–149, 151–152, 154, 160, 178, 180, 184, 186, 189, 198, 202–204, 206–208, 210, 223, 226, 262–263, 266, 275, 279, 295, 297, 309, 328, 340, 357, 362
closed loops, 04, 06
collapse, 02
column, 1, 3, 11, 71, 103–105, 109–110, 112, 114, 118–145, 151, 175, 177, 178, 181, 182, 185, 188, 191, 193, 200, 208–210, 229, 236, 241, 244–246, 249, 250, 252, 253, 256–258, 293, 295, 296, 300, 304, 307, 311, 314, 315, 317, 318, 328, 352, 353, 356, 362–364, 367–369, 372, 375, 378, 380–382, 385–388, 390, 396, 397, 399–401
column analogy, 3, 118, 119, 121–135, 137, 139, 141–145
column analogy method, 118, 119, 121–135, 137, 139, 141–145
column element, 295, 314, 315, 317, 318, 328, 362–364, 367–369, 372, 375, 378, 380–382, 385, 387–388, 390, 396–397, 399–401
compatibility, 1–4, 11, 18, 19, 32, 34, 36, 38, 39, 42, 44–52, 54, 57, 59, 64–70, 77, 78, 80, 82, 84–86, 88–90, 92, 94, 96–98, 101, 103, 106, 109, 111, 113, 115, 202, 313
compatibility conditions, 1, 3, 11, 32, 39, 52, 54, 59, 78, 82, 86, 89, 90, 92, 94, 96–98, 111, 113, 115, 313
compatibility method, 2, 313
compression, 49, 66, 69, 88, 129, 132, 133, 135, 139, 141, 195, 348, 352, 407, 408
compressive, 70, 404, 409, 410
compressive force, 70, 404
concentrated force, 75
concentrated load, 41
concentrated moment, 154, 204, 366
conjugate beam, 146–148, 202–204
consistent deformation, 3, 32–39, 41–43, 45–47, 49–55, 57, 59, 61, 63, 65–69, 71–73, 77, 78, 82, 313
continuous beam, 18–24, 26–29, 31, 33, 38–40, 42, 43, 75, 78, 80, 82, 84, 115, 116, 146, 151, 153–164, 168, 197–199, 207, 210.213, 215, 216, 218, 219, 221, 223, 225–227, 238, 263, 269–271, 274–280, 282–284, 308, 309, 314, 316, 318–321, 323, 327, 329, 356, 364, 367, 368, 370, 371, 375, 376, 410, 411
contra flexure, 152, 191, 203, 209
co-ordinates, 365
coplanar structures, 4
Cross, Hardy, 118, 202
cyclic operation, 272

D

deflected shape, 34, 38, 40, 42, 44, 47, 48, 50, 52, 54, 57, 59, 61, 65, 71–73, 78, 98, 100–103, 116, 153, 154, 156, 158, 160–162, 164, 168, 169, 171–173, 176, 180, 181, 183, 186, 187, 191, 193, 194, 197, 200, 203–206, 236, 315–321, 325–328, 330, 332–337, 339, 341, 342.344, 346, 357, 362–264, 365, 366, 368, 369, 371, 375–377, 379, 382, 384–386, 388, 391, 398, 401–403, 411, 412
deflection, 1–4, 34–36, 47, 50, 57, 72, 75, 76, 146, 191, 193, 236, 241, 245, 278, 315
deflection curve, 315
deformation, 3, 11, 32–37, 39, 41–43, 45–47, 49–55, 57, 59, 61, 63, 65–69, 71–73, 77–78, 82, 151, 168, 178, 295
degree of freedom, 11, 178, 202, 231, 266, 295
degree of static and kinematic indeterminacy, 313
degrees of freedom, 11, 146, 202, 214, 238, 266, 268, 298, 300, 304, 359
determinate, 1–8, 10, 17, 19, 21, 23, 26, 31–34, 36, 38, 43–44, 46–47, 52, 55, 57, 61, 63, 65–69, 71, 73, 75–78, 80, 86, 88, 90, 92, 96, 98, 106–110, 112, 115, 118–129, 131, 133–134, 136, 139, 140, 145–146, 201–202, 207, 223, 228, 265, 268, 285, 312–318, 325, 330–332, 334, 336–338, 341–349, 351–355, 358, 360–362, 414
determinate beam, 32–33, 36, 44, 46, 71, 75, 76, 119, 121–126, 202, 207, 228, 268, 285, 313–315, 317–318, 325, 361
determinate structure, 1–2, 4, 17, 19, 31–34, 52, 73, 75, 77, 96, 118, 120–121, 124, 134, 145, 201, 265, 313, 331–334, 341, 343, 346, 360, 362
displacement, 1–9, 11–13, 18, 21, 32–34, 36, 38, 39, 48, 50, 52, 59, 65–71, 74–76, 78, 81, 146–149, 151, 171, 175, 177, 178, 180, 184, 191–192, 202, 229, 262, 266, 293, 295–301, 303–305, 307, 308, 313–315, 317, 318, 320, 322, 326–328, 331, 332, 334, 338, 341, 343, 349, 351, 353, 355, 359–365, 367–370, 372, 375, 377, 378, 380–382, 385, 387, 390, 392, 396, 397, 399, 403–410, 413
displacement contributions, 293, 296–298, 303, 304
displacement method, 1, 3, 4, 146, 202, 262, 266, 308, 313, 360, 362
distribution factor, 204–208, 210, 211, 213, 215, 217, 219, 221, 223, 225, 228, 230–232, 235–237, 240–244, 246, 249, 251, 253, 258, 262, 268
DoF, 11, 146, 149, 151, 152, 154, 155, 157, 166, 175, 177, 191, 197.270, 354, 359, 360, 362, 366, 367, 371, 378, 379, 387, 388, 393, 395, 397, 399, 403, 406–409

E

eccentricity, 119–126
eccentric load, 118, 121, 139
elastic centre, 139
elastic curve, 18–20, 23, 31, 34, 39, 52, 55, 72, 95, 116, 117, 146, 149, 151, 152, 155, 164, 171, 173, 175, 178, 180, 183, 186, 189–191, 196, 219, 231, 256, 257, 261, 339, 378, 383, 392, 397, 412
elastic modulus, 148
elements, 328, 362, 363, 380, 396
elements of the stiffness matrix, 362–364, 367–369, 372, 375, 378, 385, 387, 390, 396, 397, 399, 406, 407, 409
elongation, 70, 405
end moment, 18, 19, 22, 24, 25, 27, 40, 104, 120–125, 128, 130, 132, 134–136, 138, 139, 144, 146–167, 169–173, 175, 178, 180, 182, 184, 186, 189, 191, 193, 195–197, 202, 203, 205–217, 219–238, 240–242, 244–256, 258–262, 266–287, 289–293, 295–301, 303–305, 307, 308, 316, 319, 321, 362, 364–370, 372–376, 378–382, 384, 385, 387, 388, 390, 391, 393, 395, 397, 403
end shears, 47, 151, 154, 155, 158, 160, 161, 178, 189, 209, 212, 218–221, 223, 228, 230, 235, 256, 261, 270, 272, 275, 276, 278, 279, 281, 283, 285, 286, 289–291, 293, 318, 320, 362, 365, 368, 370.372, 375, 378, 382
energy method, 3, 75, 77, 79, 81, 83, 85, 87, 89, 91, 93, 95–97, 99, 101, 103, 105, 107, 109, 111, 113, 115–117, 318, 330, 348
equilibrium, 1–6, 8, 9, 11, 20, 23, 25, 28, 29, 32, 34, 38, 39, 42, 44, 47, 50, 52, 54, 57, 59, 65, 77, 84, 89, 119, 146, 147, 149, 151, 152, 154, 155, 157, 158, 160, 161, 163, 165, 166, 168, 170, 172, 174, 175, 177, 179, 180, 182, 185, 188–191, 193, 197, 202, 204, 207, 209, 212, 219, 221, 223, 235, 238, 241, 245, 247, 249, 251, 253, 256, 258, 261, 266, 267, 269, 270, 276, 278, 286, 291, 296, 314, 316, 318, 320, 325, 330, 332, 333, 336, 339.341, 342, 346, 354, 360, 362, 365, 368, 370, 372, 375, 378, 382, 391, 397
equilibrium equations, 1–6, 9, 20, 23, 25, 28, 29, 34, 38, 39, 42, 44, 47, 50, 52, 54, 57, 59, 65, 77, 84, 89, 146, 149, 151, 152, 154, 155, 157, 158, 160, 161, 163, 165, 166, 170, 172, 174, 175, 177, 179, 182, 185, 188–190, 193, 197, 202, 207, 209, 212, 219, 221, 223, 235, 237, 251, 256, 261, 266, 269, 270, 276, 278, 286, 291, 314, 316, 318, 320, 325, 330, 332, 333, 336, 339, 341, 342, 346, 354, 362, 365, 368, 370, 372, 375, 378, 382, 391, 397
equilibrium method, 3, 360, 362
exact method, 266, 308
expansion, 73, 115, 351, 352, 403
expansion or contraction, 403
extensibility of the members, 11, 379
extensible, 11–13, 379
external displacements matrix, 314
external forces, 10, 70, 89, 360–362, 371, 375, 385, 390
external indeterminacy, 4, 5, 109, 111
external loads, 18–21, 31, 34, 36, 38–41, 43–48, 50–54, 56–59, 62, 65–69, 72, 73, 75–78, 80, 82, 85, 86, 89, 90, 92, 94, 98, 99, 101, 106, 107, 109, 111, 113–116, 134, 144, 147–149, 151, 207, 223, 224, 226, 268, 281, 283, 309, 313, 315, 317, 324, 327, 330–332, 334, 338, 340, 341, 343, 345, 347–352, 354–357, 362, 364, 372, 374, 378, 381, 387, 396, 397, 399, 403, 406
external stability, 8

F

fabrication errors, 2
far end has a guided, 268
far end is fixed, 202–204, 207, 267, 268, 359
far end is hinged, 204, 207, 267, 268, 360
far end is pinned, 203, 267
fictitious load, 76
final end moments, 120–125, 128, 130, 132, 134–136, 138, 139, 152, 154, 155, 158, 160, 161, 164, 166, 167, 171, 172, 175, 180, 182, 186, 189, 191, 193, 197, 202, 206, 208–212, 214, 216, 217, 219–221, 223–226, 228, 230, 232, 233, 235, 237, 238, 240–242, 245–247, 250, 251, 254–256, 258, 260–262, 266, 269, 270, 272–276, 278–281, 283, 285–287, 289–291, 293, 298–301, 304, 305, 307, 374, 375, 378, 379, 387, 391
final end reactions, 103, 332, 333, 336, 339, 342, 368
final end shears, 154, 189, 209, 212, 219–221, 223, 228, 235, 256, 261, 270, 278, 286, 289, 318, 320, 365, 375
final member end moments, 202, 207, 209, 221, 223, 235, 238, 269, 278, 281, 283, 365, 368, 370, 372, 375, 378, 381, 382, 385, 391, 393, 397
first column, 314, 315, 317, 318, 328, 362, 363, 367, 368, 372, 375, 378, 380, 385–388, 390, 396, 399
first column elements, 314, 315, 317, 318, 328, 362, 363, 367, 368, 372, 375, 378, 380, 385, 387, 388, 390, 396, 399
fixed, 1, 2, 11, 20, 23, 31–34, 57, 63, 88, 96, 97, 101, 104, 118–125, 128, 139, 142–144, 146–151, 153–158, 160, 162–166, 169–173, 178, 180, 184, 186, 189, 191, 193, 195–200, 202–211, 213, 215–227, 229–234, 236–238, 240–242, 244, 245, 247–250, 252, 253, 255, 256, 258, 259, 263, 266–273, 275, 276, 278, 279, 281, 285, 287, 290–292, 295–299, 301, 303, 304, 309, 313, 325, 328, 346, 359, 360, 362, 364, 367, 368, 371–375, 378–380, 382, 384, 387, 388, 390, 391, 393, 395, 397
fixed beam, 1, 2, 31–34, 118–124, 142–144, 147, 148, 166, 193, 197, 198, 229, 236, 313, 378

fixed end moment, 104, 119, 121, 122, 124, 125, 144, 148–151, 153–158, 160, 162–166, 169–173, 178, 180, 184, 186, 189, 191, 193, 195, 196, 222, 223, 266, 268–270, 272, 273, 275, 276, 278, 279, 281–285, 287, 290–292, 295–299, 301, 303, 304, 362, 364, 367, 368, 372–375, 378–380, 382, 384, 387, 388, 390, 391, 393, 395, 397
fixed moments, 153, 158, 210, 215, 226, 236, 244, 368
fixed support, 20, 57, 97, 144, 147, 151, 153, 191, 198, 206, 207, 210, 225, 247, 267, 268, 273, 309, 325, 328, 346
flexibility, 3, 33, 34, 38, 39, 43, 45, 46, 50–52, 54, 58–60, 66–68, 76, 110–114, 191, 313–323, 325, 327–341, 343, 345–351, 353, 355–357, 359, 407, 410
flexibility coefficient, 3, 33, 34, 38, 39, 43, 45, 46, 50–52, 54, 58–60, 64, 66–68, 76.110, 112–114, 313–315, 322, 334, 338, 340, 341, 343
flexibility matrix, 3, 313–321, 323, 325, 327–341, 343, 345–351, 353, 355–357, 359, 407, 410
flexibility matrix method, 3, 313–319, 321, 323, 325, 327, 329–333, 335–337, 339, 341, 343, 345, 347–351, 353, 355–357, 359, 407, 410
flexibility method, 3, 313
flexural rigidity, 1, 20, 23, 26, 47, 75, 88, 125, 144, 146, 197, 203, 207, 216, 219, 226, 236, 263, 266, 283, 295, 300, 362
flexural stiffness, 202–204, 262, 360
flexure, 50, 152, 191, 203, 209
force and displacement methods, 313
force-displacement, 1–3, 146, 147, 361
force-displacement equations, 1, 146
force equilibrium equations, 266
force method, 1–4, 32, 77, 313
frames, 4–16, 32, 50, 65, 75, 89, 105, 106, 118, 126, 134, 139, 146, 149, 168, 169, 175, 177, 191, 197, 202, 228, 238, 266, 268, 285, 293, 304, 313, 314, 330, 348, 359, 362, 364, 379, 403, 404, 406
frames without side sway, 168, 293
free body, 4, 20–22, 24–30, 34, 38, 40, 42, 44, 47, 48, 50, 52, 54, 55, 57, 59, 60, 63, 65, 77, 79–95, 97–104, 106, 107, 109, 111, 128, 130, 132, 134, 151–173, 175, 176, 179–183, 185–192, 195–197, 207, 209, 211, 212, 214.215, 218–225, 227, 229–231, 233, 235, 237, 239, 241–243, 245–254, 257–259, 261, 262, 269–271, 273, 275–277, 279–292, 294, 298, 300, 302, 304, 306, 308, 314, 316, 318–321, 323–328, 330, 332, 333, 335–337, 339, 340, 342, 344, 346, 365, 366, 368–373, 375–378, 380, 382–384, 388, 389, 391–395, 397, 398, 401–403
free body diagram, 4, 20–22, 24–30, 34, 38, 40, 42, 44, 47, 48, 50, 52, 54, 55, 57, 59, 60, 63, 65, 77, 79–95, 97–104, 106, 107, 109, 111, 128, 130, 132, 134, 151–173, 175, 176, 179–183, 185–192, 195, 197, 207, 209, 211, 212, 214, 215, 218–225, 227, 229–231, 233, 235, 237, 241, –243, 245–254, 257–259, 261, 262, 269–271, 273, 275–277, 279–292, 294, 298, 300, 302, 304, 306, 308, 316, 318–320, 323, 326–328, 330, 332, 333, 336, 337, 339, 340, 342, 344, 346, 364, 366, 368–373, 375, 377, 378, 380, 382–384, 388, 389, 391–395, 397, 398, 401, 402
free thermal expansion, 115, 352

G

Gauss-Seidel method, 266
general beam element, 267, 356, 362, 410
gravity load, 103, 197, 199
gravity loading, 103, 199
ground floor, 304
guided, 85, 148, 149, 158, 171, 191, 193, 198, 204, 207, 218, 219, 221, 229, 231, 235, 236, 263, 267, 268, 278, 279, 309, 320, 346, 371
guided fixed, 191, 263, 346, 367
guided fixed supports, 191, 346
guided support, 85, 149, 158, 159, 171, 191, 193, 198, 204, 207, 218, 219, 221, 229, 231, 235, 236, 267, 268, 278, 279, 309, 320, 371

H

hinge, 2, 5, 6, 26, 32, 116, 125, 136, 148, 149, 153, 162–164, 169, 198, 207, 248, 268, 272, 275, 285, 314, 316, 318, 320, 323, 341, 375, 376
hinged or pinned, 266
hinged or supported on a roller, 268
hinged support, 103, 126, 130, 132, 144, 153, 207, 210, 211, 214–216.247, 273, 309, 328, 336, 346, 349, 351, 378, 387, 409
hinge supports, 248, 320
hogging, 104, 117, 120, 122, 123
hogging moment, 104
horizontal reaction, 32, 34, 59, 65, 90, 103–105, 108, 113, 346, 349, 351
hyperstatic structures, 4

I

imaginary span, 20, 23, 24
improper constraints, 9
indeterminacy, 2–6, 11–14, 16, 20, 21, 26, 32–34, 39, 43, 45–47, 50, 52, 54, 57, 59, 65, 68, 70, 75–78, 82, 84, 86, 88–90, 92, 94, 96, 98, 101, 103, 106, 108, 109, 111, 113, 118, 146, 148, 149, 151, 153, 154, 156, 158, 160, 162, 164, 169, 171, 173, 178, 180, 184, 189, 192, 193, 214, 219, 230, 231, 236, 239, 256, 266, 268–270, 272, 275, 276, 278, 285, 290, 295, 297, 298, 304, 313, 314, 316, 320, 327, 328, 330, 332, 334, 336, 340, 341, 343, 346, 349, 353, 354, 360, 362, 364, 366, 368, 371, 373, 375, 378, 379, 382, 384, 387, 388, 391, 393, 397, 406–409
indeterminate, 1, 2, 4–8, 10, 11, 14, 16, 17, 19, 21, 23, 31–34, 36, 38, 47, 55, 57, 61, 63, 65–69, 71, 73, 75, 77, 78, 86, 88, 90, 92, 96, 98, 106–110, 112, 115, 118, 120, 121, 124, 126, 133, 134, 145, 146, 201, 202, 207, 223, 228, 265, 268, 285, 312–314, 316, 318, 325, 337, 344, 349, 351–354, 358, 360–362, 378, 414
indeterminate beam, 33, 36, 71, 75, 77, 124, 134, 202, 207, 228, 268, 285, 313, 314, 318, 361
indeterminate structures, 1, 2, 4, 17, 19, 31, 32, 73, 118, 145, 201, 265, 362
inextensibility, 11
inextensible, 11–13, 16, 17, 50, 364, 379
inflection, 191
instability, 32
interior supports, 20, 205, 207
internal forces, 1, 2, 5, 22, 33, 50, 59, 65, 74, 77, 93, 95, 98, 101, 115, 132, 134, 207, 269, 352
internal hinge, 5, 6, 12, 26, 32, 116, 162, 163, 198, 320, 323, 375, 376.
internal indeterminacy, 4, 5, 109, 111
internal moments, 18, 33, 149
internal strain energy, 75, 76

J

joint, 4–8, 10–16, 32, 50, 52, 54, 55, 57, 59, 61, 63, 65–68, 72–73, 75, 89–92, 94, 98–99, 102, 103, 105–106, 108–111, 113, 115–116, 126, 128–131, 134, 139, 144–146, 148–149, 151, 152, 154, 155, 157, 158, 160, 161, 163, 165, 166, 168–170, 172, 174, 175, 177, 179–182, 184–186, 188, 190, 191, 193, 195–197, 199, 202, 204–217, 219–221, 223–226, 228.230–233, 235–240, 242, 244, 245, 247, 249–256, 258, 260, 262, 264, 266–270, 272–274, 276, 278, 280, 281, 283, 285–287, 289–291, 293, 295–301, 303–307, 310, 311, 313, 330, 332, 334–337, 341, 343, 348, 350–354, 357, 359–362, 364, 366, 379–381, 387–390, 393, 395, 397, 399, 403–409, 412, 413
joint displacements, 11, 151, 202, 362
joint moment equilibrium, 170, 177, 267

joint or support displacements, 360
joint rotations, 202, 266, 295
joint stiffness, 204, 262

K

Kani's distribution, 297–299, 303, 306
Kani's distribution cycles, 298, 299
Kani's format, 267, 295
Kani's iterations, 273, 296
Kani's iteration table, 296
Kani's method, 4, 266–271, 273, 275–279, 281, 283, 285–287, 289–291, 293, 295, 297–301, 303–305, 307–311
Kani's rotational cycles, 276, 278, 281, 283, 285, 286, 289–291, 293
kinematic degree, 266
kinematic indeterminacy, 3, 11–13, 16, 146, 148, 149, 151, 153, 154, 156, 158, 160, 162, 164, 169, 171, 173, 178, 180, 184, 189, 192, 193, 214, 219, 230, 231, 236.239, 256, 266, 268–270, 272, 275, 276, 278, 285, 290, 295, 297, 298, 304, 313, 360, 362, 364, 366, 368, 371, 373, 375, 378, 379, 382, 384, 387, 388, 391, 393, 397, 406–409

L

lack of fit, 106, 352
lateral displacement, 266, 293, 295
lateral sway, 151, 173, 175, 186, 191, 197, 202, 228, 238.239, 247, 293, 295, 379, 388
least work, 75–87, 89–99, 101, 103, 105–109, 111, 113, 115–117
linear displacements, 146, 147
linear equations, 266
linearly elastic, 3, 75, 76, 146, 148

M

Maney, George, 146
Maxwell, 32, 39, 76
Maxwell's reciprocal, 39, 76
m1 diagram, 33, 47, 53, 56, 59, 61, 62, 64, 320, 322, 326, 328, 329, 331, 332, 334, 336, 338, 341, 343, 345, 348
member, 4, 6, 8, 10–13, 16, 17, 32, 50, 65–71, 73–75, 77, 89, 106–115, 117–118, 142, 146–148, 151, 158, 160, 161, 164, 168, 171, 172, 175, 177, 178, 180, 183, 189, 191, 197, 202, 204–214, 216, 217, 219–221, 223–226, 228, 230–233, 235, 238, 240, 242, 244, 245, 247, 248, 250, 251, 254, 255, 258, 260–276, 278–281, 283, 285–287, 289–291, 293, 296–299, 301, 303–308, 314, 348–357, 359–362, 365, 368, 370, 372, 374, 375, 378, 379, 381, 382, 385, 388, 391, 393, 397, 403–410, 413
member end forces, 71, 89, 114, 115, 164, 166, 168, 171, 172, 180, 183, 186, 189, 191, 197, 231, 236, 247, 307, 314, 350, 352, 357, 388, 403
member end moments, 146, 148, 151, 202, 207, 209, 221, 223, 235, 238, 247, 251, 260, 261, 269, 270, 276, 278, 281, 283, 291, 308, 362, 365, 368, 370, 372, 374, 375, 378, 381, 382, 385, 387, 391, 393, 397, 403
member end shear, 151, 158, 160, 161, 178, 209, 218, 219, 221, 223, 230, 235, 261, 270, 272, 275, 276, 278, 279, 281, 283, 285, 286, 289–291, 293, 362, 365, 368, 370, 372, 378, 379, 382
member force, 6, 10, 32, 65–71, 106, 108–115, 117, 352–356, 405–407, 409, 410, 413
method of joints, 66–68, 106, 108, 109, 111, 113
ML diagram, 33, 52–54, 56, 58, 59, 61, 62, 119, 121–129, 131–134, 136, 138, 140, 322, 326, 329, 331, 332, 334, 336, 338, 343, 345, 348
modified fixed end moments, 149, 156, 158, 162, 191, 272, 278, 283, 373, 379
modified stiffness, 149, 153, 163, 169, 191, 193, 210, 211, 213, 216, 219, 221, 228, 229, 231, 235, 236, 248, 272, 278, 285, 379, 387
modulus of elasticity, 2, 117, 357, 359
Mohr, Otto, 19, 32
moment - area theorem, 118
moment distribution, 4, 202, 203, 205–217, 219–233, 235–247, 249–263, 265, 266, 268, 289, 304, 308
moment distribution method, 4, 202, 203, 205–207, 209, 211, 213, 215, 217, 219, 221–223, 225–230, 233, 235–239, 241, 243, 244, 247, 249, 251–253, 255–257, 259, 261–263, 265, 266, 268, 289, 304, 308
moment equilibrium, 146, 147, 149, 152, 154, 155, 157, 160, 161, 163, 165, 166, 168, 170, 175, 177, 179, 182, 185, 188, 190, 204, 266, 267, 296
moment of area, 18, 118, 119
moment of inertia, 2, 120
Muller Breslau, Heinrich, 32
multi-story frames, 177, 266, 295, 300, 310, 311, 412

N

node, 11, 146, 166, 378
non-prismatic, 33, 46, 85, 86, 115, 122–124, 142, 143, 166, 167, 197, 198, 325, 357, 378
non-sway, 168, 228, 238, 241, 246, 247, 251, 256, 261, 285, 308

O

original beam, 3, 76, 147, 313, 315, 318, 327, 328, 332, 365, 371, 375
original structure, 34, 50, 52, 54, 313, 314, 362
other internal forces, 207, 269
other support reactions, 28, 38, 39, 42, 77, 251, 316, 320, 325, 326, 330, 336, 339, 341, 342, 362, 382, 391, 397
overhang, 21, 72, 154, 155, 197, 213, 226, 263, 275, 277, 309, 366, 367

P

partially constrained, 8
pin-jointed, 4, 6, 8, 10, 12, 13, 16, 32, 65–68, 73, 75, 106, 108, 109, 113, 313, 348, 350–354, 357, 359, 362, 364, 403, 405–409, 413
pin-jointed plane frames, 6, 8, 10, 12, 13, 16, 32, 75, 106, 313, 359, 362, 364, 406
pinned, 199, 200, 203, 210, 215, 266, 267
pinned end, 267
pinned or roller support, 267
plane frame, 4–8, 10–16, 32, 50, 52, 54, 55, 57, 59, 61, 63, 65–68, 72, 73, 75, 89, 90, 94, 98, 99, 106, 108, 109, 111, 113, 115, 116, 126, 128, 130, 139, 148, 168, 169, 175, 202, 228, 236, 238, 266, 285, 293, 310, 313, 330, 332, 336, 348, 350–354, 357, 359, 362, 364, 379, 381, 403–406, 408, 409, 412, 413
plastic hinge, 2
primary, 2, 3, 34, 37–41, 43–45, 47–50, 52–54, 56–59, 63, 64, 66–68, 77, 78, 80, 83, 85, 86, 89, 90, 92, 94, 98, 99, 111, 113, 114, 123, 125, 130, 132, 205, 266, 308, 313, 362
primary beam, 3, 34, 37–41, 43–45, 47–49, 77, 78, 80, 83, 85, 86, 123, 125, 313
primary structure, 2, 34, 50–54, 57–59, 63, 64, 89, 90, 98, 99, 130, 132
primary unknowns, 3, 50, 266, 308, 313, 362
principal axes, 126, 141
principal of parity, 105
principle of superposition, 3, 34, 50, 52, 54, 66, 67, 75, 105, 118, 146, 147, 191, 195, 202, 204, 236, 313, 314, 360
prismatic, 12–124, 142, 143, 148, 149, 166, 167, 197, 198, 202–204, 266, 325, 357, 378, 411
prismatic beam, 46, 85, 145, 148, 149, 167, 202–204, 266, 325, 357, 411
probable deflected shape, 34, 38, 40, 42, 47, 50, 52, 54, 57, 59, 61, 65, 71, 98, 100–103, 116, 153, 154, 156, 158, 160–162, 164, 169, 171–173, 176, 180, 181, 183, 186, 187, 197, 200, 262, 264, 316, 318–321, 325–328, 330, 332, 333, 335–337, 339, 341, 342, 344, 365, 366, 368, 369, 371, 375–377, 379, 382, 384–386, 388, 391, 398, 401–403, 411, 412
probable elastic curve, 31, 39, 52, 55, 95, 116, 151, 152, 155, 159, 164, 171, 173, 175, 183, 189–192, 196, 251, 256, 257, 261, 339, 378, 382, 383, 389, 392, 397, 410, 412
prop, 72
propped, 3, 36, 37, 72, 77, 125, 144, 149, 156, 216, 267
propped cantilever, 3, 36, 37, 72, 77, 125, 144, 149, 156, 216, 267
propped cantilever beam, 3, 36, 37, 77, 125, 144

R

radian, 45, 148, 155, 208, 356, 357, 411

reaction, 1–6, 8–11, 20, 23, 25, 28–36, 38, 39, 42–47, 50, 52, 54, 57, 59, 61, 63, 65, 72, 75–80, 82, 84–86, 88–90, 92, 94, 96, 98–99, 101, 103–106, 108, 109, 113, 114, 128, 132, 134, 146, 154, 155, 169, 197, 207, 214, 225, 242, 251, 263, 269, 285, 313, 314, 316, 320, 325–328, 330, 332, 333, 336, 339–343, 346, 349, 351, 356, 357, 362, 368, 382, 388, 391, 395, 397, 411–413
redistribute, 2
redundant, 2–3, 22, 28, 32–34, 36–40, 43–48, 50–52, 54, 56–59, 62–70, 75–90, 92, 94, 96–98, 101, 106–109, 111, 113–115, 118, 125–127, 130, 132, 134–135, 138–139, 141, 313–316, 318, 320–321, 325–328, 330–332, 334, 336–338, 340–341, 343, 346, 349–353, 355
redundant force matrix, 314, 315
redundant forces, 2, 32–34, 65, 69, 75, 89, 113, 313, 332, 349, 351
redundant moments, 22, 28
redundant reaction, 3, 32, 34, 43–47, 50, 52, 54, 57, 59, 63, 75–80, 86, 90, 92, 94, 96–99, 101, 103
released beam, 315–320, 322–325, 327–329
released structure, 314, 315, 318, 332, 340, 351
released truss, 351, 355
restrained beam, 361, 364, 365, 367, 369–372, 378, 379
restrained frame, 379, 381, 385, 390, 393
restrained structure, 287, 362, 366–368, 388, 392
rigid, 4, 5, 7, 10–16, 18, 32, 50, 52, 54, 55, 57, 59, 61, 63, 65, 72, 73, 75, 89–92, 94, 98, 99, 102–106, 116, 126, 128–131, 134, 139, 144–146, 148, 168.169, 172, 174, 175, 179, 181, 184, 186, 188, 190, 195, 199, 202, 228, 236–239, 241, 246, 249, 250, 252–254, 256, 257, 264, 266, 285, 287, 291, 293, 297, 300, 307, 310, 311, 313, 330, 332, 334–337, 340–343, 357, 359, 362, 364, 379, 381, 387, 390, 393, 395, 399, 412
rigid joint, 168, 202, 228, 236–239, 252, 256, 264, 266, 285, 287.291, 293, 297, 310, 311, 379
rigid-jointed structures, 11, 146, 359
rigid joint frame, 202, 228, 237–239, 252, 256, 264, 266, 287, 291, 293, 297, 310, 311
rigid joint plane frames, 202, 228, 238, 266, 285, 293
roller, 12, 59, 60, 126, 130, 132, 148, 149, 191–193, 205, 207, 229, 236, 267, 268, 346, 348, 351, 362
roller support, 59, 60, 126, 130, 132, 148, 191–193, 205, 207, 229, 236, 267, 268, 346, 349, 351
rotation, 11, 13, 19, 26, 28, 38, 45, 47, 49, 51, 57, 75, 77, 104, 125, 130, 142, 143, 146–149, 151, 153, 156, 160, 161, 163, 168–170, 173–175, 177, 178, 184, 186, 189, 191, 197, 202–206, 209, 210, 213, 214, 216, 222, 223, 226, 229–231, 236, 238–240, 244, 248, 252, 256, 266–276, 278, 279, 281–283, 285–287, 289–293, 295–301, 303–308, 321, 359, 360, 363–366, 368, 374, 379–382, 385, 387, 388, 390, 396, 397, 411
rotational, 11, 13, 47, 49, 75, 104, 142, 146–148, 151, 160, 175, 177, 191, 197, 202, 222, 223, 238, 266, 268–269, 271–273, 275–276, 278–279, 281, 283, 285–286, 289–291, 293, 295, 297–298, 304, 359–360, 365, 379–382, 385, 390, 396, 397
rotational and sway degrees of freedom, 266
rotational and sway moments, 266
rotational contributions, 266, 268, 269, 271, 272, 276, 278, 281, 283, 285, 286, 289–291, 293, 297
rotational degrees of freedom, 202, 238, 266, 268, 300, 304, 359
rotational displacement, 13, 75, 148, 151, 175, 177, 360, 363–365, 382
rotational moments, 268
rotational or flexural stiffness, 360
rotational slip, 147, 160, 161, 222, 223
rotational stiffness, 360, 379
rotation contribution, 266–268–270, 272–274, 276, 278, 280, 281, 283, 285, 287, 289, 290, 293, 296–299, 301, 303, 304, 306, 307
rotation factor, 268, 269, 273, 275, 276, 278, 281, 285, 288, 290–292, 296, 298, 301, 304, 308
rotation moment, 266

S

sagging, 18, 20, 23, 38, 103, 104, 117, 172, 194, 230
SD method, 146, 151, 154, 156, 158, 160, 162, 164, 166, 169, 171, 173, 178, 180, 183, 186, 189, 191, 193, 197–201, 214, 219, 236, 256, 269, 285, 387
second column elements, 315, 317, 318, 328, 362, 363, 367, 369, 372, 375, 378, 381, 385, 387, 388, 396, 399
second moment of area, 18, 126
self-equilibrating, 72–73, 134–135, 137, 233
settlement, 1, 2, 19, 28, 77, 144, 146–148, 175, 222–224, 226, 227, 281–284, 309, 374
settlement of supports, 2, 144, 146, 148, 222, 226, 283, 374
settles down, 31, 45, 63, 72, 82, 199, 223, 263, 264, 281, 340, 356, 411
shear force, 21–28, 30, 31, 34, 36, 38–40, 42, 44, 46–48, 50, 52, 54, 55, 57, 59–61, 63, 65, 71–73, 75, 77, 78, 80, 82, 84–86, 88–91, 93–98, 100–103, 106, 107, 116, 117, 128, 130, 132, 144–147, 151–156, 158–162, 164, 166–169, 171–173, 175, 176, 180, 181, 183, 186, 187, 189–193, 196–199, 202–204, 207, 209, 212–214, 216, 218–224, 226–231, 233, 235–239, 242, 243, 247, 248, 251, 252, 256, 257, 261–264, 269–273, 275–286, 288–294, 297, 298, 300, 302, 304, 306–311, 316, 318–321, 323, 325–328, 330, 332–337, 339, 341–344, 346, 348, 349, 356, 357, 365, 366, 368, 369, 371, 373, 375–380, 382–389, 391–394, 397, 398, 401–403, 410–412
shear force and bending moment diagrams, 21–23, 25, 26, 28, 30, 31, 34, 36, 38, 39, 42, 44, 46, 47, 50, 52, 54, 57, 59, 61, 65, 71–73, 77, 78, 80, 82, 84, 85, 88–91, 94, 96, 116, 117, 128, 130, 132, 144, 145, 151, 152, 154, 155, 158, 160, 164, 166, 168, 169, 180, 183, 186, 189, 191, 193, 197–199, 207, 212–214, 216, 218, 219, 221, 223, 225, 226, 228–231, 233, 236, 238, 242, 247, 251, 256, 261, 263, 264, 269, 270, 273, 275, 276.278, 279, 281, 283, 285, 286, 289–291, 293, 297, 298, 300, 304, 307–311, 320, 325, 326, 328, 330, 332–334, 339, 341, 342, 346, 348, 356, 357, 368, 373, 375, 378, 379, 382, 385, 387, 388, 391, 393, 397, 403, 410–412
shear force diagram, 21, 22, 24, 25, 27, 28, 30, 38, 40, 42, 44, 48, 52, 55, 57, 60, 63, 80, 82, 84, 86, 88, 91, 93, 95, 98, 100–102, 106, 107, 128, 130, 132, 153, 154, 156, 159, 161, 162, 164, 167, 168, 171, 173, 175, 176, 181, 183, 187, 190, 192, 196, 209, 212, 214, 218, 220, 222, 224, 229, 231, 233, 235, 237, 239, 243, 248, 252, 257, 262–264, 271–273, 275, 277, 279, 280, 282, 284, 286, 288–290, 292, 294, 300, 302, 306, 316, 319, 321, 325, 327, 328, 330, 335, 337, 342, 344, 346, 349, 366, 369, 371, 373, 376, 377, 380, 383, 384, 386, 389, 392, 398, 401, 402
short, 69, 73, 117–119, 191, 357, 413
short column, 118, 119
shortening, 405
side sway, 168, 171, 228, 239–241, 244, 245, 247–249, 251–256, 258, 293
simple support, 290
simply supported, 1, 2, 18–21, 23, 26, 28, 29, 36, 76, 119, 123–125, 135, 136, 153, 156, 204, 216, 314, 316, 318
simply supported beam, 1, 2, 19–21, 36, 75, 76, 119, 124, 125, 136, 204, 314, 316, 318
simultaneous equations, 18, 20, 77, 89, 106, 202
single bay, 126, 193, 194, 200, 346
single-storey, 193
sliding guided support, 320
slope, 4, 18, 19, 35, 36, 72, 85, 97, 118, 146–149, 151, 153, 155, 157, 159, 161, 163, 165, 167, 169, 171, 173, 175, 177, 179, 181, 183, 185, 187, 189, 191, 193, 195, 197, 201, 202, 205, 208–210, 213, 216, 217, 219, 223, 225, 232
slope-deflection, 146, 147, 149, 151, 153, 155, 157, 159, 161, 163, 165, 167, 169, 171, 173, 175, 177, 179, 181, 183, 185, 187, 189, 191, 193, 195, 197, 201, 202, 205, 208–210, 213, 216, 217, 223, 361, 362, 364, 365, 368, 370, 372–375, 378, 379, 381, 382, 384, 385, 387, 391
slope deflection equation, 202, 205, 266, 361, 362, 365, 368, 370, 372–375, 378, 379, 381, 382, 384, 385, 387, 391

slope deflection method, 4, 146, 147, 149, 151, 153, 155, 157, 159, 161, 163, 165, 167, 169, 171, 173, 175, 177, 179, 181, 183, 185, 187, 189, 191, 193, 195, 197, 201, 202, 208–210, 213, 216, 217, 223, 263, 266, 364, 375, 378, 379, 391
slopes, 18, 19, 35, 36, 146, 148, 202, 225, 232
space, 4–6, 10–12
space structures, 4
spans, 18–24, 26–29, 162–164, 166, 173, 221, 222, 231, 233, 314, 316, 318, 368, 375, 384
spring, 47–49, 88, 89, 111, 112, 164, 165
square matrix, 314, 362
stability, 2, 4, 6, 8, 10
stable, 9, 10, 14, 16, 32, 314
standard expressions, 8
statically determinate, 1–6, 8–10, 59, 75, 103, 134, 146, 213, 226, 275, 283, 314, 362
statically indeterminate beams, 314
statically indeterminate externally, 115, 352
static equilibrium, 1, 2, 4, 20, 23, 25, 28, 29, 38, 39, 44, 47, 50, 52, 54, 59, 65, 77, 84, 89, 146, 151, 152, 154, 155, 158, 175, 189, 208, 209, 212, 219, 221, 223, 235, 247, 251, 261, 269, 270, 276, 278, 286, 291, 314, 316, 318, 320, 325, 330, 333, 336, 339, 341, 342, 346, 362, 365, 368, 375, 382, 391, 397
static equilibrium equations, 1, 2, 4, 20, 23, 25, 28, 29, 38, 39, 44, 47, 50, 52, 54, 59, 65, 77, 84, 89, 146, 151, 152, 154, 155, 158, 175, 189, 208, 209, 212, 219, 221, 223, 235, 247, 251, 261, 269, 270, 276, 278, 286, 291, 314, 316, 318, 320, 325, 330, 333, 336, 339, 341, 342, 346, 362, 365, 368, 375, 382, 391, 397
static indeterminacy, 2–6, 14, 16, 20, 21, 26, 32–34, 39, 43, 45–47, 50, 52, 54, 57, 59, 65, 68, 70, 75, 77, 78, 82, 84, 86, 88–90, 92, 94, 96, 98, 101, 103, 106, 109, 111, 113, 313, 314, 316, 320, 327, 328, 330, 332, 334, 336, 340, 341, 343, 346, 349, 354
stiffness, 3, 4, 28, 104, 118, 125, 142–144, 148, 149, 153, 154, 163, 164, 169, 191–193, 202–208, 210, 211, 213, 215–217, 219, 221, 223, 225, 226, 228–233, 235–237, 242–244, 248, 249, 252, 256, 262, 266, 268, 269, 272, 275, 276, 278, 283, 285, 287, 290, 291, 313, 359–369, 371–373, 375, 377–397, 399–401, 403, 405–413
stiffness factor, 118, 142–144, 202–208, 210, 211, 213, 216, 217, 219, 221, 223, 225, 226, 228, 230, 232, 233, 235–237, 240, 242–244, 248, 249, 252, 256, 262, 268, 269, 287, 290
stiffness matrix, 3, 4, 313, 359–369, 371–373, 375, 377–383, 385–397, 399–401, 403, 405–413
stiffness matrix method, 3, 4, 313, 359, 361–369, 371, 373, 375, 377, 379, 381, 383, 385, 387, 389, 391, 393, 395, 397, 401, 403, 405, 407, 409–413
storey, 11, 126, 177, 193, 194, 200, 264, 266, 291, 295–301, 303–306, 308, 310, 311, 346, 347, 382, 385, 397, 399, 412
strain energy, 75–77, 79, 81, 83, 85, 87–89, 91, 93, 95–97, 99, 101, 103, 105–107, 109, 111, 113, 115, 117, 318
stresses, 1, 2, 19, 118, 126
structural element, 359
structure, 1–6, 8–11, 14, 17, 19, 31–34, 50–54, 57–59, 63, 64, 73, 75, 77, 89, 90, 96, 98, 99, 117, 118, 120, 121, 124, 126, 130, 132, 134, 136–138, 145, 146, 148, 149, 201, 202, 222, 265, 313–315, 318, 326, 331, 334, 338, 340, 341, 343, 346, 351, 358–360, 362, 366–368, 387, 388, 392, 414
superimposing, 238, 313
superposition, 3, 34, 50, 52, 54, 66–68, 75, 105, 118, 146–148, 191, 195, 202, 204, 236, 313, 314, 360
support reactions, 1, 2, 4, 9, 10, 12, 20, 23, 25, 26, 28, 31, 33, 38, 39, 42, 50, 52, 54, 57, 59, 61, 65, 72, 77, 80, 82, 84, 85, 89, 96, 103, 106, 114, 128, 132, 134, 146, 169, 197, 208, 214, 251, 263, 269.285, 314, 316, 320, 325, 326, 330, 336, 339, 341, 342, 346, 356, 357, 362, 382, 391, 397, 411–413
supports, 1, 2, 5, 8, 9, 11, 12, 18–23, 26, 28, 31, 33, 45, 57, 59, 60, 63, 65, 77, 82, 89, 103, 106, 115, 116, 130, 134, 144, 146–148, 151, 160, 168, 173, 197, 199, 200.205, 207, 222, 223, 225–227, 247, 248, 253, 275, 281, 283, 284, 309, 314, 318, 320, 325–328, 331, 332, 338, 340, 341, 343, 346, 356, 360–363, 367, 368, 372, 374, 375, 378, 380, 382, 385, 387, 388, 392, 395, 399, 409, 411
support settlement, 2, 28, 175, 281, 282
sway, 151, 158, 168, 169, 171, 173, 175, 177, 178, 186, 189, 191, 197, 202, 205, 219, 228, 238–242, 244–256, 258–261, 266, 285, 293, 295–299, 303, 304, 308, 379, 388
sway analysis, 238, 241, 242, 244–249, 251–253, 256, 258, 308
sway contribution, 266, 295, 296, 298, 304
sway degrees of freedom, 266
sway/displacement factors, 296
sway moment, 261, 266, 295, 304
sway or displacement contributions, 293
symmetrical frame, 58–60, 69, 101, 103, 105, 126, 172, 177, 191, 231, 236, 244, 288, 290, 304, 333, 382, 397, 412
symmetrical span, 268
symmetry, 57, 59, 84, 85, 96, 97, 103, 105, 126, 129, 131, 133, 134, 168, 171, 192, 200, 219, 230, 231, 233, 236, 242, 263, 286, 309, 310, 320, 322, 334, 336, 356, 382, 384, 385, 411

T

tangential deviation, 118
temperature, 1–2, 70, 73–74, 113, 115, 117, 351–352, 357
temperature stresses, 1
tension, 21, 26, 50, 66, 69, 106, 108, 119, 120, 122, 123, 128, 129, 132, 133, 135, 139, 141, 348, 379
tension coefficient, 106, 108
theorem of least work, 75–87, 89–99, 101, 103, 105–109, 111, 113, 115–117
theorem of three moments, 3, 18, 26, 31, 73
thermal expansion, 73, 115, 351, 352
three-dimensional, 4, 146
three moment, 3, 18, 19, 26, 31, 73
top and bottom chord members, 111
translation, 11, 146, 148, 149, 175, 177, 184, 191, 192, 202, 204, 238, 297, 298, 300, 304, 362, 364, 379, 390, 397, 403
translational displacement, 149, 175, 184, 191, 192, 202.297, 298, 300, 304, 362, 364, 403
translational stiffness, 379, 403
translations, 11, 148, 177
transverse displacement, 11, 75, 193, 229, 359
transverse stiffness, 360
truss, 10, 12, 16, 65–70, 73–75, 106–115, 117, 314, 348, 349, 351–355, 357, 359, 364, 403, 406, 413
truss element, 364
tunnels, 134
two-bay, 84, 264, 344, 395
two-dimenssional, 4, 146
two hinged, 102–103, 105

U

unbalanced moment, 202, 205, 207, 208, 210
unbraced, 11
unequal column heights, 244, 249, 304, 311
uniform change, 115, 352
uniformly distributed load, 1, 2, 88
unit axial, 362, 363
unit couple, 34, 39, 45, 47, 58, 59, 61, 62, 315, 317
unit displacement, 359–362, 365, 367–370, 372, 375, 377, 378, 380–382, 385, 390, 392, 396, 397, 399, 403–406
unit force, 3, 43, 48–50, 63, 66, 88, 313–315, 317, 318, 320, 322, 324, 326, 328, 329, 331, 332, 334, 336, 338, 341, 343, 349–355, 359
unit load, 3, 33, 34, 36–39, 41, 43, 45–47, 49–54, 56, 58–64, 66–68, 70, 76, 78, 142, 143, 313, 314, 317, 320–322, 326–328, 331, 332, 334, 338, 340, 341, 343, 345, 347–351, 353–355
unit load method, 34, 39, 43, 46, 50, 54, 58–60, 64, 66, 68, 70, 78, 321, 322, 326–328, 331, 332, 334, 338, 340, 341, 343, 347, 349, 351, 353, 355
unit moment, 38, 44, 48, 49, 51
unit rotation, 142, 143, 202, 206, 362, 364, 379, 387

unit rotational, 362, 364
unit translation, 362, 364, 379, 403
unit translational, 362, 364, 403
unknown reactions, 1–3, 6, 34, 38, 43, 44, 50, 89, 314, 330
unstable, 8–10, 14, 16, 17, 32
unstable structure, 9
unsymmetrical, 72, 94, 105, 126, 139, 145, 175, 177–179, 199, 239, 244, 247, 252, 256, 257, 264, 293, 297, 300, 304, 388, 390, 393, 399, 412
unsymmetrical frame, 177, 244, 304, 412
unsymmetrical load, 105, 126, 175, 178, 179, 239, 297, 300, 388, 390

V

vertical deflection, 57, 75
virtual forces, 64

W

web member, 111, 351
wind load, 295

Y

yielding, 2, 18–20, 23, 45, 63, 64, 77, 82, 89, 106, 115, 151, 160, 168, 197, 222, 223, 238, 281, 314, 318, 325–327, 331, 332, 338, 340, 341, 343, 360, 362, 367, 368, 372, 375, 378, 380, 382, 385, 387, 388, 392, 395, 399
yielding of supports, 2, 18–20, 23, 45, 63, 77, 82, 89, 106, 115, 151.160, 197, 222, 223, 281, 314, 318, 325–327, 331, 332, 338, 340, 341, 343, 362, 367, 368, 372, 375, 378, 380, 382, 385, 387, 388, 392, 395, 399

www.ingramcontent.com/pod-product-compliance
Ingram Content Group UK Ltd.
Pitfield, Milton Keynes, MK11 3LW, UK
UKHW050059200926
475702UK00019B/2919
9781032889108